Lecture Notes in Mechanical Engineering

Lecture Notes in Mechanical Engineering (LNME) publishes the latest developments in Mechanical Engineering—quickly, informally and with high quality. Original research reported in proceedings and post-proceedings represents the core of LNME. Volumes published in LNME embrace all aspects, subfields and new challenges of mechanical engineering.

To submit a proposal or request further information, please contact the Springer Editor of your location:

Europe, USA, Africa: Leontina Di Cecco at Leontina.dicecco@springer.com
China: Ella Zhang at ella.zhang@springer.com
India: Priya Vyas at priya.vyas@springer.com
Rest of Asia, Australia, New Zealand: Swati Meherishi at swati.meherishi@springer.com

Topics in the series include:

- Engineering Design
- Machinery and Machine Elements
- Mechanical Structures and Stress Analysis
- Automotive Engineering
- Engine Technology
- Aerospace Technology and Astronautics
- Nanotechnology and Microengineering
- Control, Robotics, Mechatronics
- MEMS
- Theoretical and Applied Mechanics
- Dynamical Systems, Control
- Fluid Mechanics
- Engineering Thermodynamics, Heat and Mass Transfer
- Manufacturing
- Precision Engineering, Instrumentation, Measurement
- Materials Engineering
- Tribology and Surface Technology

Indexed by SCOPUS, EI Compendex, and INSPEC

All books published in the series are evaluated by Web of Science for the Conference Proceedings Citation Index (CPCI)

To submit a proposal for a monograph, please check our Springer Tracts in Mechanical Engineering at https://link.springer.com/bookseries/11693

N. M. Sivaram · K. Sankaranarayanasamy ·
J. Paulo Davim
Editors

Advances in Manufacturing, Automation, Design and Energy Technologies

Proceedings from ICoFT 2021

Springer

Editors
N. M. Sivaram
Department of Mechanical Engineering
National Institute of Technology
Puducherry
Karaikal, Puducherry, India

K. Sankaranarayanasamy
Department of Mechanical Engineering
National Institute of Technology
Puducherry
Karaikal, Puducherry, India

J. Paulo Davim
Department of Mechanical Engineering
University of Aveiro
Aveiro, Portugal

ISSN 2195-4356 ISSN 2195-4364 (electronic)
Lecture Notes in Mechanical Engineering
ISBN 978-981-99-1287-2 ISBN 978-981-99-1288-9 (eBook)
https://doi.org/10.1007/978-981-99-1288-9

This Springer imprint is published by the registered company Springer Nature Singapore Pte Ltd.
The registered company address is: 152 Beach Road, #21-01/04 Gateway East, Singapore 189721, Singapore

Preface

The book titled ‘Advances in Manufacturing, Automation, Design and Energy Technologies’ contains the select papers presented at the Second International Conference on Future Technologies (ICOFT) on Manufacturing, Automation, Design and Energy organized by the Department of Mechanical Engineering, National Institute of Technology Puducherry, Karaikal, India, during December 16–18, 2021.

The conference witnessed seven keynote lectures in Manufacturing, Automation, Design and Energy from eminent professors and scientists. The keynote speakers are Prof. Prabhakaran Ramamurthy, Old Dominion University, USA; Prof. Wei Gao, Tohoku University, Japan; Prof. T. S. N. Sankara Narayanan, University of Madras, India, and Chonbuk National University, South Korea; Prof. Vijay Kumar Thakur, Scotland’s Rural College, UK; Prof. Rajiv Prakash, IIT (BHU) Varanasi, India; Prof. Anil Kumar Sharma, Jamia Millia Islamia University, India; and Prof. S. Balasivanandha Prabu, College of Engineering Guindy, India.

The conference received 185 manuscripts as submissions, out of which 126 were accepted for presentation at the conference. Out of this 126, 84 manuscripts were recommended for publication in this book titled ‘Advances in Manufacturing, Automation, Design and Energy Technologies,’ Lecture Notes in Mechanical Engineering, Springer. These 84 papers comprise 29 papers in the manufacturing stream, 10 papers in the automation stream, 26 papers in the design stream and 19 papers in the energy stream. All these papers published in this book are believed to support academicians, researchers and practitioners, leading to a better evolution of the mechanical engineering field.

We are highly thankful to all the reviewers and session chairs in mechanical engineering for their valuable comments and suggestions in enhancing the quality of the papers. We also take this opportunity to wholeheartedly thank all the conference participants, eminent keynote speakers, organizing secretaries and task committee members of ICOFT 2021 and the host institute, National Institute of Technology Puducherry, Karaikal, India, for their support.

Karaikal, India — N. M. Sivaram
Karaikal, India — K. Sankaranarayanasamy
Aveiro, Portugal — J. Paulo Davim

Contents

Automation Stream

Design Stream

About the Editors

Dr. N. M. Sivaram is Assistant Professor and Head of the Department, Mechanical Engineering, National Institute of Technology (NIT) Puducherry, Karaikal, India. He received his B.E. degree in Production Engineering from PSG College of Technology, India, in 2008; M.Tech. Degree in Industrial Safety Engineering from NIT, Tiruchirappalli, India, in 2010 and Ph.D. degree in Mechanical Engineering from Anna University through from PSG College of Technology, India, in 2014. His research interests include machining processes, sustainable manufacturing, industrial digitization, materials for energy applications and industrial safety. He has published more than 30 papers in journals of national and international repute and is conducted consultancy projects sponsored by Central Pollution Control Board, India.

Prof. K. Sankaranarayanasamy is presently serving as Director of NIT Puducherry, is Professor (HAG) in the Department of Mechanical Engineering at National Institute of Technology Tiruchirappalli, received his B.E. (Honours) Mechanical Engineering (Hons) in 1981 from PSG College of Technology, India. He received his M.Tech. Production Engineering in 1983 and Ph.D. in 1989; both from Indian Institute of Technology Madras, India. His research targets finite element simulation of the welding processes, ergonomics study and gear design. He has authored more than 135 papers in international and national journals and has attended conferences in Slovak Republic, China, America and Singapore. To his credit, he has completed several sponsored projects (total worth of INR 6 million) funded by BHEL Tiruchirappalli, NLC Neyveli and DST, India. He served in administrative responsibilities such as Head of the Department and Dean at NIT Tiruchirappalli and Director of NIT Puducherry.

Prof. J. Paulo Davim is Full Professor at the University of Aveiro, Portugal. He is also distinguished as Honorary Professor in several universities/colleges/institutes in China, India and Spain. He received his Ph.D. degree in Mechanical Engineering in 1997, M.Sc. degree in Mechanical Engineering (materials and manufacturing processes) in 1991, Mechanical Engineering degree (5 years) in 1986, from the

University of Porto (FEUP), the Aggregate title (Full Habilitation) from the University of Coimbra in 2005 and the D.Sc. (Higher Doctorate) from London Metropolitan University in 2013. He is Senior Chartered Engineer by the Portuguese Institution of Engineers with an MBA and Specialist titles in Engineering and Industrial Management and Metrology. He is also Eur Ing by FEANI-Brussels and Fellow (FIET) of IET-London. He has over 30 years of teaching and research experience in manufacturing, materials, mechanical and industrial engineering, with particular emphasis in machining and tribology. He also has an interest in management, engineering education and higher education for sustainability. He has guided significant numbers of postdoc, Ph.D. and master's students and has coordinated and participated in several financed research projects. He has received several scientific awards and honors. He has worked as an evaluator of projects for ERC-European Research Council and other international research agencies and examiner of Ph.D. thesis for many universities in different countries. He is Editor in Chief of several international journals, Guest Editor of journals, books Editor, book Series Editor and Scientific Advisory for many international journals and conferences. He has been listed in World's Top 2% Scientists by Stanford University study.

Manufacturing Stream

The Machining Behavior of 52100 Bearing Steel

Jai Tiwari and Kalyan Chakraborty

Abstract The work presents the machining behavior of 52100 bearing steel. The turning of the 52100 bearing steel was performed on a lathe. Subsequently, the scanning electron microscopic study was done on the machining chips. The power law equation ($\sigma = K\varepsilon^n$) was formed by the universal tensile testing for this specimen. The von Mises stresses at different experimental conditions were estimated by using the equation. The shear forces of the shear zone were evaluated using the stresses. The shear force data were used to form the regression equation. The equation correlated the shear forces with the speed, feed and depth of cut (DOC). The SEM examination of the chip surface indicated the white layer at the lowest speed, feed and depth of cut. The white layer forms because of the formation of finer martensite due to the severe plastic deformation and the dynamic recrystallization. The SEM examination for the side surface of the chip showed the presence of numerous cracks at the lowest speed, feeds and DOC. The effect of strain hardening is seen to be operating extensively at the lowest and moderate speed for all feeds and DOCs. At the highest speed, the interaction by the thermal effect is evidenced by the formation of a white layer at the chip edge. The white layer forms through the phase transformation. The reverse martensitic transformation from the austenite state occurs during deformation at the high strain rate with the air quench effect during the chip formation process.

Keywords Machining · Power law equation · Plastic deformation

1 Introduction

Hard turning refers machining of materials of hardness greater than 45HRC. This can be considered as a suitable process route to replace grinding. Thus, all adverse effects generated during grinding can be avoided. The selection of the hard turning instead of the grinding can be considered as the cost-effective step in the manufacturing of items.

J. Tiwari · K. Chakraborty (✉)
National Institute of Technology Silchar, Silchar 788010, Assam, India
e-mail: chakrabortykalyan623@gmail.com

N. M. Sivaram et al. (eds.), *Advances in Manufacturing, Automation, Design and Energy Technologies*, Lecture Notes in Mechanical Engineering,
https://doi.org/10.1007/978-981-99-1288-9_1

Therefore, the study is to be carried out to establish the behavior of the material during hard turning. The von Mises stress develops within the primary deformation zone (PDZ) because of plastic deformation at the higher strain rate during machining. The quality of the machined surface depends on the extent of von Mises stress generation in the PDZ. Further, the amount of von Mises stress depends on the built-up shear force in the PDZ. A correlation between shear force and machining parameters is necessary to study the material behavior during hard turning. The present objective is to establish a regression model to correlate the machining parameters and shear force in order to study the material behavior during hard turning.

Morehead et al. [1] studied the influence of tool wear and cutting conditions on the mechanism of chip formation during machining of 52100 bearing steel. They found saw-tooth chip formation during machining of this steel. The saw-tooth chip geometry was influenced by the tool wear and the cutting parameters. Better chip type could be possible by understanding the process. Zhang and Zhuang [2] studied the effect of tool edge on the machining performance. They also studied the wear behavior of the tool. They studied for 52100 bearing steel. They performed the machining experiments with the tool edge preparation. They established the influence of tool edge microgeometry on the surface roughness and the white layer formation. They studied the white layer formation mechanism by developing a 2D finite element model. They also found that the enlarge chamfer width could improve surface roughness and tool edge chamfer angle cannot influence the quality of the surface roughness. Caruso et al. [3] conducted experiments to determine the effects of the tool geometry, material properties, machining parameters and metallurgical features on the residual stresses in hard machining of 52100 bearing steel. They determined the surface residual stress profiles by the XRD technique. They found that the tool geometry, material property and machining parameters affected the surface residual stress. They studied the microstructural aspects. They found that metallurgical phase changes have an influence on the amount of residual stress. Azizi et al. [4] conducted the experiments on the machining. They selected 52100 bearing steel for machining. They studied the influence of cutting parameters and material hardness on surface quality and machining forces. They used the coated ceramic tools for the experimentations. They used Taguchi's L27 orthogonal array for the study. The analysis of variance enabled them to validate the regression model and to study the influences of machining parameters on the surface roughness and the cutting forces. The statistical analysis revealed that the DOC is less important in reducing the surface roughness, whereas the speed is less important to influence the forces. They developed the empirical equations to correlate the selected input parameters with the output responses. The optimality conditions were also addressed in the study. Rajarajan et al. [5] conducted the experiments considering the speed, the feed and the DOC as input parameters and the surface roughness and the tool-work temperature as the output responses. They considered the dry and the micro lubrication machining environment to conduct the turning experiments. The bearing steel (52100) was the workpiece. They selected Taguchi's L25 array as the experimental sequence. They measured the tool-work interface temperature and subsequently established a regression model using regression analysis. They also developed a surface roughness model using the

MINITAB software. Suitable machining conditions were determined and verified by the experimental trial. The machining in the micro lubrication environment was better. Umamaheswarrao et al. [6] optimized the machining responses for the hard turning of the AISI 52100 steel. They considered the speed, the feed, the DOC and the nose radius as input parameters and the surface quality, the force and the workpiece temperature as the output responses. The experiments were performed according to Taguchi L9 orthogonal array. They studied the influences of the input parameters on the machining responses using the MINITAB14 software. They used decoupled Grey Relational Analysis (GRA) with Principal Component Analysis (PCA) to optimize the process parameters. They found that the nose radius is the most influential factor for the machining responses. Umamaheswarrao et al. [7] optimized the machining parameters by using the TOPSIS methodology. They performed hard machining of the 52100-steel using the PCBN tools. They followed the central composite design methodology to conduct the experiments. They performed hard machining of the 52100 steel using the PCBN tools. They followed the central composite design methodology to conduct the experiments. They considered the speed, the feed, the depth of cut, the nose radius and the negative rake angle as the input process parameters. The surface roughness and the workpiece surface temperature were the output responses. By the ANOVA, they found the optimal setting as the speed 200 rpm, the feed 0.1 mm/rev, the depth of cut 0.7 mm, the nose radius 1.2 mm and the negative rake angle 45°. Zhao et al. [8] studied the influence of the cutting-edge radius on the surface roughness and tool wear. They performed the turning of the bearing steels (52100). Different dimensions of the CBN tool edge radii such as 20, 30 and 40 microns were used for the experimental investigation. The two input parameters and the three-level design were selected for the experimentation. They found that the edge radii significantly influenced the surface roughness and the tool wear. They also observed that an edge radius of 30 microns performed better during the hard machining of this steel. Hosseini et al. [9] studied the condition of the (Fe, Cr)3C carbide morphology in the surface during hard turning of bainitic (52100) steel. They studied the nature of carbide dissolution during the hard turning of this steel. The microstructural and analytical studies of the carbide dissolution kinetics were performed by using the DICTRA1. The carbide dissolution did not take place in the white layers up to a depth up to 20 microns from the surface. The carbide dissolution could not take place due to short contact times. They found elongated carbides on the hard turned surface due to the plastic deformation of the carbides during the hard turning.

2 Theory

The chip reduction coefficient (ζ) is given by the ratio of formed chip thickness to the uncut chip thickness.

The von Mises stress [10] can be given by

$$\sigma = 1.74\,K\,(\ln \varsigma)^n, \tag{1}$$

where σ = Von Mises stress, K = strength coefficient, ζ = chip reduction coefficient and n = strain hardening exponent.

The shear force F_s is obtained by multiplying the shear stress with area of the shear plane.

The thermal number γ is given by

$$\gamma = \frac{\partial * c * v * w}{K}, \tag{2}$$

where ∂ = density = 7827 kg/m^3, c = specific heat = 458 J/(KgK), v = cutting speed, K = thermal conductivity = 43 W/(mK).

The shear plane temperature can be determined by

$$T_s = \frac{(1 - \vartheta) * ps}{\partial * c * v * t1 * w}, \tag{3}$$

where T_s = PDZ temperature, ϑ = the part of PDZ heat conducted into the workpiece, ps = rate of heat generation, $t1$ = uncut chip thickness, w = width of cut (mm).

3 Experimental Procedures

The bearing steel (52100) rod (C1.12%, Mn0.4%, Cr1.512%) was obtained to carry out experiments. The diameter of the specimen was 60 mm and the length was 300 mm. The specimen was soft annealed in the temperature range of 780–800° C. The specimen was soft annealed. The steel was made softer by this heat treatment. The machining was done by using the lathe (speed range: 45–1000 rpm, feed range: 0.06–1.72 mm/rev.). Twenty-four experiments were done according to the central composite design (DOE). Selected speeds were 10.2, 14.2, 31.6, 70.3, 97.9 m/min, feeds were 0.098, 0.018, 0.14, 0.18, 0.19 mm/rev. and depths of cuts were 0.54, 0.66, 1, 1.5, 1.77 mm. The parameter values were coded as $-\sqrt{2}, -1, 0, 1, 2$, respectively. The tool insert was Tungaloy made (SNMG 120408) with 6° rake angle and 75° main cutting-edge angle. The chip surfaces were examined in the scanning electron microscope. The chip thickness was determined for all the samples. The material hardness was noted, and the tensile testing of the specimen was done.

4 Results and Discussions

4.1 Mechanical Tests

The engineering stress–strain diagram for the 52100 bearing steel specimen was obtained by performing the universal tensile testing of the specimen. The true stress–true strain relationship was drawn on the log–log graph paper, and "*n*" and "*K*" were determined as 0.64 and 2500, respectively. The formed power law equation was

$$\sigma = 2500 * \varepsilon^{0.64}. \quad (4)$$

5 Machining Results and Analysis

Table 1 shows the chip reduction coefficient (CRC), shear angle, shear force and shear zone temperature.

The shear force and parameter relationship was formed as

$$\begin{aligned} Y = {} & 1168.8 + 29.9x_1 + 253.1x_2 + 325.9x_3 - 288.7x_1^2 - 93.3x_2^2 \\ & - 96.2x_3^2 - 73.6x_1x_2 - 8.6x_1x_3 + 185.6x_2x_3, \end{aligned} \quad (5)$$

where Y is shear force and x_1, x_2, x_3 are codes for speed, feed and depth of cut.

Figure 1 shows the differences between experimental and predicted shear forces.

The von Mises stresses develop due to the shear force. So, the analysis of variance (AOV) was done on shear force data. The MATLAB program was developed to study the analysis of variance for the obtained data. The F-ratio was found as 5.4281 (Table 2) through AOV.

The statistical F-distribution table shows that $F_{p,\, n\text{-}p\text{-}1,1\text{-}\alpha} = F_{9,\, 14,\, 0.995} = 4.72$.

The estimated F-ratio in the present study (= 5.4281) (from AOV Table 2) is greater than the F-ratio available in the statistical F-distribution table ($F_{9,\, 14,\, 0.995}$ = 4.72, from F-distribution table).

So, Eq. (5) is found to be satisfying the condition for acceptance in the discussion.

6 Shear Force

6.1 At Lowest Speed (Code: $-\sqrt{2}$)

The shear force increases with the increase of the feed at the lowest DOC. The shear force also increases with the increase of the DOC at the lowest feed. Such increase of

Table 1 Experimental results

Sl. No.	Assigned codes (v, f, d)			CRC	Shear angle, β_o (degree)	Shear force, F_s (N)	PDZ temperature, T_s (°C)
	v	f	d				
1	−1	−1	−1	1.16	43.30	77.59	282.69
2	1	−1	−1	5.45	10.54	1384.6	389.35
3	−1	1	−1	2.24	24.97	622.17	271.22
4	1	1	−1	1.70	31.94	380.54	457.54
5	−1	−1	1	1.19	42.50	197.51	347.02
6	1	−1	1	2.01	27.56	697.56	146.23
7	−1	1	1	2.22	25.18	1381.1	246.19
8	1	1	1	2.08	26.72	1236.2	487.65
9	$-\sqrt{2}$	0	0	2.09	26.61	646.64	207.67
10	$\sqrt{2}$	0	0	2.43	23.15	829.26	557.08
11	0	$-\sqrt{2}$	0	2.90	19.58	764.79	78.59
12	0	$\sqrt{2}$	0	5.53	10.39	3908.4	369.69
13	0	0	$-\sqrt{2}$	1.97	28.06	328.44	134.77
14	0	0	$\sqrt{2}$	3.90	14.68	2998.6	316.02
15	$-\sqrt{2}$	0	0	3.12	18.25	1219.8	219.20
16	$\sqrt{2}$	0	0	1.27	40.47	218.50	340.28
17	0	$-\sqrt{2}$	0	3.23	17.65	898.13	283.72
18	0	$\sqrt{2}$	0	2.31	24.27	1085.5	342.33
19	0	0	$-\sqrt{2}$	2.56	22.05	507.46	445.65
20	0	0	$\sqrt{2}$	2.24	24.97	1288.5	518.07
21	0	0	0	3.73	15.34	1584.9	393.59
22	0	0	0	5.62	10.22	2810.7	342.26
23	0	0	0	5.38	10.68	2648.3	241.61
24	0	0	0	4.70	12.21	2197.6	296.85

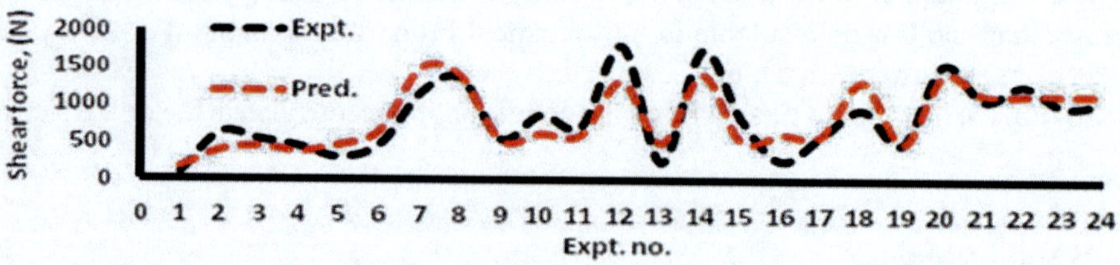

Fig. 1 Difference between experimental and predicted data

Table 2 AOV table

F-ratio	Df	ss	MS	F-ratio
Regression	9	4.0643×10^6	4.5159×10^5	5.4281
Error	14	1.1647×10^6	8.3195×10^4	
Total	23			

the shear forces in these two conditions occurs at the same rate. This shows that for these cutting conditions, the strain hardening operates similarly. At the highest DOC, the shear force increases with the increase of the feed. The shear force also increases with the increase of the DOC at the highest feed. At these cutting conditions also, the shear forces increase at the same rate. This also shows that the strain hardening operates for these cutting conditions. The strain hardening shows the predominating role to harden the material at these conditions (see Fig. 2a). The SEM examination of the chip shows that the chip formation occurs through the segmentation and the sliding at the lowest feed and the DOC. The serrations are observed at the chip top surface. The material side flow is also observed (see Fig. 2b). These features indicate improper machining responses from the process. The SEM examination of the chip surface (under) indicates the formation of the white layer at the lowest feed and the DOC. The white layer forms because of the formation of the finer martensite through the shear displacement mechanism. The martensite (finer) forms by the dynamic recrystallization. The featureless microscopic observation is revealed by the white layer appearance at the chip surface (under) (see Fig. 2c). The SEM image of the side surface of the chip shows the presence of numerous cracks (see Fig. 2d). All these observations indicate the chip formation by harder material throughout the cutting process. The effect of the strain hardening is more pronounced at the highest feed and the highest DOC.

6.2 *At Moderate Speed (Code: 0)*

The shear force increases with the increase of the feed at the lowest depth of cut. The shear force increases with the increase of DOC at the lowest feed. The shear force increases with the increase of the feed at the highest DOC. The shear force increases with the increase of the depth of cut at the highest feed. The shear force increases because of the strain hardening during the cutting process. The effect of the increase of the depth of cut is more pronounced due to strain hardening (see Fig. 3).

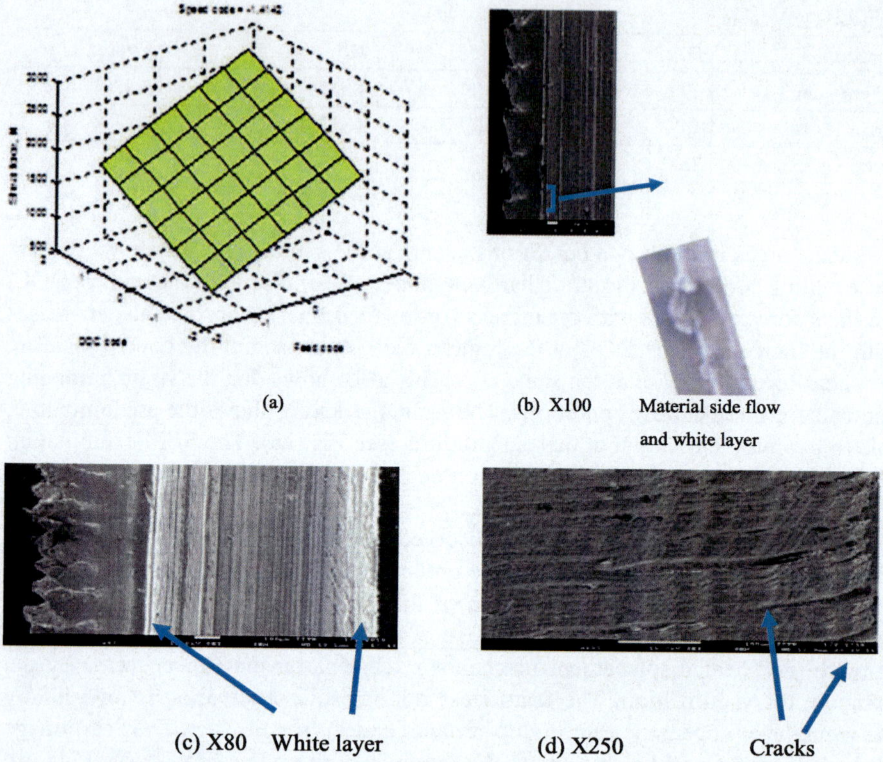

Fig. 2 **a** Variations of the shear force with respect to the feed and the DOC at the lowest speed, **b** SEM images of top surfaces of chip, **c** SEM image of undersurface of chip and **d** SEM image of side surface of chip (Condition: 10.2 m/min, 0.098 mm/rev., 0.54 mm) (Speed code: $-\sqrt{2}$)

Fig. 3 Variations of the shear forces with respect to feed and DOC for the moderate speed (Speed code: 0)

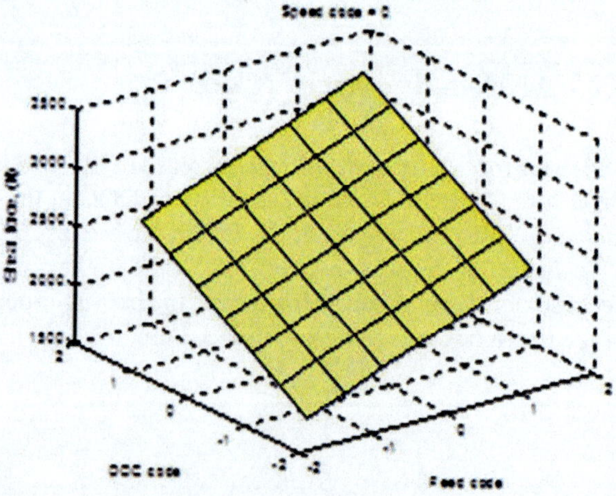

6.3 At Highest Speed (Code: √2)

The shear forces increase negligibly with the increase of the feed at the lowest DOC. The thermal softening effect is operating at this cutting condition. The shear forces increase with the increase of the DOC at the lowest feed. The strain rate hardening operates at this cutting condition. The shear force increases with the increase of the feed at the highest DOC. The shear force also increases with the increase of the DOC at the highest feed. This occurs because of the predominating strain rate hardening of the material. But, the rate of hardening is more with the increase of the DOC for the highest feed (see Fig. 4a). The thermal effect operates at the high strain rate. This causes the chip formation in a favorable mode. The continuous chips form for the high speed, feed and DOC by the successive lamellar sliding of the material (see Fig. 4b and see Fig. 4c). The interaction by the thermal effect is evidenced by the formation of a white layer at the chip edge (see Fig. 4d). The white layer forms through the phase transformation mechanism. The reverse martensitic transformation from the austenite state occurs during the deformation at the high strain rate with the air quench effect during the chip formation process (speed 97.9 m/min, feed 0.2 mm/rev., DOC 1.8 mm). The high strain rate deformation with the thermal effect forms the recrystallized martensitic (finer) grains. These appear as the featureless white layer at the chip surface (see Fig. 4d). The effect of the increase of the DOC to cause strain rate hardening is more pronounced at the highest feed.

The variations of the shear force, VMS, the shear zone temperature are shown (see Figs. 5a, b, d). The maximum temperatures are developed at (√2 0 0) (Expt. No. 10) and (0 0 √2) (Expt. No. 20). The shear forces and the von Mises stresses were also found moderately at the higher levels for these machining conditions. Figure 6 shows that the lower shear angle causes the higher shear force, the higher temperature and the higher von Mises stress.

7 Conclusions

At the lowest speed, the chip formation process is influenced by the strain hardening for all feeds and DOCs. The white layer forms at the chip surface (under) due to the formation of the martensite (finer) grains through the shear displacement and the dynamic recrystallization. At the lowest and the moderate speeds, the chip formation occurs mainly due to the predominating strain hardening of the material for all feeds and DOCs. At high speed, the chip formation process takes place by the lamellar sliding of the material over the rake face. The continuous chip forms due to the thermal effect. The white layer forms at the chip edge because of the martensite (finer) formation at the high strain rate through the reverse transformation process. The chip formation process is much influenced by the high strain rate deformation with the thermal effect. The experimental results can be reproducible for the identical

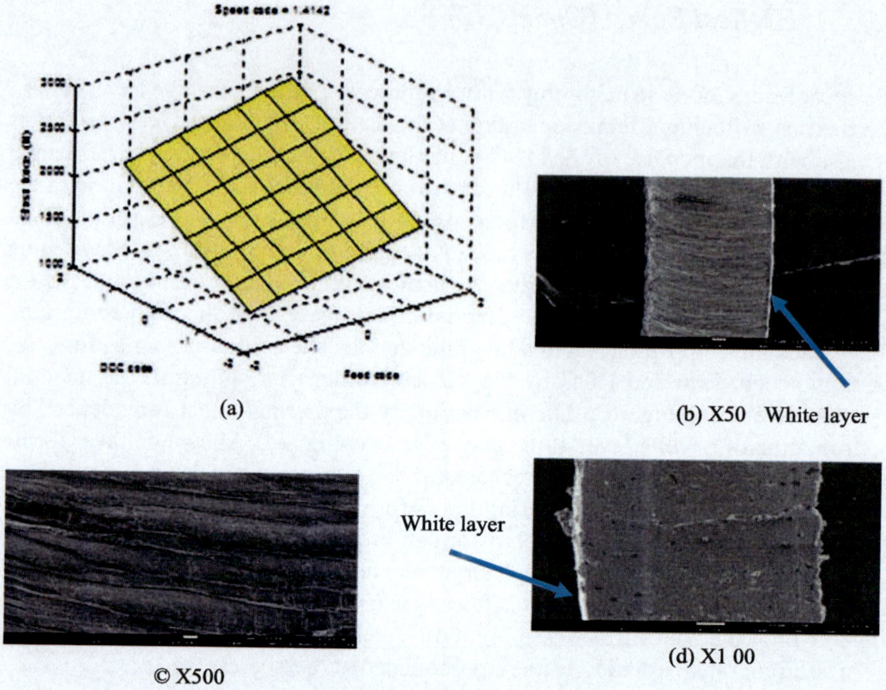

Fig. 4 **a** Variations of the shear force with respect to the feed and the DOC at the highest speed, **b**, **c** SEM images of top surfaces of chip, **d** SEM image of undersurface of chip (condition: speed 97.9 m/min, feed 0.2 mm/rev., DOC 1.8 mm) (Speed code: $\sqrt{2}$)

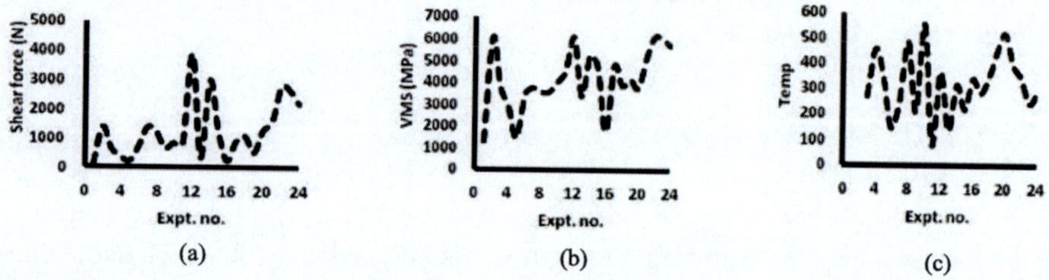

Fig. 5 Variations of **a** shear force, **b** the VMS, **c** the PDZ temperature (°C)

Fig. 6 Variation of shear angle

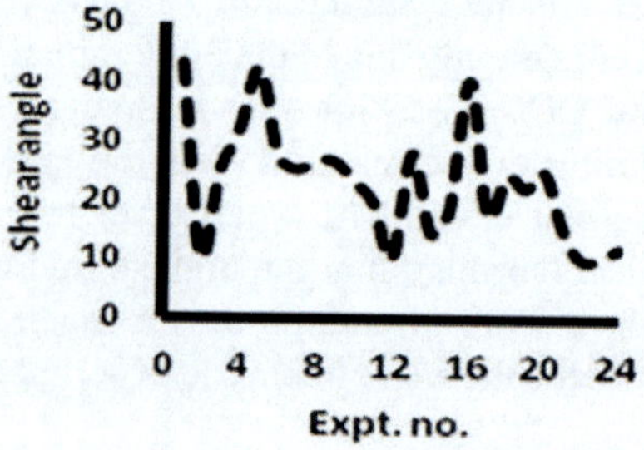

machining condition. The work is informative to develop the knowledge. The results can be incorporated for proper selection of the machining parameters.

Acknowledgements The authors are grateful to IIT Kanpur for the permission to obtain SEM images from IIT Kanpur.

References

1. Morehead MD, Huang Y, Luo I (2007) Chip morphology characterization and modelling in machining hardened 52100 steels. Mach Sci Technol 11:335–354. Taylor & Francis Group. https://doi.org/10.1080/10910340701567289
2. Zhang W, Zhuang K (2020) Effect of cutting-edge microgeometry on surface roughness and white layer in turning AISI 52100 steel. Procedia CIRP 87:53–58
3. Caruso S, Umbrelloa D, Cs R, Outeirob JC, Filicea L, Micari F (2011) An experimental investigation of residual stresses in hard machining of AISI 52100 steel. In: 1st CIRP conference on surface integrity (CSI), procedia engineering, vol 19, pp 67–72
4. Azizi MW, Belhadi S, Yallese MA, Mabrouki T, Rigal JF (2012) Surface roughness and cutting forces modelling for optimization of machining condition in finish hard turning of AISI 52100 steel. J Mech Sci Technol 26(12):4105–4114
5. Rajarajan S, Ramesh Kannana C, Denison MS (2020) A comparative study on the machining characteristics on turning AISI 52100 alloy steel in dry and micro lubrication condition. Aust J Mech Eng. https://doi.org/10.1080/14484846.2019.1710019
6. Umamaheswarrao P, Ranga Raju D, Suman KNS, Ravi Sankar B (2018) Multi objective optimization of Process parameters for hard turning of AISI 52100 steel using Hybrid GRA-PCA. Procedia Comput Sci 133:703–710
7. Umamaheswarrao P, Ranga Raju D, Suman KNS, Ravi Sankar B (2019) Topsis based optimization of process parameters while hard turning of AISI 52100 steel. Acta Mechania Malaysia (AMM) 2(2):28–31
8. Zhao T, Zhou TM, Bushlya M, Ståhl JE (2017) Effect of cutting-edge radius on surface roughness and tool wear in hard turning of AISI 52100 steel. Int J Adv Manuf Technol 91:3611–3618. https://doi.org/10.1007/s00170-017-0065-z
9. Hosseini SB, Dahlgren R, Ryttberg K, Klement U (2014) Dissolution of iron-chromium carbides during white layer formation induced by hard turning of AISI 52100 steel. In: 6th CIRP international conference on high performance cutting, HPC2014, Procedia CIRP, vol 14, pp 107–112
10. Astakhov VP, Tribology of metal cutting, Elsevier, Tribology and interface engineering series, no 52, Series Editor: Briscoe BJ

Optimization of Machining Parameters in Drilling of AA 7075 Alloys Using TOPSIS and Taguchi Method

Reddy Sreenivasulu

Abstract The selection of most suitable machining parameters is a multiple criteria decision-making problem which is based on the several qualitative and even some conflicting factors involved. In the current experimental study, technique for order of preference by similarity to ideal solution (TOPSIS) method was adopted to get best optimal selection of drilling parameters to perform holes with HSS twist drill bits on AA7075 alloy as a workpiece material. Based on the five input parameters such as rotary speed, feed rate, drill diameter, point and clearance angle and three levels of each parameter, L_{27} orthogonal array was selected according to design of experiments corresponding to Taguchi method. In this experimental investigation, the output responses burr height, thrust force and surface roughness are measured with well-calibrated instruments. Finally, concluded that rotary speed 795 rpm, feed rate 26 mm/min, drill diameter 10 mm, point angle 100°, clearance angle 8° leading to the value of optimum response variables burr height 0.164 mm, thrust force 397 N and surface roughness 1.326 μm are attained with TOPSIS.

Keywords Burr height · Thrust force · Surface roughness · TOPSIS

1 Introduction

The burrs that are produced during drilling operation may have serious problem which includes the deformation at the outer edges, complexity while assemble and further causes reduction in the dimensional accuracy [1]. One of the complexities identified in hole making on AA7075 alloy materials is generally used for structural applications. In general, formation of burrs during machining is combined with arise of plastic deformation caused by shearing action between the parent material and the cutting tool, and it is generally denoted by height and thickness. Machining of

R. Sreenivasulu (✉)
R.V.R. & J.C. College of Engineering (Autonomous), Chowdavaram, Guntur 522019, Andhra Pradesh, India
e-mail: rslu1431@gmail.com

N. M. Sivaram et al. (eds.), *Advances in Manufacturing, Automation, Design and Energy Technologies*, Lecture Notes in Mechanical Engineering,
https://doi.org/10.1007/978-981-99-1288-9_2

Table 1 Drill bit geometrical parameters and machining conditions

Parameter	Unit	Symbol	Value
Diameter	mm	d	8, 10, 12
Helix angle	degrees	δ	30° constant
Point angle	degrees	φ	100°, 110°, 118°
Chisel edge angle	degrees	γ	136° constant
Clearance angle	degrees	ψ	4°, 6°, 8°
Feed rate	mm/min	*f*	18, 20, 26
Rotary speed	rpm	n	465, 695, 795

AA 7075 alloy becomes an important research manner because of its behavior while machining. Burr formation, thrust force and surface quality are the most common problems in drilling of aluminum and its alloys. These materials adhere to cutting tool due to their smearing properties. The cutting parameters used during machining of materials are important factors affecting surface accuracy. It is noticed that there is not sufficient study about 7000 series of aluminum alloys whose usage is increasingly spreading [2, 3]. The burr minimization scheme generally depends on variations which occurs during hole making on a parent material and geometry of the drill bit [4]. In this work, HSS twist drill bits with diameters of 8, 10 and 12 mm with 118° point angle, 8° clearance angle and 30° constant helix angle (shown in Table 1) are used because of easy availability and low cost to do the experiments. To attain optimal geometry to reduce burr formation, the drill geometry was altered with three levels using Tool and Cutter grinder to do the experimentation as per Taguchi method. It is identified from previous literature, and some authors applied multi-criteria-based decision-making methods to solve multi-objective optimization wherever needed, especially in machining operations along with the design of experiments [5, 6]. The present work illustrates the application of technique for order preference by similarity to ideal solution (TOPSIS) to optimize the multi-response parameters obtained from Taguchi method with suitable orthogonal array selected based on initial machining parameters and their levels fixed at the beginning with respect to specifications of selected drilling machine.

2 Experimental Methods

According to Taguchi L_{27} orthogonal array, the holes are performed on the AA7075 workpiece with dimensions 300 mm × 50 mm × 10 mm on radial drilling machine. The Kistler dynamometer is fitted on the work table of the radial drilling machine to read the thrust forces generated during the drilling operation. The burr height is measured using tool maker's microscope after completion of drilling of 27 holes. The surface quality of the each drilled hole is measured using Mitutoyo Surftest SJ301. A Kistler type 9272 dynamometer was used to measure thrust force recorded

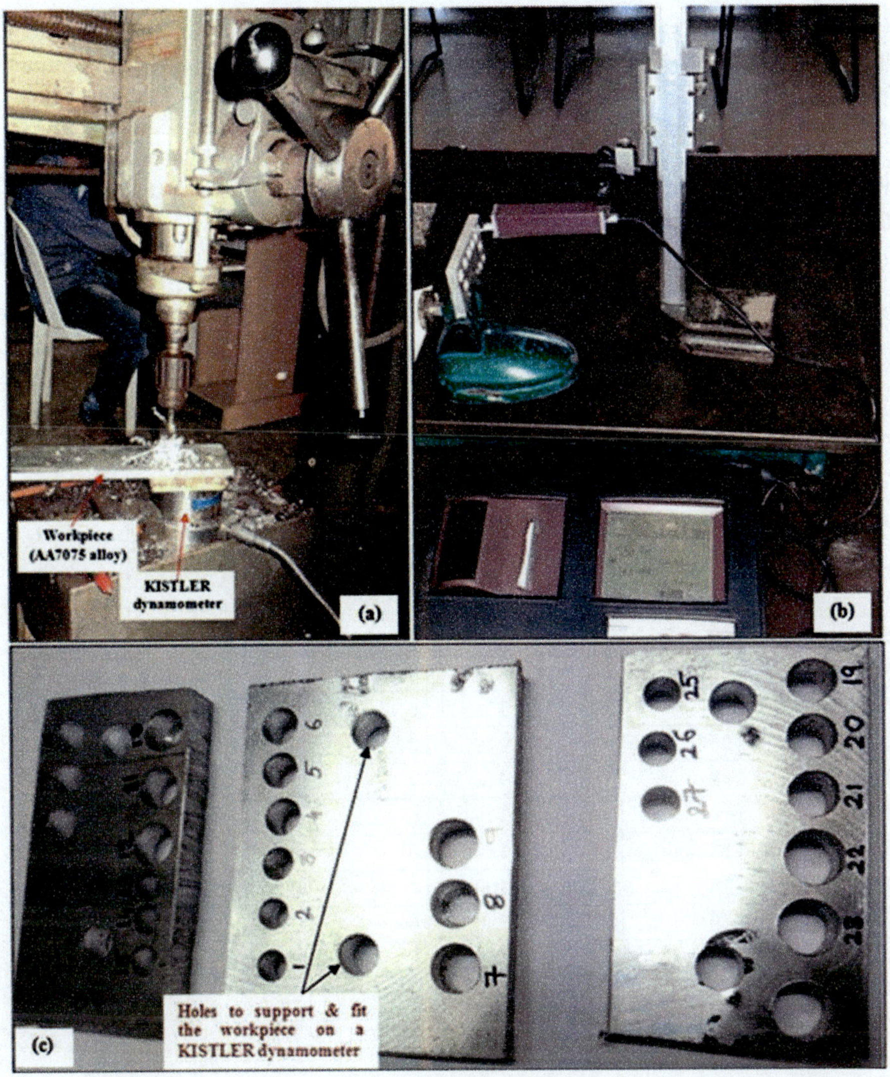

Fig. 1 **a** Radial drilling machine setup along with KISTLER dynamometer, **b** measurement of surface roughness with Surftest 301 series, **c** samples of AA7075 alloy workpieces

with DynoWare software 2825A (experimental setup shown in Fig. 1). In drilling, quality of hole primarily depends on the feed which can be applied on drill bit, also persuaded by the included angle of the drill bit at the tip and the wear rate at the outer corner chisel edge of the drill bit.

After completion of drilling, workpiece samples are placed beneath the Surftester and measured at four locations (0°, 90°, 180° and 270°) for each hole. Five measurements are taken for each location to obtain the average surface roughness value. The

Surftester was well calibrated prior to measure each sample to obtain a consistent and precise level. The sampling length is selected as 0.75 mm from the maximum range of the measuring apparatus. The output responses, viz., burr height, thrust force and surface roughnesses are recorded, and the values are depicted in Table 2. Finally, TOPSIS is used for decision-making regarding the optimal sequential selection of drilling parameters which yields the optimal burr size, thrust force and surface quality.

Table 2 Experimental runs as per Taguchi method and responses measured

Exp. No.	Rotary speed (rpm) n	Feed rate (mm/min) *f*	Drill diameter (mm) d	Point angle (deg's) φ	Clearance angle (deg's) ψ	Burr height Bh (mm)	Thrust force Ft (N)	Surface roughness Ra (μm)
1	465	18	8	100	4	0.246	281.6	1.656
2	465	18	8	100	6	0.232	235.2	2.192
3	465	18	8	100	8	0.226	395.9	1.564
4	465	20	10	110	4	0.265	232.1	1.332
5	465	20	10	110	6	0.336	291.2	1.353
6	465	20	10	110	8	0.242	265.5	1.258
7	465	26	12	118	4	0.316	336.6	1.267
8	465	26	12	118	6	0.325	283.7	1.723
9	465	26	12	118	8	0.324	252.2	2.850
10	695	18	10	118	4	0.296	241.7	1.615
11	695	18	10	118	6	0.254	237.2	2.093
12	695	18	10	118	8	0.326	395.9	1.524
13	695	20	12	100	4	0.286	262.1	1.209
14	695	20	12	100	6	0.378	208.2	1.305
15	695	20	12	100	8	0.365	268.4	1.215
16	695	26	8	110	4	0.302	346.6	1.236
17	695	26	8	110	6	0.297	280.7	1.728
18	695	26	8	110	8	0.352	252.2	2.815
19	795	18	12	110	4	0.392	246.6	1.635
20	795	18	12	110	6	0.289	236.5	2.349
21	795	18	12	110	8	0.342	305.9	1.557
22	795	20	8	118	4	0.282	272.1	1.142
23	795	20	8	118	6	0.207	298.2	1.356
24	795	20	8	118	8	0.174	345.5	1.285
25	795	26	10	100	4	0.164	397.0	1.326
26	795	26	10	100	6	0.201	286.7	1.724
27	795	26	10	100	8	0.219	362.2	2.853

3 Methodology

Technique for order of preference by similarity to ideal solution (TOPSIS) was initially introduced by Hwang and Yoon; based on their idea, the chosen input parameter has been least distance from the best feasible solution and the greatest distance from the worst feasible solution [7]. In the TOPSIS method, particular weight is given to output responses so as to rank them. The possible steps engaged in this approach are described step by step in detail.

Step 1: In the TOPSIS, units of all the parameters both input and output are eliminated and it has been converted into its normalized value (x_{ij}). The normalized value is obtained using Eq. 1, where i = number of trials, j = number of output responses, y_{ij} = actual value of the ith value of jth experimental run. The normalized performance values and weighted normalized values are shown in Table 3.

$$x_{ij} = \frac{y_{ij}}{\sqrt{\sum_{i=1}^{n} y_{ij}^2}}, i = 1, 2, 3 \ldots 27, j = 1, 2, 3. \quad (1)$$

Step 2: The weighted normalized value (β_{ij}) is computed by multiplying the normalized value by corresponding weights from Eq. 2. From the literature, it is found that numerous authors considered the weightage for output responses equally. So here, equal weightage is given for all responses [8]. Therefore, $W_{ij} = 0.33$.

$$\beta_{ij} = w_{ij} * x_{ij}, i = 1, 2, 3 \ldots 27, j = 1, 2, 3. \quad (2)$$

Step 3: PIS (S^+) and NIS (S^-) have been determined from Eqs. 3a and 3b, where J is the set of valuable attributes and J^* is a set of non-valuable attributes. The calculated values S^+ and S^- are depicted in Table 3.

$$S^+ = \{(\max(\beta_{ij}) \vdots J \epsilon J^*), \min(\beta_{ij}) \vdots J \epsilon J^*) \vdots i = 1, 2, 3 \ldots 27\}, \quad (3a)$$

$$S^- = \{(\min(\beta_{ij}) \vdots J \epsilon J^*), \max(\beta_{ij}) \vdots J \epsilon J^*) \vdots i = 1, 2, 3 \ldots 27\}. \quad (3b)$$

Step 4: Partition of each choice from PIS (S^+) and NIS (S^-) is calculated from Eqs. 4 and 5.

$$Z_i^+ = \sqrt{\sum_{i=1}^{27} (\beta_{ij} - S^+)_j^2}, \quad (4)$$

$$Z_i^- = \sqrt{\sum_{i=1}^{27} (\beta_{ij} - S^-)_j^2}. \quad (5)$$

Table 3 Normalized performance and weighted normalized values

Exp. No.	Normalized performance value			Weighted normalized value		
	Bh	Ft	Ra	Bh	Ft	Ra
1	0.040263	51.84464	0.48069	0.013287	17.10873	0.015771
2	0.035811	36.16706	0.842221	0.011818	11.93513	0.02043
3	0.033983	102.4731	0.428764	0.011214	33.81609	0.017903
4	0.046723	35.21996	0.310995	0.015419	11.62259	0.021331
5	0.075114	55.43976	0.320878	0.024788	18.29512	0.019760
6	0.038965	46.08585	0.277399	0.012858	15.20833	0.022760
7	0.066438	74.07417	0.281383	0.021925	24.44448	0.019295
8	0.070276	52.62078	0.520373	0.023191	17.36486	0.017982
9	0.069844	41.58424	1.423751	0.023049	13.72280	0.018206
10	0.058294	38.19371	0.457182	0.019237	12.60392	0.018206
11	0.042925	36.78476	0.767861	0.014165	12.13897	0.023173
12	0.070709	102.4731	0.407112	0.023334	33.81609	0.021998
13	0.054422	44.91306	0.256211	0.017959	14.82131	0.018417
14	0.095066	28.34001	0.298514	0.031372	9.352203	0.022051
15	0.088639	47.09812	0.258761	0.029251	15.54238	0.018883
16	0.060681	78.54087	0.267782	0.020025	25.91849	0.021008
17	0.058689	51.51378	0.523398	0.019367	16.99955	0.018292
18	0.082438	41.58424	1.388996	0.027205	13.72280	0.022704
19	0.102238	39.75802	0.468576	0.033739	13.12015	0.019457
20	0.055571	36.56797	0.967187	0.018338	12.06743	0.020562
21	0.077821	61.17832	0.424934	0.025681	20.18885	0.019605
22	0.052911	48.40561	0.228601	0.017461	15.97385	0.020374
23	0.028509	58.13717	0.322303	0.009408	19.18527	0.018447
24	0.020144	78.04313	0.289435	0.006648	25.75423	0.020882
25	0.017895	102.9914	0.308199	0.005905	33.98716	0.019308
26	0.026881	53.73954	0.520977	0.008871	17.73405	0.020232
27	0.031911	85.77001	1.426750	0.010531	28.30411	0.018836
(PIS) Positive ideal solution (S^{+})				**0.033739**	**33.98716**	**0.023173**
(NIS) Negative ideal solution (S^{-})				**0.008871**	**9.352203**	**0.015771**

Step 5: The value of closeness coefficient of each choice (CC_i) is found from Eq. 6 and obtained values are depicted in Table 4.

$$CC_i = \frac{Z_i^-}{(Z_i^- + Z_i^+)}. \tag{6}$$

Table 4 Separation measures, closeness coefficient value and rank

Exp. No.	Z_i^+	Z_i^-	CC_i	Rank
1	16.87844	7.756528	0.314858	14
2	22.05204	2.582933	0.104848	26
3	0.172627	24.46389	0.992993	3
4	22.36458	2.270403	0.092162	4
5	15.69204	8.942932	0.363018	11
6	18.77884	5.856133	0.237716	18
7	9.542688	15.09228	0.612636	8
8	16.62231	8.012671	0.325256	13
9	20.26436	4.370621	0.177415	21
10	21.38325	3.251734	0.131997	23
11	21.84821	2.786782	0.113123	24
12	0.171391	24.46389	0.993043	2
13	19.16586	5.469115	0.222006	19
14	24.63496	0.023362	0.000947	27
15	18.44478	6.190211	0.251277	17
16	8.068682	16.56629	0.672471	6
17	16.98762	7.647355	0.310427	15
18	20.26436	4.370641	0.177416	20
19	20.86701	3.768031	0.152954	22
20	21.91974	2.715248	0.110219	25
21	13.79831	10.83666	0.439889	9
22	18.01332	6.621654	0.268791	16
23	14.80191	9.833067	0.399151	10
24	8.232975	16.40203	0.665802	7
25	0.028101	24.63496	**0.998861**	1
26	16.25313	8.381848	0.340242	12
27	5.683109	18.9519	0.769308	5

4 Results and Discussions

From Table 4, it clears that the experimental run 25 realized the highest closeness coefficient among 27 experimental runs and the optimum condition to achieve multiple performance characteristics (rotary speed 795 rpm, feed rate 26 mm/min, drill diameter 10 mm, point angle 100°, clearance angle 8°). The sequential order of experimental runs obtained from TOPSIS was given by 14-26-3-4-11-8-13-21-23-24-2-19-27-17-6-15-20-22-25-9-16-10-7-1-12-5.

$$\eta_{opt} = \eta_m + \sum_{i=1}^{q}(\overline{\eta}_j - \eta_m). \tag{7}$$

The closeness coefficient for the obtained optimum combination of parameters was obtained from Eq. 7 and was greater than the maximum closeness coefficient corresponding to rank 1 in Table 5. Hence, the values obtained were optimal.

The main effects' plot from Fig. 2 divulges that the optimal combination of input parameters was identified at rotary speed 795 rpm, feed rate 26 mm/min, drill diameter 10 mm, point angle 100°, clearance angle 8°. The greatest value of closeness coefficient represents good performance.

Table 5 Mean response table for closeness coefficient

Level	Drilling parameters				
	Rotary speed	Feed rate	Drill diameter	Point angle	Clearance angle
1	0.357878	0.372658	0.434084	**0.443927**	0.385193
2	0.31919	0.277874	**0.44883**	0.28403	0.229692
3	**0.46058**	**0.487115**	0.254733	0.40969	**0.522762**
Max–Min	0.14139	0.209241	0.194097	0.159897	0.29307
Rank	**5**	**2**	**3**	**4**	**1**

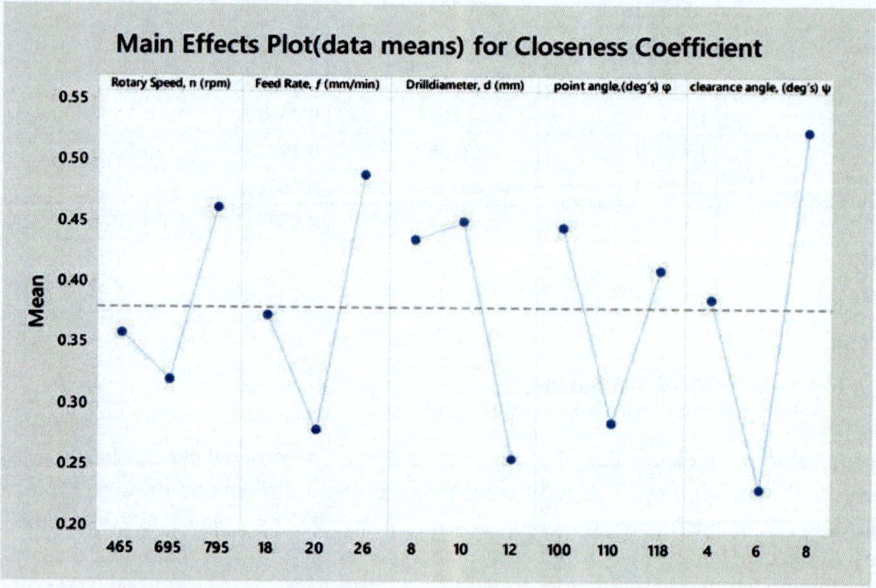

Fig. 2 Main effects' plot for closeness coefficient

5 Conclusions

In the current experimental investigation, the optimal setting of the combination of parameters was found during drilling of AA7075 alloy with HSS-R (DIN 338) twist drill bit, and the following conclusions are drawn.

- An optimal combination of input parameters attained as rotary speed of 795 rpm, feed of 26 mm/min, drill bit diameter of 10 mm, point angle 100°, clearance angle 8° foremost to the value of optimum output responses of burr height 0.164 mm, thrust force 397 N and surface roughness 1.326 μm using TOPSIS.
- From the values of closeness coefficient, the input parameters during drilling with best combination can be arranged in the order using TOPSIS was 14-26-3-4-11-8-13-21-23-24-2-19-27-17-6-15-20-22-25-9-16-10-7-1-12-5.

References

1. Reddy Sreenivasulu R, and Srinivasa Rao C (2019) Review on investigations carried out on burr formation in drilling during 1975 to 2020. Technol Eng 16(1):43–57
2. Bahçe E, Özdemir B (2019) Investigation of the burr formation during the drilling of free-form surfaces in Al 7075 alloy. J Mater Res Technol 8(5):4198–4208
3. Gopal PM, Soorya Prakash K (2018) Minimization of cutting force, temperature and surface roughness through GRA, TOPSIS and Taguchi techniques in end milling of Mg hybrid MMC. Measurement 116:178–192
4. Ali I, Quazi MM, Zalnezhad E et al (2019) Hard anodizing of aerospace AA7075-T6 aluminum alloy for improving surface properties. Trans Indian Inst Met 72:2773–2781
5. Franczyk E, Ślusarczyk Ł, Zębala W (2020) Drilling burr minimization by changing drill geometry. Materials 13:3207
6. Venkatesulu M, Rama Kotaiah K (2018) Optimization of process parameters in drilling of Al6063/B4C composites using AHP-TOPSIS method. J Eng Appl Sci 13:10461–10467
7. Gopal PM (2019) Wire electric discharge machining of silica rich E-waste CRT and BN reinforced hybrid magnesium MMC. Silicon 11:1429–1440
8. Tran Q-P, Nguyen V-N, Huang S-C (2020) Drilling process on CFRP, multi-criteria decision-making with entropy weight using grey-TOPSIS method. Appl Sci 10(20):7207

Mechanical Characterisation of a Peninsular Gneiss Rock Dust Flour Reinforced Aluminium Metal Matrix Composites

B. Prashanth, Subramanya Raghavendra, and N. B. Doddapattar

Abstract The purpose of this study is to evaluate the mechanical wear and dry sliding effects of the Al7068/rocky soil compound with varied weight fractions (2.5, 5, 7.5, and 10%) generated by the liquid metallurgy process. The average particle size of rock dust is 200 microns. Al7068 alloys were cast without stone powder reinforcement as the foundation for the research. These composites' hardness and tensile characteristics were examined. Furthermore, wear characteristics of these composites were investigated with a variety of parameters (constant and varied loads). When the weight percentage of rock dust in composites is increased to 10%, the wear and mechanical properties improve.

Keywords Rock flour · Microstructure · Dry sliding wear

1 Introduction

Peninsular gneiss-type rocks have been found throughout the Indian peninsula, as well as in rural areas of Bangalore. This type of stone is used for decorative purposes such as wall cladding, tiles, masonry bricks. The advantages of this type of stone are easy to cut and dress in shape and size as needed. Hence, it is the most demanding rock in South India. On the other hand, when cutting this type of rock, fine dust particles are produced such as wood dust when cutting wood and fly ash when burning coal. This rock powder is white in colour.

This stone flour cannot be directly used for other purposes, but as a barrier to the soil layer. Soil is made up of biotic things (living things that once lived, such as plants and insects) and abiotic materials (non-living factors such as minerals, water,

B. Prashanth · N. B. Doddapattar
Cambridge Institute of Technology North Campus, Bengaluru, Karnataka, India

S. Raghavendra (✉)
Sai Vidya Institute of Technology, Bengaluru, Karnataka, India
e-mail: rvs.sdly@gmail.com

N. M. Sivaram et al. (eds.), *Advances in Manufacturing, Automation, Design and Energy Technologies*, Lecture Notes in Mechanical Engineering,
https://doi.org/10.1007/978-981-99-1288-9_3

and air). When rock dust accumulates on the surface of the soil, it functions as an obstacle to plant development and has an impact on organism habitat.

Because of their low weight, good thermal conductivity properties, and good castability, aluminium alloys are preferred engineering materials for the aerospace and automotive industries for a broad array of applications. Aluminium 7068 alloy is the most powerful widely available. These are widely used in automobiles and military applications also as well as medical products [1]. Because of their high strength, fracture toughness, and wear resistance, Al-alloy composites have a wide range of applications.

Muthazhagan et al. [2] investigated the mechanical strength boron and graphite reinforced aluminium composites. Stir casting method was used to create Al-MMCs. The composites were made with different volume proportion rates of boron carbide and graphite. Heat treatment was used to improve the mechanical characteristics of the composites that were produced. Microstructure studies were also carried out. The inclusion of graphite to the Al matrix diminishes ductility, hardness, and ultimate tensile strength, but the addition of boron carbide enhances the composites' hardness.

The wear characteristics of hybrid MMCs with SiC, Al_2O_3, and graphite-reinforced aluminium alloy composites were investigated by Kundu et al. [3]. Stir casting manufactured the samples. Ten per cent silicon carbide and five per cent alumina were used to strengthen the Al alloy. The hybrid composites were subjected to dry sliding wear tests, and it was discovered that their wear resistance was superior to that of the Al6061-T6 alloy. The key elements impacting wear rate were load, sliding speed, and sliding distance; hence, these were the variables used for the testing. The wear strength of hybrid MMCs was investigated using a design of experiments' technique.

Nanda Kumar et al. [4] experimentally investigated hybrid aluminium MMCs reinforced with SiC and graphite. The results revealed that hybrid composites outperformed Al6016 alloy in terms of mechanical characteristics.

According to Sachin [5], aluminium and its alloys have weak tribological properties, which might lead to seizure in extreme situations. As a result, there was a tremendous push to discover new materials with superior wear resistance and tribological capabilities, which led to the creation of aluminium metal matrix composites.

Suresh et al. [6] studied the mechanical properties of Al composites, reporting that the addition of ceramic particles improved the alloy's tensile and wear resistance.

Radhika et al. [7] investigated and concluded that incorporating fly ash particles into aluminium alloy has the ability to conserve energy-intensive metal, lowering the cost of aluminium goods while also lowering of the products.

Akhlaghi et al. observed that a consistent layer of solid lubricant forms on the composites' surface during dry sliding. This greasing layer is made up of shearing Gr particles [8]. The shear stress given to the material under the contact region is reduced, and subsurface plastic deformation is reduced. The use of graphite as a solid lubricant, which causes the composite to lose strength, is a limitation of the aluminium graphite composite.

The wear behaviour of Al6061-SiC-Al_2O_3 composites was investigated by Umanath et al. [9]. The results showed that the 15% blended composite had a higher wear resistance than the 5 per cent composite. The volumetric wear loss of the Al6061-9 per cent Al_2O_3-6 per cent graphite composites, according to Nagaral et al. [10], was lower than that of the Al6061 matrix. SEM microanalysis was used to describe the worn surfaces. Development and characterisation like micro structural, mechanical, and tribological properties of Al7068/Gr are investigated.

Many attempts were made in development of aluminium matrix-based hybrid composites to increase the mechanical and thermal properties using silicon carbide, graphite, fly ash, etc., as reinforcement. But, no attempts were found in the literature about the usage of rock dust floor in aluminium. The ever-increasing demand for low-cost reinforcement sparked interest in the use of waste items like as fly ash, rock dust, and other reinforcement.

As a result, an attempt is made in this study to use rock dust as reinforcement material and aluminium as matrix material. The current study looked at the mechanical strength, such as tensile strength, hardness, and wear coefficient of friction, of a rock dust reinforced aluminium metal matrix composite made using the stir casting process.

2 Materials and Methods

The present study requires two materials to act as a matrix and reinforcement. Al7068, a matrix material with a density of 2.850 g/cm^3, was chosen as the matrix material. Tables 1 and 2 show the physical attributes and chemical composition, correspondingly.

The reinforcement is Peninsular Gneiss Rock Flour. The particle size of rock flour is 200 Microns which was measured in sleeve analyser. Density is 3.11 g/cm^3. Colour of rock flour is white. The physical properties are given in Table 3.

Table 1 Chemical properties of Al7068

Element	Weight %
Zn	7.5
Mg	2.5
Cu	2.2
Fe	0.15
Mn	0.10
Ti	0.5
Si	0.12
Others	0.15
Al	Balance

Table 2 Physical composition of Al7068

Properties	Values
Density	2.85 g/cc
Melting point	476–635 °C
Elastic modulus	73.1 GPa
Poisson ratio	0.23

Table 3 Physical properties of rock flour

Element	Weight %
SiC	59.62
Al_2O_3	15.29
MgO	6.54
Ca	6.44
Fe	10.5
K	1.25
S	0.36

3 Fabrication of Composites

Stir castings' experiment was used to develop composite samples. A crucible with a capacity of 3 kg is carried by the electric furnace. The furnace's max working temperature is 1200 °C. Aluminium ingots were heated in crucible and were heated to 800 °C. To eliminate moisture, graphite particles were preheated. Degassing tablet (C2Cl6) was applied after the molten metal had entirely melted to minimise porosity and eliminate gases trapped during melting. The stirrer was rotated at a speed of 600 rpm. During the stirring, the warmed graphite particles were gradually introduced to the molten metal. For a further 15–20 min, the stirring was continued. The liquid was then put into a mould that had been warmed to 350 °C for 45 min to ensure homogeneous solidification.

3.1 Hardness Test

Rockwell hardness test has been carried out according to ASTM E-18 standards; for each sample, an average of five measurements was collected for hardness assessment. The indenter for B scale is a 1/16″ steel ball. A 100 kgf primary load and a 10 kgf minor load are applied.

3.2 Tensile Test

Tension test specimens were prepared as per ASTM E-8 requirements and evaluated at room temperature with universal testing equipment that was connected to a computer. The yield strength and ultimate tensile strength were calculated. The average values for each alloy/condition are based on five test results.

3.3 Wear Test

A Pin-on-Disc machine was used to conduct wear test. In accordance with ASTM-G99 guidelines, a dry slide wear test was conducted using an alloy sample with dimensions of 160 mm in diameter and 8 mm in thickness. The wear rate of the specimens was calculated using the weight loss technique, which involved dividing the weight lost in sliding by the sliding distance. The average of five trials was used to calculate the wear rate.

4 Results and Discussion

4.1 Hardness Test

Figure 1 gives the hardness plot for the stir cast aluminium alloy Al7068 and its composites with different percentages of rock dust reinforcement. The specimen was prepared using 2.5, 5, 7.5, and 10 wt% of rock flour. The results indicate that the maximum hardness is seen for the Al7068/2.5% rock dust composite at an indentation distance of 0.3 mm. It also indicates that the hardness for the same range of indentation distance is highly varied for the 2.5% rock dust reinforced composite than any other composite under study. This could be due to the low percentage and a very scarce distribution of rock dust in the composite. Considering the average variation between the different specimens, it can be said that 5% rock dust in the Al7068 matrix gives the optimum results. The composite with a larger reinforcement percentage has a considerably increased hardness value in the ageing condition, even though its tensile strength may be equal to those with smaller particle fractions.

4.2 Tensile Test

Figure 2 gives the plot for the ultimate tensile strength of the as-cast aluminium alloy Al7068 and its composites with 2.5, 5, 7.5, and 10 wt% rock dust reinforcement.

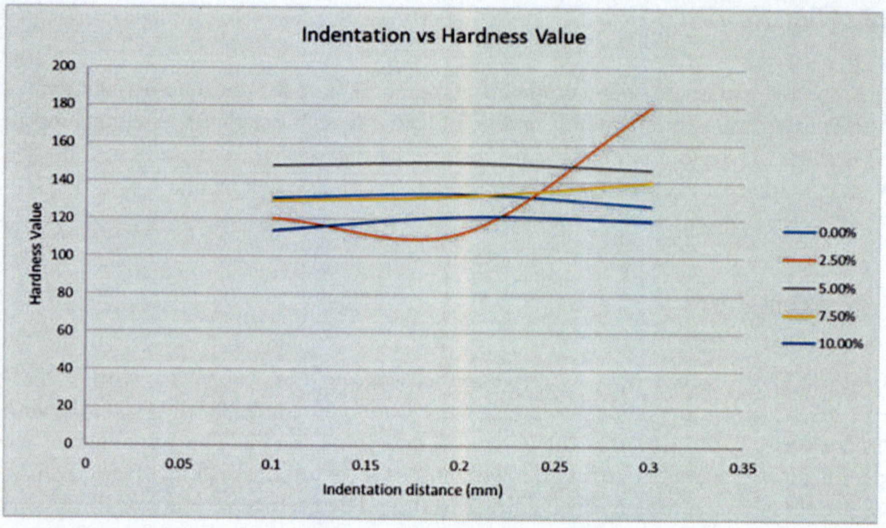

Fig. 1 Hardness values of rock floor reinforced Al7068 alloy composites

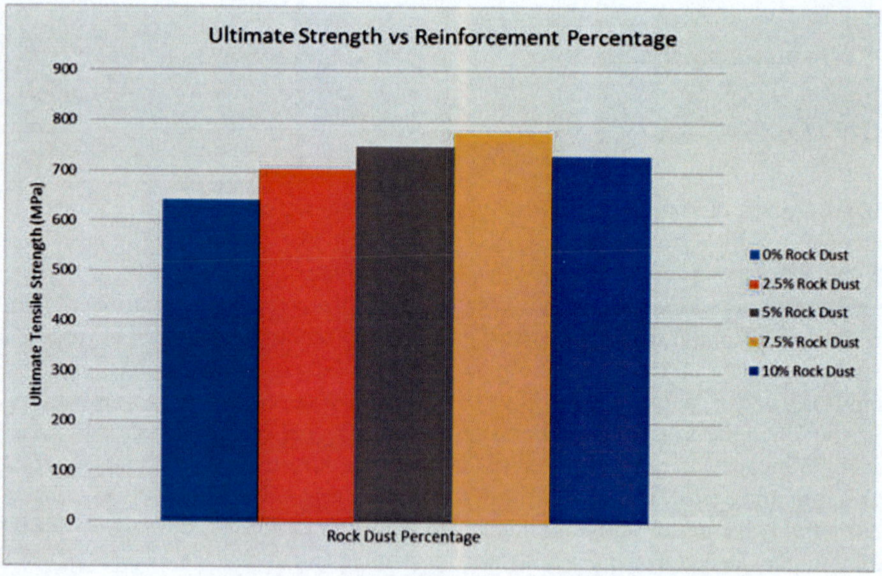

Fig. 2 UTS values of rock floor reinforced Al7068 alloy composites

The plot shows that the composite specimen prepared with rock dust reinforcement improves the strength of the aluminium alloy Al7068. The increase rock dust percentage increases the ultimate tensile strength of the material until 7.5% rock dust but then reduces for the 10% rock dust reinforcement. This could be an indication that the higher concentration of rock dust has a negative effect on the properties, by making it highly porous than the other reinforcement percentages. The change in strength of the specimen is 8.9, 14.5, 17.6, and 12.5%, respectively, for 2.5, 5, 7.5, and 10% rock dust reinforced composites. The strengthening effect of composites is induced by grain refinement at the matrix/reinforcement contact, which improves the tensile strength of composites.

Figure 3 gives the results for the % elongation of the as-cast and the rock dust reinforced composite specimen. The composite specimen prepared has the reinforcement as rock dust in the percentages 2.5, 5, 7.5, and 10. The results indicate that the lowest elongation is seen in the composite with the 2.5% reinforcement with a value of 9.83%. The highest is seen for the 10% rock dust reinforcement at a value of 15%. This means that increasing percentages of rock dust have a detrimental impact on the material's per cent elongation, where the lowest per cent elongation is desired. The presence of dimples, microfractures, and microholes created as a result of brittle fracture are the key variables impacting composite tensile strength. As illustrated in Fig. 3, adding 7.5% rock dust to an aluminium alloy reduces composite elongation.

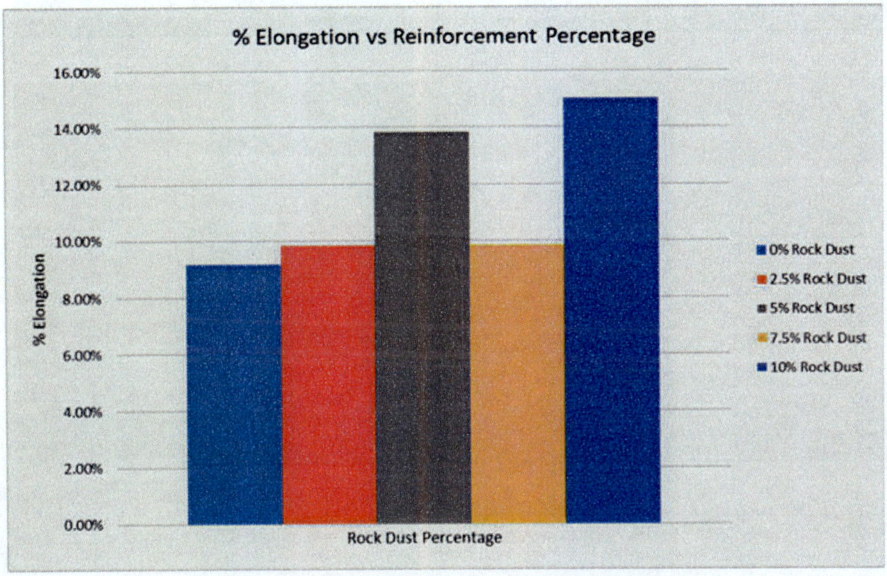

Fig. 3 % elongation values of rock floor reinforced Al7068 alloy composites

4.3 Microstructure

Scanning electron microscope images of prepared composites show the distribution on rock flour particles in aluminium 7068 matrix. Figure 4a–d show the fractured surface of tensile specimen. Surface indicates brittle fracture. The distribution of rock dust flour particles is fairly homogenous as shown in Fig. 4a, b. As the reinforcement percentage of rock flour is increased, clusters are formed in the matrix, as shown in Fig. 4c, d. The voids in the figure indicate ductile type of fracture. Also, cracks are propagated near particles [11].

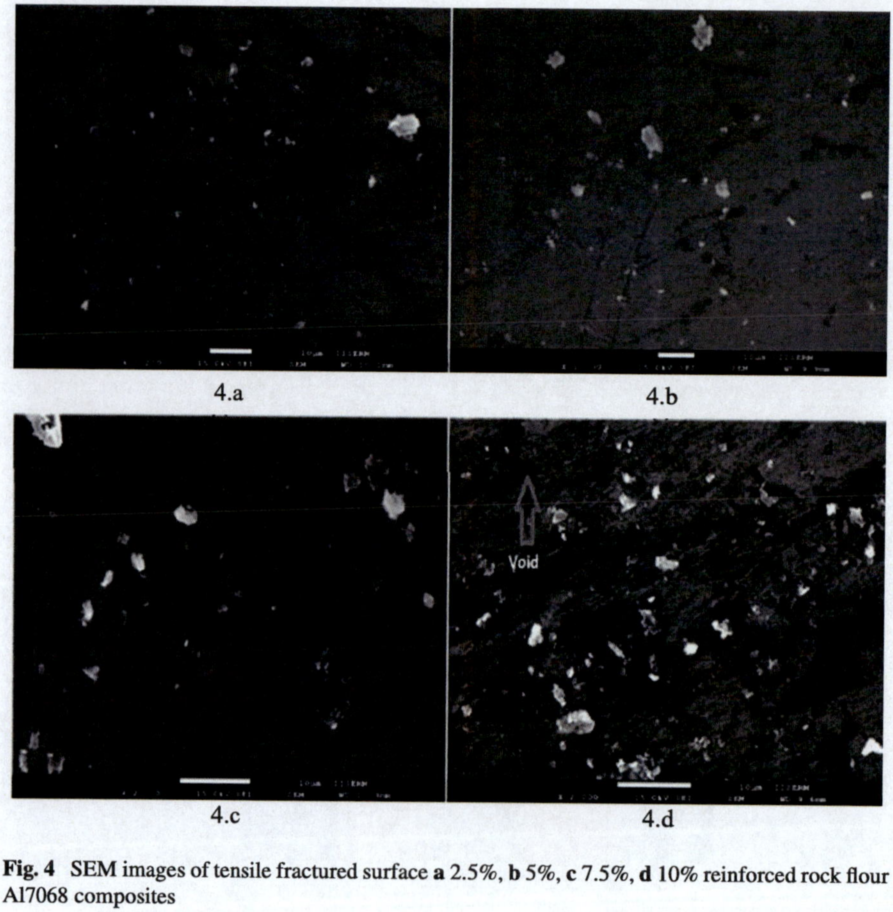

Fig. 4 SEM images of tensile fractured surface **a** 2.5%, **b** 5%, **c** 7.5%, **d** 10% reinforced rock flour Al7068 composites

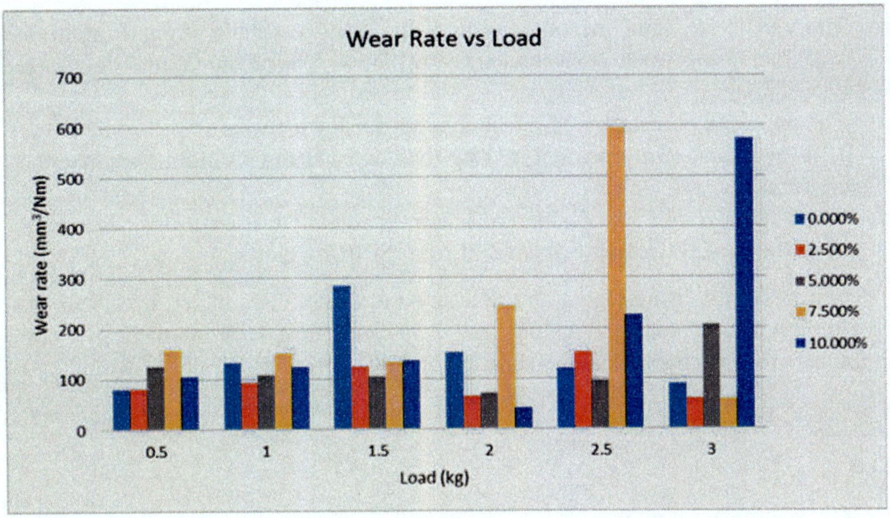

Fig. 5 Effect of load on wear rate of rock floor reinforced Al7068 alloy composites

4.4 Wear Test

Figure 5 shows the plot for the wear test results with variation in load for different percentages of rock dust reinforcement (speed: 2 m/s, load: 10–30 N). The graph indicates that the increase in reinforcement percentage decreased the wear rate. It is also seen that the increase in load has an inverse effect of lower wear rate. At 2.5 kg load, the wear rate increases drastically for the rock dust percentage of 7.5%. This is recreated for the load of 3 kg and 10% load. The lowest wear rate is seen for a load of 2 kg load and 10% rock dust reinforcement. The highest wear rate is seen for a load of 2.5 kg and reinforcement percentage of 7.5%. It is observed that the coefficient of friction varies with samples. In composites, the applied load is carried by the reinforcement, which causes a decrease. The area of contact between the pin and disc surface is covered with wear debris. When a high number of rock dust particles come in contact with disc, they begin to degrade, increasing the coefficient of friction.

5 Conclusions

From the results obtained, we can conclude that

(a) The hardness increased after the incorporation of rock dust. The optimum value of rock dust reinforcement is seen at 5% rock dust reinforcement.

(b) The tensile strength results show that the ultimate tensile strength increases with increase in rock dust percentage till 7.5% reinforcement and then starts decreasing.
(c) The optimum percentage of reinforcement for lowest wear in the composite is 10% rock dust reinforcement at 2 kg load considering all other parameters as constant.
(d) Using stir casting method, the reinforcement rock flour is uniformly distributed in Al7068 matrix which was evident from SEM Images.

Considering the results from all the tests conducted in this paper, it is clear that the aluminium composite of Al7068 and 7.5% rock dust reinforcement gives the optimum results for the mechanical and tribological characteristics of the composite.

References

1. Smith's high performance 7068 data sheet, An aluminium alloy with the strength of steel, pp 1–2 (2007)
2. Muthazhagan C, Gnanavelbabu A, Bhaskar GB, Raj Kumar K (2013) Influence of graphite reinforcement on mechanical properties of aluminium boron carbide composites. Adv Mater Res 845:398–402
3. Kundu S, Roy BK, Mishra AK (2013) Study of dry sliding wear behavior of aluminium/SiC/Al_2O_3/graphite hybrid metal matrix composite using Taguchi technique. Int J Sci Res Publ 3(8)
4. Nanda Kumar N, Kanagaraj P (2014) Study of mechanical properties of aluminium based hybrid metal matrix composites. In: International conference on advances in engineering and management (ICAEM), vol 1, pp 166–172
5. Sahin Y (2005) The prediction of wear resistance model for the metal matrix composites. Wear 258(11–12):1717–1722
6. Suresh S, Sridhara BK (2010) Effect of silicon carbide particulates on wear resistance of graphitic aluminum matrix composites. Mater Des 31:4470–4477
7. Radhika N, Subramanian R, Venkat Prasad S (2011) Tribological behavior of aluminum/alumina/graphite hybrid metal matrix composite using Taguchi's techniques. J Minerals Mater Charact Eng 10(5):427–443
8. Akhlaghi F, Zare-Bidaki A (2009) influence of graphite content on the dry sliding and oil impregnated sliding wear behavior of Al 2024–graphite composites produced by in situ powder metallurgy method. Wear 266:37–45
9. Umanath K, Palanikumar K, Selvamani ST (2013) Analysis of dry sliding wear behavior of Al6061-SiC-Al_2O_3 hybrid metal matrix composites. Composites Part-B 53:159–168
10. Nagaral M Auradi V, Parashivamurthy KI, Kori SA (2015) Wear behavior of Al_2O_3 and graphite particulates reinforced Al6061 alloy hybrid composites. Am J Mater Sci 5(3C):25–29. https://doi.org/10.5923/c.materials.201502.05
11. David Raja Selvam J, Robinson Smart DS, Dinaharan I (2013) Microstructure and some mechanical properties of fly ash particulate reinforced AA6061 aluminum alloy composites prepared by compocasting. Mater Design 49:28–34

Friction Stir Processing of Magnesium Metal Matrix Composites: A Review

Roshan V. Marode and Srinivasa Rao Pedapati

Abstract The world is more cautious about the environment and becomes aligned to eco-friendly, low-emission vehicles that are lightweight and efficient in this modern day. The lightest structural metal available on the earth is magnesium and hence is an excellent alternate material for various engineering applications where the weight of the material is a significant design parameter. Friction Stir Processing (FSP) is a solid-state severe plastic deformation (SPD) process that works on the friction stir welding (FSW) principle. Magnesium metal matrix composites (MMMCs) have gained interest in the automotive, aerospace, biomedical, and power industries. Friction Stir Processing (FSP) has recently gained favor as a method for generating composites in solid state. The review provides a thorough insight into the working operation of FSP, factors affecting tool parameters, different reinforcement particles, and their influence on a range of properties like hardness, tensile strength, corrosion, and tribological properties of Mg alloys. The current paper investigates the progress made in the field of magnesium composites manufactured through FSP since the origination of the process. This review also gives a recommendation for future research in the field of MMMCs.

Keywords Magnesium alloys · Magnesium matrix composites · Friction stir processing · Mechanical properties

1 Introduction

Lightest metals' alloys like magnesium are fetching applications broadly in a variety of disciplines, and because of their high specific strength and stiffness, great vibration absorption, electromagnetic shielding effect, superior machinability, and recyclability, they have been used in the aerospace, automobile, electronics, and biomedical

R. V. Marode (✉) · S. R. Pedapati
Department of Mechanical Engineering, Universiti Teknologi PETRONAS, 32610 Seri Iskandar, Perak, Malaysia
e-mail: roshan_20001046@utp.edu.my

N. M. Sivaram et al. (eds.), *Advances in Manufacturing, Automation, Design and Energy Technologies*, Lecture Notes in Mechanical Engineering,
https://doi.org/10.1007/978-981-99-1288-9_4

industries [1]. As compared with non-ferrous metals, for an example, titanium and aluminum alloys, the mechanical strength of Mg alloys is comparatively lower. At room temperature, crystal structure of magnesium is hexagonal close-packed, owing to which basal slip dominates plastic deformation, whereas prismatic and pyramidal slips are inhibited [2]. Therefore, to enhance the hardness, plasticity, wear resistance, and tensile strength, FSP has been used as a grain refining method. FSP is invented by Mishra in 1999, a unique SPD method [3]. FSP is a thermo-mechanical processing technology affiliated with the principle of FSW. Primarily, the tool with the shoulder and probe rotates into the workpiece. When the probe is fully inserted into the workpiece and the shoulder makes contact to the workpiece's top surface, the rotating tool travels in the required direction. The pin and shoulder work together to cause extreme plastic deformation in thermoplastic materials, producing dynamic recrystallization and considerable grain refinement [3]. The tool rotational speed (TRS) and tool traverse speed (TTS) are two key significant variables in FSP. Both of these variables are responsible for the majority of the heat essential to soften or plasticize metal through friction [4]. Moreover, with extension to the application of FSP in modification of microstructure, recently it is found to be of significant use in producing and fabricating mono and hybrid MMCs. A broad group of innovative engineered materials are MMMCs that have proven to be viable options for extensive structural applications. Adding a sufficient and specific amount of secondary phase particles to pure metals or alloys can improve their specific strength [5].

Several researchers concentrated on the subsequent features of friction stir-treated magnesium alloys: modification of the microstructure and grain size; improving the mechanical, tribological, and corrosion resistance properties of Mg alloys; and fabrication of Mg-based surface composites; Therefore, the focus of this study will be on the evolution of Mg composites processed through the FSP route in the aforementioned areas, as well as new insights into their future developments.

2 Principle of Friction Stir Processing

FSP is a solid state in nature that obeys the basic principle of FSW. FSP is used to improve grain shape and size to develop metal matrix composites with minimal greenhouse gas emissions [6]. It is also known as the green technique which modifies the microstructural qualities, which enhances mechanical properties, wear resistance, and corrosion resistance [7]. It is an innovative technique for fabricating mono and hybrid composites by solid state reinforcing different types of metal particles. FSP comprises a fast-rotating non-consumable tool that plunged into the surface with a profiled/straight probe with a greater shoulder diameter. The tool is given a traverse feed across the surface after the pin is introduced into the surface with axial load.

Figure 1 depicts the friction stir-processed composite mechanism. FSP divides the workpiece into two sides: i. Advancing side (AS) and ii. Retreating side (RS). The advancing side of the workpiece is where the tool's traverse as well as rotation directions are identical, whereas the retreating side is the side where the tool's traverse

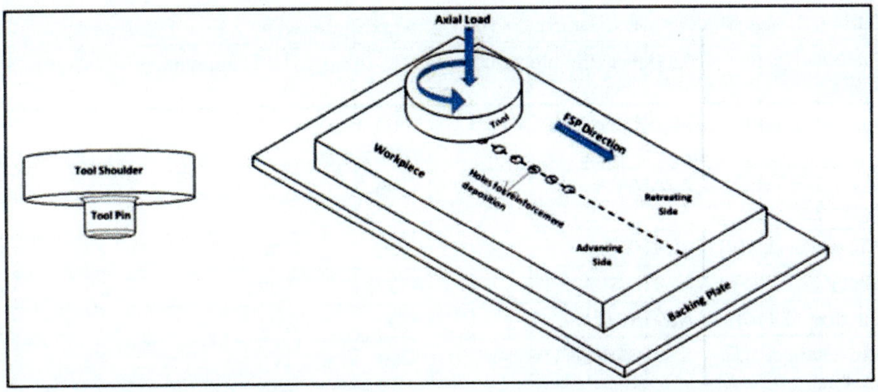

Fig. 1 Setup of friction stir processing using hole method for the fabrication of composite

and rotation directions are distinct. The heat generated due to the friction between tool's shoulder and top surface of workpiece causes plastic deformation and softens the material. The soft material which is heated is then swirled and poured in a circular motion by the probe. The flow of softened material during stirring resembles an extrusion process, filling the hole generated close to a rear side of the FSP tool. As a result, the room or cavity generated by the tool probe is routinely filled as the tool travels ahead due to the flow of softening materials through extrusion. Because the material does not melt, it eliminates the issues that come with typical molten casting [8]. This technology plays a vital role in the exploration of microstructural modification and advancement of metal matrix composites for various arrangements like blind holes, zig-zag holes, or groove methods.

3 FSP Process Parameters and Their Effect

In the field of manufacturing, FSP is considered as an attractive innovation toward microstructure refinement and property enhancement. Achieving refined microstructure and desired property requires the appropriate control of FSP parameters. FSP entails several complicated process factors related to temperature distribution, material movement, tool design, heat generation, mechanical forces, etc. [8]. The quality of the final product in FSP is determined by several factors. The quality of composite manufactured with FSP is generally determined by the settings of the FSP machine factors, the tool type utilized, and the material used. Other elements, like the number of passes and the method of strengthening the workpiece, have a major impact on surface composites. Morisada et al. [9] were the first to start developing the FSPed Mg composites. They varied the TRS and achieved the higher value of 76 Hv hardness of AZ31/MWCNT composites. Ram et al. [10] explored the influence of TRS and TTS on Mg/SiC composites and observed that beyond the TTS of 50 mm/min

Table 1 Range of tool process parameters for Mg composites used by various researchers

Authors	Workpiece/Reinforcement	TRS (rpm)	TTS (mm/min)	TTA (degree)
Ram et al. [10]	Mg/SiC	1000, 1300, and 1600	50	–
Fono-Tamo et al. [13]	AZ61/PKS	1200	20	0
Vijayan et al. [14]	AZ91/TiC	1000	30	–
Shang et al. [15]	AZ91D/SiC	450	23.5	3
Singh et al. [16]	AZ91/B_4C	900	45	–
Bhadouria et al. [17]	AZ91/MWCNT/WC	1200	60	2
Abbasi et al. [18]	AZ91/Al_2O_3	730–1800	14–80	2
Khayyamin et al. [19]	AZ91/SiO_2	1250	20, 40, and 63	3
Asadi et al. [20]	AZ91/SiC and Al_2O_3	900	63	3
Asadi et al. [21]	AZ91/SiC	710–1400	12.5–63	3
Arora et al. [22]	AZ91/TiC+ Al_2O_3	800	40	–
Morisada et al. [23]	AZ31/SiC	1500	25–200	3
Madhusudhan Reddy et al. [24]	ZM21/SiC and B_4C	1200	50	–

tunnel defects were going to form. Hence, for pure Mg with mono reinforcement, it was advice TTS below 50 mm/min. Tool tilt angle (TTA) is also the key parameter in obtaining the defect-free Mg composites via FSP; Vigneshkumar et al. [11] varied the tilt in the range of 0°–4° and exhibited the higher hardness value of 125 Hv in ZK60/SiC for 2° tilt. Recently, Dinaharan et al. [12] carried out the FSP on AZ31/Ti by altering the TTS and number of passes. They observed a significant increase in UTS and YS of 283 MPa and 193 MPa, respectively, after 5 passes and at 30 mm/min. Moreover, in the following sub-section, various FSP process factors are discussed with relation to properties. The range of process parameters employed by various researchers is illustrated in Table 1.

3.1 Microstructure Evolution

Friction stir-processed material mostly consists of three types of zones, stir zone (SZ) also known as processed zone (also known as stir zone), thermo-mechanically affected zone (TMAZ), and heat-affected zone (HAZ). Beyond these three zones, there is an unaffected zone known as base material (BM) as displayed in Fig. 2. Severe heat generated by friction in between the workpiece and tool helps in the material

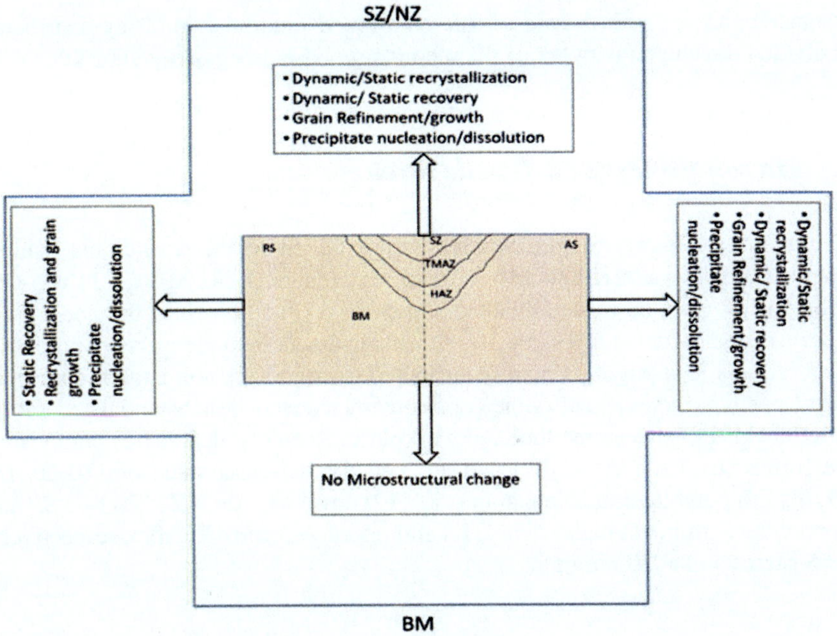

Fig. 2 Microstructural evolution in different zones of FSP

movement, and high strain distortion occurs to have in the SZ, which aids in dynamic recrystallization too. Owing to these, high strain, and dynamic recrystallization, and an equiaxed fine-refined grain structure occurs after FSP. Furthermore, the width of all zones is principally determined by the material qualities and the amount of heat generated by the tool and workpiece [25]. The effect of different pin profiles on microstructure and grain size has been studied by Patle et al. [26] and found that most refined grains of 4.08 μm were obtained for simple taper as compared to the threaded and square taper pins. Moreover, they observed the reduction in grain size of AZ91/FA composites from 166.5 μm ± 8.75 μm to 8.775 μm ± 1.3 μm after the completion of FSP [27]. When nano-SiC particles are added to the microstructure, it becomes more uniform than the micro-SiC particles which are added [20].

During the fabrication of Mg composites, the grain size of RZ5 was reduced in the range of 0.8–1.87 μm by Vedabouriswaran and Aravindan [28]. Asadi et al. [21] investigated the size of grains by varying the reinforcement particle size and noticed that the average grain size was in the range of 5 μm for 30 nm. Madhusudhan Reddy et al. [24] used a hole filling and closing approach using FSP to insert SiC and B_4C particles on a surface of ZM21 magnesium alloy. The grain size was observed to be reduced from 40 to 20 μm after FSP. Additionally, the SiC and B_4C particles were also evenly distributed, which aided in grain reduction. However, when compared to previous FSPed Mg alloys, the authors' claimed that amount of grain refinement is lower. Liu et al. [29] applied the FSP on AZ91 mg alloy and enhanced the corrosion resistance behavior with the aid of microstructural modification. They distributed and divided the cluster network of the secondary phase β-$Mg_{17}Al_{12}$ via FSP.

3.2 *Microhardness and Tensile Strength*

FSP modifies the microstructure which in turn obtained the incremental mechanical properties and the was highlighted by several researchers. AZ91/MWCNT+WC-Co-Cr composites were fabricated by Neharika et al. [17] and obtained the finest grain of 1.85 μm and hardness of 118.05 Hv. An enhancement in the property of hardness, as well as tensile strength, is mainly attributed owing to the non-agglomeration of reinforcement particles and ultrafine grains obtained due to the substantial generation of frictional heat between the tool and workpiece. Recently, defect-free composites were formed by Patle et al. [27] and attained the hardness value of 110 Hv for AZ91/fly ash combination. Dinaharan et al. [12] developed the AZ31/Ti by FSP and observed the significant increase in UTS and YS of 283 and 193 MPa, respectively, after 5 passes and at 30 mm/min.

3.3 *Superplasticity*

Dynamic recrystallization occurring during the stirring stage of FSP aids in grain reduction and texture modification. FSP attributed well in shattering up the intermetallics and precipitates with the identical division in the matrix. Superplasticity refers to a material's capacity to achieve maximum ductility and very high uniform elongation. Raja et al. [30] examined the effect of multipass FSP on AZ91, and the highest elongation of 680% was obtained with fine grain structure at the strain rate of 5×10^{-4}/s. They revealed that the dramatic change in the superplasticity is responsible, grain boundary sliding is accommodated by grain boundary migration and grain rotation, and the same was verified in microstructure and texture studies. Owing to the finer grains, high angle grain boundaries, and reasonable fractions of β-particles of FSPed GW103 alloy, the highest superplasticity of 1110% was achieved at 1×10^{-3}/s; 415 °C [31]. Harwani et al. critically reviewed the phenomena of superplasticity in Mg alloys and recommended that substantial exploitation in the field of superplastic behavior of commercial mg grades should be a future research topic to investigate [32].

3.4 *Wear and Corrosion Resistance*

FSP not only improves the mechanical strength but also improves the wear and corrosion by modifying the microstructure with the optimal parameters. In the study of

Patle et al. [27], AZ91/fly ash gives bad corrosion resistance as compared to base material AZ91; hence, fly ash composites are not favorable in corroding environments. Adetunla and Akinlabi [33] investigated the effect of various reinforcements like fly ash, palm kernel shell ash, Ti6Al4V, and SS304 on corrosion resistance. They found that out of four reinforcements, SS304 was showing high corrosion resistance 0.2 mm/yr with a high Vickers hardness value of 95 Hv. Madhusudhan Reddy et al. [24] produced ZM21/SiC and B_4C individually by hole method and compared the wear resistance for both SiC and B_4C. It was found that the lowest wear resistance with ZM21/SiC and higher value of hardness with ZM21//B_4C. The comprehensive analysis of various researchers for Mg alloys with reinforcement is depicted in Table 2.

4 Summary and Future Prospect

The approach of modifying microstructure and fabricating Mg composites through FSP has acquired much interest for improving various properties successfully. FSP's ability to encapsulate desired reinforcing particles into Mg alloys produced below melting point temperature has been proven. After reviewing the literature on the creation of FSPed Mg alloys with and without reinforcements, it is clear that the fabrication process is complicated because of the numerous process parameters involved. The selection of optimal process parameters plays a crucial role in determining the end properties of processed samples. Numerous obstacles in this sector can be addressed in future research efforts. To obtain defect-free homogeneous composites, a general relationship model between several specified tool parameters and deposition technique parameters, as well as between FSP machine parameters and deposition method parameters, must be defined. Although the FSP technique is maturing, its thermo-mechanical study must be researched, and particular benchmarking of tool design, reinforcement doping strategy, and process parameters are required to reproduce sound and defect-free mono and hybrid composites materials. There is a group of reinforcing particles obtained from the waste of industrial and agricultural, synthetic ceramic, and other metallic or non-metallic components that have yet to be studied for generating hybrid composites using FSP on various Mg alloys.

Table 2 Reported literature on Mg-based composites fabricated using FSP

Authors	BM	Deposition method and reinforcement particles	Varying parameters	Scope of study	Findings
Morisada et al. [9]	AZ31	Groove—MWCNT	• TTS	• FSP with and without MWCNTs • Microstructure analysis and hardness	• Reduced grain size in FSP/MWCNT • Higher value of hardness (78 Hv)
Sharma et al. [34]	AZ31	Zig-zag holes—MWCNT and Gr	• TRS • No. of passes	• Mechanical behavior • Morphological analysis	• Improved properties were observed at 1400 rpm • Improved hardness by 19.72% • Property enhancement verified with morphological and fracture analysis
Dinaharan et al. [12]	AZ31	Groove—Pure Ti particulate	• TTS • No. of passes • Volume fraction	• Microstructural evolution • Mechanical properties	• At 30 mm/min TTS, 5 Passes, and 21% vol, the highest UTS was observed about 283 MPa and YS 193 MPa • Different zones within SZ were seen for Ti particulates
Adetunla and Akinlabi [33]	AZ31	Groove—FA, palm kernel shell ash, Ti6Al4V, and SS304 (individually 20% vol)	• Immersion time for corrosion	• Corrosion resistance • Mechanical properties	• SS304 showed high corrosion resistance 0.2020 mm/yr (30 days) • SS304 too had high hardness 95 Hv, YS—105 MPa, and UTS—450 MPa (3 passes) with 0.26 lowest friction coefficient

(continued)

Table 2 (continued)

Authors	BM	Deposition method and reinforcement particles	Varying parameters	Scope of study	Findings
Vedabouriswaran and Aravindan [28]	RZ5	Groove—B_4C, MWCNT, ZrO_2 + Al_2O_3	• Reinforcement particles	• Microstructural investigation • Hardness	• Average grain size of 0.8–1.87 μm • Strength in between 250 and 320 MPa
Madhusudhan Reddy et al. [24]	ZM21	Holes—SiC and B_4C	• Reinforcement particles	• Metallography • Wear and hardness	• B_4C possessed high hardness and wear resistance than SiC
Vigneshkumar et al. [11]	ZK60	Holes—SiC	• TTA (0°–4°)	• Grain refinement • Particle distribution	• 2° tool tilt angle exhibited high hardness (125 Hv) • Grained size reduced and homogeneous particle distribution
Singh et al. [16]	AZ91	Holes—B_4C	• Particle size	• Microstructure • Hardness • Wear behavior	• Mass loss 0.01895 mg—particle size 50 μm • ~125 Hv—50 μm at the vertical surface below 50 μm • Significant grain refinement
Patle et al. [27]	AZ91	Groove-fly ash	–	• Wear and corrosion resistance • Hardness • Microstructure analysis	• 110 Hv for composite with 1.3 μm • Bad corrosion resistance as compared to BM; hence, composites are not favorable in corroding environment • 5.9 mg and 0.38-mass loss and COF for composites

(continued)

Table 2 (continued)

Authors	BM	Deposition method and reinforcement particles	Varying parameters	Scope of study	Findings
Bhadouria et al. [17]	AZ91D	Groove-MWCNT+WC-Co-Cr	• Mono and Hybrid • % vol fraction and 7 Passes	• Microstructure • Microhardness • Tribological property	• 1.85 μm grain size • 118.05 Hv in hybrid composites • Minimum wear in hybrid case (0.45×10^{-3} mm^3/m)

BM: Base Material; SZ-Stir Zone, YS: Yield Strength and UTS: Ultimate Tensile Strength

References

1. Mordike BL, Ebert T (2001) Magnesium properties—applications—potential. Mater Sci Eng A 302(1):37–45
2. Wang W et al (2020) Friction stir processing of magnesium alloys: a review. Acta Metall Sin (English Lett) 33(1):43–57
3. Mishra RS, Ma ZY (2005) Friction stir welding and processing. Mater Sci Eng R Rep 50(1–2):1–78
4. Węglowski MS (2018) Friction stir processing—state of the art. Arch Civ Mech Eng 18(1):114–129
5. Patil NA, Pedapati SR, Bin Mamat O (2020) A review on aluminium hybrid surface composite fabrication using friction stir processing. Arch Metall Mater 65(1):441–457
6. Srinivasan C, Karunanithi M (2015) Fabrication of surface level Cu/SiCp nanocomposites by friction stir processing route. J Nanotechnol, vol. 2015
7. Sahraeinejad S, Izadi H, Haghshenas M, Gerlich AP (2015) Fabrication of metal matrix composites by friction stir processing with different particles and processing parameters. Mater Sci Eng A 626:505–513
8. Srivastava AK et al (2021) 20th century uninterrupted growth in friction stir processing of lightweight composites and alloys. Mater Chem Phys 266(January)
9. Morisada Y, Fujii H, Nagaoka T, Fukusumi M (2006) MWCNTs/AZ31 surface composites fabricated by friction stir processing. Mater Sci Eng A 419(1–2):344–348
10. Ram B, Deepak D, Bala N (2019) Microstructural refinement and enhancement in mechanical properties of magnesium/SiC as-cast composites via friction stir processing route. Trans Indian Inst Met
11. Vigneshkumar M, Padmanaban G, Balasubramanian V (2019) Influence of tool tilt angle on the formation of friction stir processing zone in cast magnesium alloy ZK60/SiCp surface composites. Metallogr Microstruct Anal 8(1):58–66
12. Dinaharan I, Zhang S, Chen G, Shi Q (2020) Development of titanium particulate reinforced AZ31 magnesium matrix composites via friction stir processing. J Alloys Compd 820(November)
13. Fono-Tamo RS, Akinlabi ET, Tien-Chien J (2019) Experimental investigation of friction coefficient of magnesium alloy developed through friction stir processing with PKS ash powder particles R. Magnes Technol 95–99
14. Vijayan S, Gnanavel JPL, Selvakumar G, Rao SRK (2019) Study on surface characteristics of friction stir processed AZ91 with titanium carbide micro particles. Indian J Eng Mater Sci 26(3–4):205–210
15. Shang J, Ke L, Liu F, Lv F, Xing L (2019) Aging behavior of nano SiC particles reinforced AZ91D composite fabricated via friction stir processing. J Alloys Compd 797:1240–1248
16. Singh N, Singh J, Singh B, Singh N (2018) Wear behavior of B4C reinforced AZ91 matrix composite fabricated by FSP. Mater Today Proc 5(9):19976–19984
17. Bhadouria N, Kumar P, Thakur L, Dixit S, Arora N (2017) A study on micro-hardness and tribological behaviour of nano-WC–Co–Cr/multi-walled carbon nanotubes reinforced AZ91D magnesium matrix surface composites. Trans Indian Inst Met 70(9):2477–2483
18. Abbasi M, Bagheri B, Dadaei M, Omidvar HR, Rezaei M (2015) The effect of FSP on mechanical, tribological, and corrosion behavior of composite layer developed on magnesium AZ91 alloy surface. Int J Adv Manuf Technol 77(9–12):2051–2058
19. Khayyamin D, Mostafapour A, Keshmiri R (2013) The effect of process parameters on microstructural characteristics of AZ91/SiO_2 composite fabricated by FSP. Mater Sci Eng A 559:217–221
20. Asadi P, Faraji G, Masoumi A, Givi MKB (2011) Experimental investigation of magnesium-base nanocomposite produced by friction stir processing: effects of particle types and number of friction stir processing passes. Metall Mater Trans A Phys Metall Mater Sci 42(9):2820–2832

21. Asadi P, Givi MKB, Abrinia K, Taherishargh M, Salekrostam R (2011) Effects of SiC particle size and process parameters on the microstructure and hardness of AZ91/SiC composite layer fabricated by FSP. J Mater Eng Perform 20(9):1554–1562
22. Arora HS, Singh H, Dhindaw BK, Grewal HS (2012) Some investigations on friction stir processed zone of AZ91 alloy. Trans Indian Inst Met 65(6):735–739
23. Morisada Y, Fujii H, Nagaoka T, Fukusumi M (2006) Effect of friction stir processing with SiC particles on microstructure and hardness of AZ31. Mater Sci Eng A 433(1–2):50–54
24. Madhusudhan Reddy G, Sambasiva Rao A, Srinivasa Rao K (2013) Friction stir processing for enhancement of wear resistance of ZM21 magnesium alloy. Trans Indian Inst Met 66(1):13–24
25. Ma ZY (2008) Friction stir processing technology: a review. Metall Mater Trans A Phys Metall Mater Sci 39 A(3):642–658
26. Patle H, Dumpala R, Sunil BR (2018) Machining characteristics and corrosion behavior of grain refined AZ91 Mg alloy produced by friction stir processing: role of tool pin profile. Trans Indian Inst Met 71(4):951–959
27. Patle H, Sunil BR, Dumpala R (2021) Machining characteristics, wear and corrosion behavior of AZ91 magnesium alloy—fly ash composites produced by friction stir processing. Materwiss Werksttech 52(1):88–99
28. Vedabouriswaran G, Aravindan S (2018) Development and characterization studies on magnesium alloy (RZ 5) surface metal matrix composites through friction stir processing. J Magnes Alloy 6(2):145–163
29. Liu Q et al (2018) Enhanced corrosion resistance of AZ91 magnesium alloy through refinement and homogenization of surface microstructure by friction stir processing. Corros Sci 138(2010):284–296
30. Raja A, Biswas P, Pancholi V (2018) Effect of layered microstructure on the superplasticity of friction stir processed AZ91 magnesium alloy. Mater Sci Eng A 725:492–502
31. Yang Q, Xiao BL, Ma ZY (2013) Enhanced superplasticity in friction stir processed Mg-Gd-Y-Zr alloy. J Alloys Compd 551:61–66
32. Harwani D, Badheka V, Patel V, Li W, Andersson J (2021) Developing superplasticity in magnesium alloys with the help of friction stir processing and its variants—a review. J Mater Res Technol 12:2055–2075
33. Adetunla A, Akinlabi E (2018) Influence of reinforcements in friction stir processed magnesium alloys: insight in medical applications. IOP Conf Ser Mater Sci Eng
34. Sharma S, Handa A, Singh SS, Verma D (2019) Influence of tool rotation speeds on mechanical and morphological properties of friction stir processed nano hybrid composite of MWCNT-Graphene-AZ31 magnesium. J Magnes Alloy 7(3):487–500

Modeling of Volumetric Shrinkage of Nylon Parts Fabricated by 3D Printing Process

Shakti Shekhar Puri, Faheem Faroze, Sourabh Sharma, Vineet Srivastava, and Arun Tom Mathew

Abstract Three-dimensional printing (3DP) process is a very efficient manufacturing technique widely used for improving the design quality. However, 3DP process is not free from defects. Most common defect observed in 3D-printed parts is dimensional inaccuracy. One of the methods to evaluate shrinkage and accuracy of part is through the estimation of volumetric shrinkage. Here, an effort has been made to model the volumetric shrinkage of nylon parts fabricated using 3D printing. Layer height, extruder temperature, extrusion width and printing speed were the parameters selected. Thirty-one experiments have been conducted in this study. A quadratic model has been developed to estimate the volumetric shrinkage. It has been found that layer height, extruder temperature and extrusion width have major significance on the volumetric shrinkage. It was noted that volumetric shrinkage soars with increase in extrusion width and decreases with increase in layer height, extruder temperature and printing speed. The model has been validated and optimum process parameters have been identified. The effect of part orientation on volumetric shrinkage has been also analyzed, and 0° orientation gave least volumetric shrinkage.

Keywords Volumetric shrinkage · Nylon · Layer height · Extruder temperature

1 Introduction

In this competitive market, to become successful, it is mandatory to develop the new and existing product in a short interval of time to gain the advantage over the competitors. With the development of computing facilities, new technology has been developed which can begin to convert the entire production region through steeply

S. S. Puri · F. Faroze · V. Srivastava (✉)
Mechanical Engineering Department, Thapar Institute of Engineering and Technology, Patiala, India
e-mail: vineet.srivastava@thapar.edu

S. Sharma · A. T. Mathew
School of Mechanical Engineering, Vellore Institute of Technology, Vellore, India

N. M. Sivaram et al. (eds.), *Advances in Manufacturing, Automation, Design and Energy Technologies*, Lecture Notes in Mechanical Engineering,
https://doi.org/10.1007/978-981-99-1288-9_5

lowering the time and sources essential for devising, checking out and producing novel creations [1]. One such technical improvement, which targets generating speedy, correct and cost-effective viable elements, is Additive Manufacturing (AM). Three-dimensional printing is one of the forms developed to implement the concept of AM. Fused Filament modeling (FFM) is one of the most popular 3D printing techniques based on Fused Deposition Modeling (FDM) process [2]. FFM produces a three-dimensional entity by dropping a uninterrupted and steady run of liquefied work material, which is cautiously set into a layer, kicking off from the base, but still some of the problems are present like strength, dimensional accuracy, surface roughness, build time and curling of parts.

There have been quite a few efforts made to build up a relationship for dimensional accuracy of RP piece produced by diverse processes. Raghunath [3] examined the connection among shrinkage and the different cycle boundaries for SLS by utilizing Taguchi strategy. They established the connection amid the shrinkage and different cycle boundaries to be specific laser power, beam speed, hatch spacing, part bed temperature and scan length. Exact relationship for anticipating shrinkage along X, Y and Z headings was inferred utilizing relapse. Gotten results were approved and found in great concurrence with tests. Contextual investigation of seat checking part was likewise introduced to show that shrinkage model created parts that are more exact. Sood [4] contemplated the dimensional exactness of FDM-handled ABS parts considering part direction, layer height, raster point, raster width and air hole as cycle boundaries. Gray Taguchi technique is utilized to show up at a solitary connection for ideal factor settings of interaction boundaries for limiting shrinkage along length, width and height simultaneously. Mago [5] stated that the part fabrication method is dominated by various diverse factors. The choice of the diverse factors which affected the shrinkage of PLA parts was ascertained by TOPSIS. A numerical representation was evolved for the dimensional inaccuracy. It was noticed that dimensional inaccuracy amplified with addition in part length and trimmed down with rise in layer thickness and printing speed.

Schmutzler [6] considered the different imperfections created in parts manufactured by 3D printing like curling, shrinkage, shape disfigurement and so forth. They built up a numerical model to repay these imperfections depended on the information gathered. At last, they proposed to change the shape and size of the part while cutting to decrease post-manufacture abandons. Boschetto [7] proposed a mathematical model of the filament, subordinate upon the testimony point and layer height, to foresee the realistic part measurements. The model has been approved by a test crusade. The outcomes featured that this detailing can be helpful in FDM modern application. In addition, this plan was discovered to be valuable in choosing the appropriate interaction fabricating procedures since it permitted to tailor measure boundaries with the intend to upgrade expenses, time. Lee [8] investigated the merit of the printed polylactide (PLA) prototype, counting the dimensions and mechanical assets for diverse chilling air velocities. They established that the chilling air velocity had dissimilar impact on the dimensional value and mechanical strength of the printed replica.

Table 1 Physical properties of nylon

Property	Value
Density	1.12 g/cm^3
Melting temperature	217 °C
Ultimate strength	85 MPa
Coefficient of thermal expansion	95 μm/m-°C

In 3D printing, nylon is new material as compared to PLA, ABS, HIPS, PC and PET. Nylon is most appropriate to modern and designing applications, for example, for assembling elite machine parts, instruments and hinges. Nylon is an amazingly impressive and solid stuff for 3D printing as it is abrasion-resistant, heat safe and has a low friction coefficient. It is flexible when parts are thin and strong when part is thick and offers less dimensional error. With respect to the above-mentioned points, nylon has been selected as the work material to investigate volumetric shrinkage in this study. The physical properties of nylon have been provided in Table 1.

2 Planning of Experiments

In current work, the endeavor was to investigate how volumetric shrinkage is being impacted by diverse factors. From the evaluation of previous work [2–6], it was established that layer thickness, extruder temperature, extrusion width and printing speed have diverse outcome on the volumetric shrinkage. The trials were carried out maintaining these operational parameters at different ranks. The parts were developed on Creator Pro 3D printer (Flashforge). Based on the evaluation of previous work and competence of 3D printing system, the array for each of the operational parameters was chosen and is given in Table 2. Using CCRD method of design of experiments, 31 experiments have been carried out and studied based on these process parameters.

CAD model of the specimen was made in Creo 2.0, with the dimensions shown in Fig. 1a. They were then changed over into STL files. These STL files were moved to Simplify3D programming for slicing of the CAD model and make Gcode for tool path. The fabricated specimen developed after 3D printing is shown in Fig. 1b. A coordinate measuring machine was used to quantify the length, width and height measurements of the model. Then, shrinkage along length, width and height is obtained by deducting the exact prototype measurement from the initial dimension (Eqs. 1–3).

Table 2 Range of process parameters

Parameters	Range
Layer thickness (LT) (mm)	0.1, 0.2, 0.3
Extruder temperature (ET) (°C)	235, 240, 245
Extrusion width (EW) (mm)	0.32, 0.40, 0.48
Printing speed (PS) (mm/min)	2600, 3000, 3400

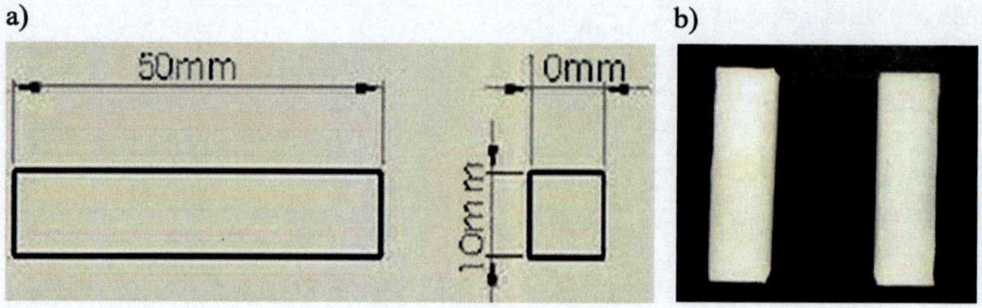

Fig. 1 **a** Dimensions of the specimen to be fabricated. **b** The fabricated specimen used

Then, the shrinkage across length, width and height is multiplied to obtain the volumetric shrinkage. The outcome is given as Eq. 4. The calculated answer resulting from each experiment is given in Fig. 2a.

$$\Delta l = \text{Original length} - \text{Measured length}, \tag{1}$$

$$\Delta w = \text{Original width} - \text{Measured width}, \tag{2}$$

$$\Delta h = \text{Original height} - \text{Measured height}, \tag{3}$$

$$\text{Volumetric Shrinkage} = \Delta l \times \Delta w \times \Delta h. \tag{4}$$

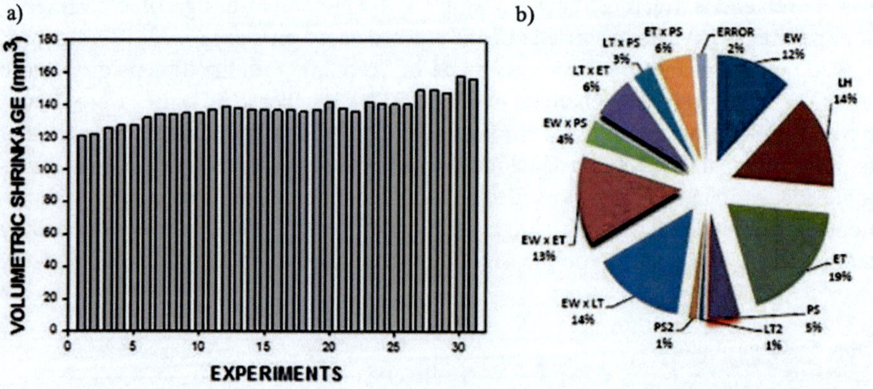

Fig. 2 **a** Histogram of the measured volumetric shrinkage. **b** Percentage contribution of the factors

3 Planning of Experiments

A representation for the volumetric shrinkage was evolved, by showing a relationship with the input considerations, namely layer thickness, extruder temperature, extrusion width and printing speed. Built on investigation of the data presented in Fig. 2a, the relationship is given below as Eq. (5).

$$\begin{aligned}\text{Volumetric Shrinkage} = {} & 202 + (1831 \times LT) + (0.416 \times ET) + (2497 \times EW) \\ & - (0.453 \times PS) - \left(211 \times LT^2\right) + \left(0.000018 \times PS^2\right) \\ & - (5.86 \times LT \times ET) - (572 \times LT \times EW) - (0.0504 \times LT \times PS) \\ & - (10.6 \times ET \times EW) + (0.00135 \times ET \times PS) + (0.0691 \times EW \times PS).\end{aligned} \tag{5}$$

The relationship sufficiency checking contains test for importance of the regression model, analysis for importance of relationship coefficients and test for deficiency of fit. ANOVA is performed for the above and is given in Table 3. The fit outline suggests that the quadratic model for volumetric shrinkage is numerically sufficient and the deficiency of fit is insignificant. The estimate of R^2 is 98.7% which demonstrates that regression model offers sturdy relationship amid discrete parameters and the response and offers excellent explanation of the connection between the discrete variables and the response. The computed F-estimate for the relationship is 87.19. The measured F-value is bigger than the assessment of $F_{0.05,14,16}$, that is 2.37, for a importance rank of $\alpha = 0.05$. This shows that the relationship is sufficient for 95% conviction level.

Proportion offerings for each parameter of the relationship are exhibited in Fig. 2b. The figure shows that extrusion temperature, layer thickness and extrusion width are the main prominent factors influencing volumetric shrinkage. The extrusion temperature is established to be the largest part swaying the volumetric shrinkage with contribution of 19% which is trailed by layer thickness and extrusion width having contribution of 14 and 12%, respectively. It is also observed from Fig. 2b that the

Table 3 ANOVA table for volumetric shrinkage

Source	DF	Seq. SS	Adj. MS	F-value	P value	R^2
Regression	14	2117.76	151.27	87.19	<0.0001	98.7
Linear	4	1065.81	130.45			
Square	4	34.03	8.51			
Interactions	6	1017.92	169.65			
Residual errors	16	27.76	1.735			
Lack of fit	10	22.26	2.23	2.43	0.144	
Pure Error	6	5.49	0.92			
Total	30	2145.52				

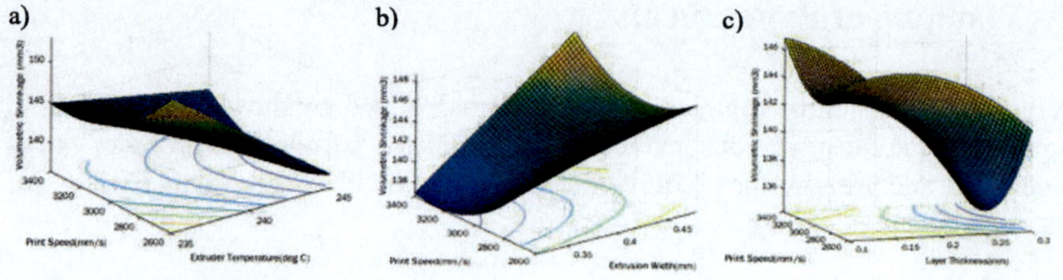

Fig. 3 Main effects for volumetric shrinkage

printing speed does not have major influence on volumetric shrinkage. Figure 3 demonstrates the key effect scheme for volumetric shrinkage. The three points in the each plot are estimated hinged on the mean of calculation of the variable enclosing the exact dealing condition. It can be noticed that volumetric shrinkage increases with raise in extrusion width. It is also seen that it reduces with raise in layer thickness and extrusion temperature.

It can be observed from Fig. 3a that the volumetric shrinkage has decreased with increase in extrusion temperature. When extrusion temperature increases, the interlayer bonding between layers also increases. An elevated extrusion temperature helps union at the crossing point between the dumping strand and the earlier dumped strand. Additionally, nylon is not as much thicker at an elevated warmth; hence, union is superior amid merged strands due to its advanced volatility [9]. Concurrently, the enhanced flow of nylon in effect reduces the extrusion opposition, conveying the volumetric nylon raw material in the course of the conduit of the extrusion head, which improves the level of union amid the filaments and reduces the volumetric shrinkage of the part.

It can be observed from Fig. 3b that as the extrusion width is amplified, volumetric shrinkage rises. This is due to extra quantity of material being dropped with raise in extrusion width resulting in increase in internal stresses. The second reason is the presence of more air voids in the part and therefore additional shrinkage. Further, it was seen that the small extrusion width across with lesser volume of material would initiate lesser heat into the part, in a predetermined time, which contrasted with higher extrusion width. Thus, the volumetric shrinkage will be more in parts made by higher extrusion width across when contrasted with parts ready by lower extrusion width.

It can be observed from Fig. 3c that the volumetric shrinkage has decreased with increase in layer height. This is due to the elevated quantity of material being dumped in the thick layers as compared to thin layer. Extra matter setting for the identical product would lead to elevated density in the completed item thereby reducing the volumetric shrinkage.

Table 4 Confirmation experiments

S. No.	LT	ET	EW	PS	Observed volumetric shrinkage	Predicted volumetric shrinkage	% error
1	0.3	245	0.48	2600	121.26	123.91	2.18
2	0.3	235	0.32	3400	128.00	134.25	4.48
3	0.2	240	0.40	3000	136.88	141.44	3.33
4	0.1	245	0.48	2600	141.38	142.84	1.04
5	0.2	240	0.48	3000	141.90	145.11	2.26
6	0.3	235	0.32	2600	149.20	150.86	1.11
7	0.1	235	0.48	2600	157.94	160.33	1.51
8	0.2	240	0.40	3000	135.90	141.44	4.08
9	0.3	235	0.48	2600	149.77	153.10	2.23
10	0.2	240	0.40	2600	140.47	145.69	3.72

4 Confirmation of the Model

The experiment conducted in the study was tested for the model developed for volumetric shrinkage given by Eq. 5. The percentage variation of volume between the printed model and the predicted model has been analyzed and reported in Table 4. As the predicted model was for 95% confidence interval, 5% threshold value for error was selected. Table 4 shows that all the error values for experimental plan lie within the threshold value with maximum deviation of 4.48% in experiment no. 2.

5 Optimization of the Model

The problem of volumetric shrinkage minimization is formulated and is given below.

$$\text{Minimize (Volumetric Shrinkage)}$$

Subjected to:

$$0.32 \leq \text{Extruding Width (mm)} \leq 0.48;\ 0.1 \leq \text{Layer Height (mm)} \leq 0.3;$$
$$235 \leq \text{Extrusion Temperature } (^\circ\text{C}) \leq 245;\ 2600 \leq \text{Printing speed (mm/sec)} \leq 3400$$

The acquired operational factors, which provide least volumetric shrinkage, are highlighted in Table 5. Experiment was performed at the optimal set of parameters, and it was found that the obtained value was within the predicted range.

Table 5 Optimum process parameters for minimizing volumetric shrinkage

LT	ET	EW	PS	Observed volumetric shrinkage	Predicted volumetric shrinkage	% error
0.3	245	0.48	2895	123.19	122.35	0.68

6 Effect of Part Deposition Orientation on Volumetric Shrinkage

The volumetric shrinkage was also analyzed with respect to the orientation of the part with respect to the bed of machine. Specimen was created using the optimized parameters given in Table 5 at diverse fabricating inclinations in the array of 0°–90° at a gap of 15° as shown in Fig. 4a. The hatching orientation was kept at 0°. The attained value for volumetric shrinkage of specimen corresponding to their orientation is given in Fig. 4b. It is visible from Fig. 4b that the minimum volumetric shrinkage was attained at 0° orientation.

Minimum volumetric shrinkage obtained was due to the fact that at 0°, the length of road was maximum as shown in Fig. 5a. Due to this, the number of void obtained was very less as material deposition over there was higher. This resulted in more compact layers, hence lower volumetric shrinkage [10]. At 45° orientation, it was noticed that the quantity of empty space at corners increases (Fig. 5b) resulting in more pour cavity and higher volumetric shrinkage.

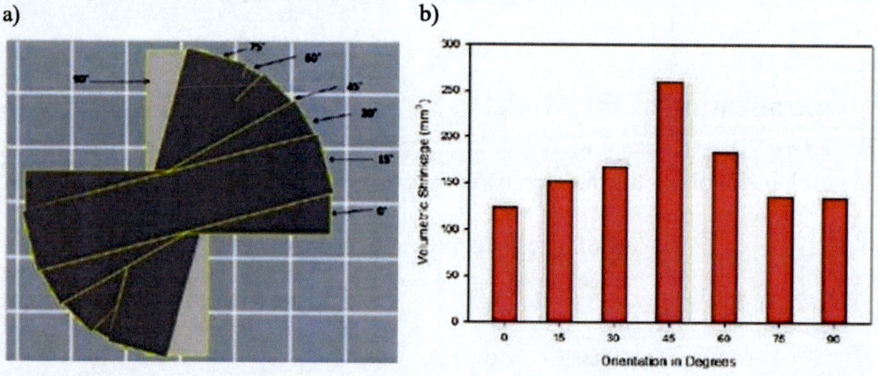

Fig. 4 **a** Arrangement of specimen on build platform. **b** Histogram of the measured volumetric shrinkage at different orientations

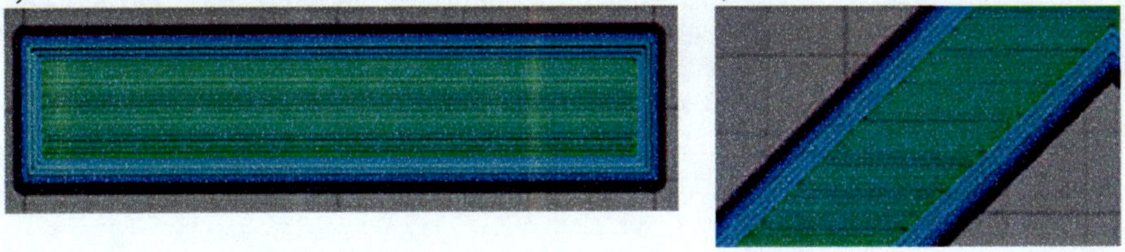

Fig. 5 **a** Part at 0° orientation. **b** Part at 45° orientation

7 Conclusion

- A relationship has been built up for forecasting volumetric shrinkage of nylon, by linking input considerations, layer thickness, extruder temperature, extrusion width and printing speed.
- The extrusion temperature was established to be the largest part swaying the volumetric shrinkage, trailed by layer thickness and extrusion width. The printing speed did not have major influence on volumetric shrinkage.
- Verification trials carried out at various examination situations demonstrate that the built-up relationship can forecast the volumetric shrinkage data precisely within 95% assurance interval and best operational factors have also been recognized.
- The effect of part orientation has been studied and it was found that 0° part orientation gave minimum volumetric shrinkage.

References

1. Chua CK, Leong KF, Lim CS (2003) Rapid prototyping, principles and applications, 2nd edn. World Scientific Publishing Co. Pte. Ltd., Singapore
2. Khatwani J, Srivastava V (2019) Effect of process parameters on mechanical properties of solidified PLA parts fabricated by 3D printing process. In: 3D printing and additive manufacturing technologies. Springer Nature Singapore, pp 95–104
3. Raghunath N (2006) Improving accuracy through shrinkage modelling by using Taguchi method in selective laser sintering. Int J Mach Tools Manuf 47(6):985–995
4. Sood AK (2009) Improving dimensional accuracy of fused deposition modelling processed part using grey Taguchi method. Mater Des 30(10):4243–4252
5. Mago J (2019) Modeling of linear shrinkage in PLA parts fabricated by 3D printing using TOPSIS method. Adv Addit Manufact Join 267–276
6. Schmutzler C (2016) Compensating warpage of 3D printed parts using free-form deformation. Procedia CIRP 41:1017–1022
7. Boschetto A (2014) Accuracy prediction in fused deposition modeling. Int J Adv Manuf Technol 73(5–8):913–928
8. Lee CY (2019) The influence of forced-air cooling on a 3D printed PLA part manufactured by fused filament fabrication. Addit Manuf 25:196–203

9. Yin B (2021) Effects of deposition speed and extrusion temperature on fusion between filaments in single-layer polymer films printed with FFF. Adv Ind Eng Polym Res 4(4):270–276
10. Singhal SK (2009) Optimum part deposition orientation for multiple objectives in SL and SLS prototyping. Int J Prod Res 47(22):6375–6396

Comprehensive Review on Mechanical Characters of Plant Particle Reinforced Polylactic Acid Biocomposites

Sachin S. Raj, Vivekanandhan Chinnasamy, and Karthik Aruchamy

Abstract Biocomposites developed in the past few decades have gained utmost attention amidst researchers among which natural fiber particulate and wood particle reinforced plastic composites commit toward versatile applications. Natural fiber powder reinforced plastics in the current market are commonly developed with petroleum-based polymers that pose environmental hazards on disposal. In order to develop an alternative resource for these, many attempts have been made using various bioplastics as matrix materials with wood and plant particle form reinforcements. Polylactic acid with natural fiber reinforcement has majorly resulted in providing good mechanical performance among which particulate or powdered form processed to micro- and nano-sized reinforcements exhibits outstanding mechanical properties when compared to long and short fiber reinforcements. This review consolidates previous research profiles that have worked with polylactic acid reinforced with various wood flour and plant fiber particulate combinations to provide a database for researchers working in the similar frame.

Keywords Polylactic acid · Wood flour · Natural plant fiber particles · Mechanical properties

S. S. Raj (✉)
Department of Mechanical Engineering, Sanjivani College of Engineering, Kopargaon 423603, Maharashtra, India
e-mail: sachinsraj1991@gmail.com

V. Chinnasamy
Department of Mechanical, Energy and Industrial Engineering, Botswana International University of Science and Technology, Palapye, Botswana

K. Aruchamy
Department of Mechatronics Engineering, Akshaya College of Engineering and Technology, Coimbatore 642109, Tamil Nadu, India

N. M. Sivaram et al. (eds.), *Advances in Manufacturing, Automation, Design and Energy Technologies*, Lecture Notes in Mechanical Engineering,
https://doi.org/10.1007/978-981-99-1288-9_7

1 Introduction

Polylactic acid (PLA) is derived through the polymerization of carbohydrates present in starchy vegetables like cassava, potato, corn, etc. [1]. It is non-toxic and is chemically represented by the formula shown in Eq. 1 [2].

$$[-CH(CH3)-CO-O-]_n \tag{1}$$

PLA is widely used for its inherent properties like high thermal stability, melt flow index and low moisture absorption [3]. It is also highly preferred for versatile applications like food packaging, 3D printing, textile industry, medical implants, etc. [1, 4, 5], due to its excellent stiffness and mechanical characteristics when compared with other bioplastics like Ecoflex, Bioflex, PHBV and PBS [6, 7]. PLA is also competitive and in some cases resulted in greater mechanical performance when compared with petroleum-based polymers like Polypropylene, Polyethylene, ABS, etc. [2, 8, 9]. The impregnation of natural plant fibers into PLA has enhanced the mechanical character of PLA in majority of the studies, while few literatures also show decrement of certain mechanical properties as an outcome. Studies also depict that the particle size plays a major role in the final properties of the specimens. Wood flour in particulate form proved to be more substantial than plant fibers' additives as long and short reinforcements [10, 11]. These factors were justified by various researches through morphological studies, where the particulate form of plant or wood fiber reinforced composites exhibited homogeneous distribution, while larger fibers exhibited scattered distribution within the PLA matrix [12–14]. Previous studies also show that processing methods are more flexible and versatile due to the rheological properties of the molten PLA composite while using smaller sized reinforcements [15]. The addition of any wood flour would increase the material viscosity during molten processing, and the complex viscosity also increases with increment in reinforcement content. Good mechanical character that is worldwide tested commonly under ASTM standards [16] is the most important strength required by a composite to perform efficiently for the application it is developed for. This review comprises such literatures dealing with the mechanical character of various wood saw dust and other plant particle reinforced polylactic acid composites.

2 Research Review

This subsection contains various literatures comprehending to PLA added with plant particle reinforcements. The method of fabricating the composites, quantity of reinforcement and other processing and testing parameters are discussed in detail along with the mechanical properties that were achieved by the researchers.

Huda et al. [8] experimented with PLA and maple wood flour having a micron particle size. The additive was added in ratios of 20, 30 and 40 wt%. Pellets were first extruded, and then, injection was molded at 183 °C. The flexural strengths of the composites at all reinforcement ratios were greater than PLA. Twenty wt% maple wood particle reinforced composite showed the highest flexural strength having an increase of 20% when compared with neat PLA. Thirty and 40 wt% reinforced composites showed consecutive lowering of flexural strength. Tensile properties showed similar pattern as of the flexural properties, and the 40 wt% reinforced composites resulted in a tensile strength lower than that of plain PLA. The increase in flexural and tensile strengths of the 20 wt% reinforced composites was due to good adhesion between maple wood flour and matrix. The impact strengths of all the maple flour reinforced PLA composites were lower than plain PLA. Higher maple wood flour content reduced the impact strength, while low wood flour had showed no significant change in the impact strength of PLA.

Lee and Wang [17] had worked with PLA and bamboo wood flour combination. Reinforcements were considered in weight ratios of 10, 20, 30, 40 and 50% and processed in a batch mixer at 180 °C, 70 rpm for 5 min. After preprocessing, the pellets were compression molded at 180 °C and 150 MPa pressure. The tensile strength had reduced with increment of bamboo wood powder into PLA. PLA showed best tensile properties, and the composite with 40% wood flour content showed a reduction of 50% of tensile strength when compared with PLA. Poor interfacial adhesion between PLA and the bamboo wood flour had resulted in reduced tensile strength. The elongation of the composites also depicted similar pattern as of the tensile strength, and this was due to the discontinuity of the polymer matrix. Tensile modulus increased with increasing fiber content due to the hard filler material. Wood fibers having greater hardness generally show better modulus values.

Pilla et al. [18] reinforced pine wood flour with PLA at 20 and 40 wt% filler ratios through compression molding process. ASTM D638 was performed to measure the tensile strength of the composites. Reinforcement of the pine wood powder showed marginal reduction in the tensile strength of plain PLA. PLA had a tensile strength of 55.5 MPa, while the 20 and 40 wt% reinforced composites showed 54.5 and 51.7 MPa, respectively. This was due to the higher stiffness of the composites obtained due to the addition of the wood powder, consequently leading to brittle fracture and low necking before fracture.

Tisserat et al. [10] analyzed the effect of paulownia wood particles in various sizes ranging from 35 μm up to 400 μm as reinforcements. A constant 25 wt% ratio was set as the analysis parameter and specimens were extruded at 170 °C. Although the addition of paulownia particles reduced the tensile strength of PLA, the tensile property emerged with an increased pattern while decreasing the size of reinforcement. Modulus value however increased drastically due to the high stiffness of the composite when compared with plain PLA. In this literature, paulownia is referred to have a high lignin content of 20% which may be the reason for the high stiffness of the resultant composite.

Petchwattana and Covavisaruch [19] developed composites using PLA and rubber wood particles with a particle size of 500 μm. The composites were processed in a twin-screw extruder at 190 °C followed by injection molding at 200 °C and pressure 70 bar. Reinforcement ratios of 5, 10, 20 and 30 wt% ratio with PLA were considered for processing the specimens. Rubber wood reinforced PLA at 30 wt% showed the best results with an increase in tensile and flexural strengths of 133% and 123%, respectively, when compared with plain PLA. Tensile and flexural moduli increased along with the addition of rubber wood, but the elongation property of the composite was vice versa. The impact strength of PLA also decreased with increasing reinforcement content. SEM images had shown agglomeration of the composites. This was the reason behind reduced mechanical characteristics.

Wan and Zhang [20] researched with PLA reinforced poplar wood flour having a size of below 100 μm. The composites were processed at a weight reinforcement ratio of 20%. Post-granulation at 195 °C using extrusion, the composite granules were injection molded into test specimens at 185 °C and pressure 0.6 MPa. Tensile strength of the poplar reinforced PLA composite was marginally higher than PLA. The flexural strength resulted in 34% higher efficiency when PLA was reinforced with the wood particle due to the high degree of crystallinity.

Guo et al. [21] developed a combination of PLA with poplar wood powder at 5 wt% reinforcement for 3D printing application. Processing was achieved through fusion deposition method at 190 °C. The tensile strength marginally reduced, flexural strength reduced from 38 MPA to 20 MPa and the impact strength reduced from 70 to 45 kJ/m^2 when poplar wood particles were added to PLA. The lowering of mechanical properties due to reinforcement was due to poor interfacial adhesion between the poplar wood particles and the matrix. Morphological images had depicted uneven particle distribution within the matrix phase.

Sachin et al. [12] processed PLA biocomposites using micro- and nano-sized wood particles extracted from the Prosopis juliflora tree. The micro and nano-sized particulates were in an average size of 10–50 μm and nanometers, respectively. The reinforcements were added to PLA in the ratio between 10 and 25 wt% in increments of 5 wt%. Fabrication of the composite was performed using a twin-screw profile extruder working with a processing temperature of 180 °C. The 20 wt% reinforced composites showed outstanding mechanical character in both the micro and nano-sized variants individually. The nano-wood particle reinforced specimens dominated the micro-wood particle reinforced specimens in terms of mechanical strength. In a couple of similar studies carried out by the author using Azadirachta indica wood flour [13] and Tamarindus indica wood flour [14] wiith PLA combinations, the mechanical property of the nano sized reinforcements depicted greater performance than the micro particulate reinforced PLA composites. Similar results trends as of the Prosopis juliflora were attained by the Azadirachta indica combined PLA composites, but the Tamarindus indica reinforcement resulted in reduction of mechanical properties of PLA when reinforced. In all the different wood species cases, the common factor was that 20 wt% reinforced composite had the best mechanical characteristics and the nano-sized reinforcement resulted in providing enhanced mechanical properties than the micro-sized particulate reinforcements.

Balart et al. [22] injection molded PLA in combination with hazelnut powdered particles at 190 °C. The powder size was 63 μm, and reinforcement ratio was maintained at 20 wt%. Comparison of the composites in this study was done with composites that were further reinforced with Epoxidized Linseed Oil plasticizer. The tensile strength of the plain composite was 45 MPa, the elastic modulus was 1019 GPa, elongation at break was 3.1%, flexural strength was 83.7 MPa, flexural modulus was 3932 MPa and hardness value was 75.6, while the plasticizer-added composites showed a tensile strength reduced to 15.3 MPa, elastic modulus reduced to 590 MPa, flexural strength reduced to 29.5 MPa, flexural modulus reduced to 2118 MPa and the hardness reduced to 59.6.

Yaacob et al. [23] used paddy straw particulates as reinforcement with PLA. The powder was reinforced in ratios of 5, 10, 15 and 20 wt%, and test specimens were processed using compression molding technique. The tensile strength of PLA was 52.4 MPa, while the reinforced composite at 20 wt% had a tensile strength of 12.97 MPa. The reduction in tensile strength was due to poor matrix-filler adhesion. Elongation of PLA also reduced when paddy powder was added. The modulus of elasticity although increased from 1445.6 MPa for plain PLA to 1824 MPa for the 20 wt% straw powder reinforced composite. The increment in the modulus was due to improved toughness offered to PLA due to the straw powder reinforcement.

Carrillo et al. [24] experimented with almond shell powder reinforced PLA composites at 30 wt% filler ratio. Injection-molded specimens that were processed at 185 °C post-pelletizing through extrusion were tested for mechanical properties. The tensile strength of plain PLA reduced from 62 to 18 MPa, tensile modulus reduced from 1900 to 1500 MPa and elongation at break reduced from 5 to 2.5% when the almond shell powder was added to it. The impact strength of PLA also reduced by 60%, post-reinforcement. Only the hardness of the PLA increased from 79 to 85 due to reinforcement.

Scaffaro et al. [25] merged Posidonia oceania leaf flour with PLA in two different particle size variants, 150–300 μm and 75–150 μm. Processing was carried out using compression molding at 190 °C. The composites were processed at 10 and 20 wt%. The tensile strength of the composites reduced irrespective of size reinforcements. Modulus of PLA increased with reinforcement and its elongation decreased with increasing leaf flour.

Table 1 summarizes the review of all the literatures, providing explanatory details about the increment or decrement of the mechanical properties of PLA when reinforced with natural plant particulates.

3 Conclusion

Some of the conclusive points through this review are summarized below:

- Not all forms of plant species lead to increment of mechanical properties when combined with PLA.
- Wood flour provides better probability of increasing the mechanical properties of PLA than other particulate matter derived for various parts of plants like, flowers, seeds, leaves, etc.

Table 1 Details of plant species, reinforcement size, quantity and result

Sl. No.	Wood/plant species	Particle size	Quantity of reinforcement (wt%)	Optimum reinforcement (wt%)	Change in properties of PLA with reinforcement of particulates	Application of the composite	References
1	Maple wood	–	20, 30, 40	20	Increment	Industrial bioplastic	[8]
2	Paulwonia wood flour	35–400 μm	25	Smaller sized particles	Decrement	Commercial products	[10]
3	Prosopis juliflora wood particles	50 μm and 50 nm	10, 15, 20, 25	20	Increment	Structural	[12]
4	Azadirachta indica	50 μm and 50 nm	10, 15, 20, 25	20	Increment	Structural	[13]
5	Tamarindus indica	50 μm and 50 nm	10, 15, 20, 25	25	Decrement	Biomedical	[14]
6	Bamboo wood flour	70 μm	10, 20, 30, 40, 50	40	Decrement	Interfacial property enhancement	[17]
7	Pine wood flour	–	20, 40	20	Decrement	Analytic model for testing	[18]
8	Rubber wood particles	500 μm	5, 10, 20, 30	30	Increment	To reduce brittleness of PLA	[19]
9	Poplar wood flour	98 μm	20	20	Increment	High performance and tolerable materials	[20]
10	Poplar wood flour	–	5	5	Decrement	3D printing	[21]

(continued)

Table 1 (continued)

Sl. No.	Wood/plant species	Particle size	Quantity of reinforcement (wt%)	Optimum reinforcement (wt%)	Change in properties of PLA with reinforcement of particulates	Application of the composite	References
11	Hazelnut powder	63 µm	20	20	Decrement	To reduce high intrinsic fragility of PLA	[22]
12	Paddy straw powder	68 µm	5, 10, 15, 20	20	Decrement	Food packaging materials	[23]
13	Almond shell flour	150 µm	30	30	Decrement	Food packaging materials	[24]
14	Oceania leaf powder	75–150–300 µm	10, 20	–	Decrement	Engineering materials	[25]

- Smaller sized particulates resulted in depicting better mechanical character than larger sized reinforcements.
- In majority of the literatures, 20 wt% reinforcement produced optimum results. Addition of reinforcement at 10 wt% and below resulted in reduced properties of PLA due to uneven distribution of particulates within the matrix.

References

1. Garlotta D (2001) A literature review of poly(lactic acid). J Polym Environ 9(2):63–84
2. Siakeng R, Jawaid M, Ariffin H, Sapuan SM, Asim M, Saba N (2018) Natural fiber reinforced polylactic acid composites: a review. Polym Compos 40(2):1–18
3. Sykacek E, Hrabalova M, Frech H, Mundigler N (2009) Extrusion of five biopolymers reinforced with increasing wood flour concentration on a production machine, injection moulding and mechanical performance. Compos: Part A 40:1272–1282
4. Shimpi NG (2018) Biodegradable and biocompatible polymer composites, processing, properties and applications. Elsevier
5. Raj SS, Kannan TK, Babu M, Vairavel M (2019) Processing and testing parameters of PLA reinforced with natural plant fiber composite materials—a brief review. Int J Mech Prod Eng Res Dev 9(2):933–940
6. Rajan KP, Thomas SP, Gopanna A, Al-Ghamdi A, Chavali M (2018) Polyblends and composites of poly (lactic acid) (PLA): a review on the state of art. J Polym Sci Eng 1(3):10–24
7. Aravind Raj S, Muthukumaran E, Jayakrishna K (2018) A case study of 3D printed PLA and its mechanical properties. Mater Today: Proc 5(5):11219–11226
8. Huda MS, Drzal LT, Misra M, Mohanty AK (2006) Wood-fiber-reinforced poly(lactic acid) composites: evaluation of the physicomechanical and morphological properties. J Appl Polym Sci 102:4856–4869
9. Haque MMU, Goda K, Ogoe S, Sunaga Y (2019) Fatigue analysis and fatigue reliability of polypropylene/wood flour composites. Adv Ind Eng Polym Res 2:136–142
10. Tisserat B, Joshee N, Mahapatra AK, Selling GW, Finkenstadt VL (2013) Physical and mechanical properties of extruded poly(lactic acid)-based Paulownia elongate biocomposites. Ind Crops Prod 44:88–96
11. Sachin SR. Wood-plastic composite processing and mechanical characteristics—a brief literature review. In: Recent advances in manufacturing, automation, design and energy technologies, Proceedings from ICoFT 2020. Edition: Lecture notes in mechanical engineering, Chapter: 32. Springer, Singapore. https://doi.org/10.1007/978-981-16-4222-7_32
12. Raj SS, Kannan TK, Rajasekar R (2020) Influence of Prosopis juliflora wood flour in poly lactic acid-developing a novel bio-wood plastic composite. Polimeros – ciencia e tecnologia 30(1):e2020012
13. Sachin SR, Kannan TK, Rajasekar R (2019) Effect of wood particulate size on the mechanical properties of PLA biocomposite. Pigm Resin Technol 49(6):465–472
14. Sachin SR, Kuzmin AM, Thiruvasagam C (2020) Mechanical, degradation and morphological investigations on Tamarindus indica reinforced poly lactic acid bio-composite for biomedical applications. J Optoelectron Biomed Mater 12(4):121–131
15. Mazzanti V, Mollica F (2017) Rheology of wood flour filled poly(lactic acid). In: Proceedings of the third international conference on natural fibers: advanced materials for a greener world 2017, pp 61–67
16. Sachin SR, Kuzmin AM, Krishnakumar S, Saravanan S, Kannan TK (2021) Philosophy of selecting ASTM standards for mechanical characterization of polymers and polymer composites. Mater Plast 58(3):247–256

17. Lee SH, Wang S (2006) Biodegradable polymers/bamboo fiber bio composite with bio-based coupling agent. Compos Part A 37:80–91
18. Pilla S, Gong S, O'Neill E, Rowell RM, Krzysik AM (2008) Polylactide-pine wood flour composites. Polym Eng Sci 48(3):578–587
19. Petchwattana N, Covavisaruch S (2014) Mechanical and morphological properties of wood plastic biocomposites prepared from toughened poly(lacticacid) and rubber wood sawdust (Hevea brasiliensis). J Bionic Eng 11:630–637
20. Wan L, Zhang Y (2018) Jointly modified mechanical properties and accelerated hydrolytic degradation of PLA by interface reinforcement of PLA-WF. J Mech Behav Biomed Mater 88:223–230
21. Guo R, Ren Z, Bi H, Song Y, Xu M (2018) Effect of toughening agents on the properties of poplar wood flour/poly (lactic acid) composites fabricated with fused deposition modeling. Eur Polymer J 107:34–45
22. Balart JF, Fombuena V, Fenollar O, Boronat T, Sanchez Nacher L (2016) Processing and characterization of high environmental efficiency composites based on PLA and hazelnut shell flour (HSF) with biobased plasticizers derived from epoxidized linseed oil (ELO). Compos B Eng 86:168–177
23. Yaacob ND, Ismail H, Ting SS (2016) Soil burial of polylactic acid/paddy straw powder biocomposite. Bioresources 11(1):1255–1269
24. Carrillo LQ, Montanes N, Sammon C, Balart R, Giner ST (2018) Compatibilization of highly sustainable polylactide/almond shell flour composites by reactive extrusion with maleinized linseed oil. Ind Crops Prod 111:878–888
25. Scaffaro R, Maio A, Lopresti F (2018) Physical properties of green composites based on poly-lactic acid or Mater-Bi filled with Pasidonnia oceanica leaves. Compos A Appl Sci Manuf 112:315–327

Experimental Research on the Hardness and Scratch Testing of Oil Hardening Non-shrinkage Material with Titanium Nitride Coating by PVD Method

M. Purusothaman, J. Britto Joseph, P. Siva, and M. Ravinth

Abstract The project experiments the impacts of heat treatment on mechanical and physical qualities, such as quenching, tempering, and TiN coating of untreated oil-hardened non-shrinking die steel which comes under tool steel grade which is widely used in manufacturing fixture and gauges. The paper contributes to better understanding of the tempering temperature up to 600 °C. Coating gets deposited at 250 °C onto the heat-treated condition, and titanium nitride coating is done by using Physical Vapor Deposition method. The paper contributes to better understanding of the tempering processes which have been conducted on uncoated substrates (specimens without TiN coating) by the specimens at austenite temperature condition slowly preheated up to 650 °C and then finally heated at 788–816 °C. These specimens were cooled via oil quench. Tempering processes were done at temperatures of 200 and 535 °C with holding time of 10–30 min.

Keywords TiN coating · Tempering · Hardness · Surface finish

1 Introduction

Oil-hardened non-shrinking die steel is economical for gauging, cutting, and blanking tools. It is composed of 0.95% carbon, 1.1% manganese, 0.6% chromium, 0.6% tungsten, and 0.1% vanadium. The in-between temperature of the hardening is 790–850 C. The (AISI O1) NSOH tool steel can be applied for blanking, punching good hardened results and produces minor dimensional changes. It provides adequate abrasion resistance and toughness for standard tool-and-die applications [1]. Hardness, wear resistance, surface polish, and other tool life properties are improved with coatings. One of the most significant achievements in the commercial use of current coating

M. Purusothaman (✉) · J. Britto Joseph · P. Siva · M. Ravinth
School of Mechanical Engineering, Sathyabama Institute of Science and Technology, Chennai, India
e-mail: purusothmani@gmail.com

N. M. Sivaram et al. (eds.), *Advances in Manufacturing, Automation, Design and Energy Technologies*, Lecture Notes in Mechanical Engineering,
https://doi.org/10.1007/978-981-99-1288-9_8

technology is the development of Physical Vapor Deposition (PVD) thin-film coatings for cutting tools in the metal cutting sector [2–5]. PACVD coatings are robust and wear resistant, and they also protect punches and forming tools from abrasive wear, allowing them to last longer. It is performed to check the good surface finish, surface hardness, and tool life. Titanium nitride (TiN) coatings are used extensively for improving wear properties of different parts due to their high hardness [6]. Chemical Vapor Deposition (CVD), Physical Vapor Deposition (PVD), and Plasma-Assisted Chemical Vapor Deposition (PACVD) are some of the coating processes that can be utilized to deposit material on a surface (Plasma-Assisted Chemical Vapor Deposition). The CVD process involves the interaction of a gas mixture with the surface of a heated substrate, resulting in the chemical breakdown of part of the gas elements and the production of a solid layer on the substrate [7]. Normal treatment care offers good toughened outcomes and minor dimensional alterations. It provides sufficient abrasion resistance and hardness for standard tool-and-die applications. Hardness, wear resistance, surface polish, and other tool life properties are improved with coatings [8, 9]. One of the most notable successes in the industrial use of current coating technology is the development of Physical Vapor Deposition (PVD) thin-film coatings for cutting tools in the metal cutting sector. PACVD coatings are robust and wear resistant, and they protect punches and forming tools from abrasive wear, allowing them to last longer. It is carried out to ensure that the surface polish, surface hardness, and tool life are all satisfactory [10, 11]. Because of their great hardness, titanium nitride (TiN) coatings are widely utilized to improve the wear qualities of many parts. Chemical Vapor Deposition (CVD), Physical Vapor Deposition (PVD), and Physical Vapor Deposition (PACVD) are some of the coating processes that can be utilized to deposit material on a surface (Plasma-Assisted Chemical Vapor Deposition). The CVD process involves a mixture of gases interacting with the surface of a heated substrate, resulting in chemical breakdown of part of the gas elements and the production of a solid layer on the substrate [12].

1.1 Scope of the Experiment

TiN is the most common and familiar coating for preventing wear on a wide range of tools and parts manufacturing. TiN is an excellent material for machining iron-based materials, die casting, tools, punches, and tooling for plastic molds. TiN is also a suitable coating for wear parts that require abrasion and adhesive wear resistance [13]. OHNS steel is a general-purpose tool steel that is commonly used in applications where alloy steels lack the necessary hardness, strength, or wear resistance. Because of the 0.90% vanadium addition, the quality is excellent, with good wear resistance and toughness properties. It has a hardness of 57–59 HRC and is normally supplied annealed [14]. In the annealed state, it is machinable. Due to its excellent alloy characteristics, DIN 2379 is better steel than D3. All molybdenum [M1–M52] and tungsten [T1–T15] class alloys are found in high-speed alloys. Hardening high-speed tool steels to 62–67 HRC is possible. Tool steel with a high carbon and chromium content is preferred for cold work. Because of the 0.90% vanadium addition, the quality is excellent, with good wear resistance and toughness properties [15]. It has a hardness of 57–59 HRC and is normally supplied annealed. In the annealed state, it is machinable. Due to its excellent alloy characteristics, DIN 2379 is better steel than D3.

2 Materials and Methods

PVD is a process for depositing thin films of one atom (or one molecule) at a time onto various surfaces, often metals, to provide a hard, lasting coating. The coating comes from a physical source, such as a solid or liquid, rather than a chemical source, as in the alternative coating technique of Chemical Vapor Deposition (CVD). PACVD processes are performed in a vacuum environment. Evaporation, Transportation, Reaction, and Deposition are the four phases in the process which is another common technique. It is used to coat cutting tools, although it is not the only one. PVD stands for vapor deposition. The coating's metal species, generated by sputtering or evaporation, occurs when a gaseous species (such as nitrogen or ammonia) interacts in the chamber and deposits on the substrate. The coating atoms and molecules only collide with the substrate a few times because PVD is a low-pressure method. PVD is thus a line-of-sight technique that necessitates the use of moving fixtures to assure consistent coating thickness.

The fundamental distinction between PVD and CVD is that PVD has a lower processing temperature (535 °C) than CVD (930 °F). PVD coatings have benefited from the decreased processing temperature in a variety of ways. Thermal cracks, which are prevalent in CVD coatings, are virtually absent in PVD coatings. PVD processing temperatures are low enough to prevent eta-phase development, allowing PVD coatings to be deposited on sharp edges. PVD coatings' relative thinness, compared to CVD coatings, improves their ability to coat sharp edges. Processing circumstances have an impact on coating microstructures. The main experimental focus areas for measuring wear with this sort of device are discussed. The coefficient of friction may also be used in full. The MDs formed during the etching stage protruded through the TiN film, resulting in a film with degraded surface characteristics, according to surface morphology. Coating thickness and indentation stresses both impacted the hardness of the deposited coatings. The coating was deposited with a thickness of 4.0–6.2 μm, according to the standard deviation as shown in Fig. 1.

3 Result and Discussion

By applying a fixed load to the material with an indenter, the Rockwell hardness test determines the permanent depth of indentation. When the material is harder,

Fig. 1 Sample specimen after TiN coating

become smaller the indentation value. The differential-depth method is used to perform the Rockwell hardness test. The indenter creates a residual depth known as the indent, which is then measured. To reduce mistakes caused by surface roughness and measurement, the complete test force is applied in two steps. It is depending on the scale C result in HRC. After TiN coating, hardness will be measured by Rockwell hardness testing machine result in before 42 HRC. Also, check the microhardness testing at different locations on the coating. Fifty g load was applied during a 10 s dwell period. The hardness of the coated and uncoated OHNS materials was discovered on their surfaces with the use of a microhardness tester as shown in Fig. 2. The values for uncoated hardness are as follows: The specimen presented is uncoated.

Optical microscopy is the most reliable method for measuring Lc. This method is useful for determining whether a failure occurred within the coating or substrate, as well as when a failure occurred at the coating–substrate border. In the ramping load scratch test, the space between the onset of the scratch and the precise break actions below examination can be with no trouble determined. Using the ratio (loading rate/sample velocity), the critical load can then be intended from the scratch length as shown in Fig. 3. Several scratch testing systems can measure acoustic emission as well as the frictional force on the stylus while scratching. A piezoelectric detector mounted near the stylus is frequently used to detect acoustic emission. With the scratch measurement, establishing the critical load, or maximum load, at which normal tin coating failure occurs is simple. It is a good tool for quantitative quality control. Failures of various loads applied to coatings are critical for examining and

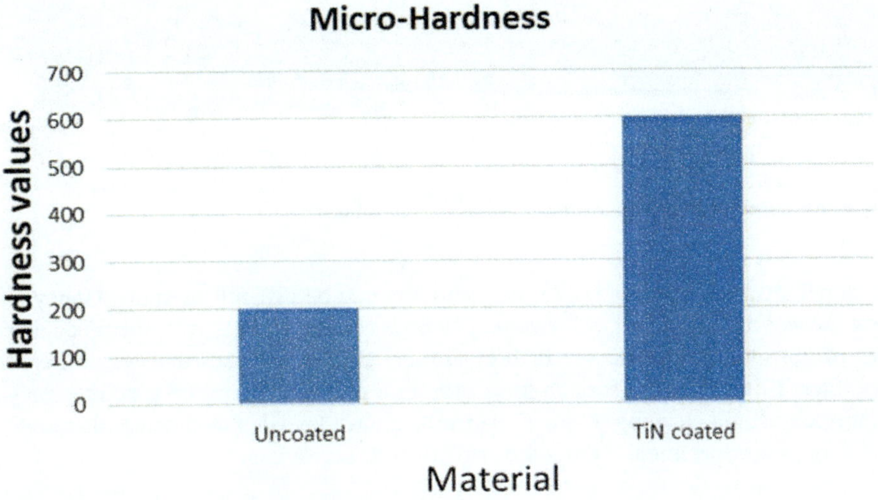

Fig. 2 Hardness test result

comparing the intrinsic quality of the coating. Large deformations can be resisted by a coating with the right interlayer. When exposed to extreme external loads and concentration stress, the coating/substrate system's cohesive and adhesive capabilities have improved, as did the coating/substrate system's strength as shown in Fig. 4.

Using an optical microscope, look at the scratch tracks. With the help of an optical microscope (magnification 200×), inspect the scratch tracks and identify the critical loads Lc1, Lc2, and Lc3 that correspond to the failure events listed in Annex. B. The onset of the regular occurrence of the linked failure event corresponds to critical failure events listed in Annex. B. The onset of the regular occurrence of the linked failure event corresponds to critical load values. The precise definition of each failure event must be carefully considered. The beginning of forward chevron fractures at

Fig. 3 Scratch tip on the TiN sample

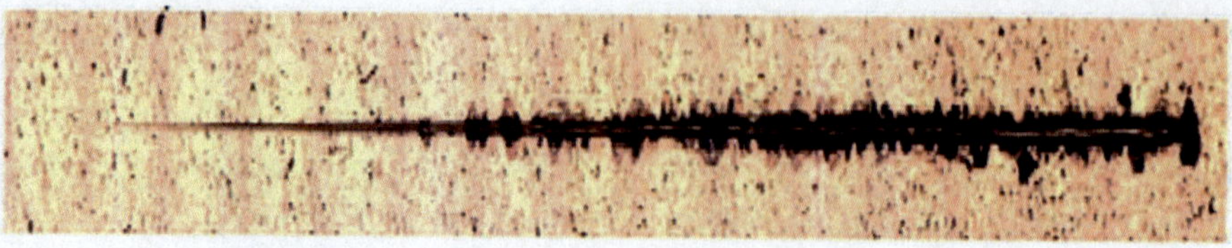

Fig. 4 After the tests, the TiN coatings were scratched completely

the scratch track's borders (Lc1), for example, can be difficult to spot. After you have located the beginning of forward chevron cracking, go back over the preceding area of the scratch track to see if there are any earlier instances that would result in a lower Lc1 value. If a scratch track crosses a coating deficit, the specimen may require ultrasonic cleaning. Critical load values must be calculated using ultrasonic cleaning of the specimen to compute critical load, Lc, values.

4 Conclusion

The hardening temperature ranges from 790 to 850 °C. The National Society for the Prevention of Cruelty to Animals Normal treatment care offers good toughened outcomes and minor dimensional alterations. It has adequate abrasion resistance and hardness for standard tool-and-die applications. The quenched specimens were heated again to temperatures of 200 and 535 °C, with each specimen held for two hours for the tempering process. Forty-two HRC is achieved with final hardness. TiN coatings are evaluated at temperatures ranging from 400 to 425 °C. The thickness of 2–6 μm on the working surfaces is achieved. The Rockwell hardness testing machine produced results of 42 HRC before and after coating. There was no difference when compared. The scratch measurements enable users to rapidly detect the load that is critical or maximum load at 3.25–8.23 N in normal tin coating failure. The conclusion is that keeping the tin coating temperature higher than the tempering temperature allows for coating. The hardness will be decreased. It is not coated because the tempering temperature is below.

Disclosure Statement No conflict of interests.

References

1. Sanders DM, Anders A (2000) Review of cathodic arc deposition technology at the start of the new millennium. Surf Coat Technol 133–134:78–90
2. Baek DK, Ko TJ, Kim HS (2001) Optimization of feed rate in a face milling operation using surface roughness model. Int J Mech Tool Manuf 41(3):451–462
3. Nouri M, List G, Girot F (2003) Experimental analysis and optimization of tool wear in dry machining of aluminium alloys. Wear 225:1359–1368
4. Dailey JH (1997) Tin coating composition. Met Finish 95. https://doi.org/10.1016/S0026-0576(97)86909-7
5. Berger M, Hogmark S (2002) Evaluation of TiB_2 coatings in sliding contact against aluminum. Surf Coat Technol 149:14
6. Thepsonthi T (2014) Modeling and optimization of micro-end milling process for micro-manufacturing. PhD thesis, Rutgers, The State University of New Jersey, Rutgers, NJ, USA
7. Thepsonthi T (2015) Modeling and optimization of micro-end milling process for micro-manufacturing
8. PhD thesis, Rutgers, The State University of New Jersey, Rutgers, NJ, USA (2014)
9. Martin PJ, Bendavid A (2001) Review of the filtered vacuum arc process and materials deposition. Thin Solid Films 394:1–15
10. Wilson S, Alpas AT (2000) Tribo-layer formation during sliding wear of TiN coatings. Wear 245:223–229
11. Lackner JM (2005) Industrially-scaled hybrid pulsed laser deposition at room temperature. Orekop, Krakow
12. Bull SJ, Berasetegui EG (2006) An overview of the potential of quantitative coating adhesion measurement by scratch testing. Tribol Int 39:99–114
13. Fernández-Abia A, Barreiro J, López de Lacalle L, Martínez S (2011) Effect of very high cutting speeds on shearing, cutting forces and roughness in dry turning of austenitic stainless steels. Int J Adv Manuf Technol 57:61–71
14. Veldhuis S, Dosbaeva G, Yamamoto K (2009) Tribological compatibility and improvement of machining productivity and surface integrity. Tribol Int 42:1004–1010
15. Settineri L, Faga M, Gautier G, Perucca M (2008) Evaluation of wear resistance of AlSiTiN and AlSiCrN nanocomposite coatings for cutting tools. CIRP Ann Manuf Technol 57:575–578

Enhancing Impact Strength of Additively Manufactured Short Carbon Fibre Reinforced Nylon Composite

Mahesh Naik, D. G. Thakur, and Sunil Chandel

Abstract Fused Deposition Modelling (FDM) is largely used additive manufacturing (AM) method because of its minimum equipment cost, variety of material availability and ease of operation. But, due to low mechanical strength, FDM has limited application in functional parts. The correct choice of printing variables can improve the strength of additively manufactured FDM components. Therefore, in this research, Izod impact strength of FDM composite specimen has been evaluated by considering three printing variables. The geometry of test specimen is fixed based upon the ASTM D256 standard. The test specimens were manufactured at various printing parameters. Based on Taguchi methodology, the experiments were conducted. This research aims to determine the optimum printing variable to get the maximum impact strength of a 3D-printed composite. ANOVA was accomplished to calculate percentage involvement of each printing variable and to get the most significant printing parameter. A confirmatory test was also carried out by considering the optimum printing parameters. Based on the result, the impact strength was affected more by layer height than infill pattern and density. The value of impact strength corresponding to the optimum printing variables of layer height, 0.2 mm; infill density, 42% and infill pattern, hexagonal was found to be 186.32 J/m.

Keywords AM · Impact strength · Taguchi method · ANOVA

1 Introduction

Additive manufacturing (AM) is evolving process which has the ability to transform digital data into real 3D parts. AM is a tool-less manufacturing technology for direct manufacturing, which leads to reduction of product development cost and enhances profitability [1]. AM has application in various manufacturing industries

M. Naik · D. G. Thakur (✉) · S. Chandel
Department of Mechanical Engineering, Defence Institute of Advanced Technology (DU), Pune 411025, India
e-mail: dinnu74@yahoo.com

N. M. Sivaram et al. (eds.), *Advances in Manufacturing, Automation, Design and Energy Technologies*, Lecture Notes in Mechanical Engineering,
https://doi.org/10.1007/978-981-99-1288-9_9

[2–9]. There are various technologies involved in AM. Fused Deposition Modelling (FDM) techniques, which come under MEX, are mostly used because of their low cost, simplicity and reliability. It has been considered as an economical and green AM technique. In 1988, Stratasys developed the FDM process [10].

Recently, some research was done to analyse the mechanical strength of FDM components by considering different printing variables. Sood and Chaturvedi [11] studied that the strength of FDM components is controlled by numerous printing variables. Rajpurohit and Dave [12] analysed the consequence of printing variables on the impact strength of FDM parts. They observed that impact strength was influenced most by the raster angle. Ramesh and Panneerselvam [13] experimentally examined the influence of printing variables such as layer height, infill density and print speed on mechanical properties of FDM components using Taguchi optimisation technique. They observed that the infill density is the significant variable that impacts the strength of FDM parts. Verma and Chaturvedi [14] have done the optimisation of printing parameters, i.e. orientation, layer height and raster angle based on multiple responses. They reported that layer height is the significant variable compared to other variables. Ahn et al. [15] experimentally studied the influence of printing factors on tensile properties of FDM samples. Based on Taguchi approach, the results found were that the air gap and raster orientation considerably influenced the tensile strength of FDM samples. Wang et al. [16] studied that at high temperature and low layer height, the impact strength of FDM parts is improved. Garg et al. [17] investigated the influence of print orientation and raster orientation on the surface finish of 3D-printed samples of ABS material. Durgun and Ertan [18] examined the consequence of printing direction and raster orientation on flexural and tensile strengths of ABS 3D-printed samples. They concluded that the printed specimen has good mechanical properties at a smaller raster angle. Lee et al. [19] scrutinised the influence of printing variables, viz., raster orientation, layer height and air gap on surface finish and elastic behaviour of the 3D-printed specimen fabricated by FDM technique using Taguchi method.

From literature survey, it was observed that the process variables influenced the strength of FDM parts. Also, some study was found on the influence of printing variables on impact properties of 3D-printed parts, but no information was found regarding 3D-printed composite specimen. Therefore, this research focuses on the effect of printing variables, viz. layer height, infill density and pattern on impact properties of FDM composite specimen using a composite 3D printer. The objective of this study is to enhance knowledge regarding the selection of printing variables so as to get good impact strength.

2 Methodology

The printing parameter optimisation of the FDM process includes designing, fabricating and testing specimens. Figure 1 shows the methodology used for the process parameters' optimisation.

Fig. 1 Methodology

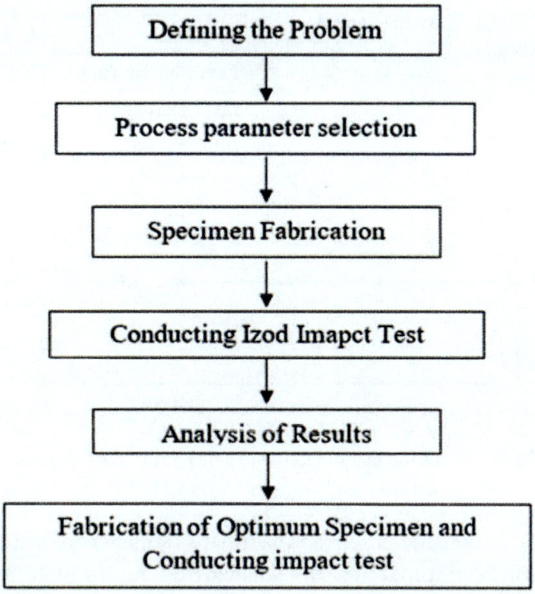

3 Experimental Setup

In this investigation, the onyx material in filament form was used, which is a combination of chopped carbon fibre and nylon. It has high mechanical strength and is tougher than ABS and PLA. An FDM technique-based composite 3D printer, named Mark Two, was used to fabricate the Izod impact specimen. The 3D printer has a print volume of $320 \times 132 \times 154$ mm^3. The 3D printer can print the minimum layer height of 100 μm and a maximum of 200 μm.

The printing variables which have effects on the strength of 3D-printed components are orientation, infill pattern and density, wall layers, layer height, raster width, raster orientation, printing speed. This study focuses on the influence of printing variables on the impact properties of 3D-printed samples. Based on the literature review, the layer height, infill density and pattern are selected as the printing parameters. The values of each level for these parameters are found based on the 3D printer and slicer software configuration. The selected variables and their levels are shown in Table 1. Using SolidWorks modelling software, a 3D model of the impact specimen is designed and saved in STL format. The geometry of specimen is set as per ASTM D256 standard, as shown in Fig. 2 [20].

For experimental design, if the full factorial method is used, then a total 27 trials of experiments would be conducted. The Taguchi method for the design of experiment, which is called the fractional factorial method, provides a simple and efficient approach for reducing the trials of experiments to be conducted. For experimental analysis, the orthogonal array is created based on Taguchi design method. Since three parameters and three levels are selected, the orthogonal array L9 is considered for

Table 1 Variables and their levels

Variables	Level-1	Level-2	Level-3
Layer height (mm)	0.1	0.125	0.20
Infill density (%)	28	42	55
Infill pattern	Triangular	Hexagonal	Rectangular

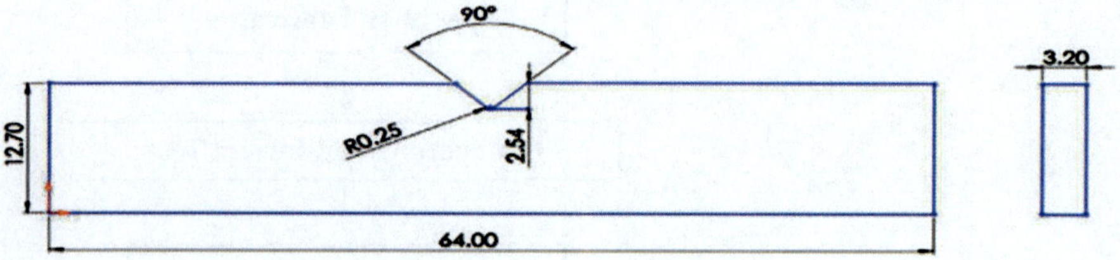

Fig. 2 Two-dimensional geometry of Izod impact test specimen

the experiments. Three specimens of each trial of L9 orthogonal array are fabricated. The Izod impact testing was carried out on IT 503 Izod impact tester as per ASTM D256 standards. Table 2 shows the L9 orthogonal array with the impact strength results. Figure 3 represents the fractured specimen of impact test.

The signal-to-noise (S/N) ratio is used to find out the effect caused by each parameter on the result. Taguchi has distributed the S/N ratio into three quality characteristics, i.e. nominal-the-best, smaller-the-best and larger-the-best. The 'larger-the-best' quality characteristic is used in this study for impact strength. Therefore, Eq. (1) has been applied to determine the S/N ratio.

$$\text{S/N ratio (larger-the-best)} = -10\log\frac{1}{n}\sum\left(\frac{1}{R}\right)^2, \tag{1}$$

where n = No. of samples, R = response value.

Table 2 L9 orthogonal array

Trial No.	Layer height (mm)	Infill density (%)	Infill pattern	Impact strength (J/m)
1	0.1	28	Triangular	136.54
2	0.1	42	Hexagonal	160.18
3	0.1	55	Rectangular	134.30
4	0.125	28	Hexagonal	156.43
5	0.125	42	Rectangular	145.57
6	0.125	55	Triangular	126.77
7	0.2	28	Rectangular	164.80
8	0.2	42	Triangular	171.22
9	0.2	55	Hexagonal	167.43

Fig. 3 Fractured specimen of the impact test

4 Results and Discussion

The impact test specimens are fabricated based on L9 orthogonal array, and the testing is done to measure impact strength. The analyses of impact strength are done to examine the influence of printing variables. Figure 4 shows S/N ratio which represents mean of S/N values for each printing parameter at each level. In this research, the larger-the-best S/N ratio specifies maximum impact strength. Based on this criteria, the optimal values of printing parameters are obtained. The optimum values of printing variables are determined from the larger-the-best S/N ratio graph, as shown in Fig. 4. The printing parameters, viz. layer height at level 3 (0.2 mm), infill density at level 2 (42%) and infill pattern at level 2 (Hexagonal), give the highest impact strength. At maximum layer height of 0.2 mm, the impact strength is high because the thickness of the layer is more which helps in resisting the impact load. Also, maximum layer height possesses more heat during deposition, which helps in forming tough bond between two layers and thus improves the strength. In the case of infill pattern, it is observed that bonding between layers is varying in each pattern. In the hexagonal pattern, each layer lays down on a similar previous layer, whereas in the triangular and rectangular patterns, the bonding between each layer corresponds to the area where the depositing layer crosses the previously deposited layer. This can be a probable explanation of why the hexagonal pattern shows maximum impact strength. The infill density at 42% leads to maximum impact strength because as the infill density increases, the stress concentration at the raster intersection increases. But, at low infill density, the distance between raster intersections is more which leads to less load absorbing capacity. Therefore, the impact strength is minimum at 28% and 55% infill densities compared to 42% infill density due to less load absorbing capacity and higher stress concentration.

Additionally, the Analysis of Variance (ANOVA) is used to determine the significant printing variables on the impact strength. Minitab software was used to carry out ANOVA. It reveals the percentage contribution of each factor on the responses. Table 3 represents the ANOVA for impact strength. If p-value of parameter is less than 0.05, then parameter is significant. From Table 3, the layer height is found to be

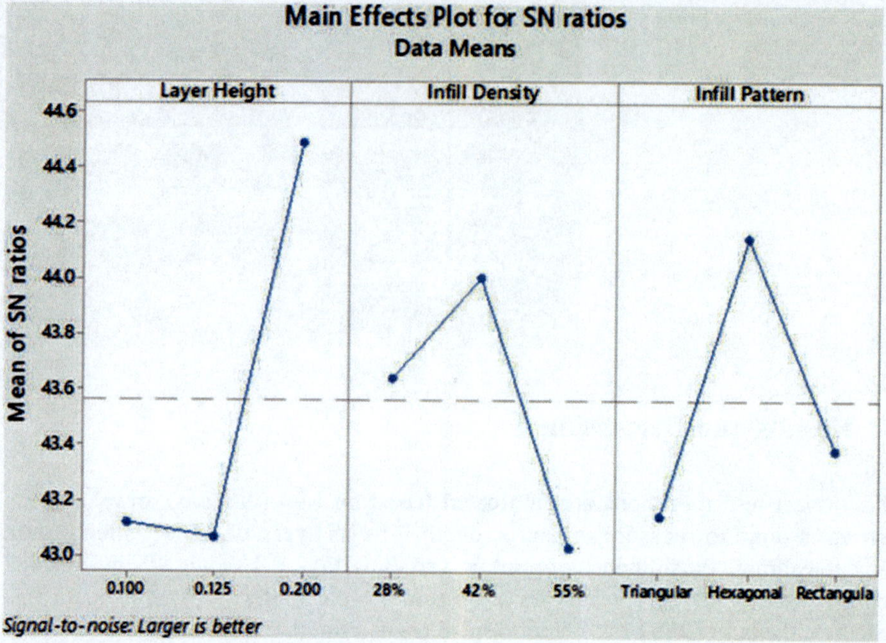

Fig. 4 S/N ratio for impact strength

significant, and the other two parameters are insignificant. The percentage contribution of layer height, infill pattern and density on impact strength was 57.69%, 21.87% and 19.04%, respectively, as shown in Table 3.

Based on the optimal values of printing parameters, a confirmation test was done. The Izod impact test of the optimal specimen was done, and the impact strength was more than that of the un-optimised specimen. The impact strength of the optimal specimen was obtained as 186.32 J/m.

Table 3 ANOVA result for impact strength

Parameter	DOF	SS	MS	F	P	Contribution (%)
Layer height	2	1203.14	601.57	41.18	0.024	57.69
Infill density	2	397.19	198.60	13.60	0.069	19.04
Infill pattern	2	456.01	228.00	15.61	0.060	21.87
Error	2	29.21	14.61	–	–	1.40
Total	8	2085.55	–	–	–	100

5 Conclusion

The aim of this study is to find out the optimal printing variables of FDM-based additive manufacturing process using Taguchi analysis. By the experimental work, the effect of variables, i.e. layer height, infill density and pattern at three levels on impact strength, is analysed. The optimum level of process parameters for impact strength is layer height—0.2 mm, infill density—42% and infill pattern—hexagonal. From ANOVA results, layer height is found to be the most significant input variable for the impact strength output response. The percentage contribution of layer height, infill pattern and density on impact strength was 57.69%, 21.87% and 19.04%, respectively. From the confirmation test, the optimal specimen has an impact strength of 186.32 J/m, which is more than that of the un-optimised specimen.

References

1. Bamiduro O, Owolabi G, Haile MA, Riddick JC (2019) The influence of load direction, microstructure, raster orientation on the quasi-static response of fused deposition modeling ABS. Rapid Prototyp J 25:462–472
2. Masood SH, Song WQ (2004) Development of new metal/polymer materials for rapid tooling using fused deposition modelling. Mater Des 25(7):587–594
3. Farina I, Fabbrocino F, Carpentieri G, Modano M, Amendola A, Goodall R, Feo L, Fraternali F (2016) On the reinforcement of cement mortars through 3D printed polymeric and metallic fibers. Compos B Eng 90:76–85
4. Kuo CC, Liu LC, Teng WF, Chang HY, Chien FM, Liao SJ, Kuo WF, Chen CM (2016) Preparation of starch/acrylonitrile-butadiene-styrene copolymers (ABS) biomass alloys and their feasible evaluation for 3D printing applications. Compos B Eng 86:36–39
5. Richter C, Lipson H (2011) Untethered hovering flapping flight of a 3D-printed mechanical insect. Artif Life 17(2):73–86
6. Rahim SL, Maidin S (2014) Feasibility study of additive manufacturing technology implementation in Malaysian automotive industry using analytic hierarchy process. Adv Mater Res 903:450–454
7. Gebler M, Uiterkamp AJS, Visser C (2014) A global sustainability perspective on 3D printing technologies. Energy Policy 74:158–167
8. Lee JY, Tan WS, An J, Chua CK, Tang CY, Fane AG, Chong TH (2016) The potential to enhance membrane module design with 3D printing technology. J Membr Sci 499:480–490
9. Mironov V, Boland T, Trusk T, Forgacs G, Markwald RR (2003) Organ printing: computer-aided jet-based 3D tissue engineering. Trends Biotechnol 21(4):157–161
10. Seitz H, Rieder W, Irsen S, Leukers B, Tille C (2005) Three-dimensional printing of porous ceramic scaffolds for bone tissue engineering. J Biomed Mater Res B Appl Biomater 74(2):782–788
11. Sood AK, Chaturvedi V, Datta S, Mahapatra SS (2011) Optimisation of process parameters in fused deposition modeling using weighted principal component analysis. J Adv Manuf Syst 10(2):241–259
12. Rajpurohit SR, Dave HK (2021) Impact strength of 3D printed PLA using open source FFF-based 3D printer. Prog Addit Manuf 6(1):119–131
13. Ramesh M, Panneerselvam K (2021) Mechanical investigation and optimisation of parameter selection for Nylon material processed by FDM. Mater Today: Proc 46:9303–9307
14. Verma S, Chaturvedi V (2018) Multiresponse optimisation of fused deposition modeling process using utility-based Taguchi approach. J Adv Manuf Syst 17(04):551–568

15. Ahn SH, Montero M, Odell D, Roundy S, Wright PK (2002) Anisotropic material properties of fused deposition modeling ABS. Rapid Prototyp J 8(4):248–257
16. Wang L, Gramlich WM, Gardner DJ (2017) Improving the impact strength of (poly lactic acid) PLA in fused layer modeling (FLM). Polymer 114:242–248
17. Garg A, Bhattacharya A, Batish A (2017) Chemical vapor treatment of ABS parts built by FDM: analysis of surface finish and mechanical strength. Int J Adv Manuf Technol 89(5–8):2175–2191
18. Durgun I, Ertan R (2014) Experimental investigation of FDM process for improvement of mechanical properties and production cost. Rapid Prototyp J 20(3):228–235
19. Lee BH, Abdullah J, Khan ZA (2005) Optimisation of rapid prototyping parameters for production of flexible ABS object. J Mater Process Technol 169(1):54–61
20. ASTM Standard D256 (2010) Standard test methods for determining the izod pendulum impact resistance of plastics. ASTM International

Hypersonic Impact, Flexural and Tensile Testing on the Natural Fiber with Influence of the Nanocompositie

S. Manigandan, M. Purusothaman, S. Venkatesh, Obuliraj, and J. Jeevanantham

Abstract The regular fiber composite because of its high strength and weariness opposition is utilized in airplane underlying application. The present proposed work centers around increment mechanical properties and framework composites by the expansion of fortifications. The principle objective of this exploration work is to make the regular fiber composite which was manufactured utilizing hand lay-up technique on epoxy sap process considering two fortifications, e.g., graphene. The natural fiber was fabricated as per ASTM standards, and mechanical tests such tensile, impact, flexural, and Sem testing are take in specimens based on the weight percentage of reinforcements. The accompanying ends are gotten from the proposed research work on concentrate on the different properties of the regular fiber under the given by the mixture composite arrangement and supporters of support and has been directed through different testing cycles to show up at the definite induction on the ideal reach for achieving the ideal presentation of the constituents in light of its properties across different areas in correlation among the diverse foreordained examples. Mechanical properties of created normal fiber composite from the SEM micrograph were cleared vulnerability our material.

Keywords Bamboo · Palm and Remy fiber · Graphene nanocomposite

1 Introduction

Normal fiber is constantly been utilized in all production of different significant designing parts in prior occasions [1]. Cost is fundamentally characteristic cycles that are changed over a refined composite material into different parts [2]. This are broadly utilized in vehicle businesses; however, cost and accessibility happen for the most part in every one of the molecules on similar created materials, harming the whole

S. Manigandan · M. Purusothaman (✉) · S. Venkatesh · Obuliraj · J. Jeevanantham
School of Mechanical Engineering, Sathyabama Institute of Science and Technology, Chennai, India
e-mail: purusothmani@gmail.com

N. M. Sivaram et al. (eds.), *Advances in Manufacturing, Automation, Design and Energy Technologies*, Lecture Notes in Mechanical Engineering,
https://doi.org/10.1007/978-981-99-1288-9_10

surface [3]. This makes picking the proper support and type of the composite regular composite materials [4]. Most strands are significantly expense and exceptionally uncommon to material in different applications and the accessibility [5]. Presently a day to build the items utilizations of the people groups in different part parts and the dependent on necessities to be increments [6]. In this way, cost of the item likewise to be increments. Then, at that point, the opportunity to be diminished the expense of the item and likewise expanded the accessibility of the necessary part to the necessary region to be utilized [7]. The most widely recognized of the car parts and the different aviation parts to be rotated by past utilized standards in the expense decrease and accessibility of restricted region [8]. We are gone through many literature surveys and cast of the component's mixture using the fabrication method [9]. We can ensure appropriate stirring parameter. Then natural fiber composite materials are tested in various testing parameter in mechanical properties using to hand lay-up method on epoxy [10]. And result and reading are obtained and plotted graphs and drawn graphs, and then SEM image is taken and its reinforcement in metal and find mixing level of graphene, bamboo fiber, palm fiber and the Remy fiber to find feasibility of the hand lay-up method on epoxy [11]. The objective of this work is to increase the strength and to decrees the cost and availability of natural fiber materials against of normal fiber materials.

2 Materials and Methods

Bamboo used in high tensile strength composite for home appliances and other small, hardened components [12]. Graphene is a blend of graphite as shown in Fig. 1. It is a solitary layer carbon foils and it is a translucent allotrope carbon the carbon molecules are thickly stuffed in a normal nuclear scale hexagonal example graphene is a two-dimensional material which draw in researcher due to high mechanical and synthetic properties [13].

Palm is the most widely recognized fiber materials utilized in as of late to substitute of the overall fiber materials [14]. It is effectively accessible into the market and to handily deal with to the handling [15]. Then, at that point, the palm and some nanoparticles consolidated materials to be utilized in many spots to a substitute course of numerous particles of the vehicle and the other significant parts [16]. These particles are exceptionally messed with amicable. Palm fiber is by and large made by the weaving system [17]. Also, these materials to be at first ready by layered particles and the layers to be joined by individually to a huge part of the materials into the roll particles [18]. After the layers arranged the lump layers to be isolated individually. In the wake of setting up the layers put away in the typical room temperature [19].

Then at that point, a similar interaction to be utilized for the bamboo fiber as shown in Fig. 2. The crude bamboo particles to be gathered [20]. The materials are to be taken out the undesirable particles of the unrefined components [21–23]. The gathered chip material is to be washed water at 90 °C for 4 h. After finishing the cleaning system, then the unrefined components to be squashed and put away in

Fig. 1 Graphene

Fig. 2 Bamboo fiber

ordinary stockpiling container to setting up the layers. The natural substance is to be moved into mechano-compound interaction [21, 24].

2.1 Hand Lay-up Method on Epoxy Resin

Then, at that point, the hand lay-up strategy is utilized to this manufacture cycle. The creation interaction generally utilized for composite trim cycle in current days. Then, at that point, the cycle is to be followed ongoing days to create the normal fiber materials and testing the normal interaction. Then, at that point, the hand lay-up strategy is exceptionally simple and least demanding technique for manufacturing

Fig. 3 Hand lay-up laminate

the materials as shown in Fig. 3. Then, at that point, this technique to be followed individually in leisurely.

Then, at that point, the readiness sheet to be set in at first and to make individually utilizing to the epoxy sap arrangement. Then, at that point, first sheet to be put in floor and the main creation sheet set in the sheet utilizing the epoxy pitch. Furthermore, the epoxy tar glued, after to be Remy sheet to be put in top of the epoxy pitch. Then, at that point, the hand lay-up apparatus to be utilized in flatted the sheet effectively and appropriately on the level surface and diminish the air bubbles. In this interaction to be handily taken care of the administrator and decreased the time. No experience to be needed to set up this interaction. Then, at that point, the roller is to be utilized for this interaction. Then, at that point, the sonication cycle is generally utilized in current days for blending the materials in readiness of nanoparticles and the primary parts of fluids. This strategy is extremely helpful to blend the part's equally in a crate.

3 Results and Discussion

Different test was led to test the mechanical, physical, synthetic properties of aluminum metal framework composite. We are taken different testing strategy like tensile test and flexural test as shown in Figs. 4 and 5.

The overlays are ready for leading test tests were done in the Met Mech Designers in Guindy. Covers were ready by hand lay-up. Every one of the filaments are slices as indicated by the measurements. In these covers, we are utilizing strands glass fiber. These covers of sap are utilizing named epoxy gum LY556 with their hardener HS951. The glass fiber and the epoxy tar are purchased from Light Mach Composites, Chennai.

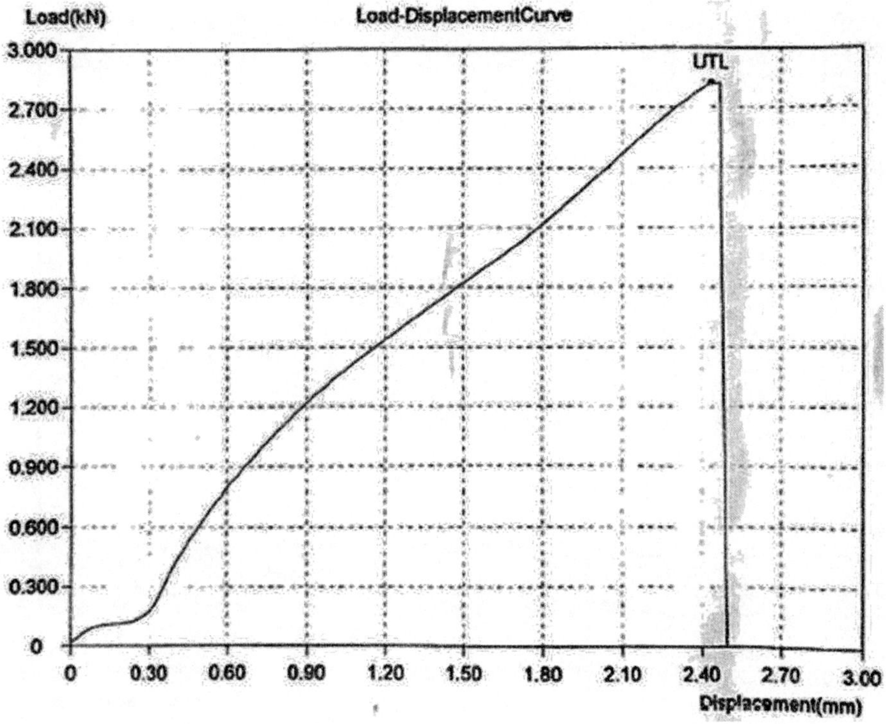

Fig. 4 Tensile test result

4 Conclusions

Then, at that point, manufactured materials tried in different tests in testing machines, identified with the test. Then, at that point, the tractable test, flexural strength, and the effect strength. The result of our venture is looked at in past normal fiber composite materials analyzed in sway test are high. The past esteem is 125 MPa, then, at that point, our worth was 535 and 561 MPa. The accompanying ends are gotten from the proposed research work on concentrate on the different properties of the regular fiber under the given by the mixture composite arrangement and supporters of support and has been directed through different testing cycles to show up at the definite induction on the ideal reach for achieving the ideal presentation of the constituents in light of its properties across different areas in correlation among the diverse foreordained examples. Mechanical properties of created normal fiber composite from the SEM micrograph were cleared vulnerability our material.

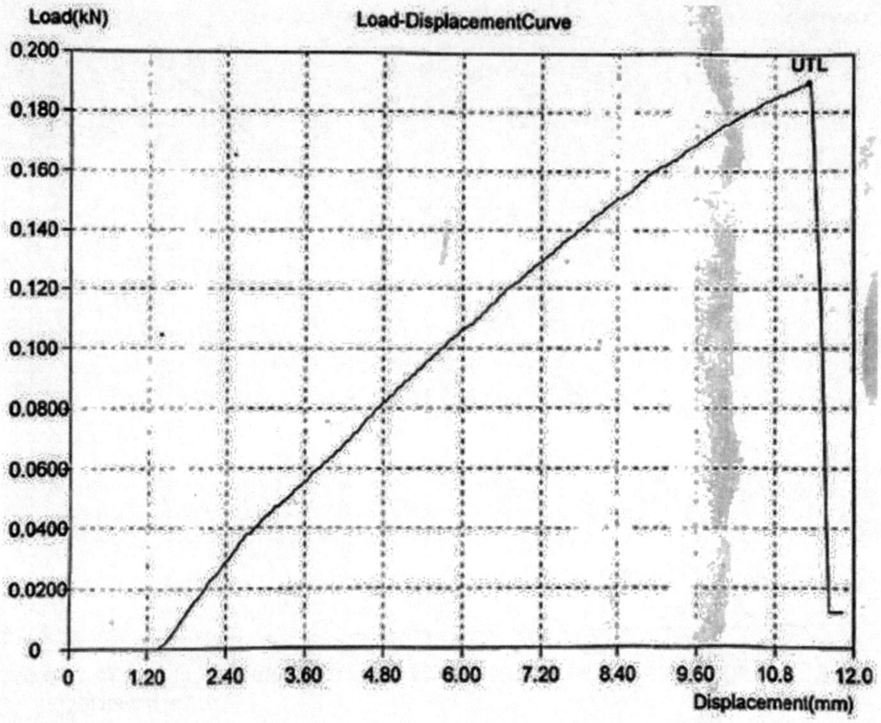

Fig. 5 Flexural test result

References

1. Sabeel Almeda K, Vijaya Ranga S (2008) Tensile, flexural and interlaminar shear properties of woven jute and jute glass fabric reinforced polyester composites. J Mater Process Technol 207:330–335
2. Sanjay MR, Yogesha B (2016) Tensile, flexural and interlaminar shear properties of woven jute and jute glass fabric reinforced polyester composites. J Miner Mater Charact Eng 4:15–25
3. Margoto OH, Morris VADS, De Paiva JMF. Mechanical properties of poly propylene matrix composites with jute fiber fabric. In: 10th International conference on composites science technology
4. Manjunath BH, Prahalad Rao K (2013) Influence of fiber particles reinforcement on epoxy composites. IJERA 3(3):1147–1151
5. Mohan TP, Kenny K (2011) Effects of synthetic and processing methods on dispersion characteristics of nonclay in polypropylene polymer matrix composites. Mater Sci Appl Sci Res 2:785–800
6. Warbhe NO, Srivastava R, Advani PS (2016) Mechanical properties of Kevlar, jute reinforced epoxy composite. IJIRSET 5(9)
7. Koo JH (2006) Polymer nanocomposite, processing, characterization and application. McGraw-Hill Companies
8. Ragosta G (2005) Epoxy-silica particulate nanocomposite: chemical interactions, reinforcement and fracture toughness. Polymer 46:10506–10516

9. Manigandan S, Gunasekar P, Nithya S, Devipriya J, Saravanan WSR, Venkatesan S (2021) Acoustic and vibration analysis of pineapple leaf fibre laminates for aircraft applications. Int J Ambient Energy 42(2)
10. Gunasekar P et al (2019) Effect of carbon nanotubes on sheet molding composite analysis for automotive application: FEA method. Int J Innov Technol Explor Eng 8(9)
11. Raja VS, Shoji T (2011) Stress corrosion cracking: theory and practice. Woodhead Publishing Ltd
12. Sim J, Park C, Moon DY (2005) Characteristics of basalt fiber as a strengthening material for concrete structures. Compos Part B Eng 36(6–7):504–512
13. Saffarian HM, Warren GW (1998) Aqueous corrosion study of α2-Ti3Al/SiC composites. Corrosion 54(11):877–886
14. Gu H, Shaw AD, Amoozgar M, Zhang J, Wang C, Friswell MI (2020) Twist morphing of a composite rotor blade using a novel metamaterial. Compos Struct 254
15. Khan TA, Kim H, Kim H-J (2019) Fatigue delamination of carbon fiber-reinforced polymer-matrix composites. In: Failure analysis in biocomposites, fibre-reinforced composites and hybrid composites. Elsevier, pp 1–28
16. Hafizal Hamidon M, Sultan MTH, Hamdan Ariffin A (2019) Investigation of mechanical testing on hybrid composite materials. In: Failure analysis in biocomposites, fibre-reinforced composites and hybrid composites, pp 133–156
17. Jawaid M, Thariq M, Saba N. Structural health monitoring of biocomposites, fibre-reinforced composites and hybrid composites
18. Abdellaoui H, Raji M, Bouhfid R, el kacem Qaiss A (2019) Investigation of the deformation behavior of epoxy-based composite materials. In: Failure analysis in biocomposites, fibre-reinforced composites and hybrid composites, pp 29–49
19. Sivashankari P, Krishnamoorthy A, Prakash S, Raja S, Venkatesh S (2018) Analysis of carbon reinforced metal matrix composite for IC engine valves. J Eng Sci Technol 13:74–82
20. Ganesan S, Hemanandh J, Sridhar Raja KS, Purusothaman M (2021) Experimental investigation and characterization of HDPE & LDPE polymer composites. In: Advances in industrial automation and smart manufacturing. Lecture notes in mechanical engineering. Springer, pp 785–799
21. Mani P, Kumar R, Kumar R, Gnanamani S, Janarthanam H, Ponnappan VS (2020) Comparative analysis of properties on leaf springs with different composite materials. AIP Conf Proc 2311(080011):1–5
22. Venkatesh S, Prakash S, Raja S, Manigandan S, Sivashankari P (2018) Effects of delamination on the critical buckling load of glass fiber reinforced epoxy composite plates. Int J Veh Struct Syst 10(4)
23. Subramanian V, Maria RV, Gunasekar P, Ponnappan VS, Jayaraman J, Joy N (2020) Investigation on the hardness characteristic of high carbonised high chromium steel with cryogenic quenching. AIP Conf Proc 2311(1):30007
24. Nirmalkumar G, Kumar CVS, Venkatesh S (2020) Experimental modal analysis of compressor rotor blades. AIP Conf Proc 2311(1):30010

Design and Fabrication of Intra Campus Mobility Electric Two-Wheeler

S. Kiranlal, V. M. Brathikan, and H. N. Lalith

Abstract There is developing interest in electric bikes in India as there will be less air contamination, lower support cost, and decreased commotion utilizing electric bikes. Since the petrol prices are increasing, there is a great market for the EV vehicles. Further, EV market in India is growing rapidly. The analysis is done to study the structure development of the product. An approach utilizing the finite model is created to rapidly examine the design. Point-by-point investigation utilizing components is done to precisely anticipate the burdens. FEA results are approved with component-level testing and stress–strain values are noted. Vibrational attributes are concentrated via completing the modular investigation with explicit dynamics on the chassis. Further, explicit dynamics are done to study how the chassis behaves with the time of the crash and also with the particular velocity. This electric two-wheeler is mainly designed for intra campus mobility for students and staff so that a lot of carbon emission will be reduced.

Keywords E-bike · FEA · Intra campus mobility

1 Introduction

E-vehicle is one of the fast-growing commodities in Indian Automotive Sectors. In the global context due to climate change, the need for advances in renewable energy, rapid urbanization, availability of data capture technologies and digital revolutions, battery advancements and security to energy requirements, the need for policies on lowering carbon emission, and an effective-cost efficient mobilizing solution.

The Indian government had established various full and semi-organizations to create policies and training necessary to establish a full-fledged EV sector in the sub-continent and to create a transition from gasoline to the electric vehicle. India, as shown in the figure, has about 79% of the total vehicle that are two-wheelers,

S. Kiranlal (✉) · V. M. Brathikan · H. N. Lalith
Kumaraguru College of Technology, Coimbatore, India
e-mail: kiranlal.s@kct.ac.in

N. M. Sivaram et al. (eds.), *Advances in Manufacturing, Automation, Design and Energy Technologies*, Lecture Notes in Mechanical Engineering,
https://doi.org/10.1007/978-981-99-1288-9_11

4% as three-wheelers, 3% as busses and large goods vehicles, 12% as economic four-wheelers and 2% as of luxury four-wheelers. With a fast-growing population and with most of the population as young and with a high-velocity market in two-wheeler, the small compact versatile electric scooter will be a solution to all regular short to medium range mobility needs.

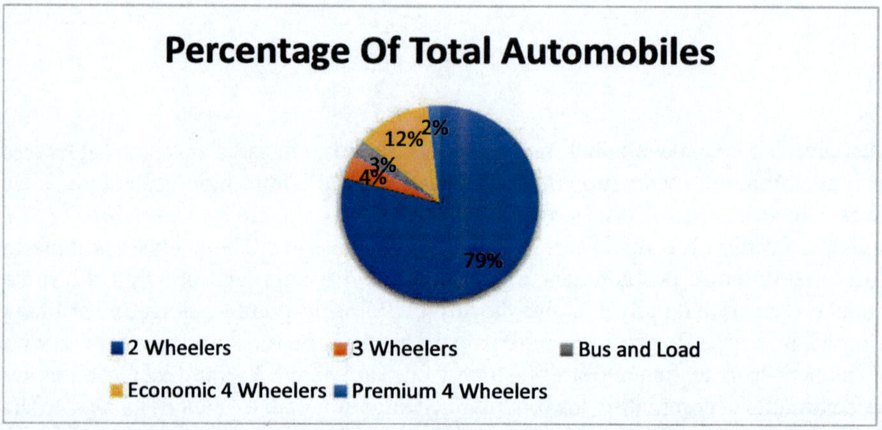

Since only 2% of the total customers are premium buyers and the electric vehicles are of high-end technology, and cost is comparatively more than the available vehicles, the market segment is relatively small, which leads to the slower adoption of this technology since 2001. But now the cost of the electric components has reduced and it is expected to reduce further. This effect along with the government policies has accelerated the EV market. The government has made a strong pitch for electric vehicle deployment in India through initiatives such as Faster Adoption and Manufacturing of E-vehicles (FAME), FAME-2, Green Mobility Fund and Make in India. In an intense move, the public authority is pushing available to be purchased EVs solely after 2030.

The electric bike is very efficient and has less pollutants as compared to the IC engine which produces more CO, NO_x and HC with lower efficiency. With an electric bike, pollutants will not be emitted into the environment [1].

The electric bike is designed in such a way that it gives assist to the rider. The motor is mounted to the pedal shaft with a total transmission ratio of 1:25. The velocity reaches 25 km/h with 100% motor assist and reaches 15 km/h with 60% motor assist [2].

The main objective of the design is to build a frame for a lightweight electric vehicle with a low cost of production and powered by a rear hub electric motor. Maximum speed is limited to 45 km/h and the weight of the entire bike is below 60 kg. The static analysis of the frame is done in ANSYS to find stress, strain and displacement in each element of the model [3].

The paper discusses the accidents that happen due to improper regulations for electric bikes. Since the motor can be of any power, speed also increases which leads to accidents. Due to this, many countries passed legislation regarding electric bikes categorized as mopeds and motorcycles which should not exceed 25 km/h and power must be below 250 W [4].

Electric bikes are known to reduce greenhouse emission impact on the ecosystem. It is estimated that 12% of carbon dioxide emission is reduced and 225 kg of carbon dioxide reduction per year from a single electric bike. These e-bikes have the potential to help the environment by reducing the impacts of the pollutants of gasoline bikes [5].

In India, electric bikes sales are gradually increasing due to various offers that government gives like subsidies for buying electric bikes and cars. Even though the sales increase, there are only a few charging stations available and that are limited to cities. The Government of India recently proposed to replace 100% of vehicles in India with EVs by 2030 [6].

The frame that we fabricated is of very simple design in order to reduce overall weight. It has been designed using rectangular tube frame since it has more resistant to bending forces. The frame is subjected to vehicle weight and rider weight with that static analysis is obtained along with that modal and explicit dynamics also has been evaluated.

This sector of two-wheeled e-vehicle lies a domain of micro-mobility which can be imagined as a hybrid of a cycle and an electric bike. These small ranged micro-e-bikes are very versatile for usage such as small to medium range mobility problems and continuous policing and roaming.

2 Computational Methodology

2.1 *Designing of CAD Model*

The entire design of the e-bike is designed in SOLIDWORKS CAD software. The complete CAD model of this e-bike is designed precisely with the required dimensions to get an accurate analysis report. The mainframe is fabricated with stainless steel and the entire assembly is built in the SOLIDWORKS software itself (Fig. 1).

3 Finite Element Analysis

Finite element analysis (FEA) is a simulation where we can find how a component behaves under various boundary conditions that act on different parts of the body. Component examination shows whether an item will break, wear out or work the way it was planned. FEA is utilized by specialists to assist with reproducing actual wonders and in this way decrease the requirement for actual models, while taking

Fig. 1 CAD model of e-bike

into consideration the enhancement of parts as a feature of the plan interaction of a venture.

3.1 Static Structural

The static structural is done on the main tubular frame with the required boundary conditions such as force, the weight of the rider and fixed support. A total force of 992.46 N is applied at the stem and a rider force of 784.8 N is also applied on the seat. The swingarm mount is given as fixed support (Figs. 2 and 3).

The total deformation in the static structural analysis is 2.0366 mm and von Mises stress is 114.79 MPa.

Table 1 gives the maximum and minimum values of the total deformation, equivalent stress, principal strain and safety factor.

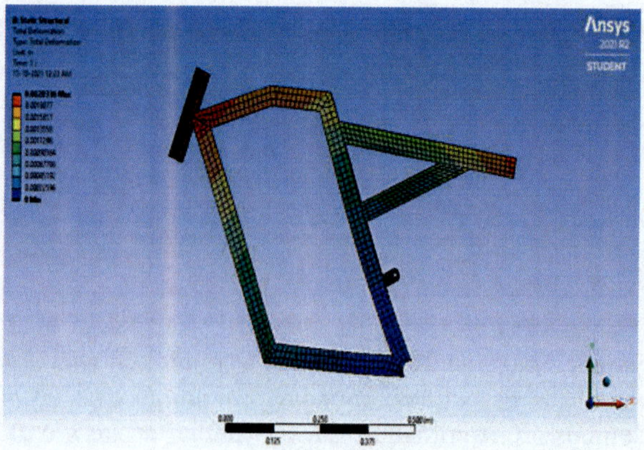

Fig. 2 Total deformation of frame in static structural

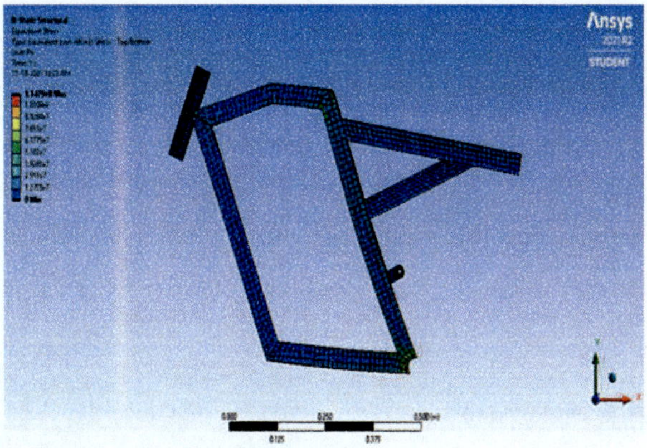

Fig. 3 Equivalent stress of frame in static structural

Table 1 Results of static structural

Results	Minimum	Maximum
Total deformation	0 m	2.0336e-3 m
Equivalent stress	0 Pa	1.1479e8 Pa
Principal strain	−1.1269e-6 m/m	4.9999e-4 m/m
Safety factor	2.1778	15

3.2 *Modal Analysis*

Modal analysis is the investigation of dynamic properties of an item or part in a frequency recurrence. It gives subtleties of how a part react to different regular frequencies. It gives a perspective on the vibration qualities of a mechanical construction or part (Table 2).

Table 2 Modal analysis with natural frequency

Mode	Frequency (Hz)
1	103.35
2	254.05
3	294.34
4	306.61
5	558.06
6	576.31
7	628.20
8	702.51
9	797.33
10	850.45

3.3 Explicit Dynamics

Explicit is utilized to ascertain the condition of a given framework at an alternate time from the current time. Interestingly, an understood examination discovers an answer by settling a condition that incorporates both the current and later conditions of the given framework (Figs. 4, 5 and 6).

From Table 3, we can get the total deformation, Tables 4 and 5 represent stress and equivalent strain, respectively, Table 6 gives the safety factor and Table 7 gives

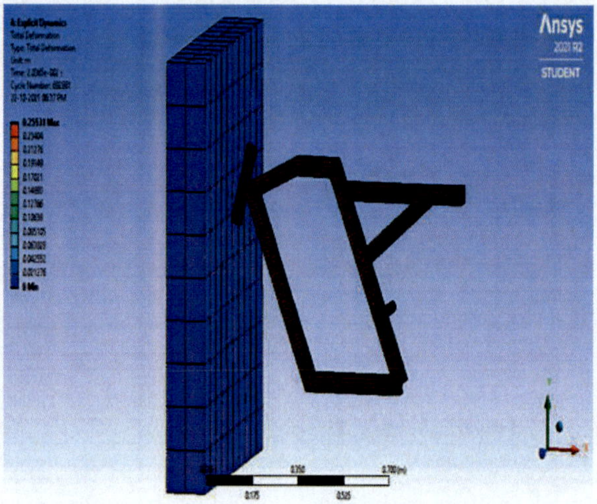

Fig. 4 Total deformation of frame in explicit dynamics

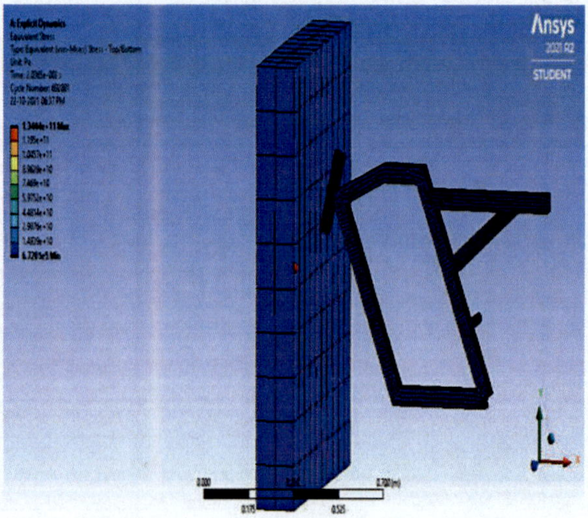

Fig. 5 Equivalent stress of frame in explicit dynamics

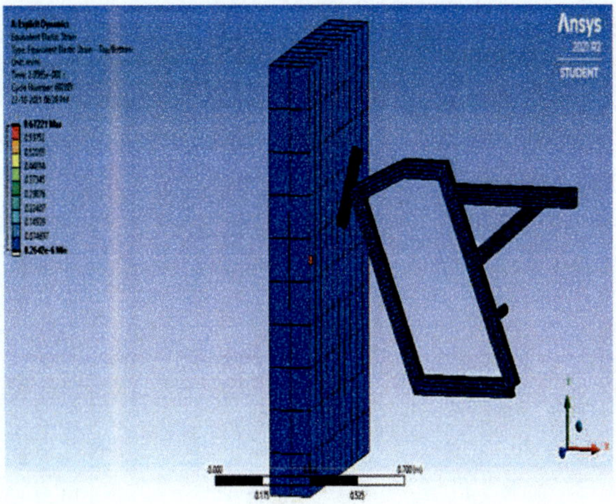

Fig. 6 Elastic strain of frame in explicit dynamics

Table 3 Deformation table of explicit dynamics

Time (s)	Minimum (m)	Maximum (m)
1.1755e−38	0	0
5e−3	0	6.2504e−2
1e−2	0	0.125
1.5e−2	0	0.18751
2e−2	0	0.25001
2.0796e−2	0	0.54514

all the results of explicit dynamics. The explicit dynamics are analyzed when the bike is moving at a velocity of 12.5 m/s and deformation of 0.25531 m, von Mises stress of 134.4e3 MPa and equivalent elastic strain of 0.67221 m/m is observed.

Table 4 Stress table of explicit dynamics

Time (s)	Minimum (Pa)	Maximum (pa)
1.1755e−38	0	0
5e−3	0	1.896e+7
1e−2	0	1.716e+7
1.5e−2	0	6.728e+6
2e−2	0	4.032e+6
2.0796e−2	0	2.135e+10

Table 5 Strain table of explicit dynamics

Time (s)	Minimum (m)	Maximum (m)
1.1755e−38	0	0
5e−3	0	1.409e−4
1e−3	0	1.121e−4
1.5e−2	0	4.514e−5
2e−2	0	3.292e−5
2.0796e−2	0	0.145

Table 6 Safety factor table

Time (s)	Minimum	Maximum
1.755e−38	15	15
5e−3	13.179	15
1e−2	14.567	15
1.5e−2	15	15
2e−2	15	15
2.079e−2	1.17e−2	15

Table 7 Results of explicit dynamics

Results	Minimum	Maximum	Time (s)
Total deformation	0	0.5451 m	2.0796e−2
Equivalent stress	0	2.1358e + 10 Pa	2.0796e−2
Equivalent strain	0	0.1458 m/m	2.0796e−2
Safety factor	1.1705e−2	15	2.0796e−2

4 Design and Development

This EV is the first version of our micro-e-vehicle. This is a scalable model with high strength in both physical and economical parameters. The vehicle can cut costs spent on fuels to a great extent. Even with industrial charges, the vehicle can be operated at a very low cost and can cover a distance of up to 25 km in one full charge. The small ergonomic design makes it weigh below 44 kg and still stiff to produce the necessary feel of driving an e-vehicle in a satisfactory environment (Fig. 7).

Specifications of E-Bike

Power	250 Watts hub motor (48 V)
Battery pack	Li-ion 48 V, 10 Ah
Range of e-bike	25 km
Top speed	30 Kmph
Time to full charge	3 h (with a 48 V and 3 Amps charger)

Fig. 7 Image of e-bike

5 Result

In this paper, a comparison of IC engine and electric bike is discussed based on the environmental issues and the e-bike has negligibly zero carbon emission along with that FEA is carried out to know the capability of the chassis to withstand the given load, and modal analysis is also done with the natural frequencies. The prototype that we developed based on the analysis weight 44 kg which is perfectly distributed as 22 kg in rear and 22 kg in front. Due to the low weight of the e-bike, the vehicle is able to achieve 25 km per charge. The results of static analysis with total deformation of 2.0366 mm, von Mises stress is 114.79 MPa and safety factor of 2.1778. In modal analysis, the reported frequencies are 103.35, 254.05, 294.34 Hz and so on. We designed in such a way that our frame will not vibrate in those frequencies. In explicit dynamic, the total deformation is 0.54514 m at 2.0796e−2 s, maximum stress is 2.135e+10 at 2.0796e−2 s and maximum equivalent strain is 0.1458 m/m at 2.0796e-2 s.

6 Conclusion

The prototype is evaluated in both software (FEA) and also in real-time testing. From the software analysis and real-time testing, it is known that the frame is very stable throughout the testing. And a lot of works is carried out by administrative staff as well as for students this will make a lot of difference, as their intra campus commute become easier. The electric two-wheeler design and developed is found to be efficient on testing. As the electric vehicle emits no pollutants, developing this micro-mobility will make a pollution free and sustainable campus.

References

1. Abagnale C, Cardone M, Iodice P, Strano S, Terzo M, Vorraro G (2015) A dynamic model for the performance and environmental analysis of an innovative e-bike. Energy Procedia 81:618–627. https://doi.org/10.1016/j.egypro.2015.12.046
2. Abagnale C, Cardone M, Iodice P, Strano S, Terzo M, Vorraro G (2015) Model-based control for an innovative power-assisted bicycle. Energy Procedia 81:606–617. https://doi.org/10.1016/j.egypro.2015.12.045
3. Godlewski T, Pawlak W (2014) Design of supporting frame for light electric motorcycle. In: Students' science conference
4. Siman-Tov M et al (2018) A look at electric bike casualties: do they differ from the mechanical bicycle? J Transp Heal 11(October):176–182. https://doi.org/10.1016/j.jth.2018.10.013
5. McQueen M, MacArthur J, Cherry C (2020) The E-bike potential: estimating regional e-bike impacts on greenhouse gas emissions. Transp Res Part D Transp Environ 87:102482. https://doi.org/10.1016/j.trd.2020.102482
6. Singh V, Singh V, Vaibhav S (2021) Analysis of electric vehicle trends, development and policies in India. Case Stud Transp Policy 9(3):1180–1197. https://doi.org/10.1016/j.cstp.2021.06.006

Experimental Investigation and Effect of Machining Parameters on Surface Roughness of AISI 1050 Steel Using Parallel Turning

Dereje Gemechu Bedane, Endalkachew Mosisa Gutema, and Mahesh Gopal

Abstract Utilizing more than one cutting tool to machine a work piece is parallel turning. But due to interaction of parallel tools at a same time an additional settings are needed to escalate the stability of the machine and tool. The current study utilizes a supplementary cutting tool, placed near to the cutting tool by using special modeled jigs. The influence of cutting parameters on surface roughness was investigated in this study utilizing AISI 1050-steel as the work piece and coated carbide cutting tools under dry machining circumstances. Surface roughness profilometer were used to assess surface roughness quality. The cutting parameters are optimized by using Taguchi L9 orthogonal array design of experiments and ANOVA. The MINITAB 17 software is used to solve the regression coefficient. The result indicates, cutting speed was the most notable surface roughness metric, followed by feed under dry circumstances. The percentage of contributions is 71.09 and 19.81% under dry turning conditions. The best surface quality achieved with optimal cutting parameters was 1.05 μm.

Keywords AISI 1050 steel · Surface roughness · ANOVA · Taguchi method

1 Introduction

Machining is an extensive term that refers to a set of procedures to remove material from a work piece. Currently, the industries are focusing not only on improving productivity, but also on improving product quality in a shorter period of time. The kinematic relationship between cutting tools and work pieces is directly related to shape, chip size, metal removal rate, and surface quality. The work can be processed with lubrication (wet machining) and non-lubricated (dry machining). During machining, tools, and work pieces generates a large amounts of heat in the

D. G. Bedane (✉) · E. M. Gutema · M. Gopal
Department of Mechanical Engineering, College of Engineering and Technology, Wollega University, Post Box No: 395, Nekemte, Ethiopia
e-mail: derejegemechu2009@gmail.com

N. M. Sivaram et al. (eds.), *Advances in Manufacturing, Automation, Design and Energy Technologies*, Lecture Notes in Mechanical Engineering,
https://doi.org/10.1007/978-981-99-1288-9_12

cutting region, affects the surface quality, life of the tool, and variation in work piece properties. Many studies have been conducted by researchers to predict surface irregularities of as the primary output parameter. The surface roughness model was developed by combining ANOVA and RSM in the turning of 42CrMo4 steel using a ceramic tool. Cutting speed, cutting depth, and feed rate are used as input variables to calculate tool vibration and work piece roughness in the radial and force created during cutting directions. The result emphases, the feed is the foremost and main factor the affects the surface roughness, but the vibration has a low effect on output Hessainia et al. [1]. Gopal [2] conducted an experimental study to predict surface roughness considering the speed, feed, depth, and nose radius of a tool while machining DSS 2205 using carbide tip tool material. Statistical techniques such as ANOVA are used for to observe the distinctiveness of performance. A quadratic model is developed to predict surface roughness. For cutting tests, cutting speed is the parameter that has the utmost influencing the cutting activity. To forecast surface roughness, a quadratic model is constructed and developed. Cutting speed is the characteristic that have greatest influence on cutting activity in cutting tests. Patel and Gandhi [3] conducted an examination to predict work piece roughness of AISI D2-Steel and CBNtool is utilized to complete turning operation, the parameters are: feed, depth, and cutting radius. Mathematical models shows the relationship between rate of feed, depth, cutting speed, and cutting edge radius to measure work piece roughness. In this study, a cutting operation is carried out by using speed, approaching angle, feed, and cutting depth as input variables to calculate force and roughness. Surface roughness (Ra) has a positive effect on feed, and tends to have a negative tendency on approaching angle, speed, and cutting depth Sharma et al. [4]. Özel et al. [5] did an experimental investigation to analyze the parameters of the cutting tool's edge shape, hardness, feed, and speed to quantify the quantity of roughness during turning of AISI H13-Steel. This research demonstrates the effects of hardness, tool cutting point shape, feed, and speed on roughness. The surface hardness of the work piece is low, and the sharpened edge form lowers forces in the direction of radial and tangential. Sundaram and Lambert [6] used tungsten carbide tools to forecast surface roughness when turning AISI 4140 steel, taking into account of speed, feed, depth, and radius in the tool.

2 Parallel (Simultaneous) Turning Processes

Previous studies focused on machining with a single tool in turning operation to optimize machining parameters. Many researchers used parallel machining tools to increase and to improve operational productivity, reduce setup time, improve component quality, and reduce floor space. Tang et al. [7] conducted an experiment using a parallel lathe process using two tools and examined both one-pass and multi-pass operations. Speed in spindle, depth and feed are considered to reduce machining time by using PSO optimization technique. By altering speeds, constant depth and constant feed, the experiments are carried out utilizing two tools to quantify cutting vibration, work surface quality, and dimensional tolerances. As a consequence, increasing the speed, the vibration is decreased and the increasing the spacing between the front and rear cutting tools Kalidasan et al. [8]. The stability of the parallel turning process was explored in this paper. Simulation models were created and compared to exploratory cutting tests to measure the effects of radial angle. The two successive cuts time between in a comparable turning procedure on the same cutting surface is affected by this angle. The simulation model is utilized to determine the best parallel turning process parameterization Brecher et al. [9]. Reith et al. [10] conducted parallel turning experiments to investigate the dynamic properties of cutting tools to eliminate chatter. Measurements were performed to validate the theoretical prediction of chip width. The results confirm that a two-cutter turning system can achieve significant improvements in cutting performance if the tool is correctly detuned. Ozturk et al. [11] performed thc two parallel turning process to study the stability of the experiment. Simulation and investigational results shows the process parameters for the high MRR rate. The author concluded that the natural frequencies of the tools are slightly different when at steady operation, and the worst stability occurs when the tools are same.

3 Developing and Modeling of Jig

In this experiment, the jig as shown in Fig. 1 was made using AISI H13 alloy steel, the hardness range of HRC 38–48, which is characterized by excellent heat resistance, high hardenability, strength toughness, and hot working dies are manufactured using this steel. The two tool holder are mounted in parallel, and designed at 83.5 mm height, 117 mm length, and 66 mm width to have a rigid supports on the tool post. The tool holder has eight holes, each of 10 mm in diameter and 15 mm in height to accommodate in correct position.

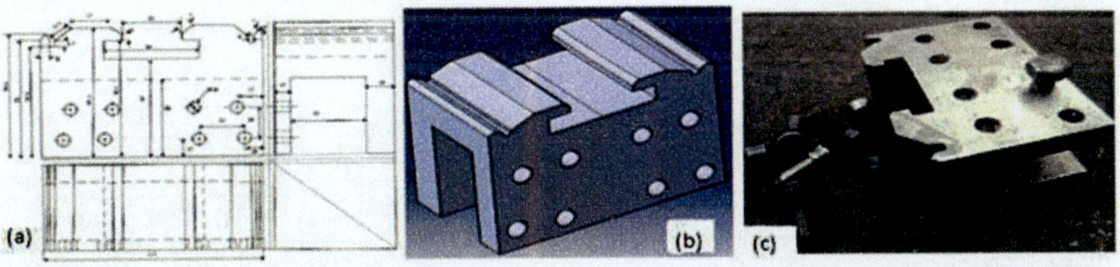

Fig. 1 Development and model of Jig

4 Taguchi Method

Taguchi methods are an efficient way to design a process. Because of its ease of adoption and application, especially for users with basic statistical understanding, the DOE employing Taguchi orthogonal array technique has grown in popularity in the manufacturing, industrial, and scientific communities. The authors Asiltürk and Akkuş [12] minimized roughness by turning the work piece to decrease roughness using the Taguchi technique. CNC lathe is employed to carry out the experiment. The feed, speed, and depth are considered during dry turning test on a hardened AISI 4140 with a carbide tool. The rate of feed has the greatest consequence on work piece. Minimum quantity lubrication (MQL) turning experiment is adapted to measure roughness, wear in the tool, and MRR. Optimization was carried out using the Taguchi and S/N ratio methods and a quantitative investigation is performed, taking into consideration feed, speed, and cutting depth, the results reveal that speed influences roughness of the material Mia et al. [13].

5 Experimental Design and Procedure

The cutting speed, feed, and depth of cut for tool 1, 2 are the parameter selected for dry cutting operation. Design of experiments is performed by Taguchi L9, analysis is performed by Analysis of Variance (ANOVA) using MINITAB 17 software. A cutting parameter is selected and its range are shown in Table 1, the highest level of all three variables are (+1) and the lowest level are (−1). The intermediate level are 0 and it is determined by the interpolation Das et al. [14]. Figure 2 shows the parallel cutting process on the lathe. A conventional medium-power lathe is used for experimental analysis. Round AISI 1050 steel is used for the processing material. Carbide insert tools are used for cutting the work piece. Surface waviness is assessed using a contact type surface roughness profilometer (model 657111). The input and output values of dry cutting conditions are tabulated in Table 2. Table 2 shows the empirically chosen four-factor design matrix with 9 coded sets or oblique criteria.

Table 1 Machining parameters and their levels of experimentation

Cutting parameter	Units	Parameter notations	Levels		
			(−1)	0	(+1)
Cutting speed	m/min	V_c	100	130	160
Rate of feed	Mm/rev	F_Z	0.1	0.2	0.3
Cutting depth (Tool 1)	Mm	a_{p1}	0.33	0.66	1
Cutting depth (Tool 2)	Mm	a_{p2}	0.25	0.5	0.75

Fig. 2 Cutting operation by using two cutting tools

6 Results and Discussion

The AISI 1050 steel is employed to conduct the experiment to assess and analyze the influence of major machining parameters on surface quality of the work piece, and the work piece after the cutting process are shown in Fig. 3.

6.1 *Analysis of Surface Roughness*

Taguchi's design of experiments methodology is used to examine the surface roughness trial findings. Taguchi (S/N) ratio approach is used to determine the best degree of response for each individual as well as to measure quality aspects that differ from the desired value. The quality characteristics were affected by uncontrollable parameters. In this method, the "Noise" is a "disagreeable value" represented as standard deviation (SD) and "signal (mean)" is represented as "desirable value". Using experimental design, the values are used to compute the average surface roughness of each factor. Table 2 displays the experimental findings for each sample's roughness value (Ra) and its average value. In this case, the "smaller is better" S/N ratio is utilized to choose the best parameter. Mean value of surface roughness of each factor would be determined using the interaction effect between the parameters. From Table 4, the mean effects are represented the expression of the smaller is better. The element that has the largest influence on the mean value under dry cutting conditions is the cutting speed (delta = 0.337 and it is ranked as 1). The depth of cut 2 has the least influence on the mean (delta = 0.053 and it is ranked as 4). These parameter levels, specifically $v_c 3$ $f_z 2$ $a_{p1} 1$ $a_{p2} 2$, with superior surface quality values for the sample, are employed as a likely combination of parameters for improved results.

The delta is calculated from the maximum and minimum averages of the effect.

$$\begin{aligned} \text{Delta} &= \text{maximum mean} - \text{minimum mean} \\ \text{Delta (A)} &= (1.540 - 1.203) = 0.337 \end{aligned} \quad (1)$$

Table 2 Output responses of experimental values

Run	Factor 1	Factor 2	Factor 3	Factor 4	Surface roughness (μm)			Average (Ra) (μm) (dry)	Signal-noise ratio
	Cutting speed (m/min)	Rate of feed (mm/rev)	Depth of cut 1 (mm)	Depth of cut 2 (mm)	Trail 1	Trail 2	Trail 3		
1	100	0.1	0.33	0.25	1.76	1.35	1.63	1.58	−3.973
2	100	0.2	0.66	0.50	1.25	1.39	1.47	1.37	−2.734
3	100	0.3	1.00	0.75	1.65	1.34	1.87	1.62	−4.190
4	130	0.1	0.66	0.75	1.18	2.16	1.73	1.69	−4.557
5	130	0.2	1.00	0.25	1.27	1.13	1.80	1.40	−2.922
6	130	0.3	0.33	0.50	1.26	1.43	1.90	1.53	−3.693
7	160	0.1	1.00	0.50	1.54	0.97	1.39	1.30	−2.278
8	160	0.2	0.33	0.75	1.05	1.03	1.07	1.05	−0.423
9	160	0.3	0.66	0.25	0.92	1.48	1.38	1.26	−2.007
Average performance mean value								1.42	−2.975

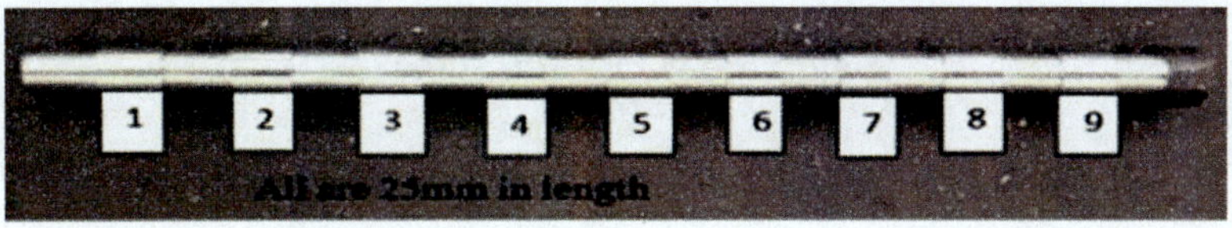

Fig. 3 Sample of workpiece after turning process

6.2 Response Analysis of Mean Value

The delta value indicates that the surface roughness ranges from high impact to low impact, and Fig. 3 shows that the control coefficients for the variation in surface roughness value are plotted. As observed from Fig. 3, the cutting speed is in level 3 (160 m/min), the rate of feed level 2 (0.2 min/rev), the cutting depth 1 is level 1 (0.33 mm), and the depth of cut 2 is 4 as shown in Table 3. The velocities are on plane 2 (0.50 mm). This is because the optimal turning parameters identified were high cutting speed, medium feed, low cutting depth 1, and medium cutting depth 2. This indicates that the optimal machining parameter selected is $v_c3\ f_z2\ a_{p1}1\ a_{p2}2$. The mean effects of parameters are shown in Fig. 4.

Table 3 Response table of mean effects

Levels	Parameters (mean)			
	Cutting speed	Rate of feed	Cutting depth 1	Cutting depth 2
1	1.523	1.523	1.387	1.413
2	1.540	1.273	1.440	1.400
3	1.203	1.470	1.440	1.453
Delta	0.337	0.250	0.053	0.053
Rank	1	2	3	4

Table 4 S/N ratio response table

Levels	Cutting parameters (S/N ratio)			
	Cutting speed	Rate of feed	Cutting depth 1	Cutting depth 2
1	−3.633	−3.603	−2.697	−2.968
2	−3.725	−2.027	−3.100	−2.902
3	−1.570	−3.297	−3.131	−3.057
Delta	2.155	1.576	0.434	0.155
Rank	1	2	3	4

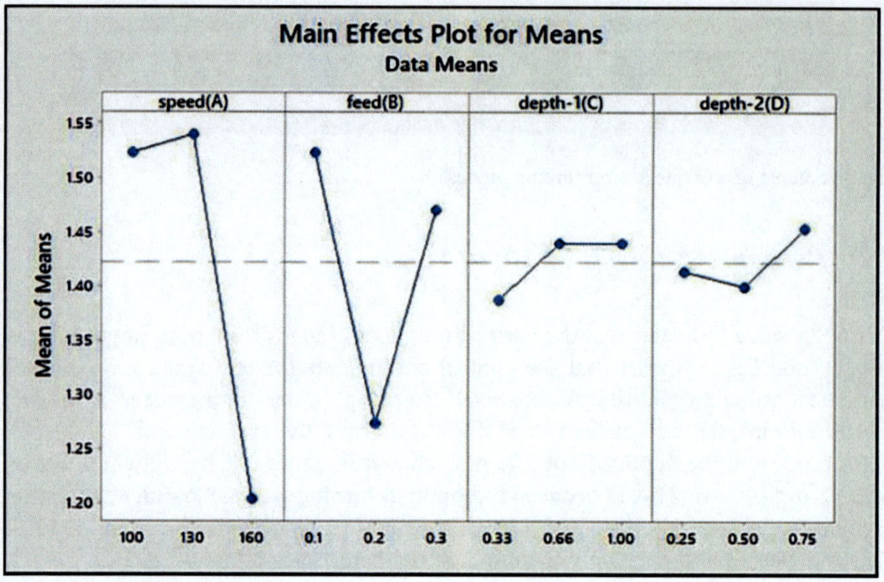

Fig. 4 Plot of mean's main effect

6.3 S/N Ratio Response Analysis

Figure 4 shows how the S/N ratio approach was utilized to quantify the change in roughness values during the experimental design. Based on the response characteristics, an equation was devised to determine the S/N ratio. Smaller of the response, the better for this study principle is adopted. Table 4 shows the response to a lower S/N ratio. The S/N effects of parameters are shown in Fig. 5.

The cutting speed S/N ratio of the weight percentage has the greatest influence with the Delta = 2.155 and is ranked as 1, and cutting depth 2 is the factor with the slightest effect on the S/N ratio and is ranked as 4 with delta value 0.155, the better surface quality characteristics, the optimal parameters are $v_c3\ f_z2\ a_{p1}1\ a_{p2}2$, and there are possible parameter combinations to improve the surface quality of AISI 1050 steel.

6.4 Prediction of Performance at Optimal Levels for Mean Value During Dry Condition

Four parameter coefficients are selected by a mean value to predict the best response. $v_c3\ f_z2\ a_{p1}1\ a_{p2}2$ is the average optimum parameter levels for the surface roughness values. To find the responses of the surface roughness predicted value the regression equation is used.

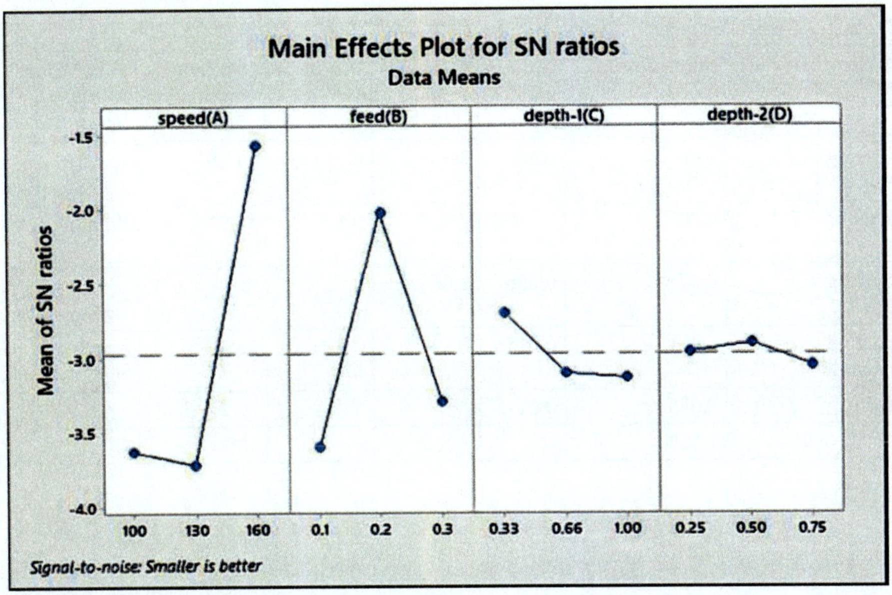

Fig. 5 Plot of S/N main effect

$$Yopt = m + (mv_C3 - m) + (mf_Z2 - m) + \left(ma_{p1}1 - m\right) + \left(ma_{p2}2 - m\right) \quad (2)$$

where $ma_{p1}1$—optimal parameter level value of a_{p1},
$ma_{p2}2$—optimal parameter level value of a_{p2},
mv_C3—optimal parameter level value of v_C,
mf_Z2—optimal parameter level value of f_Z,
m–roughness average performance mean value.

$$Yopt = 1.42 + (1.203-1.42) + (1.273-1.42) + (1.387-1.42) + (1.40-1.42)$$

$$Yopt = 1.003 \text{ (surface roughness predicted mean value)}$$

6.5 *Prediction of Performance at Optimal Levels for S/N Value During Dry Condition*

A low S/N ratio value selected for four parameter coefficients to predict the optimal response. The optimum value for the S/N ratio parameter of the roughness value is $v_c3\ f_z2\ a_{p1}1\ a_{p2}2$. The result was predicted using a regression Eq. (2).

Table 5 Variance analysis for surface roughness value

Source	Degrees of freedom	Sum-of-squares	Model sum of squares	F-statistic value	P-statistic value	Percentage contribution (%)	Significance E (α = 0.05)
Cutting speed	2	1.08	0.54	30.9	0.032	71.09	Highly significant
Rate of feed	2	0.301	0.1505	8.6	0.102	19.81	Significant
Cutting depth 1	2	0.072	0.036	2.06	0.637	4.7	Medium significant
Cutting depth 2	2	0.031	0.0155	0.89	0.920	2.04	Less significant
Error	2	0.035	0.0175			2.32	
Total	10	1.519				100	

$$
\begin{aligned}
Yopt &= -2.975 + (-1.57 + 2.975) + (-2.027 + 2.975) \\
&\quad + (-2.697 + 2.975) + (-2.902 + 2.975) \\
Yopt &= -0.271 (\text{predict S/N value for surface roughness})
\end{aligned}
$$

6.6 *ANOVA for Surface Roughness During Dry Condition*

ANOVA analysis is used to identify the most noteworthy elements influencing the roughness of a work material. Table 5 displays the computed and reported values. Table 5 depicts the variations in the contribution variation of the surface roughness of each element as a function of cutting speed (v_c), rate of feed (f_z), cutting depth 1 (a_p1), and cutting depth 2 (a_p2). The cutting speed factor was discovered to be an important factor that influences the roughness of AISI 1050 steel material; during the machining with a 95% confidence level. The weight percentage of contribution during ANOVA analysis that effect workpiece are cutting speed (71.09%), rate of feed (19.81%), cutting depth 1 (4.71%), cutting depth 2 (2.04%). As a result, the cutting speed during dry hard turning has a substantial influence on AISI 1050 steel roughness.

6.7 *Determining the Optimum Condition During Dry Condition*

Tables 3 and 4 provide the conditions for dry cutting. It is obvious that the ideal cutting speed optimal weight percent value is 160 m/min, optimal rate of feed value

Table 6 The comparison of Expected (Predicted) and Trail (Experimental) values

Response	Optimal Best parameter	Forecast (predicted) value		Trail value (Experimental)	
Roughness of work piece (μm)	$v_c3\ f_z2\ a_{p1}1\ a_{p2}2$	Mean	S/N	Mean	S/N
		1.003	−0.271	1.05	−0.424

0.2 mm/rev, cutting depth 1 0.33 mm, cutting depth 2 optimal value 0.50 mm. The S/N ratio has a lower value. As a consequence, the S/N ratio of the optimum processing parameter was chosen for the surface roughness determined by the trial findings. In this investigation, the ideal settings for $v_c3\ f_z2\ a_{p1}1\ a_{p2}2$ were superior cutting speed, average rate of feed, minimum cutting depth 1, and medium cutting depth 2.

6.8 *Confirmation Test for Dry Cutting Condition*

Physical experiments are used to validate Taguchi's projected values. Cutting parameters following optimum combinations are as follows: cutting speed (160 m/min), feed rate (0.2 mm/rev), cutting depth 1 (0.33 mm), and cutting depth 2 (0.33 mm) (0.50 mm). Table 6 summarizes the comparison between expected (predicted) and trailed values (experimental).

The value experimentally obtained was approximately near to predicted by Taguchi design and the percentage of error was 4.76%.

7 Conclusion

Current research uses Taguchi L9 orthogonal arrays and S/N ratio methods are used to optimize machining parameters of AISI 1050 steel materials with multi-tool turning operation.

- At the cutting speed (160 m/min), rate of feed (0.2 mm/rev), cutting depth 1 (0.33 mm), and depth of cut 2 (0.33 mm), a low surface roughness of 1.05 μm was attained in the dry cutting condition (0.50 mm).
- According to the ANOVA results, cutting speed is the greatest noteworthy factor that influence surface roughness, next to that rate of feed is the most influential factor. The percentage of contributions is 71.09 and 19.81% under dry turning conditions, respectively.

References

1. Hessainia Z, Belbah A, Yallese MA, Mabrouki T, Rigal J-F (2013) On the prediction of surface roughness in the hard turning based on cutting parameters and tool vibrations. Measurement 46(5):1971–1681
2. Gopal M (2021) Prediction of surface roughness in turning of duplex stainless steel (DSS) using response surface methodology (RSM) and artificial neural network (ANN). Mater Today Proc
3. Patel Vallabh D, Anish Gandhi H (2019) Analysis and modeling of surface roughness based on cutting parameters and tool nose radius in turning of AISI D2 steel using CBN tool. Measurement 138:34–38
4. Sharma Vishal S, Dhiman S, Sehgal R, Sharma SK (2008) Estimation of cutting forces and surface roughness for hard turning using neural networks. J Intell Manuf 19(4):473–483
5. Özel T, Hsu T-K, Zeren E (2005) Effects of cutting edge geometry, workpiece hardness, feed rate and cutting speed on surface roughness and forces in finish turning of hardened AISI H13 steel. Int J Adv Manuf Technol 25(3):262–269
6. Sundaram RM, Lambert BK (1981) Mathematical models to predict surface finish in fine turning of steel. Part I. Int J Prod Res 19(5):547–556
7. Tang L, Landers RG, Balakrishnan SN (2008) Parallel turning process parameter optimization based on a novel heuristic approach. J Manuf Sci Eng 130(3)
8. Kalidasan R, Ramanuj V, Sarma DK, Senthilvelan S (2014) Influence of cutting speed and offset distance over cutting tool vibration in multi-tool turning process. Adv Mater Res 984:100–105
9. Brecher C, Epple A, Neus S, Fey M (2015) Optimal process parameters for parallel turning operations on shared cutting surfaces. Int J Mach Tools Manuf 95:13–19
10. Reith Marta J, Bachrathy D, Stepan G (2017) Optimal detuning of a parallel turning system—theory and experiments. J Dyn Syst Meas Control 139(1)
11. Ozturk E, Comak A, Budak E (2016) Tuning of tool dynamics for increased stability of parallel (simultaneous) turning processes. J Sound Vib 360:17–30
12. Asiltürk I, Akkuş H (2011) Determining the effect of cutting parameters on surface roughness in hard turning using the Taguchi method. Measurement 44(9):1697–1704
13. Das SR, Panda A, Dhupal D (2018) Hard turning of AISI 4340 steel using coated carbide insert: Surface roughness, tool wear, chip morphology and cost estimation. Mater Today Proc 5(2):6560–6569
14. Mia M, Dey PR, Hossain MS, Arafat MT, Asaduzzaman M, Ullah MS, Zobaer ST (2018) Taguchi S/N based optimization of machining parameters for surface roughness, tool wear and material removal rate in hard turning under MQL cutting condition. Measurement 122:380–391

Recent Advancements in the Fabrication of Ceramic Matrix Composite: A Critical Review

Prasoon Choudhary and Gurinder Singh Brar

Abstract This article is a comprehensive review of the material property and different processing methods to enhance the properties of ceramic matrix composites, e.g., toughness, stiffness, high-temperature resistance, and wear resistance. Ceramic matrix composites (CMCs) have better industrial importance due to their better tribological properties as compared to single phase ceramic material. Due to wide use of ceramic material as both for matrix and reinforcement, various properties of CMC can be improved such as thermal conductivity, mechanical strength, and hardness. It has been observed that different manufacturing process also affects the properties of ceramic matrix composites. The advantages and disadvantages of various existing processing methods for the CMCs are also discussed. CMCs performances are affected by the nature of the matrix, reinforcement, and different fabrication process. It is important to select proper matrix and reinforcement material of ceramic matrix composites with suitable fabrication processes for better performance. Also, it is important to improve the mechanical strength at the terminals between the matrix and the reinforcement of ceramic matrix composite. By proper selection of materials and processing by appropriate methods, CMCs are used for various applications, e.g., automobile industries, aircraft, marine, gas turbine, nuclear reactors, and other structural applications. This review on the fabrications of ceramic matrix composites is expected to furnish suitable guidance for the development of various CMCs for industrial and structural applications.

Keywords CMC · Matrix · Reinforcement

1 Introduction

Ceramics are made from non-metallic inorganic solid materials having wide range of compositions [1]. These are considered as the best material for components,

P. Choudhary (✉) · G. S. Brar
Department of Mechanical Engineering, National Institute of Technology, Uttarakhand, India
e-mail: prasoon.phd2020@nituk.ac.in

N. M. Sivaram et al. (eds.), *Advances in Manufacturing, Automation, Design and Energy Technologies*, Lecture Notes in Mechanical Engineering,
https://doi.org/10.1007/978-981-99-1288-9_13

e.g., bearings, IC engine components, and tools because of high hardness, young's modulus, and creep resistance. Ceramic matrix composites (CMCs) are composites which contains ceramic fibers reinforced in ceramic matrix. In this composite, both matrix and reinforcement contains any ceramic material. CMCs have high toughness, better strength, and wear resistant under high temperature [2]. Generally, ceramic matrix composites reinforcement is composed of oxide fibers or non-oxide fibers. The non-oxide fibers have lower density and resistance at higher temperature than oxide fibers [3–6]. Some oxide fibers (Alumina, Mullite, Zirconia)-based CMCs performs poor high-temperature mechanical property due to grain growth and poor creep resistance [7, 8]. Ceramic matrix composites have gained its importance in military and civil fields due to light weight, oxidation resistance, and good microwave absorption properties. Now-a-days CMCs are being developed to overcome with the problem that arise with the conventional ceramic, e.g., SiC, Alumina, and Zirconia [9]. It is being produced using several fabrication processes, e.g., chemical vapor or liquid phase infiltration, powder metallurgy techniques, polymer infiltration, electrophoretic deposition, and sol–gel process.

2 Material

2.1 Mullite

Mullite, found in nature is a combination of alumina and silicon oxide has chemical formulae $3Al_2O_3{\cdot}2SiO_2$. Figure 1a shows mullite ceramic powder that is being used in various fields due to its tremendous properties, e.g., high thermal stability, excellent thermal shock behavior, low density, and high thermal conductivity with excellent strength. Mullite has better room temperature strength at higher temperature and thus it has higher creep resistance and high-temperature shock resistance. These properties make mullite a suitable matrix material for CMC [10–14].

Rashad [12] developed mullite-alumina composites from clay and aluminum fluoride trihydrate ($AlF_3{\cdot}3H_2O$) by adding Al_2O_3 powder in different quantities and

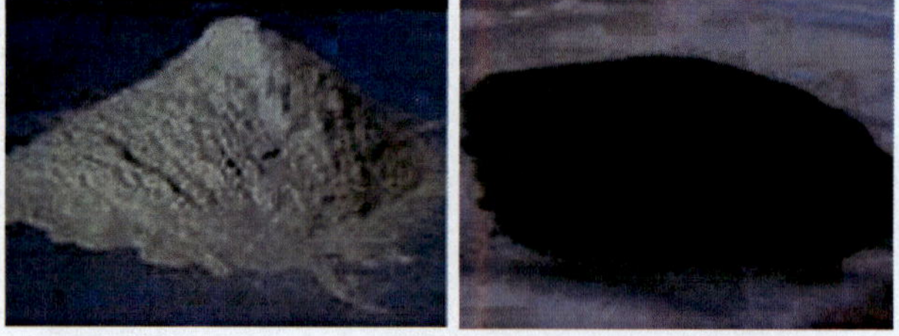

Fig. 1 **a** Mullite powder **b** SiC powder (400 mesh) [21]

sintered at different temperatures. It has been observed that the developed composites added with 10 wt% alumina and sintered at 1400 °C has high porosity and flexural strength. The alumina powder in mullite increased the tribological properties of the sample.

Wang [15] developed SiC/Mullite porous ceramic matrix composite from recycled mud and it performed an increase in flexural strength and decrease in porosity. Alkali metal oxide in recycled mud acts as sintering aid and it decreases the sintering temperature.

Mullite matrix composites have good corrosion resistance and high thermal resistance than alumina. It also produces very tough oxide composites by interface engineering. Coating of BN or SiC–BN coating for mullite matrix composites provides a non-brittle fracture and thus improves mechanical property of composites [5, 16, 17].

2.1.1 Improving the Properties of Mullite Matrix Composite

Mullite ($3Al_2O_3{\cdot}2SiO_2$) is easily purified and shows a low formation temperature, which makes it suitable to prepare an oxide matrix for SiC/oxide composites. The property of mullite can be improved by the adding alumina in proper ratio because it shows enhanced thermal shock resistance [18]. It has been found that a mixture of clay and commercial alumina performs better mullitization at a temperature of 1200 °C. Mullite obtained from clay has excess silica which deteriorates its property because of its glassy form [19].

Also with an increase of sintering temperature, more mullite is integrated. At a temperature above 1400 °C, alumina particles in mullite help in sintering at a temperature of 1600 °C. The fracture toughness of mullite can be improved by the introduction of high-strength ceramic.

With an increase in mullite fibers, the porosity of ceramic matrix composite increases below 3 wt% and it gradually increases at 4 wt%. At room temperature, flexural strength increases at 3 wt% mullite fibers and after that, it decreases. Mullite in combination with SiC has excellent high-temperature strength due to the same thermal expansion of both materials [20].

2.2 Silicon Carbide

Silicon carbide are ceramic materials that have strength and chemical inactivity at high temperatures. Figure 1b shows SiC powder of 400 mesh particle size. SiC materials in crystalline forms have stability in high-temperature atmosphere and making them particularly attractive for use in difficult environments [21].

Ceramic matrix composites with SiC matrix and SiC fibers reinforcement have excellent reliability with better mechanical properties and increased damage tolerance. SiC matrix composites are the best materials suitable for application at high temperatures. It is best composites where the working conditions are extremely unfavorable, e.g., aircraft, gas turbines, nuclear reactors, etc. Silicon carbide as a reinforcement in form of particulates shows high thermal, chemical, mechanical stability. SiC ceramics with little grain boundary maintain their strength up to 1600 °C [22–24].

SiC fiber reinforced ceramic matrix composites (CMCs) have excellent properties, thermal resistance, and low density. They are now generally used in high-temperature applications, e.g., turbine engines, aerospace, etc. [25, 26].

2.2.1 Improving the Property of SiC-Based CMC

By increasing the density of SiC, CMCs improves their toughness and reduces porosity. If carbon content in SiC CMCs increases, then residual silicon in composite material decreases and thus, CMCs obtained has better flexural strength at high temperatures. Also, due to enhanced thermal conductivity of the SiC, CMCs performs uniform heat distribution and thus reduces thermal stress. Various literatures have shown that increasing the sintering temperature and hydraulic pressure can increase the density and refine the grain size of SiC-based CMC. There is a slight increase in porosity of composites after the addition of 10–20 wt% of SiC and a slight increase in thermal conductivity at 30–40 wt% of SiC. The effects of SiC, sintering temperature, and Al_2O_3 on the microstructure and behavior of SiC-reinforced ceramics were studied and figured out that the composite ceramics have excellent properties when SiC content is 20 vol%, and sintering temperature is 1530 °C [27].

2.3 Tungsten and Molybdenum Fiber Reinforcement

Tungsten and molybdenum in the wire form can be used as reinforcement in ceramic matrix composites. Both in fiber form contain potassium precipitates which improve the microstructure of ceramic matrix composites at high temperatures. In general, Mo and W fibers perform necking and show same tensile behavior. The tungsten fibers reinforcement shows better tensile strength and bending strength than molybdenum fibers and Mo fibers have lower young modulus than W-SiCN. The open porosity of Mo-SiCN composites was better than its tungsten counterpart. Also, W-SiCN composites have brittle fracture behavior with low fracture strength [28].

2.4 Carbon Fiber Reinforcement

Carbon fibers are accepted as reinforcement in composite materials because of their better mechanical property and low cost. However, carbon fibers in ceramic matrix composites are rarely used because carbon fibers perform deterioration by corrosion, oxidation, and thermal stress during fabrication and its application. A carbon fiber reinforced oxide matrix composites fabricated by sol–gel techniques has performed crack deflection and thus increases fracture toughness [29–31]. It also increases interfacial properties of CMC at higher temperature and thus improves better load transfers between matrix and reinforcement.

Ceramic matrix composite with carbon–silicon carbide as matrix and carbon fiber as reinforcement have been esteemed as high implementation materials due to their good tribological, chemical, mechanical, and thermal properties [17]. The SiC in the dual ceramic matrix of C/C–SiC has an important role during the sliding friction between two surfaces. Due to the abrasive nature of SiC, mating surfaces during wear are increased which improves coefficient of friction of C/C–SiC composites [32, 33].

3 Methods

3.1 Electrophoretic Deposition

EPD is used for the manufacturing of CMC with various types of ceramic matrix and fiber. Figure 2 shows the flow chart describing the various steps involved in EPD. This technique was developed for the manufacturing of the alumina matrix composite with a high fraction of fiber reinforcement. It is a process in which charged particles in a solvent move under a potential toward oppositely charged electrodes and form a suitable deposit. Ethanol was used as a solvent to reduce any porosity formation in the deposits due to gas evolution. The solvent suspension remains at rest for 24 h before EPD. The CMCs produced after EPD were dried in air at room temperature and after that pressure less sintering process was operated to improve the density of composites. The microstructure control and reduced processing times are the main benefits of EPD [13, 34].

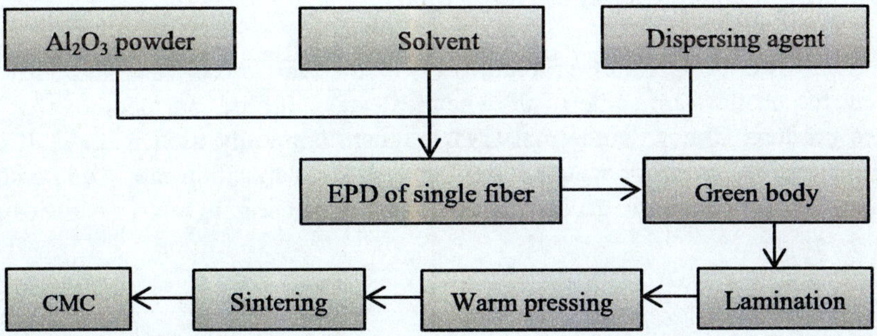

Fig. 2 Electrophoretic deposition [34]

3.2 Powder Metallurgy Process

In order to make ceramic matrix composite, both ceramic matrix and reinforcement powders will be mixed with binder polyvinyl alcohol (PVA) in a planetary ball mill. Initially, ceramic matrix material and ceramic reinforcement in the powdered form will be mixed as per composition required with PVA binder in ball mill for appropriate duration for making a homogeneous mixture. The homogeneous mixture prepared will be pressed in die with the hydraulic press as per the required load for making compacted samples. The compacted ceramic composites will be processed further for sintering to improve its property. The samples will be sintered by controlling temperature and atmosphere using a microwave sintering furnace. Based upon different conditions, it will produce a finally procured sample [35–38].

3.3 Chemical Vapor Infiltration

The chemical vapor infiltration (CVI) method is one of the important method used in industry to produce fiber reinforced ceramic matrix composites. It is a process where a porous long continuous fiber is placed in an atmosphere by use of a reactive gas mixture which disintegrates and yields a solid deposit that fills the pores inside the fiber. CMCs that give good results from the CVI process is silicon carbide (SiC) matrix reinforced with SiC fibers and carbon (C) ceramic reinforced with C-fibers [34].

SiC_f/SiC composites and C_f/C composites fabricated by CVI have excellent high-temperature properties, high strength, corrosion resistance, modulus of elasticity, shock resistance, fatigue resistance, chemical stability, and higher fracture toughness relative to unreinforced SiC [36].

3.3.1 Type of Chemical Vapor Infiltration

Isothermal CVI, Temperature gradient-CVI, Isothermal forced flow, and Thermal gradient forced flow are some of the chemical vapor infiltration methods that are used to produce CMCs. Isothermal CVI is most commonly used method. It is a flexible and slow process because it performs at low deposition rate. Temperature gradient CVI processing is used where the density of ceramic matrix is necessary [23].

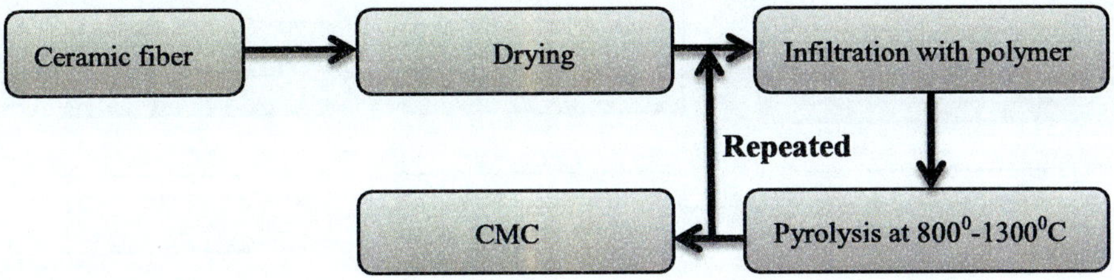

Fig. 3 PIP process [25]

3.4 *Polymer Infiltration and Pyrolysis*

The PIP process can manufacture ceramic matrix composites with complex structures and low thermal damage. Figure 3 shows a flow chart describing various steps involved in the process. It is the method to produce ceramic matrix composites which consists of an infiltration with polymers into the ceramic reinforcement followed by pyrolysis. Infiltration is a process by capillary action where pores of fiber are filled with low viscosity polymers. Pyrolysis is the process in which polymer is heated in argon atmosphere when it decomposes and converts into ceramic. Alumina silicate fiber reinforced ceramic matrix composite prepared by polymer infiltration and pyrolysis method and the mechanical and microstructural property was investigated. It was found that density and mechanical properties increase with increasing no. of PIP Cycle [3, 7, 22].

The CMC laminate produced using PIP has a good microstructure, where re-infiltration and pyrolysis were performed five times. In PIP, fibers damage is reduced due to processing of CMCs at relatively low temperature [31, 32].

3.5 *Sol–Gel Process*

Sol–gel methods are used to produce a various range of ceramic matrix composites. It is generally used to manufacture different high-temperature ceramic matrix composites, e.g., Ni:Al_2O_3, Fe:Al_2O_3, Ni:ZrO_2, SiC:Al_2O_3 as well as advanced variety of some of these composites. In all types, the ceramic matrix composites have better microstructures with uniform distribution between matrix and reinforcement phase and thus have high toughness and machinability. SiC-reinforced alumina composites materials produced by the sol–gel process (Fig. 4) were lighter, stiffer, and tougher than pure alumina [26, 36, 37].

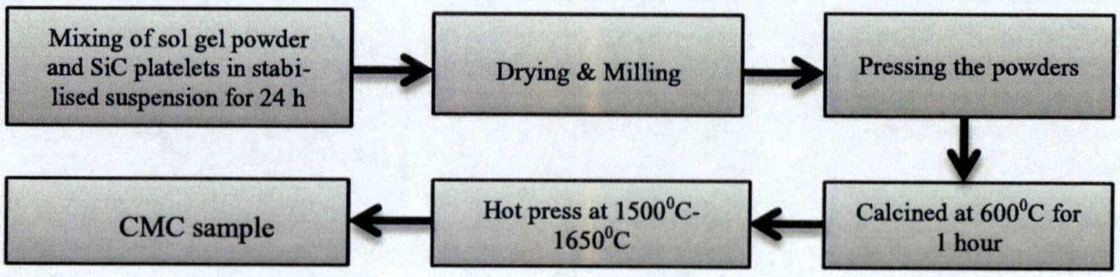

Fig. 4 Sol–gel processing [26]

4 Conclusions

Ceramic matrix composites have shown various improvement over other ceramic constituents, which makes them suitable materials for various applications, e.g., high-speed turbines, bearings, nuclear reactor, aircraft, etc. After studying various research papers, the recent advancements in the matrix material, fiber reinforcement, and different processing methods of CMCs are evaluated and the following effects are observed on CMCs:

1. Amalgamation of various ceramic constituents improves mechanical and tribological behaviors of CMCs.
2. The porosity of Mullite/SiC composite reduces with increase of sintering temperature and it increases with increase of Si content.
3. Strength and toughness of CMC can be improved by increasing the performance of matrix and reinforcement
4. CVI technology is the best method to produce C-SiC/SiC_f–SiC ceramic composites.
5. PIP processing is used for CMCs with complex structure while EPD process fabricates CMCs with various ranges of matrix and fiber reinforcement.

A Future area of research with CMC's is in various applications working under high load with low lubrication, higher temperature, tribological behavior under high speed.

References

1. Affatato S, Ruggiero A (2015) Advanced biomaterials in hip joint arthroplasty: a review on polymer and ceramics composites as alternative bearings. Composites 83:276–283
2. Jiancheng A, Tiezhu G, Enxia X (2020) Preparation and properties of mullite-SiC-O-SiAlON composites for application in cement kiln. Ceram Int 46:15456–15463
3. Peters PWM, Daniels B, Clemens F (2000) Mechanical characterisation of mullite based ceramic matrix composites at test temperature upto1200⁰C. J Eur Ceram Soc 20:531–535
4. Boccaccini AR, Atiq S, Boccaccini DN (2005) Fracture behaviour of mullite fibre reinforced mullite matrix composites under quasi-static and ballistic impact loading. Compos Sci Technol 65:325–333

5. Ha JS, Chawla KK (1995) Mechanical behaviour of mullite composites reinforced with mullite fibres. Mater Sci Eng 203:171–176
6. Chawla KK, Xu ZR, Ha JS (1996) Processing, structure, and properties of mullite fiber/mullite matrix composites. J Eur Ceram Soc 16:293–299
7. Zhaowen R, Wancheng Z, Yuchang Q (2020) Microwave absorption and mechanical properties of SiC_f/SiOC composites with SiO_2 fillers. Ceram Int 47:8478–8485
8. Win XY, Cheng LF, Zhang LT (2016) Fibre-reinforced multifunctional SiC matrix composite materials. Int Mater Rev 62(3):117–172
9. Xiulei W, Zhenghe Z (2021) Advances in modification and high temperature application of silicon carbide ceramic matrix composites in aerospace: a focused review. J Eur Ceram Soc 41:4671–4688
10. Yani J, Jianbao L (2014) Fabrication and properties of SiC/mullite composite porous ceramics. Ceram Int 40:1329–1334
11. Yongfa K, Zhengfang Y, Guangyin Z (2002) Sliding friction and wear of alumina-reinforced zirconia-toughened mullite composites. Wear 252:607–613
12. Rashad M, Sabu U, Logesh U, Balasubramanian M (2019) Development of porous mullite and mullite-Al_2O_3 composite for microfiltration membrane applications. Sep Purif Technol 219:74–81
13. Kaya C, Butlera EG, Selcuk EG (2002) Mullite (Nextel TM 720) fibre-reinforced mullite matrix composites exhibiting favourable thermomechanical properties. J Eur Ceram Soc 22:2333–2342
14. Cascales A, Tabares N (2015) Processing and mechanical properties of mullite and mullite-alumina composites reinforced with carbon nanofibers. J Eur Ceram Soc 35:3613–3621
15. Wang W, Chen W, Liu H (2019) Recycling of waste red mud for fabrication of SiC/mullite composites porous ceramics. Ceram Int 45:9852–9857
16. Chawla KK (2008) Interface engineering in mullite fiber/mullite matrix composite. J Eur Ceram Soc 28:447–453
17. Samuel JF, Timothy PC (2018) Carbon fiber reinforced ceramic matrix composites with an oxidation resistance BN interface coating. Ceram Int 44:15310–15316
18. Broutelle M, Frederic L (2020) Bearing damage identification in oxide/oxide ceramic matrix composite with a new test design. Compos Struct 236:111902
19. Yahua B, Patrick SN (2008) $AlPO_4$ coated mullite/alumina fiber reinforced reaction bonded mullite composites. J Eur Ceram Soc 28:3041–3048
20. Shuxin N, Xiqing X, Xin L, Xin C, Yushi L (2020) Microstructure evolution and properties of silica-based ceramic cores reinforced by mullite fibers. J Alloy Compd 829:154494
21. Naslain RR, Pomeroy MR (2015) Ceramic matrix composites: matrices and processing. In: Reference module in materials science and materials engineering, pp 1060–1066
22. Yang L, Zheng SL, Shu HM (2018) Mullite bonded SiC reinforced SiC matrix composites, preparation, characterisation and toughening mechanism. J Eur Ceram Soc 38:5282–8293
23. Naslain R (2004) Design, preparation and properties of non-oxide CMCs for application in engines and nuclear reactors: an overview. Compos Sci Technol 64:155–170
24. Hui G, Luo F, Qinlong W (2019) Effect of preparation conditions on mechanical, dielectric and microwave absorption properties of SiC fiber/mullite matrix composite. Ceram Int 45:11625–11632
25. Gao H, Luo F (2019) Fabrication of SiC_f/SiC-mullite composite with improved pretreatment condition via precursor infiltration-sintering combined with infiltration-pyrolysis process. Ceram Int 45:16062–16069
26. Han S, Yang LW, Liu HT (2017) Micro-mechanical properties and interfacial engineering of SiC fiber reinforced sol-gel fabricated mullite matrix composites. Mater Des 131:265–272
27. Shatumbu TA (2020) Microstructure and mechanical properties of alumina matrix composites reinforced with SiC by spark plasma sintering. Mater Res 1–9
28. Mainzer B, Chaorong L, Riedel R (2021) Novel ceramic matrix composites with tungsten and molybdenum fibre reinforcement. J Eur Ceram Soc 41:3030–3036

29. Yang LW, Liu HT, Jiang R (2017) Weak interface dominated high temperature fracture strength of carbon fiber reinforced mullite matrix composites. J Eur Ceram Soc 37:2991–2996
30. Zhang R, Qingsong M, Kuanhong Z (2021) Microstructure and properties evolution of C/Mullite composites during fabrication process. Ceram Int 47:6623–6632
31. Yutaro A, Ryo I, Ken G (2019) Carbon fiber reinforced ultra-high temperature ceramic matrix composites: a review. Ceram Int 45:14481–14489
32. Kumar P, Shrivastava VK (2019) Reciprocating sliding tribology of brake oil treated carbon fibre reinforced ceramic matrix composite. Science Press 29:1903–1913
33. Duan J, Menghang Z, Pengju C (2021) Tribological behaviour and applications of carbon fiber reinforced ceramic composites as high performance frictional materials. Ceram Int 47:19271–19281
34. Andrea L, Stoll L, Mahr P (2006) Fabrication technology for oxide-oxide ceramic matrix composites based on electrophoretic deposition. J Eur Ceram Soc 26:1567–1576
35. Mittal G, Singh R, Brar GS (2018) Investigation and performance analysis of ceramic matrix composite for bearing application. In: Reference module in material science and materials engineering, pp 1–6
36. Syed SA (2021) A critical review on self-lubricating ceramic-composite cutting tools. Ceram Int 47:20745–20767
37. Omole SO, Barnabas AA, Akinfolarin JF (2015) Production and evaluation of ceramic and metal matrix composite by powder metallurgy. Res Eng Struct Mater 1:73–79
38. Rezaie HR, Rainforth WM (1999) Fabrication and mechanical properties of SiC reinforced mullite matrix composites. J Eur Ceram Soc 19:1777–1787

Effect of MIG Welding Process on Hardfacing of Inconel 718 Over Stainless Steel 347

A. Karpagaraj, R. Sarala, L. R. Shobin, Hemant Priyadarshi, and Aswin C. Gowda

Abstract The welding techniques used for cladding and hard facing are a cost-effective process. Nowadays, Metal Inert Gas Welding (MIG) is very popular to deposit wear-resistant metal over the surface. This study uses MIG welding to deposit Inconel 718 (filler with 1.2 mm diameter) on 5 mm thick stainless steel 347. The layer deposition parameter is identified based on the Bead on Plate (BoP) trials. An optimal parameter of 120 A, 20 V, travel speed of 250 mm/min, wire feed rate of 4.5 mm/min, and gas flow rate of 20 L/min are recognized. Overlap of 50% is selected to make the layer thickness of 3 mm deposited in the SS 347 surfaces. The effect of heat input on the microstructural changes and mechanical properties are studied. At the top portion of the Inconel 718, coarse grains are identified. The tensile test performed at the interface and deposit is 686 and 798 MPa. Metal deposition strength is about 20.17% higher than the base metal strength. The three-point bend test shows that not much more brittleness in the samples. Microhardness in the top portion is 218 Hv, at interface 180 Hv, and base metal 164 Hv.

Keywords Hardfacing · MIG welding · Inconel 718 · Mechanical behavior

---

A. Karpagaraj (✉)
National Institute of Technology Puducherry, Karaikal 609609, India
e-mail: karpaganit@gmail.com

R. Sarala
Alagappa Chettiar Government College of Engineering and Technology, Karaikudi, Tamil Nadu 630004, India

L. R. Shobin
SRM Institute of Science and Technology Tiruchirappalli Campus, Tiruchirappalli, Tamil Nadu 621105, India

H. Priyadarshi
National Institute of Technology Patna, Bihar 800005, India

A. C. Gowda
Visvesvaraya Technological University, Bangalore, Karnataka 560091, India

N. M. Sivaram et al. (eds.), *Advances in Manufacturing, Automation, Design and Energy Technologies*, Lecture Notes in Mechanical Engineering,
https://doi.org/10.1007/978-981-99-1288-9_14

1 Introduction

Hardfacing is a technique that usually deposits high hardness/strength materials over the base substrate. It will increase the lifetime of the base substrate by providing it from wear. Many techniques were available to deposit the metals over the base substrate [1–4]. Welding is one of the techniques used for making layer deposition. Filler metals are fused on the surface where Hardfacing is needed. The most suitable and economical arc welding is the Metal Inert Gas (MIG) welding process. In this process, the filler metal is directly fused by applying heat and deposit on the base substrate [5]. Suresha et al. perform the Hardfacing operation using the MIG welding process. The Hastelloy is deposited over the stainless steel 316L sample. It was reported that the hardness of 27% is higher than the base metal [6]. Rajkumar et al. studied the Hardfacing of Inconel 625 on stainless steel 347 pipes. It was found that the hardness was more at the heat-affected zone (HAZ). The wear behavior is increased because of the Inconel 625 layers [7]. Rajeev et al. studied the effect of cold metal transfer and plasma arc welding on the Hardfacing of the steel. The finer microstructure was present in the samples made through the CMT process compared to the plasma arc process. It was found that oxidation occurred at higher temperatures, and it improved the wear performance at elevated temperatures [8]. Pradeep et al. analyzed the various welding process and checked the wear performance at various levels. It was found that the TIG welding was performed at a lower speed. The arc welding samples showed promising results [9]. Sakiru et al. performed the Hardfacing using the TIG welding process. It was reported that dendritic structure is formed due to the nature of the TIG welding process. The maximum hardness value is reached on the layers compared to the base metal. The hard-faced Cp-Ti is more significant than its base metal due to the complex weld thermal cycle [10].

2 Experimental Setup

The welding setup used for this study is the Welbee 500P low spatter MIG welding machine with three-axis controllers (Fig. 1). The filler metal used is an Inconel 718 wire spool having a diameter of 1.2 mm. Argon gas is selected (20 L/min) for providing the shielding to the Hardfacing area. A customized fixture is used for holding the stainless steel plates (SS347) of 5 mm thickness.

The major welding parameters and their levels are listed in Table 1. The mode used for depositing the layer is DC pulsed with the asynchronous option. The frequency is 0.5 Hz, and the back current is 50% of the peak current.

Based on the previous experience, the layers are mode with 50% offset. The layer thickness of 3 mm is maintained after the machining made by a vertical milling machine. The samples are mounted with Bakelite powder and moved for polishing, etching purpose. The standard etchants from the ASTM handbook reveal the macro and microstructure at the Inconel and steel portions. Tensile test and three-point

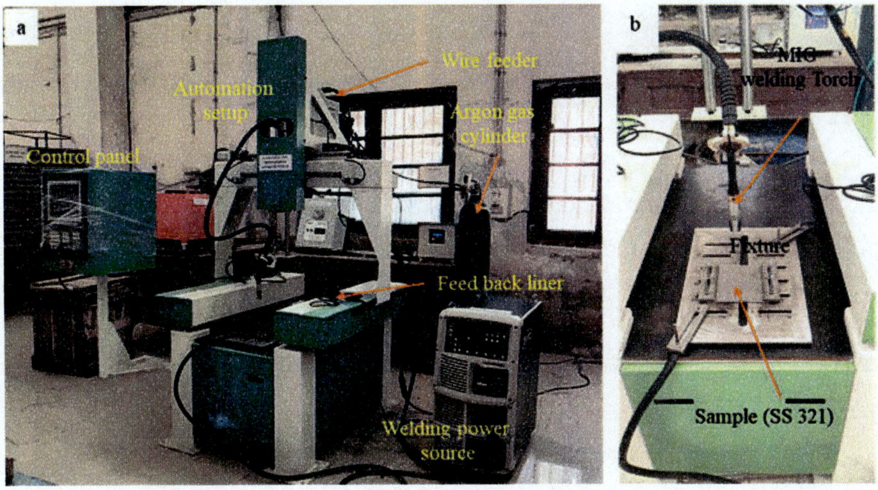

Fig. 1 The overall experimental setup **a** welding setup **b** closure view of fixture and torch setup

Table 1 Welding parameters and their ranges

S. no	Parameter	Range
1	Welding current (A)	90–120
2	Travel speed (mm/min)	200–300
3	Wire feed rate (mm/min)	3.5–5.5

bending test are performed with Zwick Roell universal testing machine. Optical microscope images are captured from the DM750 Leica with a 5× to 100× magnification range. Microhardness is measured at the top, interface, and bottom side using the Tinius Olsen with a load of 500 g for 30 s. Later, heat treatment is done using the furnace having the range up to 900 °C. The 375, 575, and 775 °C are maintained for 8 h after the air-cooled method is used.

3 Result and Discussion

The significant findings from the research are discussed as follows. The welding parameters are optimized from experience gained from the preliminary works. The welding current of 120 A, wire feed rate of 5.5 mm/min, welding/travel speed of 250 mm/min are identified as optimal parameters. The metal deposit of Inconel 718 is made on the SS347 with an overlap of 50%. Later the surface is milled for further analysis. The results gathered from the various characterization methods are submitted as follows.

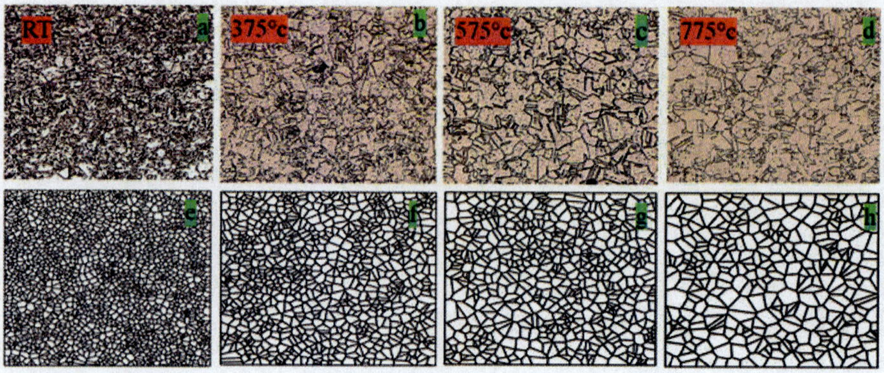

Fig. 2 Grain size comparison at room temperature to elevated temperature **a–d** experimental results **e–h** predicted results

3.1 Microstructural Analysis

Microstructure at different locations like Inconel deposit (top side), the interface of Inconel 718–SS347 (middle), and the base metal SS347 (bottom) are focused. In the middle portion, the interface clearly showed an excellent bonding exists between Inconel 718 to SS347. There are no defects present in the metal deposits on the examined part. A slender level of Heat Affected Zone (HAZ) is identified.

Heat treatment is done in the induction furnace to identify the grain coarsening. The grain coarsening is started at 375 °C. The grain coarsening at 775 °C is nearly matching with HAZ zone microstructure [11, 12]. The heat-treated base metal grains are compared with simulated results to determine the grain coarsening level. A simulation (neper software * [open source]) study is done to predict the grain coarsening. This microstructural profile comparison is shown in Fig. 2. Because this study will be correlated with temperature history for accurate prediction of HAZ, this prediction and the heat treatment results are needed for future studies related to heat transfer analysis.

At the top portion (Inconel deposition), grain coarsening occurs. Inconel 718 has a needle-like structure. Dentaries with different orientations are identified in Fig. 3. This grain coarsening type typically occurs while performing the metal deposit using welding [6, 8, 10, 13].

3.2 Tensile Test

The tensile test is performed with 0.5 mm/s cross-head speed for all the samples. Three specimens are tested for each category, and the average values are presented in Table 2. The ultimate strength for the Inconel 718 is about 798 MPa (refer Table 2) and for the interface 686 MPa. The Inconel 718 is 14% higher than the interface

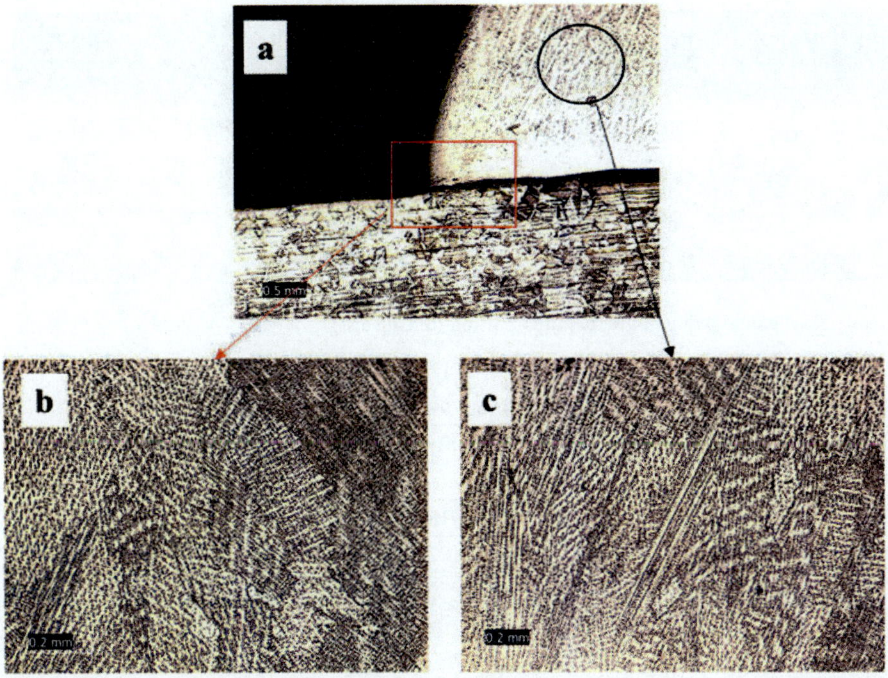

Fig. 3 Microstructure of hardfaced sample **a** macro graph **b** interface, and **c** Inconel deposit

and 20% higher than SS347 (base metal). At the interface, the fusion is perfect, and there are no defects present (refer Fig. 4).

The arrow marks show the line of the interface in Fig. 4a and the circle indicating the macro voids present at the fractured portion. In Fig. 4b, the closed view shows the boundary region, proving no defects at the interface zone. The square part marked in the area represents the existence of the micro-voids. These micro-voids and the fractured pattern absorbed at the Inconel 718 fractured side are the reason for less strength than other welding processes. But the strength produced by the MIG welding is comparable with the other works presented in the area of arc welding [14].

Table 2 Tensile test results obtained from the samples

S. no	Location	UTS (MPa)	Displacement (mm)
1	Base (SS347)	637	22.78
2	Interface (SS347-Inconel 718)	686	19.29
3	Metal deposit (Inconel 718)	798	16.40

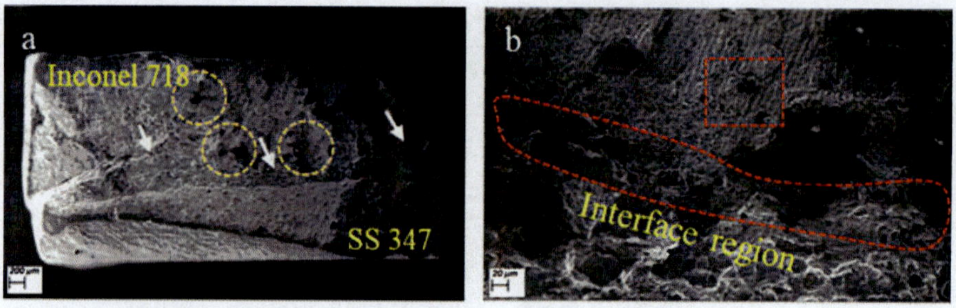

Fig. 4 SEM examination at the fractured portion (tensile test)

Table 3 Microhardness measurement at various locations

S. no	Location	Microhardness (Hv 0.5)	Average hardness (Hv 0.5)
1	Base (SS347)	161 + 168 + 164	164.3
2	Interface (SS347-Inconel 718)	174 + 186 + 182	180.6
3	Metal deposit (Inconel 718)	213 + 224 + 219	218.3

3.3 Microhardness Test

The tensile test and optical microstructure present in the weld deposit and the interface are not created complex microstructure. So, it is necessary to read the microhardness at various locations. The microhardness indirectly gives the surface's behavior against the wear environment. The microhardness is 218 Hv at the deposited layer (Inconel 718), as displayed in Table 3. This reading is about 24.7% higher, which is needed to make necessary wear resistance against the working environment [6, 7].

Hardness is increased due to the changes of microstructural, alloying elements, and weld thermal cycles. Also, the tensile test results agree with the microhardness measurements made at various locations. The microhardness at the interface is 17.26% higher than the base metal, while the strength/toughness is also higher than the base metal.

3.4 Three-Point Bend Test

A three-point bend test is conducted on the samples at various locations like base (SS347), interface (SS347-Inconel 718), and deposit (Inconel 718). All the specimens are passed without showing any discontinuity under the loading condition. The samples after the test are shown in Fig. 5. Since the Inconel 718 portions do not offer any stern faces, the Inconel samples give minor deformation while performing the test. Since the material is quite stranger than the SS347, extra care is taken while performing the three-point bend test. There is no sign of cracks or flows in the Inconel

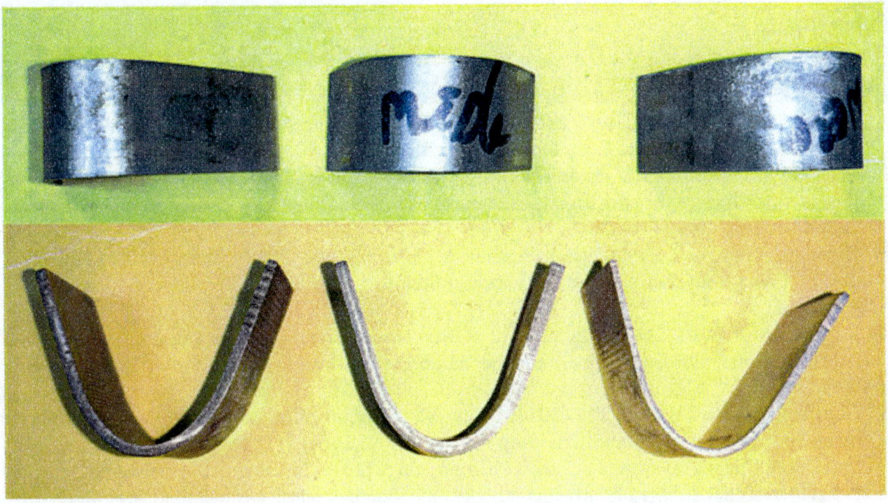

Fig. 5 Three-point bend test sample after the test

718 and interface. But for the interface region, it is 19.2 mm, and for the base metal, it is about 22.3 mm. Because in the preliminary stage, the wet angle and dilution calculation are carefully analyzed. This initial analysis is helped a lot while performing the tensile and three-point bend test.

4 Conclusion

The optimal welding parameter (120A/5.5 mm/min/250 mm/min, 0.5 Hz, 50% peak current) for the pulsed mode is identified for depositing the Inconel 718 over the SS347 surface. A single layer of Inconel 718 metal deposition is made over the SS 347 plates. The overlap of 50% is taken to achieve the hard layer thickness of 3 mm over the base substrate. Layer deposition is performed without defects (spatter, crack, and porosity). Tensile (14% higher), microhardness (24.7% higher), and bending tests (no cracks in Inconel 718 depositions) are given satisfactory results as compared with a base substrate of SS347. Grain coarsening is occurred due to the complex weld thermal cycle caused by the MIG welding process. The possibility of using the software to forecast the grains for macro analysis is given positive results only.

Acknowledgements The author thanks SERB/EEQ/file no 2018/000676 for providing the fund to carry out the research. The authors thank the procurement section (Prof. Prakash Chandra–team) NIT-Patna for their support and guidelines towards purchasing the equipment. The authors also thank the National Institute of Technology–Puducherry for their current supports.

References

1. Shahroozi A, Afsari A, Khakan B, Khalifeh AR (2018) Microstructure and mechanical properties investigation of Stellite 6 and Stellite 6/TiC coating on ASTM A105 steel produced by TIG welding process. Surf Coat Technol 350:648–658
2. Vimalraj S, Varahamoorthi R, Umesh Bala A, Karthikeyan R (2020) Modeling and optimizing the laser parameters for corrosion resistance in 316 SS laser hardfaced surface using tungsten carbide. Mater Today Proc 26:2485–2490
3. Brownlie F, Anene C, Hodgkiess T, Pearson A, Galloway AM (2018) Comparison of Hot Wire TIG Stellite 6 weld cladding and lost wax cast Stellite 6 under corrosive wear conditions. Wear 404:71–81
4. Kaszuba M (2020) The application of a new, innovative, hybrid technology combining hardfacing and nitriding to increase the durability of forging tools. Arch Civ Mech Eng 20(4):1–13
5. Chavda SP, Desai JV, Patel TM (2014) A review on optimization of MIG welding parameters using Taguchi's DOE method. Int J Eng Manag Res (IJEMR) 4(1):16–21
6. Suresha B, Channabasavanna SG, Siva Shanmugam N (2019) Microstructure and abrasive wear behaviour of nickel based hardfacing stainless steel deposited by gas metal arc welding. Appl Mech Mater 895:278–283
7. Rajkumar V, Arjunan TV, Rajesh Kannan A (2020) Investigations on hardfacing and wear characteristics of nickel-based Inconel 625 overlaid welds over AISI 347 pipe. J Braz Soc Mech Sci Eng 42(1):1–12
8. Rajeev GP, Kamaraj M, Bakshi SR (2019) Comparison of microstructure, dilution and wear behavior of Stellite 21 hardfacing on H13 steel using cold metal transfer and plasma transferred arc welding processes. Surf Coat Technol 375:383–394
9. Pradeep GRC, Ramesh A, Durga Prasad B (2013) Comparative study of hardfacing of AISI 1020 steel by three different welding processes. Glob J Res Eng 13(4-A):11–14
10. Sakiru A, Shahjahan M (2015) Thin surface layers of iron-based alloys deposited by TIG hardfacing. Tribology Online 10(6):434–440
11. Chandra K, Kain V, Tewari R (2013) Microstructural and electrochemical characterisation of heat-treated 347 stainless steel with different phases. Corros Sci 67:118–129
12. Gonzaga AC, Barbosa C, Tavares SSM, Zeemann A, Payão JC (2020) Influence of post welding heat treatments on sensitization of AISI 347 stainless steel welded joints. J Market Res 9(1):908–921
13. Ram GDJ, Venugopal Reddy A, Prasad Rao K, Madhusudhan Reddy G, Sarin Sundar JK (2005) Microstructure and tensile properties of Inconel 718 pulsed Nd-YAG laser welds. J Mater Process Technol 167(1):73–82
14. Ramkumar T, Selvakumar M, Narayanasamy P, Ayisha Begam A, Mathavan P, Arun Raj A (2017) Studies on the structural property, mechanical relationships and corrosion behaviour of Inconel 718 and SS 316L dissimilar joints by TIG welding without using activated flux. J Manuf Proc 30:290–298

Effect of B_4C on Mechanical Properties of AZ91C Metal Matrix Composite

Nagallapati Jaya Krishna and Dega Nagaraju

Abstract Metal matrix composites are advantageous due to their excellent stiffness and high specific strength in aerospace and automobile industries. Magnesium-based composites are the predominant choice due to their low density. As strength, machinability, surface texture and corrosion resistance are important factors for the application. Magnesium matrix composites are less corrosion resistance and strength and to improve the present work focused on reinforcing the AZ91C alloy with Boron carbide (B4C) in two different proportions. Magnesium composite material is manufactured by casting process. The samples are prepared in accordance with the ASTM standards and tests like hardness, tensile and compressive strength are performed on the sample specimens with the proportions varying 2% of B4C with the rest AZ91C and 4% B4C with rest AZ91C. It is observed that the tensile strength is increased due to the addition of B4C while the hardness and the compressive strength decreases.

Keywords AZ91C · B_4C · Tensile strength · Compressive strength · Hardness

1 Introduction

Magnesium is one of the lightest structural metals which is 35% lighter than Al and 65% lighter than T. It has a density of 1.74 g/cm^3. Due to its low density, stiffness and high specific strength the magnesium-based materials found many applications in automobile, military and aerospace industries to reduce energy cost and improve the performance. The fabrication of magnesium composites would also be difficult due to the metal's porosity. Due to the addition of ceramic reinforcement particles to the base magnesium metal will tend to change its properties to a great extent. Nie et al. [1], Esmaily et al. [2], Muley [3], Deng [4], Zhou [5] and Deng [6] was manufactured Mg-matrix composites by using solid processing, liquid processing or combination

N. Jaya Krishna · D. Nagaraju (✉)
School of Mechanical Engineering (SMEC), Vellore Institute of Technology (VIT), Vellore, Tamil Nadu 632014, India
e-mail: deganagarajulc@gmail.com

N. M. Sivaram et al. (eds.), *Advances in Manufacturing, Automation, Design and Energy Technologies*, Lecture Notes in Mechanical Engineering,
https://doi.org/10.1007/978-981-99-1288-9_15

techniques fortified with SiC particles of various sizes such as micron, submicron and nano particles with respective volumes for AZ91 alloy. In this exploration, the mechanical properties were compiled as a function with some amount of reinforcement. In general, some of the mechanical properties like ultimate tensile strength, the yield strength and the elastic modulus are enhanced with increasing amount of SiC. These effects are found to be more pronounced as the size of SiC particles is reduced. For example, in this paper the author said that the ultimate tensile strength and elongation to fracture was decreased due to more SiC clusters along with the grain boundary of matrix compared with the 1 vol% SiCp/AZ91 composite for particle size of 1 μm [1]. The improvement of the composite (AZ91/SiC) quality is ascribed to the refinement of Mg grains, better dispersion of the $Mg_{17}Al_{12}$ stage and expansion of SiC particles. Esmaily et al. [2] announced that the MMC delivered utilizing bigger SiC particles showed better corrosion resistance over smaller particles. The hardness and ultimate compressive quality have expanded just about 90% and the wear conduct additionally progressed [3]. The bimodal particles increase in the refining grain assess and the yield nature of Mg-based composite with bimodal molecule is more than single size help. The nature of the MMC extended with sub-micron particles at 2% of volume divide however higher than will be gained by using micron particles growing volume division from 5 to 10% [1]. The yield quality and extreme rigidity of the bimodal size AZ91/Sicp composite is more than the solid micron molecule reinforce composite. The best volume proportion of Sicp enemy mechanical properties is $S - 1.5 + 10 - 8.5$. The $\Delta\sigma$Hall–Petch and $\Delta\sigma$CTE increment with the expansion of volume proportion [5]. Tian [7], Bhingole [8], was manufactured Mg-matrix composites fortified with various fortifications like Al_2O_3–SiO_2, Al_2O_3 particles and their individual volumes or weights for AZ91 compound. Jun Tian [7] discovered the results that the composite and the matrix have a similar genuine pressure type and genuine initiation vitality for crawl, which are 3 and 144.63 kJ/mol. Bhingole [8] revealed that the mechanical properties of MMC with its hardness, yield quality and strain hardening are extended by 64, 43 and 115% independently when differentiated and AZ91 amalgam. MMC's also gets more wear safe when differentiated and AZ91 amalgam. Park et al. [9], Zhao et al. [10], Selvamani et al. [11] and Yuan et al. [12] was manufactured Mg-composites fortified with various fortifications like CNT, MWCNT and so forth., Zhao et al. [10] revealed that the yield quality, extreme rigidity and stretching of the 0.5 wt% CNTs/AZ91D composite were upgraded altogether when compared to the strong AZ91D amalgam. Yongha Park et al. [9] discovered the change in elasticity, wettability, dissemination, holding quality by fortifying Si covered CNT to AZ91 magnesium composites. Selvamani et al. [11] discovered that the composites with 4% reinforcement of CNT demonstrate high hardness when compared with 3% fortification show better malleable and yield. Yuan et al. [12] was fabricated MWCNTs/magnesium compound composites by powder smaller laser sintering and watched that MWCNTs can expand the laser retention of the MWCNTs/AZ91D composites contrast and solid AZ91D and give high temperature and cooling rate to make the composites extreme. Govindaraju [13] and Zheng et al. [14] have done experimentation on the effect of heat treatment

Table 1 Chemical composition of AZ91C alloy

Element	Percentage of an amount	Element	Percentage of an amount
Magnesium	88.6–91.4	Silicon	0–0.3
Aluminum	8.1–9.3	Copper	0–0.1
Zinc	0.4–1.0	Nickel	0–0.010
Manganese	0.13–0.35	Others	1–0.3

process on magnesium alloys. The rigidity of homogenized-moved AZ91 combinations increments drastically with extension declining marginally as opposed to homogenized amalgams. The break surface of homogenized-moved example shows more malleable crack with the sign of a lot of dimples appropriating higher thickness in framework. However, the effect of B_4C particles has not been studied yet according to literature. Therefore, the objective of this present paper is to study the mechanical properties of the AZ91C/B_4C composites. To achieve this objective the B_4C particles are reinforced with AZ91C by conventional casting processes, i.e., induction furnace and then the effect of the B_4C on mechanical properties such tensile properties and hardness of the cast magnesium matrix composites were studied.

2 Materials and Experiments

2.1 Material

To fabricate the magnesium metal matrix composite, the AZ91C was selected as matrix and Boron carbide particulates are used as reinforcement. The chemical composition is shown in Table 1.

2.2 Processing of Composite Material

The metal and ceramic composite materials are fabricated by liquid state processing. In liquid state processing there are stir casting, infiltration, gas pressure, squeeze casting and pressure die are the conventional manufacturing methods. In the present study the synthesis of magnesium alloy matrix composites was carried out by casting process. The AZ91C was placed in a graphite crucible and heated up to 650 °C and method and preheated B_4C particulates was poured in molten metal the stirring is continued for 15 min under open atmosphere in an induction furnace. Therefore for 1 kg of composite melts containing X g wt% B_4C particles was poured in a mold to fabricate ingots of 150 mm × 150 mm × 15 mm. To study the effects of B_4C particles

Table 2 Composition for metal matrix composites

Sample No.	AZ91C (%wt)	B_4C (%wt)
1	100	0
2	98	2
3	96	4

Fig. 1 Test sample for tensile test

on the mechanical properties of the composite ingots, the unreinforced AZ91 alloy ingot was also cast under the same conditions (Table 2).

2.3 *Tensile Strength*

The tensile strength was carried out under tensile loading with the Universal Testing Machine (UTM) and the samples are prepared as per ASTM B 557:2010 as shown in Fig. 1. The testing machine has a capacity of 1000 KN and loading rate was 15 KN/min. The tensile test is carried out for three model samples with different percentage of compositions namely AZ91C (100%) + B_4C (0%), AZ91C (98%) + B_4C (2%) and AZ91C (96%) + B_4C (4%).

2.4 *Compressive Strength*

The compressive strength was carried out under compressive loading with the Universal Testing Machine (UTM) and the samples are prepared as per Standards. The testing machine has a capacity of 1000 KN and loading rate was 15 KN/min. The tensile test is carried out for three model samples with different percentage of compositions namely AZ91C (100%) + B_4C (0%), AZ91C (98%) + B_4C (2%) and AZ91C (96%) + B_4C (4%).

2.5 *Hardness*

In present work the Brinell hardness tester is used to determine the hardness of the MMC specimens. The load applied is 500 kg on the surface of the MMC specimens. For each specimen 3 readings are taken and then their mean is taken as a result as BHN. The hardness test is carried out for three samples.

3 Results and Discussion

3.1 Tensile Strength

The obtained results from the tensile test for the three specimens are shown in Table 3. After conducting the tensile test for AZ91C/B_4C specimens, the AZ91 (96%) and B_4C (4%) had found ultimate load, ultimate strength is increased than the other two specimens.

3.2 Compressive Strength

The obtained results from the tensile test for the three specimens are shown in Table 3. After conducting the compressive test for AZ91C/B_4C specimens, the AZ91 (100%) and B_4C (0%) had found compressive load, compressive strength has higher strength than the other two specimens.

3.3 Hardness

The obtained results from the hardness test for the sample specimens are shown in Table 3. The experimental and referenced values for each metal were compared. From observed data, it is noticed that third specimen is having the better hardness than other two specimens.

4 Conclusions

The need of composite materials in every field is an-evidence to wide research in this field, because of their key characteristics like high tensile strength, high corrosion resistance, good toughness, good surface finish, good weldability. In support of this work involved in development of magnesium metal matrix composite material of AZ91C/B_4C was fabricated. The investigation of the prepared Mg MMCs using various tests like tensile test and hardness test have revealed the various key characteristics. The empirical findings from this work are made into the following conclusions.

- The tensile strength of the material increases more when it is mixed with the B_4C of 4% than when mixed with the B_4C of 2%.
- The compressive strength of the material decreases more when it is mixed with the B_4C of 4%, 2% than when mixed with the B_4C of 0%.

Table 3 Results of tensile, compressive and hardness test

S. No.	Materials		Hardness test (BHN)	Tensile test		Compressive	
	AZ91C (wt%)	B_4C (wt%)		Ultimate load (KN)	Ultimate strength (N/mm^2)	Load (KN)	Strength (N/mm^2)
1	100	0	76.67	7.0	35.450	247.29	535.816
2	98	2	48.67	8.0	40.873	197.82	362.886
3	96	4	64.00	9.48	47.244	185.22	212.823

- The hardness of the material is increased slightly when it is mixed with the B_4C of 4% than when mixed with the B_4C of 2% but it has decreased when compared with AZ91C.
- The increase in the percentage of tensile strength of the hybrid composite materials by increasing the composition of the B_4C is 15.58%.
- The decrease in the percentage of compressive strength of the hybrid composite materials by increasing the composition of the B_4C is 32.27% for 2% and 60.28% for 4%.
- The increase in the percentage of hardness of the hybrid composite materials by increasing the composition of the B_4C is 31.49% but it has decreased when compared with AZ91C.

Future scope

This work can be further extended by conducting some machining operations on the prepared magnesium based MMCs which can yield the machinability, surface roughness, cutting forces, material removal rate etc.

References

1. Nie KB, Wang XJ, Wu K, Xu L, Zheng MY, Hu XS (2012) Fabrication of SiC particles-reinforced magnesium matrix composite by ultrasonic vibration. J Mater Sci 47:138–144
2. Esmaily M, Mortazavi N, Svensson JE, Halvarsson M, Jarfors AEW, Wessen M, Arrabal R, Johansson LG (2016) On the microstructure and corrosion behavior of AZ91/SiC composites produced by rheocasting. Mater Chem Phys 180:29–37
3. Muley SV, Singh SP, Sinha P, Bhingole PP, Chaudhari GP (2014) Microstructural evolution in ultrasonically processed in situ AZ91 matrix composites and their mechanical and wear behaviour. Mater Des 53:475–481
4. Deng K, Shi J, Wang C, Wang X, Wu Y, Nie K, Wu K (2012) Microstructure and strengthening mechanism of bimodal size particle reinforced magnesium matrix composite. Compos Part A 43:1280–1284
5. Zhou S-S, Deng K-K, Li J-C, Shang S-J, Liang W, Fan J-F (2014) Effects of volume ratio on the microstructure and mechanical properties of particle reinforced magnesium matrix composite. Mater Des 63:672–677
6. Deng KK, Wang XJ, Wu YW, Hu XS, Wu K, Gan WM (2012) Effect of particle size on microstructure and mechanical properties of SiCp/AZ91 magnesium matrix composite. Mater Sci Eng A 543:158–163
7. Tian J, Shi Z (2014) Creep mechanism and creep constitutive model of aluminum silicate short-fiber-reinforced magnesium matrix composite. Trans Nonferrous Met Soc China 24:632–640
8. Bhingole PP, Chaudhari GP, Nath SK (2014) Processing, microstructure and properties of ultrasonically processed in situ MgO–Al2O3–MgAl2O4 dispersed magnesium alloy composites. Compos Part A 66:209–217
9. Park Y, Cho K, Park I, Park Y (2011) Fabrication and mechanical properties of magnesium matrix composite reinforced with Si coated carbon nanotubes. Procedia Eng 10:1446–1450
10. Zhao F-Z, Feng X-H, Yang Y-S (2016) Microstructure and mechanical properties of CNT-reinforced AZ91D composites fabricated by ultrasonic processing. Acta Metall Sin (Engl Lett) 29(7):652–660

11. Selvamani ST, Premkumar S, Vigneshwar M, Hariprasath P, Palanikumar K (2017) Influence of carbon nano tubes on mechanical, metallurgical and tribological behavior of magnesium nanocomposites. J Magnes Alloys 5:326–335
12. Yuan X, Huang S (2015) Microstructural characterization of MWCNTs/magnesium alloy composites fabricated by powder compact laser sintering. J Alloy Compd 620:80–86
13. Govindaraju M, Balasubramanian K, Chackingal U, Prasad Rao K (2014) Effect of severe plastic deformation and heat treatment on toughness of magnesium alloys. In: 3rd international conference on materials processing and characterisation (ICMPC 2014). Procedia Mater Sci 6:37–42
14. Zheng L, Nie H, Liang W, Wang H, Wang Y (2016) Effect of pre-homogenizing treatment on microstructure and mechanical properties of hot-rolled AZ91 magnesium alloys. J Magnes Alloys 4:115–122

Mechanical, Microstructural Characterization and Tribological Response of Al 5050/B_4C/SiC Stir Cast Hybrid Metal Matrix Composite

P. Sri Sukesh, Y. Vasu, M. Venkateswara Rao, and V. Damodara Naidu

Abstract Al alloy matrix composites are popping up as modern engineering materials on account of their high stiffness, good wear opposition and corrosive resistance properties. The utilization of Al 5050 alloy has been expanded remarkably in Naval and space research. Now the present effort is concerned with the Mechanical, Microstructural characterization and Tribological behavior of Al 5050 fortified with SiC and B4C by assorting weight percentages of particles by stir casting approach. The properties such as hardness, the microstructural properties such as SEM, XRD and Porosity, and the tribological properties such as wear behavior are studied in the present investigation. The results yield that on an average the hardness of the composite upsurges with rise of B4C proportion. It was also found that the resistance to wear is sky-high for the composite with Al 5050-91%, B4C-6%, SiC-3%. For the detection of crystalline phases in the composites X-Ray diffraction technique was used. Diffraction peak positions obtained from the XRD test can also be used to calculate unit cell dimensions. For all the casted composites, density and porosity performance was pin pointed. It was observed that the porosity of all the combinations are under 3%.

Keywords Aluminum metal matrix composites · Al 5050 · B_4C · SiC · MMC · XRD

1 Introduction

For aviation and vehicle manufacturing industries metal matrix composite materials are pivotal materials. These days, lightweight sheet metal fabrication has achieved at most importance. Aluminum alloys are patented as high strength to weight fraction alloys. In addition, Al–MMCs can also be prepared with less effort using variety of materials and can be applied in multiple domains of design applications [1, 2].

P. Sri Sukesh (✉) · Y. Vasu · M. Venkateswara Rao · V. Damodara Naidu
Department of Mechanical Engineering, Bapatla Engineering College, Bapatla, A.P, India
e-mail: srisukesh99@gmail.com

N. M. Sivaram et al. (eds.), *Advances in Manufacturing, Automation, Design and Energy Technologies*, Lecture Notes in Mechanical Engineering,
https://doi.org/10.1007/978-981-99-1288-9_16

AMMC can be flourished by various methodologies such as stir casting, squeeze casting, powder metallurgy and liquid infiltration. Tribological along with mechanical properties of AMMCs can be upgraded by prepending of ceramic particles in aluminum/aluminum alloy matrix materials [3]. Even though MMCs offer matchless properties than monolithic materials, the properties always depend on the type of reinforcement, matrix-reinforcement interactions along with manufacturing procedures [4].

SiC, B_4C particles with less micron size as possible, fortified aluminum composites have acquired more application-oriented recognition than any other kinds of MMCs on account of their high concert, viz., high hardness, high wear resistance and high thermal conductivity. However, to improve the strength at elevated temperatures, particulate reinforcement is observed as one of the best options [5]. Chennakesava Reddy [6] worked on Al 5050 alloy and proposed that on account of increase in wettability, strength and modulus of elasticity were improved. Grain boundary strengthening and precipitation strengthening are the additional mechanisms which could intensify the properties of SiC/5050 Al alloy composites. Bhujanga and Manohara [7] observed that the fusion of B4C particulates in aluminum matrix leads to the development of thin titanium rich interfacial reaction layer.

Further it has been observed that the particle sizes and weight fractions of the reinforcement are crucial on the perfunctory properties of the composites. Stir casting process is the preferred process because of its sundry dominance like simplicity, flexibility and help full in the production of large quantities from all liquid state processings [8].

From review of literature, it was observed that no researchers used two different reinforced particles such as B4C and SiC with the Al 5050 as matrix material. In this outstanding research, an effort has been made to emerge $\mathrm{Al\,5050/B_4C/SiC}$ composite material by varying weight ratios of its constituents. The mechanical, microstructural characterization and tribological behavior were scrutinized to note the influence of B_4C and SiC in the casted composites.

2 Experimental Details

2.1 Materials

Aluminum alloys especially for 5xxx series having excellent corrosion resistance even in salt water and equipped with very high toughness even at subfreezing temperatures [6]. In many engineering sectors including automotive and Naval fields Al 5050 alloy is widely used. On that note Al 5050 is chosen as matrix material owing to its wide application. Good strength and formability are the added properties of this alloy. Table 1 shows the material composition and chemical composition of Al 5050 alloy.

Table 1 Chemical composition of Al 5050 alloy as matrix material (wt.%)

Constituents of Al 5050	Al	Mn	Si	Mg	Cu	Zn	Ni	Ti	Cr
Wt. (%)	97.6	0.224	0.129	1.72	0.043	< 0.008	< 0.005	0.018	0.01

Boron carbide is the third hardest material in the universe, after diamond and cubic boron nitride. B4C exhibits the properties such as high temperature safe, high hardness, high granulating proficiency, high versatile modulus and great hostile to radiation execution [9].

Spiffing mechanical, thermal, electrical and chemical properties possessed by silicon carbide particles SiC has high hardness, high wear resistance and good self-lubricating capacity. Moreover silicon carbide holds high thermal conductivity, low thermal expansion coefficient and high temperature strength characteristics.

Boron carbide (B_4C) with mass density 2.52 gm/cm^3 and silicon carbide (SiC) particles with mass density 3.21 gm/cm^3 in the range of 20–30 μm were used as the reinforcement phase in this study.

2.2 *Composite Preparation*

Stir casting technique was the promising route employed to fabricate Al 5050 – B_4C – SiC composites. The crucible used and the furnace to heat the raw materials are shown in Fig. 1.

2.3 *Stir Casting method*

Stir casting is the fabricating technique used in this present investigation to cast the composites. Stir casting is generally accepted as a promising route among the variety of manufacturing processes available for casting metal matrix composites. The key element in this process is the mechanical stirring. The reinforcing particles which are in powder form are scattered uniformly into molten aluminum by mechanical stirring. In order to melt the constituents inside the furnace graphite crucible was generally used in this process. Upto 30% volume fraction of the reinforcements, stir casting is the satisfactory method for manufacturing composites [10–12]. The parameters that influence the uniform distribution of the ceramic particles include stirring parameters, melting temperature, geometry of the mechanical stirrer, characteristics of the particles added and the placement of the mechanical stirrer in the melt [13, 14]. Eddy currents and inert gases are passed to the molten material through the circumference and at the bottom of the crucible in order to reduce porosity and getting homogeneous

Fig. 1 Heating furnace

distribution of the composite. Nine castings of Al 5050/B_4C/SiC composites were prepared as per the following weight fractions.

MMC 1: Al 5050 – 97% + B_4C – 2% + SiC – 1%
MMC 2: Al 5050 – 95% + B_4C – 2% + SiC – 3%
MMC 3: Al 5050 – 93% + B_4C – 2% + SiC – 5%
MMC 4: Al 5050 – 95% + B_4C – 4% + SiC – 1%
MMC 5: Al 5050 – 93% + B_4C – 4% + SiC – 3%
MMC 6: Al 5050 – 91% + B_4C – 4% + SiC – 5%
MMC 7: Al 5050 – 93% + B_4C – 6% + SiC – 1%
MMC 8: Al 5050 – 91% + B_4C – 6% + SiC – 3%

2.4 Die Preparation and Specimens

The metallic mold used in preparing the cast ingots is shown in Fig. 2. It was made from a low-grade metal and was surface finished by using emery papers. Wire EDM machine shown in Figs. 3, 4 and 5 represents the cast ingots before cutting and after cutting the specimens as per the ASTM standards, respectively.

Fig. 2 Metallic mold

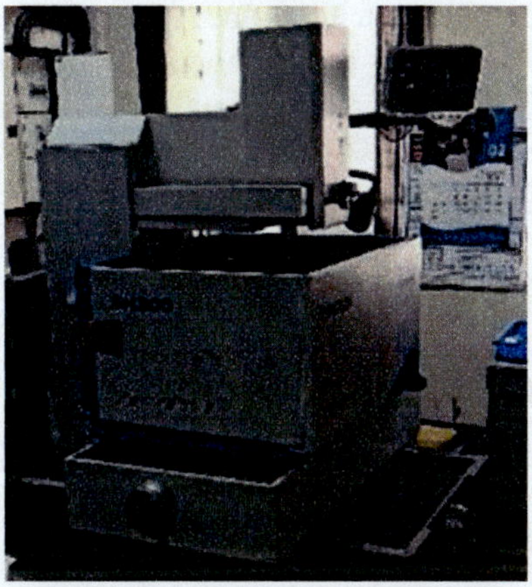

Fig. 3 EDM cutting machine

Fig. 4 Cast ingots before cutting

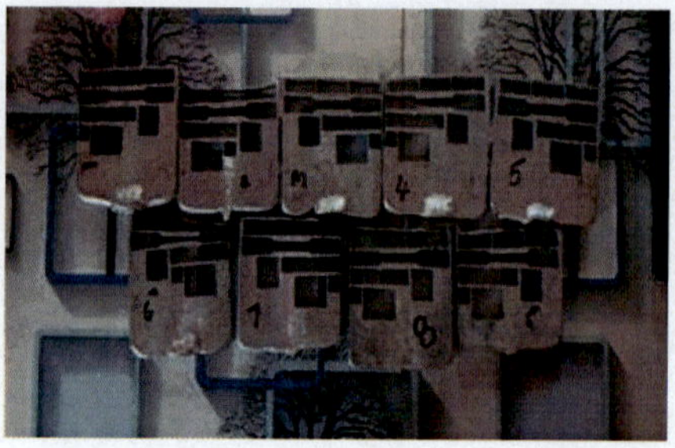

Fig. 5 Cast ingots after cutting

2.5 *Microstructural analysis*

SEM analysis. Figure 6 shows the scanning electron microscopic apparatus. A focused beam of electrons urged to fall on the surface of the sample. The interaction of electrons with atoms within the sample, producing various signals that suppress details about the surface topography and composition of the sample or the composite. The signals generated during SEM analysis produce a two-dimensional image and reveal information about the sample, including external morphology (texture), chemical composition. SEM images of various composites are identified and studied. The test performed for different composites are under

ASTM E766-14 standards. The SEM test samples are exhibited in the Fig. 7. Each sample shown below, used for testing of SEM are of 10 * 10 * 10 mm^3.

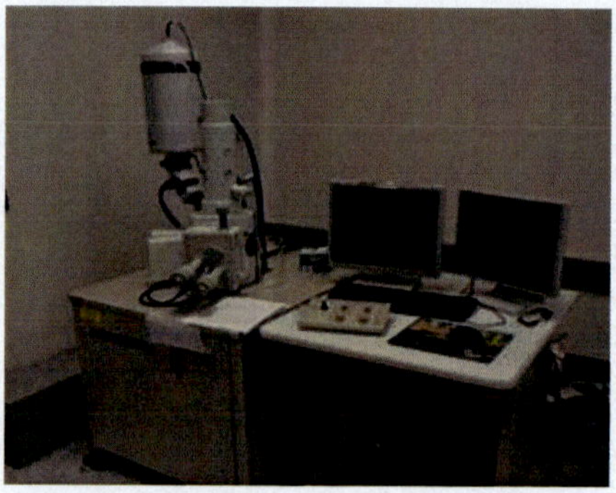

Fig. 6 Scanning electron microscope

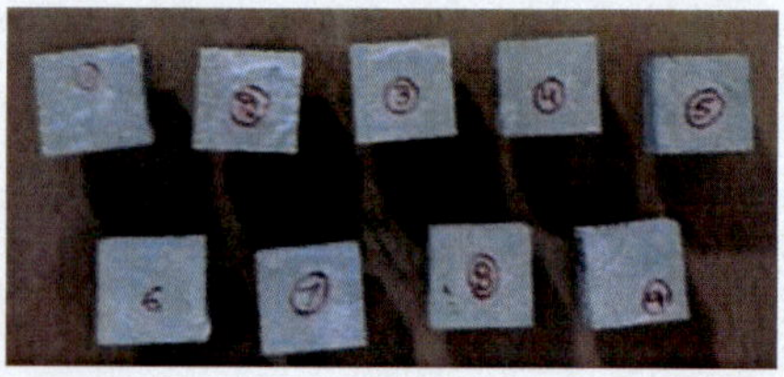

Fig. 7 SEM test samples

Fig. 8 XRD apparatus

Fig. 9 XRD test samples

X-Ray Diffraction analysis. XRD is the well-known non-destructive processes and is useful to analyze the crystalline structure of materials. XRD analysis is employed to spot the crystalline phases present in an exceedingly material, by the way of the study of the crystal structure and thereby it gives the information about chemical composition.

Crystalline structure of various compositions were observed using X-Ray Diffraction apparatus shown in Fig. 8.

All the tests were performed on the report of the ASTM D3906-19 standards and the samples before testing are shown in Fig. 9. Each sample shown below, used for testing of XRD are of 30 * 20 * 10 mm^3.

2.6 *Porosity and Density of composites*

Rule of mixtures was the method used to find out the theoretical density of the composite,

$$\rho_c = \rho_m \times v_m + \rho_{r_1} \times v_{r_1} + \rho_{r_2} \times v_{r_2} \tag{1}$$

and the experimental density from the Archimedes Principle or the Physical law of Buoyancy. The porosity content in the composite can be obtained by

Table 2 Porosity percentage in different composites

S.No	Theoretical density (gm/c.c)	Experimental density (gm/c.c)	Porosity in %
1	2.69	2.6893	0.026
2	2.7223	2.703	0.715
3	2.71	2.693	0.63
4	2.69	2.6823	0.286
5	2.706	2.7	0.222
6	2.71	2.6593	1.87
7	2.746	2.687	2.196
8	2.751	2.697	1.962
9	2.7546	2.7076	1.706

$$P = (1 - \rho_{\text{Exp}}/\rho_{\text{The}}) \times 100\%$$

ρ_{The} Theoretical density (gm/cm^3)
ρ_{Exp} Experimental density (gm/cm^3)

Table 2 represents the percentage of porosity in 9 different weighted composites. This was done using above porosity content formula which was discussed above.

2.7 *Wear Test*

Wear analysis is performed on a pin-on-disc wear tester is as shown in Fig. 10. As per ASTM G99-17 standards all the tests were conducted. A circular wear path is described by the machine by loading the pin against a flat rotating disc in the pin-on-disc wear apparatus. Dead weight loading system was used to perform the wear test by loading the pin against the disc. Weight loss is recorded carefully after the test. Acetone is used to clean the pins and wear track before and after the test.

2.8 *Wear Measurement/Quantification*

Wear measurement is carried out to determine the amount of material removed (or worn away) after a wear test. The material worn away can be expressed either as weight (mass) loss, volume loss, or linear dimension change depending on the purpose of the test, the type of wear, the geometry and size of the test specimens.

Common techniques of wear quantification include using a precision balance to estimate the mass loss or using a microscope to measure the wear depth or cross-sectional area of a wear track, in order to determine the wear volume loss. Three

Fig. 10 Pin on disc wear apparatus

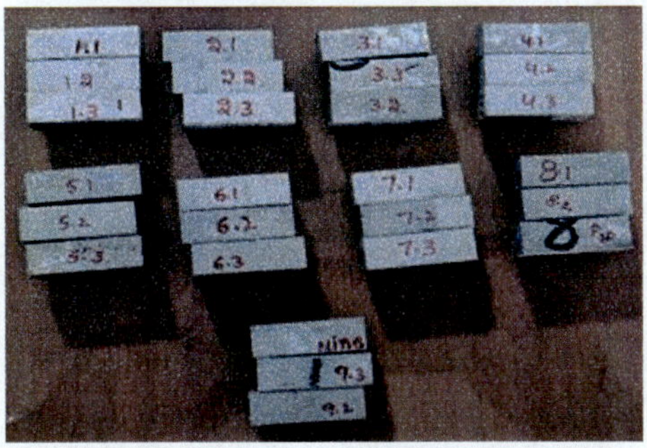

Fig. 11 Wear test samples before testing

different sliding velocities that is at 1, 2&3m/s is performed to study the wear properties at different velocities. The load and sliding distance on the pin is maintained constant that is 10N and 2000 m, respectively. Dry sliding wear test is done at room temperature. The wear test specimens before testing and after testing are shown in Figs. 11 and 12, respectively. Parameters maintained during wear test is mentioned in below Table 3.

2.9 Hardness

Hardness is the opposition offered by the composite for controlled deformation. This controlled deformation of the composites can be achieved by indentation or scratching or bending. The hardness of various composites were measured using the Micro Vickers hardness testing apparatus with diamond indentation shown in Fig. 13.

The tests were instructed according to the ASTM E92-17 standards with a diamond indenter with a load of 2N. The Micro Vickers hardness testing samples are shown

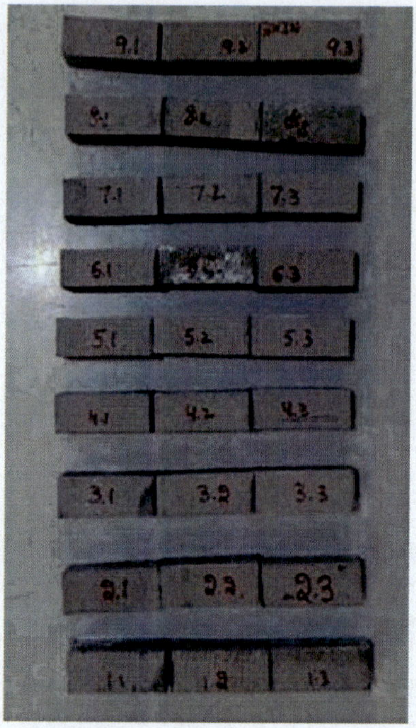

Fig. 12 Wear test samples after testing

in Fig. 14. Each sample shown below, used for testing of hardness values are of 20 * 20 * 10 mm^3 in dimension.

The wear resistance and wear rate for varying compositions of Al 5050, B_4C, SiC at 1, 2, 3m/s are calculated and shown in Table 4.

3 Results and Their Discussion

3.1 Hardness

The fluctuation of the hardness of cast composites with variation of B_4C is conceived on the graph in Fig. 15. It was clearly catched the sight of that with increase of wt.% B_4C, hardness value upsurged. Reinforcing particles, i.e., B_4C and SiC are denser and stronger than the matrix material, i.e., Al 5050, which focusses on the increment of limiting the plastic deformation of matrix in the time of test. As long as the wettability increase the dispersal of particles in the composite, the hardness quality is intensified [7].

Table 3 Wear treating parameters

Sample No	Composition (%)			Sliding distance (m)	Velocity (m/s)	DiscSpeed (Rpm)
	Al 5050	B_4C	SiC			
1.1	97	2	1	2000	1	191
1.2	97	2	1	2000	2	382
1.3	97	2	1	2000	3	573
2.1	95	2	3	2000	1	191
2.2	95	2	3	2000	2	382
2.3	95	2	3	2000	3	573
3.1	93	2	5	2000	1	191
3.2	93	2	5	2000	2	382
3.3	93	2	5	2000	3	573
4.1	95	4	1	2000	1	191
4.2	95	4	1	2000	2	382
4.3	95	4	1	2000	3	573
5.1	93	4	3	2000	1	191
5.2	93	4	3	2000	2	382
5.3	93	4	3	2000	3	573
6.1	91	4	5	2000	1	191
6.2	91	4	5	2000	2	382
6.3	91	4	5	2000	3	573
7.1	93	6	1	2000	1	191
7.2	93	6	1	2000	2	382
7.3	93	6	1	2000	3	573
8.1	91	6	3	2000	1	191
8.2	91	6	3	2000	2	382
8.3	91	6	3	2000	3	573
9.1	89	6	5	2000	1	191
9.2	89	6	5	2000	2	382
9.3	89	6	5	2000	3	573

3.2 Density Comparison

The rule of mixtures is the principle used to assess the theoretical density of the casted composites and the experimental density was achieved using the method of Archimedes displacement [15] which were performed as per the ASTM Standards D 792-66. Theoretical and measured densities of prepared composites are shown in the graph below. The outcome of density measurement advised that inclusion of ceramic B4C and SiC particulates to Al 5050 matrix shows the enhancement in the overall density of composites [16]. Moreover in distinction to Fig. 16, it is evident

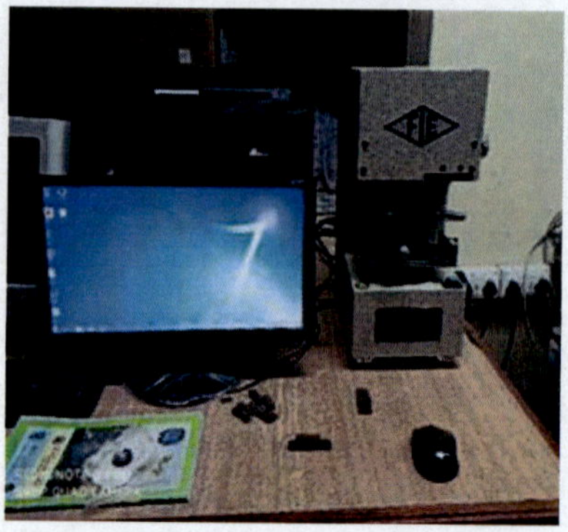

Fig. 13 Micro Vickers hardness tester

Fig. 14 Hardness test samples

that experimental density of composites 4, 5 and 7 seems to be under estimated theoretical density which in turn shows the porosity development in the respective cast ingots.

3.3 Micro Structural Characterization

XRD analysis was carried out to determine the existing phases in the composites. The XRD pattern of matrix material and the reinforced particles is shown in Figs. 17, 18 and 19, respectively. The X-ray diffraction pattern of the composite Al 5050 – 97% + B_4C – 2% + SiC – 1% is shown in Fig. 20 and also indicated the presence of the alloys Al 5050, B4C and SiC. As can be seen, only Al 5050, B_4C and SiC peaks are observable in diffraction pattern of the composite.

Scanning Electron Microscopic images at 100, 50, 10 μm obtained for the composite with 97% Al 5050 + 2% B4C + 1% SiC is shown in Figs. 21, 22, 23 and 24 at X100, X500, X1000, X2000 magnifications, respectively. From the below images it was clear that there were no agglomerations or clustering's visible in the fabricated composite.

Table 4 Wear resistance and wear rate of composites at various velocities

Composition (%)			Sliding distance (m)	Velocity (m/s)	Rpm	Initial weight (gms)	Final weight (gms)	Mass loss (gms)	Density (gm/c.c)	Volume loss (mm^3)	Wear resistance (mm^3/m)	Wear rate (m/mm^3)
Al 5050	B_4C	SiC										
97	2	1	2000	1	191	8.098	8.068	0.03	2.69	11.152	179.34	0.006
97	2	1	2000	2	382	8.063	8.018	0.045	2.69	16.729	119.552	0.008
97	2	1	2000	3	573	8.232	8.198	0.034	2.69	12.639	158.24	0.006
95	2	3	2000	1	191	8.167	8.098	0.069	2.703	25.527	78.348	0.013
95	2	3	2000	2	382	8.131	8.088	0.043	2.703	15.908	125.723	0.008
95	2	3	2000	3	573	8.176	8.138	0.038	2.703	14.058	142.267	0.007
93	2	5	2000	1	191	8.079	8.028	0.051	2.71	18.819	106.275	0.009
93	2	5	2000	2	382	8.205	8.158	0.047	2.71	17.343	115.320	0.008
93	2	5	2000	3	573	8.170	8.128	0.042	2.71	15.498	129.049	0.007
95	4	1	2000	1	191	8.047	7.988	0.049	2.69	18.215	109.799	0.009
95	4	1	2000	2	382	8.092	8.058	0.034	2.69	12.639	158.24	0.006
95	4	1	2000	3	573	8.220	8.188	0.032	2.69	11.896	168.124	0.005
93	4	3	2000	1	191	8.118	8.058	0.06	2.7	22.222	90.001	0.011
93	4	3	2000	2	382	8.248	8.218	0.03	2.7	11.111	180.001	0.005
93	4	3	2000	3	573	8.199	8.168	0.031	2.7	11.481	174.201	0.006
91	4	5	2000	1	191	7.978	7.918	0.06	2.71	22.140	90.334	0.011
91	4	5	2000	2	382	7.969	7.918	0.051	2.71	18.819	106.275	0.009
91	4	5	2000	3	573	7.967	7.938	0.029	2.71	10.701	186.898	0.005
93	6	1	2000	1	191	8.238	8.18	0.058	2.687	21.585	92.657	0.011
93	6	1	2000	2	382	8.074	8.028	0.046	2.687	17.119	116.829	0.008

(continued)

Table 4 (continued)

Composition (%)			Sliding distance (m)	Velocity (m/s)	Rpm	Initial weight (gms)	Final weight (gms)	Mass loss (gms)	Density (gm/c.c)	Volume loss (mm^3)	Wear resistance (mm^3/m)	Wear rate (m/mm^3)
Al 5050	B_4C	SiC										
93	6	1	2000	3	573	8.128	8.088	0.04	2.687	14.886	134.354	0.007
91	6	3	2000	1	191	8.254	8.208	0.046	2.697	17.055	117.268	0.008
91	6	3	2000	2	382	8.133	8.098	0.035	2.697	12.977	154.118	0.006
91	6	3	2000	3	573	8.242	8.218	0.024	2.697	8.899	224.744	0.004
89	6	5	2000	1	191	8.264	8.018	0.246	2.7076	90.855	22.013	0.045
89	6	5	2000	2	382	8.148	8.108	0.04	2.7076	14.773	135.382	0.007
89	6	5	2000	3	573	8.186	8.118	0.068	2.7076	25.114	79.637	0.012

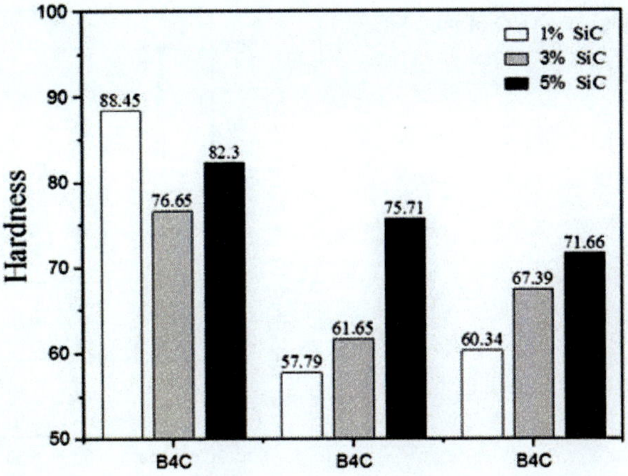

Fig. 15 Consequence of B4C on Vickers hardness of composites

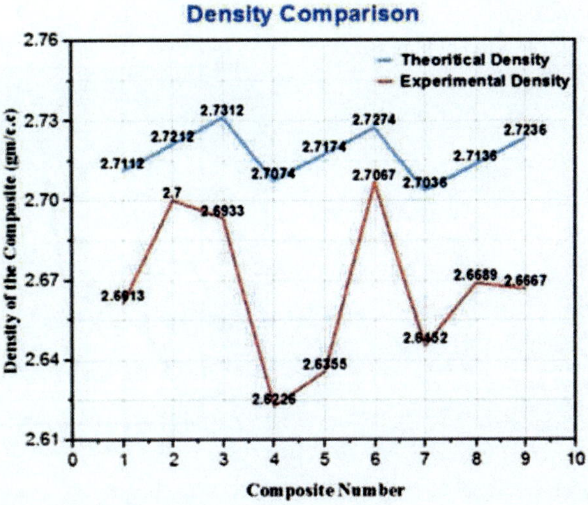

Fig. 16 Comparison between theoretical and experimental density of composites

3.4 *Effect of Sliding Velocity on Wear*

The Fig. 27 explains the effect of sliding velocity on the wear loss of composites. On indeep observation, an upsurge in sliding velocity from 1 to 2 m/s, a climb in wear loss was occurred for the composite and at high sliding velocity, i.e., 3 m/s, decrease in wear loss was observed clearly for the composite shown in Fig. 27. From this we can clear that on higher sliding velocities, there is a slight increase in wear resistance. It is observed from the Fig. 25 that the drop in wear rate can be seen for all the composites at high sliding velocity, i.e., at 3 m/s. so it can be declared that wear rate was minimized with an increase in sliding velocity. On the other hand Fig. 26 shows that wear resistance at 3m/s tops on comparison with the other two velocities.

Fig. 17 XRD of pure SiC

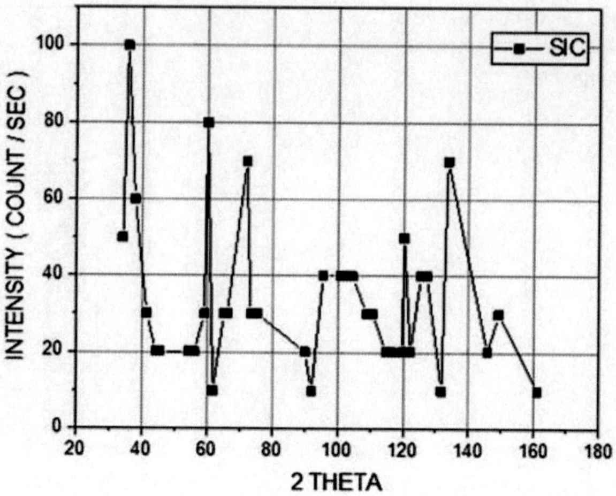

Fig. 18 XRD of pure B_4C

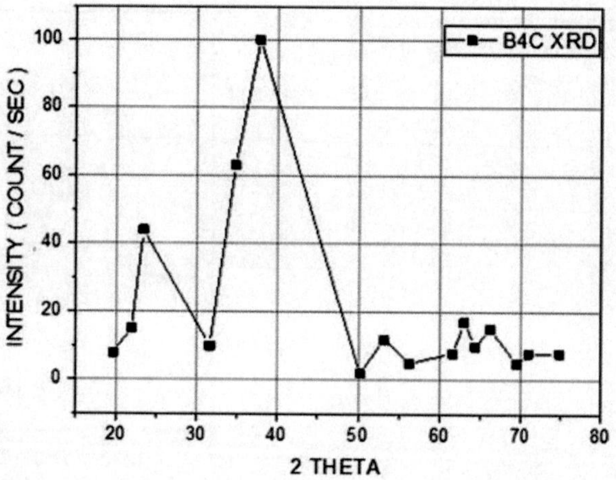

This may be the reason because of lower contact time while increasing the sliding velocity. When sliding velocity increased, the travel time was minimized to complete the required sliding distance (2000 m) that may cause the generation of wear debris particles. Therefore, results in reduction of wear rate of the composite.

Uthayakumar et al. [17] reported that B4C easily creates a boron oxide (B_2O_3) layer on the surface of a sample. This layer reduces the wear rate significantly. At the tribo contact, the wear of composite at higher speeds leads to increase in temperature. The improved high temperature strength due to ceramic reinforcement also contributes to improve the wear resistance at higher speeds. The ceramic reinforcement also hand out the improvement of wear resistance at higher speeds due to high temperature strength. Volume fraction and the type of reinforcement are contributing factors on which the wear resistance of the composite may depend.

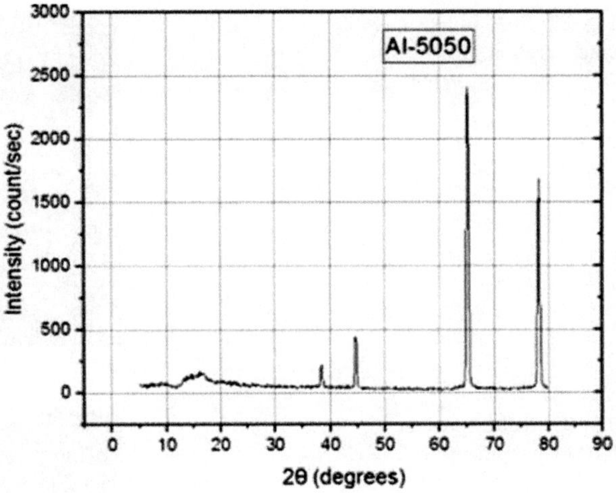

Fig. 19 XRD of pure Al 5050

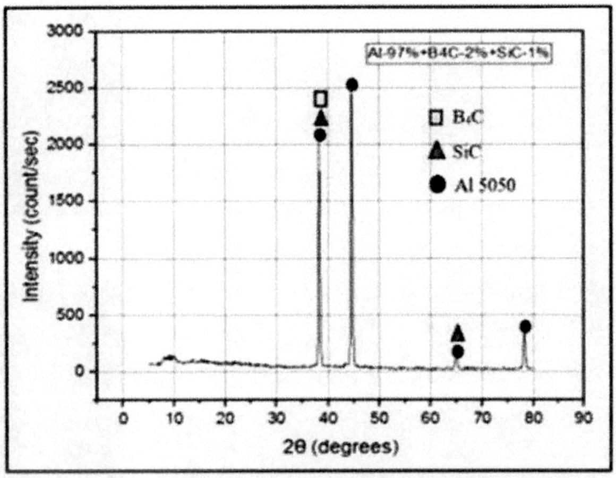

Fig.20 XRD plot of the composite

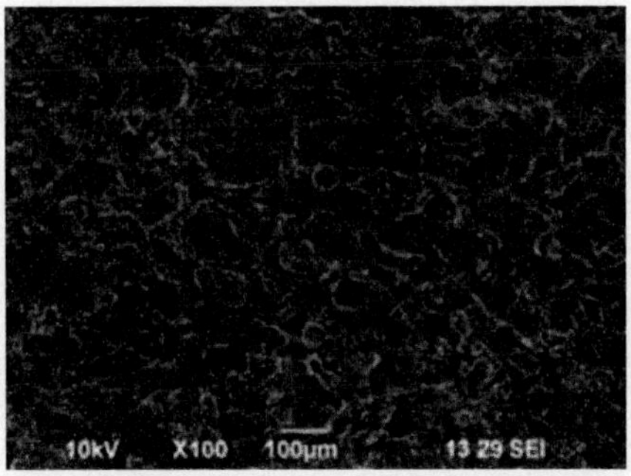

Fig. 21 SEM image at X100 magnification

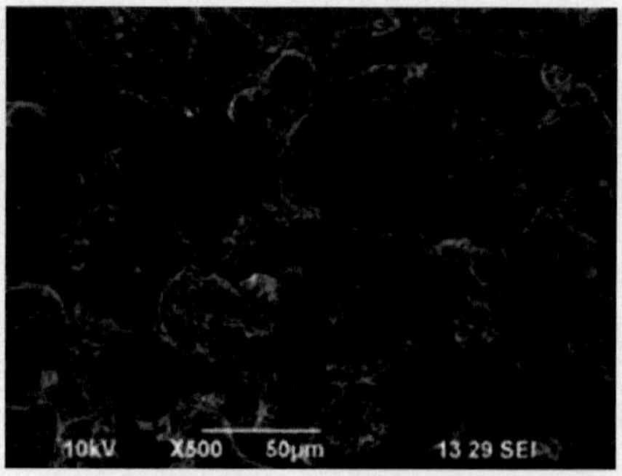

Fig. 22 SEM image at X500 magnification

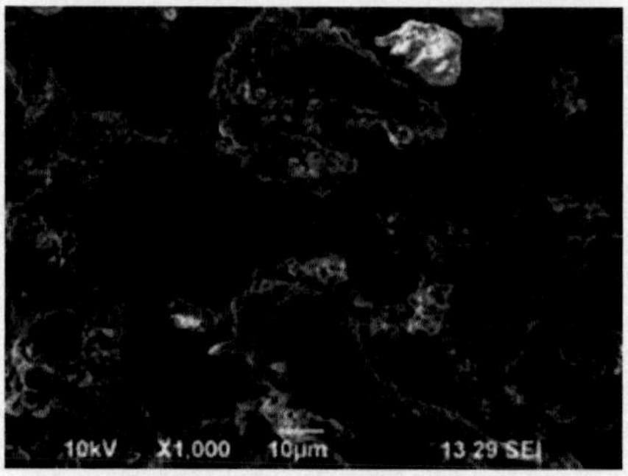

Fig. 23 SEM image at X1000 magnification

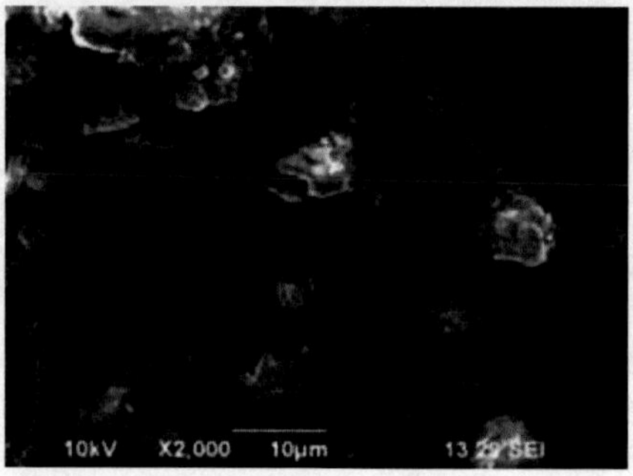

Fig. 24 SEM image at X2000 magnification

4 Conclusions

In this probing investigation, MMCs of differing B_4C and SiCWt.% were developed using stir casting technique. Microstructural aspects such as grain size, uniform distribution of the prepared composites were deliberated. Based on the experimental evaluation, following establishments were drawn:

- Al 5050 alloy can be fruitfully fabricated by using boron carbide and silicon carbide as reinforcement material through stir casting technique apparatus.

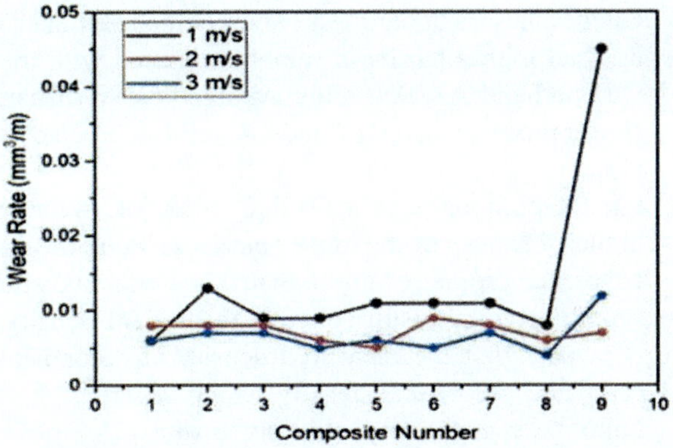

Fig. 25 Wear rates of composites with variation of velocity

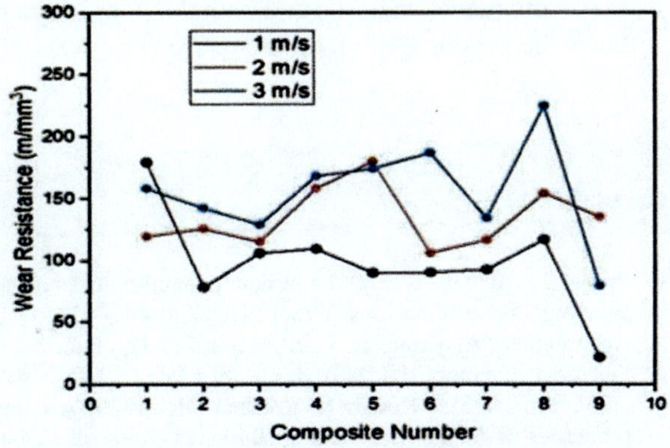

Fig. 26 Wear resistances of composites variation of velocity

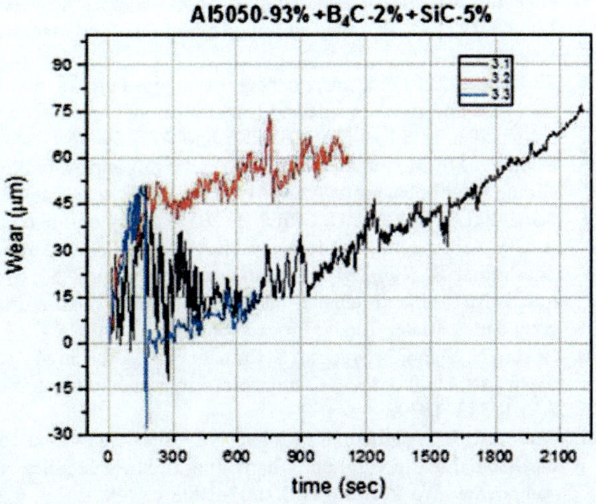

Fig. 27 Wear vs Time plot of composite with variation of velocity

- On the analysis of hardness values of the composites, found that fortified composites had higher hardness values compared with the unreinforced Al 5050 alloy. On one hand Al 5050 alloy average 46 HV of hardness value, but on the other side composite with B_4C and SiC reinforced displayed the highest value of 88.45 HV.
- For the composite with 6%B_4C + 3%SiC wear resistance was recorded to be highest because of the higher hardness value of individual B4C and the superior adherence properties of the matrix material and silicon carbide.
- From the comparison between theoretical density and experimental density it is evident that the measured density of composites 4, 5 and 7 underneath the estimated theoretical density which specifies the porosity development in the minor parts of the respective composites.
- As per the XRD analysis, the peaks obtained in every composite reflects the presence of the matrix material and both the reinforcements in all the composites. This shows the presence of all the materials, i.e., Al 5050, B_4C, SiC in the composite.

References

1. Prasad SV, Asthana R (2004) Aluminum metal-matrix composites for automotive applications: tribological considerations. Tribol Lett 17(3):445–453
2. Natarajan N, Vijayarangan S, Rajendran I (2007) Fabrication, testing and thermal analysis of metal matrix composite brake drum. Int J Veh Des 44(3–4):339–359
3. Nirala A, Soren S, Kumar N, Kaushal DR (2020) A comprehensive review on mechanical properties of Al-B4C stir casting fabricated composite. Mater Today: Proceed 21:1432–1435
4. Kumar DS, Suman KNS, Poddar P (2017) Effect of particle morphology of Ni on the mechanical behavior of AZ91E-Ni coated nano Al2O3 composites. Mater Res Express 4(6):066505
5. Lloyd DJ (1994) Particle reinforced aluminium and magnesium matrix composites. Int Mater Rev 39(1):1–23
6. Reddy AC (2017) Consequences of magnesium in 5050 aluminum alloy on wettability, strengthening mechanisms and fracture behavior of silicon carbide nanoparticle metal matrix composites. Int J Res Mech Eng Technol 7(1):89–96
7. Bhujanga DP, Manohara HR (2018) Processing and evaluation of mechanical properties and dry sliding wear behavior of AA6061-B4C composites. Mater Today: Proceed 5(9):19773–19782
8. Karabulut Ş, Gökmen U, Cinici H (2016) Study on the mechanical and drilling properties of AA7039 composites reinforced with Al_2O_3/B_4C/SiC particles. Compos B Eng 93:43–55
9. Shashidhar S, Kumar PV, Shantharaja M, Nagaral M, Shivananda BK (2018) Processing, microstructure and tensile behavior of nano B_4C particulates reinforced Al2219 alloy composites. Mater Today: Proceed 5(11):25158–25164
10. Lokesh N, Manoj B, Srikanth K, Ramanayya PKV, Rao MV (2018) Mechanical characterization of stir cast Al 6063 TiO_2-Cu reinforced hybrid metal matrix composites. Mater Today: Proceed 5(9):18383–18392
11. Sharma SS, Jagannath K, Mohan NS, Bhat C, Prabhu PR, Chadaga RC (2010) Thermo-mechanical and precipitation hardening of stir-cast silica reinforced aluminium alloy matrix composites. World J Eng 7(3):1007–1008
12. Sharma SS, Prabhu PR, Jayashree PK, Shankar G, Porwal S, Singh H (2013) The tensile and wear property enhancement of Al–Zn–Mg alloys by precipitation hardening and thermomechanical treatments. Int J Curr Eng Technol 3(4):1510–1515

13. Sharma SS, Jagannath K, Prabhu PR (2013) Microstructure and mechanical characterization of heat treated stir cast silica (sea sand) reinforced 7XXX Al alloy MMCs. Int J Mech Mechatron Eng 7(1):171–176
14. Jagannath K, Sharma S, Mohan N, Kini A, Prabhu P (2014) Characteristic study for low melting point MMC using stir casting technique. World J Eng
15. Bharath V, Ajawan SS, Nagaral M, Auradi V, Kori SA (2019) Characterization and mechanical properties of 2014 aluminum alloy reinforced with Al 2 O 3p composite produced by two-stage stir casting route. J Inst Eng (India): Ser C 100(2):277–282
16. Srivatsan TS (1996) Microstructure, tensile properties and fracture behaviour of Al 2 O 3 particulate-reinforced aluminium alloy metal matrix composites. J Mater Sci 31(5):1375–1388
17. Uthayakumar M, Aravindan S, Rajkumar K (2013) Wear performance of Al–SiC–B_4C hybrid composites under dry sliding conditions. Mater Des 47:456–464

An Evaluation of Density and Compression Properties of AZ91D–SiC Metal Matrix Composite Produced Through Powder Metallurgy

A. Packia Antony Amalan, P. Arun Kumar, P. Balasundaram, and N. M. Sivaram

Abstract This study reported that the physical and compression properties of silicon carbide reinforced AZ91D magnesium alloy composite. Silicon carbide powders were added as reinforcement, 1–3 weight percent with magnesium alloy AZ91D. The samples were prepared through powder metallurgy method. Ball milling process was carried out for one hour at a rotational speed of 315rpm. Compaction was performed at 10 ton hydraulic press. Sintering temperature was maintained at 380 °C for 90 min. The samples were studied with two geometrical shapes such as cuboid and cylindrical shapes. The cuboid shape AZ91D–1SiC wt.% composite sample possessed density and porosity are 1.54 g/cc and 15.6%, respectively. Similarly, the cylindrical shaped composite had 1.62 g/cc, 10.96%, respectively. Maximum obtained Young's modulus and compressive strength of cylindrical shaped composite are 249.4, and 9.9 MPa, respectively. The possible reason of lower values of compressive strength and Young's modulus were due to compaction pressure during fabrication and more porosity in the samples.

Keywords Magnesium · Compressive strength · Powder metallurgy

1 Introduction

The demand for light weight high strength materials is increasing rapidly in the field of automobile and aerospace engineering. Magnesium (Mg) has a density of 1.74 g/cm^3 [1]. However, pure magnesium does not have good strength as compared to aluminum. Thus, the use of magnesium alloys is increasing in recent days. The most commonly used Mg alloy in the automotive and aerospace industry is the

A. Packia Antony Amalan (✉)
PSN College of Engineering and Technology, Tirunelveli 627 152, India
e-mail: amalan234@gmail.com

P. Arun Kumar · P. Balasundaram · N. M. Sivaram
National Institute of Technology Puducherry, Karaikal 609 609, India

N. M. Sivaram et al. (eds.), *Advances in Manufacturing, Automation, Design and Energy Technologies*, Lecture Notes in Mechanical Engineering,
https://doi.org/10.1007/978-981-99-1288-9_17

AZ91D (nominal composition 9 wt.% Al, 1 wt.% Zn (Mg) and balance [2]. Hardness of AZ91D is 71 HV and ultimate tensile strength of 264MPa [3]. Apart from light weight and high strength AZ91D has other properties such as good heat dissipation, good damping, readily available, good machinability, good electromagnetic shielding [4, 5]. Thus, enhancing the mechanical properties strength will be able to widen the applications of AZ91D in various fields. Magnesium based materials successfully produced through power metallurgy method [1, 6, 7]. In addition magnesium based materials produced through powder metallurgy method possesses higher wear resistance [1]. Addition of silicon carbide (SiC) particles as reinforcements improves mechanical properties of AZ91D magnesium alloy composites [8–10]. Laurent et al. [11] reported that metal matrix have high specific strength and stiffness. Kipouros et al. [12] reported that dimensional control could be achieved by powder metallurgy method.

From the literature survey it was found that addition of silicon carbide as reinforcement can increase the mechanical property of AZ91D. In this present work, analysis was done for the failure of magnesium alloy AZ91D reinforced with silicon carbide fabricated for compression test. The samples used were containing 1–3 wt.% of SiC with AZ91D which were subjected to different tests to obtain the optimum sample. The compression test was carried out and the results depicted relations with porosity of respective samples.

2 Experimental Method

2.1 *Blending of Metal Powders*

The powders of AZ91D and SiC were supplied from Parshwamani metals, India. AZ91D and SiC powders for different composition sample were taken and mixed using planetary mill for one hour at a rotational speed of 315 rpm. After taking out the powder mixture from the mill, Polyvinyl alcohol (PVA) was added in the powder mixture to increase the strength and bonding between the particles of powder mixture. The composition of AZ91D magnesium alloy used for the work is listed in the Table 1. Figure 1 represents the fabricated cuboid and cylindrical shape composites.

Table 1 Composition of AZ91D

Sample	Al	Zn	Cu	Ni	Mn	Si	Fe	Mg
AZ91D	8.5	0.5	0.03	0.002	0.25	0.1	0.005	Bal.

Fig. 1 Fabricated cylindrical and cuboid samples

2.2 Compaction and Sintering

Compaction process was carried out using 10 ton hydraulic press using both cylindrical and cuboid dies. Sintering was carried out in a box furnace. Normal atmospheric condition was used for sintering rather than inert gas atmosphere (argon atmosphere, controlled sintering). The sintering temperature (maximum temperature limit) was fixed to 380 °C and the time was set for one hour and thirty minutes. The rate of temperature rise inside furnace was approximately 5 °C per minute.

2.3 Physical Property Evaluation

Theoretical density (ρ_{th}) was calculated by using the rule of mixtures principle (Eq. 1). Porosity for different samples was calculated using Eq. 2.

$$\text{Density}(\rho\text{th}) = (\text{wt.\%} \times \rho(\text{AZ91D})) + (\text{wt.\%} \times \rho(\text{SiC})) \tag{1}$$

$$\text{Porosity} = 1 - \frac{\text{Actual density}}{\text{Theoretical density}} \tag{2}$$

3 Results and Discussion

3.1 Physical Properties of Specimen

The results of physical properties density, porosity and mechanical properties such as compressive strength and Young's modulus are discussed in detail in this chapter.

Table 2 represents the density of samples. Figure 2a showed actual density of cylindrical and cuboid samples. It can be observed that the actual density of the cuboid samples was lesser than the cylindrical samples insisting that cuboid samples has more empty spaces between the particles. The minimum density is for 2% cuboid sample (1.384 g/cc) and the maximum density (1.624 g/cc) is for 1% cylindrical sample. It is observed that with addition of SiC powder, the theoretical density increase.

Figure 2b represented that porosity for 1% cylinder sample is the least (10.96%) as compared to 2% cuboid sample which has the maximum porosity (24.7%). The surface area to volume ratio of cylindrical shape is less than cubical shape for unit dimension, hence obtained more compaction and the porosity of cylindrical sample was less compared to cubical sample (Table 2). The maximum density (1% cylinder

Table 2 Porosity of produced samples

Geometrical shape of sample	Sample	Actual density (g/cc)	Theoretical density (g/cc)	Porosity (%)
Cuboid	AZ91D alloy	1.478	1.81	18.34
	AZ91D–1SiC	1.538	1.824	15.6
	AZ91D–2SiC	1.384	1.838	24.7
	AZ91D–3SiC	1.459	1.852	21.22
Cylindrical	AZ91D alloy	1.565	1.81	13.48
	AZ91D–1SiC	1.624	1.824	10.96
	AZ91D–2SiC	1.432	1.838	22.075
	AZ91D–3SiC	1.526	1.852	17.6

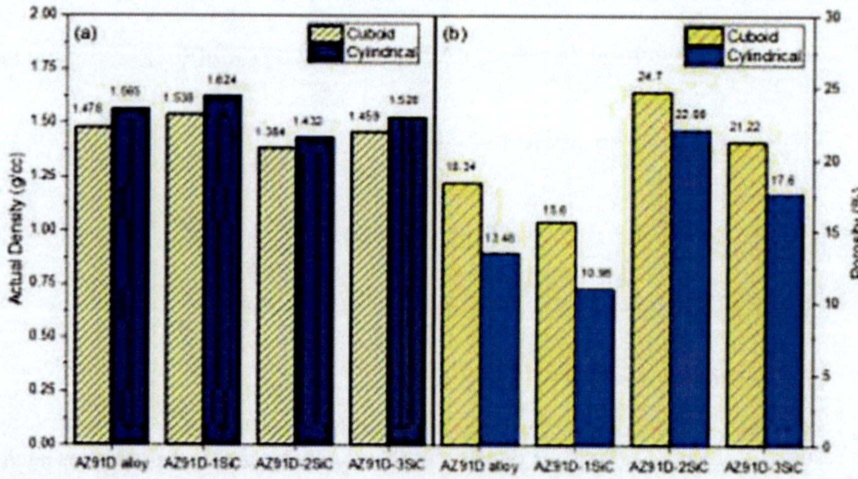

Fig. 2 Comparison of **a** actual density, **b** porosity of samples

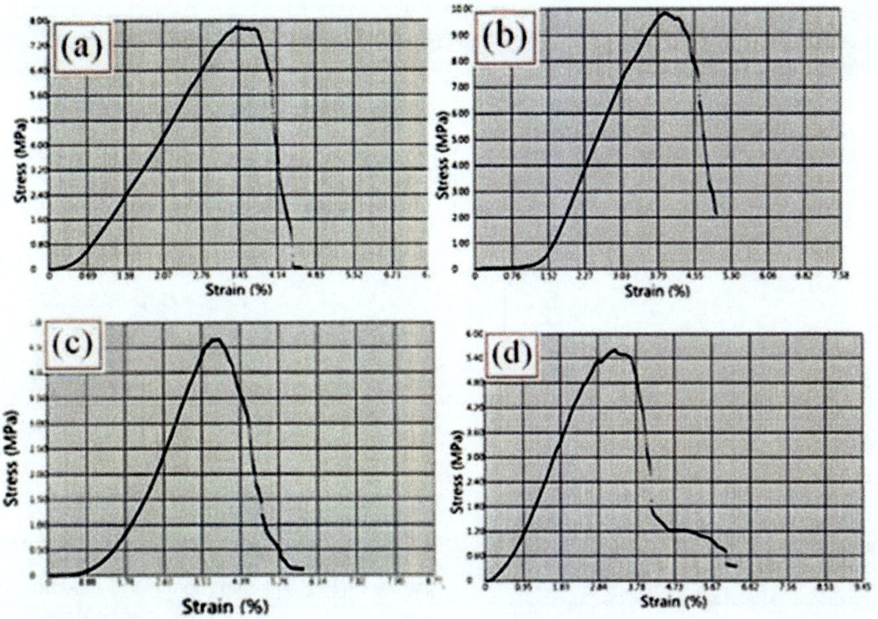

Fig. 3 Compressive stress–strain graph of **a** AZ91D alloy **b** AZ91D–1SiC (wt.%) **c** AZ91D–2SiC (wt.%) **d** AZ91D–3SiC (wt.%)

sample) and the minimum density (2% cuboid sample) clearly indicates that the presence of empty spaces or voids between the particles which might have occurred due to inappropriate fabrication of the samples. Further, it can be inferred that the cylindrical samples had lesser voids as compared to their respective cuboid samples.

3.2 Compression Property

The compression test was carried out using micro compression testing machine. The results of compression test were recorded and plotted in a Fig. 3a–d for different samples. Figure 4 represents that comparison of ultimate compressive stress and porosity of produced alloy and composites; it is observed that the cylindrical shape AZ91D–1SiC sample has the maximum ultimate compressive stress values of 9.85 MPa and porosity 10.96%. The AZ91D–2SiC sample has the minimum ultimate stress values of 4.66 MPa and porosity 22.08%. The low values of ultimate compressive stress can be attributed to the lower strength of the samples causing mainly due to inadequate compaction pressure and poor sintering parameters. The change in values of compressive stress can be ascribed to the porosity of the samples. It is clearly observed that as the porosity increases, the ultimate compressive stress tends to decrease. This is due to increased void between the particles which lead

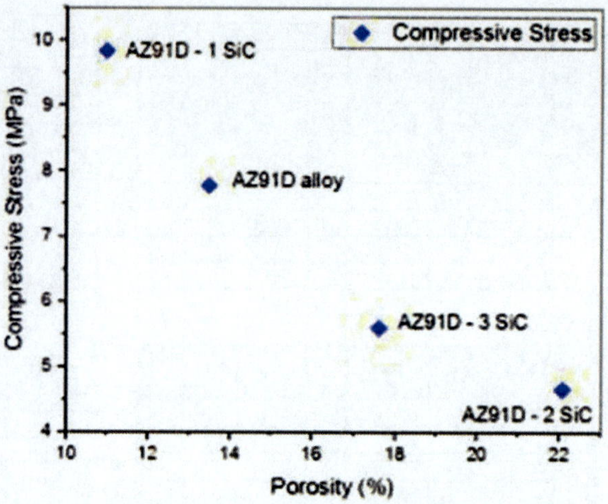

Fig. 4 Comparison of ultimate compressive stress with porosity of samples

Table 3 Compressive properties of produced cylindrical shaped samples

Sample	Compressive stress (MPa)	Compressive strain (no unit)	Young's modulus (MPa)
AZ91D alloy	7.79	0.0343	227.11
AZ91D–1SiC	9.85	0.0395	249.36
AZ91D–2SiC	4.66	0.0377	123.60
AZ91D–3SiC	5.60	0.0328	170.73

to lower strength of bonds between the particles thus decreasing the force required to break the samples. Table 3 indicated the compression properties and Young's modulus of samples. A point is taken on the stress–strain graph of compression test and the corresponding values of stress and strain is used to calculate the Young's modulus of different samples. The maximum and minimum values of the Young's modulus were observed to be 249.36 MPa for 1% SiC and 123.60 MPa½ for 2% SiC cylindrical sample, respectively. The compaction pressure used was less than the ideal compaction pressure (140 MPa). The sintering temperature was taken below the ideal sintering temperature 421 °C, 80% of melting point).

4 Conclusion

Magnesium alloy AZ91D was reinforced with 1, 2 and 3% (wt.%) SiC were produced by using powder metallurgy process. The theoretical and actual density, porosity and compression properties were evaluated.

Maximum actual density 1.624 g/cc of the cylindrical shaped AZ91D–1SiC (wt.%) sample which was 3.76% higher than cylindrical shaped base alloy AZ91D. Maximum theoretical density 1.852 g/cc of the AZ91D–3SiC (wt.%) sample (both cylindrical and cuboid shape) which was 2.32% higher than base alloy AZ91D. Maximum porosity 24.7% was reported for AZ91D–2SiC (wt.%) cuboid sample.

The compression tests resulted that the maximum compressive stress for 1% SiC reinforced AZ91D cylindrical sample was 26.44% higher than pure AZ91D sample. Minimum compressive strength was 4.66 MPa for AZ91D–2SiC (wt.%) sample. These results can be related to the porosity of the respective samples. As the porosity of sample increases the ultimate compressive stress was decreases. The low values of compression test results can be described to the low bonding strength between the particles due to poor fabrication of the samples.

References

1. Soorya Prakash K, Balasundar P, Nagaraja S, Gopal PM, Kavimani V (2016) Mechanical and wear behaviour of Mg–SiC–Gr hybrid composites. J Magnes Alloy 4(3):197–206
2. Kumar DS, Sasanka CT, Ravindra K, Suman KNS (2015) Magnesium and its alloys in automotive applications—a review. Am J Mater Sci Technol 4(1):12–30
3. Selvamani ST, Premkumar S, Vigneshwar M, Hariprasath P, Palanikumar K (2017) Influence of carbon nano tubes on mechanical, metallurgical and tribological behavior of magnesium nanocomposites. J Magnes Alloy 5(3):326–335
4. Wang J et al (2015) Optimization of mechanical and damping properties of Mg-0.6Zr alloy by different extrusion processing. J Magnes Alloy 3(1):79–85
5. Sklenička V, Kuchařová K, Kvapilová M, Svoboda M (2015) Factors influencing creep resistance in discontinuously reinforced magnesium metal matrix composites. Kov Mater 53(4):221–229
6. Turan ME, Sun Y, Aydin F, Zengin H, Turen Y, Ahlatci H (2018) Effects of carbonaceous reinforcements on microstructure and corrosion properties of magnesium matrix composites. Mater Chem Phys 218(July):182–188
7. Turan ME, Sun Y, Akgul Y (2018) Mechanical, tribological and corrosion properties of fullerene reinforced magnesium matrix composites fabricated by semi powder metallurgy. J Alloys Compd 740:1149–1158
8. Aravindan S, Rao PV, Ponappa K (2015) Evaluation of physical and mechanical properties of AZ91D/SiC composites by two step stir casting process. J Magnes Alloy 3(1):52–62
9. Packia Antony Amalan A, Sivaram NM, Bavatharani C, Ragupathy D (2021) A study on the effect of ageing heat treatment on hardness, tensile and corrosion behaviour of stir-cast AZ91D–5SiC–1Gr hybrid magnesium composite. Int J Metalcast 1–14 (2021)
10. Hassan HA, Lewandowski JJ (2014) Effects of particulate volume fraction on cyclic stress response and fatigue life of AZ91D magnesium alloy metal matrix composites. Mater Sci Eng A 600:188–194
11. Laurent V, Jarry P, Regazzoni G, Apelian D (1992) Processing-microstructure relationships in compocast magnesium/SiC. J Mater Sci 27(16):4447–4459
12. Kipouros GJ, Caley WF, Bishop DP (2006) On the advantages of using powder metallurgy in new light metal alloy design. Metall Mater Trans A Phys Metall Mater Sci 37(12):3429–3436

Incorporating Six Sigma in e-Learning Platform During COVID-19 Pandemic

S. Pratheesh Kumar, V. Nithin, S. Akash, and N. Sheik Musthaq Ahamed

Abstract The COVID-19 pandemic has caused various changes in the routine lives of people all across the world. In terms of education, the pandemic resulted in the closure of educational institutions. Because students were unable to attend courses and lectures in person, the institutions resorted to using the Internet to provide classes in order to continue their education. Currently the Six Sigma implementation in this whole system might help to analyze the quality of the education which is being delivered through online mode. The data collection through surveys helps to find shortcomings and the advantages. The methodology used in this paper is DMAIC which undergoes define, measure, analyze, improve and control phases in which the data collected will be analyzed using critical to quality tree (CTQ), bar charts, cause and effect diagram, weighted score matrices, and statistical testing to control the process. From this review paper the quality of the online education and statistical data of the students and faculty undergoing or conducting the online classes are consolidated and discussed. The data collected can be used in the educational research domain if one intends to make research on online education and this can be used to assess the quality of education.

Keywords Online education · Six Sigma · DMAIC · COVID-19 pandemic

1 Introduction

A pandemic is an infectious disease outbreak that has spread across a large area, such as multiple continents or the entire world, and has impacted a large number of people. Since its discovery in Wuhan, China's Hubei province capital, in December 2019, the coronavirus illness 2019 (COVID-19) has spread worldwide, culminating in the ongoing 2019–20 coronavirus pandemic. India confirmed its first COVID-19

S. Pratheesh Kumar (✉) · V. Nithin · S. Akash · N. Sheik Musthaq Ahamed
Department of Production Engineering, PSG College of Technology, Coimbatore, Tamil Nadu, India
e-mail: spratheeshkumarth@gmail.com

N. M. Sivaram et al. (eds.), *Advances in Manufacturing, Automation, Design and Energy Technologies*, Lecture Notes in Mechanical Engineering,
https://doi.org/10.1007/978-981-99-1288-9_18

case on January 27, 2020, and has since adopted a number of harsh steps to contain the outbreak, including imposing a 24-h curfew and closing schools and institutions. Residents were only allowed to leave their homes for necessary requirements between 6 a.m. and 3 p.m. inside their residential area during the 24-h curfew, which went into place practically immediately in numerous cities. Temperature checks have been performed on a regular basis. The suspension of all colleges and educational institutions in order to limit the COVID-19 epidemic was immediately followed by the activation of online education during that time. Children and students had to rely more on their own resources to continue studying remotely via the Internet, television, or radio during this time, while the educational community worked hard to ensure learning continuity. Teachers had to adjust to new pedagogical concepts and instructional methods for which they had not been trained. Learners from the most marginalized groups, in particular, are at risk of falling behind because they lack access to digital learning resources as well as the resilience and engagement to learn on their own. The COVID-19 outbreak has had a tremendous impact on higher education, with colleges and countries closing their doors in response to lockdown measures. Despite higher education institutions' eagerness to replace face-to-face lectures with online learning, the closures had an influence on learning and examinations, as well as the safety and legal status of international students in their host country. Perhaps most importantly, the problem raises doubts about the value of a university education, which includes networking and social opportunities in addition to academic content. To be relevant, universities will have to reinvent their learning environments such that digitalization broadens and complements student–teacher and other connections. The global epidemic created by Coronavirus has wrecked most of the aspects relating to the education system [1]. The normal mode of teaching had changed completely to an online meeting system. This mode was hard to implement in the beginning, since most of the students and faculties had to afford Internet facilities and electronic devices of their own. Interactions between students and staff had gone low [2]. Almost 32 crore youngsters have been denied an education as a result of the disease. This is being referred to as a national crisis, which has resulted in a rise in unemployment. Almost 65 percent of young people got affected during the lockdown due to transition from classroom to online [3]. Countries worked to provide alternate learning opportunities to reduce the loss of learning when schools were closed. The major e-learning platforms include Google Hangouts Meet, Zoom, WebEx Meetings, Microsoft Teams, ConnectWise Control, GoToMeeting, Blue Jeans Meetings etc., e-learning is a learning approach that is based on formalized instruction but uses electronic resources. The use of computers and the Internet to attend meetings and classes with colleagues and classmates to create a classroom atmosphere is a fundamental component of e-learning [4]. The rapid progress in technology and current COVID-19 pandemic, e-learning is now embraced by the masses. However, there are lots of difficulties for students as well as faculty.

The Six Sigma approach to e-learning necessitates a large amount of time, effort, and energy. It necessitates the e-learning professional's acquisition of new skills. Six Sigma's analyses, efforts, and talents, on the other hand, are currently being applied to almost every other business process in almost every industry [5]. Business procedures must be followed if e-learning is to be regarded a business process. Investing in Six Sigma for e-learning ensures that the business gets what it needs, the customers get what they want, the language is a common business language, and e-learning has corporate credibility, maybe most importantly.

2 Six Sigma Implementations

The conventional method of education was replaced with the new online mode owing to the effect of COVID-19 pandemic. At the same time, e-Learning has received multiple concerns from professors and students concerning the shift from the traditional system to the online system. The necessity for a research to analyze the current state of education was realized by e-learning. With the goal of boosting staff and student satisfaction in the institution, it was decided to apply a Six Sigma Define, Measure, Analyze, Improve, and Control (DMAIC) approach to assess and measure the performance of e-learning. Six Sigma is a relatively new quality improvement strategy that is gaining traction in a variety of businesses throughout the world [6]. The Define-Measure-Analyze-Improve-Control (DMAIC) cycle is at the heart of the Six Sigma quality program. It is a crucial component of Six Sigma, as it offers a framework for executing Six Sigma initiatives [7]. The following sections explain how the DMAIC cycle was implemented to the project.

2.1 Define Phase

Process flow in a conventional education process: The traditional setup makes use of a range of elements, like as infrastructure and study materials, to make education more accessible to students (textbooks, journals). Traditional learning is more organized than online learning and allows for a free flow of communication and open contact between students and teachers [8]. Figure 1 summarizes the conventional or offline education process.

Process/Information flow in a e-learning: Fig. 2 represents the process flow chart for Process/Information flow in e-learning.

Critical to Quality tree (CTQ): Starting with identifying the requirement, drivers, and all of the critical to quality characteristics (CTQs) that have been found, the critical to quality (CTQ) tree is structured as a flow chart [9]. In Fig. 3 the numbers 1, 2, 3, 4, 5 represent the need, drivers, CTQ characteristics, measurement, and target, respectively.

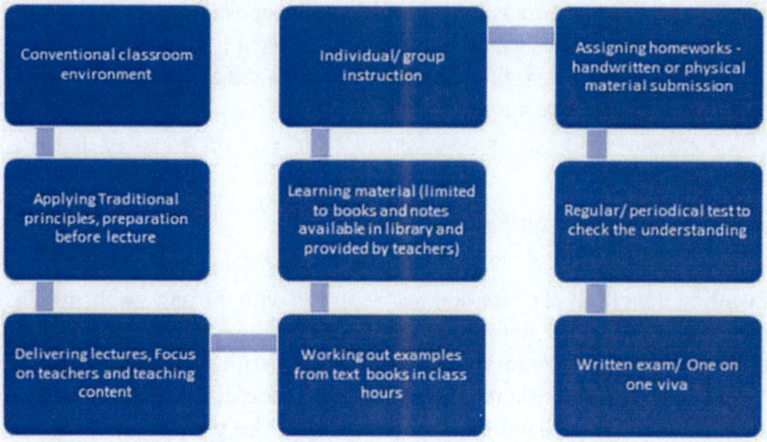

Fig. 1 Process flowchart for conventional/offline education

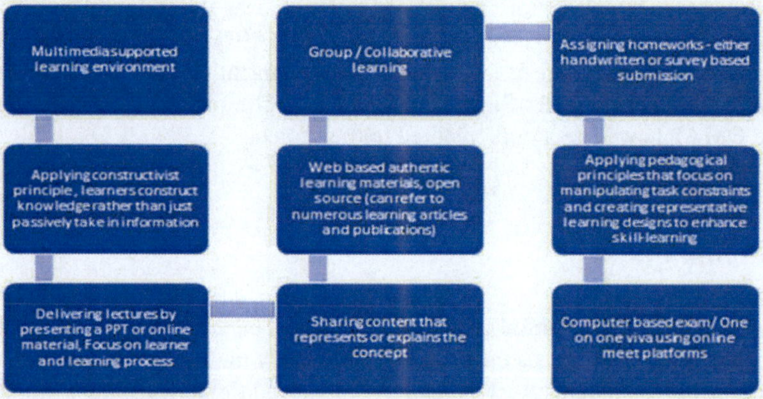

Fig. 2 Process flow chart for information flow in a e-learning

2.2 Measure Phase

The COVID-19 epidemic had a significant impact on education. Students had to not only adjust to online learning but also maintain a social distance from their classmates and family. Some children found it difficult to adjust to the "new normal," and they missed having one-on-one contact with their teachers. Some folks simply desired more time spent with their parents. Education institutions must understand more about how students feel about remote education and how they have experienced it. We can collect data by sending out a survey to students about remote learning. So, we will be able to determine what students enjoy about the current setup and what they would like to change once we obtain the results [10]. To measure the experience

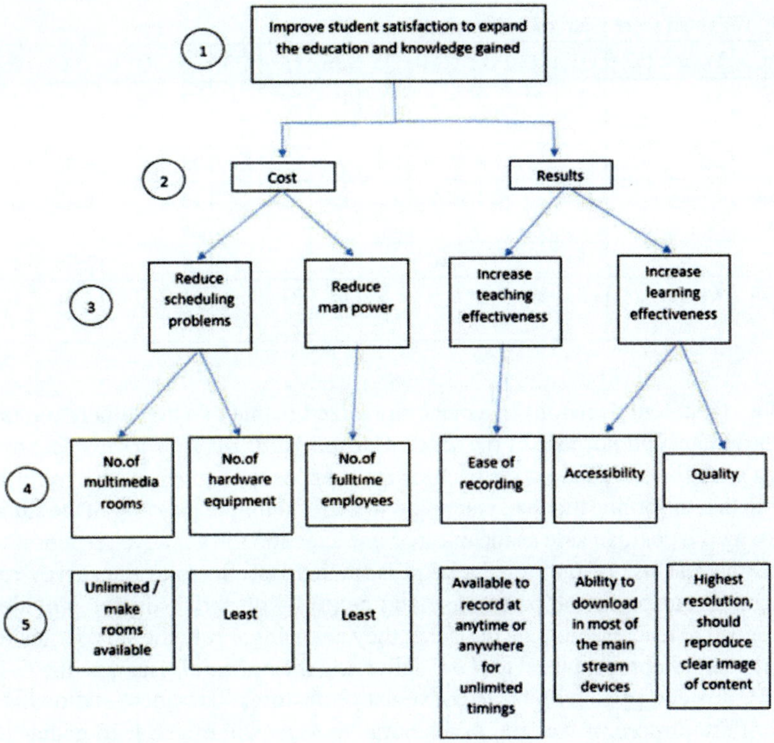

Fig. 3 Critical to quality (CTQ) tree

of online education in students' point of view the following survey questions are created. Because of the global disruption caused by the COVID-19 epidemic, most schools and institutions have switched to online instruction. If there is a shift in generally used standards and process models, it is critical to track progress to ensure that you are on the right track. While there has been widespread disdain for the remote learning and work environment, people have no choice but to accept the change. Teachers must be at ease and content in their work in order to succeed in distance learning. Schools, universities, and all academic institutions must provide support to instructors in their time of need so that they can provide their best effort [11]. To measure the experience of online education from a faculty point of view the following survey questions are created.

2.3 Analysis Phase

For students survey: Making decisions are both necessary and challenging. One must make decisions that are both justified and supported by all parties. It's also

Table 1 Weighted score matrix for student survey

Criteria	Weight (%)	Q1	Q2	Q3	Q4	Q5	Q6	Q7	Q8	Q9	Q10	Total	Rating
Extreme (YES)	5	2	39	2	2	11	17	29	4	7	36	7.45	1
Very	4.5	11		8	12	20			21	25		4.365	3
Moderate (Maybe)	4	29		23	23	16	19		27	16		6.12	2
Slight	3.5	8		18	10	2			1	4		1.505	5
Not at all (No)	3	3	13	2	6	4	17	24		1	17	2.61	4

crucial to document decisions in a clear, organized manner so that others can understand the reason and purpose of the score. Consolidating all the survey questions for students with the split scores for each of the responses as given in the Table 1, we found out that the online mode of education was useful for the majority of the students. So, ranking the positive side as the number one spot and the negative response which explains the less use of online education is ranked last. So as of the survey results students find online education so easy and helpful with the resources provided by their respective teachers and the tools that they are using. From the survey, it is found that the most reliable and used tool for delivering the online classes was the G-Meet tool [12]. It is equipped with basic conversation features like video, audio, and chat options. Furthermore, it has got a collaborative tool (white board) to gather ideas, caption is another excellent feature which helps the students to understand the audio if the Internet connection is not good at times. To make the consolidation easier, the answers were combined and considered like the answer–extreme also denotes YES as another answer, moderate also denotes MAY BE as another answer, and for NO it is Not at all.

For faculty survey: From the survey responded by the faculties, the most important polls are taken into consideration. Figure 4 shows one of the survey results. The interaction between students and the faculty in this new mode of online learning was not as satisfied as in the traditional offline mode. So, there is a huge drawback since the faculty might not know how understandable the students are when the lecture is being delivered through online meetings [13]. However, the faculties are satisfied with the online mode of delivery in concern with the pandemic, some find it less qualitative. As like the reviews from the student side the faculties also feel that the online education gives more space and open resources for students to work with. Since there is no direct eye to eye contact with the students, faculties find it difficult to involve everyone present in the online meeting in the tasks allocated. So, to sum up the faculty survey, the online mode of education is not as satisfying and qualitative as the traditional education. Evaluating and reviewing the activities of students still remains difficult. The qualitative analysis of each student is not possible in this mode.

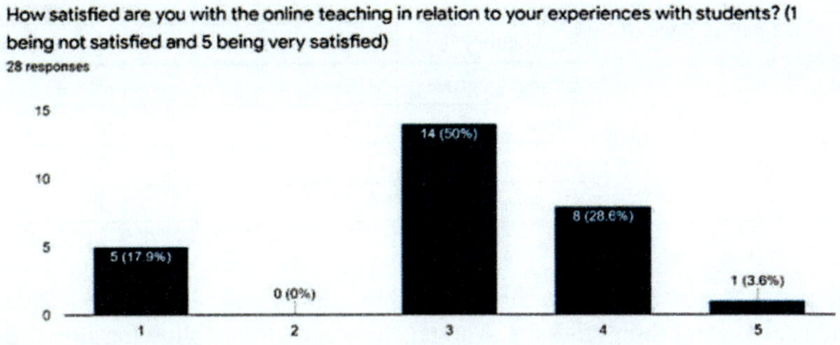

Fig. 4 Faculty survey

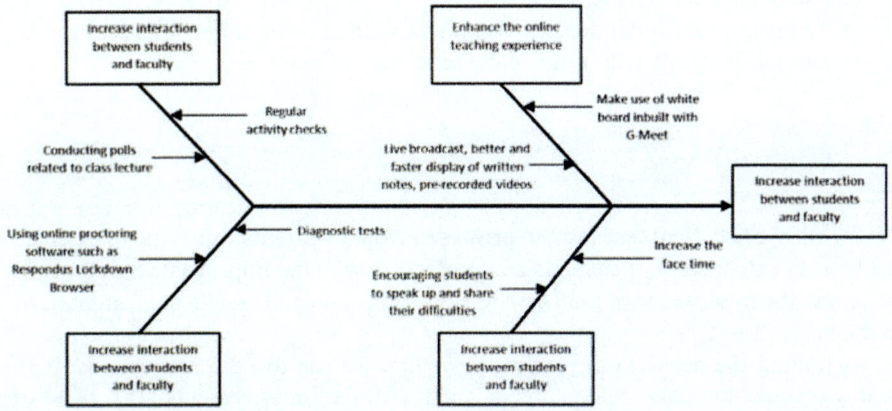

Fig. 5 Cause and effect diagram

Cause and Effect Diagram for Quality of Online Education: The information gathered was utilized to determine the characteristics that influence the quality of the new online course delivery system, which is depicted in Fig. 5 as a cause and effect diagram.

2.4 *Improve Phase*

List of improvements needed: While students believe that online education is a step forward over the traditional method of delivering lectures, they believe that it still needs to be improved. Instructors, on the other hand, loved being able to record a lesson outside of the classroom and not having to be physically present in a classroom to teach an online class. The findings also show that some teachers are unaware of this recording option [14]. More training on the new system's functionality is required as

Table 2 Satisfactory level of improvement in online education

Satisfactory level	Response percentage (%)	No. of responses
Strongly agree	51.2	86
Agree	34.5	58
Neutral	11.9	20
Disagree	2.37	4

a result of this. Instructors think that improving quality and resolution is critical for both themselves and their students. The teachers noted the following opportunities for improvement in the existing system:

i. Include live broadcasts and conversation to create a classroom-like experience for students who view lessons online.
ii. When using the digital pen, the display of written notes is better and faster.
iii. Smart whiteboards will improve the efficiency of teaching.

2.5 *Control Phase*

Performing Statistical test for the proposed improvements: It is vital to evaluate the level of satisfaction of students and professors with the improvement in the new system in the final stages of the DMAIC approach. The poll results are summarized in the below Table 2.

By polling the new learning delivery system, we can test the premise that both professors and students are equally satisfied with the improvements [15]. Most of the students gave positive feedback for the improvement measures that can be implemented to the online education system. From the student brainstorming session, it is found that adding new features and activities to the regular online classes might help to increase the interest to get involved in the lecture. The total number of responses were taken as size of the sample and negative feedbacks were considered as defects. Considering the number of defects in total responses, the sigma level of the improved phase was found to be 2.6. 142,857 is the DPMO. The table shows that there is ample evidence that there is a very modest difference in teacher and student satisfaction as a result of the new online education system's improvements.

3 Conclusion

The DMAIC technique was used to evaluate the success of the present online lecture delivery/education system, which was implemented in response to the COVID-19 epidemic, at colleges and institutions. The findings revealed that the institutions have upgraded their current distant learning delivery infrastructure. The advances made possible by the new system, as well as other gains that can be made, will allow schools and universities to grow and scale their present online offerings in order to provide an excellent online teaching experience. The new enhancement recommendations allow teachers to develop instructive content without having to be physically present in the classroom, allowing students to participate more actively.

References

1. Alghamdi AA (2021) Impact of the COVID-19 pandemic on the social and educational aspects of Saudi university students' lives. PLoS ONE 16(4):e0250026
2. Moreno V, Cavazotte F, Alves I (2017) Explaining university students' effective use of e-learning platforms. Br J Edu Technol 48(4):995–1009. https://doi.org/10.1111/bjet.12469
3. Schooling disrupted, schooling rethought: How the Covid-19 pandemic is changing education (2020) OECD policy responses to coronavirus (COVID-19). https://doi.org/10.1787/68b11faf-en
4. Bourne JR, Brodersen AJ, Ccampbell JO, Dawant MM, Shiavi RG (1996) A model for on-line learning networks in engineering education. J Eng Educ 85(3):253–262. https://doi.org/10.1002/j.2168-9830.1996.tb00241.x
5. Antony J, Banuelas R (2002) Key ingredients for the effective implementation of Six Sigma program. Meas Bus Excell 6(4):20–27. https://doi.org/10.1108/13683040210451679
6. Mehrabi J (2012) Application of six-sigma in educational quality management. Procedia Soc Behav Sci 47:1358–1362. https://doi.org/10.1016/j.sbspro.2012.06.826
7. Narayanamurthy G, Gurumurthy A, Chockalingam R (2017) Applying lean thinking in an educational institute—an action research. Int J Product Perform Manag 66(5):598–629. https://doi.org/10.1108/ijppm-07-2016-0144
8. Ahmad WFW (2006) Development of a multimedia courseware for visualization on teaching and learning: area and volume
9. Sangwan A, Sangwan A, Punia P (2021) Development and validation of an attitude scale towards online teaching and learning for higher education teachers. TechTrends 65(2):187–195. https://doi.org/10.1007/s11528-020-00561-w
10. Zalat MM, Hamed MS, Bolbol SA (2021) The experiences, challenges, and acceptance of e-learning as a tool for teaching during the COVID-19 pandemic among university medical staff. PLoS ONE 16(3):e0248758. https://doi.org/10.1371/journal.pone.0248758
11. Kuwaiti AA, Al Kuwaiti A (2020) Use of six sigma and lean six sigma methodologies in higher education institutions: a systematic review. Lean Six Sigma for Higher Educ 1–20. https://doi.org/10.1142/9781786348500_0001
12. Meishar-Tal H, Levenberg A (2021) In times of trouble: higher education lecturers' emotional reaction to online instruction during COVID-19 outbreak. Educ Inf Technol 26(6):7145–7161. https://doi.org/10.1007/s10639-021-10569-1
13. Son C, Hegde S, Smith A, Wang X, Sasangohar F (2020) Effects of COVID-19 on college students' mental health in the United States: interview survey study. J Med Internet Res 22(9):e21279. https://doi.org/10.2196/21279
14. Arbaugh JB (2000) How classroom environment and student engagement affect learning in internet-based MBA courses. Bus Commun Q 63(4):9–26. https://doi.org/10.1177/108056990006300402
15. e. Yaylak (2019) The attitudes and opinions of prospective teachers towards the use of technology. Int J Educ Technol Sci Res 4(9):149–175. https://doi.org/10.35826/ijetsar.20

Improving the Efficiency of the Vehicle Service Sector Using CPM and PERT

S. Pratheesh Kumar, E. Selvavignesh, Jonathan Cecil Fernando, S. Sree Kannan, and R. Shishanth

Abstract The automobile service industry is vital to our economy and lifestyle. The demand for automobile services continues to outpace supply. According to a Bain & Company study, gaining a new customer costs 6–7 times more than keeping one. This creates a significant difference with huge implications. It is also the world's most valuable sector. The main reason is to research service industry procedures and strive to implement a time-saving, error-proof solution while maintaining quality. The steps of the total maintenance procedure are grouped according to the customer's needs. So, in this study, the service procedure is broken down into two diagnosis modes: forward and backward. The time for each step is taken from a functioning service centre. Project scheduling strategies like the critical path method and project evaluation and review technique are applied to achieve the best results in this study. The outcomes of the project scheduling methodologies are compared to create an efficient method of vehicle servicing. Using project scheduling approaches like the critical path method and project evaluation and review technique saved processing time. The outcomes directly improve efficiency and customer happiness. The best approach can be employed in vehicle repair shops and is more efficient than current methods.

Keywords Vehicle service sector · Critical path method · Project evaluation and review technique · Network analysis

1 Introduction

Scheduling is the process of allocating resources, which in theory are time and energy, but in practice are time and energy. A project is made up of occupations, activities, functions, or tasks that are all connected in some way and must all be accomplished

S. Pratheesh Kumar (✉) · E. Selvavignesh · J. C. Fernando · S. Sree Kannan · R. Shishanth
Department of Production Engineering, PSG College of Technology, Coimbatore, Tamil Nadu, India
e-mail: spratheeshkumarth@gmail.com

N. M. Sivaram et al. (eds.), *Advances in Manufacturing, Automation, Design and Energy Technologies*, Lecture Notes in Mechanical Engineering,
https://doi.org/10.1007/978-981-99-1288-9_19

in order for the project to be finished. Services trade has grown at the same rate as goods trade over the 1990s (about 6.5%), and its share in total trade has reached around 24% [1]. The service sector is the third layer of the economy's three tiers. This industry produces services such as maintenance and repairs, training, and consulting rather than products.

Authorized service centres provide servicing and maintenance for a certain automotive brand, whether they are company-owned or franchised. Unorganized local garages are low-cost, bare-bones improvised auto garages that provide car services at the lowest feasible prices. Multi-brand vehicle service centres offer a wide range of services, parts, and support for a variety of automobile manufacturers. On Indian roadways, there are over 60 million automobiles. As a result, there is a huge need for after-sale servicing and maintenance. To keep operating costs low, local garages operate on a small budget. Multi-brand automobile workshops are structured in the same way as authorized service centres are.

1.1 Project Scheduling and Its Methods

The practice of utilizing information, skills, tools, and strategies to execute a project according to precise specifications is known as project management. The main goal of the project schedule is to identify the issue, develop a solution, and then implement it until the issue is resolved [2]. PERT and CPM are the two most commonly used techniques to calculate the period of a project.

1.2 Critical Path Method

The critical path method (CPM) determines the maximum and minimum time required to complete a project. It facilitates in the separation of independent variables and their identification [3]. CPM can also help you identify critical tasks that must be covered. The timeline is independent of changes in delivery timings. You will need the project's scope and a list of activities to utilize CPM.

1.3 Program Evaluation and Review Technique (PERT)

PERT is a strategy for calculating the total time necessary to complete a project by planning the flow of activities inside it. PERT is used to organize a project, identify activities, arrange them in a logical order, and set milestones. The PERT system employs a network diagram that illustrates the events that must occur to achieve project goals. Calculating the time necessary for each action leads to the critical path (the longest path connecting all the occurrences). PERT and CPM vary in that the former stresses the significance of time, meaning that if time is reduced, costs will be reduced as well.

In this project, PERT and CPM methods are used, hence past research papers are reviewed. Ambekar et al. [1] aimed to investigate the discrepancy between predicted service quality metrics and actual performance in service centres deemed to be of high quality. The SERVQUAL model has long been the most widely used tool for assessing consumer perceptions of service quality. Tatterson et al. [2] describe when implementing PERT techniques to the export process and examining the transportation process by calculating how to accelerate each task. Agboola et al. [3] describe the CNC machine tools used are more extensive and become more essential. Two approaches were used to diagnose a defective LC-195V5 CNC milling machine. Shreeranga et al. [4] have discovered issues in automotive repair centres and found the best remedies to increase overall performance. Wallace Agyei et al. [5] studied the key factor in estimating project's duration and cost. The information on the cost and duration of activities was gathered. Both CPM and PERT were used for the analysis (Muhammad Khalil et al. [6]). The goal of this study is to determine the influence of network planning in boosting efficiency in home construction projects. This study employs a descriptive technique, in which the data is gathered through the company's firsthand observation. (Natcha Thawesaengs kulthai et al. [7]) The car service system was made more efficient. The research used lean six sigma approaches, including the DMAIC process and the 5D stages for new service development (Mazlum et al. [8]). Classic project management approaches such as PERT and CPM will be employed in this study, as well as fuzzy PERT and fuzzy CPM, which are used in fuzzy project management, to develop an online Internet branch and plan the project of an online Internet branch (Mudita Dave Nagar et al. [9]). PERT and CPM were used to compare the cookie baking process. This study employs a descriptive approach, in which the data is gathered by direct observation by the firm. CPM and PERT were investigated, and it was discovered that CPM is a strategy utilized in projects with predictable activities and tasks, such as building projects (Jadhao et al. [10]). The issues that an Indian car service sector has were investigated, and many gaps that would directly or indirectly effect service quality and factors that affect service were discovered. The report continues with a discussion of the gaps that impact the four-wheeler service industries' current position and market share in the vehicle service industries.

2 Workflow

The workflow employed in the service centre are.

1. Arrival and problem discussion
2. Cleaning the tyres
3. Troubleshooting the vehicle
4. Disassembly of the vehicle
5. Spare part procurement

6. Engine component repair
7. Wheel balancing
8. Repairing electrical and computer systems
9. Reassembly
10. Wheel alignment
11. Vehicle final inspection
12. Testing of a vehicle on the road
13. Cleaning and dispatching the vehicle.

3 Modes of Diagnostics

To find out the effectiveness of the service centre, we have to prepare two different approaches to find out which one has less time to complete the whole process. The two types of approaches are the forward approach and the backward approach [3, 4].

3.1 *Forward Approach*

In this approach, we select the standard process which is commonly used in service centres. It starts with receiving the vehicle from the customer, cleaning, troubleshooting, engine component repair, repairing electrical and computer systems, etc. The detailed approach is shown in Table 1.

Table 1 Processes in forwarding and backward approaches

Sl. No.	Forward approach	Backward approach
1.	Arrival and problem discussion	Arrival and problem discussion
2.	Cleaning the tyres	Cleaning the tyres
3.	Troubleshooting the vehicle	Troubleshooting the vehicle
4.	Disassembly of vehicle	Repairing electrical and computer systems
5.	Spare part procurement	Disassembly of vehicle
6.	Engine component repair	Spare part procurement
7.	Wheel balancing	Engine component repair
8.	Repairing electrical and computer systems	Wheel balancing
9.	Reassembly	Reassembly
10.	Wheel alignment	Wheel alignment
11.	Vehicle final inspection	Vehicle final inspection
12.	Testing of the vehicle in the road	Testing of the vehicle in the road
13.	Cleaning and dispatch of the vehicle	Cleaning and dispatch of the vehicle

3.2 *Backward Approach*

In this approach, the majority process is the same as the forward approach but the change of repairing electrical and computer systems then engine component repair, wheel balancing, etc. The forward approach and backward approach processes are shown in Table 1.

4 Data Processing by PERT and CPM

4.1 *Using CPM for Project Scheduling*

A time study was conducted in an automotive service centre, and it was tabulated along with the activities according to the precedence order. Both PERT and CPM are conducted for two different approaches that were mentioned in the above sections. Cleaning the tyres activity B does not need a predecessor; hence, it is an independent activity that can be done before or after activity A. The list of activities in order of precedence and duration are shown in Table 2.

After arranging the activities in order of precedence, a network diagram for the same is developed as shown in Fig. 1.

After the completion of the network diagram, early start (Es), early finish (Ef), late start (Ls), late finish (Lf), and slack are determined and the activities with slack zero fall in the critical path. The total duration without using the project scheduling

Table 2 Activities in order of precedence and duration

Activity	Index	Predecessor	Duration (mins)
Arrival and problem discussion	*A*	–	5
Cleaning the tyres	*B*	–	3
Troubleshooting the vehicle	*C*	*A*	10
Disassembly of vehicle	*D*	*B, C*	25
Spare part procurement	*E*	*C, D*	5
Engine component repair	*F*	*E*	25
Wheel balancing	*G*	*D*	4
Repairing electrical and computer system	*H*	*A*	2
Reassembly	*I*	*F, G*	5
Wheel alignment	*J*	*I*	45
Vehicle final inspection	*K*	*I, J*	10
Testing of the vehicle in the road	*L*	*K*	20
Cleaning and dispatch of the vehicle	*M*	*L, H*	5

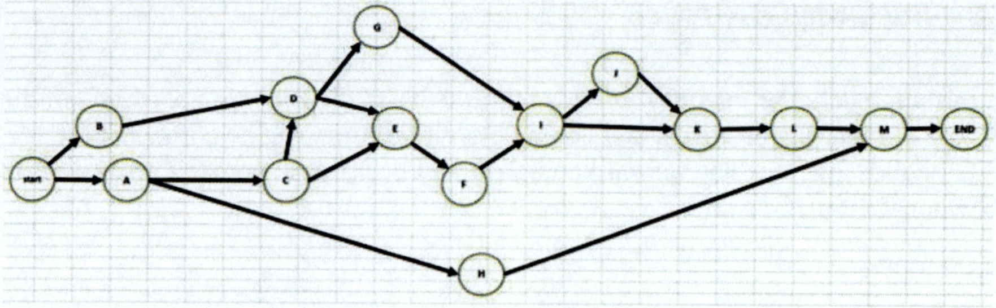

Fig. 1 Network diagram for a forward approach

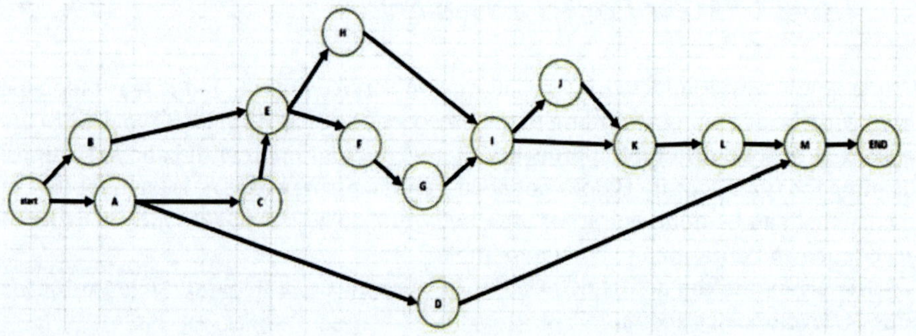

Fig. 2 Network diagram for a backward approach

techniques is estimated to be 164 min. The CPM of the forward pass of activities shows that the project can be completed in just 143 min.

Start–A–B–C–D–E–F–G–H–I–J–K–L–M = 3 + 25 + 5 + 25 + 5 + 45 + 10 + 20 + 5 = 143 min.

Similarly, the network diagram for the backward pass is shown in Fig. 2.

After the development of the network diagram, Es, Ef, Ls, Lf, and slack are determined and the activities for which the slack is 0 are the activities that should not be delayed and these activities lie on the critical path. The critical path is found to be.

Start–A–B–C–D–E–F–G–H–I–J–K–L–M–End = 3 + 25 + 5 + 25 + 5 + 45 + 10 + 20 + 5 = 143 min.

It is found that the critical path for both approaches is 143 min.

4.2 *Using PERT for Project Scheduling*

PERT is used to manage the project because the time taken to accomplish each activity is not constant and may vary for different vehicles. Now, we will utilize the same forward and backward techniques, but this time the PERT method will be employed to determine the critical path with the shortest potential time of the project completion. The precedence table, as well as the three-time estimations time, is divided into three categories: optimistic, most likely, and pessimistic are shown in Table 3. Calculate and tabulate the total anticipated time, variance, and standard deviation. To calculate the total expected time, the formula is shown in Eq. 1.

$$\text{Expected time} = (\text{Optimistic} + 4 \times \text{Most likely} + \text{Pessimistic})/6 \tag{1}$$

The formula for the variance is.

$$\sigma^2 = ((t_p - t_o)/6)^2 \tag{2}$$

where

Table 3 Precedence of activities for the forwarding approach

Activity	Index	Predecessor	Optimistic time	Most likely time	Pessimistic time
Arrival and problem discussion	*A*	–	3	5	10
Cleaning the tyres	*B*	–	2	3	6
Troubleshooting the vehicle	*C*	*A*	5	10	15
Disassembly of vehicle	*D*	*B*, *C*	20	25	30
Spare part procurement	*E*	*C*, *D*	5	5	15
Engine component repair	*F*	*E*	20	25	90
Wheel balancing	*G*	*D*	3	4	10
Repairing electrical and computer system	*H*	*A*	2	2	15
Reassembly	*I*	*F*, *G*	5	5	30
Wheel alignment	*J*	*I*	35	45	60
Vehicle final inspection	*K*	*I*, *J*	5	10	10
Testing of the vehicle in the road	*L*	*K*	15	20	30
Cleaning and dispatch of the vehicle	*M*	*L*, *H*	3	5	10

Table 4 Result

Time without project scheduling	CPM forward (mins)	CPM backward (mins)	PERT forward (mins)	PERT backward (mins)
164	143	143	160.5	160.5

σ^2= variance.
t_p= pessimistic time.
t_o= optimistic time.
Following the determination of the expected time, T_e, variance, and standard deviation. Es, Ef, Ls, Lf, and slack are found, and the activities with slack zero fall in the critical path. *Start–B–D–E–F–I–J–K–L–M–END* = 160.5 min is the critical path for the forward approach. *Start–B–E–F–G–I–J–K–L–M–end* = 160.5 is the critical path for the backward approach.

5 Result and Discussion

After all the research, we found out that if we follow the CPM method to manage this project, then the estimated time for the completion of the project will be 143 min in both cases while through the PERT method, it will take 160.5 min to complete. Hence, for this project, CPM method will be the right choice.

It is evident from the above table that, about 21 min are saved when the project is scheduled using CPM while only 3.5 min are saved while using PERT and it is found that the probability of completing the project within the target time is 92% in forwarding approach and in the case of backward approach it is 94%. Since CPM saves more time for completing the project CPM is considered as most efficient in these cases as shown in Table 4.

6 Summary

The car servicing sector contributes significantly to the global economy by providing value. During their initial servicing offerings, more cars will be queued up in a queue for service, and this will continue for additional services as well. It is thus vital to table out a less time-consuming and speedier technique to complete the service operation. We conducted a time study utilizing project scheduling methodologies and compared them to many features depending on the techniques used in this work, which are detailed in the findings section. The major reason we chose the PERT and CPM methods are that they may decrease work delays and provide better outcomes when it comes to reducing interruptions and conflicts. Hence, the probability of project completion in the backward approach is greater than the forward approach. The implementation of a backward approach in the workflow can service more vehicles and reduce the waiting time for the customers and provide good customer satisfaction.

7 Conclusion

Projects are the fundamentals of every business, organization, industry, or service centre. As a result, project management is an essential component of any business. We discovered that the CPM is more successful in this scenario after completing the investigation. CPM is a project management approach that is utilized in projects with predictable activities and tasks, such as a building. CPM is a deterministic technique that estimates the amount of time it will take to finish a project. PERT, on the other hand, is a probabilistic tool utilized in projects with unpredictably uncertain tasks and activities, such as research and development. For additional calculations to complete the project, this approach depends on three-time estimates: the most probable (most likely time), the most promising (optimistic time), and the most unfavourable (unfavourable time) (pessimistic time).

References

1. Ambekar SS (2013) Service quality gap analysis of automobile service centres. Ind J Res Manage Bus Soc Sci 1(1):38–41
2. Tatterson JW, Wood DF (1974) PERT, CPM and the export process. Omega 2(3):421–426
3. Agboola OO, Kareem B, Akinnuli BO (2016) Development of a diagnostic schedule for a defective LC-195V5 CNC milling machine using PERT. Leonardo Electronic J Pract Technol 28:107–118
4. Bhat S, Jnanesh NA, Kamath PP, Nayak RR, D'souza NM (2016) Enhancing the performance of an automobile service industry: lean thinking approach. J Mech Eng Autom 6(5):40–48
5. Agyei W (2015) Project planning and scheduling using PERT and CPM techniques with linear programming: a case study. Int J Sci Technol Res 4(8):222–227
6. Kholil M, Alfa BN, Hariadi M (2010) Scheduling of house development projects with CPM and PERT method for time efficiency 1755–1315
7. Thawesaengs kulthai N, Jarumanee P (2016) Improvement of vehicle service system by the application of lean Six Sigma. Eng J 20(2):159–178
8. Mazlum M, Guneri AF (2015) CPM, PERT and project management with fuzzy logic technique and implementation on a business. Procedia Soc Behav Sci 210:348–357
9. Nagar MD (2017) Project management techniques: PERT and CPM. Asian J Comput Sci Eng 2(2):28–31
10. Jadhao MMD, Kedar AP (2016) Service quality challenges in an Indian automobile service industries. Int J Eng Res Appl 6(12):39–43

Enhancing Efficiency in Microcircuit Manufacturing Using Six Sigma

S. Pratheesh Kumar, S. Shanchai Kumar, S. Shiriraam, and J. Sowdhariya Kumar

Abstract Six sigma is a powerful technique for increasing the quality of any process that can be utilized in both the manufacturing and service industries. Printed circuit boards (PCBs) have evolved through time into a basic tool for enhancing the manufacturing of electrical devices. Factories all around the world confront hidden and common challenges such as low productivity, low production rates, high defect rates, high waste percentages, excessive material utilization, and so on. During the testing cycle, the existing process was unable to give a good first pass of the built PCB. This case study focuses on defect detection and control in the manufacturing process. The data was collected in predicting the key causes of frequent manufacturing process errors. A Pareto investigation diagram was utilized to dissect the framework's concerns. Efficient information investigation joined with process boundary examination takes into account the right ID of the issue and afterward the slow landing in the ideal answer for the issue, bringing about an impressive improvement in the first time pass (FTP) of built PCB following testing. The main impact of six sigma is its application in the plan of new item improvement. This makes design engineers acutely aware of the significance of creating and specifying products that can be built with six sigma qualities at a reasonable cost. Accurate and flexible estimation models are consequently required for electronic products, particularly printed circuit boards, which comprise the majority of electronic product costs.

Keywords Six sigma · Printed circuit board · DMAIC · Pareto chart · Quality

1 Introduction

One of the most regularly used words in today's industrial and manufacturing contexts is "competitiveness." Manufacturing enterprises are becoming more competitive year

S. Pratheesh Kumar (✉) · S. Shanchai Kumar · S. Shiriraam · J. Sowdhariya Kumar
Department of Production Engineering, PSG College of Technology, Coimbatore, Tamil Nadu, India
e-mail: spratheeshkumarth@gmail.com

N. M. Sivaram et al. (eds.), *Advances in Manufacturing, Automation, Design and Energy Technologies*, Lecture Notes in Mechanical Engineering,
https://doi.org/10.1007/978-981-99-1288-9_20

after year, and a manufacturer's competitive edge must be maintained to stay in business. To put it another way, manufacturers must strive to gratify their customers by providing high-quality goods and services at the lowest possible price. As a result of these challenges, manufacturers lose hundreds of thousands, if not millions, of dollars each year. This money might have gone into expenditures that would help the firm maintain its competitive edges, such as the acquisition of new, up-to-date machinery or the hire of new, well-trained employees. Among other things, this includes the creation of new factories and the deployment of new work systems.

Surface-mount innovation permits electrical parts to be put straightforwardly on the outer layer of a printed circuit board. An electrical part that is surface-mounted is alluded to as a surface-mount gadget. Due to enhanced production automation that lowers costs and improves quality, SMT has virtually replaced the through-hole technology construction approach of fitting components in the industry. It also allows for the placement of additional components on a given substrate area [1, 2].

Meeting the Demand for Smart Electronic Devices.

i. The popularity of smart devices is skyrocketing. Smartphones have been around for a while, and smart TVs are quickly becoming a popular form of entertainment, but people are demanding more and more "smart" technology.
ii. Researchers and engineers are now looking for new technologies to meet those expectations, but they are running into a new customer demand challenge: consumer electronics are decreasing.
iii. The electronics assembly industry has been hit the most by job losses. Smartwatches, cellphones, remote controls, and fitness monitors all need the assembly of minuscule components. Micro-equipment tiny enough to construct smart gadgets by microdispensing adhesives would be required to assemble such minute parts. This is the source of the issues.
iv. Only a few manufacturing companies provide microdispensing technologies, yet they are critical to the electronics manufacturing industry's long-term viability. Without the little and precise amounts of fluids (adhesives and silicones) keeping these gadgets together to be able to use, consumers would have no idea how to pick the right size TV or have a watch that can receive phone calls and measure their heart rate [3].

2 Six Sigma Implementation

Six sigma may offer a quantitative description of the performance of a company or industrial process. The description is presented as a statistical representation. To achieve six sigma, a process must have fewer than 3.4 faults per million chances. Defects are defined as any standard that deviates from the customer's standards. Six sigma is a strategy that furnishes organizations with instruments to upgrade the capacities of their business processes. This improvement in execution and lessening in process fluctuation adds to abscond decrease and upgrades in productivity, staff resolve, and the item or administration quality. Previously, soldering by hand was

referred to as hand soldering, whereas soldering by machine was referred to as wave soldering. With the introduction of surface-mounted devices, the soldering technique has been dubbed the reflow soldering process [4].

The process engineer's task is to arrive at the right soldering profile during the reflow process such that the soldering connection is flawless, with no dry solder joints, sorting, or other types of soldering faults. There are components in the SMD family that have solder joints beneath the component, such as ball grid arrays and lead less grid arrays. Soldering these components is a significant problem for process engineers. In addition to this, the components are divided into two types: (a) lead-free components and (b) lead-containing components because of the restriction on hazardous chemicals legislation, the majority of European component producers have converted to lead-free substances. The temperature required for soldering varies depending on the composition. However, components utilized in space, military, and medical applications are usually led due to dependability concerns. These components are primarily produced on continents other than Europe. It is difficult to determine the best reflow soldering profile for components that include both lead and lead-free solder. The eutectic point of lead-containing components is usually 187 °C centigrade, while the eutectic point of lead-free components is roughly 217 °C centigrade. When both types of components are present, the peak temperature must be at least 5–10 °C higher than 217 °C centigrade. However, there are a few essential components, such as BGA and FPGA that cannot survive temperatures beyond 230 °C centigrade. Components may be damaged if they are exposed to temperatures higher than those specified. To get the best peak soldering profile, process engineers must carefully regulate the temperature between 220 and 228 °C centigrade [5]. Efficiency in the sense to get more microcircuit board from the manufacturing unit with help of best peak soldering profile.

3 DMAIC an Overview

The majority of the industry is attempting to use six sigma methodologies in their processes to achieve defect-free throughput and customer satisfaction. To meet consumer expectations, the industry aims to produce defect-free throughput at the lowest possible cost. Because the majority of the electronics industry's operations are automated, there is a push to get as near to 100% defect-free production as possible. As a result, businesses are using six sigma methodologies. In one case study, six sigma methodologies were effectively used to improve defect-free production in the reputable Indian electronics industry. Four different types of PCBs are utilized in a hand-held radio communication set. An automated assembly line is used to build these three types of PCBs with surface-mounted components. These PCBs do not have a good first-time pass (FTP). Because repairing damaged PCBs takes a long time, it was determined to apply six sigma approaches to enhance first-time pass yield.

3.1 Define

i. The radio set's controller PCB contains 107 components. Surface mounted devices account for 104 of the total.
ii. Reflow defects are a type of defect that occurs when a process is repeated and soldering techniques are used to repair soldered PCBs, which takes a long time and lowers the life of solder connections.
iii. Because multilayer PCBs are so complex, rework frequently fails to provide satisfactory results, and faults such as interlayer soldering, delamination, and other flaws emerge when the boards are tested.
iv. As a result, the production line is delayed.
v. Defect-free PCBs flow through the manufacturing process, allowing quality production to go to the next step.
vi. The project was chosen with all of these considerations in mind [4].

Pareto analysis of defects:

A Pareto diagram is utilized to show the most pervasive sorts of deficiencies, just as the most well-known reasons for imperfections and client protests. The analysis and implementation of corrective and preventative measures are part of the prioritizing of evaluated problems. It may use the Pareto analysis to enhance the quality of industrial goods, helping to keep faults under control, and improve production processes. There are several benefits to using them, including:

i. Identifying and prioritizing important reasons in order of significance to them to solve problems effectively.
ii. Many practical applications, such as efforts to enhance a process, customer demands, suppliers, and investment prospects, need the establishment of priorities.
iii. Shows us where our efforts should be focused to enhance the process/product.

The case study focuses on major fault detection and control in the manufacturing process. The information was obtained to pinpoint the major difficulties that lead to frequent manufacturing process mistakes. As illustrated in Fig. 1, a Pareto analysis chart was produced utilizing the sorts of problems detected [6].

Analyzing Fig. 1 makes it straightforward to form conclusions regarding the major sources of faults. The most prominent of them have "dry solder" and "shorting" as their causes, as can be observed. Faults in the first group account for half of all defects. The cumulative percentage is shown by the line segment that extends from the bottom left corner to the top right corner. Corrective activities were taken to reduce flaws and enhance the quality of industrial goods.

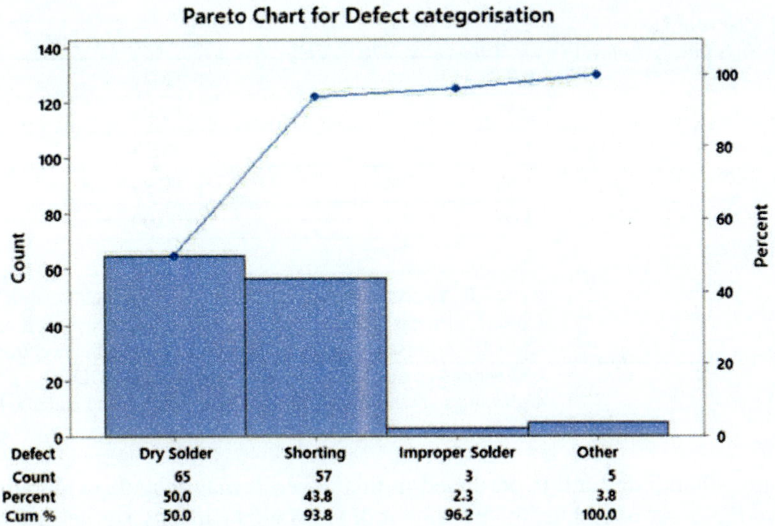

Fig. 1 Pareto chart for defect categorization

3.2 *Measure*

The action stage is the second stage in the DMAIC approach. In this stage, information on the quality trait of interest is gathered to:

i. Understand the current presentation of the cycle.
ii. Determine the gauge of the current interaction execution.

Process capability analysis:

Process limit is portrayed as a factual appraisal of a trademark's intrinsic interaction inconstancy. A cycle ability study can be utilized to assess an interaction's ability to fulfill determinations. An ability gauge is now and again made toward the beginning and fulfillment of a quality improvement program, like six sigma, to demonstrate the degree of progress that occurred.

Several capability estimates are in widespread use, including:

i. Processability assessments incorporate possible capacity (Cp) and genuine capacity during creation (Cpk). When joined with ceaseless information, Cp and Cpk delineate how competent a cycle is of arriving at its particular limitations. They help decide a section's and interaction's underlying and proceed with limit.
ii. As indicated in Table 1, "sigma" is a capacity estimate that is commonly employed with attribute data (i.e., with defect rates).

These kinds of ability gauges viably address the nonconformance pace of a cycle by expressing it as a solitary number. This generally involves deciding a proportion between as far as possible and the interaction spread. A deliberate method for

Table 1 Value of Cp signifies with the sigma level

CpK	Sigma level	Process yield (%)
0.33	1	68.27
0.67	2	95.45
1	3	99.73
1.33	4	99.99
1.67	5	100.00
2	6	100.00

$Cp < 1.0$ Process is not capable of meeting its specification
$Cp = 1.0$ Process is marginally capable of meeting specifications
$Cp > 1.0$ Process is capable of meeting the specification limits
Cpk values greater than 1.33 are considered GOOD
Cpk values greater than 1.33 and equal to Cp are considered BEST

surveying a handling limit is portrayed in this paper. It may include exploration to improve the processor, thus, the cycling ability. Process capacity studies are much of the time completed as a feature of an interaction advancement project.

Process Flow diagram:

The procedure begins with a bare PCB that is ready to assemble. If there is an SMD component on at least one side of the board, go to the next step: pick and place SMD components → reflow soldering → solder paste screen printing. If there are SMD components on the other side of the board, SMT adhesive is used, and all of the processes from step 2 are repeated. SMT inspection is done with an X-ray inspection machine if there is no SMD component on the opposite side. Wave soldering is required if there are through-hole components. If there are no through-hole components, it is determined whether hand soldering is required. If hand soldering is required, it is completed, and the board is ready for the final inspection. If no hand soldering is required, the board is sent straight to final inspection. There must be through-hole components to assemble if there are no SMD components on either side of the board. This is accomplished by placing through-hole components on the circuit board and then wave soldering them. It is now being determined whether or not hand soldering is required. If affirmative, hand soldering is performed, after which the board is subjected to a final inspection. If no hand soldering is required, the board is sent straight to final inspection [7].

3.3 *Analysis*

Cause and Effect diagram:

The fishbone chart provides comprehensive information on all probable causes, allowing you to pinpoint the main cause of the problem. After deciding on all the

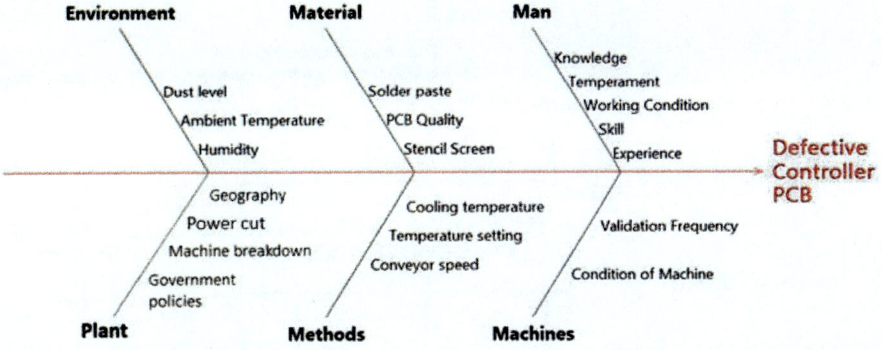

Fig. 2 Cause and effect diagram

possible root causes of the problem, it was decided to focus only on the main issues that would contribute most to improving the system. Focusing on 20% of the problems can lead to an 80% improvement in the whole system (Pareto principle). It also offers potential solutions for eradicating those fundamental problems to some extent. After examining all of the flaws using the Pareto chart, we discovered that dry soldering is the most common problem, followed by shorting, with maximum rejection rates of roughly 50 and 43.8%, respectively. As a result, we concentrate primarily on soldering defects, attempting to identify, and reduce all probable root causes using a fishbone diagram to limit section rejection as shown in Fig. 2. The cause and effect diagram was developed using Minitab software [8].

CNX Diagram:

Based on the cause and effect diagram, a control, noise, and experimental (CNX) diagram was created. Figure 3 depicts a CNX diagram.

New CNX Diagram:

After reviewing some literature, a new CNX diagram was created. The main difference between the old CNX diagram and new CNX diagram is only two variables in the experimental factor (heating temperature and conveyor speed) are dependent on the manufacturing process. Figure 4 is a diagram that gives the pictorial representation of variables in the CNX diagram.

3.4 *Improve*

Design of Experiments (DOE):

The strategies for solving reflow soldering challenges utilizing design of experiment (DOE). It researches the utilization of three unique ways of upgrading the warm

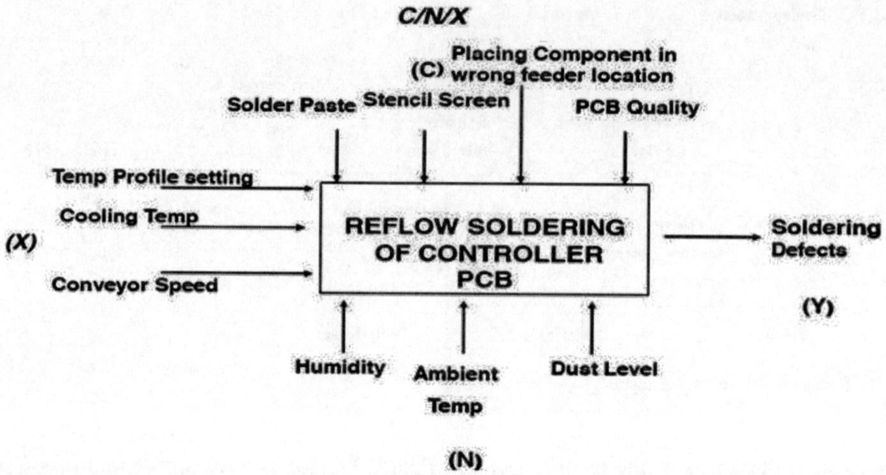

Fig. 3 CNX diagram

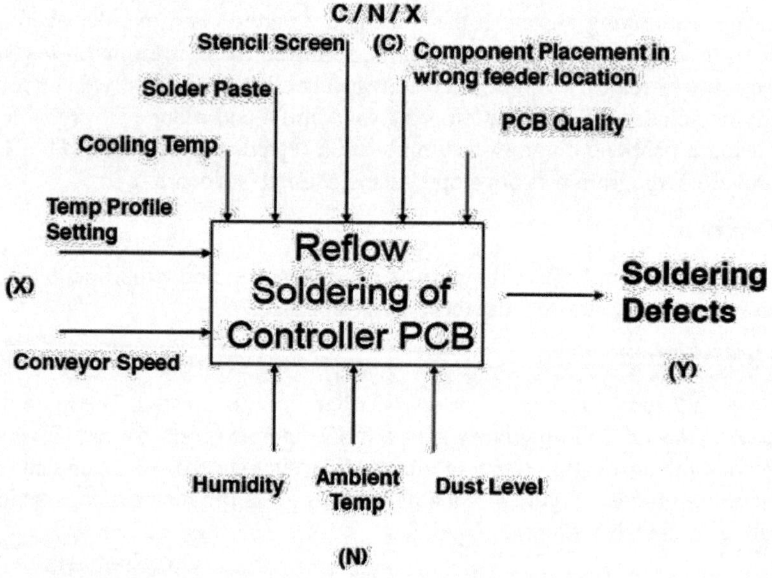

Fig. 4 New CNX diagram

boundaries of the reflow fastening process (regular reaction surface technique (RS), nonlinear programming (NLP), and a half and half AI methodology). To display and address the warm boundary streamlining difficulties for the reflow patching process in PCB, three unmistakable procedures were applied. Reflow soldering is a nonlinear technique with a variety of performance characteristics. As a result, the thermal reflow profile was employed to manage the impacts of heating on the board assembly and adjust the process parameters. It is an experimental concept for reflow thermal profiling that uses eight-factor levels (input) and eleven answers (output). As a consequence, all three approaches produced satisfactory soldering results. However, the hybrid AI strategy outperformed the other two methods in terms of constructing nonlinear mapping and solving optimization issues, as well as providing higher optimization performance [8, 9].

A DOE model can be created to examine the boundaries that impact heat misfortunes at high and low levels. Their goal is to uncover the components that play a major impact in heat misfortune while additionally further developing stove plans to help efficiency. There are eight components in the experiment, including flap design, conveyor belt speed, blower speed, and insulation, which all contribute to heat losses. At the levels investigated in the experiment, the DOE technique assisted designers in identifying important elements and linkages between them. The folding plan and the blower speed were the main components prompting heat misfortune in stoves, as per the Statistical Analysis Software (SAS) factual program [9].

We established a new categorization of factors in the design of experiments while focusing on the thermal profile. The two orders of parts used in try configuration are controllable and uncontrolled, albeit the two orders of information factors may not forever be fruitful in displaying the "noticed construction" of certain examinations. We remember semi-controllable information factors for the general cycle model structure since specific parts classed as controllable are semi-controllable. He demonstrated the assembling climate utilizing the three interaction input factors. The semi-controllable info variable guides process execution and helps an expert in fostering a proper model for foreseeing mean reaction and reaction variety through pre-arranged investigations [10].

They employed a different approach in this investigation, even though they used the design of an experiment. To work on the reflow patching temperature profile, he utilized the FEM reproduction model. Since diminishing the most extreme warm pressure altogether affects patch association toughness, the temperature and stress conveyances of a particular BGA-contained electronic gathering during reflow were demonstrated. I researched specific basic reflow settings, for example, the most noteworthy reflow temperature, stay time above fluids, splash times, slope rate, and transport speed, to diminish the greatest warm pressure in the whole get-together. The greatest warm pressure can be diminished by upgrading the previously mentioned reflow profile boundaries as an outcome of the reenactment model used [8–11].

Examining the reflow broiler's temperature profile for ideal welding quality is comparative. For this, he utilized four rules (top temperature, reflow time, cooling slant, and douse time). There were three levels in every part. Taguchi's L9 symmetrical exhibit was utilized [12].

We investigated to work on the reflow binding procedure and gauge heat factors. They sorted out some way to change the warming zone in the profile to guarantee reflow for every single bind intersection, which was a major concern. The major goal of their study is to improve solder joint quality and dependability. They identified the following factors that influence the soldering process and product quality [13, 14]:

Table 2 Possible thermal problem and cause

Problem	Possible thermal profile root cause
Cracked chip capacitors	Extreme ascent rate in the preheat zone
Solder balls	Fragmented drying before reflows The dry-out area is excessively cool and too short a length Extreme drying temperature Ill-advised gas air: air versus nitrogen
Cold solder joints	Inadequate time over reflow temperature
Solder not wetting to leads	Unnecessary drying time makes the transition weaken Unreasonable reflow temperature/time causing oxidation
Solder not wet on the pad	Lead is warming quicker than board(too much wind current
Components/board burning	Over the top reflow temperature

1. PCB	2. Oven	3. Target profile
• PCP thickness • PCB number • PCB type • PCB size • PCB materials and structure	• Cooling capacity • Heat transfer capacity • Total heating length • The total heating zone number	• The upper limit of the heating rate • The upper limit of the cooling rate • Upper heat factor • Lower heat factor • Conveyor speed

Reflow soldering of electrical components is the process of heating a board with solder paste until the paste flows around the components of the printed circuit board, then cooling the board before removing it from the oven to avoid injury to the worker as reflected in Table 2.

3.5 Control

The control phase is the final step of the DMAIC. The major purpose of this phase is to keep the system in its better form and prevent the problem from recurring. Following a discussion of the phase goals, it was determined that the best way to monitor and maintain the benefits of our project is to develop standard operating procedures that will provide clear steps to be followed whenever the waste begins to evolve beyond the proposed limits as an immediate action to deal with it at the source before it worsens.

The initial step of the control stage is to report and normalize the enhancements that were carried out during improve. This takes a few structures. The interaction guide of the new cycle that was made during improve ought to be investigated and refreshed as important to mirror any adjustments that might have happened during rollout. It will be utilized for preparing and reference so the new interaction will be clear. Assuming numerous people or gatherings are associated with the cycle, an arrangement flowchart ought to likewise be created to explain jobs and errands. While a cycle map is a critical part of the documentation of the new interaction, it is normally likewise useful to have a client guide that explains the means of the interaction and gives reasoning. This is especially significant assuming various upgrades are made and on the off chance that the new interaction is considerably not quite the same as the first. Following the implementation of the modifications, they should be monitored to see if they have a good impact on the manufacturing process and provide any revenues for the firm. This may be accomplished by developing a control strategy that specifies what data should be controlled, how it should be controlled, and who should control it. If a nonconformance is discovered, advice for corrective action should be supplied as well. Such a strategy should be adjusted over time based on post-implementation reviews.

4 Conclusion

The purpose of all of the aforementioned areas is to save money while delivering high-quality items to customers on time. As a result of all of these efforts, the printed circuit board yield has been optimized for continual improvement; improving productivity while maintaining high quality, while lowering costs, and shortening the time to fulfill client demands. For the electronic firm to be competitive in the market, it must continue to improve its production process to provide high-quality goods within realistic budgets. The familiar research allows us to examine how past study was conducted, what tactics researchers employed, and what may be done better or differently from what they did. Of course, each company has its approach to optimizing printed circuit boards for continuous improvement; however, most PCB assembly line manufacturing processes, such as loading PCB, solder paste, and screen printer, placing components, reflow solder (oven), washer machine for cleaning, inspection, and rework station, are shared by all companies. As a result, there is a distinction between an assembly line configuration and the number of machines in each line, as well as the machine's brand and capacity. Furthermore, the company's martial such as solder paste or the chemical that will be utilized in washer machines. The primary issue regarding cleaning PCB in the washer machine is which chemical will be utilized or which one is already accessible on the market. Additionally, the company's inspection methods such as solder paste measurement, AOI, and automatic test electronics are to be standardised. As a consequence, in addition to the data acquired from the actual assembly line and identifying certain bottlenecks to be rectified, all of the earlier studies were beneficial.

References

1. Meenakshi Sundaram R (1990) Chi-Ching Yu: a scheduling strategy in the manufacture of printed circuit boards (PCBs) using Surface Mount Technology (SMT). Comput Ind Eng 19:47–52
2. Derman G (1986) The impact of surface mount technology on electronics manufacturing. Microelectronics J 17:5–11
3. Strauss R (1998) SMT soldering handbook. Elsevier
4. Dey S, Sharma S, Dutt S (2013) Applications of Six Sigma in electronics industry—a case study. Int J Eng Sci Innov Technol 2:302–315
5. Tong JPC (2004) Fugee Tsung, Benjamin Yen: a DMAIC approach to printed circuit board quality improvement. Int J Adv Manuf Technol 23:523–531
6. Dunford R (2014) The pareto principle
7. Balazs HI, Krammer O, Geczy A (2020) Reflow soldering: apparatus and heat transfer processes. Elsevier
8. Tsai T-N (2012) Thermal parameters optimization of a reflow soldering profile in printed circuit board assembly: a comparative study. Appl Soft Comput 12:2601–2613
9. Mapa LB, Vancha AR (2012) Design of experiments modeling of a heat tunnel. In: 2012 ASEE annual conference & exposition
10. Flaig JJ (2006) A new classification of variables in design of experiments. Qual Technol Quant Manag 3:103–110
11. Gong Y, Quanyong L, Yang DG (2006) The optimization of reflow soldering temperature profile based on simulation. In: 2006 7th international conference on electronic packaging technology, pp 1–4
12. Shu M-H, Hsu B-M, Hu M-C (2012) Optimal combination of soldering conditions of BGA for halogen-free and lead-free SMT-green processes. Microelectron Reliab 52:2690–2700
13. Gao J, Wu Y, Ding H (2007) Optimization of a reflow soldering process based on the heating factor. Soldering Surf Mount Technol 19:28–33
14. Jin GG (2008) Thermal profiling: a reflow process based on the heating factor. Soldering Surf Mount Technol 20:20–27

Delta 3D Printer—A Review on Electrical Components

S. Kiranlal, V. M. Brathikan, C. S. Harish, A. Asfaq Moideen, and B. Anandh

Abstract Delta 3D printers are the most speed and efficient 3D printer compared to Cartesian model 3D printers. Delta printers can work faster and smoothly due to less movement in actuators resulting in vast pointing of nozzles and less energy consumption. For prototyping a table top delta 3D printer, the selection and usage of electronic components like controllers, sensors and actuators are more important for required accurate and precise 3D models.

Keywords 3D printer · Cartesian model · Additive manufacturing

1 Introduction

In this modern era, 3D printers have become a revolutionary thing on the concept of rapid prototyping. Nowadays students, professionals and even in industries do rapid prototypes to ensure one mechanical model's efficiency and make it into a

S. Kiranlal (✉)
Department of Mechanical Engineering, Kumaraguru College of Technology, Coimbatore, India
e-mail: kiranlal.s@kct.ac.in

V. M. Brathikan
Center for Exploratory Research, Kumaraguru College of Technology, Coimbatore, India
e-mail: brathikan.vm.admin@kct.ac.in

C. S. Harish · B. Anandh
Department of Electrical and Electronics Engineering, Kumaraguru College of Technology, Coimbatore, India
e-mail: harish.18ee@kct.ac.in

B. Anandh
e-mail: anandh.20ee@kct.ac.in

A. A. Moideen
Department of Electronics and Communication Engineering, Kumaraguru College of Technology, Coimbatore, India
e-mail: asfaqmoideen.20ec@kct.ac.in

N. M. Sivaram et al. (eds.), *Advances in Manufacturing, Automation, Design and Energy Technologies*, Lecture Notes in Mechanical Engineering,
https://doi.org/10.1007/978-981-99-1288-9_21

small-scale prototype. The delta 3D printer is a type of fusion deposition modeling 3D printer which is quite different from the CNC type 3D printer. Electricity and electronics play a vital role in 3D printing. This review report talks about the influence of electrical components of delta 3D printers in detail. A good mechanical structure and an errorless controller circuit makes the printer sensible. Consider a Cartesian model 3D printer.

1.1 Background

The first successful model of 3D printer was made in the 1980's which is a stereolithography printer. That is quite different from fusion deposition modeling printers. The first product of FDM printers was in the 1990's. The demand and usage of 3D printing technology has vastly increased till date. Many of the design circuits and processing technology became open source so people are ready to assemble 3D printers on their own. Not only looking forward to a single company which will give the entire product. On FDM printers, there are many types namely Cartesian model, delta robot, robotic arm 3D printer, etc. [1]. The delta model 3D printer is more efficient and faster than FDM printers compared to normal Cartesian model printers. Delta 3D printers are derived.

1.2 Scope

The field of additive manufacturing is a revolutionary change for prototyping a computer aided design model [4, 5]. Rather than conventional manufacturing for small-scale production 3D printers is more viable and efficient both in cost and structure [7]. The delta 3D printers can replace the Cartesian model printers due to many main factors. The motor or any other actuators are not placed as the moving parts except the hot end so it is easier to move the nozzle in an efficient way. Hundreds of calculations are done every second for pointing the nozzle by just moving the hot end in a three-dimensional way by only using three vertical movement actuators [8, 9]. The speed and time taken for a single print. The delta model only needs three drives for three sides [10]. But in the Cartesian model, we need at least 4 or 5 drives.

2 Methodology

Most of the working is the same compared to the Cartesian model, but the delta model only has 3 upward and downward motions which will result in 3D dimensional coverage on the heat bed.

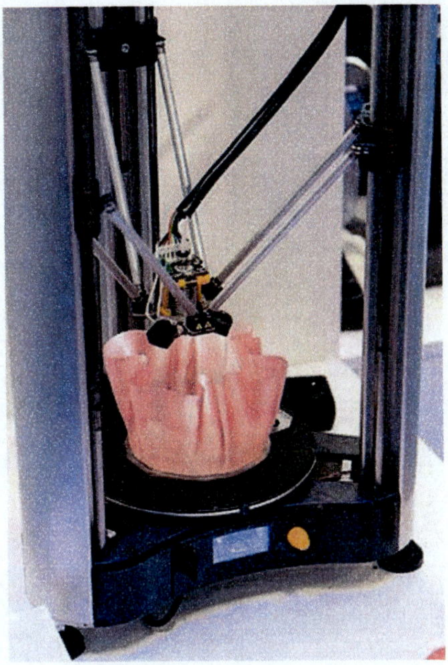

Fig. 1 Delta model 3D printer

Three stepper motors are placed on the base of the 3D printers as three Y axes and three belts with pulleys are connected to the top of the frame for each motor. A supporting rod is connected between three pillars and the hot end frame. Delta 3D printers work in the method of reverse kinematics. It calculates 100 kinematic equations in a second which helps the stepper motor to move in either direction to point the nozzle at the correct position. A high-speed microcontroller should be given for faster calculations. Data is given to the stepper motor as field excitation from the controller board; the motor will run either direction and make three-dimensional coverage (Fig. 1 and Table 1).

2.1 Materials

2.2 Detailed Specifications

Power Supply

The power supply used here is DC 12 v 10 A. It uses a 230 V 5A AC supply as input. This switch mode power supply works on the S-60-12 transformer for output power of 51–100 W. The constant and uninterrupted power supply is most important for 3D printing.

Table 1 Materials and quantity of the components used in the delta model 3D printer

S. No	Device	Specification	Quantity
1	Power supply (AC–DC convertor) (LM100-10C051212-35)	S-60–12 220 V AC–12 V–5A DC	1
2	Micro controller (Arduino mega 2560)	Mega 2560 Rev-3	1
3	Controller board (RAMPS v 1.4)	32-bit 120 MHz ARM CPU	1
4	LCD 12,864 smart controller	128 × 64	1
5	NEMA17 4.8 kg stepper motor (bipolar)	Single shaft input DC—3.1 V Step angle—1.8°	4
6	MK-8 Bowden extruder kit	17.5 mm filament	1
7	End stops Ramps 1.4 i3 CNC	Operating voltage—3.3 V/2A	3
8	Thermistor (4 pin digital thermal thermistor module)	Operating voltage—3.3 V–5 V	2
9	Cooling fan (3010 fan model)	Input voltage—12 V current—0.08A	1
10	Hot end (RepRap MakerBot J-Head Hot end)	Power—40W Nozzle diameter—0.4 mm	1

Arduino Mega

The Arduino mega 2650 is used as the microcontroller which has an AVR RISC-based microcontroller and flash program memory (flash) of 256 KB. And the reason to choose this microcontroller is I/O pins. This is the only microcontroller which has this amount of input output pins, 54 digital I/O pins (of which 14 provide PWM output) and 16 analog input pins. So, it will be easier to connect many pins with RAMPS 1.4 which is designed to be the same size as an Arduino mega 2560. So, we can insert the RAMPS axially upward on the Arduino.

RAMPS v 1.4

The main controller used here is RAMPS which is an Arduino supported 3D printer controller and specially designed for small-scale 3D printers. Which takes every single of the 3D printers from extruder to end stop. Like every stepper motor, cooler fan, heat resistor, thermal sensor, LCD display and SD card interface and every single part is connected to this controller which is placed above the Arduino [2].

Stepper Motor

A 3D printer's efficiency can be achieved if the three axes are moved precisely on exact time. For that we are using a 17 4.8 Kg stepper motor. Which did not need any external motor controller. The four wires from the stepper motor are directly connected to the main controller board. For this delta 3D printer, we use only a 3-stepper motor which is moved only upward and downward [3].

LCD with Smart controller

LCD 12864 smart controller which comes with a 128 × 64 LCD display and a microSD slot for giving .dodge files as input. This is connected to the RAMPS by 16 pins which are personalized pins for both controller and display. For interfacing a variable, resistor with a button is used.

Extruder kit

For this 3D printer, we are using MK-8 Bowden extruder kit which has a 17.5 mm aluminum nozzle with (4 pin digital thermal thermistor module) for getting heat signal as analog output and it comes with a cooler fan (3010 Fan model) for maintaining constant temperature [6]. The extruder kit with nozzle is the moving part which runs on the entire process so the extruder must be attached firmly to the moving glide rail. RepRap MakerBot J-Head Hot end is used as the hot end which works on 12 v (40 W) and directly connected to the RAMPS controller. The approximate heat dissipation of the extruder to heat the filament is 200–250 °C.

End stops

End stops are like a normal push button placed on the end of the glide rail which sends a signal as high as the stepper motor reaches the end of that axis so the controller will stop sending signals to the motor. Ramps 1.4 i3 CNC is a specialized end stop for CNC machines so it is used as the default end stop.

3 Procedure

The RAMPS v 1.4 is Arduino aided controller board which is axially placed on the Arduino MEGA 2560 the controller board like Arduino shields. The required digital input/output pins will be connected while placing them correctly on the Arduino. The power supply is given to both the Arduino and the main controller board. To avoid errors, use an uninterrupted power supply. The data >dodge file is given as input via SD card. The SD card slot is placed on the smart graphical LCD user interface which in turn is connected to the RAMPS controller the data and led input is given via 16 pins connected between them. The stepper motors are connected directly to the controller as 4 pins each. The end stops are connected on the top of each pillar. When the arm of the on pillar reaches the top end, it gives a signal to the controller to the stop stepper motor. The extruder stepper motor is placed on the center top of the delta frame. The slide format of the CAD model (.dodge file) is given as an input to the comptroller and the controller gives signals to the motors and hot end. The feedback from the sensors is also taken by the controller and processed to get a final output as a 3D model (Figs. 2 and 3).

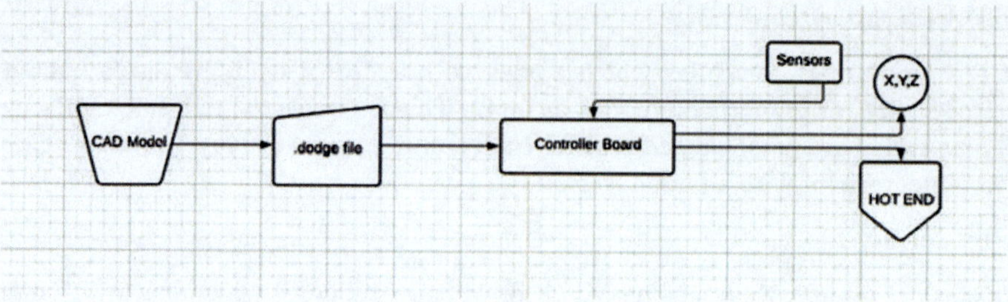

Fig. 2 Procedure diagram

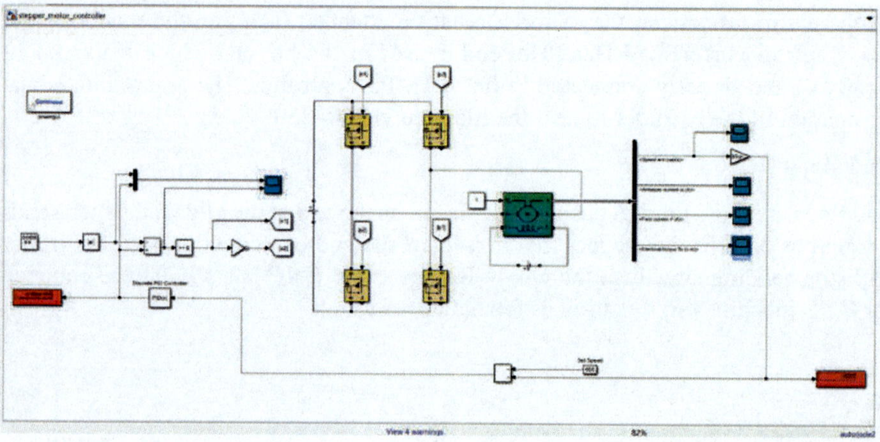

Fig. 3 MATLAB simulation power

4 Results and Findings

$$(\mathrm{W}) = \text{Voltage (V)} \times \text{Current (A)}$$

$$\begin{aligned}\text{for Hot End} &= 12\ \mathrm{v} \times 3.3\ \mathrm{A}\\ &= 40\ \mathrm{W}\end{aligned}$$

$$\begin{aligned}\text{for Stepper motor} &= 12\ \mathrm{v} \times 0.4\ \mathrm{A}\\ &= 3.6\ \mathrm{W}\end{aligned}$$

$$\text{for 4 motors} = 14.2\ \mathrm{W}$$

for controller and other components

$$= 5 - 6\ \text{v} \times 1\ \text{A}$$
$$= 5.8\ \text{v}$$

Total power

Hot end	: 40 W
Stepper motor $\times 4$	: 14.2 W
Other Components	: 5.8 W
Total	: 60 W(approx.)

The power consumption and power ratings of every electronic component is particularly important. The power consumed by every single actuator like stepper motor is calculated using MATLAB Simulink simulation, and calculated power consumptions are discussed above. For a single hot end, it consumes 3.3 A approximately and for stepper motor it consumes 0.4 A. After calculating all power requirements, the delta 3D printer will consume 60 W of power for one hour of printing.

5 Discussions

The electrical circuit diagram of the delta 3D printer with all components is mentioned below as a schematic diagram. The errorless connections are more important. There will be no response or signal from the controller if the main components are not connected correctly. The selection of a power supply with the required amount of power, using appropriate gauge wires and avoiding wire joints are preferred for sending and receiving signals accurately and avoiding soldering points instead of using a new wire joint entirely. Use Dc pins for connecting required cut joints. The uninterrupted power supply should be given to the printer (Fig. 4).

6 Conclusion

Influence of electrical circuits in 3D printing, background, selection of components, methodology, procedure and influence of embedded C programs are analyzed and discussed in this review paper. The main objective is to have a clear idea about the circuital parts of 3D printing. The project prototype of "delta 3D printer" was successfully completed and working perfectly. The required components are selected and assembled perfectly, according to the circuit given by the product owner. A minute error in the circuit might result in a greater loss if ignored. All the sub components are used as recommended by the controller manufacturer. The embedded code plays a vital role in interfacing with the Arduino. The manufacturer gave different codes for

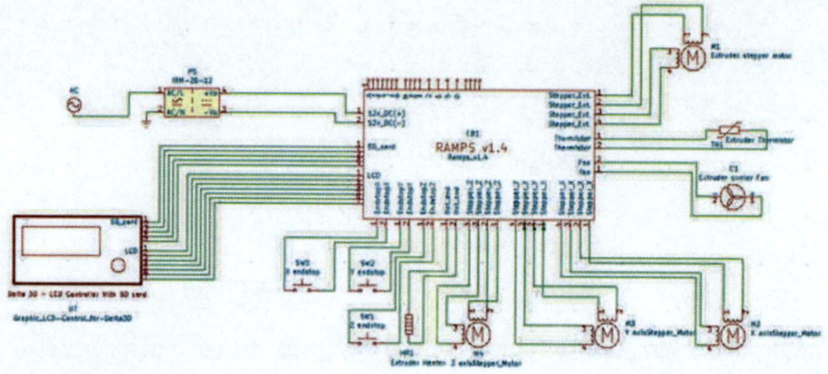

Fig. 4 Circuit diagram of delta 3D printer

delta and Cartesian type 3D printers. The review report was done after the completion of the prototype.

References

1. Celi R, Sempertegui A, Morocho D, Loza D, Alulema D, Proano M (CHILEAN conference on electrical, electronics engineering, information and communication technologies (CHILECON)). Study, design and construction of a 3D printer implemented through a Delta Robot
2. RAMPS 1.4 Assembly Guide
3. Simons A, Avignon KLM, Addy C (2019) Design and development of a delta 3D printer using salvaged E-waste materials. J Eng
4. MacDonald E, Salas R, Espalin D, Perez M, Aguilera E, Muse D, Wicker RB (2014) 3D printing for the rapid prototyping of structural electronics. IEEE Access 2:234–242
5. Haith G (2001) Apparatus and method for three-dimensional model printing. U.S. Patent 6259962
6. Landry T (2017) Extruders 101: a crash course on an essential component of your 3D printer Matter Hackers. [online] Matter Hackers
7. Scott C (2017) New delta wasp pellet 3D printer print without big cost
8. Wang M, Sun Z (2020) Dongguan university of technology, intelligent fault diagnosis of delta 3D printers using local support vector machine by a cheap attitude multi-sensor. In: 2020 prognostics and health management conference. PHM-Besançon
9. Wu C, Yi R, Liu Y-J, He Y, Wang CCL (2016) Delta DLP 3D printing with large size. Conference Paper
10. Schmitt BM, Zirbes CF, Bonina C, Lohmann D, Lenina DC, da Costa Sabino Netto A. A comparative study of cartesian and delta 3D printers on producing PLA parts

Fabrication and Compressive Strength of Functionally Graded Dual Filler Polymer Composite Materials

Vasavi Boggarapu, Raghavendra Gujjala, Shakuntla Ojha, L. Ruthik, Venkateswara Babu Peddakondigalla, and Satish Jain

Abstract Functionally graded materials (FGMs) are a class of composites that show a variation in properties along the graded direction. In general, FGMs comprise one type of filler particle having specific properties. To achieve the advantage of different fillers simultaneously in a single material, the dual filler concept was introduced. In this regard, the present work focuses on the fabrication of functionally graded polymer composites (FGPCs) using aluminum and copper as dual fillers and epoxy as matrix material. Two distinct curing conditions were employed in the development of samples; no curing time (C1) and providing curing time (C2) between the graded layers. The results showed an improvement in compressive strengths of C2 by 3.87% as compared to C1. Furthermore, the failure mechanism was due to the formation of microcracks and kink bands.

Keywords Functionally graded polymer composites · Dual filler · Compressive strength

1 Introduction

Composites are the class of high-performance engineering materials alongside metals, ceramics and alloys. These materials comprise two distinct phases, i.e.,

V. Boggarapu
School of Mechanical Engineering, VIT-AP University, Amaravati 522237, Andhra Pradesh, India

R. Gujjala (✉) · L. Ruthik · S. Jain
Department of Mechanical Engineering, National Institute of Technology, Warangal 506004, India
e-mail: raghavendra.gujjala@nitw.ac.in

S. Ojha
Department of Mechanical Engineering, Kakatiya Institute of Technology & Science, Warangal 506015, India

V. B. Peddakondigalla
Department of Mechanical Engineering, Vasireddy Venkatadri Institute of Technology, Guntur 522508, India

N. M. Sivaram et al. (eds.), *Advances in Manufacturing, Automation, Design and Energy Technologies*, Lecture Notes in Mechanical Engineering,
https://doi.org/10.1007/978-981-99-1288-9_22

matrix and reinforcement (or) filler which are connected by interfacial layer. The final characteristics of composites depend on properties of phases, distribution of filler in matrix and adhesion among them. Composites exhibit diverse properties due to which they were successfully applied in various engineering sectors like aerospace, automotive, electrical and electronic, building industry, etc. [1, 2]. The inclusion of fillers into the matrix affects mechanical, thermal, electrical and optical properties of composite materials [3]. These filler particles in the form of powders, whiskers, fibers and fabrics [4, 5] can be added to the matrix materials. A further step that evolved in the development of composites was to differentiate filler content in any of the geometrical dimensions. The resultant material displays a variation in properties along the graded direction. These new classes of composite materials were termed functionally graded materials (FGMs). Due to varied filler distribution, FGMs can prevent the drawbacks of traditional composites like sharp boundaries and internal stress concentration zones [6]. FGMs exhibit a transition of one material to another in step-wise gradient layers by eliminating the discrete interfaces. The linear or non-linear variation in the chemical content of phases results in the distribution of properties from one side to the other.

Engineering materials like metals, polymers and ceramics were utilized as matrices to develop FGMs. If a polymer is used as a matrix, then materials were referred to as functionally graded polymer composites (FGPCs). Polymeric graded materials show unique properties such as thermal conductivity, scratch and wear-resistant, toughness, strength, electrical conductivity and dielectric strength [7, 8]. A few techniques involved in the fabrication of FGPCs were casting, selective laser sintering, hand lay-up lamination, corona discharge, hot isostatic pressing and compression molding, as reviewed by Naebe and Shirvanimoghaddam [9]. Among the above-mentioned techniques, hand lay-up is popularly used methodology owing to its least cost and ease of fabrication. But obtaining the gradation at each layer with uniform variation of filler particles in the matrix was difficult in conventional hand lay-up process due to accumulation of denser particles at bottom surface. To overcome this drawback, a layer-wise curing approach was introduced. Concerning this, the present work focuses on the development of FGPCs through two fabrication conditions (traditional hand lay-up and layer-wise curing) and a comparison was presented in terms of compressive strength.

Among the different thermoset polymers, epoxy is widely used as a composite matrix in several applications [10]. Yet, it possesses limited mechanical properties due to its brittle nature. Moreover, epoxy has low thermal and electrical conductivity. To overcome this, researchers have added metallic fillers like silver, copper, gold, aluminum, etc., to the epoxy resin [11–13]. Polymer materials with the inclusion of conductive particles find their applications in electrical and biosensors. Adequate studies were reported on the electrical properties of FGPCs with the addition of metal particles [7, 14]. But limited literature was available on mechanical properties. In terms of distribution, different filler particles have distinct effects on the surface and volumetric properties of FGMs. However, a dual filler concept was presented to accomplish the benefit of two types of reinforcement in a single material [15]. Thus, the objective of the present work is to fabricate epoxy-based functionally graded composites reinforced with aluminum (Al) and copper (Cu) as dual fillers through two distinct curing conditions and evaluate their compressive strength.

2 Experimentation

2.1 *Raw Materials*

Gradient composite materials comprise aluminum and copper particles of 325 mesh size each as filler materials which were added to epoxy resin (LY 556). To ensure appropriate curing, polymer matrix is added with amine-based hardener of Aradur (HY951) in the 10:1 ratio. Four different layers with varied volume fraction of filler particles as shown in Table 1 were chosen in the study. As copper is denser than aluminum, it is graded at the bottom to prevent diffusion into subsequent layers. Thickness of each layer was kept constant as 6 mm to ensure layered homogeneity. Each layer of constituents was mixed properly in the individual glass containers to avoid entrapment of air bubbles. Moreover, pure epoxy samples were also prepared.

2.2 *Fabrication of Graded Composites*

Samples were prepared through hand lay-up technique in cylindrical molds at two distinct curing conditions. In the first condition (C1), constituents were poured in layer upon layer simultaneously. There was no waiting time between the depositions of each layer, whereas in second condition (C2), curing time was provided for infusing the layers. The second layer was poured on to preceding (first layer) at its gel time of 30 min. Similarly, subsequent layers were impregnated on the previous layers. After casting four layers in the molds by different conditions, specimens were finally cured at room temperature for 24 h. Distribution of elements at the interface was obtained using electron-dispersive X-ray spectroscope (EDX). Figure 1b represents the variation of filler content corresponding to different layers of sample C2. An

Table 1 Volume fraction of filler particles and epoxy at different layers of FGPCs

Number of layer	Vol. % of Al	Vol. % of Cu	Vol. % of epoxy
First (top)	10	0	90
Second (middle)	20	0	80
Third (middle)	0	20	80
Fourth (bottom)	0	10	90

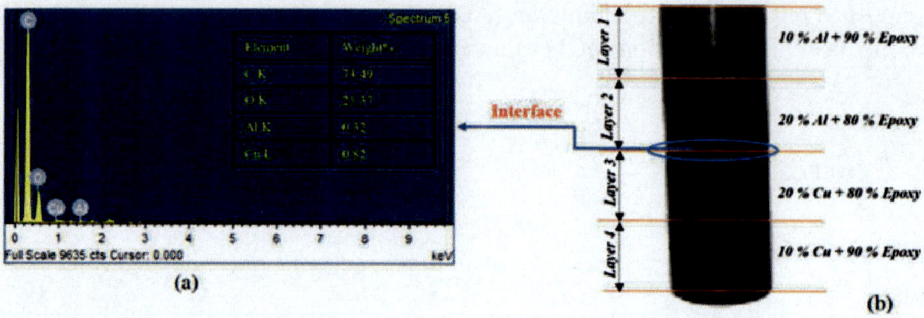

Fig. 1 **a** EDX spectrum at the interface and **b** distribution of fillers at different layers of sample C2

interface was identified between layer 2 and layer 3 from EDX analysis which confirm the presence of both Al and Cu in the epoxy matrix as shown in Fig. 1a.

2.3 Compressive Strength Testing

Following ASTM D 695 standards, the compressive strength of FGPCs was evaluated on universal testing machine (UTM) maintaining 2 mm min^{-1} as crosshead speed. The testing sample is of cylindrical shape having diameter 10 mm and thickness (height) of 24 mm. Five identical samples for each curing condition were prepared and tested to obtain the mean value. To study the failure mechanism, samples were analyzed using TESCAN VEGA3 scanning electron microscopy (SEM) equipped with EDX.

3 Results and Discussion

Figure 2a represents the compressive strength of epoxy, and FGPCs fabricated at two different conditions (C1 and C2). Pure epoxy being brittle, its failure is catastrophic. Thus, its strength was reported to be lower among all the samples, which was then enhanced with addition of metallic fillers. Sample C2 has shown 3.87% and 46.6% greater compressive strength as compared to C1 and epoxy, respectively. Different curing conditions among the graded layers result in strength difference between C1 and C2. In sample C2, the provision of curing time resulted in layer homogeneity and good interfacial bonding between the layers. Hence, the strength obtained for sample C2 was higher among others, whereas in C1, all the layers were poured in layer upon layer without providing curing time. This results in diffusion of denser particles toward the bottom from previous layers leaving more amount of epoxy at top most layer. Due to this, the compressive load was not transferred within the graded layers, resulting in early failure of C1 than C2. In FGPCs, two different failure modes were observed: first mode was identified as longitudinal splitting followed by crushing. Presence of low filler content at layers 1 and 4, crack initiation takes place at these zones. On increasing the load further, crushing of sample takes place which may also decrease its height. These cracks were then propagated toward the succeeding layers with a raise in stresses. The failure sample of C2 reveals the partial deformation of layer 1 and layer 4, while cracks were only observed at layer 2 and layer 3 owing to high filler volume content. At the interface of layers 2 and 3, cracks were not formed due to the presence of Al and Cu particles. Thus, the presence of dual fillers provides more resistance toward compressive loading which enhance the strength. It was apparent from above discussion that, the curing time shows a significant effect on compressive strength of FGPCs.

After compressive testing, the surface of sample C2 was observed under SEM to understand the failure mechanisms. Although Al and Cu are ductile materials, the failure mechanism does not indicate ductility; instead, the micrographs show a brittle fracture. Initially, a crack was observed at the top layer having low filler content under the application of compressive load as shown in Fig. 3a. Morphology reveals the development of microcracks surrounding the crack tip which represents the stepped fracturing. On further increasing the load, a new crack was developed at the bottom layer without extension of the previous crack. Later on, the external surface of the sample separates from their interface layers as shown in Fig. 2b.

Formation of kink band was observed from the crack-initiated region toward the succeeding layers (Fig. 3b). Kinking phenomenon occurs at the final stage of the fracturing process, and it does not have any effect on FGPCs strength. The fracturing of filler particles can be observed at the boundary of the kink band under the influence of shear stresses (Fig. 3c). During compressive loading, materials fragment into several pieces along their weakest direction. But in the case of FGPCs, sample was divided into two pieces at their low volume content layers. This difference

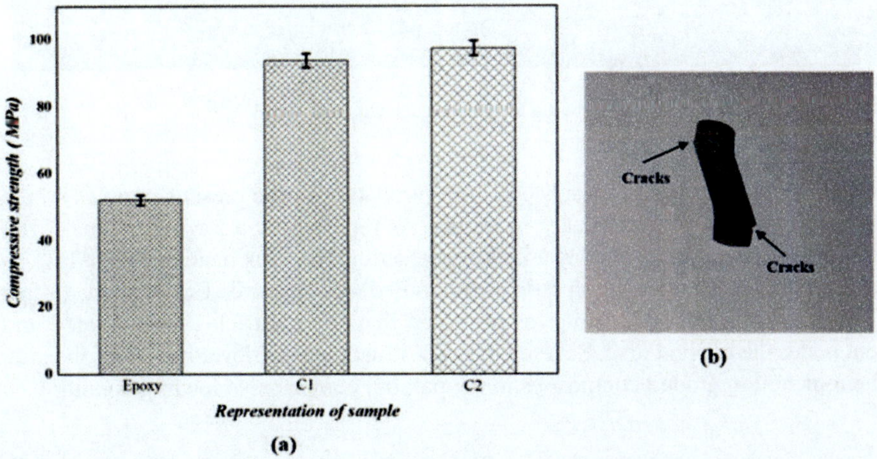

Fig. 2 **a** Compressive strength of epoxy and FGPCs and **b** failure sample

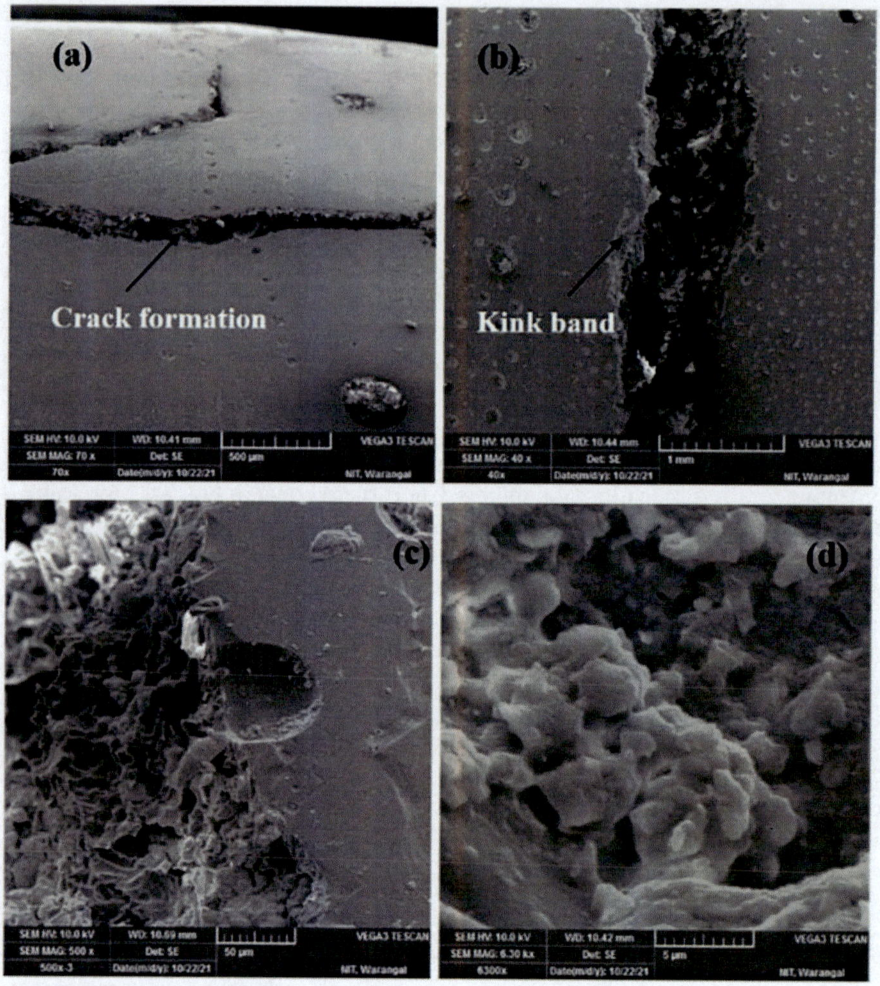

Fig. 3 **a–d** SEM micrographs showing the fractured surface of sample C2

was attributed due to the presence of filler gradation in the present samples where fracturing occurs only at weaker locations. As reported by a few researchers [16–18], layered composites display a kinking behavior. The kink band shown in Fig. 3b was formed by a row of filler particles that failed subsequently. From the magnified SEM micrograph (Fig. 3d), it was observed that filler particles were rotated and bent under the applied load. Schematic Fig. 4 illustrates the development of the kink phenomenon in graded composites in the parallel compressive loading condition.

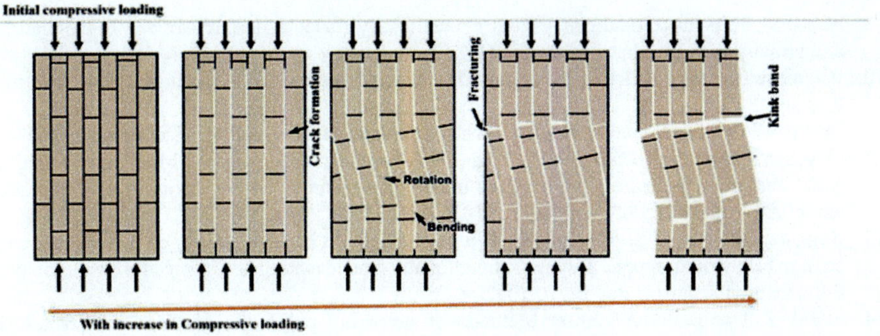

Fig. 4 Schematic illustration of fracture mechanism in FGPCs during parallel compressive loading

4 Conclusions

Functionally graded polymer composites (FGPCs) with dual filler were successfully fabricated with the layer sequencing method and cured at two distinct conditions. Curing time plays a vital role in evaluating the compressive strength of samples. In this context, sample C2 which was provided with curing time between the graded layers display a 3.87% greater strength than C1. Brittle type of failure mechanism was identified in the fractured samples. Furthermore, the formation of microcracks and kink bands along the parallel compressive loading occurs in FGPCs. The presence of dual fillers at the interface of the sample provides resistance toward compression.

References

1. Van Den Einde L, Zhao L, Seible F (2003) Use of FRP composites in civil structural applications. Constr Build Mater 17(6–7):389–403
2. Boggarapu V, Gujjala R, Ojha S (2020) A critical review on erosion wear characteristics of polymer matrix composites. Mater Res Express 7(2):022002
3. Stabik J, Dybowska A, Chomiak M (2010) Polymer composites filled with powders as polymer graded materials. J Achievements Mater Manuf Eng 43(1):153–161
4. Khatkar V, Vijayalakshmi AS, Manjunath RN, Olhan S, Behera BK (2020) Experimental investigation into the mechanical behavior of textile composites with various fiber reinforcement architectures. Mech Compos Mater 56(3):367–378
5. Kumar A, Kumar S, Mukhopadhyay NK, Yadav A, Kumar V, Winczek J (2021) Effect of variation of SiC reinforcement on wear behaviour of AZ91 alloy composites. Materials 14(4):990
6. Boggarapu V, Gujjala R, Ojha S, Acharya S, Chowdary S, Kumar Gara D (2021) State of the art in functionally graded materials. Compos Struct 113596
7. Stabik J, Dybowska A (2017) Epoxy-copper composites with gradation of filler content. Compos B Eng 127:36–43
8. Xu Y, Chung DDL, Mroz C (2001) Thermally conducting aluminum nitride polymer-matrix composites. Compos A Appl Sci Manuf 32(12):1749–1757

9. Naebe M, Shirvanimoghaddam K (2016) Functionally graded materials: a review of fabrication and properties. Appl Mater Today 5:223–245
10. Parameswaranpillai J, Hameed N, Pionteck J, Woo EM (eds) (2017). Springer, New York
11. Kanzow J, Horn PS, Kirschmann M, Zaporojtchenko V, Dolgner K, Faupel F, Wehlack C, Possart W (2005) Formation of a metal/epoxy resin interface. Appl Surf Sci 239(2):227–236
12. Goyanes S, Rubiolo G, Marzocca A, Salgueiro W, Somoza A, Consolati G, Mondragon I (2003) Yield and internal stresses in aluminum filled epoxy resin. A compression test and positron annihilation analysis. Polymer 44(11):3193–3199
13. Boggarapu V, Adapa S, Kanakam R, Ojha S, Gujjala R (2022) Research on mechanical and erosion behavior of stepped polymeric functionally graded materials reinforced with aluminum. Polym Compos. https://doi.org/10.1002/pc.26433
14. Stabik J, Dybowska A (2007) Methods of preparing polymeric gradient composites. J Achievements Mater Manuf Eng 25(1):67–70
15. Farahnakian M, Elhami Joosheghan S, Moradi M (2021) Dual filler functionally graded polymermaterials–manufacturing process and characteristics. Mater Manuf Process 36(3):301–307
16. Chau CC, Blackson J, Im J (1995) Kink bands and shear deformation in polybenzobisoxazole fibres. Polymer 36(13):2511–2516
17. Ji HM, Liang SM, Li XW, Chen DL (2020) Kinking and cracking behavior in nacre under stepwise compressive loading. Mater Sci Eng, C 108:110364
18. Nizolek TJ, Begley MR, McCabe RJ, Avallone JT, Mara NA, Beyerlein IJ, Pollock TM (2017) Strain fields induced by kink band propagation in Cu-Nb nanolaminate composites. Acta Mater 133:303–315

Analysis of MRR, TWR and Surface Roughness in EDM Using Artificial Neural Network Technique

C Veera Ajay, K Karthik Kumar, A S Kamaraja, C T Justus Panicker, C Arun Sudhan, and S Ashok Kumar

Abstract In this work, optimisation for electric discharge machining parameters using ANN technique was carried out. Artificial neural network is used for analysing the material removal rate (MRR), tool wear rate (TWR) and surface roughness (SR) of EDM. An artificial neural network-based prediction model was developed to evaluate MRR, TWR and surface roughness while electric discharge machining of aluminium alloy (AA 6063) material. The ANNs (4-n-2, 4-n-1) were generated by introducing different combinations of transfer functions and a number of neurons. The optimum 4-n-2 model was built with overall *R*-value of 0.99987 for MRR and TWR. The optimum 4-n-1 model was built with overall *R*-value of 0.98774 for SR.

Keywords Artificial neural network (ANN) · Electric discharge machining (EDM) · Surface roughness · Material removal rate (MRR)

1 Introduction

Electric discharge machining process stands an advanced manufacturing method which do not use sharp cutting tools like as traditional machining processes such as drilling, turning and shaping. EDM is one of the best method aimed at machine exotic, more power, conductive and heat resistive tools for generating complex profiles. For machining composites, EDM is a well-known method [1]. Balasubramaniam et al. [2] machined the Al-SiCp metal matrix composite using different conductor resources such as brass, tungsten and copper were used. The performance measures

C. V. Ajay (✉) · K. K. Kumar · C. A. Sudhan · S. A. Kumar
National Engineering College, Kovilpatti, India
e-mail: ajayveera2@gmail.com

A. S. Kamaraja
Kalasalingam Academy of Research and Education, Krishnan Koil, India

C. T. J. Panicker
National Institute of Technology, Trichy, India

N. M. Sivaram et al. (eds.), *Advances in Manufacturing, Automation, Design and Energy Technologies*, Lecture Notes in Mechanical Engineering,
https://doi.org/10.1007/978-981-99-1288-9_23

are MRR, TWR and circularity. The machining factors such as current, pulse and flushing pressure are optimised using an artificial neural network. According to research, the current is the most important factor that affects the selected response. Copper performs better than the other two electrodes. With improved performance, machining time is reduced. Ali and Nejad [3] used the neural network to enhance the machining settings of an EDM for silicon carbide. For maximal MRR and minimal SR, 81 trials were carried out using various machining parameters. The goal of this research is to optimise the inflexibility of the exterior formed in die tumbling EDM by taking into account the effects of many contribution factors at the same time. The neuro solutions package was used to create multiperceptron neural network models. The network's weighting factors were optimised using the genetic algorithm approach [4].

Rao and Rao [5] performed on cold working tool steel manganese alloy. The outcomes specify that inherited programme design technique offers a marginally minor deviance from the dignified values of the model from fuzzy logic plus neural network [6]. An ideal cutting condition for cable-EDMed K460 apparatus toughen finishing cuts in terms of measurements precision and exterior irregularity. Critical speed, topmost current and offset distance were among the cutting terms evaluated in this study. The experimental approach was Box–Behnken, and the desirability function was used to do many reply optimisations on measurements precision and exterior irregularity. The outcomes revealed which maximum current and counter-balance distance have a considerable impact on the specimen's measurement, but peak current alone has an impact on the surface roughness.

2 Resources and Technique

Figure 1 depicts the prototype model. For testing, an aluminium alloy AA 6063 work piece with parameters of 10 mm × 10 mm × 4.1 mm is being used. The equipment is made of brass electrodes with a diameter of 2 mm, and the dielectric fluid is standard EDM oil.

In these machining process, considered parameters are discharge current (I), pulse-on-time (T_{on}), voltage (V) and flushing pressure (P_f). The set of try-outs was generated built on RSM and considered factors besides by their stages and output parameters are shown in Table 1.

The ANN is a statistical-based modelling method that does not need somewhat mathematical model. The neural network construction encloses of the input layer, hidden layer and an output layer. The ANN modelling technique is employed on the combined response MRR and TWR followed by individual response SR.

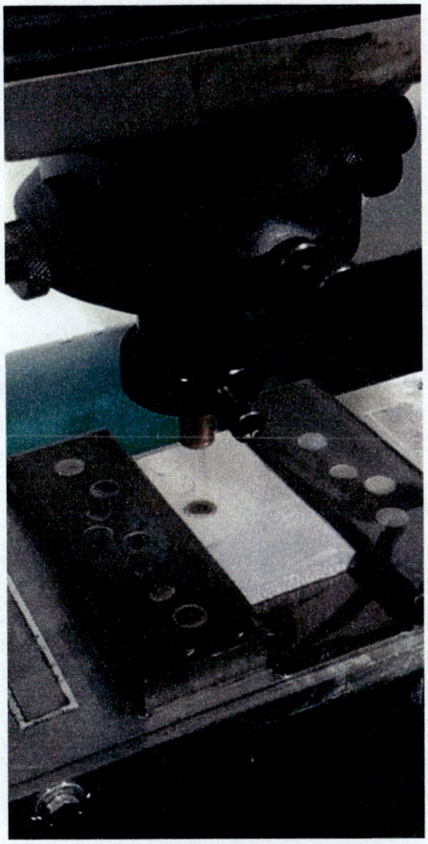

Fig. 1 RELIABLE-type EDM instrument

3 Results and Discussions

3.1 Typical ANN Aimed at the Combined Characteristics (MRR and TWR)

Table 2 reveals the *R* and MSE for MRR and TWR estimate model '4-n-2' once accomplished by expending trial-and-error method. The system 4-7-2 provides maximum *R* rate of 0.99989 and minimum MSE rate of 0.01210 available among all the ANN models, then the effects concluded that the neural setup by seven neurons with a tangent sigmoid transfer function in the unknown layer can forecast MRR and TWR nearer towards the experiential targeted result. In the optimal ANN model aimed at the MRR and TWR, the tansig and purelin transfer functions remained engaged in the concealed and output layer, respectively. The consistent *R*-values for preparation, analysis and validation were 0.99989, 0.99984 and 0.99996, respectively.

Table 1 Input and output parameters

S. no	Input parameter				Output parameter		
	I (A)	V (V)	T_{on} (μs)	P_f (MPa)	MRR (g/min)	TWR (g/min)	SR (micrometre)
1	5	40	300	0.25	0.014678	0.004573	6.41265
2	5	40	450	0.5	0.02371	0.00524	6.82945
3	10	40	450	0.75	0.104118	0.005843	7.3012
4	10	20	450	0.5	0.124935	0.006035	7.31605
5	15	40	300	0.75	0.18549	0.007778	7.35615
6	15	20	300	0.5	0.16631	0.00597	7.371
7	5	40	300	0.75	0.037843	0.003508	6.11615
8	10	60	450	0.5	0.070135	0.006715	7.58285
9	10	40	300	0.5	0.085085	0.005175	6.8844
10	10	60	300	0.75	0.064268	0.004983	6.86955
11	10	60	150	0.5	0.035235	0.004315	6.45275
12	15	40	300	0.25	0.132328	0.006843	7.65265
13	10	40	150	0.25	0.066053	0.004508	6.4676
14	15	60	300	0.5	0.11151	0.00665	7.6378
15	15	40	150	0.5	0.12646	0.00511	6.93935
16	10	40	300	0.5	0.085085	0.005175	6.8844
17	10	40	300	0.5	0.085085	0.005175	6.8844
18	10	40	300	0.5	0.085085	0.005175	6.8844
19	10	20	150	0.5	0.100035	0.003635	6.18595
20	10	60	300	0.25	0.051103	0.006048	7.16605
21	10	40	300	0.5	0.085085	0.005175	6.8844
22	10	20	300	0.25	0.105903	0.005368	6.89925
23	10	20	300	0.75	0.119068	0.004303	6.60275
24	15	40	450	0.5	0.15136	0.00751	8.06945
25	5	60	300	0.5	0.01386	0.00438	6.3978
26	5	20	300	0.5	0.02866	0.0037	6.131
27	10	40	450	0.25	0.090953	0.006908	7.5977
28	5	40	150	0.5	0.04881	0.00284	5.69935
29	10	40	150	0.75	0.099218	0.003443	6.1711

3.2 *ANN Typical for the Separate Response (SR)*

The R result and MSE again for SR estimation method '4-n-1' while developed utilising trial-and-error technique are shown in Table 3. Among all the ANN models, the system 4-11-2 has the greatest R-value of 0.98982 and the minimum MSE significance of 0.0247, as well as the data shows that the neuron to eleven hidden units and just a log sigmoid carrier signal inside the hidden units can anticipate SR nearest

Table 2 Overview of trial-and-error scheme for estimate of MRR and TWR with 4-n-2 ANN model

Hidden layer neurons	Transfer function		R standards				MSE
	Layer 1	Layer 2	Training	Testing	Validation	All	
4	Tansig	Purelin	0.98146	0.97284	0.99763	0.98225	0.421
5			0.9991	0.9948	0.99983	0.99912	0.0198
6			0.99769	0.99894	0.99971	0.99798	0.0427
7			0.99989	0.99984	0.99996	0.99987	0.01210
8			0.98789	0.99409	0.99743	0.98994	0.229
9			0.91866	0.9692	0.9957	0.93229	0.1276
10			0.99838	0.99523	0.99742	0.99738	0.0361
11			0.99235	0.99192	0.9964	0.99256	0.1881
12			0.98663	0,095,453	0.9872	0.97651	0.8763
4	Logsig	Purelin	0.99448	0.99875	0.99768	0.99302	0.14164
5			0.99197	0.99169	0.99974	0.99263	0.66127
6			0.9855	0.98525	0.98747	0.985	0.45827
7			0.99557	0.99287	0.99666	0.99429	0.3746
8			0.99144	0.98884	0.9884	0.99042	1.1746
9			0.98448	0.99561	0.98507	0.98605	0.75704
10			0.98276	0.9937	0.99276	0.98381	0.5132
11			0.97378	0.97365	0.96486	0.968	0.4715
12			0.99362	0.98702	0.99822	0.99341	0.65412
4	Purelin	Purelin	0.97141	0.97788	0.98803	0.97237	0.23966
5			0.97236	0.98124	0.97289	0.97353	0.60736
6			0.9488	0.95202	0.98474	0.95115	0.47043
7			0.98343	0.99938	0.99371	0.98619	0.55676
8			0.9902	0.33193	0.9942	0.98883	0.14386
9			0.96565	0.99872	0.96944	0.97077	0.6466
10			1	1	0.9999	1	0.57421
11			0.95825	099,391	0.94247	0.95799	0.78496
12			0.95919	0.99194	0.97219	0.96495	0.5552

towards the measured set point. The logsig and purelin model parameters were used in the hidden and output layers, appropriately, in the best ANN model again for SR. For training, testing and validation, the values of r were 0.98982, 0.98332 and 0.99053, respectively.

Table 3 Overview of trial-and-error technique for estimate of SR by 4-n-1 ANN model

Hidden layer neurons	Transfer function		*R*-values				MSE
	Layer 1	Layer 2	Training	Testing	Validation	All	
4	Tansig	Purelin	0.99148	0.9908	0.99886	0.99231	1.2049
5			0.98079	0.97244	0.97916	0.97501	0.9621
6			0.99368	0.99816	0.99054	0.99368	0.21657
7			0.97488	0.98636	0.95727	0.97334	1.0507
8			0.99361	0.9953	0.99864	0.99269	0.937
9			0.99507	0.99439	0.97259	0.99215	0.55458
10			0.98835	0.99368	0.9724	0.98707	0.68706
11			0.98794	0.99531	0.99258	0.98951	0.3793
12			0.98135	0.99775	0.99774	0.98436	0.2745
4	Logsig	Purelin	0.99326	0.99075	0.99863	0.99331	0.11674
5			0.9912	0.99609	0.99856	0.9923	0.44287
6			0.99282	0.99721	0.98845	0.99145	0.30223
7			0.96704	0.97257	0.9926	0.97112	0.2789
8			0.99502	0.99244	0.99338	0.99394	0.16625
9			0.99497	0.97515	0.99311	0.99171	0.2133
10			0.99331	0.99243	0.99974	0.99396	0.14433
11			0.98982	0.98332	0.99053	0.98774	0.0247
12			0.98398	0.99251	0.95897	0.97979	0.87203
4	Purelin	Purelin	0.97255	0.98019	0.98953	0.97573	0.22247
5			0.99204	0.98912	0.97681	0.98907	0.50431
6			0.98421	0.98873	0.99564	0.98597	1.2243
7			0.95253	0.97555	0.96668	0.92767	0.86038
8			0.97799	0.97571	0.98112	0.97646	0.52921
9			0.99239	0.99291	0.99729	0.99301	0.71915
10			0.94906	0.93765	0.9638	0.94936	0.82676
11			0.98252	0.99235	0.98598	0.97985	1.50348
12			0.95274	0.98143	0.9378	0.95472	1.4138

4 Conclusion

1. Among all the ANN models, the system 4-7-2 has the greatest rating of 0.99989 and the minimum MSE number of 0.01210, indicating that perhaps the neural network with seven number of neurons and a tangent sigmoid transfer function can estimate MRR and TWR near to the measured predicted values.
2. Among all the ANN models, the system 4-11-2 seems to have the greatest *R* rating of 0.98982 and the minimum MSE level of 0.0247, and the findings show

that the neural network using eleven neurons in the hidden layer and just a log sigmoid transfer function can estimate SR near to the measured set point.

References

1. Pellicer N, Ciurana J, Delgado J (2011) Tool electrode geometry and process parameters influence on different feature geometry and surface quality in electrical discharge machining of AISI H13 steel. J Intell Manuf 22:575–584
2. Balasubramaniam V, Baskar N, Sathiya Narayanan C (2014) Optimization of electrical discharge machining parameters using artificial neural network with different electrodes. In: 5th international & 26th all India manufacturing technology, design and research conference
3. Ramezan A, Mahdavi N (2011) Modeling and optimization of electrical discharge machining of SiC parameters using neural network and non-dominating genetic algorithm. Mater Sci Appl 2:669–675
4. Krishna Mohana Rao G, Hanumantha Rao D (2010) Hybrid modeling and optimization of hardness of surface produced by electric discharge machining using artificial neural networks and genetic algorithm. ARPN J Eng Appl Sci 5:72–81
5. Rodic D, Gostimirovic M, Kovac P, Mankova I, Pucovsky V (2014) Predicting of machining quality in electric discharge machining using intelligent optimization techniques. Int J Recent Adv Mech Eng 3(2):1–9
6. Kanlayasiri K, Jattakul P (2013) Simultaneous optimization of dimensional accuracy and surface roughness for finishing cut of wire-EDMed K460 tool steel. Precision Engineering, Article in Press

Wear Characteristics of Hard Coatings on Austenitic Stainless Steels Using Detonation Spray Process

Jhansi Jadav, U. S. Jyothi, S. Shanti, and P. V. S. L. Narayana

Abstract In hydroelectric power plants, severe damage to steam turbine components is caused by solid particle erosion. Hence to protect the base material, thermal spray coatings are used. In this study, austenitic stainless steel is used as substratc material and 86WC-10Co-4Cr, $75Cr_3C_2$ and 100% Al_2O_3 are the coating materials employed using detonation gun method. The erosion test is conducted with the help of air jet sand erosion tester with silica as erodent particles at a velocity of 60 m/s at 90° impact angle. Stepwise erosion and cumulative weight loss was calculated at different time intervals. It is observed that the erosion damage is less for austenitic stainless steel (ASS). Also 86WC-10Co-4Cr coating on ASS has shown better resistance to erosive wear compared to Al_2O_3 coating and $Cr_{23}C_6$-Ni Cr coating. An attempt has been done to understand the mechanisms of erosion with microscopic analysis.

Keywords Austenitic stainless steel coatings · Detonation spray process · Thermal spray process

1 Introduction

Thermal spray coating processes are industrially significant due to the combination of thermal energies, kinetic energies and wide variety of choice of selection of materials [1]. Applications of the thermal spray coatings include chemical processing equipment, paper processing rolls, aero-engine components against corrosion and corrosive wear and wear resistance [2–4]. Detonation gun process (DG) is one of the best thermal spray coating techniques to get low porosity, dense and hard coatings [5]. Generally ceramic coatings and hard metallic coatings are used for wear and oxidation resistance based on the component-design, service and environmental

J. Jadav (✉) · S. Shanti · P. V. S. L. Narayana
Metallurgical and Materials Engineering Department, MGIT, Hyderabad, India
e-mail: jjhansibai_mme@mgit.ac.in

U. S. Jyothi
Mechanical Engineering Department, GRIT, Hyderabad, India

N. M. Sivaram et al. (eds.), *Advances in Manufacturing, Automation, Design and Energy Technologies*, Lecture Notes in Mechanical Engineering,
https://doi.org/10.1007/978-981-99-1288-9_24

conditions. Different ceramic powders used in and multi-layered composite coatings are detailed in [6, 7]. Hard metallic coatings (WC–CO) with different thickness are coated using DG process, where residual stresses have transformed from tensile nature to compressive nature with increase in thickness up to 365 μm [8]. Wang et al. [9] reported the continuous caster drive roll life was doubled with Cr_3C_2-NiCr coatings with DG process. In the current research work, erosion resistance of austenitic stainless steel with several coating materials Al_2O_3, WC–Co and Cr_3C_2-NiCr is studied using air jet sand erosion tester.

2 Materials and Methods

100%Al_2O_3, 86WC-10Co-4Cr and 75Cr_3C_2-25NiCr are the coating materials used on substrate (austenitic stainless steel) using detonation gun (DG) process. The elemental constituents of ASS are given in Table 1.

The optical microstructures of coating materials are given in Fig. 1. In Fig. 1a, dark gray areas represent the CO-Cr binder matrix and within the brighter phases are W and W_2C are distributed. In Fig. 1b, the Cr_2O_3 phase appears darker and the ductile metallic matrix appears brighter. Alumina coatings are observed to be porous when compared to DG sprayed carbide coatings and the α-Al_2O_3 and γ-Al_2O_3 phases appeared in Fig. 1c.

2.1 *Detonation Gun Spray Process*

The detonation gun is filled with mixture of fuel gases along with N_2 carrier gas. The details of coating materials, pressure of fuel-gas mixture and carrier gas along with feed rate of gases are given in Table 2.

Austenite samples are sand blasted at room temperature, and spray distance maintained is 160 mm for carbide particles and 200 mm for Al_2O_3 particles, during coating. Spraying spot diameter maintained is 2 mm. The fuel mixture got ignited in air atmosphere with spark plug, and the energy released from each detonation will melt the powder material, accelerated and propelled with carrier gas at a speed of 900 m/s, toward the substrate. 5–6 μm thickness is obtained for each shot, and 3-shots/sec are

Table 1 Elemental constituents present in ASS

Element	Carbon	Silicon	Phosphorous	Sulfur	Chromium
wt%	0.0218	0.688	0.018	0.0132	26.703
Element	Nickel	Boron	Copper	Molybdenum	Iron (Bal.)
wt%	9.165	0.0033	0.07	0.048	63.07

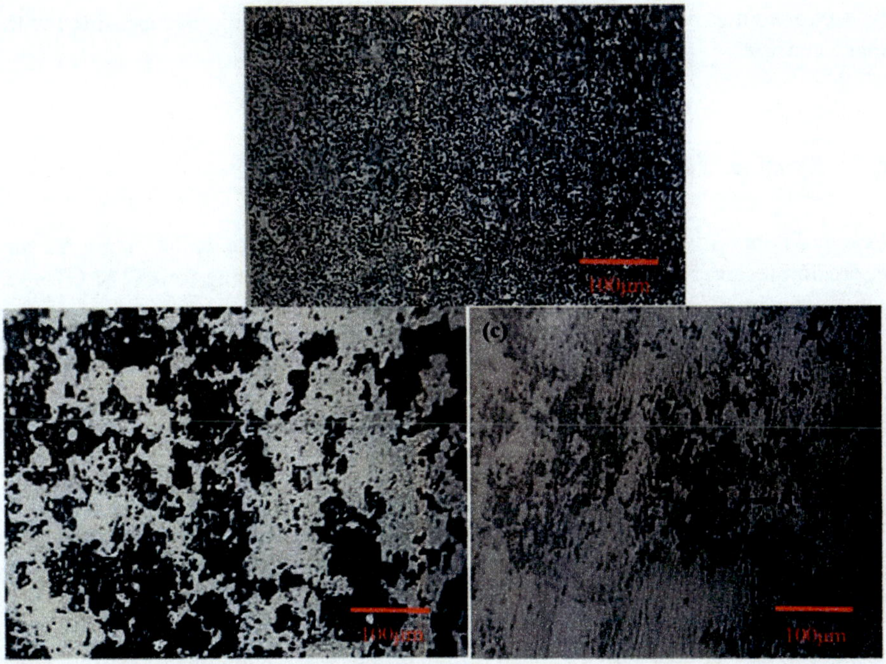

Fig. 1 Optical microstructures of **a** tungsten carbide coating, **b** chromium carbide coating and **c** 100% alumina on ASS substrate

Table 2 Details of feed rate of gases [10]

Material	Pressure bar (C_2H_2, O_2, N_2)	Feed rate of gases [SLPH], (C_2H_2, O_2, N_2)
Cr_3C_2-NiCr	1, 2, 3	2640, 2320, 800
WC-Co-Cr	1, 2, 3	2960, 2400, 720
Al_2O_3	1, 2, 3	1960, 3200, 800

employed. The process parameters greatly influence the bonding characteristics of the coating like hardness, porosity, residual stresses, wear resistance, etc. [11].

2.2 Hardness and Porosity

The ASS substrate and the coating materials hardness are determined with Vickers hardness tester. 300 gm load for 10 s is applied to measure the microhardness of the coatings. Macrohardness of the substrate material is determined at a load of 30 kgf load with a dwell time of 10 s, then unloaded. Porosity measurements are done on the sectioned and polished surface of coatings using optical microscope, fitted with image analyzer.

2.3 *Erosion Test*

Wear resistance of the coatings and substrate samples are conducted using the air jet erosion tester (Fig. 2). Erosion test was conducted according to ASTM G76-95 standards using SiO_2 erodants [12].

In air jet erosion tester, the silica erodent material (250–350 μm size) is supplied by particle feeder, at a 45 g/min feed rate, into mixing space and continual flow rate of air with silica erodants are maintained with a compressor. Generally angular and round powder particles are accelerated with compressed air jet through a stainless steel converging nozzle of ($\Phi = 10$ mm) and hit the surface of the sample kept fixed at sample holder. The distance maintained between the convergent nozzle and the samples to be tested is 10 cm. The impact angle and impact velocity maintained throughout the experiment are 90 °C and 60 m/s, for a period of 2 h. Before the erosion test, the samples are cleaned with acetone, dehydrated and then weights are recorded using an electron balance having a precision of 0.01 mg. After the 2 h test, the samples are again cleaned in acetone and dried, weights are recorded to find the weight loss.

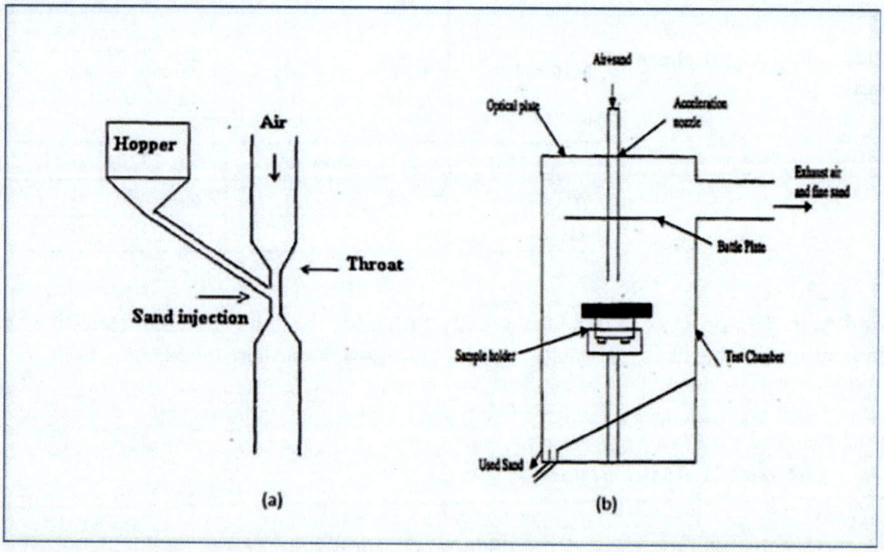

Fig. 2 Air jet erosion tester **a** compressed air and sand injection system and **b** vertical erosion chamber [10]

Table 3 Porosity and hardness of wear resistance coatings [10]

Coating material	Microhardness (HV0.3 kg)	Porosity (%)
86WC-10Co-4Cr	1075	0.41
$75Cr_3C_2$-25NiCr	961	0.44
Al_2O_3	1100	2.19

3 Results and Discussion

3.1 *Hardness and Porosity*

The high velocity of the metal coating powders (>900 m/s) and high temperatures (~4000 °C) employed during spraying have resulted with good dense and hard coatings of WC, Al_2O_3 and Cr_3C_2 of materials on ASS substrate.

The macrohardness value of austenitic stainless steel substrate is 280 HV30 kg, and the microhardness and porosity values of wear resistance coatings are given in Table 3.

In general, a coating that contains non-melted powders may exhibit lower microhardness because of poor interlamellar bonding. It was reported in Table 3 that the hardness of the Cr_3C_2 coating was the lower one and the main reason for the finding of hardness changes is the variation in their chemical composition. The parameters employed in the DG spray coating process, such as particle velocity, nozzle to substrate distance, ratio of fuel gas to oxygen mixture, temperature, surface roughness of the substrate will dictates the hardness and porosity of coated substrates [11]. The measurement of coated samples porosity is done using microprobe analyzer, tailored to optical microscope and is reported in Table 3.

It is observed from Table 3, the porosity value is low (<1%) for 86WC-10Co-4Cr and Cr_3C_2-NiCr coated samples. The porosity is low due to the combination of thermal energy and kinetic energies employed in the spray process, to get melted and join with another powder particle without leaving pores in between them. Lower porosity is expected for stream of completely molten particles with high velocities and shorter spray distance. In the current study, slightly higher porosity values are reported in Al_2O_3 coating may be due to improper process parameters employed like spray distance and fuel mixture to oxygen ratio.

3.2 *Erosion Test*

Cumulative Weight Loss

The collective weight loss of all the coated samples with 30, 60, 90 and 120 min time interval is given in Table 4. The erosion behavior of detonation gun sprayed-three coating materials at four intervals on ASS substrate are shown in Table 4 and Fig. 3.

Table 4 Weight loss of coated and uncoated samples for two-hour test duration

Coating material	Cumulative weight loss (gm) after			
	30 min	60 min	90 min	120 min
Uncoated	0.0022	0.0042	0.0046	0.0050
86WC-10Co-4Cr	0.0064	0.0201	0.0213	0.0366
$75Cr_3C_2$-25NiCr	0.0126	0.0209	0.0250	0.0410
Al_2O_3	0.0038	0.0051	0.0056	0.0412

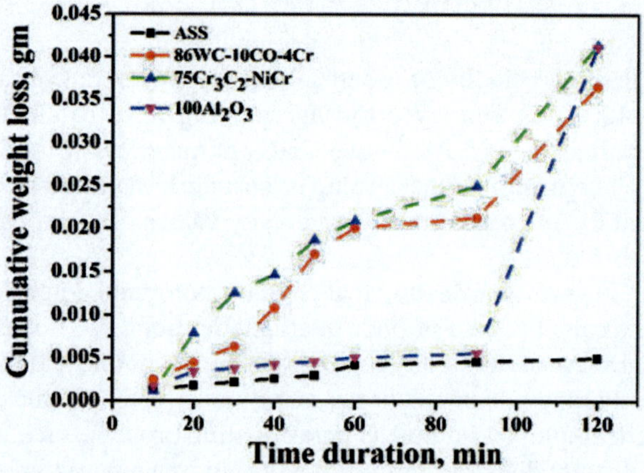

Fig. 3 Cumulative weight loss of different coating using DG spray process

Wear rate of ductile and brittle materials are function of impact angle of erodants. Erosion rate is maximum between 15° and 30° for ductile materials and for brittle materials, it is maximum at 90° [13]. Hence, in the current study, the cumulative weight loss reported is low for uncoated austenite stainless steel (Table 4). The erosion rate is nearly uniform when the substrate is ductile in nature and fracture is characterized by substantial plastic deformation before and during propagation of crack. The crack is stable, and it resisting further propagation of crack unless there is a rise in the applied stress. A minor erosion mechanism for ductile material is micro-cutting and plowing at low angles of erodent particle impact.

Comparatively, the coated substrates have shown higher weight loss to that of ASS substrate. Here, the erosion is due to anisotropic microstructure of the wear resistant coating and preferred crack propagation corresponding to the wear resistant coating surface followed by crack interlocking and spalling [14].

It is observed from the results that the weight loss is highest for alumina coating and minimum for WC-coating. 86WC-10Co-4Cr shows better erosion resistant behavior on this steel due to better adhesion of the coating with low porosity. A combination of optimum hardness and toughness increases the erosion resistance [15]. The binder phase Co-Cr provides the toughness to the matrix, and WC particles contribute to the hardness. Alumina coating has shown higher erosion rate may be due to high porosity and improper process parameters. Also erosion rate of alumina crystal strongly depends on the grain size [16].

3.3 Scanning Electron Microscopic Analysis of Eroded Samples

The coated samples are subjected erosion test with SiO_2 particles (250–350 microns) as erodent, shown in Fig. 4a. The eroded microstructure of substrate along with entrapped sand particle is seen in Fig. 4b. After 2 h of erosion test, the eroded surfaces of 86WC-10Co-4Cr, $75Cr_3C_2$-25NiCr and Al_2O_3 coatings are depicted in Fig. 5a–c.

Microstructures obtained through scanning electron microscopy (SEM) of uncoated ASS are shown in Fig. 4b, and it is observed that there is a shallow crater with the sand grain center and material dislodge with finer cavities in the surroundings. This is because of the good toughness and shock absorbing nature of the ASS. Figure 5a displays the microstructure of WC-Co-Cr coating on ASS, shows a cavity at the center, due to pull out of WC particles. Surface deformation, cracks and plastic deformation are observed. Less erosion rate is observed in this coating because of uniform distribution of tungsten carbides in the coating and good adhesion of coating with the substrate. Figure 5b shows less amount of plastic deformation of chromium carbide coatings on ASS. Large amount of deep cracks are observed, and subsurface material is distorted in severe manner along with deep micro cavities. Cracks have propagated deeply due to de-carburization and poor bond of the coating with the substrate. Figure 5c shows more erosion because of large amount of plastic deformation is observed for alumina coating on ASS. Ductile flakes and bulging of coating are seen and erosion has occurred in ductile mode.

Fig. 4 SEM micrographs of **a** SiO_2 erodent particles and **b** substrate

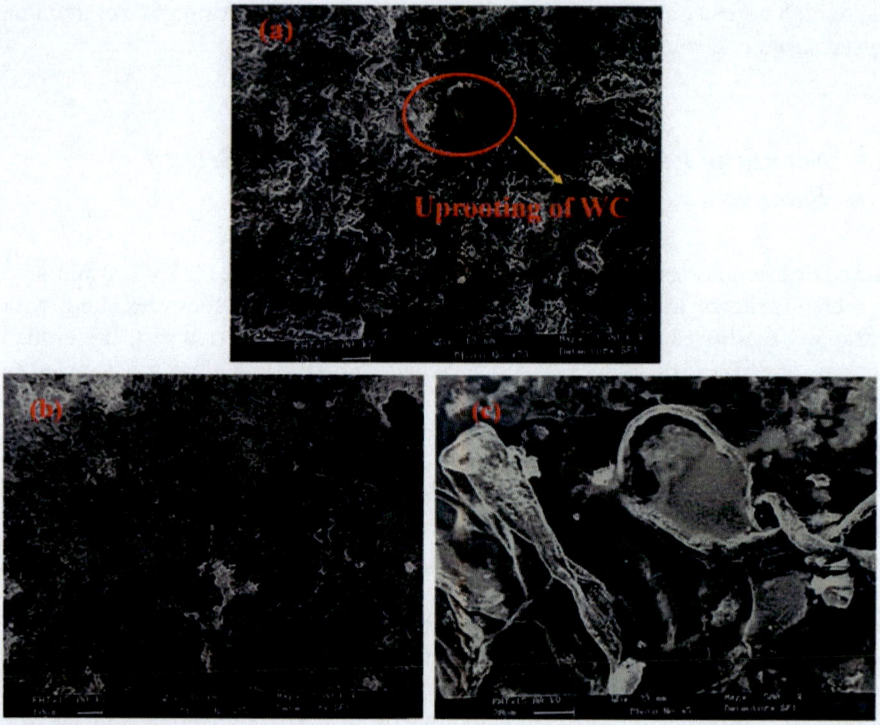

Fig. 5 Eroded microstructures of coating of **a** 86WC-10Co-4Cr, **b** $75Cr_3C_2$-25NiCr and **c** Al_2O_3

4 Conclusions

Sand particle erosion is conducted on austenitic stainless steel and DG spray coated substrates with air jet sand erosion tester for two hours.

- Better erosion resistance is observed for 86WC-10Co-4Cr coatings on ASS is due to low porosity, better hardness and toughness.
- Cumulative weight is less for WC-coated ASS when compared $75Cr_3C_2$-25NiCr and 100% Al_2O_3 coatings.

References

1. Jonda E, Latka L, Pakiela W (2021) Comparison of different cermet coatings sprayed on magnesium alloy by HVOF. Materials 14(7):1594
2. Wilson S (2013) Abradable thermal spray applications and technology. ASM handbook, vol 5A: thermal spray technology. ASM International, pp 287–295

3. Fauchais P, Vardelle A (2012) Thermal sprayed coatings used against corrosion and corrosive wear. In: Salimi Jazi H (ed) Advanced plasma spray applications. InTech, pp 3–39. https://doi.org/10.5772/34448
4. Vikas C, Singh SB, Puri D, Prakash S (2008) Performance of plasma sprayed nanostructured and conventional coatings. J Aust Ceramic Soc 44: 56–62
5. Rakesh G, Singh SB, Grewal JS (2010) Surface engineering and detonation GunSpray coating. Int J Eng Stud 2:351–357
6. Sova PD, Smurov I (2010) Development of multimaterial coatings by cold spray and gas detonation spraying. Surf Coat Technol 205(4):1108–1114
7. Wang L, Yan D, Dong Y, Zhang J, Chen X (2013) Nanostructured ceramic composite coating prepared by reactive plasma spraying micro-sized Al–Fe_2O_3 composite powders. Ceram Int 39(3):2437–2442
8. Tie-Gang W, Sheng-Sheng Z, Wei-Gang H, Jia-Bao L, Jun G, Chao S (2010) Estimation of residual stress and its effects on the mechanical properties of detonation gun sprayed WC-Co coatings. Mater Sci Eng A 527(3):454–461
9. Jun W, Li Z, Baode S, Yaohe Z (2000) Study of the Cr_3C_2-NiCr detonation spray coating. Surf Coat Technol 130(1):69–73
10. Jhansi J et al (2022) Detonation gun spray coatings on martensitic stainless steels. Lecture notes in mechanical engineering series. Recent advances in manufacturing, automation, design and energy technologies. Proceedings from ICoFT. Springer Science and Business Media LLC, pp 85–93
11. Singh L, Chawla V, Grewal JS (2012) A review on detonation gun sprayed coatings. J Miner Mater Charact Eng 11(3):243–265
12. ASTM G76–95 (2000) Standard test method for conducting erosion tests by solid particle impingement using gas jets
13. Bharat B (1999) Principles and applications of tribology. Wiley, pp 852–854
14. Wood RJK, Mellor BG, Binfield ML (1997) Sand erosion performance of detonation gun applied tungsten carbide/cobalt-chromium coatings. Wear 211(1):70–83
15. Suzhenleio Y, Jinli W (2003) Effect of curing degree and fillers on slurry and erosion behaviour of fusion bonded epoxy powder coatings. Wear 24:292–297
16. Franco A, Roberts SG (1998) The effect of impact angle on the erosion rate of poly crystalline α-Al_2O_3. J Eur Soc 18:269–274

Correlation Between Microstructure and Hardness in 5356Al Fabricated by Wire-Arc Additive Manufacturing Process

Tagore Kumar Pasupula, V. Jaya Prasad, and Satish Kumar

Abstract This paper is about the investigation of the hardness and metallurgical properties of 5356Al metal with wire-based arc additive manufacturing technology, which is the chemical addition of metal parts production process. In this work, the choosing of all process parameters is based on the availability of machine by using cold metal transfer (CMT), the bead formation in primary level is more in thickness as compared to increases the wall height, it will depend upon the voltage, wire feed rate, and weld speed, etc., and they fixed after getting the successive bead of the metal 5356Al on 6062Al as the base metal. Along with that, the cooling rate of the metal bead is a normal cooling process to be taken to room temperature, after getting the bead with the required height and length, samples for both hardness and metallurgy are taken. The microstructures are obtained and calculated grain size numbers with the Vickers hardness.

Keywords Grain orientation · Grain size · Cold metal transfer · 5356Al

1 Introduction

Additive manufacturing (AM) techniques are most extensively used nowadays to nest the layer-by-layer deposition as compared to other traditional process techniques [1]. AM is the process substitute technique for manufacturing complex, non-complex products, and expensive materials. They are various types of methods being as laser-based melting and deposition [1], direct energy process [2], electron beam-based melting [3], shape metal deposition [4], and wire-based arc additive manufacturing [5–7]. As compared to all these processes, cheap and the best is wire arc additive

T. K. Pasupula (✉) · S. Kumar
Department of Mechanical Engineering, QIS College of Engineering and Technology, Ongole 523001, India
e-mail: tagore.mech63@gmail.com

V. J. Prasad
Department of Mechanical Engineering, JNTU, Kakinada 533003, India

N. M. Sivaram et al. (eds.), *Advances in Manufacturing, Automation, Design and Energy Technologies*, Lecture Notes in Mechanical Engineering,
https://doi.org/10.1007/978-981-99-1288-9_25

manufacturing (WAAM) due to low maintains. In addition to other advantages of WAAM is higher in deposition, which means in laser or electron beam welding the deposition rate is 3–10 g/min, but compared to the arc technology the deposition rate is 60–135 g/min [8–10]. In the WAAM process, the layer is nested layer by layer, finally it shows as a single bead with a proper cooling rate [11].

Aluminum alloys have a wide range of applications in different areas like aerospace, shop constructions, and automobile applications, because of their properties like extraordinary strength, lightweight, and greatest correction resistance [12, 13]. So that, to increase the deposition rate with a low investment, several welding techniques are converted to additive manufacturing processes such as Gas metal or gas tungsten or plasma-based arc welding are used [14–17]. Fronius Company was investigated by modifying in gas metal welding process titled as cold metal transfer (CMT) [18]. Due to the back and forth movement of the filler wire and the no current during the dip transfer, the CMT process is treated as low heat input process [19].

2 Experimentation

The experimental promotion for the study and characteristics of metal 5356Al obtained by the cold metal transfer process consists of the following:

1. The micrographic observation.
2. Grain size analysis.
3. Different heat treatment process.
4. Vickers micro-hardness.

The experimentation starts with washing in alkaline water and using hot air-dried the substrate material as 6062Al, the dimensions of the plate are lengthwise 150 mm, width-wise 70 mm, and thickness-wise 20 mm. The fabricated material is 5356Al of wire of diameter is 1.6 mm with the weight of 10 kgs, these metals are procured from Ramu metal company, Chennai (Table 1).

The set-up of wire arc additive manufacturing which shows in Fig. 1a consists basic cold metal transform (CMT) system with an automatic wire feed rate system which is available in Sri Venkateswara College of Engineering Chennai. The torch is fixed at one position and the worktable moves forward and backward direction with numerical controls, the voltage, feed rate, and other process parameters are fixed based on pilot experiments.

Table 1 Nominal composition of 5356Al and 6062Al (%wt.)

Material	Cr	Cu	Fe	Mg	Mn	Si	Ti	Zn	Be	Al
6062Al	≤0.25	≤0.1	≤0.5	0.6–1.2	0.4–1.0	0.7–1.3	≤0.1	≤0.2		Bal
5356Al	0.05–0.2	0.1	0.4	4.5–5.5	0.05–0.2	0.25	0.06–0.2	0.1	0.008	Bal

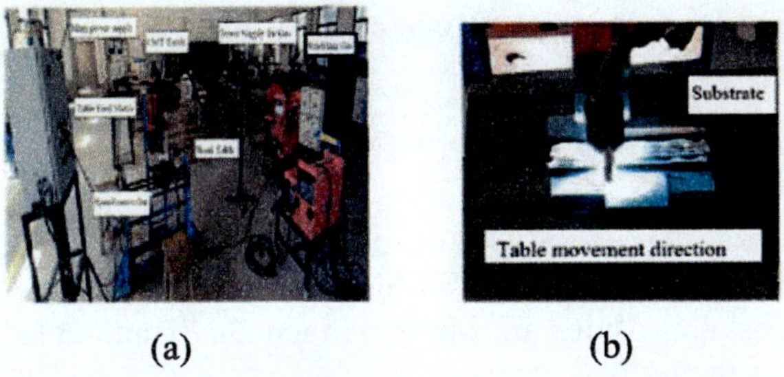

Fig. 1 **a** Cold metal transform machine and **b** nesting formation

The process parameters are wire feed rate 70 mm/sec, weld speed is 0.2 cm/min, the gas flow rate is 12 bar (lit/min, Ar), voltage is 60 V, and arc length correction is 0 are kept constant. Apart from them, the main weld speed plays the main role because of work table moves only in one direction shown in Fig. 1b.

During the process of manufacturing, the arc dowsed after completing each layer in the span of 45 s., then the next layer is initiated, and also maintain the distance between the nozzle of the torch and the workpiece is 7 mm. Due to this, the layer nested efficiently as layer by layer is shown in Fig. 2. The nested bead observed that bottom to up length was decreased, this is due to metal contraction. The specimens are taken from the bead after complete cooling to the room temperature, for heat treatment and hardness process.

Each sample was cut-off up to substrate level, and the middle of the piece was taken for the observing and testing bead. A little bit of machining is required for removing the non-uniform material from the bead then by using emery paper for polishing and to close getting of surface cleaned and apply the Kellers reagent 600 s., to remove the presence of impurities. The samples are detected at low magnification to high magnification level and considered high level only with an OLYMPUS-BX53M. The micro-hardness is calculated for each layer with a METCO-Economet VH-1MDX, a load of 100 g and 500 g for 20 s. is used. The heat treatment process was done in a control atmospheric electric furnace with an SV HEAT ENGINE, while cooling based on the availability of saltwater (1:10), normal coconut oil, and a fast air process are taken.

Fig. 2 Nested bead layers by layers

3 Results and Discussion

3.1 *Macroscopic Inspection*

The wall is nested, and the dimensions are 145 mm in length, 80 mm in height, and 15 mm in width. For finishing operation, it requires machining and polishing, the final dimensions are 140 mm in length, 75 mm in height, and 12 mm in width laid out in Fig. 3a, b.

3.2 *Microstructure Evaluation*

Microstructure of 5356Al alloy specimen taken from the bead at center shown in Fig. 3b. It displays a dendritic structure [20] with intermetallic components of Al alloys due to improper heat distribution of layers, long that porosity, stringers, and segregation were not observed shown in Fig. 4a. This dendritic structure is removed by the simple technique is the heat treatment process. So that the specimen is heated to 480 °C temperature and held at the same temperature of 90 min, then cooled room temperature in a different medium like coconut oil, brain solution, and fast air cooling process, that process depicted in Fig. 5.

It is noticed that after heat treatment properly done for the workpiece depicted in Fig. 6a–d and observed that the α-aluminum boundary [15] in microstructure depicted in Fig. 4b–d. Under oil quenching the microstructure shows Fig. 4d. Completely removed dendritic structure and forms fine coarse grains [21]. Similarly in salt bath observed that uniform coarse grains [21] are formed and in fast air cooling non-uniform coarse grain [21] structures are formed.

The line intercept method is used to measure grain size according to ASTM E112 [22]. It observed that the grain size number various a little bit only but in oil quenching it is more and that grain numbers are given in Table 2.

Fig. 3 **a** Schematic diagram of finished bead. **b** Finished bead

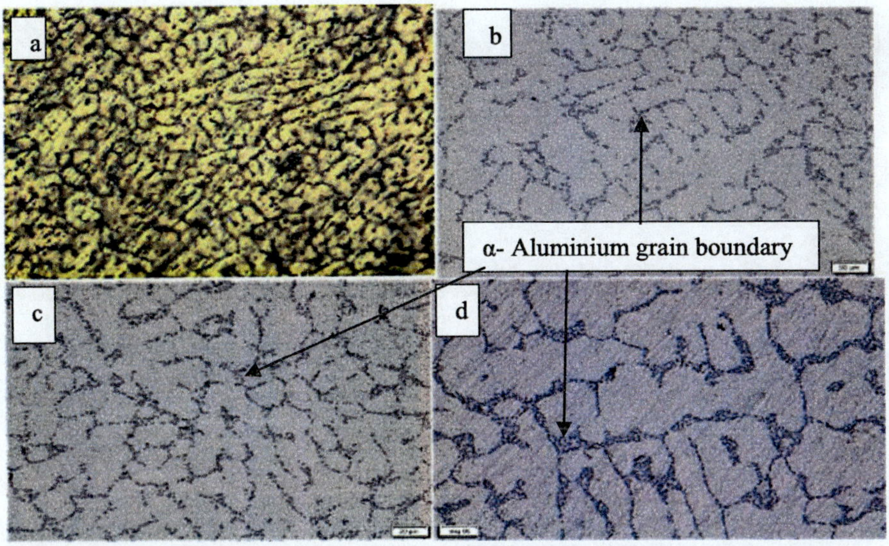

Fig. 4 Microstructure of 5056Al **a** without heat treatment, **b** cooled in brain solution treatment, **c** cooled in forced air cooling, and **d** cooled in coconut oil

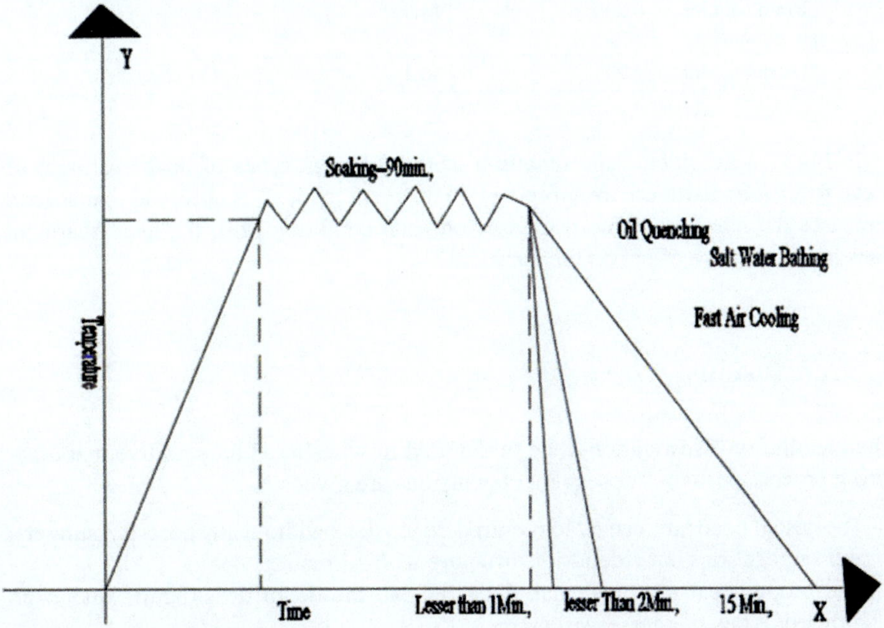

Fig. 5 Heat treatment process

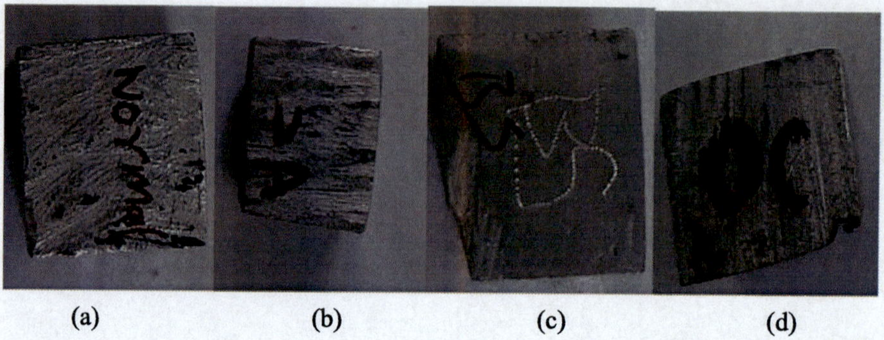

Fig. 6 **a** Without treatment. **b** Brain solution treatment. **c** Fast air cooling. **d** Oil cooling

Table 2 Grain number and lengths for deposited samples

S. no	Type of cooling process	Grain size number	Mean intercept length (micro meter)	Number of intercepts per unit length (1/mm)
1	Without treatment	8.44	13.77	65.74
2	Fast air cooling	8.69	15.74	63.53
3	Brain solution treatment	8.83	14.99	66.7
4	Oil quenching	9.99	10.3	157

In Fig. 7 a graphical representation of the different types of heat treatment of metal to Vickers hardness by applying two different loads. It is observed that as load increases the hardness value decreases on each condition [23], the heat treatment does not have more effect on the hardness.

4 Conclusions

The metallurgy investigation done on 5356Al in wire-based arc additive manufacturing process with the subsequent conclusions are given:

- The nested bead are formatted required dimensions without any porosity, stingers, and segregations but a dendritic structure with Al matrix.
- The proper heat treatment has done remove the dendritic structure and from different types of courses structure.
- The grain structure is calculated by each treatment with line intercept method accurately.
- The hardness calculated by varying load, two different load different types and average valve are considered and also load increases the hardness value decreases.

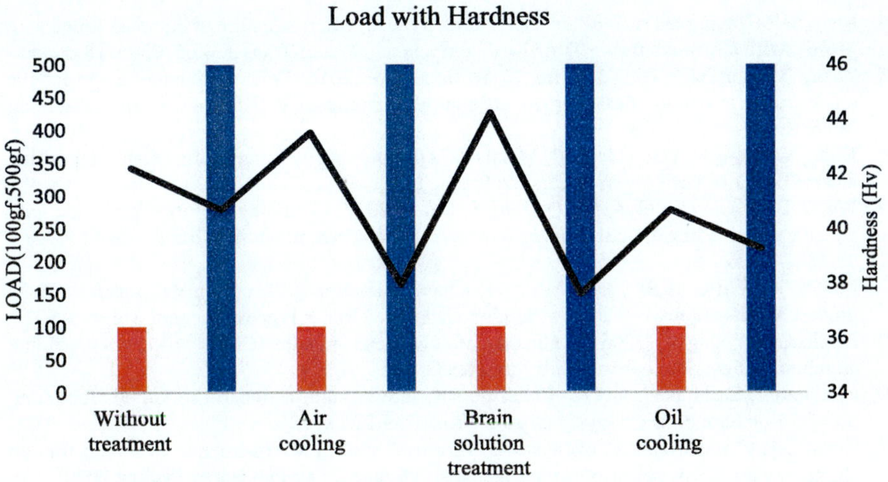

Fig. 7 Different types of heat treatment process with hardness

References

1. Agarwala M, Bourell D, Beaman J, Marcus H, Barlow J (1995) Direct selective laser sintering of metals. Rapid Prototyp J 1:26–36
2. Lewis GK, Schlienger E (2000) Practical considerations and capabilities for laser assisted direct metal deposition. Mater Des 21:417–423
3. Taminger KM, Hafley RA (2003) Electron beam freeform fabrication: a rapid metal deposition process. In: Proceedings of the 3rd annual automotive composites conference, pp 9–10
4. Merz R, Prinz FB, Ramaswami K, Terk M, Weiss L (1994) Shape deposition manufacturing: engineering design research center. Carnegie Mellon University, Stanford, California, USA
5. Almeida PS, Williams S (2010) Innovative process model of Ti–6Al–4V additive layer manufacturing using cold metal transfer (CMT). In: Proceedings of the twenty-first annual international solid freeform fabrication symposium. University of Texas, Austin, TX, USA
6. Ding J, Colegrove P, Mehnen J, Ganguly S, Sequeira PM, Wang F et al (2011) Thermo-mechanical analysis of wire and arc additive layer manufacturing process on large multi-layer parts. Comput Mater Sci 50:3315–3322
7. Wang F, Williams S, Rush M (2011) Morphology investigation on direct current pulsed gas tungsten arc welded additive layer manufactured Ti6Al4V alloy. Int J Adv Manuf Technol 57:597–603
8. Brandl E, Michailov V, Viehweger B (2011) Deposition of Ti–6Al–4V using laser and wire, part I: microstructural properties of single beads. Surf Coat Technol 206:1120–1129
9. Zhang Y, Wei Z, Shi L, Xi M (2008) Characterization of laser powder deposited Ti–TiC composites and functional gradient materials. J Mater Process Technol 206:438–444
10. Karunakaran KP, Suryakumar S, Pushpa V, Akula S (2010) Low cost integration of additive and subtractive processes for hybrid layered manufacturing. Robot Comput-Integr Manuf 26:490–499
11. Mazumder J, Dutta D, Kikuchi N, Ghosh A (2000) Closed loop direct metal deposition: art to part. Opt Lasers Eng 34:397–414
12. Kaibyshev R, Musin F, Lesuer DR, Nieh TG (2003) Superplastic behavior of an Al–Mg alloy at elevated temperatures. J Mater Sci Eng A 342:169–177
13. Engler O, Miller-Jupp S (2016) Control of second-phase particles in the Al-Mg-Mn alloy AA 5083. J Alloys Compd 689:998–1010

14. Martina F, Colegrove PA, Williams SW, Meyer J (2015) Microstructure of interpass rolled wire + arc additive manufacturing Ti-6Al-4V components. Metall Trans A 46:6103–6118
15. Zhang X, Martina F, Ding J, Wang X, Williams SW (2016) Fracture toughness and fatigue crack growth rate properties in wire + arc additive manufactured Ti-6Al-4V. Fatigue Fract Eng Mater Struct 40:790–803
16. Xiong X, Zhang H, Wang G (2009) Metal direct prototyping by using hybrid plasma deposition and milling. J Mater Process Technol 209:124–130
17. Wu Q, Lu J, Liu C, Shi X, Ma Q, Tang S, Ma S (2017) Obtaining uniform deposition with variable wire feeding direction during wire-feed additive manufacturing. Mater Manuf Process 32:1881–1886
18. Pickin CG, Williams SW, Lunt M (2011) Characterization of the cold metal transfer (CMT) process and its application for low dilution cladding. J Mater Process Technol 211:496–502
19. Pickin CG, Young K (2006) Evaluation of cold metal transfer (CMT) process for welding aluminium alloy. Sci Technol Weld Join 11:583–585
20. Okamoto T, Kishitake K (1975) Dendritic structure in unidirectionally solidified aluminum, tin, and zinc base binary alloys. J Crystal Growth 29(2):137–146
21. Zykova A, Chumaevskii A, Kolubaev E, Tarasov S (2020) Microstructure of in-situ friction stir processed Al-Cu transition zone. Accepted: 16 June 2020; Published: 18 June 2020
22. ASM Handbook (2014) Metallography and microstructures, vol 9. ASM International
23. Petrík J, Palfy P (2011) The influence of the load on the hardness. Metrol Measure Syst 330930. ISSN 0860–8229

Mechanical Performance Analysis of Ecofriendly Fiber Composites

J. Ronald Aseer, K. Sankaranarayanasamy, S. Renold Elsen, Amit Kumar Thakur, and R. Prema Latha

Abstract Today world, the rising environmental impact related with the production, removal and recycling of synthetic fiber reinforced polymer composites has encouraged the progress of ecologically friendly composites for automotive, construction, sporting goods, etc. In this context, natural fibers such sisal, kenaf, jute, banana, hemp, flax are considered as an alternate for synthetic fibers. Among various natural fibers, production of banana fiber is increasing due to the extensive scope of application, abundantly available and cheaper. In this work, acid treated banana fiber/phenol formaldehyde composites were produced by using hydraulic press and their tensile, flexural and impact strength properties were investigated. The results indicated that acid treated banana reinforced phenol formaldehyde composites have shown optimum strength values as compared to control samples. Also, mechanical properties of the of acid treated banana reinforced composites are higher at 40% V_f of fiber. This obtained composites could be an effective alternative for ceiling, partition wall in construction industries.

Keywords Mechanical properties · Composites · Phenol formaldehyde · Banana fiber

J. R. Aseer (✉) · K. Sankaranarayanasamy · R. Prema Latha
Department of Mechanical Engineering, National Institute of Technology Puducherry, Karaikal, India
e-mail: ronaldaseer.j@nitpy.ac.in

S. R. Elsen
School of Mechanical Engineering, VIT University, Vellore, Tamil Nadu, India

A. K. Thakur
Department of Mechanical Engineering, Lovely Professional University, Punjab, India

N. M. Sivaram et al. (eds.), *Advances in Manufacturing, Automation, Design and Energy Technologies*, Lecture Notes in Mechanical Engineering,
https://doi.org/10.1007/978-981-99-1288-9_26

1 Introduction

Due to their abundance in nature, researchers are concentrating on lignocellulosic fiber materials. Many researchers disseminated their results of fiber characterization using various chemical treatment. Characterization results indicated that natural fibers derived from plants, tree can be a good substitute for making bio-composites [1–4]. The composites properties were analyzed by varying the natural fiber ratio, length of fiber for different proportion of resin content. Strength and modulus values are high as compared to the control samples [5, 6]. The thermal properties are analyzed, and results are reported for various fiber materials by changing the volume fraction. Composites with fibers have shown better conductivity and diffusivity values as compared to resin. Nowadays, windshield wiper blade is mostly manufactured from E-glass fiber reinforced polymer composite using traditional methods [7, 8]. Also, carbon fiber is used to manufacture the windshield wiper blade which provides the required strength and stiffness. Also, manufacturing cost of carbon fiber and fiber misalignment, fiber waviness leads to less usage in the field of wiper blade manufacturing [9]. Glass fiber has some issues of high density, poor recyclability, hazard to health, high tooling cost, effective handling and disposal problems [10]. Natural plant fibers are having unique characteristics of low density, less cost, low weight, easy recyclability and biodegradable in nature which makes them suitable for polymer reinforced composite production. It will not release harmful gases during the production and handling. Also, the specific strength and stiffness of natural fiber is good which compared to glass fiber. It is understanding that manufacturing of polymer composites using natural fiber and proper use of fiber is essential. Aseer et al. investigated the effect of $NaClO/H_2O$ treated banana fiber on various properties which has shown superior properties that raw fibers [11]. Natural fiber composite provides rigidity to modify the properties of materials depends on the necessities [12]. Mechanical properties of pineapple and sisal fiber reinforced composites were investigated by Mishra et al. [13]. Pine apple and sisal hybrid fiber composites have shown good properties. The obtained composite can be used in automobile application. The mechanical properties of banana fiber and glass fibers PF composites were reported [14]. The composite reinforced with glass fiber has given optimum results. The aim of this work is to analyze the properties of acid treated banana fiber/phenol formaldehyde composites static loading condition for the production of ceiling panels.

2 Methodology

2.1 Materials

Banana fiber were collected from Tamilnadu, India. The hardener (NH4CL) used for the investigation is added as 1% with the phenol formaldehyde resin. Banana

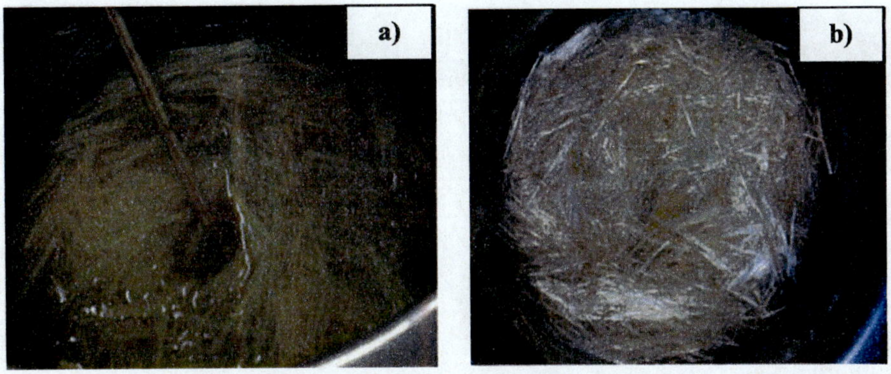

Fig. 1 Fiber extraction (**a**) raw and (**b**) acid treated fiber

fiber extraction and preparation have been reported by Aseer et al. [11]. The raw and treated fiber (Fig. 1a, b) was collected and combined with phenol formaldehyde for the fiber mat manufacturing.

2.2 *Fabrication of Composites*

Composite panels are prepared using traditional methods using various fiber and resin ratio. A layer of gel was glued on the metallic mold, and the fibers were arranged in the mold by combing. It was pressed with 60kN and kept to cure at 30° C for four hrs. The mat made with fibers were cut as per the required dimensions. The final sample was fabricated using press. The phenol formaldehyde and hardener were mixed in the ratio of 1:1.

Then, a thin coat of the mixed resin was distributed into the mold and mats were arranged by layer. Mats made of fiber were pressed with hand roller (Fig. 2a). The mold was then positioned in a press under a load of 60kN to make the resin to set [14]. After six hours, the specimens were ejected and prepared as per the ASTM standards (Fig. 2b).

3 Results and Discussion

3.1 *Fiber Effect on Tensile and Flexural Properties*

The tensile and flexural strength properties of banana/PF composites are given in Table 1. Neat PF resin has shown lower value of tensile, flexural and impact strength. Fiber addition to the phenol formaldehyde increases the tensile, flexural strength

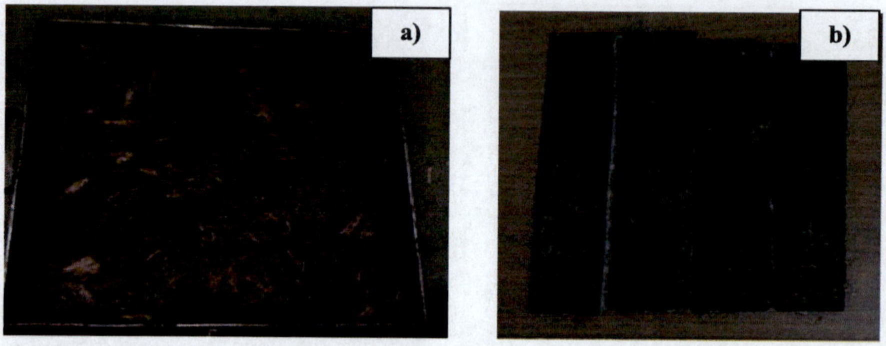

Fig. 2 Composite with (**a**) 40% V_f of fiber and (**b**) various combination of fiber

of composites. This is mainly due the fiber matrix adhesion which increases the properties of composites [14]. Mechanical properties of composites are reaching high vales when fiber is 40% V_f. Further, addition of fiber decreases the properties due to the less resin the fiber matrix region (Fig. 3a, b).

Table 1 Properties of the composites

Volume fraction (%V_f)	Tensile strength (Mpa)	Young's modulus (Gpa)	Flexural strength (Mpa)	Flexural modulus (Gpa)	Impact strength (J/cm^2)
Phenol formaldehyde	47 ± 1.2	4.6 ± 1.2	76 ± 1.1	2.6 ± 1.2	0.4 ± 0.1
10	58 ± 1.2	6.2 ± 2.2	85 ± 1.5	3.1 ± 1.1	6 ± 1.3
20	69 ± 1.1	7.1 ± 1.1	96 ± 1.4	4.3 ± 1.3	10 ± 1.1
30	79 ± 1.3	8.3 ± 1.2	101 ± 1.2	5.2 ± 1.2	16 ± 1.4
40	**88 ± 1.2**	**9.1 ± 1.4**	**112 ± 1.3**	**6.1 ± 1.6**	**24 ± 1.2**
50	77 ± 1.3	8.5 ± 1.3	102 ± 1.2	5.6 ± 1.7	20 ± 1.6

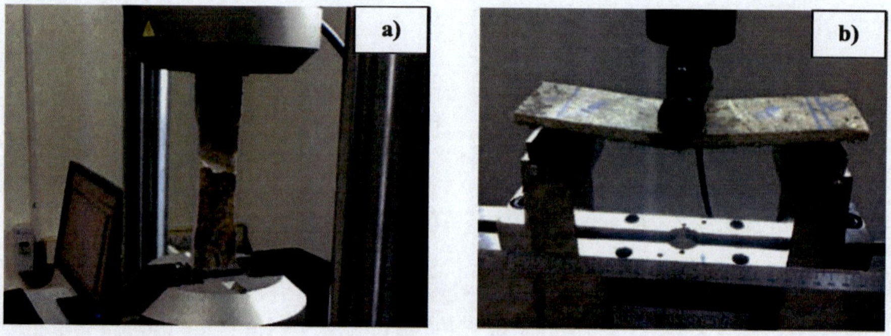

Fig. 3 Composite sample after (**a**) tensile and (**b**) bending test

3.2 *Fiber Effect on Impact Strength*

The impact strength of the composites with various fiber is given in Table 1. The impact strength of resin is low as compared to fiber composites. When the fiber incorporation increases, impact strength increased which indicates high adhesion with phenol formaldehyde resin. It increases up to 40 vol%) due to more energy dissipation at the fiber and matrix region by pull-out of fibers. The poor adhesion of phenol formaldehyde is a due to its polar character. This damages the fiber and matrix region which leads to failure of composites [15, 16].

4 Conclusions

Composite prepared with traditional methods have produced reliable results. The static properties of banana fiber/phenol formaldehyde composites were analyzed. Composite with 40% has shown higher mechanical properties as compared to neat phenol formaldehyde resin. It is understood from results that banana fiber can be an alternate for synthetic fibers. Impact strength of the fiber composites also produced the good results which makes them suitable for using in construction industries.

References

1. Lee C, Greenhalgh ES, Panesar A (2021) Optimization of patch-wise laminated composite panels for enhanced dynamic characteristics. Compos Struct 269
2. Ekoi EJ, Dickson AN, Dowling DP (2021) Investigating the fatigue and mechanical behaviour of 3D printed woven and nonwoven continuous carbon fibre reinforced polymer (CFRP) composites. Composites Part B 212
3. Sharma SP, Lakkad SC (2020) Comparative study of the effect of fiber surface treatments on the flexural and interlaminar shear strength of carbon fiber-reinforced composites. Mater Today Commun 24
4. Singh JIP, Singh S, Dhawan V (2020) Influence of fiber volume fraction and curing temperature on mechanical properties of jute/PLA green composites. Polym Polym Compos 28(4)
5. Kumar U, Rathi R, Sharma S (2020) Carbon nano-tube reinforced nylon 6,6 composites: A molecular dynamics approach. Eng Solid Mech 8(4)
6. Mann GS, Singh LP, Kumar P (2019) Effect of volume fraction and heating temperature on hybrid natural fibre composites developed through the die moulding process. Int J Eng Adv Technol 9(1)
7. Saba N, Jawaid M, Alothman OY Parida HMT (2016) A review on dynamic mechanical properties of natural fibre reinforced polymer composites. Constr Build Mater 106, 149–159
8. Sreenivasan VS, Rajini N, Alavudeen A, Arumugaprabu AV (2015) Dynamic mechanical and thermo-gravimetric analysis of Sansevieria cylindrica/polyester composite: Effect of fibre length, fibre loading and chemical treatment. Compos Part B: Eng 69, 76–86
9. Pickering KL, Efendy MGA, Le TM (2016) A review of recent developments in natural fibre composites and their mechanical performance. Composites: Part A 83, 98–112
10. Jawaid M, Abdul Khalil HPS, Alattas OS (2012) Woven hybrid biocomposites: Dynamic mechanical and thermal properties. Composites: Part A 43, 288–293

11. Ronald Aseer J, Sankaranarayanasamy K, Jayabalan P, Natarajan R, Priya Dasan K (2013) Morphological, physical, and thermal properties of chemically treated banana fiber. J Nat Fibers 10(4), 365–380
12. Velmurugan R, Manikandan V (2007) Mechanical properties of palmyra/glass fiber hybrid composites. Compos Part A-Appl Sci 38:2216–2226
13. Mishra S, Mohanty AK, Drzal LT, Misra M, Parija S, Nayak SK (2003) Studies on mechanical performance of biofibre/glass reinforced polyester hybrid composites. Compos Sci Technol 63, 1377–85
14. Joseph S, Sreekala MS, Oommen Z, Koshy P, Thomas S (2002) A comparison of the mechanical properties of phenol formaldehyde composites reinforced with banana fibres and glass fibres. Compos Sci Technol 62,1857–1868
15. Ornaghi Jr, HR, Silva HSP, Zattera AJ, Amico SC (2011) Hybridization effect on the mechanical and dynamic mechanical properties of curaua composites. Mater Sci Eng A 528, 7285–9
16. Gregorova A, Hrabalova M, Kovalcik R, Wimmer R (2011) Surface modification of spruce wood flour and effects on the dynamic fragility of PLA/wood composites. Polym Eng Sci 51, 143–50

A Novel Design and Development of Low-Cost Electro-Chemical Machining Unit with Optimized Parameters

M. Armstrong, M. Sivaprakash, M. Sivasubramanian, J. T. Winowlin Jappes, C. Rajaganapathy, and S. Ram Kumar

Abstract The electrochemical machining technique paved the way for a new approach of material removal mechanism through electrochemical reactions between the workpiece and the tool. It is a method for creating a variety of profiles without using mechanical forces between the work and the tool, using optimized input parameters. Since the size and expense of the conventional electrochemical machining equipment are so high, it could not be made available to everyone who wants to work under it. As a result, with fewer components and low-cost electrolytes, a novel design and fabrication of the electrochemical machining unit was produced in this work. The device is small and can conduct machining operations such as blanking, surface polishing, and generating various profiles by adjusting various parameters. To develop a square blank on the surface of the work piece, the unique ECM unit was used with high-speed steel (HSS) as a tool, an aluminium block as a workpiece, and salt water as an electrolyte. It functioned similarly to standard ECM machines in that it created the desired dimension on the work piece with an excellent surface quality. It is primarily intended for low-level machining and laboratory applications.

Keywords Electrochemical machining unit · Design and fabrication · Compact · Low cost

M. Armstrong (✉) · M. Sivaprakash · J. T. Winowlin Jappes
Department of Mechanical Engineering, Kalasalingam Academy of Research and Education, Krishnankoil—626126, Srivilliputhur, Tamilnadu, India
e-mail: mecharmstrong@gmail.com

M. Sivasubramanian
Department of Automobile Engineering, Kalasalingam Academy of Research and Education, Krishnankoil—626126, Srivilliputhur, Tamilnadu, India
e-mail: m.sivasubramanian@klu.ac.in

C. Rajaganapathy
Department of Mechanical Engineering, Government College of Engineering Srirangam, Trichy—620012, Tamilnadu, India

S. Ram Kumar
Department of Mechanical Engineering, Sri Krishna College of Technology, Kovaipudur, Coimbatore, Tamilnadu, India

N. M. Sivaram et al. (eds.), *Advances in Manufacturing, Automation, Design and Energy Technologies*, Lecture Notes in Mechanical Engineering,
https://doi.org/10.1007/978-981-99-1288-9_27

1 Introduction

Electrochemical machining is a technique for removing material using electrochemical reactions. This type of machining technique is used to machine hard materials that cannot be machined using traditional methods. At the micro-level, this approach can be used to create small pits or cavities in intriguing materials. In this technique, the tool serves as a cathode, while the workpiece serves as an anode. The material removal process occurs as a result of the energy suspension through the anode, and the workpiece gets cut according to the topographies of the cathode [1]. Electrolytes such as sodium chloride and sodium nitrate aqueous solutions, which are non-toxic and ecologically friendly, are used in the electrochemical machining process [2]. High current is supplied through the tool apparatus and the workpiece when an electrolyte is passed through the gap known as standoff distance between the tool apparatus and the workpiece.

The tool and the workpiece are not in direct contact, rather they are kept apart by a millimetre space, which is called the standoff distance (SOD). As electrons pass between the workpiece and the substrate, it is chemically broken down. The electrolyte is given a controlled stream. Due to the speed of the electrolyte, the material being eliminated is flushed away. This maintains an assortment of eliminated materials from accumulation. In this manner, the tool encounters no wear or dimensional changes. The electrolyte has again flow back to the electrolyte storage tank. In the current scenario, researchers are focusing their efforts on constructing a compact ECM machine with smaller electrodes and smaller space between the electrodes than the traditional design. Because it allows for the machining of complicated features and these advancements might shape the future of ECM design technology [3]. ECM machining technology is becoming increasingly used in a variety of fields, including electronics, automotive, medicine, and biomedicine [2–8]. Advancement in ECM also very much needed to improve the machining ability of ECM machine [10]. The effects on electrolyte on machining process play a vital role in performing varied machining operation [11]. The electrochemical machining (ECM) unit is not compact and is not readily available for machining work in all locations. The installation of conventionally accessible ECM devices necessitates significant financial expenditure. Due to the use of metal panels, replacing and repairing existing traditional ECM machines require a larger expenditure. In traditional ECM machines, high-cost electrolyte and tool materials were used. The recent development on electro-machining unit with incorporating and varying many factors like easily available and low-cost electrolytes, electrodes, micro-level machining in extreme conditions, micro-drilling was studied [12–17]. In this work, the new design and construction of a low-cost ECM machine with the goal of solving the above challenges were studied.

2 Materials and Methods

The electrochemical machine unit comprises hydraulic systems, mechanical feed systems, and electrical connections. The following is the discussion on the materials, methods, and design process in considerations with various factors.

2.1 *Hydraulic System*

A hydraulic system is comprised of a flushing and pumping system. To eliminate the metal hydroxides that arise during the reaction, the metal electrolyte must be given at a certain rate. If the electrolyte does not move at a fast enough rate, the produced sludge will build up on the tool's surface, causing dimensional changes. The electrolyte is pumped into the system with the help of two diaphragm pumps. The machine contains two distinct jets for circulating the electrolyte between the tool and the electrolyte, as well as two suction tubes connected to the pumps.

Each jet has the same inner diameter and can be operated independently. These two jets each have their own velocity valves for adjusting the electrolyte velocity at the jet's exit. The hydraulic and flush systems are schematically depicted in Fig. 1. The pump's inlet and outflow are separated into two sections. The suction tubes and flush tank are attached to the pump's input, while the velocity valve (which is connected to velocity controller tubes) and jets are connected to the pump's outlet. A flush valve regulates the flushing system. The flush tank is filled with normal tap water. Whenever the machining of different materials is to be achieved, the pumping system should be flushed by opening the flush valve, which could remove the stains of the previous electrolyte used and exist inside the walls of the tube. By adjusting the opening of the velocity valves, six different values of velocity can be obtained. From Table 1, the velocity can also be increased and decreased by increasing and decreasing the diameter of the jet.

2.2 *Velocity and TDS of the Electrolyte*

Another major controlling boundary is the speed of the electrolyte, which controls the clearance rate of the material from the surface; if the velocity is slowed down, the machined material will get deposited on the tool apparatus, which decreases the life of the tool. Stored material will likewise influence the response, subsequently bringing about no uniform material removal on the surface of the work piece during the surface machining process.

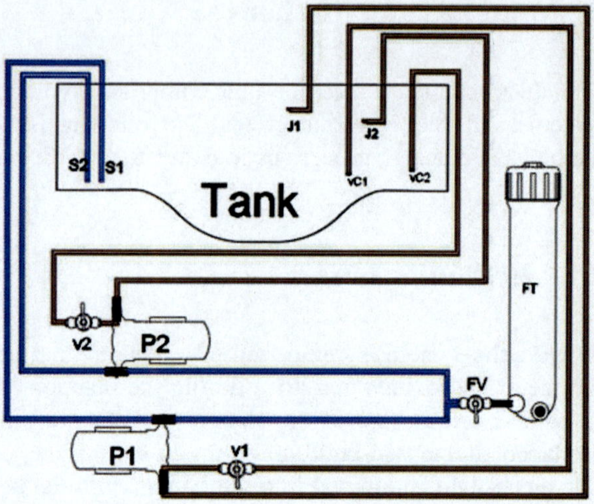

Fig. 1 Schematic layout of the hydraulic and flush system
P1—pump 1, 2. P2—pump 2, v1—velocity valve 1, v2—velocity valve 2, FV—flush valve, FT—flush tank, S1—suction tube 1, S2—suction tube 2, J1—jet 1, J2—jet 2, vC1—velocity changing tube 1, vC2—velocity changing tube 2, blue—inlet connections, and brown—outlet connections.

Table 1 Hydraulic system specifications

Inlet pump	Velocity [mm/s]
Maximum pressure 150 Psi Inlet velocity 36 V DC Diameter of jet 4 mm	Velocity 1—15.287 Velocity 2—13.9246 Velocity 3—12.607 Velocity 4—11.17 Velocity 5—10.17 Velocity 6—8.69 Velocity 7—8.299
Outlet pump	**Velocity [mm/s]**
Maximum pressure 150 Psi Inlet velocity 36 V DC Diameter of jet 4 mm	Velocity 1—16.69 Velocity 2—13.227 Velocity 3—12.607 Velocity 4—10.977 Velocity 5—8.66 Velocity 6—7.9189

2.3 *Standoff Distance (SOD)*

SOD is the distance between the tool apparatus and the workpiece. SOD decides the material evacuation rate, regularly known as the MRR. If the SOD is ideal, then the MRR will be in an ideal way. The range of SOD typically lies between 1 and 6 mm, where if the SOD is less than 0.9 mm, the circuit shorts and results in dark spots, whereas assuming the SOD is more than 6 mm, the material removal rate rates down to nothing and influences undesirable zones. In any case, the ideal value of SOD fluctuates for various materials. The ideal value of SOD is determined by the experimentation technique.

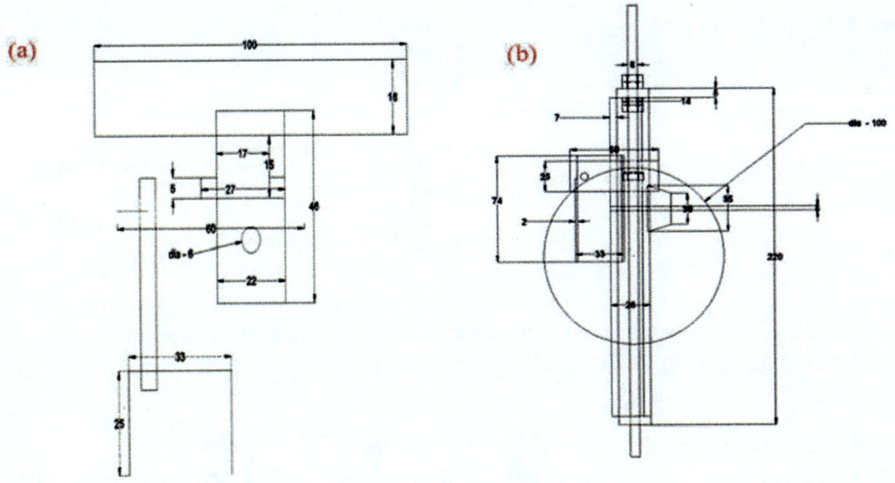

Fig. 2 (**a**) Top view and (**b**) front view of the feed system

2.4 *Feed System*

The feed system is controlled manually, and it can be utilized for inclined profiles. It is controlled by a stainless-steel threaded rod which prevents corrosion in the feed system. Figure 2 depicts the top and front views of the feed system.

2.5 *Semiautomatic Feed System*

A servo motor is connected to the lead screw of the tool feed mechanism. It is controlled by a microcontroller and is programmed to rotate the shaft of the servo motor at a very slow rate. This rotation is connected to the feed system, coupled using a gear locking mechanism. Since, it can rotate only up to 180°, the return motion of the servo motor should not be transmitted to the feed system. To achieve this, a gear locking mechanism is used as coupling between the servo motor shaft and the feed screw. After completion of 180 degrees, the shaft of the motor will come to 0-degree anti-clockwise. The lock gear mechanism is designed to transmit the anti-clockwise rotation of the servo motor and at its centre, connected with the base plate, will lock the anti-clockwise rotation of the servo motor with the lead screw of the feed system. When the servo motor rotates in the clockwise direction, the locking arm will release the connection between the lead screw and the shaft of the servo motor (Fig. 3).

A rubber band is used to ensure the locking arm returns to its position. The microcontroller is uploaded with the programme by default, and the delay time is given as the input to the microcontroller. Different delay values will control the shaft at a different speed, so the movement of the tool can be controlled at different

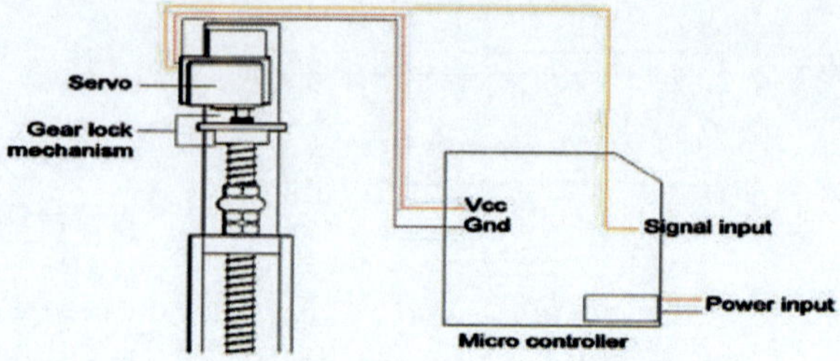

Fig. 3 Semiautomatic feed system

speeds. Since this system is semiautomatic, the tool must be placed in the home position manually (Fig. 4, Table 2).

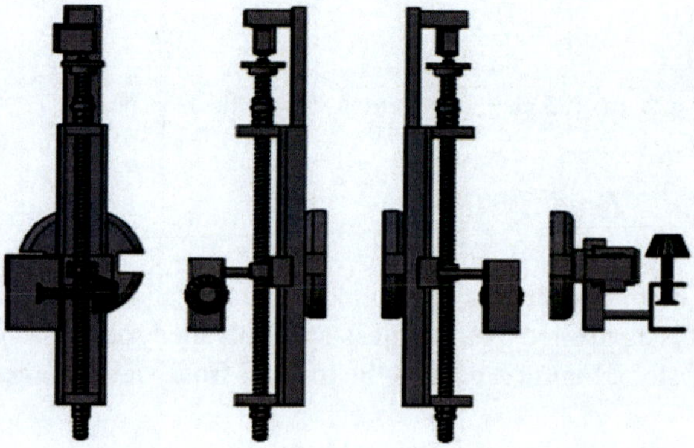

Fig. 4 3D models of the feed system

Table 2 Input delay vs time taken for the tool to travel 5 mm

Input delay value (microseconds)	Time taken for the tool to travel 5 mm
15	40[s]
30	1[min]
50	1[min] 30[s]
100	2[min] 30[s]
200	4[min] 30[s]
300	7[min]
400	8[min] 30[s]
500	10[min]
600	11[min] 30[s]
700	13[min]
800	17[min]
900	20[min]

2.6 Tank Design

The tank is made of transparent plastic material, resistive to any reaction to the electrolyte incorporates on the surface. The total volume of the tank is 1.4 L (Figs. 5, 6, 7, 8).

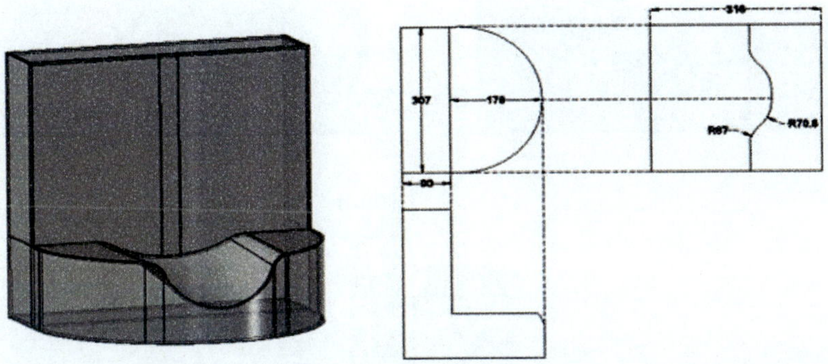

Fig. 5 Container design

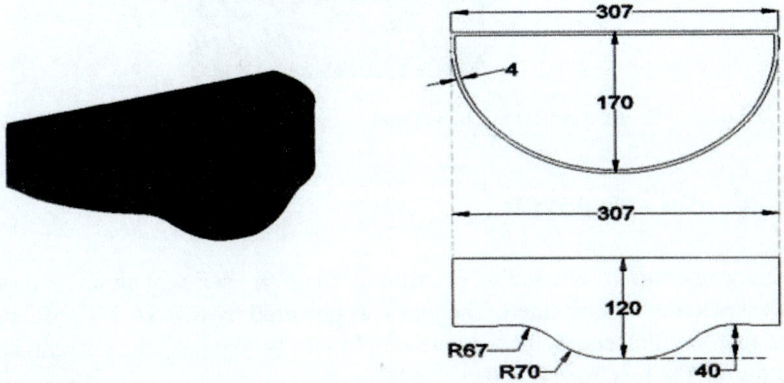

Fig. 6 Tank design

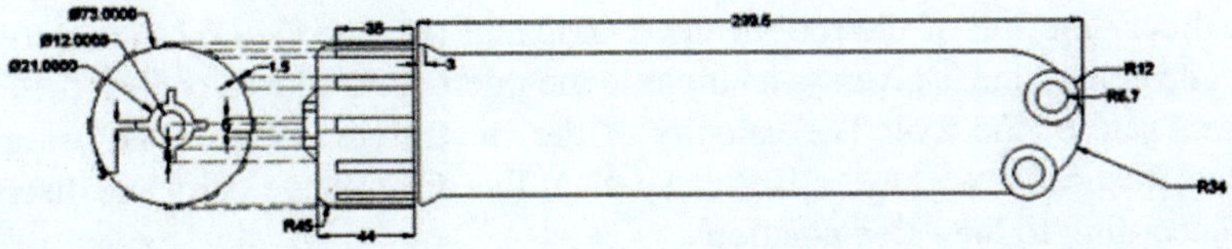

Fig. 7 Flush tank

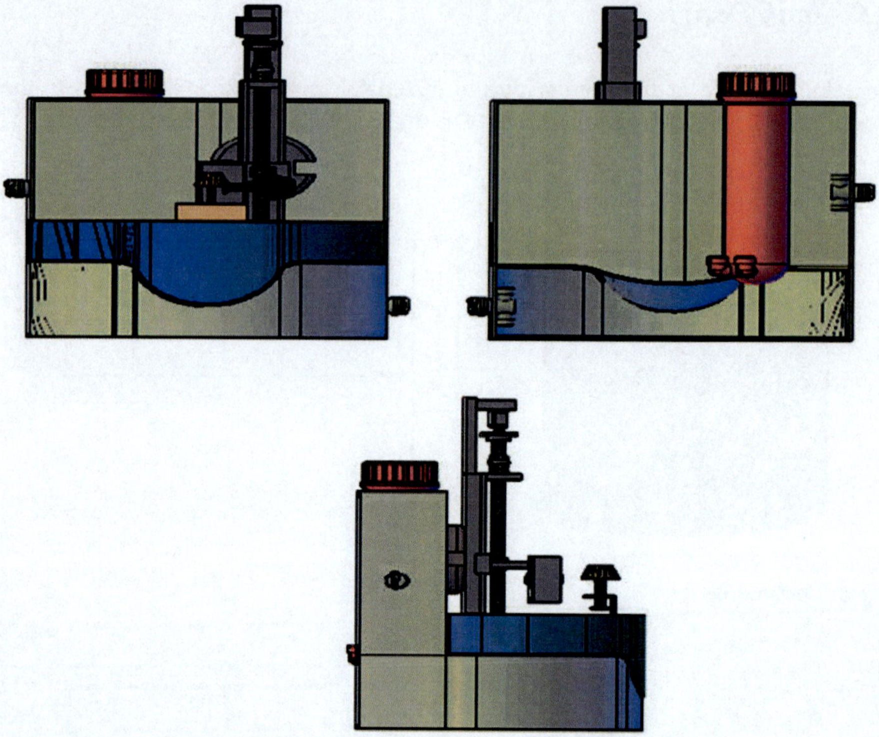

Fig. 8 3D model of electrochemical machining unit

3 Result and Discussion

The machine operates by circulating electrolytes through a series of pumps and valves to the workpiece to be machined. The unit was powered by two 5A 24 V AC to DC adapters that simultaneously power two diaphragm pumps and the current supply utilized during the reaction (Fig. 9).

The flush tank supplies desalinated water to clean the internals of the machine after the completion of every reaction to ensure the hydraulic circuit is clean from electrolyte and oxide depositions. The tool is fixed with a lead screw mechanism to achieve linear motion of the tool in the Y-axis and is controlled by the microcontroller. The delay time is given as an input to the microcontroller to control the linear movement rate of the tool. The velocity of the electrolyte is controlled using two individual flush valves. These valves can deliver five different velocity and discharge values according to the valve positions.

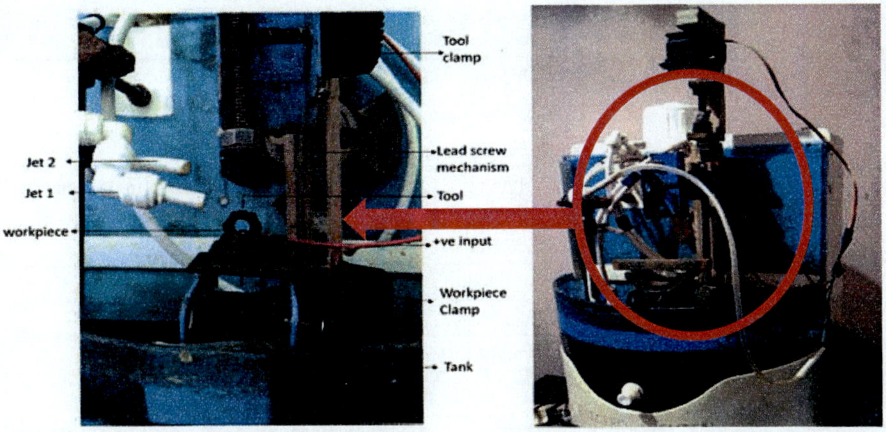

Fig. 9 Electrochemical machining unit

3.1 Input Parameters

The input parameters that had been fed in the low-cost ECM unit were.

- Velocity
- The temperature of the electrolyte
- The concentration of the solute in solvent—electrolyte
- Standoff distance
- Number of jets
- Diameter of the jets (maximum 4 mm)
- Input voltage
- Direct and reverse polarity

3.2 Electrochemical Machining of Aluminium Block

The aluminium block is taken as a workpiece, and it is machined using the novel designed ECM unit. The tool material was taken as high-speed steel (HSS). The workpiece is masked using cello tape. The area exposed to the tool is left unmasked. The aluminium workpiece and the tool are connected to a positive charge and negative charge, respectively. This ensures the material removal rate (MRR) on the tool is negligible compared to the MRR of the workpiece. The electrolyte is prepared with a mixture of salt and water (Table 3).

The electrolyte is incorporated into the space between the tool and workpiece, and a DC supply is given. Due to electrochemical reaction, the unmasked portion of the workpiece is removed and the surface had been machined and a blank was formed.

Figure 10a, b depicts the aluminium block without and with blank machined using Novel ECM Unit. The following are the parameters involved in the blanking

Table 3 Specifications of the experiment performed

Content	Specifications
Workpiece material	Aluminium
Time of machining	12 min
Power source	24 V DC
Electrolyte used	NaCl salt 3.0 g in 450 ml tap water
Tool material	High-speed steel
Stand of distance	2 mm
Initial weight	13.7 g
Final weight (after machining)	13 g

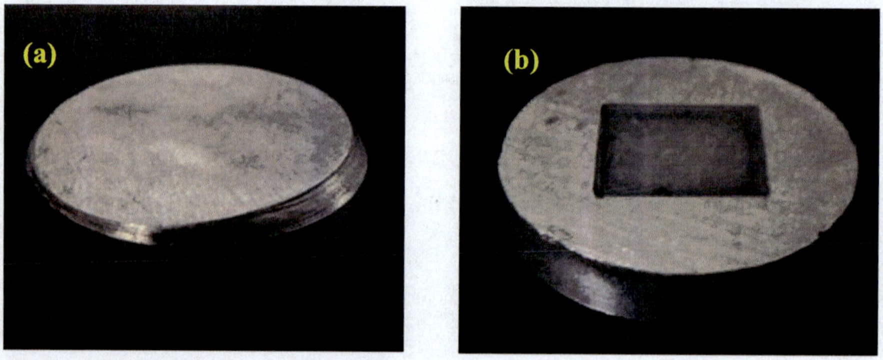

Fig. 10 (**a**) Workpiece before machining and (**b**) after machining using ECM unit

of aluminium block. The process of blanking a piece of aluminium block showed its best result in the electrolyte velocity range of 6–8 mm/s which took around 15 min to complete the required blank. Figure 11 shows the microscopic view of the before and after blanked surface.

4 Conclusion

The new compact electrochemical surface machining setup has been designed and fabricated with diverse components that are designed for handling dimensionally varying workpieces. Many modifications were made to make the machine flexible enough to surface finish workpieces that required different clamping methods. This unit mainly focused on reducing the use of metals for the structures to avoid corrosion issues. The feed system will be operated manually by precalculated rpm values. SOD will be constant and controlled using the thickness of the acrylic covers used in front of the tool. The unit is designed to be portable, with three jets with varying areas of the

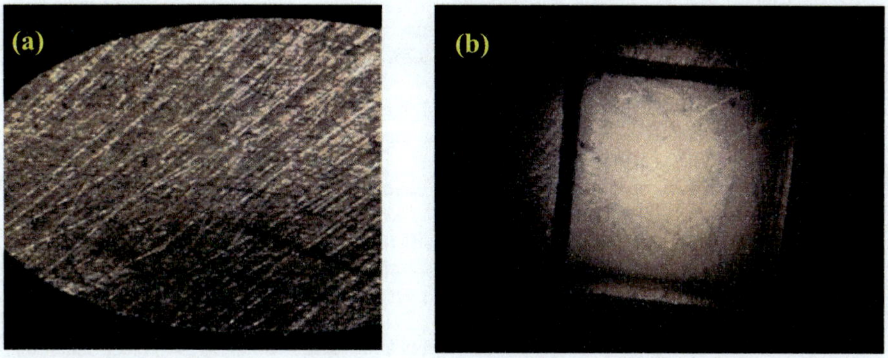

Fig. 11 Microscopic view of before (**a**) and after (**b**) blanked work surface

cross-section to adjust the velocity of the electrolyte. The surface finish made in the aluminium block shows the accuracy of the material removal with required dimension in less time without much manpower. Hence, this work provides a significant application for both industrial and academic purposes for machining small objects at a lower cost through electrochemical reaction process.

Acknowledgements The authors felt very much grateful for the Management and Department of Mechanical Engineering, Kalasalingam Academy of Research and Education for providing the fullest support in utilizing the department laboratories for fabrication and testing of the work.

Conflict of Interest The authors confirm that there is no conflict of interest.

References

1. Datta M, Harris D (1997) Electrochemical micromachining: An environmentally friendly highspeed processing technology. Electrochim Acta 42(20–22):3007–3013
2. Wei B (1994) Modeling and analysis of pulse electrochemical machining. The University of Nebraska– Lincoln
3. McGeough JA (1974) Principles of electrochemical machining. Chapman and Hall, London
4. Li Z, Ji H (2009) Machining accuracy prediction of aero-engine blade in electrochemical machining based on BP neural network. Proc Int Work Inf Secur Appl Academy Publisher, Qingdao, China, 9–12. ISBN 978-952-5726-06-0
5. Wang M, Zhu D, Qu NS, Zhang CY (2007) Preparation of turbulated cooling hole for gas turbine blade using electrochemical machining. Key Engg. Materials. 329:699–704. https://doi.org/10.4028/www.scientific.net/KEM.329.699
6. Datta M, Landolt D (2000) Fundamental aspects and applications of electrochemical microfabrication. Electrochim Acta 45:2535–2558
7. Kamaraj AB, Sundaram MM, Mathew R (2013) Ultra high aspect ratio penetrating metal microelectrodes for biomedical applications. Microsyst Technol 19:179–186. https://doi.org/10.1007/s00542-012-1653-3

8. Lu X, Leng Y (2005) Electrochemical micromachining of titanium surfaces for biomedical applications. J Mater Process Technol 169:173–178. https://doi.org/10.1016/j.jmatprotec.2005.04.040
9. Kalaimathi M, Venkatachalam G, Sivakumar M, Ayyappan S (2020) Multi-response optimisation of electrochemical machining process parameters by harmony search-desirability function optimiser. Int J Manuf Technol Manag. 34:4
10. Chan Jia Yuan, Ahmad Shahrun Hijrie Abu Bakar, Muhamad Najib Roslan, Cheah Weng Cheng, Muhammad Nadzmie Syafiq Mohd Rosekhizam, Jaharah A. Ghani, Zaliha Wahid (2021) Electrochemical machining (ECM) and its recent development. J Tribol. 28, 20–31
11. Saravanan KG, Prabu R, Venkataramanan AR (2021) Eden Tekle Beyessa: Impact of different electrolytes on the machining rate in ECM process. Adv Mater Sci Eng. https://doi.org/10.1155/2021/1432300
12. Spieser A, Ivanov A (2013) Recent developments and research challenges in electrochemical micromachining (μECM). Int J Adv Manuf Technol 69:563–581. https://doi.org/10.1007/s00170-013-5024-8
13. Fan ZW, Hourng LW, Lin MY (2012) Experimental investigation on the influence of electrochemical micro drilling by short pulsed voltage. Int J Adv Manuf Technol 16:957–966. https://doi.org/10.1007/s00170-011-3778-4
14. Newman J, Thomas-Alyea KE (2004) Electrochemical systems. 3rd ed. John Wiley & Sons, Inc., Hoboken, New Jersey
15. Spieser A, Ivanov A (2015) Design of an electrochemical micromachining machine. Int J Adv Manuf Technol 78:737–752. https://doi.org/10.1007/s00170-014-6332-311
16. Dharmalingam S, Florence Prasad, Ragupathy V, Satish Kumar P (2015) Machinablity study on Electrochemical Machining—A Review. Int J Chem Tech Res. 7(6), 2596–2600
17. Armstrong M, Sivasubramanian M, Selvapalam N (2021) Experimental investigation on the heat transfer performance analysis in silver nano-coated double pipe heat exchanger using displacement reaction. Materials Today: Proceedings. 45(2):2482–2490. https://doi.org/10.1016/j.matpr.2020.11.100

Influence of Turning Parameters on the Surface Roughness and Cutting Force of the Aluminum Matrix Hybrid Composites

P. Shanmughasundaram, S Thirumurugaveerakumar, and R Kalaiselvi

Abstract Turning operation was performed on Al LM 25/ /Fly ash (FA)/Steel powder (SP) hybrid composite specimens to measure the surface roughness (SR) and cutting force (CF). The tests were performed in the conventional lathe using tungsten carbide plain and textured tool inserts to study the performance of the micro-textured turning tool on the machining of Al /FA/SP hybrid composites. Spindle speed, depth of cut, and feed were chosen as operating parameters. The surface roughness of the composites deteriorates more rapidly at the higher feed rate. The cutting force increases with feed, cutting speed, and depth of cut. An increase in depth of cut tends to increase the chip area which causes additional cutting force thereby increasing the surface roughness of the surfaces.

Keywords Turning · Composite · Plain and textured tool inserts · Surface roughness · Cutting force

1 Introduction

Precision machining is gaining importance which has an influence on operational attributes like fatigue strength, creep strength, heat transfer rate corrosion resistance, and wear. Cutting velocity is to be chosen by considering the various parameters

P. Shanmughasundaram (✉)
Department of Automobile Engineering, Karpagam Academy of Higher Education, Coimbatore-641021, India
e-mail: sunramlec@gmail.com

S. Thirumurugaveerakumar
Department of Mechanical Engineering, Kumaraguru College of Technology, Coimbatore-641049, Tamilnadu, India

R. Kalaiselvi
Department of Information Science and Engineering, Kumaraguru College of Technology, Coimbatore-641049, Tamilnadu, India
e-mail: kalaiselvi.r.cse@kct.ac.in

N. M. Sivaram et al. (eds.), *Advances in Manufacturing, Automation, Design and Energy Technologies*, Lecture Notes in Mechanical Engineering,
https://doi.org/10.1007/978-981-99-1288-9_28

which include surface quality, tool life, productivity, and machine capacity in turning operation. Generally, higher velocity enhances material removal rate and productivity with relatively lesser operating time. Conversely, vibration, cutting force, and temperature increase with velocity. A larger depth of cut and higher feed rate enhance material removal rate which increases productivity and reduces the machining time and cost. But it increases vibration and degrades the surface finish. It also affects the tool life. Recent research findings concerned with the machining of materials explored, that textured tools have the ability to influence the tribological characteristics between the tool and workpiece interfaces. Various works have been undertaken by authors to enhance the machining behavior of the tools by making textures on the rake surface of the tool [1]. Developed the cutting model for turning of MMCs. Correlation between the chip compression ratio and the deformation of the chip or shear plane angle was obtained [2]. Optimized the level of parameters in turning of MMCs employing gray-fuzzy algorithm. They recommended the levels of machining parameters to enhance the MMR and drop the surface roughness and flank wear. Sekhar and Singh [3] reviewed the mechanism involved in the turning of MMCs. Aurich [4] analyzed the influence of the reinforcement phase on the roughness as well as the surface layer of the of Al-MMCs experimentally [5]. Studied the machinability of AA356-SiC MMCs using medium-grade PCD 1500 inserts during turning. They reported that higher cutting speeds enhance the surface finish [6]. Reviewed the machining of aluminum MMCs. They reported that the desired properties can be obtained by optimizing process parameters [7]. Investigated the squeeze casted hybrid nano composite under different machining environments [8]. Optimized the level of turning factors of GFRP composites [9]. Evaluated the surface roughness using ANN and RSM methods in turning of SiC and graphite-reinforced hybrid AMCs.

Analyzed the influence of textured tools in turning of Ti–Al alloy [10]. It was reported that tool wear was found to be relatively lower with perpendicular texture orientation at a higher cutting speed [11]. Reported that tool life tends to increase by patterning on the tool rake portion and it reduces the both the time and cost. Lower cutting force and CoF caused less tool wear [12]. Reported the performance of the tools which have dimple-textured surfaces on the rake surface in the machining of alloys. They reported that micro dimples efficiently reduce the tendency of chip adhesion. Hence, the machining of MMCs is a key area of research, but limited works have been made on SR and CF in the machining of hybrid MMCs. The type of cutting tool and tool angle has a noteworthy effect on SR and CF in the turning of the materials. The selection of the cutting tool is an important criterion for obtaining better-machined parts. Proper care must be taken while selecting the cutting tools. In particular, while machining the metal matrix composites, the selection of tools depends on several factors like cutting conditions, matrix material, reinforcement, and properties of materials.

An attempt is made to get a better surface finish by optimizing the operational factors, feed, speed, and Depth of Cut (DoC) through the Taguchi technique in the machining of MMCs. The turning operation was performed on Al alloy LM 25–5wt% FA and 5 wt% steel particles (SP) hybrid composites specimens to measure the SR and CF.

2 Materials and Experimentation

AA LM25–5wt. %FA and 5 wt. % SP composites were chosen as the workpiece material. Squeeze casting was employed to produce composite materials in order to enhance wettability. Specimens were machined in the 50 mm diameter and 350 mm length cylindrical bars. Sumitomo, 13^0rake angle and 0.8 mm nose radius turning tool were used. The experimental study was made using a lathe which is shown in Fig. 1. Turning insert (plain) and turning insert (textured) are portrayed in Fig. 2 and Fig. 3, respectively. Micro textures were formed on the rake side of the insert through the laser machining process. An abstract of testing conditions is presented in Table 1. Mitutoyo SJ-210 was employed to measure the surface finish of the surface.

The thrust was measured employing Kistler piezoelectric dynamometer. Kistler piezoelectric dynamometer with data acquisition is shown in Fig. 4.

The operational parameters and their levels used for the turning operation were chosen after performing an extensive literature survey and trial operations. Turning operational factors and their values at three levels are given in Table 2. The cutting speed was computed from the measured speed of the spindle and the workpiece diameter.

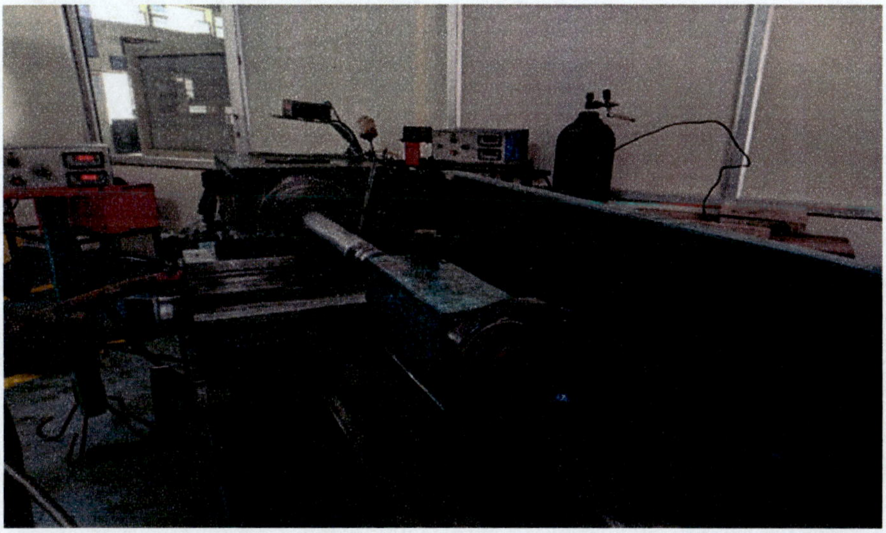

Fig. 1 Experimental setup

Fig. 2 Turning insert (Plain)

Fig. 3 Turning insert (textured)

3 Results and Discussion

Figure 5 infers that the surface roughness of the composites deteriorates more rapidly at the higher feed rate. A little quantity of the reinforcement particles which are present on the surface of the workpiece is partly or fully removed during turning and increases the surface roughness. Moreover, reinforcement particles have an opportunity to attach themselves at the cutting edge of the insert due to higher interface temperature. It leads to built-up edge formation and increases the roughness. When

Table 1 Abstract of experimental conditions for turning

Item	Specification
Lathe	Center lathe
Model	Kirloskar, Turnmaster -35
Workpiece	Bar size: Φ 50 mm × Machining length 350 mm
Spindle motor	3 hp DC Compound motor
Feed motor	1 hp DC Compound motor
Tool holder	PCLNL -2020 K12Square shape (length 120 mm, 20 mm width, 20 mm thick)
Tool inserts used	Tin-Coated Tungsten carbide (Plain & Textured) Tool radius 0.8 mm
Spindle speed (rpm)	60–5000
Dynamometer	Kistler Multi Component dynamometer (Type 5297B), Charge amplifier 5070A

Fig. 4 Kistler piezoelectric dynamometer with data acquisition

Table 2 Turning operational parameters and their values at three levels

Level	Feed rate (mm/rev) (A)	Depth of cut (mm) (B)	Cutting speed (m/min) (C)
I	0.075	0.2	54.95
II	0.11	0.6	109.9
III	0.15	1.0	164.85

the speed increases beyond 109.90 m/min, roughness increases, caused by higher interfacial temperature between the tool and workpiece and vibrations.

The cutting force of particulate composites during turning is affected by machining factors. Figure 6 shows the cutting force obtained in the turning operation of composite material for three different variants of feed, speed, and depth of cut. Figure 6 depicts that the cutting force increases as the feed, cutting speed, and depth of cut increases.

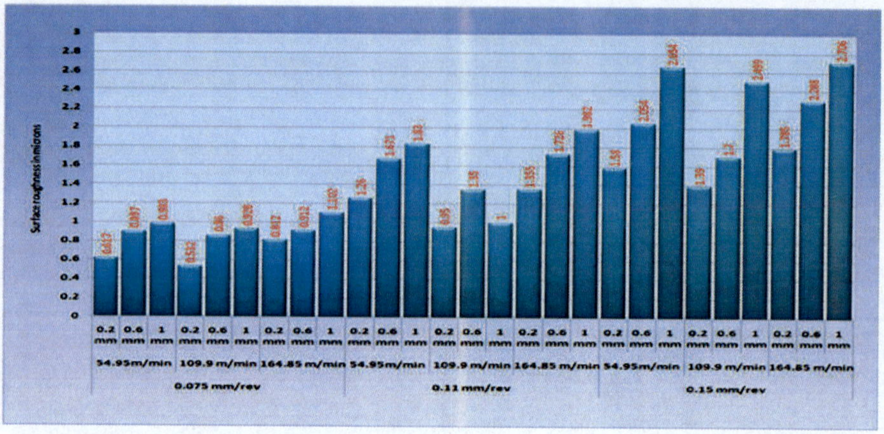

Fig. 5 Surface roughness values of composite material during turning operation for three different variants of feed rate, depth of cut, and cutting speed

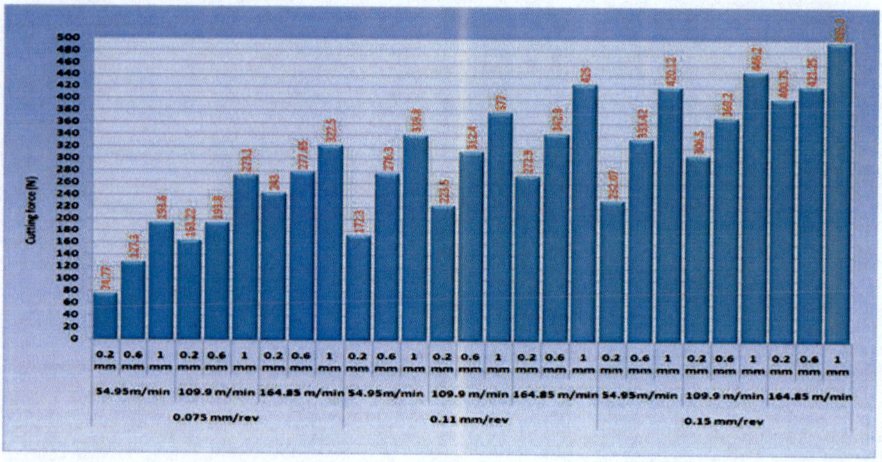

Fig. 6 Cutting force values of composite material during turning operation for three different variants of feed, depth of cut and cutting speed

3.1 Taguchi Design of Experiments

The “smaller is better” S/N ratio is chosen to foresee the optimum parameters as lower surface roughness and cutting force of the specimens were needed.

The impact of factors such as feed rate, depth of cut, and cutting speed on the SR and CF has been analyzed using Taguchi L27 orthogonal array. It can be inferred that the feed rate is a foremost parameter followed by depth of cut, and finally cutting speed on the SR and CF of the tested specimens.

Response diagram of S/N ratio for SR and cutting force of the composite are shown in Figs. 7 and 8, respectively. Figure 7 showed that 109.90 m/min cutting speed is the optimal speed in obtaining the minimum surface roughness. When the speed increases beyond 109.90 m/min, roughness tends to increase due to higher interfacial temperature between the tool and workpiece, vibrations. The optimum depth of cut and feed rate was 0.2 mm and 0.075 mm/rev, respectively. Feed rate is the main factor followed by the DoC and cutting speed. Similar results were in obtaining the lowest cutting force during the turning of the material (Fig. 8).

The impact of textured and non-textured turning tool on the surface finish, SR values were plotted by keeping the optimum feed of 0.075 mm/rev, DoC 0.2 mm, and speed of 109.90 m/min.

Figure 9 displays the variation of SR of the surface produced by textured and non-textured in turning of composite. SR observed with textured tool produced was much lesser than that of the non-textured insert. The improvement was noted to be approximately 25%. Textures that are formed on the tool rake prevent adhering the chips on the tool by diffusion during turning. It also prevents the built-up edge formation which enhances the surface finish. It also facilitates lubrication at the

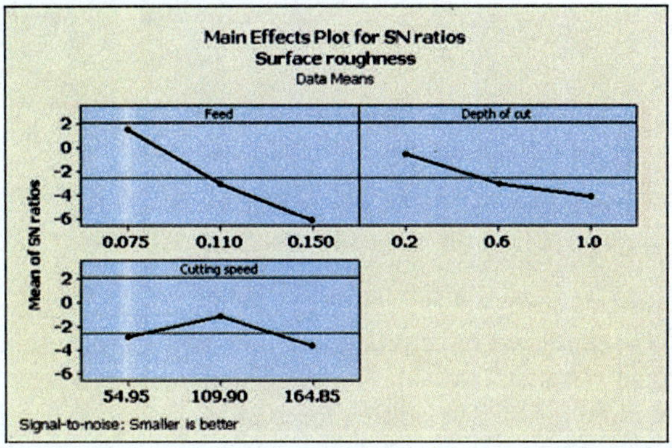

Fig. 7 Response diagram of S/N ratio for surface roughness of composite

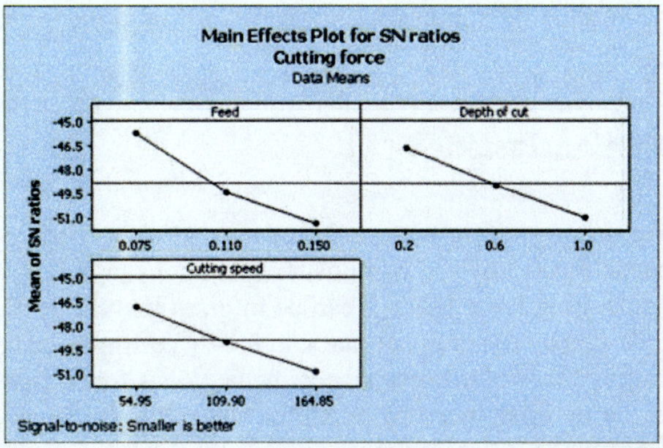

Fig. 8 Response diagram of S/N ratio for cutting force of composite

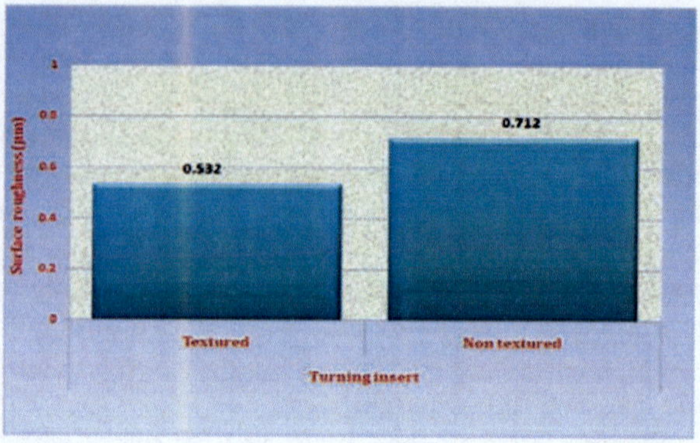

Fig. 9 Effect of turning insert on SR

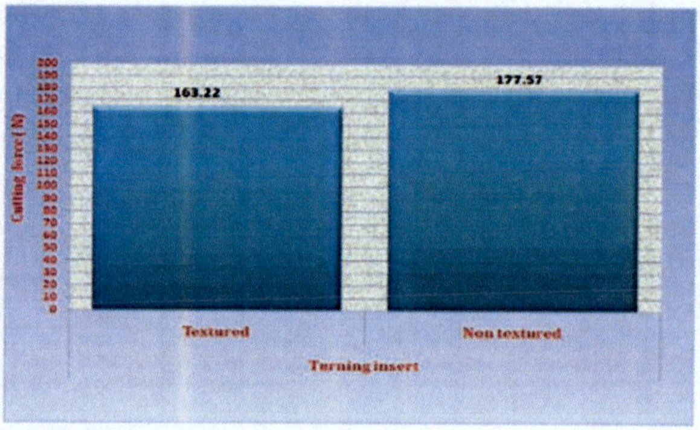

Fig. 10 Effect of turning insert on cutting force

interface and reduces friction drastically. Figure 10 illustrates the variation of cutting force produced by textured and non-textured in turning of composite. The cutting force observed with the textured tool produced was lesser than that of the non-textured insert. The cutting force produced by textured and non-textured in turning of the composite are 163.22N and 177.57 N, respectively. The cutting force produced by the non-textured tool is increased by 8.79% compared to the textured tool during the turning operation which was performed at the optimal conditions.

4 Conclusions

Turning operation was performed on Al alloy LM 25–5wt% FA and 5 wt% steel particles (SP) hybrid composites specimens to measure the surface roughness (SR) and cutting force (CF). It can be inferred that the feed rate is a foremost parameter followed by the depth of cut, and finally cutting speed on the surface roughness and cutting force of the tested specimens. It was found that 109.90 m/min cutting speed is the optimal speed in obtaining the minimum surface roughness. The optimum feed rate is 0.075 mm/rev. The lowest depth of cut (0.2 mm) reduces the surface roughness. SR observed with textured tool produced was much lower than that of the non-textured insert. The improvement was noted to be approximately 25%. Textures that are formed on the tool rake prevent in adhering the chips on the tool by diffusion during turning. It also prevents the built-up edge formation which enhances the surface finish. It also facilitates lubrication at the interface and reduces friction drastically. The cutting force observed with the textured tool produced was lower than that of the non-textured insert.

References

1. Davim JP, Conceição António C (2001) Optimal drilling of particulate metal matrix composites based on experimental and numerical procedures. Int J Mach Tools Manuf 41(1), 21–31
2. Suresh P, Marimuthu K, Ranganathan S, Rajmohan T (2014) Optimization of machining parameters in turning of Al-SiC-Gr hybrid metal matrix composites using grey-fuzzy algorithm. Trans Nonferrous Metals Soc China 24(9):2805–2814
3. Sekhar R, Singh TP (2015) Mechanisms in turning of metal matrix composites: a review. J Market Res 4(2):197–207
4. Aurich JC, Zimmermann M, Schindler S, Steinmann P (2016) Turning of aluminum metal matrix composites: influence of the reinforcement and the cutting condition on the surface layer of the workpiece. Adv Manuf 4(3):225–236
5. Muthukrishnan N, Murugan M, Prahlada Rao K (2007) Machinability issues in turning of Al-SiC (10p) metal matrix composites. Int J Adv Manuf Technol 39 (3-4), 211–218
6. Suthar J, Patel KM (2017) Processing issues, machining, and applications of aluminum metal matrix composites. Mater Manuf Process 33(5):499–527
7. Kannan C, Ramanujam R, Balan ASS (2017) Machinability studies on Al 7075/BN/Al2O3 squeeze cast hybrid nanocomposite under different machining environments. Mater Manuf Processes 33(5):587–595
8. Hussain SA, Pandurangadu V, Kumar KP (2015) Machining parameters optimisation in turning of GFRP composites by desirability function analysis embedded with Taguchi method. Int J Mach Mach Mater 17(2):95–107
9. Suresh P, Poongodi T (2018) Evaluation of surface roughness during turning of Al-SiC and Al-SiC-Gr composites. Multidiscip Model Mater Struct 14(5):874–890
10. Arulkirubakaran D, Senthilkumar V, Kumawat V (2016) Effect of micro-textured tools on machining of Ti–6Al–4V alloy: An experimental and numerical approach. Int J Refract Metal Hard Mater 54:165–177
11. Kim DM, Lee I, Kim SK, Kim BH, Park HW (2016) Influence of a micropatterned insert on characteristics of the tool-workpiece interface in a hard turning process. J Mater Process Technol 229:160–171
12. Sugihara T, Singh P, Enomoto T (2017) Development of novel cutting tools with dimple textured surfaces for dry machining of aluminum alloys. Procedia Manuf 14:111–117

Investigations on Modified Friction Welding Samples of SA-Grade Materials Used in Box-Type Heat Exchangers

A. Daniel Das and K. Thirunavukkarasu

Abstract This research investigation states that seamless ferritic and austenitic alloy (SA) boiler grade material were joined by a friction welding process implementing an external tool. This process was further investigated to blooming the advantages in the mechanical and metallurgical areas. In this process, capital joint strength was analyzed by taking the tensile strength test from nine welding samples. Three input parameters, such as the rotational tool speed, the extrusion of the tube, and the depth impression on the tube plate, were investigated in the study of capital weld interface strength. Meanwhile, the output constrain is considered as pull strength. Previously, the joint is made by welding by employing a supporting block at interference fit conditions. Finally, the optimum collective strength characteristics from this research area segregate 1073.5 N/m^2.

Keywords Friction Welding of Tube-to-Tube Plate using an External Tool · Taguchi Method · ANOVA · SA 213 · SA 387

1 Introduction

Even though there were many joining metals, the welding process carves its way in this innovative world. Industrial sectors such as power plants, automobiles, aerospace, and welding techniques were mainly valuable members as everyone knows that its applications categorized welding. Among those, friction welding plays an essential role in metal joining creatures [1]. In the twenty-first century, the solid-state welding process was prevalent in the engineering industries. The friction welding process by employing an external tool was discovered in 2006 [2]. After lots and lots of struggle and many changes, the process was patented in 2008. After this valuable invention, this process overcomes many disadvantages caused by the fusion welding process.

A. Daniel Das (✉) · K. Thirunavukkarasu
Department of Mechanical Engineering, Karpagam Academy of Higher Education, Coimbatore, Tamilnadu, India
e-mail: danielamaladas@gmail.com

N. M. Sivaram et al. (eds.), *Advances in Manufacturing, Automation, Design and Energy Technologies*, Lecture Notes in Mechanical Engineering,
https://doi.org/10.1007/978-981-99-1288-9_29

The solid-state welding process delivers no solidification cracks and porosity on the welded samples [3–7]. The comparison of optimization techniques such as genetic algorithm and artificial neural networks was compared by Senthil Kumaran et al. in 2011. At 2012, mechanical and metallurgical properties are being discussed by joining steel tube with Al tube plate. This method is most eco-friendly, and this was confirmed by comparing the TIG welding process introducing Al filler plate. He also explained the welding abilities by preparing different types of types, pin diameters, and tube diameters in 2013. Meanwhile, the projection with the top part was concentrated in 2014 because projection was the most important parameter for finding capital attachment strength by joining the aluminum plate and steel tube by this welding process for studying the feasibility of the welding parameters. Further, the research was concentrated on different projections. Senthil Kumaran et al., investigations were all about this solid-state welding process [8, 9]. This article was the first and foremost research work that concentrate tensile test and tube length on the steel tubes.

2 Experimental Investigation

2.1 Selection of Base Metals and Tool

In this study, SA213 as tube and SA 387 as tube plate are investigated for metal joining ability and microstructural characteristics, where SA means seamless ferritic and austenitic alloy a boiler grade material. An external tool is used to weld both the tube and the tube plate which have high heat resistant due to rich tungsten carbide [10–12]. The pin diameter was 15.9 mm. Hence, it was concluded as interference fit because the inner tube diameter was 16 mm. No mistakes happen when welding because the tool made up of tungsten carbide has very good mechanical properties that withstand high temperatures.

2.2 FWTPET Process

Friction welding using an external tool was done by the machine which was developed in-house. The tube and tube plate were arranged in a rigid position which was very easy for welding purpose. The workpiece fitted inside the cast iron cube, and the cube was firmed in-between the male and female vice. Tungsten carbide tool has a tool pin which acts as “ankle supported by cuff”, which means the material movement avoided during friction welding process. The tool was rotated at different levels of rpm. When it hits the workpiece, the rubbing force takes place and heat generated on the base metal. The timing of the welding was all about 150–180 secs, and finally, the tool rushed back to its own original position. After the welding process done

Fig. 1 Modified tensile testing machine

the samples are allowed for quenching. Figure 1 shows that the modified UTM for tensile testing of specimen.

3 Results and Discussions

Research values of this convincing parameters helps to detect optimized result. Among nine samples, the load was applied until the tube getting failure also considering the length of the tube at the time of pull tests. Figure 2 shows the tube failure due to the tensile test take from the UTM.

3.1 Significance of Taguchi Analysis

Tensile strength was considered as a resultant parameter. Experimented values were shown in Table 1. Also, the deformation of tube also considered practically to

Fig. 2 Fractured Sample after pull test

Table 1 Input and Output parameters of nine samples

Runs	Speed (rpm)	Projection (mm)	Depth (mm)	Tensile strength (N/m^2)	Tube length (Initially 30 mm)
1	750	0.5	0.3	384.5	30
2	750	1.0	0.6	529.5	30
3	750	1.5	0.9	674.5	30
4	1000	0.5	0.6	584.0	31
5	1000	1.0	0.9	729.0	31
6	1000	1.5	0.3	874.0	31
7	1250	0.5	0.9	783.5	32
8	1250	1.0	0.3	928.5	33
9	1250	1.5	0.6	1073.5	34.5

conclude the welding strength. By the experimental investigation, the most consequential parameter 1250 rpm tool rotational speed, 1.5 mm tube projection, and 0.6 mm depth of cut produces all about 1073.5 N/m^2 of tensile strength [12]. The deformation of the tube from the fixed end was 34.5 mm. Change in length of the tube was about 4.5 mm from its original position. Figure 3 shows the graph plot of response effects of signal-to-noise ratio. Already this signal was drawn with correspondence with the noise produced at the background. Table 2 indicates the ranking process of all the three convincing parameters. Tube projection helps the composition of tube mixed more with the tube plate. Table 3 shows the comparison between the theoretical calculation and the experimental investigation. Hence, there must be slight variation between the two methods. Machine errors were added into account on experimental investigation [13].

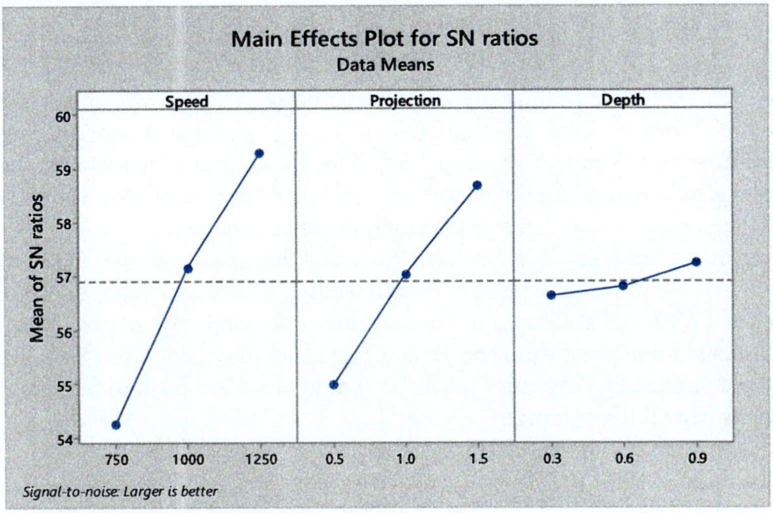

Fig. 3 Graph plot of Signal-to-Noise ratio

Table 2 S/N ratio ranking process

Level	Speed	Projection	Depth
1	54.25	54.97	56.63
2	57.14	57.03	56.81
3	59.28	58.68	57.24
Delta	5.03	3.71	0.61
Rank	1	2	3

Table 3 The comparison statement along with theoretical and experimental values

Method	Input parameters			Output parameter
	Speed mm	Projection mm	Depth mm ara>	Tensile strength N/m^2
Taguchi	1250	1.5	0.6	1073.5
Experimental	1250	1.5	0.6	1050.78

3.2 Regression Analysis

To violate machine errors regression analysis were used as a theoretical investigation. According to Darwin theory, the generations were better than their previous. So by analysis with the chromosome continuously, the output values are refined and give better and better results. For this experimental investigation, regression equation predicts the retarding output strength.

Tensile Strength = –359.0 + 0.7980 Speed + 290.0 Projection + 0.0 Depth

3.3 Statistical Analysis (ANOVA)

The statistical software performing the calculation of ANOVA which was used to find the inheritance percentage of output constrains [14]. Table 4 shows the details of the inheritance percentage by comparing all the three input values used to find out the capital joint strength. Figure 4 explains about the interaction plot which clearly denotes the graphical visualization of each process parameters.

Here, the percentage of inheritance denotes the speed of the tool rotation contributes more for being done a successful weld of 65.4%, whereas the projection contributes 33.8% and the depth of cut contributes 0.016%. By graphical preview, the inheritance for of error was zero. Hence, this concludes that perfect welding was done by the parameter 1250 rpm speed, 1.5 mm tube projection, and 0.6 mm depth of cut gives capital joint strength.

Table 4 Analysis of Variance for Clearance fit method

Source	DF	Adj.SS	Adj.MS	Percentage of contribution
Speed, rpm	1	238,801	238,801	65.4
Projection, mm	1	120,001	120,001	33.8
Depth, mm	1	6149	6149	0.016
Error	5	0	0	0.00
Total	8	364,951		100

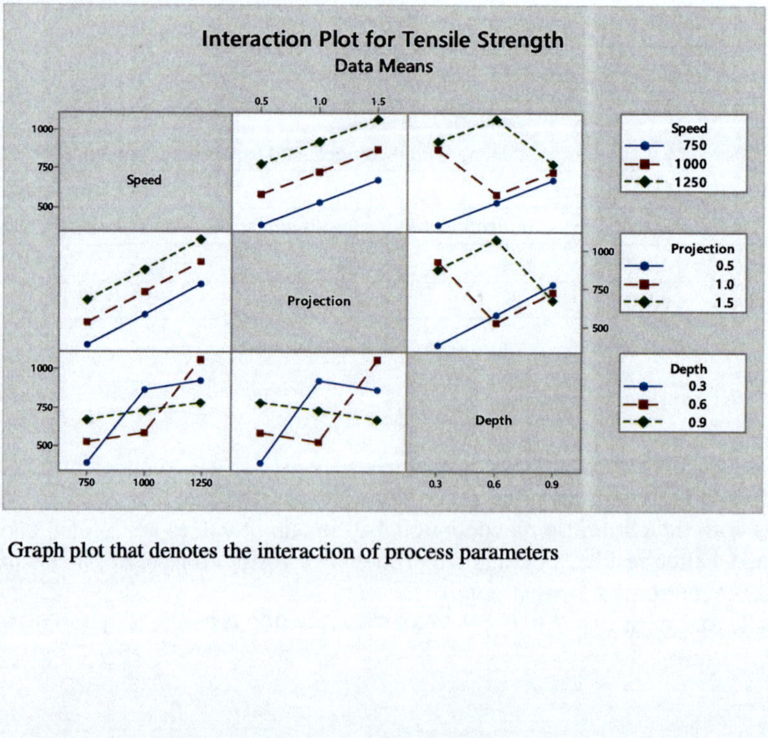

Fig. 4 Graph plot that denotes the interaction of process parameters

3.4 Mechanical and Metallurgical Investigations

Impact and Peel Test. After statistical variations, the samples were introduced to special characterization to prove the welding strength by both mechanical and metallurgical aspects. For mechanical testing, impact test was the test which analyze by applying mechanical load with a help of hammer. Table 5 shows the values calculated by impact izod test on the weld interface [15].

By impact test, the tool rotational speed of 750 rpm will produce maximum impact strength, which peel test was also another pull test which the test carries at the weld interface. The peel test shows that the tube failure at 1000 rpm rotational speed condition. Figure 5 shows the peel test samples and the tube failure at 1000 rpm condition.

Metal Bonding Analysis at Weld Interface. The microscopic analysis was considered as a metallurgical analysis which will help to find out the optimum joint characteristics. Figure 6 denotes the metallurgical investigation at 750 rpm rotational speed of WC tool.

The grains are refined at the weld interface to produce capital joint strength, whereas the base metal grain size was all about 36.9 micron at tube and 52.1 micron at the tube plate. At the weld interface, the average grain diameter was about 19.9 micron [16]. This shows better joint strength on the weld interface.

Scanning Electron Microscope with EDS. The scanning electron microscope image shows the intermolecular compound present inside the weld interface. Intermolecular compound was denoting as tungsten carbide. The compound helps in better bonding strength.

Table 5 Impact test tables

S.No	Conditions (rpm)	Interference fit energy lost/area (J/m^2)
1	750	59
2	1000	55
3	1250	50

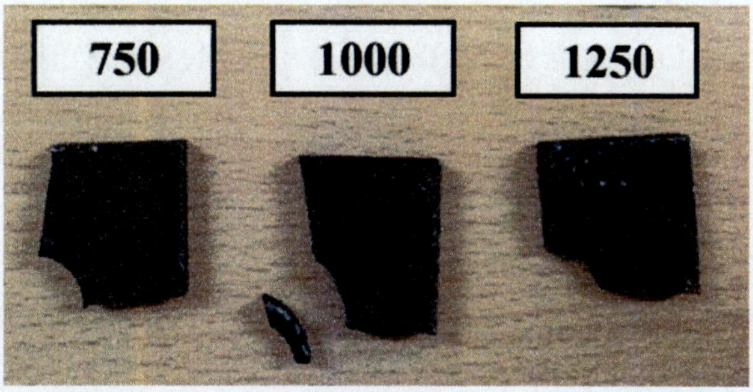

Fig. 5 Peel test samples with tube failure at 1000 rpm

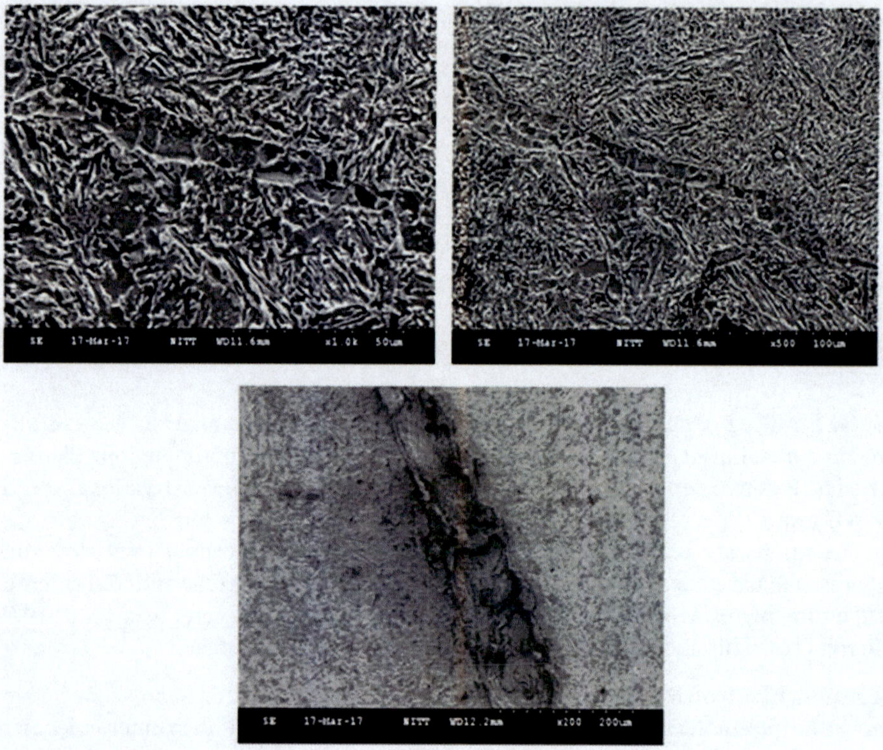

Fig. 6 SEM images with various magnifications on weld interface at 750 rpm condition

Moreover, at 750 rpm, the speed of the tool was slow, and this results poor generation of heat. So the bonding at the weld interface was little low. Same case follows at the 1000 rpm tool rotational speed. Hence, at 1250 rpm, enormous heat produced, and the base metal turns into plastic form. Finally, this achieves better bonding. Figure 7a, b, and c show the EDS along with SEM for 750 rpm, 1000 rpm, and 1250 rpm, respectively. From Fig. 7d, the sample confirms that level of compounds like chromium, magnesium, etc., are high and that provides maximum corrosion resistance.

Metal Analysis by XRD. Figure 8 clearly shows that the XRD analysis graph taken from the weld interface and there was no speck detected at welding samples clearly give solid-state welding produced optimum joint strength characteristics [17].

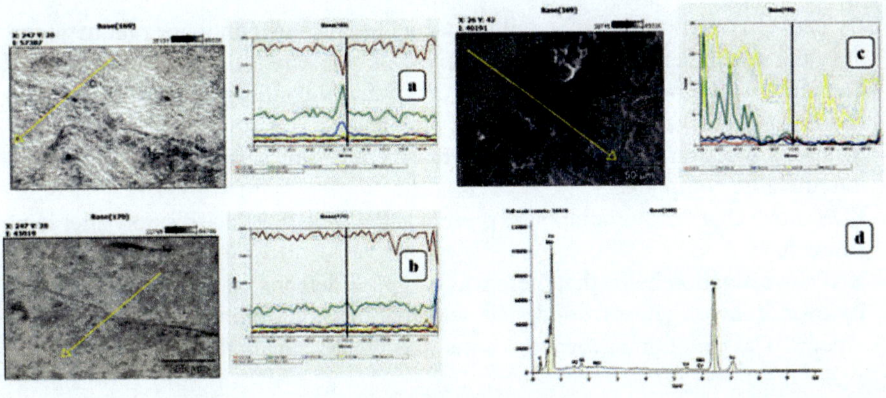

Fig. 7 Scanning Electron Microscope with EDS at **a** 750 rpm **b** 1000 rpm (**c**) 1250 rpm **d** EDS of sample provides maximum tensile strength

Fig. 8 XRD graph plot at 1250 rpm tool rotational speed condition

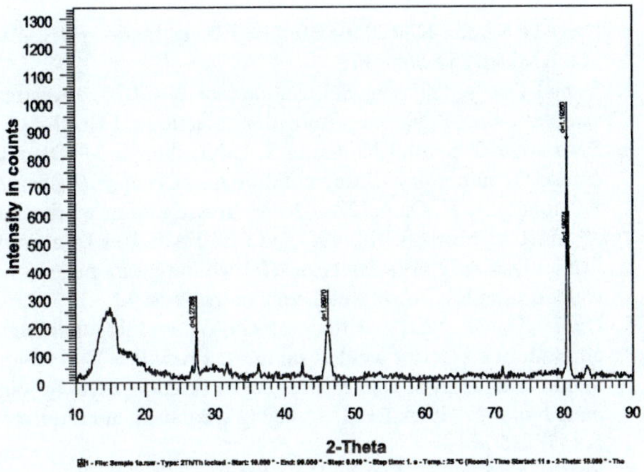

4 Conclusions

New implementation in this research article was tensile strength was considered as an output parameter.

- As a result, the convincing parameters (Speed 1250 rpm, Projection 1.5 mm, and Depth 0.6 mm) produced almost 1073.5 N/m^2. Hence, it was compared with an experimental investigation value was about 1050.78 N/m^2. There must be slight variation because of machine errors.
- Among all the three-input parameter, tool rotational speed contributes more than other two constrains.
- Impact strength produced 59 J by izod method at 750 rpm condition than the other two, because at tool rotation disturbed the molecules at minimum level

- By peel test, at 1000 rpm tube failure occurs because by continues production of heat, the molecules disturbed moderately.
- Grain size at the weld interface proves optimum joint strength, because the grain is refined at the weld interface than the base metals.
- Scanning electron microscope shows that better joint happens and there is no intermetallic compound.
- EDS shows that ferrous compound present at the weld interface results better joint strength.
- XRD investigation helps that there was no speck detected at the weld interface.
- By overall investigations, solid-state welding gives optimum joint strength in all aspects. The metal flows towards center axis results a capital joint strength.

References

1. Das AD, Senthil KS (2016) FWTPET Investigation on SA213 Tube to SA387 Tube Plate. Appl Mech Mater 852:355–361
2. Daniel Das A, Vijayan SN, Subramani N (2020) Investigation on welding strength of fsw samples using Taguchi optimization technique. J Crit Rev 7:179–182
3. Sivakumar S, Senthil Kumaran S, Uthayakumar M, Daniel DA (2018) Garnet and Al-flyash composite under dry sliding conditions. J Compos Mater 52:2281–2288
4. Karuppasamy R, Daniel DA (2019) Investigations on mechanical properties of squeeze casted Al MMC reinforced with TIC and BN. J Adv Res Dyn Control Syst 11:1086–1092
5. Daniel Das A (2019) Effect of TIG welding parameters on mechanical properties of Al6063 welded samples. J Adv Res Dyn Control Syst 11
6. Daniel Das A (2019) Effect of GTAW welding parameters on mechanical properties of aluminium six-series welded samples. J Adv Res Dyn Control Syst, 11
7. Senthil Kumaran S, Daniel DA (2018) Friction welding joints of SA 213 tube to SA 387 tube plate boiler grade materials by using clearance and interference fit method. Mater Today Proc 5:8557–8566
8. Senthil Kumaran S, Daniel DA (2018) An investigation of Boiler Grade Tube and Tube Plate without block by using friction welding process. Mater Today Proc 5:8567–8576
9. Senthil Kumaran S, Daniel DA (2018) An examination of seamless ferritic tube and austenitic alloy tube plate joining by friction welding process. Mater Today Proc 5:8539–8546
10. Daniel Das A TK (2020) Multi criteria decision making investigation on friction welded samples of boiler grade materials using topsis method. Test Eng Manag, 83.
11. Das A, Daniel, Thirunavukkarasu K (2021) Parametric optimization of FWTPET for optimum tensile strength by genetic algorithm approach. In: Materials Today: Proceedings 37: 1571–1577
12. Elango D, Daniel Das A, Kumaresh Babu SP, Natarajan S, Yeshitla A (2021) Electrochemical studies of WC-Flyash HVOF coating interface on SA209-T1 Steel under 3.5 NaCl Solution. Adv Mater Sci Eng 2021
13. Das A Daniel, Karuppasamy R (2021) Optimization on hardness values of FSW samples for Al 7–Series & AA 6–Series samples using Taguchi method. In: Materials Today: Proceedings 37: 596–599
14. Sathishkumar D, Daniel Das A (2021) Investigations on effect of process parameters on GTAW of aluminium alloy welding using full factorial design technique. In: Materials Today: Proceedings 37: 621–626

15. Das A Daniel, Thirunavukkarasu K (2020) Metal joining technique by FWTPET process of seamless ferritic and austenitic alloy grade materials with a peripheral support. In AIP Conference Proceedings, 2270(1), p 130002. AIP Publishing LLC
16. Thirunavukkarasu K, Kavimani V, Gopal PM, Daniel Das A (2020) Recovery and recycling silica flux in submerged Arc Welding–Acceptable properties and economical correlation. Silicon 1–10
17. Das A Daniel, Thirunavukkarasu K (2021) Investigation on boiler grade tubes and tube plate using FWTPET process assisted with PCA. In IOP Conference Series: Materials Science and Engineering, 1059(1), p 012066. IOP Publishing

Sustainable Turning of 6063 Aluminum Alloy in Dry Condition Using Gray Relational Analysis

A. Kannan, S. Sivakumar, P. Balasundaram, and N. M. Sivaram

Abstract The aim of this work was finding the best input parameters while turning of 6063 aluminum alloy for minimizing the machining time and also for maximizing the material removal rate simultaneously. L_9 design was accomplished to develop target functions to be improved within the selected design of experiments. Turning process was done in dry condition to ensure the sustainability. The multi-objective optimization was done using grey relational analysis. In this method, two objective functions were transformed into only one objective function in the form of GRG, thus simplifying the complex engineering problems into simple one. Optimum results were identified as $V_3F_3D_3$ using GRA for better MRR and less machining time. Optimal results were verified through additional experiments. Comparing the initial parameter settings, the optimal machining parameters resulted in, 93.75% improvement in MRR and 88.28% reduction in machining time were found using GRA optimization. From the results, it was understood that, the correct selection of machining factor produce the enhanced chip removal and reduction in machining time.

Keywords Sustainable machining · Material removal rate (MRR) · GRA

1 Introduction

Making a required product into desired shape and size, we have different manufacturing practices. Out of which, machining plays an important role to produce many engineering components. Turning process is very significant and fundamental

A. Kannan (✉) · S. Sivakumar · P. Balasundaram · N. M. Sivaram
Department of Mechanical Engineering, National Institute of Technology Puducherry, Karaikal, India
e-mail: kannanmeed@gmail.com

S. Sivakumar
e-mail: shivalesprit@gmail.com

P. Balasundaram
e-mail: pbalasundaram@ymail.com

N. M. Sivaram et al. (eds.), *Advances in Manufacturing, Automation, Design and Energy Technologies*, Lecture Notes in Mechanical Engineering,
https://doi.org/10.1007/978-981-99-1288-9_30

Fig. 1 Experimental work-turning process

machining process. Unwanted material is removed from the original material using single point cutting tool in the condition of metal chips from a turning cylindrical component to achieve the desired shape and size [1, 2]. Many parameters are there in turning process which is the reason for concluding performance. The main and known factors are: Speed, feed, depth of cut, cutting tool, cutting tool nomenclature, and workpiece hardness. The authors mentioned that turning may be used in machining of most of the metals, non-metals also [2, 3]. Turning is the main machining method in most of the manufacturing premises and has been applied in the creation of many cylindrical components [2, 4, 5]. The cutting speed (v) is specified as the peripheral speed of the component. The depth of cut (d) is the advancement of cutting insert tool perpendicular to the axis of the workpiece. The feed rate is the advancement of the insert tool along the length of the component, or advancement of the tool similar to the axis of revolution of the workpiece [1, 2]. Figure 1 shows the practical work of the turning process. In lathe machine, an electric power is converted into mechanical power to do the machining operation in terms of chip removal [6, 7].

Usually, the anticipated machining factors are considered based on familiarity or by handbook [6]. Nevertheless, it may not guarantee for the designated machining factors ensure best or near optimum machining condition for a selected machine or an atmosphere. The final output of a turning process could be represented by cutting tool life, machining duration, expenses, chip removal, and surface finish [2, 8]. These outputs can be classified into three groupings: Product quality, production, and financial responses. Since, 6063 aluminum alloys are widely used in small-scale industries for producing varies components through turning operation, productivity is the main concern for them in achieving their daily target. The product quality has encompassed by the surface finish. But, production has characterized mainly through the material removal rate and machining time [8, 9]. So the aim of this study was estimating productivity and machining time while machining 6063 aluminum alloy for higher productivity with less machining time using gray relational analysis.

2 Practical Machining Work

All machining trails have done on an experimental lathe setup Kirloskar turnmaster-35 all geared lathe as we see in Fig. 1. The capacity of the motor is 2.2 kW. Cutting tool used was uncoated tungsten carbide THN SNMG08 inserts made by Sandvik. Cutting tool frame of a description of DBSNR was used to clamp the tool. 6063 aluminum alloy cylindrical bar of 320 mm length and 60 mm diameter have used for component material. Response machining time was calculated using stopwatch for all experiments. The chip removal rate was computed using Eq. (1) [2]. Cutting speed in (m/min), feed rate in (mm/rev) and depth of cut in (mm). MRR was obtained in (cm^3/min) unit here.

$$MRR = vfd \tag{1}$$

When the number of factors and levels are increases, the number of trial turns also increases drastically. If the practical trail turns are more it will increase the research cost and will consume more time. So, for negotiation of these antagonistic effects we have considered only the restricted amount of trial runs of design of experiments (DOE) consisting of nine collections of statistics were designated to find the best conditions in the turning of 6063 aluminum alloy [7]. Trials were directed with the turning factors as per Table 1, to get the better material removal rate in 6063 aluminum alloy with reduction in machining time. All trails were done considering dry machining settings in line with a framework of sustainable machining approach [10]. The initial cutting parameters [11] were selected for the reference was trail number one in Table 2. The factors and their ranges for doing the turning process were designated as per the appropriate ISO normal, endorsements of the cutting insert tool and machine tool enterprise, also the capability of the machine. It may also be selected based on the previous experience and from the proper literature survey [12–14]. Three trails were performed for each combination of machining factor. New cutting tool insert was introduced for every trail to evade lifelessness in insert tool and to have precise value of all machining characteristics [7].

Table 2 displays the opted design of experiments and their response values of chip removal rate and cutting duration. All estimated data was investigated to find the expected value within the design domain.

Table 1 Turning factors and their ranges

Turning Factor	Notation	Unit	Factors ranges		
			Low	Medium	High
Speed	V	m/min	100	150	200
Feed	F	mm/rev	0.05	0.075	0.1
Depth	D	mm	0.25	0.5	1

Table 2 Experimental tracks and response outcomes

Trail No	Factor level			Response data	
	Speed	Feed	Depth	MRR (cm^3/min)	Machining time (min)
1	1	1	1	1.25	1.11
2	1	2	2	3.75	0.70
3	1	3	3	10	0.46
4	2	1	2	3.75	0.61
5	2	2	3	11.25	0.35
6	2	3	1	3.75	0.27
7	3	1	3	10	0.33
8	3	2	1	3.75	0.23
9	3	3	2	10	0.15

3 Gray Relational Analysis

In GRA normally, trial statistics, that is, measured features of quality appearances, are first standardized between the ranges from 0 to 1 [2, 7, 15]. The method is known as standardization. Then, as per the standardized value, GRC was designed for characterizing the association among the anticipated plus the real trial statistics. Afterwards, total gray relational grade was computed using all GRC'c matching to the designated response values. The total enhancement of the response values depending upon the calculated GRG [2, 7, 15]. GRA changes the several objective problems into an individual objective problem by converting the values into total GRG. In gray relational generation, for regularization of machining time, minimum-the-better (MB) criterion is considered. This may be computed as Eq. (2)

$$x_i * (k) = \frac{xi(k) - \min xi(k)}{\max xi(k) - \min xi(k)} \tag{2}$$

MRR calculation higher-the-better (HB) criterion, which may be computed as Eq. (3)

$$x_i * (k) = \frac{\max xi(k) - xi(k)}{\max xi(k) - \min xi(k)} \tag{3}$$

Here, xi*(k) and xi (k) are the regularized and experiential value, respectively, for ith number of trial through the kth value of response [2, 7, 15] which is presented in Table 3. The succeeding phase is computing the gray GRC value through the subsequent Eq. (4)

Table 3 Regularization of each performance appearances

Run No	MRR	Machining Time
	HB criterion	MB Criterion
1	0.000	0.000
2	0.250	0.427
3	0.875	0.677
4	0.250	0.521
5	1.000	0.792
6	0.250	0.875
7	0.875	0.813
8	0.250	0.917
9	0.875	1.000

$$\xi \mathrm{i(k)} = \frac{\Delta \mathrm{min} + \zeta \Delta \mathrm{max}}{\Delta \mathrm{i(k)} + \zeta \Delta \mathrm{max}} \quad (4)$$

Here, $\Delta i\ (k) = |\ xi^*(k)\ \text{-}\ xi^0\ (k)\ |$ is the variations of absolute data among xi^0 (k) and xi^*(k). Δ_{max} and Δ_{min} are the total supreme and total least values in dissimilar value series, correspondingly [2, 7, 15]. The exclusive distinguish co-efficient (ζ) is placed among zero and one in order to escalating or for compacting the range of GRC. Commonly, $(\zeta) = 0.5$ is considered if all process parameters have equal weightage [13] which is presented in Table 4. In the final phase, accomplishment of the average value of gray relational co-efficient was done. Here, GRG (γi) may be computed using Eq. (5) and which is given in Table 5.

$$\gamma \mathrm{i} = \frac{1}{n} \sum_{k=1}^{n} \xi i(k) \quad (5)$$

Table 4 Gray relational co-efficient (GRC)

Run No	MRR	Machining Time
	HB criterion	MB Criterion
1	0.333	0.333
2	0.400	0.466
3	0.800	0.608
4	0.400	0.511
5	1.000	0.706
6	0.400	0.800
7	0.800	0.727
8	0.400	0.857
9	0.800	1.000

Table 5 Gray relational grade (GRG)

Run No	GRG	RANK
1	0.3333	9
2	0.4330	8
3	0.7038	4
4	0.4553	7
5	0.8529	2
6	0.6000	6
7	0.7636	3
8	0.6286	5
9	**0.9000**	**1**

As we can see from Fig. 2, the best settings for turning 6063 aluminum alloy for better material removal rate (MRR) and lesser machining time becomes $V_3F_3D_3$. Table 6 displays the average GRG ratio for every range of the turning factor. From Table 6, cutting speed finds the rank one which is most dominant independent parameter which affects the turning process, followed by DOC, and feed, respectively.

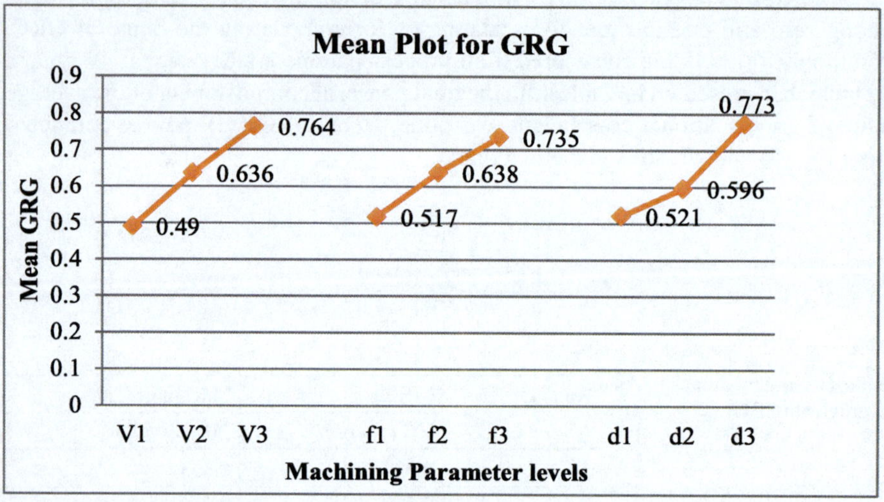

Fig. 2 Mean plot for GRG

Table 6 Results of mean GRG

Factors	GRG values				
	Low	Medium	High	Range	Rank
V	0.490	0.636	**0.764***	0.274	1
F	0.517	0.638	**0.735***	0.217	3
D	0.521	0.596	**0.773***	0.253	2

Overall average GRG = 0.630
*Shows optimum range of the turning parameter in each factor

4 Validation Test

Once calculating the best factor combinations, subsequent phase is to forecast and confirm the performance improvement of the machining appearances using the optimal factor settings. The valued GRG through the best factors can be computed as Eq. (6)

$$\gamma = \gamma m + \sum_{i=1}^{0} (\gamma i - \gamma m) \tag{6}$$

where γ_m is the overall average GRG [2, 7, 15], γ_i is the average GRG at best range, and 0 is the amount of the core plan factors that disturb the excellence appearances. Table 7 indicates the results of confirmation test.

Table 7 indicates the association between the estimated material removal rate and machining time with the test trail through the best machining settings. Decent promise among the test rail and the estimated outcomes were seen (Improvement in the total GRG has understood as 0.6667).

Table 7 Validation test

Factor ranges	Reference trail	Best trial settings	
		Estimation	Test trial
	$V_1f_1d_1$	$V_3f_3d_3$	$V_3f_3d_3$
MRR (cm^3/min)	1.25		20
Machining time (min)	1.11		0.13
Overall GRG	0.3333	1.0000	1.0000

Improvement in GRG = **0.6667**

5 Conclusions

- The optimum cutting parameter for better MRR and less turning time has seen as V3F3D3 using gray relational analysis.
- Since, the optimum combinations of cutting parameter fall out of the L_9 experimental domain, an experiment has been conducted again for the optimum machining combinations to calculate chip removal rate and cutting duration.
- Comparing the reference parameter and best turning circumstance, 93.75% improvement was there in MRR and 88.288% reduction in machining time was achieved using grey relational optimization techniques.
- The enhancement in total GRG has understood as 0.6667 which displays the performance improvement in the turning process for the desired objective functions.

References

1. Groover MP (2010) Fundamentals of modern manufacturing. John Wiley & Sons, Bethlehem
2. Luiz G.Cardoso, Deise S.Madeira, Thulio E.P.A.Ricomini, Ruben A.Miranda, Tarcisio G.Brito, Emerson J.Paiva, Optimization of machining parameters using response surface methodology with desirability function in turning duplex stainless steel UNS S32760. The international journal of advanced manufacturing technology. https://doi.org/10.1007/s00170-021-07690-3 (2021)
3. Saini M, Yadav RN, Kumar S (2015) An overview on turning process. In: Proceedings of 1st international conference on advancements and recent innovations in mechanical, production and industrial engineering (ARIMPIE-2015), vol2. ITS Engineering college greater Noida, India, pp 377–386
4. Davim JP, Gaitonde VN, Karnik SR, (2007) Investigations into the effect of cutting conditions on surface roughness in turning of free machining steel by ANN models. J Mater Process Technol 205:16–23. https://doi.org/10.1016/j.jmatprotec.2007.11.082
5. Grzesik W, Influence of tool wear on surface roughness in hard turning using differently shaped ceramic tools. Wear 265:327–335. https://doi.org/10.1016/j.wear.2007.11.001 (2008)
6. Yang WH, Tarng YS (1988) Design Optimization of cutting parameters for turning operations based on the Taguchi method. J Mater Process Technol 84:122–129
7. Kazancoglu Y, Esme U, Bayramoglu M, Guven O, Ozgun S (2011) Multi-objective optimization of the cutting forces in turning operations using the grey-based Taghuchi method. Mater Technol 45(2):105–110
8. Rastee DK, Heisel U, Schmauder S, Eisseler R (2014) Experimental investigation and multiobjective optimization of turning duplex stainless steels. Int J Manuf Eng 2014:921081–921013. https://doi.org/10.1155/2014/921081
9. Gomes JHF, Paiva AP, Ferreira JR, Costa SC, Paiva JP (2011) Modeling and optimization of multiple characteristics in the AISI52100 hardened steel turning. Adv Mater Res 223:545–553
10. Jaffery S, Mativenga P (2009) Assessment of the machinability of Ti-6Al-4V alloy using the wear map approach. Int J Adv Manuf Technol 40:687–696
11. Bariş Buldum, Uğur Eşme,Mustafa Kemal Külekci, Tarsun-Mersin, Aydin Şik, Ankara, Yiğit Kazançoğlu, Balcova-Izmir (2012) Use of Grey-Taguchi Method for the Optimization of ObliqueTurning Process of AZ91D Magnesium Alloy, Materials Testing, surface roughness measurements. https://doi.org/10.3139/120.110392

12. Sandvik Coromant (2015) Turning tools
13. ISO 3685: tool-life testing with single-point turning tools. International organization for standardization (ISO): Geneva, Switzerland. 1993. (1993)
14. Salman Sagheer Warsi, Mujtaba Agha, Riaz Ahmad, Syed Husain Imran Jaffery, Mushtaq Khan (2018) Sustainable turning using multi-objective optimization: a study of Al 6061 T6 at high cutting speeds. Int J Adv Manuf Technol. https://doi.org/10.1007/s00170-018-2759-2
15. Biswajit Das, Susmita Roy, RN Rai, SC Saha, Multiobjective optimization of in situ process parameters in preparation of Al-4.5%Cu–TiC MMC using a grey relation based teaching–learning-based optimization algorithm, J Process Mechanical Engineering, 0(0) 1–15 https://doi.org/10.1177/0954408917710555,(2017)

Automation Stream

ANN and Fuzzy Logic Based Direct Instantaneous Torque Control for 8/6 Switched Reluctance Motor

G. Jegadeeswari, D. Lakshmi, and B. Kirubadurai

Abstract Due to its simplicity of construction, robustness and high dependability, low-cost production capabilities, and high torque-to-mass ratio, the engine receives considerable reactions from industry during the past decade. SRM drive consists of a salient, concentrated coil polar stator and a salient rotor that has no wires or magnets. The double-defining structure of the engine is extremely nonlinear in magnet features. However, due to a large torque rib, which results in noise and vibration in the motor, the use of SRM is limited. A MPC-based ANN Instant Direct Torque Control (IDTC) (MPC) method for control of SRM drive torque is employed to resolve these difficulties. With this approach, the torque output of the motor may be controlled by the current hysteresis band. The simulation of the IDTC with ANN and the IDTC using fuzzy logic controllers is conducted to decrease the periodic speed ripple in the SRM drives. Comparative analysis is also given.

Keywords Switched Reluctance Motor (SRM) · BR converter · Adaptive Neural Network (ANN) · Torque ripple

1 Introduction

Switched reticence motor functionalities have been known for over 150 years, but only extensive enhancements in power electronics drive technology have led to significant success with the switched reticence motor in adjustable speed drives. The conventional reluctance machine was created due to huge requirement for variable drives and power semiconductor development and is known as the switched

G. Jegadeeswari (✉) · D. Lakshmi
Department of Electrical and Electronics Engineering, AMET Deemed to Be University, Chennai, Tamil Nadu, India
e-mail: jegadeeswari.dharan@gmail.com

B. Kirubadurai
Department of Aeronautical Engineering, Vel Tech Dr. Rangarajan Dr. Sagunthala R&D Institute of Science & Technology, Chennai, India

N. M. Sivaram et al. (eds.), *Advances in Manufacturing, Automation, Design and Energy Technologies*, Lecture Notes in Mechanical Engineering,
https://doi.org/10.1007/978-981-99-1288-9_31

reluctance machine. The first name used by one of the writers is "Switched Reluctance" that describes the two functions of a configuration machine (a) switched, (b) reticence. The term switched is shown because this machine may be used in the continuous mode of switching. In the scenario, both the stator and the rotor consist of variable reticence magnets, or it may be said that their structure is double salt. A SRM has both the stator and the rotor salient pillars [1]. The rotor does not have any type of permanent winding or magnet in every stator pole. It consists of a laminated steel soft magnet substance. In order to produce motor phases, two diametrically opposed windings are linked. A single regulated switch circuit is adequate to provide a one-way current for every phase throughout the rotor rotation. When the rate of change of phase induction is positive, the stator phase winding has to be excited for forward motor running. Otherwise, a breaking time or no torque will be produced on the machine [2]. Since SRM has a simple, robust design, low-production costs, fault tolerance, and excellent efficiency, the electric drive is becoming more recognized. There are also some drawbacks that it requires an electronic control system and a sensor of the shaft position. SRMs are usually built to make effective use of the converter ratings.

Torque control and torque control making it more complicated to analyze the machine than in other common machinery [3]. In this report, torque-sharing features are suggested to regulate the SRM instantaneous torque and to decrease the ripple of torque during operation fixed or within limits. The proposal to decrease ohimic losses was made to anticipate pulse width modulation with DITC. The fundamental flaw of this investigation is that two phases SRM were swapped to extend any number of phases without the need for couple control or pre-calculated commutation current, increasing the benefit of time consumption.

Bearingless switched reluctance motor of the rotor has proposed this author without using of any mechanical support structure. Objective of this author was Bearingless of rotor magnetically hover in air without a mechanical support provided of single layer winding [4]. Number winding in this paper was reduced from 8 to 6 further with reduce the cost converter topologies. The speed regulation technique with the driving theory was proposed. Main drawback is applicable for medium- and low-speed applications, whereas, the field harmonics in the stator and rotor increase the higher value of hysteresis and eddy current losses in these slots for the high-speed applications.

The results of the instantaneous torque control approach are presented in this study. The critical angle or switching angle of the two adjacent phases is the same torque and current condition or the same flow connection [5]. The improved current reference switching was then intended to decrease the torque drop. This article has not examined the influence of the winding resistor on the supply voltage [6].

The basic aim behind this article is to operate smoothly. The optimization of the current profile may be achieved using this article torque ripple minimization. The current waveform is controlled with a relatively small number of optimization variables with a simple procedure and the genetic algorithm [7, 8]. The new adjusted process was to decrease the maximum value of the present phase by affordable solution quality and a less time's The major disadvantages are the present expression of reference picked with just a few variables to determine the maximum STO range. It is challenging the universal SRM mathematical model to represent the specified mathematical formula [9].

The energy efficiency of SRMs using a zero-voltage switching system [10]. Null-voltage loop switching can significantly improve its performance in SRM operation by reducing the flow connection peaks. The harmonic magnitudes within the SRM core are reduced by the magnetic flux densities [11].

The current controller has to closely monitor the modulated reference current and track it even during dynamic activities. The current control unit was utilized to monitor the reference phase present and peak current technique for optimizing torque torsion hysteresis. In this study, the hysteresis current controller works at a constant frequency of switching. Two successive samples were preserved inside the range limit for the positive voltage. The strong voltage cuts were used to decrease switch losses and increase the life of the condenser. Even though, outgoing switch hard switching method the noise created higher than the soft-chopping mode [12].

2 Proposed Work

The system block diagram is shown in Fig. 1. The torque generation depends on the reticence principle, where the phase functions independently and in sequence in the event of the switching reluctance engine.

Due to the nonlinear characteristics of the magnetic circuit, the expression for the phase torque is given by,

$$T(\theta, i) = i\frac{\partial\psi(\theta, i)}{\partial\theta} \tag{1}$$

where 'teta' is the corner location of the rotor, I is the current of the phase. Thus, the preceding equation shows that the 'T (teta, I torque is precisely proportional to In

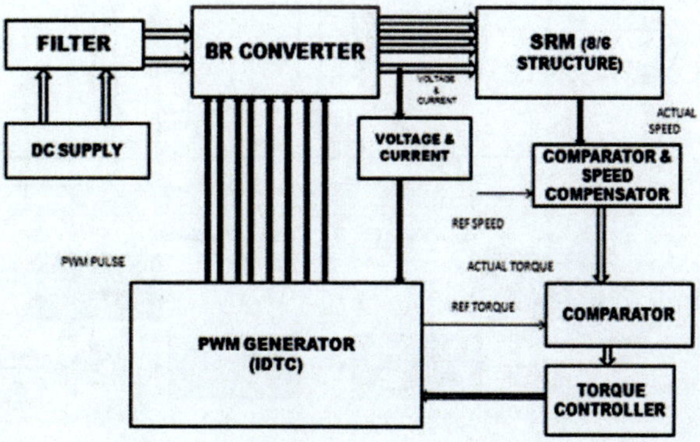

Fig. 1 Proposed system block diagram

order for a positive torque, there must be a widening of the statistical flux amplitude with regard to the rotor position and a decrease of the stator flux amplitude with regard to the rotor position to generate a negative torque change. Figure 2 shows the block diagram representing the technique of direct torque control. This torque control approach includes three key functions: torque and flux control hysteresis; optimized vector search table switching; and an engine design. In this approach, the actual or estimated velocity is compared to the referral velocity, the result of which is called an error signal. This includes the speed controller which is nothing but the speed controller, the output of which is the reference value of the T ref. In this situation, the torque and flux reference value may be compared with its current value, and a torque and flux hysteresis control technique can be used to generate the monitor signal. Input of the vector search table has been provided via a hysteresis band controller output. The best selection of the switching vectors is presented at all feasible stator fluid link locations.

The angle of the computed stream, which defines the area where the stream vector is stimulated, then passes the output signal via the switching table. The pulse of gate to the reverse circuit is given by the switching table signals. We may thus infer that the inverter is dependent mostly on the three variables.

i. Flow control signal hysteresis.
ii. Control signal for torque hysteresis.
iii. The vector flow angle and vector flow direction.

In the case of direct control of the coupled motor reticence, our major goal is to directly regulate the flow connection and electromagnetic torque through the selection of the correct inverter switching status. This allows us to decrease losses due to switches and harmonic distortions of stators. We require two control loops that are a flux-hysteresis control loop and a torque hysteresis control loop to regulate the torque and flux of the switched reactivation motor on their own.

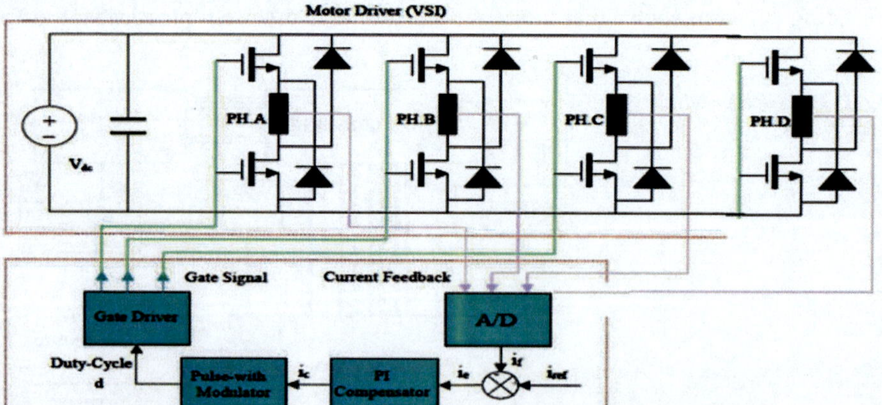

Fig. 2 Proposed system circuit diagram

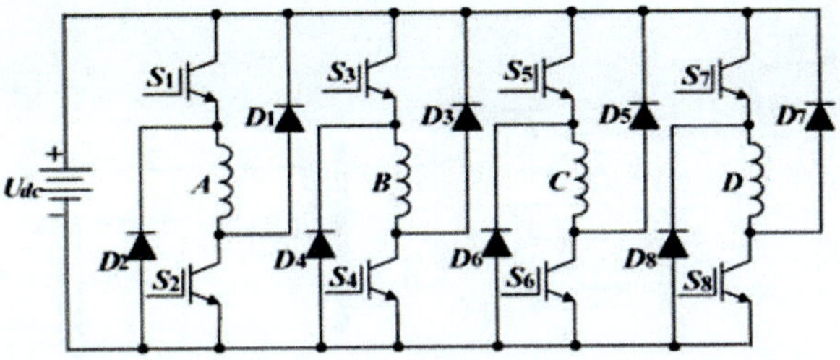

Fig. 3 $(n + 1)$ power semi-conductor switching devices and (n + 1) diodes

2.1 BR Converter

Figure 3 illustrates the number of phase windings $(n+1)$ of the Power Semi-conductor switching systems, and $(n + 1)$, per phase. Powered devices $T\&T_1$ connct the phase winding 'A' to the power. When the rotor position is used to disconnect this winding, the T_1 device is switched on by turning ON $T_1\&T$. This winding is activated. Winding A now re-save (feedback) energy to the source via the $D_1\&D$ diodes (i.e.) when the energy is transferred back to the source, the phase demagnetizes through the common diode and the phase diode.

Until the output has been fully removed, the following step cannot be switched ON. The B-winding stage is activated when devices T and T_2 are turned on. The stored energy is supplied to the main power by $D\&D_2$ in the phase B winding. The C winding phase is also turned on and off from the power supply. Through the activation and disabled of the corresponding equipment, the windings B and C are linked to and connected from the supply. This repeats the cycle.

2.2 Direct Instantaneous Torque Control of SR Motor

As the SRM drive has an outstanding construction, it has significant torque ripples and an issue with acoustic noise. Different approaches are being employed to minimize the ripple torque. One way is to decrease the torque ripple by skewing the rotor. The SRM direct torque control technique is similarly another way. The vector control technique is advanced DTC. This approach is used to regulate SRM's torque by controlling the flow linkage magnitude and changing the speed of the stator flux vector (acceleration or deceleration).

3 Simulation and Results

The model SR Simulink utilizing the adaptive neural network (ANN) method is illustrated in Fig. 4 Instantaneous direct torque management method reduces the engine torque ripples.

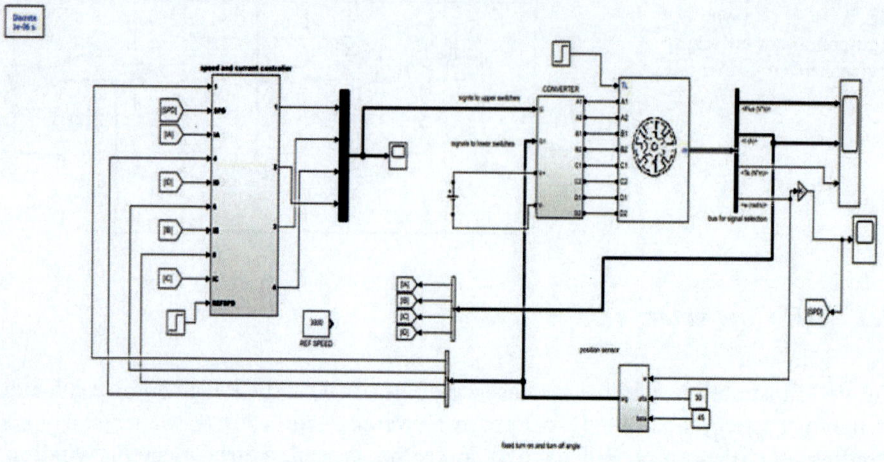

Fig. 4 Matlab Simulink model of SR motor using ANN algorithm

The SR motor model Simulink with fuzzy logic algorithms is shown in Fig. 5. Instantaneous direct torque control technology reduces the motor torque ribs.

The input voltage of the BR converter in Fig. 6, and the BR converter voltage is shown in Fig. 7. It converts the DC tension to AC tension. This voltage then powers the four-phase SR motor.

The pulses of the PWM to the BR converter are shown in Fig. 7. The pulses directly control the torque instantaneously. The rate of pulse switching is 25 kHz.

The flux, current, and torque waveform of the SR engine with fuzzy logic is shown in Fig. 8. Due to the nature of the material, the engine initially takes high power. This causes noise in the torque shape also with higher orders of ripple currents.

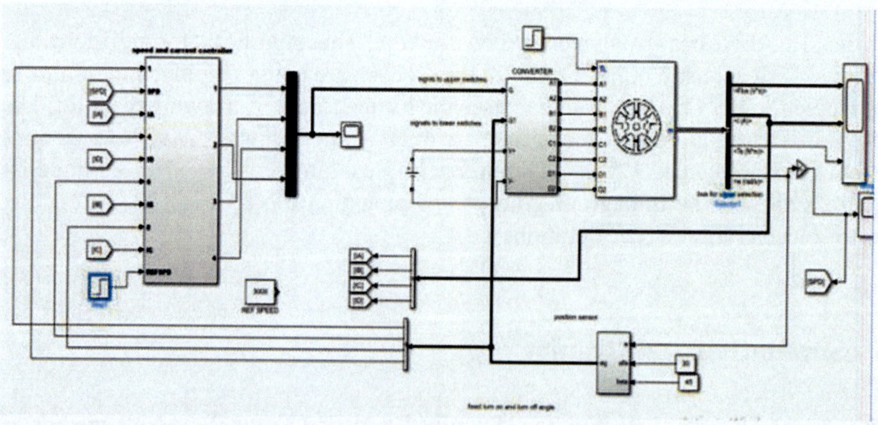

Fig. 5 Matlab Simulink model of SR motor using fuzzy logic controller

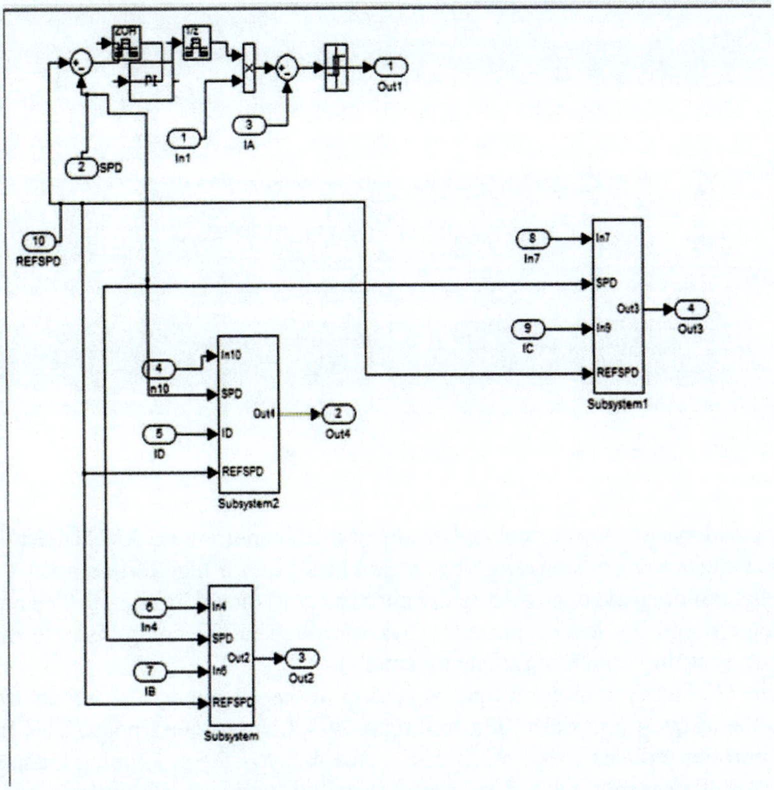

Fig. 6 Hysteresis current controller Simulink

Fig. 7 PWM pulses to the BR converter

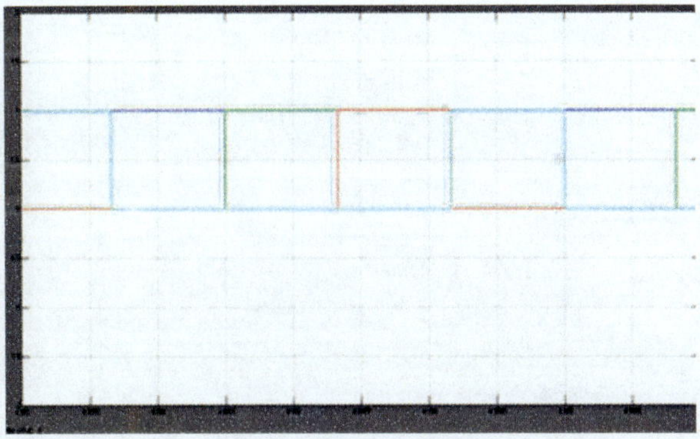

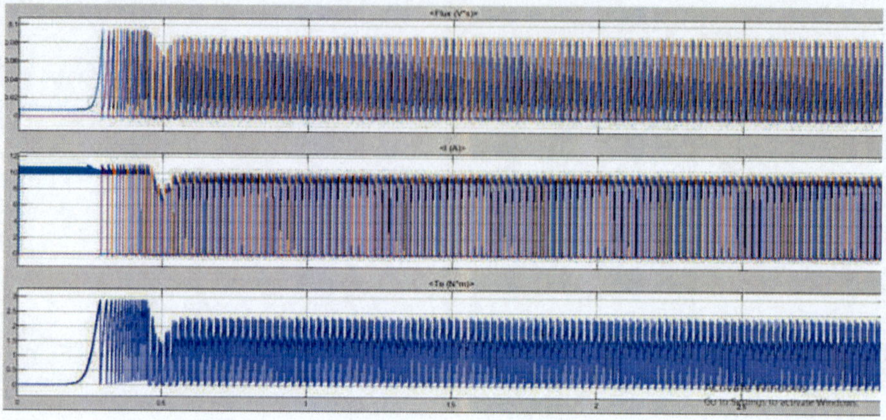

Fig. 8 Motor flux, current and torque waveform of fuzzy logic

The waveform of flux, current and torque of the SR motor with ANN is displayed like Fig. 9 because of the existing ribs created in SR motor high torque ribs.

The SR motor speed controller with neuro fuzzy is shown in Fig. 10. The neuro-fuzzy logic controller lowers the motor waveform speed. In comparison to the PI control, the settling time is significantly smaller.

Figure 11 shows the motor torque waveform of the SR motor. The NF controller reduces the torque ripples also. It's coming only 1.7 Nm torque ripples. This fuzzy logic controller reduces speed oscillation, since fuzzy is the self-tuning technique. Figure 12 illustrates motor speed and torque waveform utilizing a fuzzy logic control. The fluctuations in the torque are 1.75 Nm.

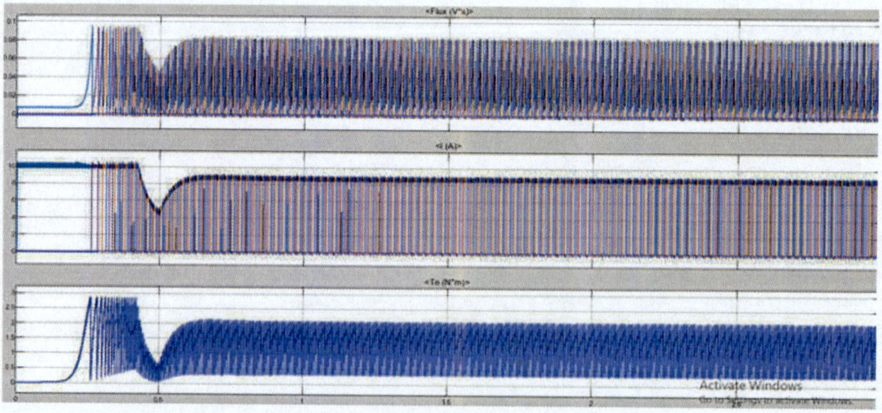

Fig. 9 Motor flux, current and torque waveform of PI-fuzzy logic

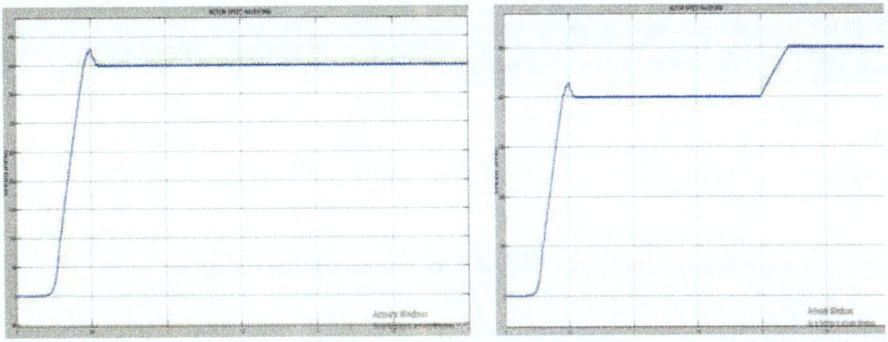

Fig. 10 Motor speed waveform without and with step change using fuzzy logic controller

Fig. 11 Motor speed waveform with step change using fuzzy logic controller

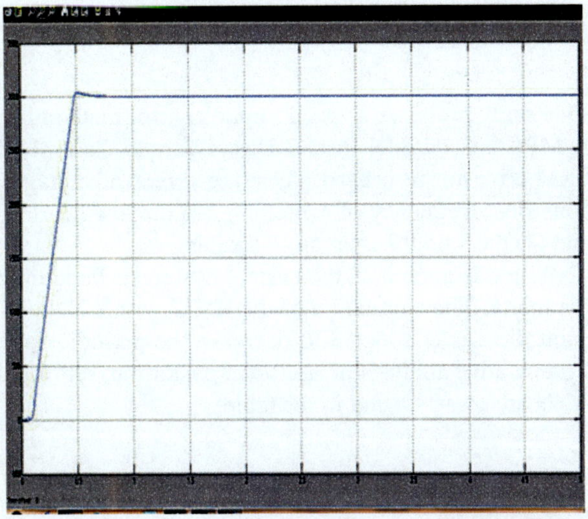

Fig. 12 Motor speed waveform with step change using ANN

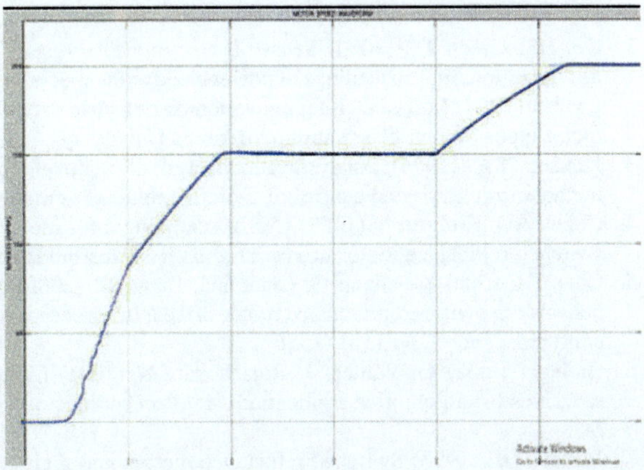

Table 1 Comparison of various controller

Algorithm	Rise time (t_{r1})	% overshoot	Settling time (t_s)	Peak time (t_P)
PI controller	0.36	0.5	0.8	0.5
Fuzzy	0.25	0.48	0.62	0.48
ANN	0.21	0.40	0.5	0.47

3.1 Comparison Table of Different Control Techniques

See Table 1.

4 Conclusion

This study proposes a speed torque control method based on fuzzy logic and ANN. A MPC-based ANN Instant Direct Torque Control (IDTC) method for control of SRM drive torque is used. There are several advantages of IDTC control like: (1) the consistent frequency of switching, (2) the low hearing level, (3) the low power rip, and (4) the smooth creation of torque.

With this approach, the current hysteresis band may control the torque output of the motor. The simulation of the IDTC with ANN and the IDTC using fuzzy logic controllers is conducted to decrease the periodic speed ripple in the SRM drives. Comparative analysis of rise time, peak time, settling time for both fuzzy logic and ANN are clearly listed in the table.

References

1. Ooi HS, Green TC (2000) Sensor less switched reluctance motor drive with torque ripple minimization. In: Proceedings of power electronics specialists conference, vol 3, pp 1538–1543
2. Lai J-S (1994) Soft-switching converters for electric propulsion drives with consideration of motor types. D.Ing. dissertation, Afrikaans University
3. Jackson TW (1996) Analysis and design of a novel controller architecture and design methodology for speed control of switched reluctance motors. Master's thesis, VT
4. Cheok AD, Ertugrul N (1999) Use of fuzzy logic for modeling, estimation, and prediction in switched reluctance motor drives. Proc IEEE Trans Ind Electron 46:1207–1224
5. Chen JH, Chau KT, Jiang Q, Chan CC, Jiang SZ (2000) Modeling and analysis of chaotic behavior in switched reluctance motor drives. In: Proceedings of power electronics specialists conference, vol 3, pp 1551–1556
6. Gallegos-Lopez G, Walters J, Rajashekara K (2001) Switched reluctance machine control strategies for automotive applications. In: Proceedings of SAE 2001 world congress, Detroit, MI
7. Miller THE (1993) Switched reluctance motors and their control. Magna Physics Publishing and Clarendon Press Oxford

8. Lai J-S (1996) Resonant snubber based soft-switching inverters for electric propulsion drives. In: Proceedings of IECON 22nd international conference on, vol 1, pp 47–52
9. Song B-M, Lai J-S (2000) A novel two-quadrant soft-switching converter with one auxiliary switch for high power applications. IEEE Trans Ind Appl 36(5):1388–1395
10. Jegadeeswari G, Kirubadurai B, Lakshmi D (2020) Multi carrier based new random pulse width modulation for three phase inverters. Int J Sci Technol Res (IJSTR) 9(2):5425–5432. ISSN 2277-8616. Impact factor: 3.03
11. Jegadeeswari G (2017) Design and implementation of THD reduction for cascaded multilevel H-bridge inverter. J Adv Res Dyn Control Syst (JARDCS) 11(638–644). ISSN 1943-023X. Impact factor: 3.514
12. Jegadeeswari G, Kirubadurai B (2021) A fuzzy logic controller based SRM with torque ripple suppression for EV applications. Int J Aquatic Sci 12(3)

Design and Fabrication of Automatic Screw Gage Calibrator and Component Tester Using IoT

V. Arunkumar, S. Sathiyavathi, A. Tonythomas, A. P. Parameswaran, M. Megavarthini, B. Kishor Kumar, and R. Gokul Raj

Abstract In metrology laboratories, the vogue practice of calibration is done with the help of instruments like master calibrators, and the cost for calibration of such devices is high. Hence, calibration of such tools is also time-consuming. This proposed project is named an automatic screw gage calibrator. By implementing this project, time is substantially reduced as encoders are used and mini slip gages for calibration. The project is developed as a prototype that can only calibrate the screw gage at the initial stage. After upgrading, it can calibrate instruments like Vernier caliper and dial gage. This instrument is not only used for calibration but also for measuring components. Hence, it acts as a multipurpose machine. In industries, small tiny components are manufactured, and our instrument can also measure the parameters of these components. The added advantage of our project is that the manufacturer doesn't want to consult the calibrating industries as our calibrator tests on its own. Thus, qualifying and testing departments are integrated with the sector as our project does both the action and, hence, is automated.

Keywords Micrometer · Calibrator · Tester · Slip gage

1 Introduction

Advancements in technology have played a significant role in replacing many manual operations with machines. In the industry, the calibration of the screw gage is done

V. Arunkumar (✉)
School of Computer science and Engineering, Vellore Institute of Technology (VIT University) Chennai Campus, Chennai, Tamil Nadu, India
e-mail: arunkumar.v@vit.ac.in

S. Sathiyavathi · A. P. Parameswaran · M. Megavarthini · B. Kishor Kumar · R. Gokul Raj
Department of Mechatronics Engineering, Kongu Engineering College Perundurai, Erode, India

A. Tonythomas
Department of Mechatronics Engineering, KCG College of Technology, Chennai, Tamil Nadu, India

N. M. Sivaram et al. (eds.), *Advances in Manufacturing, Automation, Design and Energy Technologies*, Lecture Notes in Mechanical Engineering,
https://doi.org/10.1007/978-981-99-1288-9_32

manually by placing the slip gages one after the other. In the wake of this shortfall, a new innovative project has proposed reducing the human effort and automating it. In addition to this, the industry also calibrates Vernier calipers, dial gage, pressure gage, multimeter, etc. Most of these devices are calibrated manually, which is more liable to human errors, and time-consuming. Initially, the project was developed only for screw gages, making the calibration process for small industries simple without consulting the calibrating industries.

Moreover, the calibration for handheld devices is often more expensive than the price of a new one. It is not only used for calibration purposes but also quality checking in industries. The work done by machines is more precise and also saves a lot of time. Hence, this project serves all the requisites as mentioned above.

2 Literature Survey

Micrometer heads have different measuring ranges, out of which British Standards [1] relate to specific measuring ranges. Micrometer heads have rotatable and non-rotatable spindles and are classified into three types based on degrees of thimble diameter. Each micrometer head will be provided with means to adjust the zero setting and compensate for wear between screw and nut. The dimensions and maximum permissible errors for all three types of micrometer heads were tested and tabulated according to the British Standards.

Micrometer for external measurements gives the dimensional quantity of an external feature of a workpiece. Micrometers are designed and manufactured according to the International Standards formulated by International Organization for Standardization (ISO). The International Standard of a product can be regarded as Geometrical Product Specification (GPS) [2]. The GPS specifies the most critical dimensional, functional, and quality characteristics of micrometer calipers for external measurement. These standards are applicable for micrometers having both analog and digital indications.

The project is based on automated calibration of micrometer using machine vision which has high accuracy [3]. It is used to measure micrometer screw error and forces. The project has two motorized stages, one for moving the micrometer and another for moving the camera. The project also consists of a CCD camera with various zoom options. The micrometer is manually checked and given to the system. The micrometer error and the direction of camera motion are measured. The camera focuses on the points marked on the thimble and scale. The thimble is coupled to a motor that rotates the thimble at regular intervals of 0.05 mm. The camera is moved using another motor. The images are captured and are processed using the software. The time needed for the measurement of one point is 18 s. Here one point refers to rotation of thimble, movement of measurement stage and camera, reading the values, and processing it. This project is time-consuming, say for calibrating 400 points, it takes around 2 h even though its accuracy is higher than manual calibration.

A project for controlling snake robots using a mobile app for rescue operations was developed [4]. The user can give commands through the app, interfacing with the robot through a Wi-Fi or a Bluetooth module to control its movement. Similarly, the control of each link/leg through Arduino and servo motors was proposed to define the motion of the octopod [5].

An optical encoder is used to measure load angle, angular velocity, and acceleration [6]. This method helps control the motor and measure the rotation and position of the motor shaft. An optical encoder gives accurate results for transient speed and load angles. But, the measurement of acceleration is complex with some errors. These errors are rectified by increasing the resolution of the encoder.

The speed of the DC motor is controlled by using the ATmega16 microcontroller. This method uses the pulse width modulation (PWM) technique [7]. Varying the PWM duty cycle will control motor terminal voltage, which changes the speed of the motor. A TV remote control is used as a transmitter to send data to the microcontroller with an IR receiver. The ATmega16 is connected to an L293D driver IC which controls the direction and speed of the DC motor. A photodiode is used to measure the speed of the motor. The speed and direction can be set with a TV remote control, and the process is executed using the PWM technique using a microcontroller. The results are displayed on an LCD.

The LCD screen is interfaced with the Arduino to display the characters as well [8]. LCD can operate in both 8-bit and 4-bit modes. The interfacing consists of two parts: one is wiring and the other is programming. Once all the wires are properly connected, the Arduino uses an LCD library function to display the output on the screen.

A review paper on the working of pitch screw gage, which discusses a screw check, is used to estimate the breadth or thickness of a material [9]. The micrometer has two types of scales: circular scale and pitch scale. There are usually two types of errors: positive zero error and negative zero error. This paper provides the fundamentals and importance of screw gages.

3 Proposed Method

The objective of this project is to reduce time and cost consumption for calibration. Secondly, accuracy is maintained. Future updates to automated calibrators can calibrate other instruments like Vernier calipers, dial gages, and pressure gages. In this automatic calibrator, we can also test the component without using any other external device. The micrometer enclosed for calibration is taken into account for testing purposes. Arduino Mega 2560 plays a pivotal role in our project. The Arduino programming is done for rotation of the motor and digitally getting the output of the encoder. Thus, it simplifies the major work in this project. The output of Arduino will be 5 V which will not be sufficient for the rotation of motors. The motor requires 12 V for its process, which could be provided by the L293D module, which converts

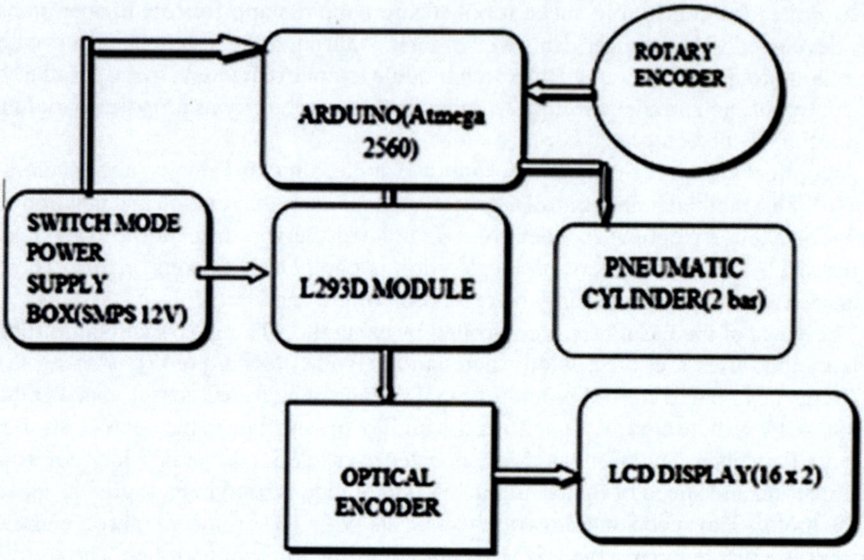

Fig. 1 Flowchart of proposed system

digital to analog. Hence, the L293D module is used for the 12 V supply and rotation of the motor. The relay module plays an essential role in moving the slip gage or component to the ends of the screw gage. To pick the component or slip gage electromagnet is used. It is energized with the help of Arduino programming.

Figure 1 shows a clear view of how the system works with the encoder feedback with the help of programmed Arduino. Here two encoders were used, one for mode selection which is to operate. There are two modes that the system is configured: calibration mode and testing mode, which helps the user select the operation to be done. Another encoder is used to measure the rotation and position of the motor with the help of a feedback system. This feedback is controlled by Arduino [10]. Mainly, the output from the encoder is in the form of a digital one, so by using Arduino, the digital form is converted into degrees. The encoder helps in the measurement of the screw gage by calculating the number of rotations moved. Typically, the screw gage moves 0.5 mm for one rotation of the motor, so the linear movement is determined by calculating the rotation. By calculating the value of the screw gage, the output is displayed in the LCD at corresponding modes.

4 Working Principle

Micrometers of any type can be used. The micrometer is fixed inside the chamber. There is a stand provided for fixing the micrometers. Depending on size, the screws are adjusted so the micrometer can improve tightly to it. A shaft with an encoder is fixed in-between screw gage and motor to couple each other. Here in this project, a photoelectric encoder disk as a model for calibration is used. Then the chamber is closed to select the mode by using a rotary encoder connected to the LCD. There are two modes: calibrating mode (for calibration) and testing mode (for testing). When calibration is selected, the machine changes the way to calibration, and the process starts. Initially, the home position of the screw gage is checked. It should be at a maximum position. If not, it automatically moves to the home position with the help of the encoder.

Then the conveyor belt is moved near the screw gage. By using a servo motor, the slip ring is fixed to the center of the screw gage. The screw gage is rotated to get the desired value. This process is repeated to get the desired average output of the error factor. This calibrated compensation factor is displayed on the LCD screen. For testing a component, the testing mode should be selected. In the testing mode, the component is placed in the conveyor belt. A servo motor controls the electromagnetic hand to place the component in the center of the screw gage. Using the same process, the measurement is calculated and compensated with an error factor calculated in the calibration mode.

Figure 2 shows the project flow diagram. The process starts with selecting the modes; when the modes are selected, the conveyor is filled with either a component or slip gauge one by one, depending on the mode selected by the user. When a user selects a particular mode, such as calibration mode, the corresponding slip gage is picked with the help of electromagnet and activators. It is placed in the position between two ends of the screw gage. When it is placed, the screw gage is moved to the position, and the measurement is taken. After the output is taken, the values are displayed on the LCD. This process is the same while we are testing the component in the screw gage.

A mobile app was developed for the calibration of screw gages and identifying the errors. All the error values of each screw gage are sent to a database which can be viewed in the same app. The app contains all the details of the previous screw gages calibrated so that it can be used for future reference.

5 Design and Implementations

Figure 3 shows the CAD modeling of the project designed using AutoCAD 2018. The CAD modeling shows how all components are aligned, and the type of material is also selected according to the project implementations.

Figure 4 shows the circuit design of the project designed using Proteus V8.9. Proteus contains designs of almost all microcontrollers. The components are wired and connected based on this design version CBR1.0.

Figures 5 and 6 show the complete product version of the Automatic Screw Gage Calibrator and Component Tester CBR1.0, which is done by assembling all the components according to the designs mentioned above.

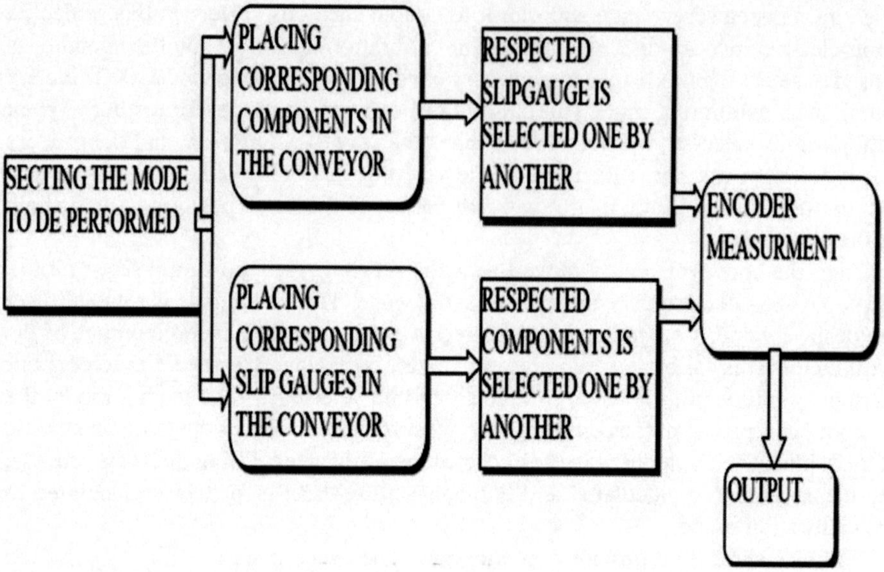

Fig. 2 Flow diagram

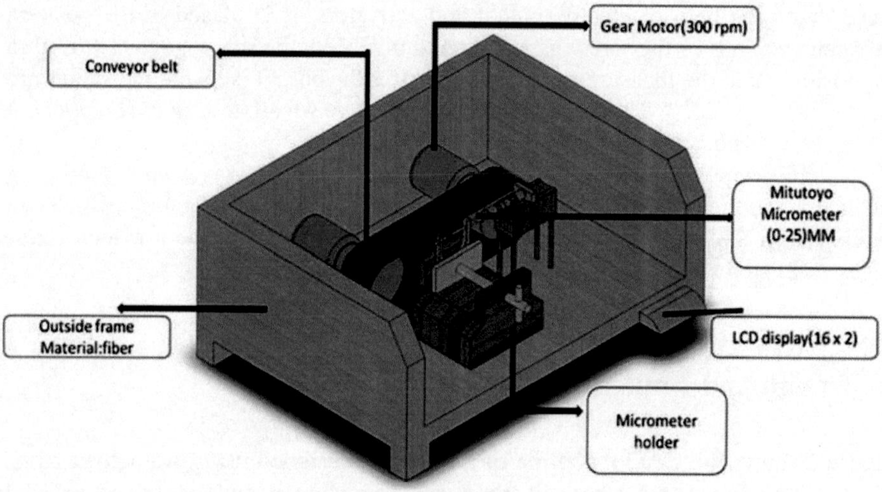

Fig. 3 CAD model of the screw gage calibrator

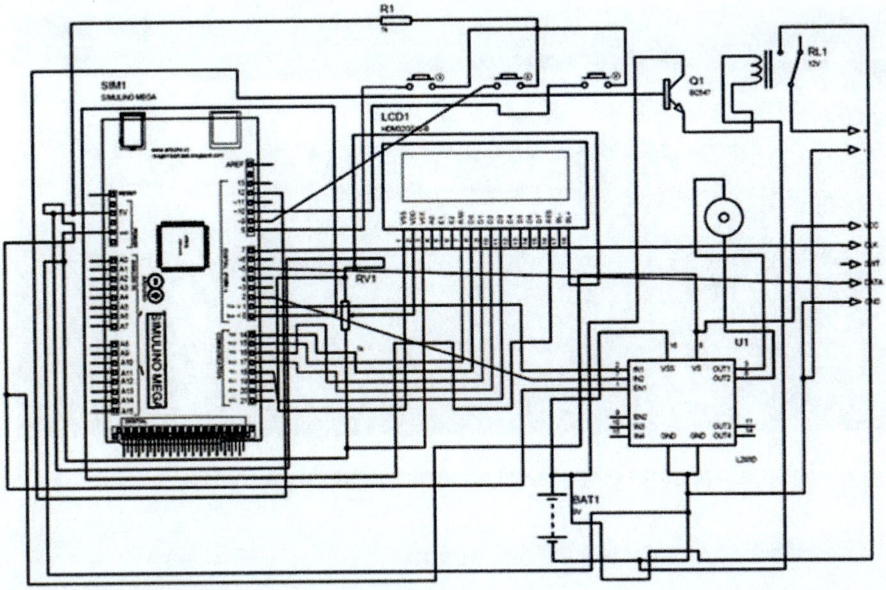

Fig. 4 Circuit diagram of screw gage calibrator

Fig. 5 Automatic screw gage calibrator and component tester

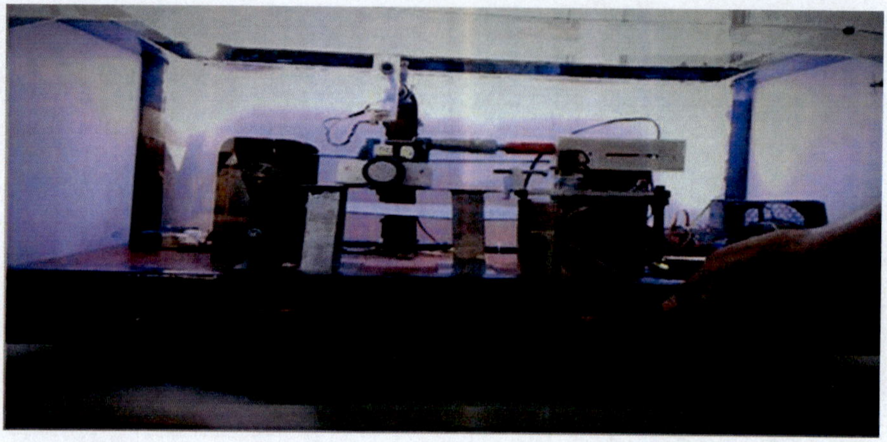

Fig. 6 Automatic screw gage calibrator and component tester version CBR1.0

6 Feasibility Study and Conclusion

6.1 Feasibility Study

This project aims to minimize the human resources in calibrating industries, i.e., to eliminate manual use of slip gage in calibrating the screw gage. The proposed project can be implemented in all calibrating and metrological industries, thus serving the prime aim of the sectors. It can be further developed for all measuring instruments, like components used in computerized numerical control, casting process, etc. It is simple in all sorts like handling, calibrating, and maintenance. Thus, the proposed project has the probability of serving all industrial needs.

6.2 Conclusion

This work concludes that this automatic calibrator provides a good calibration, reduction in time, and cost with high accuracy of output. The limitation of this calibrator version CBR1.0 is that it can measure only one micrometer at a time, and storage of data is also limited. This limitation is overcome in the next version of the calibrator. This project can also be extended to calibrate other instruments like Vernier calipers, dial gages, pressure gages, and the components can also be tested.

References

1. BS 1734:1951—specification for micrometer heads
2. ISO 3611:2010 Geometrical product specifications (GPS)—dimensional measuring equipment: micrometers for external measurements—design and metrological characteristics
3. Hemming B, Fagerlund A, Lassila A (2007) High-accuracy automatic machine-vision based calibration of micrometers. Measurement Science and Technology
4. Arun Kumar V, Adithya B, Bijoy Antony PT (2021) Snake robots for rescue operation. IOP Conf Ser Mater Sci Eng 1055:012001
5. Arun Kumar V, Meenakshipriya B, Bijoy Antony PT, Adithya B, Hari Vikinesh M (2021) Design and fabrication of octopod for survey and rescue operation. IOP Conf Ser Mater Sci Eng 1055:012021
6. Kadhim AH, Babu TKM, O'Kelly D (1992) Measurement of steady-state and transient load-angle, angular velocity, and acceleration using an optical encoder. IEEE Trans Instrum Meas 41(4)
7. Raka Agung IGAP, Huda S, Arta Wijaya IW (2014) Speed control for DC motor with pulse width modulation (PWM) method using infrared remote control based on ATmega16 microcontroller. In: International conference on smart green technology in electrical and information systems (ICSGTEIS), Kuta, Indonesia
8. Clarry M (2015) Interfacing to an LCD screen using an Arduino
9. Chauhan V, Chauhan M, Verma S, Kumar R (2018) A review paper on pitch screw gauge working. Pramana Res J 8(8)
10. Chu Y, Park JH (2020) Efficient learning modules for embedded system. Int J Electr Eng Educ

Multiple Regression Analysis of Performance Indicators in the Tertiary Food Processing Industry

S. Pratheesh Kumar, V. Sathya Nandhana, R. Akash, and A. R. Kamalesh Krishna

Abstract Almost all food is processed in some way before it is ingested. Food is processed to make it more tasty, pleasurable, and safe to consume, as well as to preserve it for consumption outside of the harvest season. Food processing is also a technique that allows for greater variety in meals, giving customers additional options. Any technology that involves transforming fresh components into edible goods is referred to as food processing. Food processing also includes adding ingredients such as vitamins and minerals to increase nutritional content or preservatives to enhance shelf life. On the other hand, a failure to investigate the relationship between system characteristics and performance indicators severely limits profitability. The goal of this study is to analyze and quantify the impact of financial variables on the overall sales of the tertiary food processing business. A fishbone diagram is used to depict the potential causes and consequences of an input variable on output variables. Based on the coefficient correlation coefficient, several regression algorithms are applied to a variety of financial parameters to discover which parameter has the most significant impact on the performance indicator. As a result, the current study determines the most relevant parameter influencing tertiary food processing industry sales and also selects an appropriate regression model by comparing the two models using mean squared error. The found regression model can be used to forecast sales in the future.

Keywords Tertiary food processing industry · Performance indicator · Regression analysis · Correlation coefficient

S. Pratheesh Kumar (✉) · V. Sathya Nandhana · R. Akash · A. R. Kamalesh Krishna
Department of Production Engineering, PSG College of Technology, Coimbatore, Tamil Nadu, India
e-mail: spratheeshkumarth@gmail.com

N. M. Sivaram et al. (eds.), *Advances in Manufacturing, Automation, Design and Energy Technologies*, Lecture Notes in Mechanical Engineering,
https://doi.org/10.1007/978-981-99-1288-9_33

1 Introduction

Food processing, food preservation, food production, and food manufacture are all phrases that refer to the methods that are used to transform raw agricultural resources and in certain cases, wild-harvested items into consumable foods. These methods are used to make meals safer and more appealing to humans. Food processing and production on a large scale continue to play an essential role in the twenty-first century. Consumers throughout the world would be limited to food varieties produced locally if these systems were not in place. For the majority of the population, especially those living in cities, these would be significant limiting considerations. Figure 1 depicts the share of India's food processing market toward the total economy in 2020 and 2025 [1].

Food processing methods fall into three categories:

- Primary food processing industries
- Secondary food processing industries and
- Tertiary food processing industries (Ready–to-eat foods).

Tertiary food processing involves the production of ready-to-eat foods and includes industries that produce macaroni, noodles, processed meat, starches, bakery products, dairy products, etc. The assumptions of multilinear regression analysis and missing value analysis were investigated, and the KPSS was used to estimate lessons measuring and evaluation, instructional strategies, counseling, program creation, and educational psychology [2]. Collinearity between independent factors via diagnosis was identified and multiple regression forecasting models were used to contrast the predicted results [3]. The influence of independent factors such as the size of the company, maturity, debt level, cash ratio, inventories, sales development, and asset liquidity ratio, on the dependent variable which was the accountancy rate of return on assets (ROA) was studied using a regression model [4]. Meta-regression analysis was conducted on the causes for the unpredictability of past climate change impact research findings [5]. According to corporate statistics, food processing companies' profitability has been either greater or equivalent to that of other companies. Food processing enterprises, on the other hand, were found to have a smaller value-added

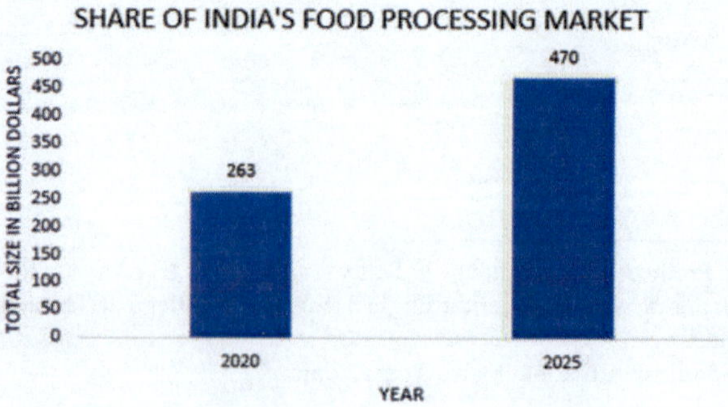

Fig. 1 Size of India's food processing market

component than other industrial firms [6]. Research studies and interviews were used to create the customer happiness model, and tested the model using factor analysis and multiple linear regression [7]. The effect of firm-specific and macroeconomic variables on the viability of Pakistan's food sector was explored using multiple regression analysis under strong correlation conditions [8]. The impact of firm-specific variables on the financial performance of enterprises in the Czech Republic was investigated using cumulative sum chart and panel pass time series techniques [9]. The performance of coffee processors and the coffee market in Serbia using capacity utilization study, financial accounting, and efficiency factors assessment [10]. An investigation was carried out to identify whether and how much of a link existed between alleged important success factors and revenue, and what function crucial factors serve in profit forecasting using correlation and regression analysis [11].

2 Performance Parameters

2.1 Major Influencing Factors in the Tertiary Food Processing Industry

Figure 2 shows the cause-and-effect diagram, which illustrates the factors that affect the 'Sales'.

Major influencing factors in the tertiary food processing industry were referred from the yearly edition of RBI Bulletin related to food processing industries, published by the Reserve Bank of India [6]. Profit before tax evaluates the earnings of a company before corporate income tax is paid.

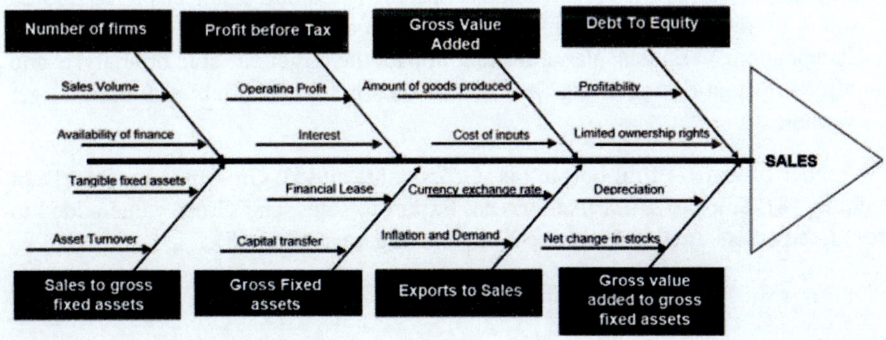

Fig. 2 Cause and effect diagram used for sales

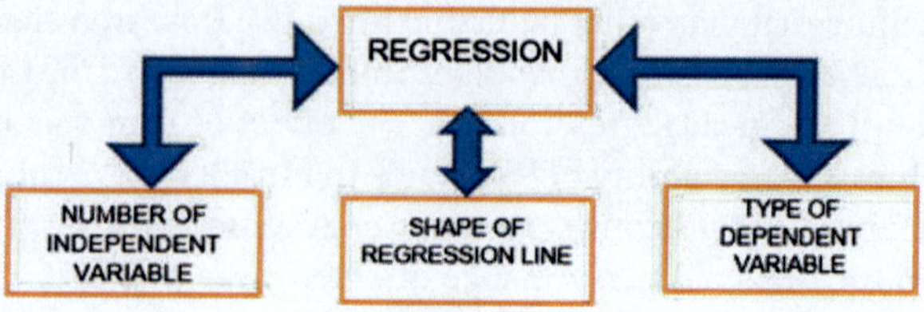

Fig. 3 Primary attributes for regression algorithm

2.2 *Identification of Key Performance Indicators*

The firm's sales were selected as the study's major performance measure. Sales are a metric used to assess an organization's efficiency. The level of profit-generating capability may be used to assess the organization's performance. The data for this research was obtained from the official website of the Reserve Bank of India. The dataset includes the following attributes: number of businesses, sales, debt to equity, and profit before tax, and so on. Some of the terms that may be used to describe gross value added to gross fixed assets are gross fixed assets, gross value added, exports to sales, and sales to gross fixed assets.

3 Multiple Linear Regression Analysis

3.1 *Attributes of Multiple Linear Regression Analysis*

The primary attributes of the regression model, as shown in Fig. 3, include:

i. Dependent Variables: The dependent variable is the one that needs to be figured out or predicted. Sales is considered as the dependent variable.
ii. Independent Variables: Variables that impact the target variable or analysis and offer information regarding the relationships between variables and the target variable.

Number of firms, Profit before tax, Gross value added, Gross fixed assets, Debt to equity [8], Sales to gross fixed assets, Export to sales, and Gross value added to gross fixed assets are considered to be independent variables [5].

3.2 *Regression Model Development*

Based on data obtained from the website of the Reserve Bank of India, a multiple regression analysis was conducted. The analysis was completed in a Google Collab using the Python programming language. As indicated in Fig. 4, the libraries needed

```
#importing required libraries
import pandas as pd
import matplotlib.pyplot as plt
import numpy as np
import seaborn as sns
```

Fig. 4 Importing the required libraries

to execute regression analysis must first be imported. The data to be examined must be imported once the relevant libraries have been loaded.

Figures 5 and 6 depict the development of multiple linear regression and elastic net regression models. The Scikit-Learn library was used to create these models. Elastic Net combines the features of lasso and ridge and reduces the impact of several features.

```
from sklearn.metrics import mean_squared_error
#importing linear regression model from scikit learn packages
from sklearn.linear_model import LinearRegression
#instantsiating the linear regression model
linear_reg = LinearRegression()
#fitting the linear regresssion model with train datasets
linear_reg.fit(x_train,y_train)
#predicting the test datas with linear regression model
y_pred = linear_reg.predict(x_test)
```

Fig. 5 Model development-multiple linear regression

```
#importing elastic net regression model from scikit learn packages
from sklearn.linear_model import ElasticNet
#importing GridSearchCV hyper parameter optimization technique from scikit learn packages
from sklearn.model_selection import GridSearchCV
#instantsiating elastic net regression model
elastic_net = ElasticNet()
#storing different hyper parameter values of in a dictionary
parameters = {'alpha':[0.0001,0.0025,0.001,0.01,0.1,1,10,100],
              'l1_ratio':[0.0001,0.0025,0.001,0.01,0.1,0.2,0.5,1,1.5]}
#instantsiating the GridSearchCV with elastic net regression model
elastic_reg = GridSearchCV(elastic_net, parameters, cv=4)
#fitting the GridSearchCV based elastic net regresssion model with train datasets
elastic_reg.fit(x_train,y_train)
```

Fig. 6 Model development-elastic net regression

```
#importing train-test split package
from sklearn.model_selection import train_test_split
#creating train-test split with 80% train and 20% test datas
x_train,x_test,y_train,y_test=
train_test_split(x,y,test_size=0.20,random_state=42)
#printing the shapes of test and train datas
print ('Train set:', x_train.shape,  y_train.shape)
print ('Test set:', x_test.shape,  y_test.shape)
```

Fig. 7 Train and test split evaluation used in both regression models

The train test split is an approach to evaluate a machine learning system's performance. They include two datasets: (1) Dataset for Training: The dataset that is utilized to fit the machine learning model is called the train dataset. (2) Dataset for Testing: This dataset is used to determine how well a machine learning model fits a given situation. Figure 7 demonstrates the train and test split evaluations used in the Multiple Linear and Elastic Net Regression models, respectively.

4 Influence of Factors on Key Performance Indicator

4.1 Identification of Critical Factors

The degree to which the variables are related is measured using correlation analysis. The correlation coefficient, which describes how much one variable changes when the other does, is calculated using correlation analysis. The correlation between feature and target variables was evaluated. From the obtained results, a correlation heat map between the input and the output variables was plotted using visualization libraries, as illustrated in Fig. 8.

4.2 Graphical Representation of Critical Characters

From the results, it was inferred that Profit before tax and Gross value added had a profound effect on the sales of the organization. Graphic visualization of the relationship between profit before tax, Gross value added, and sales was plotted using a line plot, as shown in Fig. 9.

The line plots elucidate that with an increase in Profit before tax and Gross value added, Sales also increase, and hence, these two independent variables have a significant effect on Sales as shown in Fig. 9a, b.

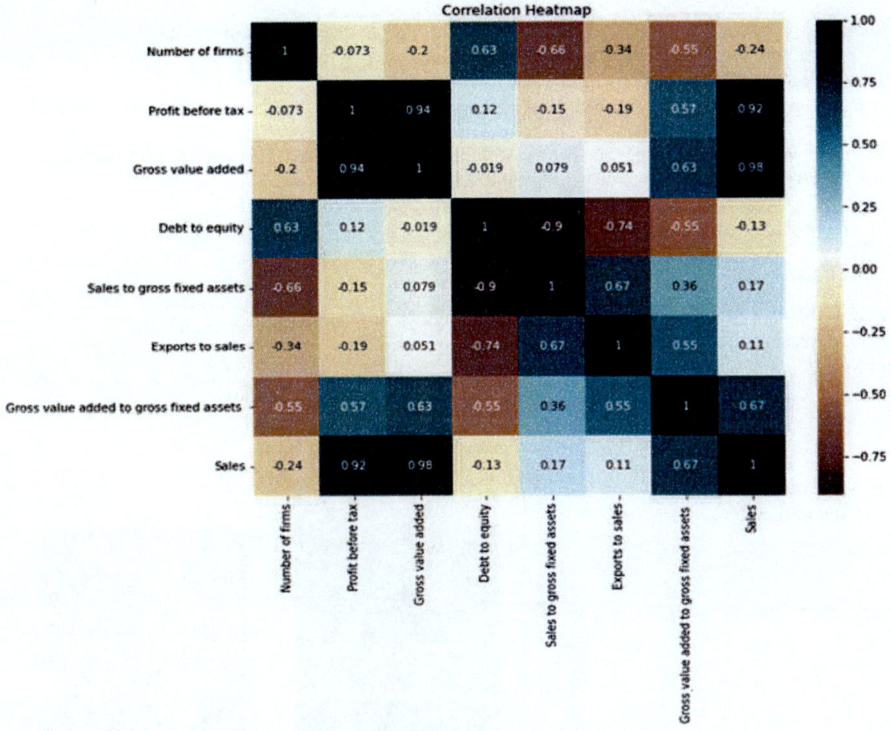

Fig. 8 Correlation between the input and output variables

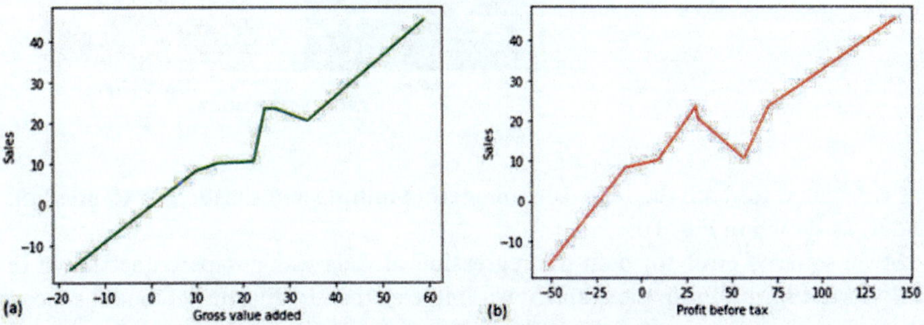

Fig. 9 **a** Gross value added versus sales. **b** Profit before tax versus sales

4.3 *Comparison of Regression Models*

The mean squared error (MSE) of a regression line indicates how near it is to a set of points. It is accomplished by squaring the distances between the points and the regression line (these distances are referred to as the "errors"). The mean squared error

```
print("The root mean squared error of the tuned linear regression model is:"
      ,np.sqrt(mean_squared_error(y_test,y_pred)))
linear_regressor_R2_score = linear_reg.score(x_train,y_train)
print("The R2 Score of the tuned linear  regresion model is:",linear_regressor_R2_score)

The root mean squared error of the tuned linear regression model is: 11.165478745097822
The R2 Score of the tuned linear  regresion model is: 1.0
```

Fig. 10 Mean squared error for regression models

Table 1 Mean squared error of the regression models employed

Regression model	Mean squared error
Linear	11.165479
Elastic net	10.181988

Fig. 11 Comparison of mean squared error

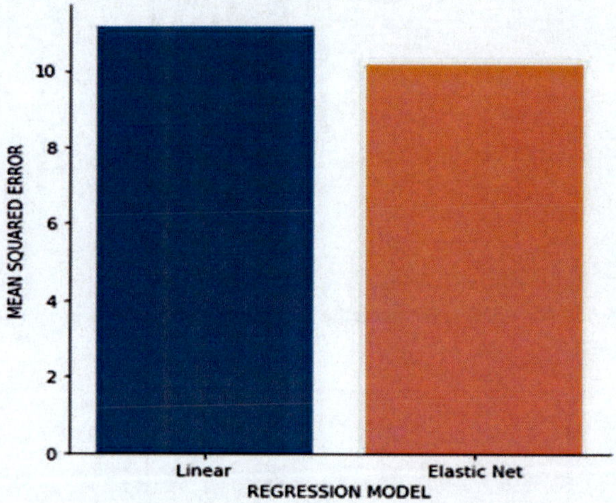

was calculated for both the regression models (Multiple and Elastic Net Regression model) as shown in Fig. 10.

Mean squared error for both the regression models was compared as shown in Table 1 and Fig. 11. From the graph, it was inferred that the Elastic net model's mean squared error was lesser than the multiple regression model.

5 Conclusion

The impact of financial variables on overall sales in the tertiary food processing industry was studied in this paper. The fishbone diagram was used to show the input variable's probable causes and effects on the output variables. Multiple regression algorithms are applied to identify the impact of parameters on the performance indicator based on correlation analysis. It was found that Profit before tax and Gross value added had the most significant impact on the sales of the organization. The graphical representation of the relationship between the parameters showed the existence of a linear relationship between them. On comparison of mean squared errors of both multiple linear regression and elastic net regression models, it was observed that the mean squared error of the elastic net regression model was lesser than that of the multiple linear regression model, making the elastic net regression model better than the multiple linear regression model in predicting results. By training these machine learning models with the data obtained during the pandemic period, these models can be applied to a pandemic situation and can make accurate predictions.

References

1. Invest India. https://www.investindia.gov.in/sector/food-processing. Last Accessed 26 Nov 2021
2. Uyanik GK, Guler N (2013) A study on multiple linear regression analysis. Proc Soc Behav Sci 106:234–240. Elsevier
3. Li Z, Cao X, Ding X, Chen H (2015) Prediction model of multiple linear regression analysis in grain production. In: 5th international conference on information engineering for mechanics and materials. Atlantis Press, Huhhot, pp 1290–1293
4. Dakić S, Mijić K (2020) Regression analysis of the impact of internal factors on return on assets: a case of meat processing enterprises in Serbia. Strat Manag 25:29–34. Center for evaluation in education and science
5. Rathgeber AW, Stockl S, Gaugler T (2020) Global climate impacts of agriculture: a meta-regression analysis of food production. J Clean Prod 276:122575. Elsevier
6. RBI Bulletin. https://www.rbi.org.in/Scripts/BS_ViewBulletin.aspx?Id=18823. Last Accessed 26 Nov 2021
7. Andaleeb SS, Conway C (2006) Customer satisfaction in the restaurant industry: an examination of the transaction-specific model. J Serv Mark 39:88. Academia
8. Bhutta NT, Hasan A (2013) Impact of firm specific factors on profitability of firms in food sector. Open J Account 2(2):19–25. Scientific Research Publishing
9. Chandrapala P, Knapkova A (2013) Firm-specific factors and financial performance of firms in the Czech Republic. Acta Universitatis Agriculturae et Silviculturae Mendelianae Brunensis 61(7):2183–2190. Mendel University Press
10. Nuševa D, Mijić K, Jakšić D (2017) The performances of coffee processors and coffee market in the Republic of Serbia. Ekonomika Poljoprivrede 64(1):307–322. Center for Evaluation in Education and Science
11. Tadić J, Jevtić J, Jančev N (2019) Modeling of critical profitability factors: empirical research from the food industry in Serbia. Ekonomika Poljoprivrede 66(2):411–422. Center for Evaluation in Education and Science

A Multipurpose Agribot

E. Sai Bhavinya, K. Vijaya Lakshmi, and P. Srinivas

Abstract Agriculture is the major source of food in India as it accounts for 17% of the total Gross Domestic Product (GDP). The world is currently witnessing a significant increase in population growth which demands for an increase in food production. But there is scarcity of labor in agriculture due to the National Rural Employment Generation Scheme (NREGS). This has an impact on the productivity, which in turn affects the GDP. Automation in agriculture can be considered as one of the solutions to this problem. It not only helps in improving the efficiency of the crop production but also helps in developing devices for performing various mechanical works in the fields. So, a solar powered multipurpose agribot that can perform certain operations like plowing land, sowing the seeds, and sprinkling of water is designed.

Keywords Agribot · Solar · Arduino · Automation

1 Introduction

More than 60% of Indian population depends on agriculture. India's agriculture sector accounts only for around 17% of the country's economy as depicted in Fig. 1 because the agricultural productivity is decreasing over the years [1].

One reason is that the farmers are declaring crop holidays due to deficiency of labor for farming and their increasing demand for wages. Other reasons are farmer suicides due to inadequate price for their crops and decrease in the agricultural land due to urbanization.

As seen in Fig. 2, agricultural land has continued to decrease because of rapid urbanization. Approximately 45% land across India is cultivated area, whereas 22% is forest land. Over the years, the urban expansion to the boundaries has lowered the agricultural area by 16.31% [2].

E. Sai Bhavinya (✉) · K. Vijaya Lakshmi · P. Srinivas
Velagapudi Ramakrishna Siddhartha Engineering College, Vijayawada, India
e-mail: bhavinya1999@gmail.com

N. M. Sivaram et al. (eds.), *Advances in Manufacturing, Automation, Design and Energy Technologies*, Lecture Notes in Mechanical Engineering,
https://doi.org/10.1007/978-981-99-1288-9_34

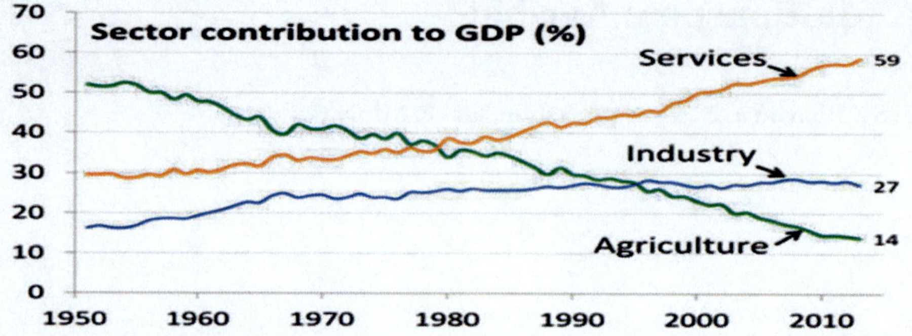

Fig. 1 Agriculture sector in India's economy

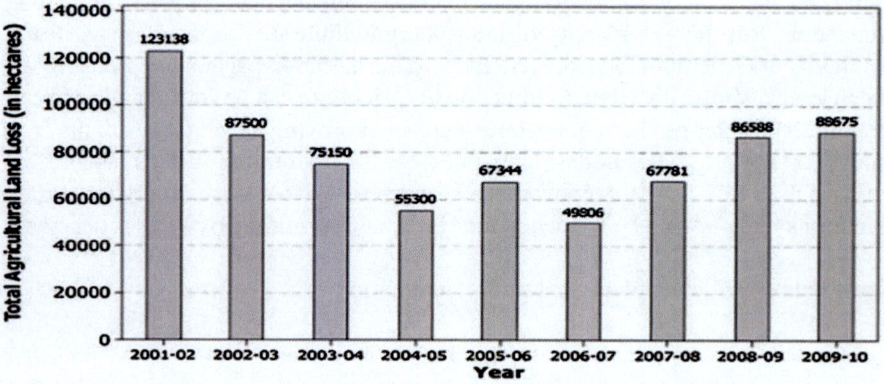

Fig. 2 Loss of agricultural land due to urbanization over years

According to the NCRB, the total number of farmer suicides reported in India each year is shown in Fig. 3. Farmer suicides account for 11.2% of all suicides. More than 23,000 farmers have committed suicide between 2009 and 2016 [3]. High debt burdens, poor government policies and corruption in subsidies, crop failure, personal issues, and family problems are all possible causes.

Scenarios like this results in scarcity of food as our country is highly populated. Drones and robots are gaining popularity in improving crop productivity due to the advantages of mechanization and precision agriculture [4, 5]. They also can be used for spraying pesticides in difficult to reach areas. This paper aims to develop a robot that performs farming activities such as plowing, sowing seeds, and spraying [6]. This type of devices operates 24/7, 365 days a year. Using solar power can decrease pollution [7]. It increases productivity at an affordable cost and reduces the need of manpower.

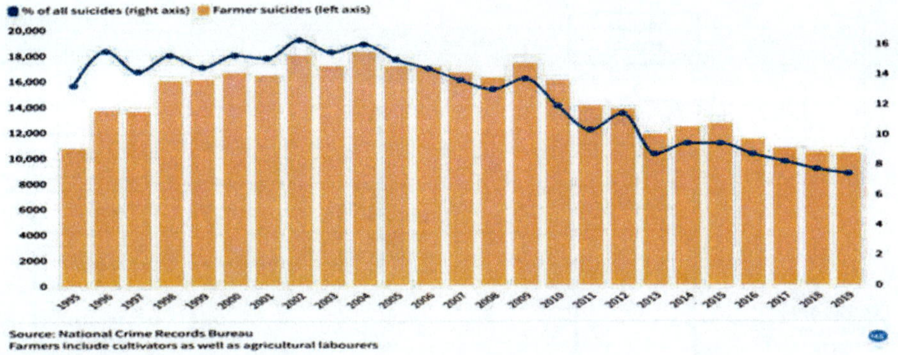

Fig. 3 Statistics of farmer suicides

2 Block Diagram of Proposed Model

Arduino Mega 2560 is the controller used for proposed multipurpose agricultural robot [8]. Solar panel is used as an energy source for robot. The solar panel specifications are shown in Table 1. The energy transferred from the solar panel is stored in the battery and used to power other components. Bluetooth HC-05 is interfaced to Arduino and Android Smartphone which sends commands for plowing, sowing, and spraying [7, 9]. Mobile application gives commands to Arduino via Bluetooth module as shown in Fig. 4. It instructs the robot to move forward and backward. L298N and L293D motor driver runs the DC motors which include plowing and sowing motors. Spraying motors are controlled by a relay as shown in [10].

Robot base is made up of aluminum frame of dimensions 16 × 12 × 9cm. This base relies on tires which run by DC motors [11]. Specifications of DC motor are speed: 10–1000 rpm, load current: 700 mA and shaft diameter: 6 mm. L298N motor drives four dc motors which run the robot. Relay is switch which is used for running sprinkling motor [12]. L293D motor driver simultaneously controls the speed and direction of two motors. L293D motor drives plowing and sowing motors [13]. Soil moisture sensor is used to sense the moisture content in soil [14]. Ultrasonic sensor measures distance and detects obstacle while the robot is moving.

DHT11 senses the temperature and humidity in surroundings [15, 16]. These three layers are attached to an IR sensor which intimates the level of seeds that is if the

Table 1 Solar panel specifications

Specifications	Range
Output power	75 W
Operating voltage	12 V
Cell conversion efficiency	>19%
Voltage at maximum power	20.0 V
Current at maximum power	3.75 Amps

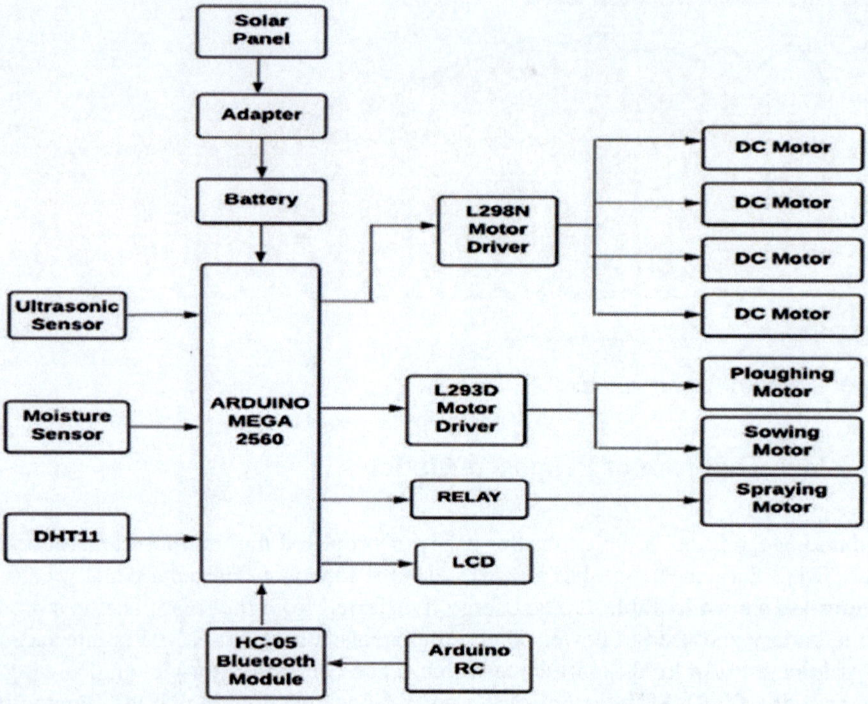

Fig. 4 Schematic diagram of proposed system

level is low, it gives a buzzer [17]. Sprinkling is done with the help of motor and relay. When the command is given, the relay gets on and the operation is performed [18].

3 Hardware and Software Implementation

Arduino software is downloaded from the web for programming purpose. Bluetooth HC-05 module is connected to the mobile phone. Serial commands are sent to the module. Arduino RC is the mobile application. In app, scan for devices and tap on HC-05. Bluetooth connection transmits the commands that is if "1"–robot moves front, "0"–stop all operations, "2"–seeding, "3"–Plowing down, "4"–Plowing up, and "W"–watering. Based on the command in the app, the corresponding operation is performed.

Ultrasonic sensor is interfaced to detect obstacle or end. If an end comes in the farm, the robot takes right and then turns right again and goes back into another way. Moisture sensor is used to measure the soil moisture content. DHT11 sensor is used to know the temperature and humidity of surroundings as shown in Fig. 5.

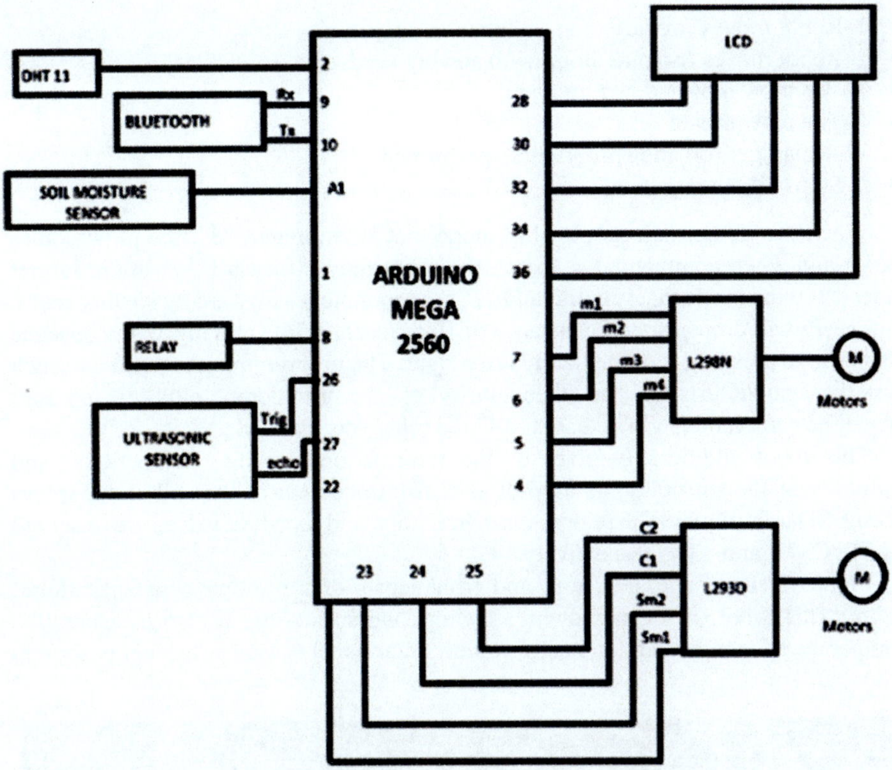

Fig. 5 Connection diagram of proposed system

It is programmed in such a way that if the robot is moving forward, it measure front temperature, if moving backwards it measures back temperature and displays. LM317 variable voltage regulator is selected for charging the battery from solar energy [19].

The energy from solar is transfer to 12 V/1.3Ah lead acid battery via adapter which is a connector. A diode provides reverse voltage protection to the battery when it is not charging. Arduino RC mobile application gives commands to robot.

4 Results and Discussion

A multipurpose agricultural robot is designed to perform operations like plowing, seed sowing, and sprinkling water. These actions are carried out in accordance with the commands passed by the Arduino RC App which are transmitted through Bluetooth. Initially, the farmer needs to select the mode of operation to be performed. The commands are as follows:

- 1–Robot moves forward
- 2–Robot moves forward along with sowing seeds
- 3–The plow moves down into land
- 4–The plow moves up
- W–Relay gets on and sprinkling is performed
- 0–Stops all operations.

When the farmer selects plowing mode that is command "3", the plow comes down and when command 1 is given, the robot moves forward. When the farmer selects sowing mode that is command "2" the robot moves forward by sowing seeds. The seeds will drop at specific distance of 10 cm each. While moving, if any obstacle comes in to the path, it immediately turns right. The programming is written in such way that again it turns right and then it turns left. The sprinkler motor is kept in water, when the command is given that is "W" the relay gets enabled.

This robot additionally gives us the information regarding temperature and humidity of the surroundings as well as moisture content of the soil and displays it on LCD as in Fig. 6. The temperature, humidity, and moisture values are observed as 31 °C, 73, and 15%, respectively.

Figure 7 shows the side, top, and front views of a multipurpose agricultural robot. This robot performs plowing, sowing, and sprinkling. It also measures the temperature, humidity, and moisture values. Solar panel is used as an energy source.

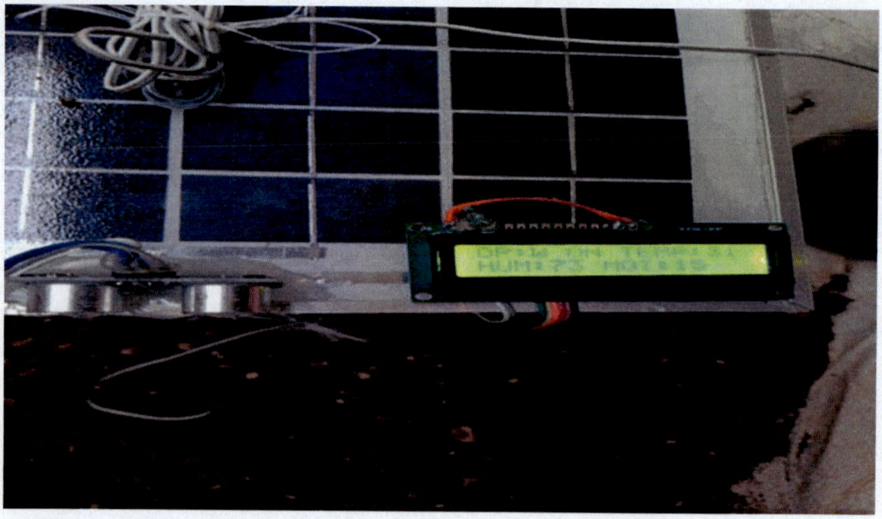

Fig. 6 LCD display of measured values

Fig. 7 Solar powered multipurpose agricultural robot

5 Conclusion

An autonomous multipurpose agricultural robot is developed in this paper to perform various complex farming tasks like seed sowing, plowing, and pesticide spraying. The operations are performed using the android app. The advantages of this robot are reduced human intervention and efficient resources utilization. Wireless communication between the robot and mobile using Bluetooth ensures the safety of the operator as there is no direct contact. This mechanism has a significant influence on agriculture as the farmer can get the crop in less time and with reduced labor cost. This work can be extended further using artificial intelligence or machine learning to make the device performance more efficient and accurate.

References

1. https://commons.wikimedia.org/wiki/File:1951_to_2013_Trend_Chart_of_Sector_Share_of_Total_GDP_for_each_year,_India.png
2. Pandey B, Seto KC (2015) Urbanization and agricultural land loss in India: comparing satellite estimates with census data. J Environ Manage 148:53–66
3. Opindia Homepage. https://www.opindia.com/2020/09/ncrb-data-reveals-farmer-suicides-the-lowest-in-india-since-1995-ncrb-report/
4. Mogili UR, Deepak BBVL (2021) Influence of drone rotors over droplet distribution in precision agriculture. Advanced manufacturing systems and innovative product design: select proceedings of IPDIMS 2020. Springer Singapore
5. Inkulu AK et al (2021) Challenges and opportunities in human robot collaboration context of industry 4.0-a state of the art review. Ind Robot Int J Robot Res Appl
6. Mogili UMR, Deepak BBVL (2018) Review on application of drone systems in precision agriculture. Procedia Comput Sci 133:502–509
7. Ramalingam D, Renitha P, Vignesh P, Karthick Sories MR (2018) A study about the design and fabrication of automatic seed sowing and fertilizer spraying. Int J Manag Technol Eng 8

8. Sujon MDI, Nasir R et al (2018) Agribot: Arduino controlled autonomous multi-purpose farm machinery robot for small to medium scale cultivation. In: International conference on intelligent autonomous systems (ICoIAS), Singapore
9. Senthilnathan N, Gupta S, Pureha K, Verma S (2018) Fabrication and automation of seed sowing machine using IOT. Int J Mech Eng Technol
10. Gan H, Lee WS (2018) Development of a navigation system for a smart farm. IFAC-Papers Online 51(17)
11. Berenstein R, Edan Y (2018) Automatic adjustable spraying device for site-specific agricultural application. IEEE Trans Autom Sci Eng 15(2):641–650
12. Ahmed BK, Prakash A et al (2020) An approach for digital farming using mobile robot. In: Second international conference on inventive research in computing applications (ICIRCA), Coimbatore, India
13. Ragavi B, Pavithra L, Sandhiyadevi P et al (2020) Smart agriculture with ai sensor by using agrobot. In: Fourth international conference on computing methodologies and communication (ICCMC), Erode, India
14. Ramesh kumar S, Sriram kalyan H et al (2018) Design and fabrication of autonomous robot for precision agriculture. Int J Mech Prod Eng Res Dev (IJMPERD)
15. Madiwalar S, Patil S et al (2020) A survey on solar powered autonomous multipurpose agricultural robot. In: 2nd international conference on innovative mechanisms for industry applications (ICIMIA), Bangalore, India
16. Amer G, Mudassir SMM, Malik MA (2015) Design and operation of Wi-Fi agribot integrated system. In: International conference on industrial instrumentation and control (ICIC), Pune, India
17. Kumar P, Ashok G (2021) Design and fabrication of smart seed sowing robot. Mater Today Proc 39
18. Cantelli L, Bonaccorso F, Longo D et al (2019) A small versatile electrical robot for autonomous spraying in agriculture. Agri Eng
19. Ranjitha B, Nikhitha MN, Aruna K, Afreen, Murthy BTV (2019) Solar powered autonomous multipurpose agricultural robot using Bluetooth/Android App. In: 3rd international conference on electronics, communication and aerospace technology (ICECA), Coimbatore, India

AI-Based Automated Surface Inspection of Steel Sheets

V. V. N. Satya Suresh, C. Ankith Kumar, and Y. Kalyani

Abstract Surface defects identified after rolling process in the manufacture of steel sheets are one of the most important factors affecting the quality. Though an inspector could identify some of the defects, but the process of manual inspection reduces productivity and also leads to inspector's fatigue. Commercially available software is prohibitively high in cost and cannot be used during online inspection. Hence, an automated system which is fast and identifies all the defects is the need of the hour. In this paper, machine vision process related to CAD/CAM is used to collect the images. An artificial intelligence (AI)-based technique is used to categorize the nature of defects through convolution neural network model under deep learning techniques. Coding is carried out using Python programming language. Samples of few numbers were tested which proved the efficacy of the algorithm.

Keywords Machine vision · Artificial intelligence · Deep learning · Surface defects

Abbreviations

AI	Artificial intelligence
CAD	Computer aided design
CAM	Computer aided manufacturing
CNN	Convolutional neural network
VGG	Visual geometry group
ANN	Artificial neural network
CIFAR	Canadian institute for advanced research
MNIST	Modified national institute of standards and technology
FERET	Face recognition technology
GPU	Graphics processing unit

V. V. N. Satya Suresh · C. Ankith Kumar (✉) · Y. Kalyani
Mahatma Gandhi Institute of Technology, Gandipet, Hyderabad, India
e-mail: ankithkumar.c@gmail.com

N. M. Sivaram et al. (eds.), *Advances in Manufacturing, Automation, Design and Energy Technologies*, Lecture Notes in Mechanical Engineering,
https://doi.org/10.1007/978-981-99-1288-9_35

SIFT	Scale invariant feature transform
SYM	Support vector machine
NORB	Nyu object recognition benchmark
SVHN	Street view house numbers
CPU	Central processing unit
NEU	North eastern university
CCD	Charge coupled device
ROI	Region of interest
RELU	Rectified linear unit

1 Introduction

Steel is the foremost principal metal of all metals particularly for its quantum and sort of use. Surface defects of steel sheets noticed after rolling process are an important quality aspect to be considered. Traditional surface inspection done by human inspectors is far from satisfactory. Manual inspectors were unable to identify all the surface due to fatigue, thus many defects get unnoticed. Many types of surface defects, viz., cracks, inclusion, scratch, pitches, and hole, may occur during rolling process. These damages not solely affect the physical appearance of the steel, additionally they decrease the corrosion resistance, wear withstanding capacity, and many other properties. The fact is that several times these defects not visible through naked eye until the rolling operation is over.

The surface quality of the steel which is rounded in coil shape is tested manually by cutting a random part of the steel coil of about 0.05% of the entire coil. Hence, major part is thus ignored. Manual inspectors cannot inspect all the damages and defects in a steel; thus, a huge number of defects are neglected by the human inspectors. Thus, the main objective of this work is to develop and incorporate an automated arrangement to the existing manufacturing process to identify all types of surface defects in the steel sheet.

2 Literature Review

In this section, different object detection techniques and surface inspection techniques have been briefly discussed. Traditional machine vision involves the use of stroboscopic method which works on the principle of light radiated from a source at high frequency so that a human operator identifies the defects on a belt conveyor moving at high speed. Often, it leads to optic fatigue and also unnoticed defects. There are other non-destructive detecting techniques, viz., infrared detection, laser detection, eddy current testing, and magnetic flux leakage detection. All these techniques cannot function effectively since the images require high resolution. Some non-contact inspection techniques, viz., electromagnetic testing and ultrasonic scanning use electromagnetic signals or ultrasound frequencies which get converted into optical signals. The results obtained need to be judged by professionals since they are not intuitive.

Neogi et al. [1] discussed automated steel inspection techniques with use of image processing and observed that steel surface pictures contain great amount of noise because of surface scale, etc., and defects are of non-uniform shapes, type and varies from one mill to a different. Sun et al. [2] discussed the software and hardware components required for visual inspection of steel and iron using deep learning-based network framework for intelligent manufacturing. They recommended that a camera whose frame rate is greater than that of speed of object should be adapted. Tang et al. [3] discussed in detail about the automated surface inspection system with its software and image processing algorithm which was able to detect and classify the defect of steel strip (cold rolling steel strip). Their experiment showed that additional analysis is to be carried out to meet the feature extraction of the defected image in real-time-detection which can further have to be expanded and optimized.

In recent years, deep learning based on object detection techniques has become the research topic. Zhao et al. [4] gave a review on deep learning based on the various object detection frameworks with a short introduction on the history of CNN and deep learning. They proposed many promising future ways to achieve an intensive understanding of the object detection concepts through the use of neural networks and connected learning systems. Balan et al. [5] surveyed on the feature extraction techniques in image processing. They mentioned some applications of the feature extraction, viz., data compression, data decomposition, and latent semantic analysis through pattern recognition and projection. In the same manner, Kumar and Bhatia [6] discussed various types of the extraction techniques and its application that might be useful in pattern recognition systems, as patterns shall have different orientations, style, etc.

Liu [7] proposed a technical approach for image recognition by involving an algorithm called CNN and described its structure, layers and working process. The thorough recognition method of CNN such as weight sharing features, layers, and characteristics of CNN were discussed. Krishna et al. [8] proposed a deep learning model that demonstrates the feasibility of adapting CNN with the VGG 16 architecture to detect fake currency notes, which is relatively better than traditional image processing methods. Although their dataset was small and failed to represent the real-world problem of fake currency, it was helpful to detect fake notes quickly under the data trained. Further, Neha et al. [9] evaluated the performance of the popular CNN for detecting objects in the real time by considering three CNN's for the object detection, viz., ResNet50, GoogleNet, and AlexNets with the CIFAR10 and CIFAR100 datasets. The result indicated that the accuracy of GoogleNet and ResNet50 were higher compared to AlexNet.

Ramprasath et al. [10] used a deep learning algorithm to attain the expected development in areas such as computer vision. They mentioned that by adding more layers to the CNN model and training it with large data using GPU higher accuracy could be achieved. Chauhan et al. [11] built a CNN model to estimate its performance on the detection datasets and image recognition on the MNIST and CIFAR-10 datasets using the central processing unit only. Krishna et al. [12] discussed image classification using the deep learning techniques. In this case, CNN alone was employed in AlexNet design for the classification purpose. Experimentally, it was noticed that the sample images are classified accurately, showing the effectiveness of the deep learning techniques.

Alex et al. [13] developed a deep CNN to categorize about one million images and classified them into one thousand categories. Their results showed that a huge, deep CNN has the ability to achieve an extremely difficult dataset using supervised learning. Similarly, Browne et al. [14] developed a CNN architecture with an application of practical image processing on a mobile robot to detect cracks on an autonomous robot. CNN was trained by employing a dataset of 48 $\times$ 48 sub regions drowned from thirty still images of 320 $\times$ 240 pixels frame taken from pre-recorded video. Akash et al. [15] developed a model by combining the CNN with Scale Invariant Feature Transform (SIFT) as CNN needs a huge dataset, whereas SIFT needs smaller datasets for object detection. They found that the integrated model has performed better than the traditional state of art model. Yim et al. [16] developed a two-track deep learning neural network algorithm which was composed of pre-trained CNN and the fully connected layer algorithm. In the majority of cases, features from the top layers of CNN are utilized for the classification but those features might not contain helpful information to predict an image accurately, whereas in few cases, features from the lower layers possess more power compared to top layers. Their model improved the pre-trained CNN models accuracy by making use of the employed intermediate layer and its features. Nitesh et al. [17] developed a CNN model with eight convolutional layers, three fully connected layers which is a modified version of VGG architecture to classify and find objects like knives, water bottles, and utensils in visual pictures. The training accuracy of previous AlexNet and ZfNet models were around 85% and testing accuracy is around 80%. Their proposed model attained an accuracy of 98% in training and an accuracy of 92.2% in testing, and performed better than other ZfNet, AlexNet, and VGG 13 layers and could be used for sharp object detection.

From the literature, it has been observed that very little work has been carried out to detect the surface defects in steel sheets and thus it is necessary to develop an automated arrangement to identify defects in the steel sheet. The occurrence of these defects depends on various factors such as foreign particles sticking to rolls, dirt on strip, lack of strip flatness, carbon soot, oxide layer formation, and chemical attacks. Manual inspection is hardly satisfactory since the process need high accuracy at high conveyor line speed. Alternatively, a method of collecting reliable and accurate images during the online process is very important. Since a permanent solution is aimed to detect the nature of defects, a large collection of image data is needed.

3 Problem Statement

Processing of cold and hot rolled steel sheets is carried out by cutting and slitting operations to produce blanks of across-the-board sizes, measuring up to 1600 mm × 2500 mm. During manual inspection, the operator manually inspects each sheet at the offline inspection stations. It takes nearly 20 s at the offline inspection stations to inspect each side of the sheet viz. bottom and top surfaces.

Surface defects, viz., Dent Marks, Roll Mark, SPM Pinch Mark, SPM Matte less Mark, Gauge Mark, Oxidation mark, Rubbing Mark, Sliver, Rust patch, crazing, pitted surface, rolled-in-scale, inclusion, patches, scratches Lamination/Flying scab, Carbon soot mark, Coils slips, Black/brown patches, Holes, Scratch, Fold marks, etc., require a minimum size of 5 mm to be detected. The proposed system should be able to integrate with the existing manufacturing process. It should be able to analyze the severity and the occurrence of defects based on the identification and classification of defects.

4 Dataset

In this work, NEU-DET dataset [18] which was developed by North-Eastern University, USA, has been used. The dataset composed of six common surface defects of a metal such as crazing, patches, pitted surface, inclusion, rolled-in-scale, scratches, and each class of defect contains 300 image samples. Therefore, a total number of 1800 of grayscale images were considered in the NEU-DET Dataset. Additionally, another class of 118 non-defect images were added to the 1800 grayscale images which were captured in a steel industry using a mobile device to classify whether the sample is a defected one or not. If it contains defects, they shall be classified into the type of defect. Initially, the model was tested on a defect-free steel sheet to validate the accuracy of the model. The results showed an accuracy of 99% in predicting the images generated from the model. Later, the model was extended to predict and categorize the six types of defects nature of defects, viz., Crazing, Inclusion, Patches, Pitted surface, Rolled-in-scale and Scratches.

The developed model performed well in predicting the defected images which contain sample test images as predicted by the model. All the images in the model were able to predict the type of defect, and the probability of defection is around 99%. The accuracy of the model was found to be 99.77% under training and 97.40% under validation.

Therefore, a total number of 1918 images of six different classes were used as a dataset for training the model. Figure 1 shows six different classes of defected images.

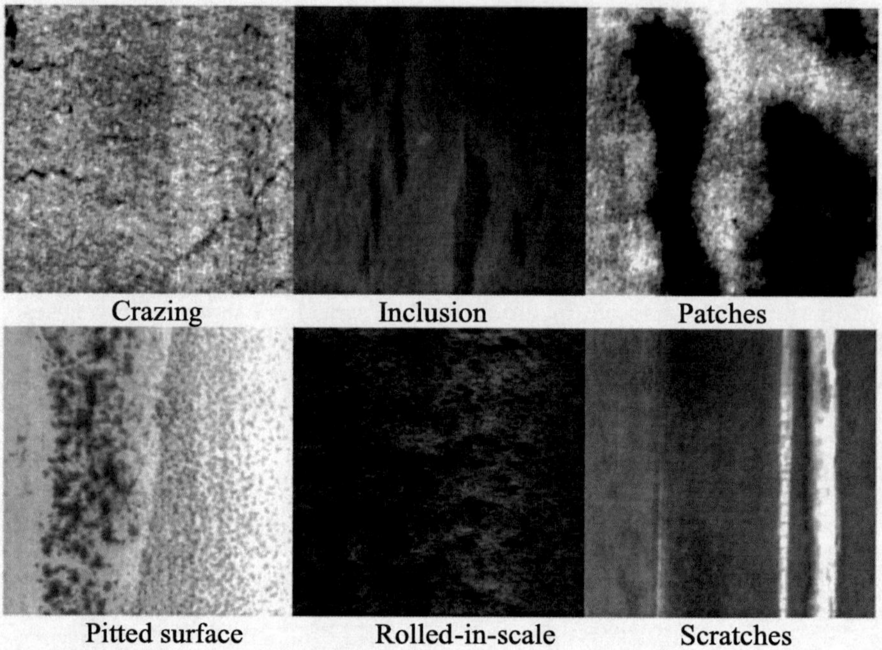

Fig. 1 Dataset containing the nature of defects

5 Methodology

The proposed methodology involves the use of the latest artificial intelligence (AI) technologies to detect and classify various defects in rolled steel sheets. AI with deep learning techniques that mimic human behavior helps to identify defects in steel sheets by visual symptoms and appearance.

Initially, the image captured by a CCD camera is considered to be the region of interest (ROI) to identify and classify the nature of surface defect. It is necessary to use smoothing technology and filter to reduce the noise contained in the images. The image is compared with the stored image in the ROI. The image is segmented to find the defects region through region splitting. Later, it is characterized to categorize into gray ones to correlate as area. Finally, the classification of defects is carried out by using CNN architecture on a server with a large backup memory.

Coding part was performed using Google Colaboratory by importing libraries, viz., TensorFlow, Matplotlib, OpenCV, Keras, NumPy, Scikit-Learn, etc., into the drive. Images were loaded by giving a path towards the images directory and later pre-processed the images by reshaping them and converting them into NumPy arrays. Training and testing data were split into 90% for training and 10% for testing. The CNN has three convolutional layers followed by max pooling layers and a fully connected layer. The output layer has seven neurons for seven different classes, all of these layers use Rectified Linear Unit (RELU) as its activation function while the last layer uses SoftMax as its activation function. The mathematical equation of RELU is

$$f(x) = \begin{cases} 0 forx < 0 \\ xforx \geq 0 \end{cases} \text{and Range} = 0 \text{ to } \infty. \quad (1)$$

The mathematical equation of SoftMax is

$$(x_i) = \frac{\exp(x_i)}{\sum_j \exp(x_j)} \text{and Range} = 0 \text{ to } 1. \quad (2)$$

The CNN model was trained with 30 epochs and a batch size of 64; however, these values may be changed accordingly in order to make changes in model performance. For compiling the model, Categorical Cross Entropy loss function and Adam optimizer were used. The performance of other CNN models such as VGG16 and Alex net was were also carried out but they failed to classify the images correctly as they require large data for training.

6 Results and Discussion

Initially, the model was tested on a defect-free steel sheet to validate the accuracy of the model. The results showed an accuracy of 99% in predicting the images generated from the model. Later, the model was extended to predict and categorize the nature of defects. The developed model performed well in predicting the defected images as shown in the Fig. 2 which contains sample test images as predicted by the model. All the images in the model were able to predict the type of defect, and the probability of defection is around 99%.

The model performed well in predicting the images of the testing dataset. The loss of the model was found to be 1.75% under training and 9.47% under validation as shown in Fig. 3. The model was saved in hierarchical data format which contained the weights and other parameters.

7 Conclusion

In this paper, a methodology was proposed to detect defects in a steel sheet using deep learning of artificial intelligence. A CNN model was trained on 1918 images of six classes of defects and a class of non-defect images. Results showed that the model is very accurate in predicting the class of steel defect. Coding part and the model were accurate in predicting the images from the testing set which were obtained through graphs generated for training and validation loss.

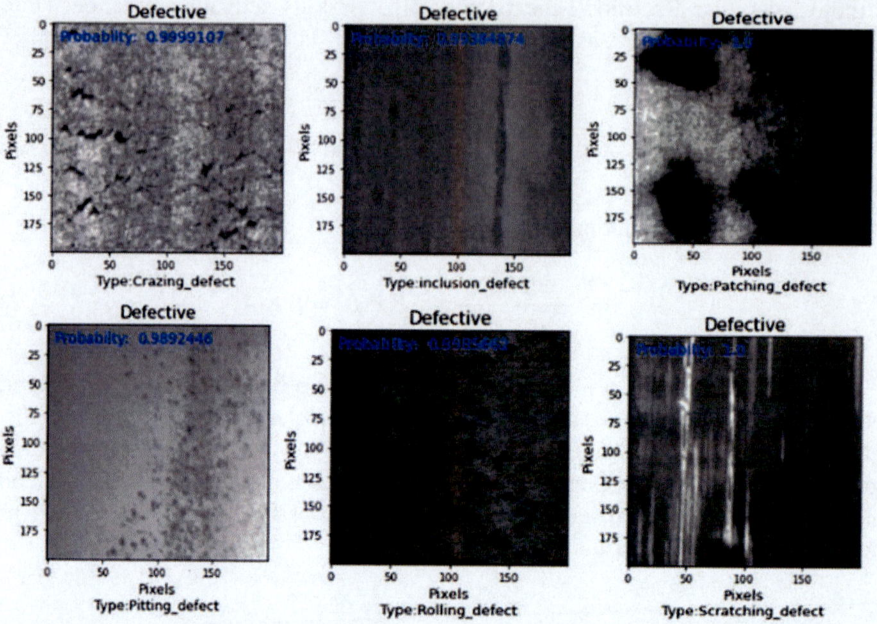

Fig. 2 Defects as predicted by the model

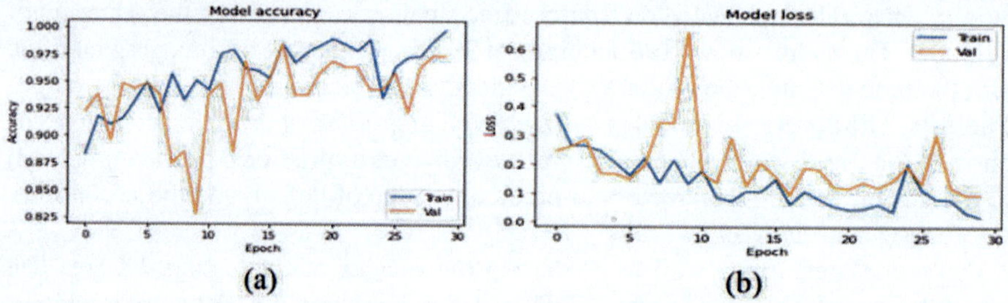

Fig. 3 Training and validation curves for checking **a** model accuracy, **b** model loss

8 Future Work

Further work can be extended to make the system work in real time by using Raspberry Pi microcontroller and Raspberry Pi camera, where the Raspberry Pi camera will capture the images to be loaded in the Raspberry Pi microcontroller. It can be extended to establish human–robot collaboration to pass the signals from human to robot [19, 20].

References

1. Neogi N, Mohanta DK, Dutta PK (2014) Review of vision-based steel surface inspection systems. EURASIP J Image Video Process 1–19. Springer
2. Sun X, Gu J, Tang S, Li J (2018) Research progress of visual inspection technology of steel products—a review. Appl Sci 8(2195):1–25
3. Tang B, Kong J, Wang X, Chen L (2009) Surface inspection system of steel strip based on machine vision. In: 1st international workshop on database technology and applications. IEEE, Wuhan, China, pp 359–362
4. Zhao Z, Zheng P, Xu S, Wu X (2019) Object detection with deep learning: a review. IEEE Trans Neural Netw Learn Syst 30(11):3212–3232
5. Balan SP, Sunny EL (2018) Survey on feature extraction techniques in image processing. Int J Res Appl Sci Eng Technol 6(3):217–222
6. Kumar G, Bhatia PK (2014) A detailed review of feature extraction in image processing systems. In: 4th international conference on advanced computing and communication technologies. IEEE, Rohtak, pp 5–12
7. Liu YH (2018) Feature extraction and image recognition with convolutional neural networks. In: 1st international conference on advanced algorithms and control engineering. IOP Publishing, pp 1–7
8. Krishna G, Pooja G, Ram B, Radha V, Rajarajeswari P (2019) Recognition of fake currency note using convolutional neural networks. Int J Innov Technol Explor Eng 8(5):58–63
9. Neha S, Vibhor J, Anju M (2018) An analysis of convolutional neural networks for image classification. In: International conference on computational intelligence and data science. Elsevier, The Northcap University, India, pp 377–384
10. Ramprasath M, Anand VM, Hariharan S (2018) Image classification using convolutional neural networks. Int J Pure Appl Math 119(17):1307–1319
11. Chauhan R, Ghanshala KK, Joshi RC (2018) Convolutional neural network (CNN) for image detection and recognition. In: 1st international conference on secure cyber computing and communication. IEEE, Jalandhar, India, pp 278–282
12. Krishna MM, Neelima M, Harshali M, Rao MVG (2018) Image classification using deep learning. Int J Eng Technol 7(2.7):614–617
13. Alex K, Ilya S, Hinton GE (2017) ImageNet classification with deep convolutional neural networks. 60(6):84–90. Association for Computing Machinery
14. Browne M, Ghidary SS (2003) Convolutional neural networks for image processing: an application in robot vision. In: Gedeon TD, Fung LCC (eds) 16th Australian conference on artificial intelligence, AI 2003 LNCS, vol 2903. Springer, Heidelberg, pp 641–652
15. Akash T, Ajay TV, Tarunkanth D, Selva J (2018) Real time object detection using CNN. Int J Eng Technol 7(2.24):33–36
16. Yim J, Ju J, Jung H, Kim J (2015) Image classification using convolutional neural networks with multi-stage feature. In: Kim JH, Yang W, Jo J, Sincak P, Myung H (eds) 3rd international conference on robot intelligent technology and applications, AISC, vol 345. Springer, Switzerland, pp 587–594
17. Nitesh R, Anandhanarayanan K, Chelliah B, Govindaraj R (2019) Computer vision framework for visual sharp object detection using deep learning model. Int J Eng Adv Technol 477–481. Blue Eyes Intelligence Engineering and Sciences Publication
18. NEU Dataset. http://faculty.neu.edu.cn/yunhyan/NEU_surface_defect_database.html. Last Accessed 08 Oct 2020
19. Oudah M, Al-Naji A, Chahl J (2020) Hand gesture recognition based on computer vision: a review of techniques. J Imaging 6(8):73
20. Inkulu AK, Bahubalendruni MR, Dara A, Sankaranarayanasamy K (2021) Challenges and opportunities in human robot collaboration context of industry 4.0-a state of the art review. Ind Robot. https://doi.org/10.1108/IR-04-2021-0077

Design of Smart Glove for Sign Language Interpretation Using NLP and RNN

Akshay V Nayak, B. S. Karthik, L. C. Sudhanva, Akshay A Ganger, K. S. Rekha, and K. R. Prakash

Abstract The sign language interpreter aims to build a communication bridge between a speech- and hearing-impaired person and an able person. It has been a challenging task to understand sign language which uses finger shapes and hand movements. There is a need to build an interpreter to convey the ideas of deaf and mute to able people. The proposed system consists of a pair of wearable gloves with sensors embedded in them, which convert the hand gestures to text and speech. It uses technologies like natural language processing (NLP) and recurrent neural networks (RNN) for processing signals. The hand glove consists of sensors that track the movement of fingers and hands. The sensed values are sent to the processing unit using a wireless module. The dataset is prepared using simulation software. The machine learning model is built based on the training set, to help in identifying the hand gesture. A mobile phone application runs this machine learning model. The phone processes the data received from the glove using the proposed model to identify the letter or word. These words and letters are interpreted as a sentence using NLP. The speaker outputs the natural voice from the mobile phone.

Keywords Sign language interpreter · Natural language processing (NLP) · Recurrent neural networks (RNN) · Machine learning

1 Introduction

It has been a major challenge to establish efficient communication between mute people and normal people for several decades. Presently, non-speakers use American Sign Language (ASL) or Indian Sign Language (ISL) to communicate; however, most of this does not give a completely viable solution for Indian languages, therefore there is a need for a system to bridge the gap in the communication. The system should

Akshay V Nayak (✉) · Karthik B S · Sudhanva L C · Akshay A Ganger · Rekha K.S · K. R. Prakash
The National Institute of Engineering, Mysuru 570008, India
e-mail: vakshaynayak@gmail.com

N. M. Sivaram et al. (eds.), *Advances in Manufacturing, Automation, Design and Energy Technologies*, Lecture Notes in Mechanical Engineering,
https://doi.org/10.1007/978-981-99-1288-9_36

capture the movement of the hand through portable gloves fitted with sensors. The smart gloves prepared is fitted with flex sensors, an accelerometer, and controllers to capture the bend and rotation of the hand. The data captured is predicted by the ML model. Input/Output is processed using a mobile phone, and the output from the system is either text or speech.

2 Related Works

In [1], Khan et al have developed a product called "Talking Hands". The American Sign Language is used to capture the sign of the mute person. The sensor gloves are designed to capture the signs. The sensor values are recognized by artificial neural networks and the captured values are categorized in 24 alphabets. The system can be used in public places as it is very dynamic in the recognition of hand gestures.

In [2], Ahmed et al have done a detailed survey and A review on systems-based sensory gloves for sign language recognition. A survey is made on the sign language approaches like sign language based on sensor, vision, and hybrid model. The different stages of the collection of datasets, recognition of sign language gestures using hand gloves are discussed. The major components of the glove-based recognition system are hardware components and software components. The hardware components of the glove-based recognition system are composed of three main units: input, processing, output, and tools.

In [3], Vijayakumar et al have presented an electronic system to help mute people to converse with normal people. The system will identify the hand gestures and convert the identified gestures into speech and text. The prototype is built using Raspberry Pi3 and sensors. The authors have tried to reduce the number of sensors to make the system lightweight and robust. The Raspbian operating system is mounted on the Raspberry Pi.

In [4], Shen et al have designed two types of data gloves using soft bending sensors. The sensors are built using electrical and mechanical components with good sensitivity. The sensors are flexible and stretchable. The soft bending structure will accurately capture the motion of the hand when a large amount of skin stretch is required. The two sensor gloves designed can be used for various applications. The results obtained from the two gloves were evaluated with some parameters.

In [5], Bairagi et al have proposed a system to identify the hand gesture. The gloves are designed with an accelerometer sensor and resistance sensor which are very easy to wear, and it can measure the bend and movement of the hand accurately. The microcontroller is used for the recognition of hand movement. An Android App is developed to convert the signal to speech. The system produces more accurate results than image processing-based hand gesture recognition.

In [6], Nagavkar et al have proposed a communication system for mute people to translate sign language into computer text using a hand glove. The hand gloves were constructed with flex sensors to monitor finger movements. The neural network was used to train and classify the 26 letters of the alphabet.

In [7], Manjunatha et al have proposed a non-verbal sign language for deaf people. As the language interpretation was difficult as the language was not from a common source of origin. A device called a Deaf-mute Language Interpreter is used to convert from hand gestures including a facial expression to speech. The developed system would identify ten characters from the American Sign Language (ASL).

In [8], Mohan et al have designed smart hand gloves to ease the communication of disabled people with society. The smart gloves help the disabled to convert the hand signals to text or pre-recorded voice. The gloves designed will help disabled people to control their home appliances too. The gesture datasets stored will help to identify the hand gestures and express the same as voice.

In [9], Abhilasha et al have proposed a robotic gadget that tries to bridge the gap between dumb and normal people. The robotic gadget converts the American Sign Language into appropriate text or speech. The glove is designed with flex sensors, accelerometer and microcontroller for input, processing, and output. The proposed glove is a robotic gadget that interprets American Sign Language Standard into text or speech to evacuate the information transmission gap between the mute and the ordinary public. This glove has been actualized with the assistance of flex sensors, accelerometer, microcontroller (Arduino Nano), and the Bluetooth chip.

In [10], Katti et al have proposed a system to make normal people understand the sign language of dumb people. The gloves are built using a set of sensors, Arduino Mega Board, MPU6050, and Raspberry Pi. The flex sensors which are attached to every finger will capture the motion and fingers bend and process the sign and produce the speech output.

In [11], Karibasappa have proposed a device to bridge the gap between dumb people and normal people to minimize this barrier a wireless glove was designed using flex sensors and an accelerometer. The speech synthesizer circuit converts the movement of hand gestures into speech output.

In [12], Shareef et al have proposed a system that helps to reduce the communication gap in the communication systems. The proposed system is built using Raspberry Pi and Arduino nano as an A/D converter. The input is taken through flex sensors and an accelerometer. The sensed values are stored in the database. The sign made is compared with the stored value in the database and suitable text and speech messages are produced.

In [13], Kamble et al proposed a system for people with mobility issues and mute persons. The designed system will control the wheelchair based on finger movement thereby providing the speech output. The flex sensor will control the wheelchair and provide audio output. The wheelchair is controlled with hand gestures.

3 Existing System

Sign language recognition systems are used to translate sign language into text or speech form which is understood by normal people. Many efforts have been made to establish the communication between individuals and deaf-mute. There are three approaches: vision based, sensor based, hybrid based, to capture configurations of hand and recognize the corresponding gestures.

The vision-based approach uses cameras to capture the gestures. Various existing systems use the vision-based approach because it removes the need for sensors and reduces expenses. Although it only requires a camera, the disadvantage is that the limited field of view of capturing device and the requirement of another individual to hold the camera.

The sensor-based approach uses different sensors like Inertial Measurement Units (IMUs), However, mounting many sensors to capture the data of hands and fingers was a tedious task, and the hand glove so developed will be complex and managing the device is an issue.

The hybrid device will have a camera and sensors both used to effectively capture the data and correlate the dataset with each other and develop an efficient system, but again, the device is more complex. The existing systems are mostly based on American sign language and only for recognition of alphabets and numbers.

4 Proposed System

A pair of gloves is built to sense the accurate pose of each finger and the orientation of the hand. Sensors will fetch the data, and the values of the sensor inputs are sent to the mobile phone through a bluetooth module. A dataset is prepared, and a deep learning model is trained on these datasets. The model consists of techniques such as recurrent neural networks and natural language processing. A mobile application is built to run this machine learning model. The output of the model would be a text which is equivalent to the sign language performed. The output text can be read using the text-to-speech system.

5 Comparision of Existing System with Proposed System

The proposed sign language interpreter aims to overcome the problems faced by the three existing systems. The proposed system uses flex sensors for finger movement capture and IMU sensor to capture hand and wrist rotation. With new techniques in gloves building, reduction in the number of hardware mounted on the gloves is achieved. Also flex sensors are embedded to the glove in form of layers which is closely fit to cloth/leather of the gloves making it more rigid and maintenance-free system. A small single control unit sits on gloves making total systems elegant and the advantage of these gloves based system over vision-based systems is that gloves directly give the required data in terms of voltage values which can be easily used and computed as required rather than processing raw image data into meaningful values.

6 Design and Implementation

The sign language interpreter is designed to support the mute-dumb person. There are several stages in the interpreter such as an input phase, processing phase, and deployment phase.

- **Input phase**:

  The input phase will fetch the data with the help of gloves. A pair of gloves is built to sense the accurate pose of each finger and the orientation of the hand. Both the gloves should be integrated with the accelerometer, gyroscope, and a flexible sensor. The obtained value is filtered to reduce noise. The values of the sensor inputs are sent to the mobile phone through a BLE module.
- **Processing phase**:

  A 3D model of the hand with gloves is simulated in software. This software helps in debugging the errors in the glove and its implementation. The software also helps in preparing datasets. A deep learning model is trained on these datasets. This model consists of techniques such as recurrent neural networks(RNN), natural language processing(NLP), and long-short term memory(LSTM). The output of the model would be a text which is equivalent to the sign language performed. This output text can be further processed using the Text-To-Speech (TTS) system.
- **Deployment phase**:

  A mobile phone application is built to run this machine learning model. The application will collaborate with the hand glove using a wireless module and is tested on the ML model.

The processing of the fetched data will be done based on the predictive model designed, and the output will be in the form of text or speech through a mobile application. Figure 1 shows the flowchart of the working of the system.

6.1 Algorithm of Working of the System

Step 1: Start.

Step 2: If the device is off go to step 10.

Step 3: Input flex and IMU sensor values.

Step 4: Process the data received.

Step 5: Send the values to the mobile phone.

Step 6: The values received is passed to the trained model to get the gesture values.

Step 7: Natural language text is extracted using the gesture value.

Fig. 1 Flowchart of the working of the system

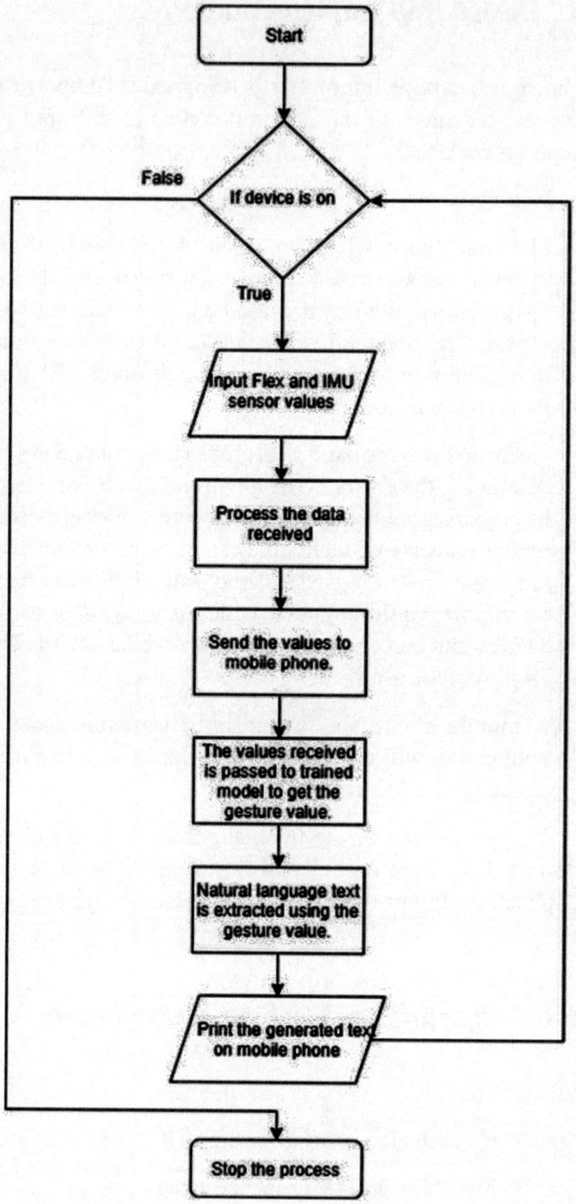

Step 8: Print the generated text on the mobile phone.

Step 9: Go to step 2.

Step 10: Stop.

7 Results

The testbed is set up and able to monitor the movement of the hand gestures. The input is sensed and processed with the patterns stored and appropriate results are displayed. Figure 2. shows the glove prototype for hand gestures.

The hand glove will sense the hand gestures based on the movement of the hand. Figure 3 shows the gesture for number 2 and the corresponding output.

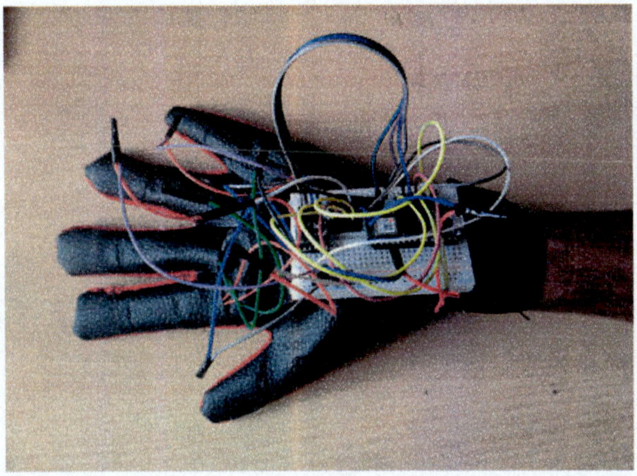

Fig. 2 Design of the smart glove for sign language interpretation

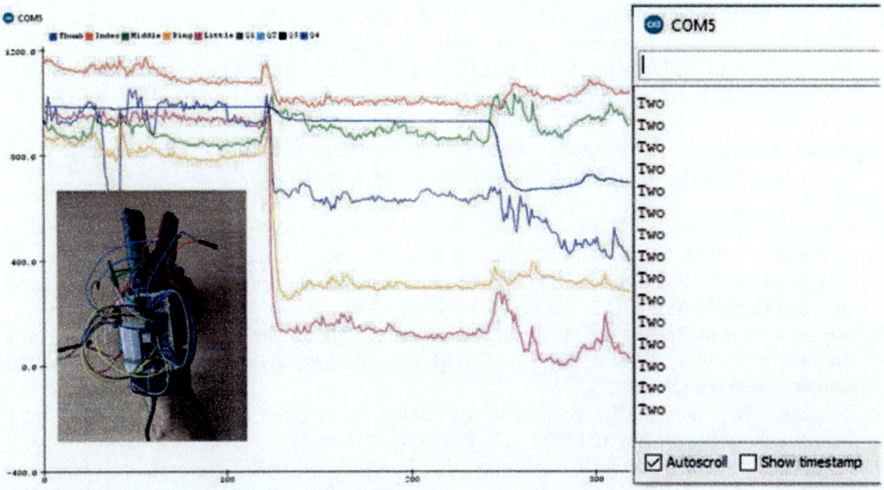

Fig. 3 Output for number 2 and the corresponding output

8 Conclusion

The glove prepared can meet the needs of the mute people in establishing communication with society. The Indian Sign Language has several variations, and the work carried by us concentrate on the Indian standard gestures. The Sign Language Interpreter is supported by the concepts of RNN and NLP and robust enough to catch the dynamic changes in the hand gestures. The system is lightweight and smart enough to cater for the needs and fill the gaps in understanding the hand gestures of the Indian Sign Language. Future work may be carried to identify the advanced sign gestures of Indian Sign Language. The system can be extended to establish human–robot collaboration to pass the signals from human to robot.

Acknowledgements This work was supported by the ARTPARK, AI & Robotics Technology Park, promoted by IISC. We are grateful to our project mentor Dr.Ipsita Biswas Mahapatra, Lead–Academic Initiative AI & Robotics Technology Park (ARTPARK), Indian Institute of Science, Bangalore.

References

1. Khan YN, Mehdi SA, Sign language recognition using sensor gloves. Res Gate. https://doi.org/10.1109/ICONIP.2002.1201884. Source: IEEE Xplore, Research Gate
2. Ahmed MA (2018) A review on systems-based sensory gloves for sign language recognition state of the art between 2007 and 2017. Sensors 18:2208. www.mdpi.com/journal/sensors; https://doi.org/10.3390/s18072208
3. Vijayakumar KP, Nair A, Tomar N (2020) Hand gesture to speech and text conversion device. Int J Innov Technol Explor Eng (IJITEE) 9(5). ISSN: 2278-3075
4. Shen Z, Yi J, Li X, Lo MHP, Chen MZQ, Hu Y, Wang Z (2016) A soft stretchable bending sensor and data glove applications. Robot Biomim 3:22. https://doi.org/10.1186/s40638-016-0051-1
5. Bairagi VK (2017) Gloves based hand gesture recognition using Indıan sign language. Int J Latest Trends Eng Technol 8(1–4):131–137. https://doi.org/10.21172/1.841.23, e-ISSN:2278-621X
6. Nagavkar G, Potdar K (2017) A neural network classifier based American sign language recognition system. IJSRD—Int J Sci Res & Dev 5(7). ISSN (online): 2321-0613
7. Manjunatha KN, Raghu N (2020) Design of sign language interpreter for specially challenged community. Eur J Mol & Clin Med 7(7) ISSN 2515-8260
8. Mohan P, Mohan Raj M, Kathirvel M (2020) Smart speaking glove for deaf and dumb. Int J Eng Res & Technol (IJERT). ISSN: 2278-0181. Published by, www.ijert.org NCFTET—2020 Conference Proceedings
9. Chougule AC, Sannakki SS, Rajpurohit VS (2019) Smart glove for hearing-impaired. Int J Innov Technol Explor Eng (IJITEE) 8(6S4). ISSN: 2278-3075
10. Katti SN, Sirsi SS, Prakash V, Vinith Raj BU, Jha V (2021) Talking gloves: sign language to speech conversion for deaf and mute person. J Univ Shanghai Sci Technol 23:(9) ISSN: 1007-6735

11. Karibasappa R, Choodarathnakara AL (2018) Smart glove based gesture vocalizer for deaf and dumb. Int J Manag, Technol Eng 8(XI):81. ISSN NO: 2249-7455
12. Shareef SR, Hussain MM, Gupta A, Aslam HA (2020) Hand gesture recognition system for deaf and dumb. Int J Multidiscip Curr Educ Res (IJMCER) 2(4):82–88. ISSN: 2581-7027
13. Kamble S, Naukudkar N, Suresh A, Bansode A (2020) Flex sensor based glove to control wheel chair and sign language translator for speech impaired people. Int J Scı Eng Res 11(6). ISSN 2229-5518

A Smartphone-Based Digital Image Colorimetry Model for Identifying Fuel Types in Downstream Petroleum Sector

S. Hemachandiran, G. Aghila, and R. Siddharth

Abstract The industry 4.0 in petroleum sector integrates the advancement of Internet of Things (IoT) application to increase the automation and digitization. The fuel adulteration is the major problem faced by the consumer and retailer in the downstream sector. The sophisticated laboratory-based testing is not a feasible solution to identify fuel adulteration. In the process of automating the identification of fuel adulteration, characterizing the fuel types is the first and foremost step. In this research, a smartphone-based colorimetry model is proposed to identify the fuel types. The fuel samples are collected and digitized using a custom designed imaging box. This research highlights the importance of data acquisition and its different characterizations such as color space and lighting conditions. The proposed research also identifies Delta E is comparatively better than Euclidean distance to identify the subtle color changes in the fuel types.

Keywords Color identification · Delta E · Euclidean distance

1 Introduction

Adulteration is one of the major issues in the downstream petroleum sector. In petroleum industry, adulteration is defined as the illegal or unauthorized introduction of foreign substance into the quality petroleum products [1]. This results shortfall in meeting the requirements and deteriorating the quality of the petroleum products. Foreign substances like kerosene or lubricants are called adulterants which alter and degrade the quality of the qualified petroleum product. Adulteration leads to malfunctioning/seizure of the engine, failure of vehicle components, increased emission of toxic substances like carcinogenic pollutants and financial losses to national economy where the lower taxed kerosene is mixed with higher taxed petrol/diesel. The procedures for quality assurance and control (QA/QC) are critical parameters

S. Hemachandiran (✉) · G. Aghila · R. . Siddharth
National Institute of Technology Puducherry, Karaikal, India
e-mail: spahhema@gmail.com

N. M. Sivaram et al. (eds.), *Advances in Manufacturing, Automation, Design and Energy Technologies*, Lecture Notes in Mechanical Engineering,
https://doi.org/10.1007/978-981-99-1288-9_37

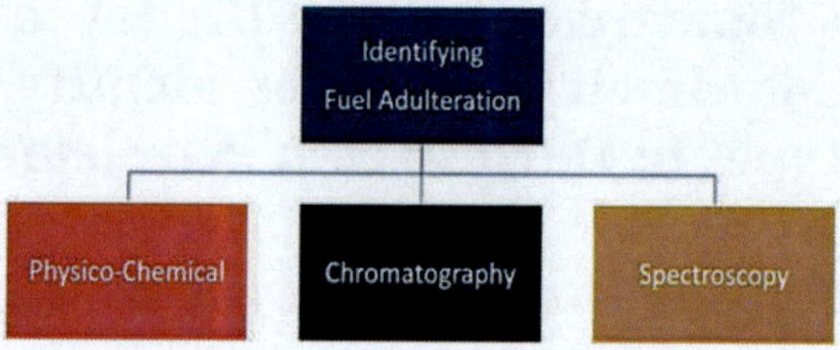

Fig. 1 Existing approaches to identify petroleum adulteration

in ensuring the quality of petroleum fuels. The QA/QC of the final product, what the consumers are consuming should strictly complied to the standards and protocols formed by some of the popular organization like International organization for standardization (ISO), Bureau of Indian standards (BIS) and American society for testing and materials (ASTM) [2]. These standards emphasize the major parameters density, sulfur content, flash points, kinematic viscosity and aromatic are to be evaluated. Currently, the laboratory-based analytical method is used to identify petroleum adulteration by evaluating any one or more listed parameters. The widely adapted methods are grouped into three categories (as represented in Fig. 1) based on the analytical approach that requires sophisticated infrastructure: Physico-chemical properties, Chromatography and Spectroscopy.

2 Literature Review

2.1 *Physico-Chemical Properties*

Barbeira et al. [3] studied the adulteration in petrol by evaluating the physicochemical properties like density, distillation, octane number (research octane number (RON) and motor octane number (MON)) and cetane number after testing with varying ethanol concentration. Onojake et al. [4] compared the adulterated and unadulterated petroleum product samples by comparing the physicochemical properties. From this study, it is identified that there is a significant variation in the octane number, Reid vapor pressure, specific gravity and distillation. Generally, kerosene is used as the primary adulterant in diesel. The studies in [5, 6] use kinematic viscosity measurement to identify the varying amount of kerosene adulteration in diesel.

2.2 *Chromatography*

Chromatography is an analytical approach most commonly used for separating a mixture into individual components. The most commonly used chromatography methods for finding petroleum adulteration are gas chromatography and liquid chromatography. Gas chromatography is an efficient technique that reveals molecular level information with high accuracy. Because of the likelihood of earlier presence, detecting solvent adulteration in the petroleum product is difficult; nevertheless, their disproportionate presence may be easily monitored using gas chromatography with flame ionization detector (GC-FID) [7].

2.3 *Spectroscopy*

Spectroscopy is a study of interaction (in terms of absorption and emission) between the petroleum product and electromagnetic radiation. Spectroscopy is broadly classified into four types: Mass spectrometry, Raman spectroscopy, NMR spectroscopy and Colorimetry spectroscopy.

Mass spectrometry: Mass spectrometry detects and separates ions of different masses and it is coupled with gas chromatography (GC–MS) to find petroleum adulteration. In the literature, Quach et al. [8] utilized GC–MS method for quantitative analysis to estimate methyl tert-butyl ether (MTBE) and benzene in petrol for adulteration monitoring. Alberici et al. [9] introduced "easy ambient sonic-spray ionization mass spectrometry" for checking quality and adulteration of different petroleum products.

Raman spectroscopy: Raman spectroscopy or infrared spectroscopy is typically used to determine vibrational, rotational and other states in a chemical system. In petroleum adulteration, Raman spectroscopy used along with Fourier transform infrared spectroscopy (FTIR). FTIR-based multivariate analysis is proposed by Pereira et al. successfully discriminated the adulterants with the petrol [10]. De Souza et al. [11] investigated the diesel adulterant using FTIR method along with statistical based partial least square method.

NMR spectroscopy: Nuclear magnetic resonance (NMR) spectroscopy is an analytical technique to observe the magnetic field around the sample. ASTM D5292 and D4808 are standard test procedures to find the amount of aromatic carbon and hydrogen content respectively by NMR spectroscopy. These standards do not provide solution to the adulteration issue but provides additional information to assist other analytical methods. Cunha et al. [12] use NMR time domain method for identifying kerosene as an adulterant in diesel.

Colorimetry spectroscopy: Colorimetry is used to describe the perception of human color to the system. Generally, colorimetry is classified into visual and photoelectric colorimetry. Visual colorimetry is observing the changes in the color through naked eyes. It is difficult to interpret subtle changes through naked eyes [13]. Photoelectric colorimetry uses spectrophotometer to identify the intricate changes in the color difference. Santos et al. [14] investigated various approaches to identify the petroleum product quality using color detection method.

2.4 Research Gap

The existing analytical methods and standards are verified and maintained by the oil companies. But most of the petroleum adulteration happened during the downstream process, i.e., either in transporting or retailing. The intermediate quality checks are carried out in the retail outlet to prevent the adulteration using density measurement by hydrometer method. This method is simple and inexpensive that also gives quick measurement. However, its small measuring range and the need of large sample volume makes this method less effective in many scenarios [15]. The additional learning curve to understand the working principle of hydrometer makes it difficult to adapt for the end-customer verification. The laboratory-based digital density estimation and various analytical methods listed above are alternative approach to identify the adulteration. But the listed methods are expensive and time consuming due to laborious process and the requirement of a sophisticated laboratory infrastructure.

3 Methodology

The digital image colorimetry (DIC) is one of the efficient and alternative approaches to analyze the samples using digital images. As a part of the initial study in this research, the petroleum samples are collected from the professional digital single lens reflex camera in the daytime with various illuminating condition: direct sunlight, natural lighting inside the room and artificial lighting using fluorescent bulb. The sample images are represented in Fig. 2. In the process of finding petroleum adulteration, the foremost task is to identify the petroleum product type whether it is petrol or diesel. From the samples, it is easy to identify from the human perspective by comparing the color and classify the petroleum products. The process of making the system to identify the color difference among the petroleum type is a complex and difficult task. The color difference is commonly measured using two different metrics:

- **Euclidean distance between RGB values**: In the RGB color space, Euclidean distance metrics finds the color difference between the image samples as images are represented as n-dimensional vector. The distance between two points in the Euclidean space represents the difference between the color information. For the RGB image, the Euclidean distance is calculated using Eq. 1. The samples in Fig. 2 are evaluated using the Euclidean distance measure to identify the color difference. Figure 2 (a), (b) and (c) shows the diesel samples; (d), (e) and (f) denote the petrol samples. In this measurement, the resultant value 0.0 denotes that both the images are similar else the average distance measurement between the RGB values are given as output. The distance metric is directly proportional to the color difference. The larger value denotes the two colors in the two DIC sample images are more different from each other. The overall results of the samples are presented in the Table 1. In this experiment, the Euclidean distance between the

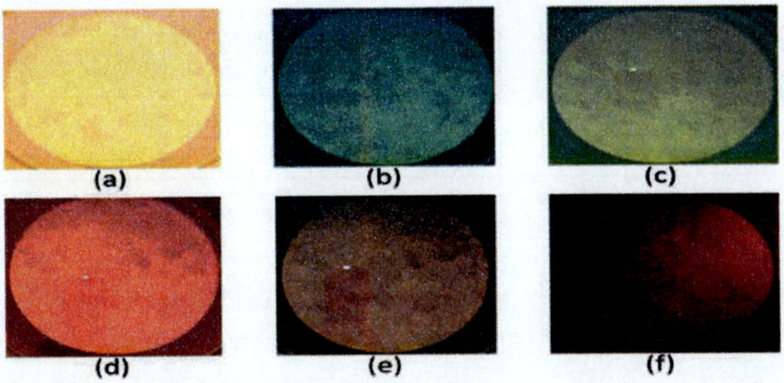

Fig. 2 (**a**), (**b**) and (**c**) are the samples of diesel in direct sunlight, artificial lighting and natural lighting, respectively. Similarly (**d**), (**e**) and (**f**) are the sample of petrol in direct sunlight, artificial lighting and natural lighting, respectively

Table 1 Comparison chart of Euclidean distance metric over samples represented in Fig. 2

	(a)	(b)	(c)	(d)	(e)	(f)
(a)	0.0	34,870.2	25,694.1	28,843.5	33,844.5	34,936.6
(b)	34,870.2	0.0	34,214.6	17,480.0	26,229.3	31,550.5
(c)	25,694.1	34,214.6	0.0	33,712.8	31,431.4	25,219.5
(d)	28,843.5	17,480.0	33,712.8	0.0	34,676.0	18,981.5
(e)	33,844.5	26,229.3	31,431.4	34,676.0	0.0	21,207.7
(f)	34,936.6	31,550.5	25,219.5	18,981.5	21,207.7	0.0

diesel sample (a) and (b) is more than the diesel sample (a) with petrol sample (d). It implies that the diesel and petrol samples are closer to each other compared with another diesel sample. Similarly, there are many samples which shows incorrect measurement. Hence, it is identified that the minor difference in the DIC samples are not efficiently differentiated using the Euclidean distance metric.

$$distance = \sqrt{(R_2 - R_1)^2 + (G_2 - G_1)^2 + (B_2 - B_1)^2} \quad (1)$$

– **Delta E:** This measurement is introduced by international commission on illumination to quantify the color difference for matching accuracy and it is widely adopted in the uniform color space. In this experiment, Delta E is compared between the samples represented in the Fig. 2 to identify the color difference in petrol and diesel. The value of Delta E is between the ranges of 0 to 100, where the value 0 represents there is no color difference. Delta E is calculated using Eq. 2 where L* represents the lightness, a* and b* represent the chromatic components which show the color from green to red and blue to yellow, respectively. Delta E measurement provides significantly better result than Euclidean distance approach and the overall results are represented in Table 2. However, the Delta E

Table 2 Comparison chart of Delta E metric over samples represented in Fig. 2

	(a)	(b)	(c)	(d)	(e)	(f)
(a)	0.0	42.7	29.8	47.2	45.8	77.8
(b)	42.7	0.0	17.9	53.3	38.2	60.7
(c)	29.8	17.9	0.0	40.3	28.3	52.9
(d)	47.2	53.3	40.3	0.0	15.4	26.4
(e)	45.8	38.2	28.3	15.4	0.0	27.4
(f)	77.8	60.7	52.9	26.4	27.4	0.0

value of diesel sample (a) and (b) is more than the diesel sample (b) with petrol sample (e).

$$DeltaE = \sqrt{(L*_2 - L*_1)^2 + (a*_2 - a*_1)^2 + (b*_2 - b*_1)^2} \tag{2}$$

The traditional color difference method for comparing DIC samples to identify the petrol and diesel is complex and inefficient. To improve the accuracy of the DIC system to identify petroleum samples by color information, the proposed research shifted the focus to change the data acquisition methodology. The next section discusses the proposed data acquisition approach.

4 Proposed Data Acquisition

In this research, the samples are digitally photographed using the CMOS camera sensor from the smartphone. There are three important characteristics to be considered while performing the data acquisition.

1. **Color space**—To provide standard specification of color information, various color spaces are used in the literature, Example: Red, Green and Blue (RGB), Cyan, Magenta, Yellow and black (CMYK), Hue, Saturation and Brightness/Light/Luminosity (HSB/HSL) and L*a*b color model where L* represents the luminance, a* and b* are two chromatic components. In this research, all this popular color spaces are studied and L*a*b color model is taken into consideration because it covers entire range of human color perception.
2. **Lighting condition**—The lighting condition influences the color appearances of the collected samples. The quality of DIC image samples typically gets affected by the non-uniform and non-reproducible lighting. Hence to capture the sample images, the ambient lighting is completely eliminated by using the imaging box. The imaging box is a custom printed three-dimensional structure used to place the smartphone to avoid ambient lighting. The flash light LED present in the smartphone is used for uniform illumination. A flash diffuser is used to avoid the

shadows created using the direct flash light. The sample prototype of the imaging box is represented in Fig. 3.

3. **Color quantification software**—The role of color quantification software is to convert the color information to the system understandable digital form. This research utilized the advantage of huge open source community supported OpenCV and its application for quantifying the color information.

Figure 4 shows the new samples collected using the proposed smartphone-based data acquisition model. The samples are taken using the imaging box with natural sunlight and flash light using the smartphone.

After the data is collected in the digital format, the RGB color space of the model is converted into L*a*b color space for covering the human perception of different color spectrum. Finally, the color difference is evaluated using Euclidean distance and Delta E metrics. Tables 3 and 4 represent the comparison of samples in Fig. 4 using Euclidean distance and Delta E, respectively. The Euclidean distance between the diesel sample (a) and (b) is more than the distance between diesel sample (a) and petrol sample (c) and (d). From these results, it is evident that Euclidean distance is not an effective approach to identify the fuel samples. The Delta E metric is much more robust against the Euclidean distance and clearly differentiates the petrol and diesel samples.

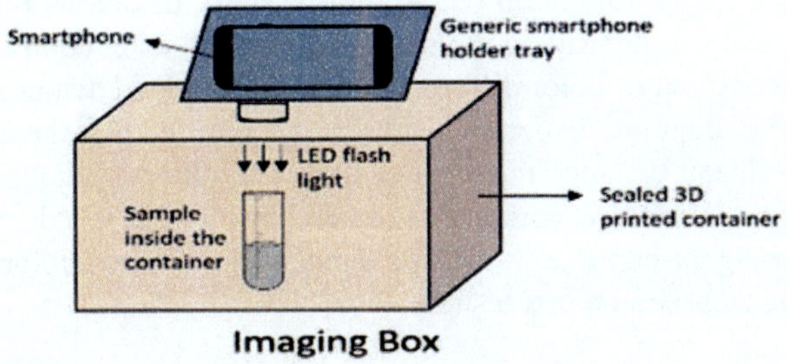

Fig. 3 Prototype of the 3D imaging box for reference

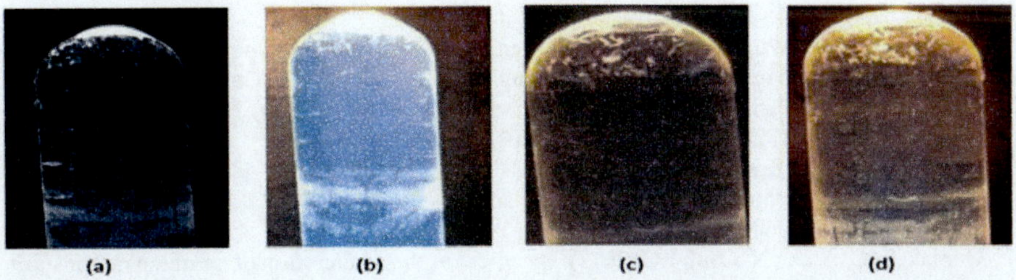

Fig. 4 **a** and **b** are the samples of diesel in natural sunlight and flash light using imaging box. Similarly, **c** and **d** are the sample of petrol in natural sunlight, and flash light using imaging box, respectively

Table 3 Comparison chart of Euclidean distance metric over samples represented in Fig. 4

	(a)	(b)	(c)	(d)
(a)	0.0	14,040.7	12,080.3	11,486.1
(b)	14,040.7	0.0	9815.0	11,729.5
(c)	12,080.3	9815.0	0.0	5274.1
(d)	11,486.1	11,729.5	5274.1	0.0

Table 4 Comparison chart of Delta E metric over samples represented in Fig. 4

	(a)	(b)	(c)	(d)
(a)	0.0	18.4	28.6	32.3
(b)	18.4	0.0	42.2	40.7
(c)	28.6	42.2	0.0	7.0
(d)	32.3	40.7	7.0	0.0

5 Conclusion

This article highlights the challenges faced in identifying the fuel types for colorimetry based fuel adulteration recognition system. In this work, L*a*b color space has been used for identifying the subtle color changes in contrast to the RGB color space. The two major color difference metric Euclidean distance and Delta E is considered for evaluation. From the results, it is identified that the Delta E metric outperforms Euclidean distance in terms of differentiating the fuel samples in the complex scenarios. The future work in this research is planned in a direction to introduce a deep learning based model which uses smartphone-based colorimetry to find the fuel adulteration based on this results.

References

1. Frederick, Halim, Winda, Astuti, Iwan Solihin, Mahmud: Automatic petrol and diesel engine sound identification based on machine learning approaches. E3S Web Conf. 130, 01011 (2019)
2. Han, Z., Wan, J., Deng, L., Liu, K.: Oil adulteration identification by hyperspectral imaging using qhm and ica. PLOS ONE 11(1), 1–13 (01 2016)
3. Barbeira PJS, Pereira RCC, Corgozinho CNC (2007) Identification of gasoline origin by physical and chemical properties and multivariate analysis. Energy Fuels 21(4):2212–2215. https://doi.org/10.1021/ef060436l
4. Onojake MC, Atako N, Osuji LC (2013) The effect of the adulteration of premium motor spirit (pms) on automotive engines. Pet Sci Technol 31(1):1–6. https://doi.org/10.1080/10916466.2010.524466
5. Shinde, D.B.: Analysis of adulterant kerosene in diesel by kinematic viscosity measurement (2012)

6. Yadav SR, Murthy VK, Mishra D, Baral B (2005) Estimation of petrol and diesel adulteration with kerosene and assessment of usefulness of selected automobile fuel quality test parameters. Int J Environ Sci Technol 1(4):253–255. https://doi.org/10.1007/BF03325839
7. Moreira LS, d'Avila LA, Azevedo DA (2003) Automotive gasoline quality analysis by gas chromatography: Study of adulteration. Chromatographia 58(7):501–505
8. Quach DT, Ciszkowski NA, Finlayson-Pitts BJ (1998) A new gc-ms experiment for the undergraduate instrumental analysis laboratory in environmental chemistry: Methyl-t-butyl ether and benzene in gasoline. J Chem Educ 75(12):1595. https://doi.org/10.1021/ed075p1595
9. Alberici, R.M., Simas, R.C., de Souza, V., de S´a, G.F., Daroda, R.J., Eberlin, M.N.: Analysis of fuels via easy ambient sonic-spray ionization mass spectrometry. Analytica Chimica Acta 659(1), 15–22 (2010), https://www.sciencedirect.com/science/article/pii/S0003267009015189
10. Pereira RCC, Skrobot VL, Castro EVR, Fortes ICP, Pasa VMD (2006) Determination of gasoline adulteration by principal components analysis-linear discriminant analysis applied to ftir spectra. Energy Fuels 20(3):1097–1102. https://doi.org/10.1021/ef050203e
11. Dc Souza, L.M., Mitsutake, H., Gontijo, L.C., Borges Neto, W.: Quantification of residual automotive lubricant oil as an adulterant in brazilian s-10 diesel using mir spectroscopy and pls. Fuel 130, 257–262 (2014), https://www.sciencedirect.com/ science/article/pii/S0016236114003019
12. Cunha, D.A., Montes, L.F., Castro, E.V., Barbosa, L.L.: Nmr in the time domain: A new methodology to detect adulteration of diesel oil with kerosene. Fuel 166, 79– 85 (2016), https://www.sciencedirect.com/science/article/pii/S0016236115010832
13. Fan, Y., Li, J., Guo, Y., Xie, L., Zhang, G.: Digital image colorimetry on smartphone for chemical analysis: A review. Measurement 171, 108829 (2021), https: //www.sciencedirect.com/science/article/pii/S0263224120313221
14. Dos Santos, A.P.F., da Silva, K.K., Borges, G.A., d'Avila, L.A.: Fuel quality monitoring by color detection. In: Zeng, L.W., Cao, S.L. (eds.) Color Detection, chap. 5. IntechOpen, Rijeka (2020), https://doi.org/10.5772/intechopen.86531
15. Babu, V., Krishna, R., Mani, N.: Review on the detection of adulteration in fuels through computational techniques. Materials Today: Proceedings 4(2, Part A), 1723 – 1729 (2017), 5th International Conference of Materials Processing and Characterization (ICMPC 2016)

Real-Time Detection of Edge Defects on a Rolled Steel Sheet Using Transfer Learning Technique

V. V. N. Satya Suresh, Y. Kalyani, and C. Ankith Kumar

Abstract Automation of the manufacturing processes especially in the rolling process of steel needs better quality assessment methods so that proper inspections are carried out before the end product is released. In order to achieve higher productivity through non-contact inspection, it is important to identify edge defects during the rolling process itself with the help of computer vision and machine learning techniques. In this work, a new approach consisting of an improved convolutional neural network (CNN) as classifier using Visual Geometry Group-19(VGG-19) algorithm as feature extractor is used to detect the edge defects under the framework of tensor flow. About 4000 images have been collected for model testing and training. By adopting the recently developed VGG-19 trained CNN network model, 3 Fully Connected layers were reduced into one flat layer and 2 Fully Connected Layers with reduced parameters. Testing accuracy of the model was 87% and the model achieved 98% training accuracy.

Keywords VGG-19 algorithm · Convolutional neural network · Edge defects · Transfer learning

1 Introduction

Iron and steel industry has expanded many folds in recent years and have considerable effect having a major impact on economic growth. Defects have been on the rise during the rolling process in steel production which may result in significant financial losses for companies as well and damage the company's reputation. Since manual inspection techniques are not reliable due to operator's fatigue and it is also not feasible from the manufacturing point of view in terms of time, cost, and accuracy, a study is critical to understand and categorize various detects.

V. V. N. S. Suresh · Y. Kalyani (✉) · C. A. Kumar
Mahatma Gandhi Institute of Technology, Gandipet, Hyderabad, India
e-mail: ykalyani_pg19mct8008@mgit.ac.in

N. M. Sivaram et al. (eds.), *Advances in Manufacturing, Automation, Design and Energy Technologies*, Lecture Notes in Mechanical Engineering,
https://doi.org/10.1007/978-981-99-1288-9_38

In the past few decades, the process of identifying surface and edge defects was carried out through magnetic flux leakage testing, artificial visual inspection, etc. [1]. Under the influence of miscalculated inspection rate and particular factor, the above mentioned methods fail to meet the expectations under real-time conditions [2]. Song and Yan [3] adopted the adjacent evaluation completed local binary pattern (AECLBP)-based feature extraction method and achieved good accuracy by using support vector machine (SVM) algorithm as a classifier to classify various edge defects in steel sheets. However, it uses the knowledge of a precise field to extract features, thus makes it difficult to identify all the defects. In the year 2016, Lei Wang et al. [4] adopted a new method for detecting the edge defects in rolled steel strips. It worked well, but it needs more information which is time consuming. Later, S Zhou et al. [5] developed a classification based on CNN, which performed very well. However, it also consumes more time to arrive at the solution since it needs lot of training data sets.

In the year 2019, Zheng Liu et al. [6] used the Inception Dual Network which is a deep neural network subsidiary to detect the defects. They were able to achieve high accuracy. The disadvantage is that it also requires huge data for training the neural network. Pan and Yang [7] published a review article on transfer learning in the year 2009. They concluded that the transfer learning technology can utilize the knowledge in which neural network learns on other data sets, thereby reduces the demand for neural networks to train the datasets. Kasthurirangan et al. [8] used the first 15 layers of VGG-16 network model of CNN as feature extractors and interchanged the last few fully connected layers of VGG-16 with machine learning techniques such as support vector machines, random forests, and multilayer perceptron. They achieved an accuracy of 90% by using a small dataset consisting of 760 images. However, in their work, the data enhancement method has not been used and they have also not analyzed principle of the neural network in detail, which is vital to resolve practical problems.

Scime et al. [9] adopted the well-known AlexNet pre-trained technique keeping all the parameters fixed and achieved detection and classification by re-trained the final fully connected layer. They have copied the grayscale image to three times to adopt the grayscale image to the shape of the RGB color image input. They also added various scaling features in diverse channels. By conducting a thorough literature survey, it has been noticed that the usage of deep learning, viz., transfer learning technology in manufacturing operations is limited thus become a research topic at present. In order to overcome the shortcomings of large dataset and minimize the computation time, it is suggested that the image feature extraction model is trained initially and apply this model to identify the defects on strips based on their classification. In this work, the VGG-19 model which is developed using transfer learning technique is used to build image net data set for extraction of image feature. This is used to construct a classification model connected to a CNN.

After carrying out thorough literature review, it was found that many researchers have performed experiments on surface steel defects rather than on edges of the steel strip [10]. The proposed work involves detection of edge defects, viz., burrs, cuts, etc., in a steel strip by changing the dataset images unlike the method used by Spinola et al. [11] which used to measure the width of the sheet. This study used the VGG-19 model and convolutional neural network with the application of AI techniques.

2 Proposed Work

Based on the Tensor Flow framework, this work focused on the identification of the edge defects. Firstly, the number of edge defects are classified into regions of interest and few defects are categorized based on understanding and experience of the operator based on the collected images. They are used for training and testing set of the CNN model. Since the number of pixels of the collected image burden the processing time on a CPU, therefore, they are required to be preprocessed to meet the image recognition necessities of the Tensor Flow framework. The data is then loaded using the VGG-19 framework under the aegis of CNN to construct the model and fine tune the model parameters according to model training result and model testing result. VGG-19 network model framework extracts the low- and high-level features of images by reviewing layer after layer and classify the defects through image classification, to achieve a desired accuracy. In this paper, identification of defects is carried out through the process of collection of data and its preprocessing, model training and model testing. The main edge defects considered in this work are sheared edge, edge crack, scratch edge, burnt edge, damaged strip edge and chipped edge.

3 VGG-19 Network Configuration

CNN's VGG-19 framework as a preprocessing model is used in this work. It has six unique structures. Each structure comprises various sets of convolutional layers and fully connected layers. In this work, a size of 3*3 convolutional kernel is used with the input size taken as 256*1600*3. The number of layers considered is between 16 and 19. Compared to earlier CNN models, VGG-19 has a better in-network depth since, it uses associate discontinuous structure of various convolutional layers and nonlinear activation layers, that leads to one convolution. The VGG-19 has three main functions, viz., rectified linear unit (ReLU) function, activation function and sigmoid function. The layer structure will extract image options by means of max-pooling for down sampling. It modifies the ReLU function as well as the activation function. The down sampling layer is especially accustomed to improve the anti-distortion ability of the image so that unwanted noise in the image can be overcome. The mathematical expression for down sampling layer is given in Eq. (1).

$$X_{pj}^{(n)} = f\left(T_j^n down\left(X_j^{(n-1)}\right) + b_j^{(n)}\right) \tag{1}$$

where $down\left(X_j^{(n-1)}\right)$- is the most pooling sampling operate, T_j^n-is the constant akin to the j-th feature map of the n-th layer, and $f\left(T_j^n down\left(X_j^{(n-1)}\right) + b_j^{(n)}\right)$- is the ReLU activation function.

4 Edge Defect Identification with an Upgraded VGG-19 Network Model

Due to the variability of defects identified by various processing industries, a large number of defect images are to be classified, hence, an improved version of VGG-19 network can easily identifies the defects and have the capacity to allow more number of images. In this work, transfer learning which is a subset of deep learning CNN model is used to target detection. It proposes a trained VGG-19 model which uses a pre-trained model. By refining the transfer learning methodology, the model parameters of the pre-trained VGG model of the convolution layer gets optimized thus able to solve the classification of edge defects.

The model parameters of VGG-19 are contained in 3 FC layers which in turn form a network. This model has the capability of detecting thousand classifications, however, this work mainly focused on four classifications. Each classification undergoes rigorous scrutiny for defect free objects. At each juncture, the image is passed through three connected layers. Sometimes, it can also be one Flatten layer and 2 totally connected layers. Further analysis is carried out until the image converges into a single convolution. A flatten layer acts like an intermediate layer because the fully connected layers cannot be merged with the convolution layer. The developed model uses the refined transfer learning technique to transfer the model parameters of VGG-19 pre-trained model into pooling layer, convolution layer and also the fully connected layer of the model. Within the fully connected layer, a tool named as 2-label Softmax classifier is used which measures thin options, viz. Max pooling, Dropout, etc., with good accuracy.

5 Results and Discussion

Installation of deep learning technology platform is carried out on Windows 10 operating system along with Python 3.6 / 3.7 language and Tensor Flow 2.0 software framework. It aids in constructing model framework for detection of edge defects based on improved VGG-19 model. It was trained and tested using Python language. In this paper, the dataset related to SEVERESTAL: STEEL DEFECT DATABASE is used as training samples.

Altogether 4520 training samples and 1128 test samples are contained in the above mentioned dataset. After the standard training and testing of the datasets, edge defects were identified in the collected images. Sample set are manually labeled in the keras network with the labeling tool to categorize the defects. Later, the training sample data is used as a validation set. This validation set contained all the training set samples of 4520 and test samples of 1128 without any exclusions. The training and testing sets are put in the proportion of 8: 2. This process is performed on the transfer learning model of VGG-19. Under this model, 10 epochs were designed at a learning rate of 0.005. Each epoch was analyzed for the loss value. It was noticed that the learning rate is inversely proportional to the loss value as shown in the Fig. 1. Approximately, it took about 3 and ½ hours to process all the images until the loss value was stabilized to a minimum value of 0.0217. An accuracy of 97.4% has been achieved after running 10 epochs.

Figure 2 shows the plot between the accuracy of the model and the number of repetitions of each epoch. The x-axis shows the number of epochs and the y-axis reflects the accuracy achieved. It can observed that the accuracy increases with the number of epochs within a certain region.

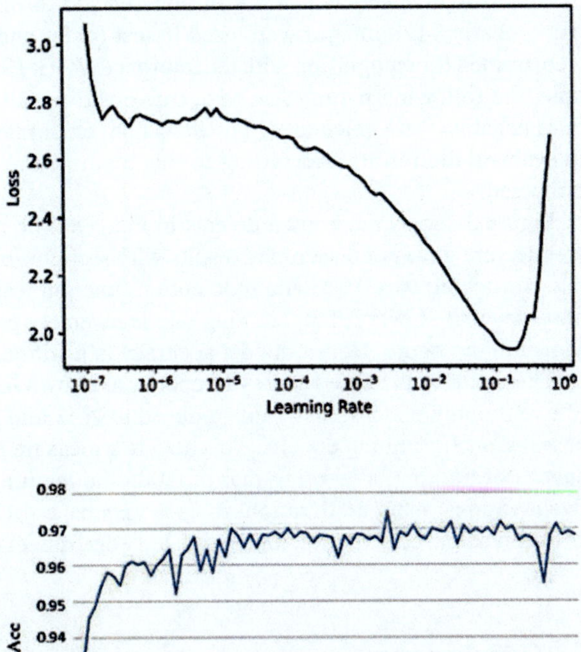

Fig. 1 Relation between loss and learning rate

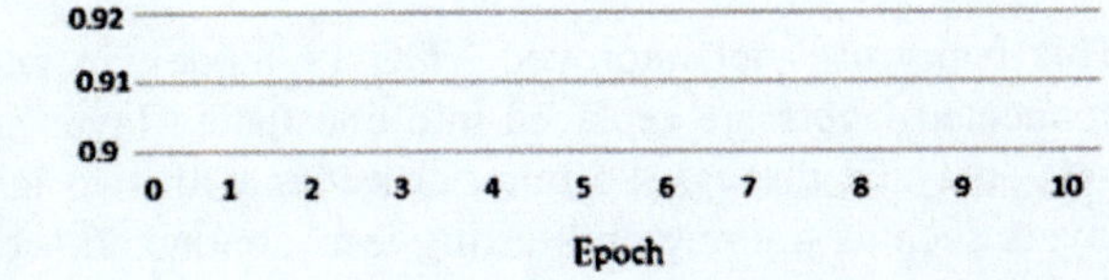

Fig. 2 Accuracy of the model

Fig. 3 Loss of the classification model

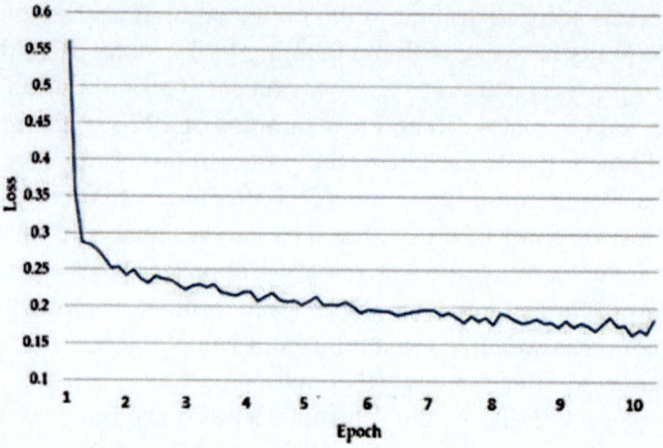

The upgraded VGG-19 network model constitutes two components, viz. machine vision as well as human vision. VGG-19 network model recognizes the defects within the rolled steel sheets which are further verified through visual inspection. In this work, nearly 1128 images were used in test folder and the remaining 4520 images were trained for recognition with the improved VGG-19 network model. At any given time, the following parameters, viz., true positive, false positive, true negative and false negative were calculated. The precision, error rate and accuracy are calculated to evaluate the results. According to this method, the undetected edge defects are evaluated.

Figure 3 shows the known defects in images. The results of improved VGG-19 architecture gave good accuracy results with recognition. In this model, the learning rate considered was 0.005 and momentum function was chosen as optimizer. It was observed that at every step, learning rate is inversely proportional to the loss rate as shown in the figure. Hence, model accuracy is justified.

The model was trained across 10 epochs as shown in the Fig. 4. After completion of five training epochs, loss value reduced to 20% and further reduced to 15% after completion 10 training epochs. This also is a measure of model accuracy. Using 10 epochs of training, a learning rate of 0.005 the learning rate doubled by 0.25 after five iterations. After 10 iterations of each epochs, an accuracy of the model is 0.974 has been achieved. Original and simulated edge defects are shown in Fig. 5.

6 Conclusions

This paper uses an improved VGG-19 model, in which the original three fully connected layers are replaced into one flatten layer and two FC layers. It further gets analyzed through Softmax classifier with two labeled Softmax classification layers such as training and testing tests conducted for edge defects resulted in the following:

Fig. 4 Accuracy of the classification model

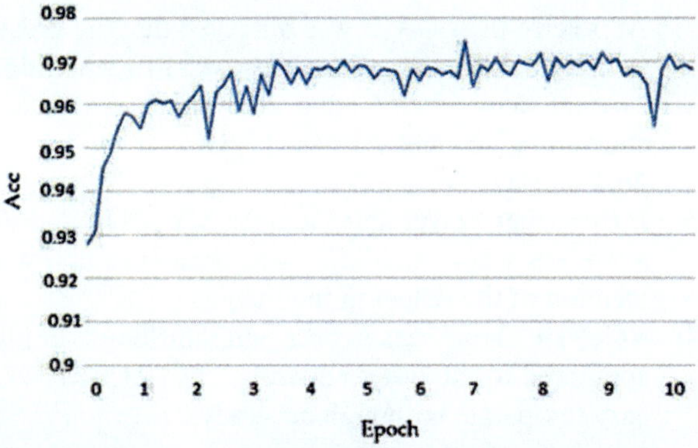

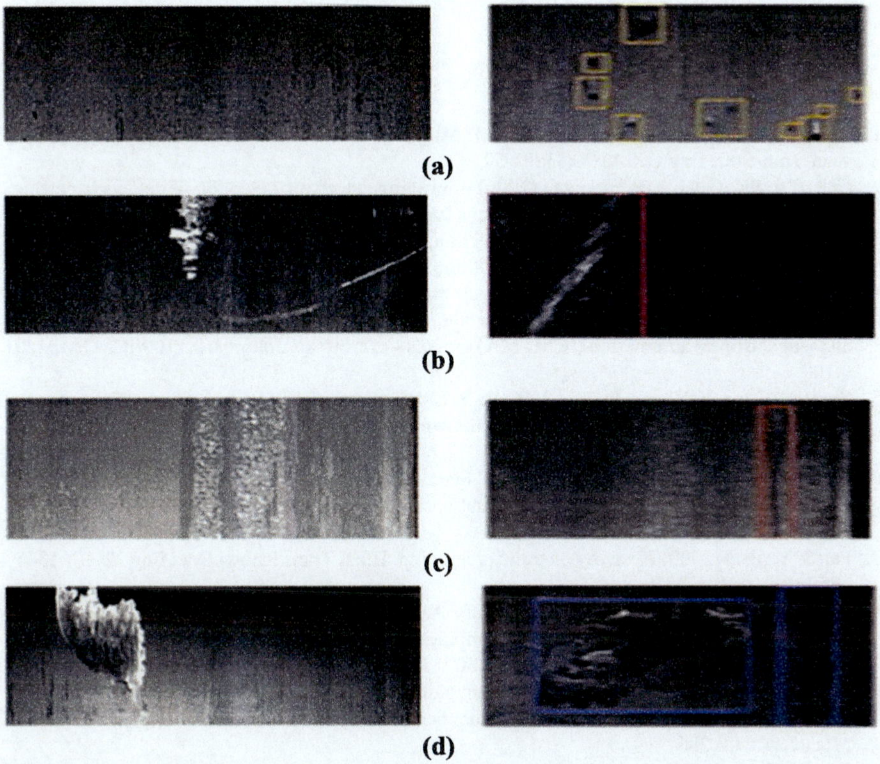

Fig. 5 Original and simulated edge defects **a** Edge crack **b** Scratches **c** Busted edge **d** Burnt edge

(1) At precise intervals, it was observed that the greater the number of epochs in a training period, the greater the model's recognition of detection rate.
(2) Manual detection of edge defects in steel sheets is cumbersome and time consuming, hence, AI-based VGG-19 model is effective and speed up the productivity.
(3) In this paper, two unique methods, viz., CNN and VGG-19 were employed. CNN addresses whether a defect was present or not, whereas VGG-19 predicted the location of the defect in the image.
(4) VGG-19 algorithms works with different kinds of defects simultaneously compared to the other networks, viz. ALEXNET and RESNET, and it is also easy to operate with high accuracy.

References

1. Maki H, Tsunozaki Y, Matsufuji Y (1993) Magnetic online defect inspection system for strip steel. Iron Steel Eng (USA) 70(1):56–59
2. Li X, Tso SK, Guan X-P, Huang Q (2006) Improving automatic detection of defects in castings by applying wavelet technique. IEEE Trans Ind Electron 53(6):1927–1934
3. Song K, Yan Y (2013) A noise-robust technique supported completed native binary patterns for hot-rolled steel strip surface defects. Appl Surf Sci 285(21):858–864
4. Wang K, Xu K, Zhou P (2016) Online detection technique of 3d defects for steel strips supported the mensuration stereo. In: 2016 eighth international conference on measuring technology and mechatronics automation (ICMTMA), pp 428–432. https://doi.org/10.1109/ICMTMA.2016.109
5. Zhou S, Chen Y, Zhang D, Xie J, Zhou Y (2017) Classification of surface defects on steel sheet victimization convolutional neural networks. Mater Technol 51(1):123–131. https://doi.org/10.17222/mit.2015.335
6. Liu Z, Wang X, Chen X (2019) Inception dual network for steel strip defect detection. In: 2019 IEEE 16th international conference on networking, sensing and control (ICNSC). IEEE, pp 409–414
7. Pan S, Yang Q (2009) A survey on transfer learning. IEEE Trans Knowl Data Eng 22(10):1345–1359
8. Gopalakrishnan K, Khaitan SK, Choudhary A, Agrawal A (2017) Deep convolutional neural networks with transfer learning for computer vision-based data-driven pavement distress detection. Constr Build Mater 157:322–330
9. Scime L, Beuth J (2018) A multi-scale convolutional neural network for autonomous anomaly detection and classification in a laser powder bed fusion additive manufacturing process. Addit Manuf 24:273–286
10. Sun X, Gu J, Tang S, Li J (2018) Research progress of visual inspection technology of steel products—a review, MDPI. Appl Sci 8(11). https://doi.org/10.3390/app8112195
11. Spinola CG, Bonelo JM, Canero-Nieto M, Martin Vazquez MJ (2018) Real-time image processing for edge inspection and defect detection in stainless steel production lines. Appl Sci 8(11):2195. https://doi.org/10.3390/app8112195

Semi-Automation in Chilli Pulverization

S. Shankar, R. Nithyaprakash, R. Naveenkumar, S. Kulasekaran, C. S. Kavinganesh, and R. Gokulraj

Abstract Chillies are an important and most commonly used ingredient in many cuisines around the world. It is used both in raw form and as powdered form. The pulverization of chillies to chilli powder is done using machines like plate mills. These mills have several disadvantages, and one amongst them is the task of refeeding the hopper with the chilli powder. The chillies are turned into fine powder by pulverizing them by using 3–4 cycles. This cannot be achieved in a single cycle as it will produce a lot of heat, and thus, it will modify the taste and quality of the powder. In this work, a cost-efficient low-pressure suction method has been adopted as an alternative for the conventional human work force in the refeeding process of the chilli powder after each cycle to minimize human eye irritation. This method utilizes an electric motor with high RPM to create enough suction to carry the powder from the bottom and transfer it back to the hopper. With the usage of distance sensors, wiper motor, and microcontroller, the automatic opening and closing of the bottom of the container are also implemented.

Keywords Chilli · grinding · Pulverization · Semi-automation

1 Introduction

The chilli pepper belongs to the genus capsicum, a member of the nightshade family, Solanaceae. Chillies are used to add pungency and flavour to culinary dishes. It is used in many different cuisines—American, Chinese, Indian, Korean, Mexican, Portuguese, and Thai. The substances which give chilli peppers, its intensity when ingested inside or applied topically are called capsaicin, and its related compounds are known as capsaicinoids. Chilli powder is also blended with other products like

S. Shankar (✉) · R. Nithyaprakash · S. Kulasekaran · C. S. Kavinganesh · R. Gokulraj
Department of Mechatronics Engineering, Kongu Engineering College, Erode, India
e-mail: shankariitm@gmail.com; shankar@kongu.ac.in

R. Naveenkumar
Department of Mechanical Engineering, Kongu Engineering College, Erode, India

N. M. Sivaram et al. (eds.), *Advances in Manufacturing, Automation, Design and Energy Technologies*, Lecture Notes in Mechanical Engineering,
https://doi.org/10.1007/978-981-99-1288-9_39

coriander, cumin, onion, garlic powder, and salt. India has come to be known as "land of spices". India is a primary producer and consumer of spices with a production of around 947,790 tonnes (60% of world production) [1]. But spice grinding is not an easy task which can be done in all households and hotels. It comes at a high cost and a lot of human effort. Packaged spice powders are mixed with preservative which result in a lot of health problems. So, most of the people rely on the spice mill shops to grind the chilli into fine powdered form. These spice mills consume a lot of power, and also the workers amongst the mills are facing a lot of difficulties in the processing of chillies.

The chillies that have been dried in the sun are fed into the hopper of the plate mill machine shown in Fig. .1. The plate mill machine is driven by a belt driven shaft connected 4 to a 7.5 hp to 10 hp single-phase AC motor. The motor rotates at 700–900 rpm. In some region, chillies are first fed into a hammer mill machine where it cuts chillies in small fragments, and then, they are fed into the plate mill for efficient pulverizing and reduced processing time. The chillies/chilli fragments travel from the hopper to the processing hull via a vibrating plate. The processing hull consists of two plates for pulverizing. One plate is fixed, and the other is movable. The fixed plate is connected to the main shaft and so it rotates with the shaft at 700–900 rpm. The chillies from the vibrating plate enter into this processing hull and get pulverized in the gap between these two plates and come out as powder via the output pipe, where it gets collected in a container. This process is repeated 3–4 times until the customer demanded fineness of the chilli powder is reached. To pulverize the chillies into smaller and even smaller particles, for each round, the adjustable knob is rotated to some extent. This in-turn reduces the gap between the two plates and thus increasing friction and increasing the pulverization effect [2]. During grinding process, the heat in the chamber boosts the temperature from 30 degrees up to 90 degrees. But substances like volatile components, etheric oil, and heat sensitive compounds of spices boils off at temperatures about 50 to 60 degrees. So too much friction causes burn out of the chillies and destroy its properties which modifies the taste. The whole process takes 15–20 min depending on the amount of chilli that was fed and the sharpness of the pulverizing plates.

A performance evaluation of drying of red chillies in Vietnam was carried out [3]. A comparison between double pass solar drier, normal cabinet drier, and conventional open-air sun drying was studied. It took 32 h under DPSD and 73 h under cabinet drier to achieve moisture content of 10%. Whereas under open-air sun drying, the desired moisture content was not able to be achieved even after 93 h. The American Spice Trade Association (ASTA) colour values were also higher in DPSD as compared to the other two methods. Aflatoxin B1 contamination was lower in DPSD. The drying cost was also 39% lower (0.077US$/kg). The grinding characteristics of cinnamon and turmeric under cryogenic and ambient grinding conditions were investigated in literature [3]. In this study, it was found that the average particle size of the ground turmeric and cinnamon was 0.336 mm and 0.356 mm under cryogenic condition, whereas under ambient grinding, they were 0.407 and 0.454. The energy constants and the energy consumptions were lower as well in cryogenic grinding when compared with ambient grinding. The colour values retained better in cryogenic grinding. A study conducted [6] communicates about the importance of temperature control in case of grinding black pepper for the retainment of volatile oil. In this study, the black pepper was subjected to grinding at 4 different temperatures which is 40 °C, 24 °C (ambient), 10 °C, and −120 °C by the usage of hot/chilled water/liquid nitrogen through the grinder jacket along with the feed. This experiment showed that the volatile contents as well powder fineness were better under cooled conditions than hot or ambient conditions.

A research on occupational exposure to spice dust and its health effects was carried out in spice packing section of a company in Gauteng, South Africa [4–6]. The research collected complaints from the employees in the spice packing department, and they stated sneezing, rhinitis, and skin symptoms such as dry skin and pruritus as most frequently occurring health effects. The research also revealed that the efficiency of usage of dust masks to protect from spice dust inhalation is pretty low as in 95% of total respondents who worn it, 83.8% of them still complained of the upper respiratory tract (URT) symptoms of sneezing, runny, or blocked nose. A study with comparison between paprika splitters and chilli grinders was conducted in the central chest clinic located in Colombo, Sri Lanka. The workers stated symptoms like sneezing, cough, and watery nose at the starting stage of their employment. But these symptoms disappeared after an average of 2 ½ months. The major symptoms amongst chilli workers are coughing, sneezing, running nose weight loss, and haemoptysis. The data about the workers who experienced those symptoms were reported in literature [7].

Another study [8] investigated the degradation in the quality of chilli during comminuting with the help of modelling. Two methods were considered in this study: grinding and high-speed cutting. Colour retention, concentration of pungent compounds, and moisture were taken as the performance metrics for the study. The result of this study was that although the quality of chilli powder due to grinding was lower than high-speed cutting, the rate of degradation in quality was lower as well in grinding as compared to high-speed cutting. Several studies in past [9–12] suggest repetitive works which cause work related to musculoskeletal disorders for the workers. Redesigning the workplace with appropriate modification may be results with reduced stresses to workers.

In this work, to tackle the problem of human workload, the refeeding process is taken care of by the implementation of a suction motor which creates a low-pressure system in a container placed above the hopper of the existing machine. As fluids tend to move from high pressure to low pressure naturally, the chilli powder output from the bottom can be easily sucked and filled into the container. After all the chillies in the hopper are pulverized, the bottom of the container is opened automatically to allow the collected powder to fall back into the hopper for the next round of pulverization. This automatic opening and closing of the bottom of the container are controlled by ultrasonic sensor and Arduino microcontroller and are being achieved by a rack and pinion mechanism.

Table 1 Survey amongst the spice mill owners

S. No	Title	Shop 1	Shop 2	Shop 3	Shop 4	Shop 5	Shop 6	Shop 7
1	Num. of workers (male, female)	(2,0)	(2,1)	(2,0)	(1,1)	(3,0)	(1,2)	(2,0)
2	Avg. age of workers	40–45	40–45	> 50	> 40	40–50	> 40	> 40
3	Salary per worker per day (in Rs.)	300	300	350	300	300	350	350
4	Avg. customer count	25	15	10–20	10–20	12	15	15–20
5	Avg. amount of chilli brought by a customer (in kg)	0.5	1	1	1	0.5	1	1
6	Avg. amount of chilli processed per day	12.5	10–15	15	14	15	10–13	15
7	Avg. time to grind 1 kg of chilli (in min)	15	15–20	15	20	15	15	20
8	Power of the electric motor (in Hp)	10	7.5	10	10	10	7.5	7.5

2 Data Collection

A survey regarding the problems faced by the spice mill workers was conducted in 7 spice mill shops around the Villupuram district in Tamil Nadu, India. The data gathered in this survey is tabulated and shown in the Table 1. From the data of Table 1, average of avg. customer count is 16.25 which can be considered as 16 (approx.). Similarly, average of avg. quantity of chilli brought per customer is 0.8125 which can be approximated to 1 kg. Letus assume that the chilli pulverizing blade is new, and it takes only 3 rounds to grind the chillies to the required fineness. Then, a worker will have to refeed the hopper 48 times a day excluding attending to other machines in the shop. The issues faced in the mills are as follows: excessive workload, blowback of chilli powder in the face of workers, noise, and nose and skin irritation.

3 Proposed Design

Designing is one of the crucial stages of any product. The machine is designed using solid works software. Figure 1 shows the complete assembly of the proposed solution on the existing machine. It consists of motor housing, motor tube, container, transferring tube, hopper, and other accessories as shown below. Properties used for the design are provided below.

- Material (Container)—High density polyethylene
- Density (Container)—930–970 kg/m^2
- Material (Filter plate)—Commercial wood

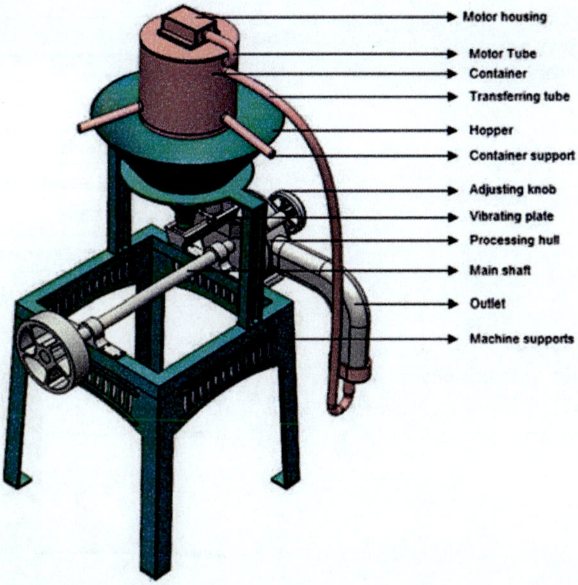

Fig. 1 Solid works model of the proposed solution

- Density (Filter plate)—680–760 kg/m^3
- Material (Filter sheet)—Sponge

Table 2 shows the specifications of the suction motor. Table 3 shows the specifications of the wiper motor used for pinion.

The mechanical section consists of the plastic bucket container, its supporting PVC pipes, wooden filter plate, rack, rack holder, and pinion. Figure 2 shows the fabricated prototype model. The electrical section consists of the circuitry responsible for the automatic opening/closing mechanism. It consists of the ATMEGA328—Arduino Nano controller, L298N motor driver, 12 V geared DC motor, and ultrasonic sensor. The Arduino Nano microcontroller is powered from the 5 V port of the motor driver,

Table 2 Specifications of AC suction motor

Sr. No	Title	Description
1	Power type	Single phase AC 230 V
2	Power (watts)	1000
3	Current rating	4.34
4	Torque (Nm)	0.477
5	Speed (RPM)	>20,000
6	Motor dimensions (mm)	66*130*130
7	Fan blade diameter (inch)	4.9
8	Fan blade material	Aluminium
9	Fan blade weight (Kg)	0.15

Table 3 Specifications of DC Wiper motor

Sr. No	Title	Description
1	Power type	12 V DC 2A
2	Power (watts)	24 ~ 48W
3	Current rating	1A–2A
4	Torque (kg-cm)	15
5	Speed (RPM)	30
6	Motor mass (grammes)	1500
7	Motor dimensions (cm)	21.08 × 19.05 × 18.29

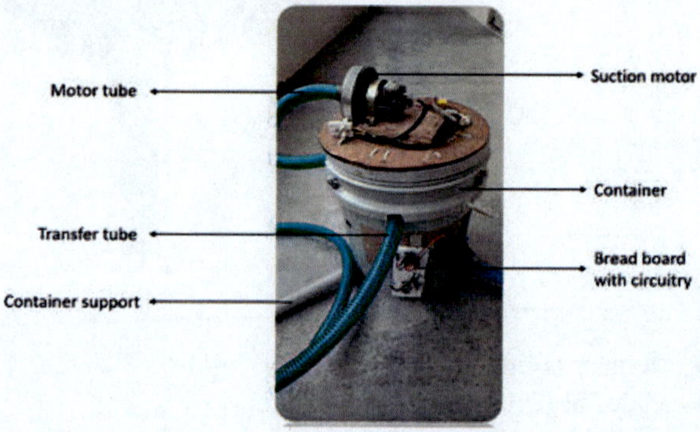

Fig. 2 Fabricated setup

and the driver is powered by a 12 V DC supply. Ultrasonic sensor has 4 pins: VCC, ground, trigger, and echo. The sensor is powered by Arduino Nano pins. The digital pins 2 and 3 are connected to the echo and the trig pin of the sensor. The digital pins 6 and 7 of the Arduino are connected to IN1 and IN2 of the motor driver.

4 Conclusion

This semi-automation in chilly pulverization process thus reduces the workload in manual refeeding of the chilly powder after each pass of pulverization and therefore also prevents the irritation of nose and skin due to chilly particles coming into contact whilst the conventional refeeding. For further improvement in this system, drawing power for the suction motor from the main shaft drive of the machine and meshing it with suitable gears to achieve higher speeds that is required for enough suction capacity can be worked out. Noise reduction can be worked out further to achieve a noise free working environment and thereby increasing the health of the workers.

References

1. Bharathi V, Raja P, Harikrishnan M (2018) Smart multi utility spice pulverizer machine. Int J Appl Innov Eng & Manag (IJAIEM) 7(11):040–047. ISSN 2319-4847
2. Banout J, Ehl P, Havlik J, Lojka B, Polesny Z, Verner V (2011) Design and performance evaluation of a Double-pass solar drier for drying of red chilli (Capsicum annum L.). Sol Energy 85(3):506–515
3. Barnwal, P., Ashish Mohite, K. K. Singh, P. Kumar, T. JOHN Zachariah, and S. N. Saxena. "Effect of cryogenic and ambient grinding on grinding characteristics of cinnamon and turmeric." *International Journal of Seed Spices* 4, no. 2: 26–31,(2014).
4. Goring D (2003) Health effects of exposure to spices in a spice packing factory. Occup Health South Afr 9(2):19–23
5. Mallappa JM, Sharankumar H, Roopa Bai RS (2015) Effect of milling methods and its temperature on quality parameters of ByadagiChilli: with emphasis on cryogenic grinding. Res J Eng Sci ISSN 2278: 9472
6. Murthy CT, Krishnamurthy N, Ramesh T, Srinivasa Rao PN (1996) Effect of grinding methods on the retention of black pepper volatiles. J Food Sci Technol 33(4):299–301
7. Uragoda CC (1983) A comparative study of chilli grinders with paprika splitters. Occup Med 33(3):145–147
8. Zhou W, Goh J (2005) Investigation of the effect of different comminution methods on the quality of chilli powder. Dev Chem Eng Miner Process 13(5, 6):709–718
9. Shankar S, Kumar N, Hariharan CPS (2021) Ergonomic evaluation of ergonomically designed chalkboard erasers on shoulder and hand-arm muscle activity among college professors. Int J Ind Ergon 84:103170
10. Kumar RN, Saravanan A, Shankar S, Nithyaprakash R, Jaikumar B, Mohanraj A, Mohanprasath K (2021) Investigating the impact of the hoe handle length and angle on the shoulder and arm muscle activity during manual farming activities. J Inst Eng (India): Ser A 102(4):1053–1060
11. Shankar S, Naveenkumar R, Karthick J (2019) Management of musculoskeletal shoulder and neck pain through ergonomic intervention: a pre-post design analysis in hand screen printing industry. Int J Bus Innov Res 18(3):392–409
12. Shankar S, Maheswari C, Gowtham R, Kiruba P, Mohansrinivas K (2019) Design and fabrication of portable sugarcane harvesting machine. Int J Sci Technol Res 8:2059–2062

2.4 GHz Microstrip MIMO Antenna Design

Reeya Agrawal

Abstract This study explores low-cost and fast 2.4 Hz multiple in multiple out antenna manufacturing. This paper briefly introduces an antenna and its use in today's era. Then, it describes different parameters such as array gain, spatial diversity gain, spatial multiplexing gain, interference reduction, and avoidance. A small literature survey over multiple in multiple out antenna by different authors is also a small literature survey. Furthermore, the introduction of multiple in multiple out wireless communication system, microstrip antenna technology, multiple in multiple out antenna, and different S-parameters is simulated such as S11, S22, S21, S12, return loss, and envelope correlation coefficient in dB concerning frequency (GHz). At last, the conclusion describes those multiple in multiple out antennas as the future of the upcoming wireless antenna society. So much research is going on over multiple in multiple out antenna. In the end, this work can be done in the form of an array.

Keywords Multiple input multiple output (MIMO) · Single input single output (SISO) · Ultra-wide band (UWB) · Microstrip antenna (MSA) · Bandwidth (BW) · Signal to noise ratio (SNR)

1 Introduction

Spectral efficiency and interference are urgent issues. The available spectrum will be squeezed [1], owing to the prominence of wireless Internet, mobile video, and data transmission. This will raise the possibility of signal interference with other services by expanding the utilization of the spectrum. The ultra-wideband is an outstanding example of interference in both directions. Because it is now recognized that spectrum efficiency is significantly enhanced when interference reduction technologies

R. Agrawal (✉)
GLA University, Mathura, Bharthia, India
e-mail: Reeyaagrawal0304@gmail.com

Microelectronics & VLSI Lab, National Institute of Technology, Patna 800005, India

N. M. Sivaram et al. (eds.), *Advances in Manufacturing, Automation, Design and Energy Technologies*, Lecture Notes in Mechanical Engineering,
https://doi.org/10.1007/978-981-99-1288-9_40

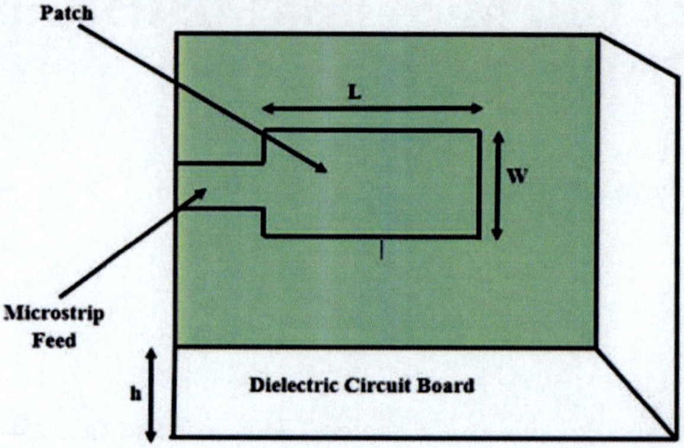

Fig. 1 MIMO reconfigurable antenna for Wi-Fi at 2.4 GHz [6]

are applied, the primary objective for communications scientists is to improve spectral efficiency [2]. Everything will be altered with intelligent antenna technologies developing in the communications sector. Until recently, the communication systems solely utilized the spectrum, time, and code domain. Code and restrictions on the range are tight. Spatial domain antennas, such as smart antennas, provide significant benefits over conventional antennas [3]. Two antennas on either end of the communication link, one of which transmits multiplexed data streams and the other receives multiplexed data streams, are required to multiplex data streams in both the spatial and frequency domains. With multiple input multiple output (MIMO) systems, data rates are much higher, and the overall spectrum efficiency is thus greater [4]. The primary source of wireless channel communication impairments is multipath fading. The phrase "multipath" refers to electromagnetic wave dispersion in the environment, which leads to various time delays, frequency shifts, and angles of arrival for a broadcast signal to its intended recipient. It is also challenging to construct reliable wireless communication systems due to unlimited power and infinite frequency bandwidth [5].

Wireless communication system design has reached a watershed moment with MIMO technology. A unique MIMO wireless system may employ time and frequency dimensions, which were not feasible on ordinary single-antenna wireless systems. Figure 1 displays a 2.4 GHz MIMO reconfigurable Wi-Fi antenna. MIMO technique is favorable for arraying gains, spatial diversity gains, spatial multiplexing gains, and interference reduction.

1.1 Array Gain

This is the increase in SNR received due to the coherent combining effect of the wireless signals at the receiver. Spatial processing at the antenna array on the receiver's side, or spatial pre-processing on the antenna array on the transmitter's side, is an effective technique for achieving coherent combining. Wireless network coverage and range improve as the number of gains rises [7].

1.2 *Spatial Diversity Gain*

The signal strength of a wireless system varies or fades. Fading can be minimized by several copies of a broadcast signal (preferably independent). The variety sequence has led to the overall chance that at least one of the copies has become profoundly faded, improving reception quality and reliability [8].

1.3 *Spatial Multiplexing Gain*

Spatial multiplexing or transmission into the operating spectrum of several independent data streams leads to linear data rate increases with MIMO systems. The recipient can split the data streams under particular channel circumstances, such as dense dispersion in the surroundings. Capacity rises by a ratio commensurate with the number of streams [9].

1.4 *Interference Reduction and Avoidance*

Another reason spatial dimensions may be utilized is to avoid interference [10]. Pointing the signal in the desired direction while diverting the energy away from unwanted users reduces interference. Coverage and range are improved when interference is reduced and avoided. It may not be possible to use all of the benefits simultaneously since they are vying for available spatial degrees of freedom. While many of these features will provide more capacity, coverage, and reliability for a wireless network, they will come at the cost of more equipment [11].

1.5 *Literature Review*

MIMO often traces research articles from the 1970s that explore cable pair interference in the multi-channel digital transmission systems Kaye and George (1970), Branderburg and Wyner (1974), and van Etten (1975, 1976). While they are not multi-way propagation cases for multiple information streams, some mathematical approaches have effective MIMO. In the Bell Laboratories in the mid-1980s, Jack

Salz continued this investigation into the Bell Laboratories by looking at multi-user system operations utilizing mutually interconnected linear and additive noise networks [12].

- Chandan Kumar et al. describe that the four-channel MIMO microstrip and the U-shaped microstrip resonator, plus a pair of line resonators, were integrated into producing a single 5.25 GHz band resonator (WLAN band). Microstrip resonators were used to decrease mutual interaction between antenna components closely located. To retain the compactness of the MIMO structure, a distance of 0.5 between neighboring patch components is maintained. The length, breadth, and distance between the inverted U-shaped resonator arms have been tuned to provide the lowest possible mutual coupling [13].
- Wen et al. describe a stacking patch antenna that can be utilized in multiple input multiple output systems which has been designed. The proposed stacked patch antenna consists of two microstrip patches printed on the top and bottom of the substrate. We discover that eight desirable modes must resonate in the field by examining the different unique modes. Four distinct emission patterns appear in resonant modes on the bottom patch. To combine parallel resonances with a broader working band, three resonance frequencies are placed within the desired frequency range, including two similar resonances and one series resonance [14].
- Gao et al. author, have suggested, simulated, and built a proposal for future wireless communications using a conceptual design of a multiple input (MIMO) antenna system packed inside a metallic telephone. This MIMO antenna design employs four identical magnetic dipoles. The whole antenna system is within a mobile device that imitates the appearance and feel of a mobile phone. Some prototypes of antennas are built and measured to validate the conceptual design. In contrast with traditional single-input wireless communication systems (SISO), a single-input single-output telephone (MIMO) may almost double the channel capacity [15].
- Mohamed et al. describe that using a tiny four-element multiple input–output (MIMO) antenna, the authors explain and suggest a practical approach with good performance across an ultra-wide bandwidth. This is done by intrinsic directional radiation features and their asymmetrical configuration. The concept for a compact MIMO antenna system is confirmed by an adequate coefficient of envelope correlation [16].
- Bhavarthe et al. describe that author has designed the new type of multilayer meander strip line step-by-step (MLSV-EBG) structure to create the compact multilayer multi-input multi-output (MIMO) antenna with a mutual decrease in connections. This indicates a multi-layered cell, a unique meander line, and a step-by-step idea. The suggested EBG is evaluated utilizing an LC model parallel approach. The step-through architecture of the MLSV-EBG design results in a higher route length and cell compactness. This program also offers parametric research. The MLSV-EBG is recommended to illustrate how it may decrease the reciprocal coupling between two MIMO multilayer antennas. The study reported in this statement makes an essential addition by demonstrating that the

compact multilayer EBG structure is more efficient and less often in a multilayer environment [17].

- Behera et al. describe that UWB MIMO portable wireless, both compact and efficient, might be equipped with a UWB MIMO fractal antenna diametrically fed. Conventional mode analysis is used in conducting organisms to examine electromagnetic activity (CMA). The inherent properties are studied for these devices, I modal distribution of surface current, (ii) narrow/comprehensive bandwidth capability, and (iii) radiation potential. There are profound studies of the existence of fundamental and non-essential geometric modes. More specifically, CMA may demonstrate where the antenna radiation is the most likely to arrive on the body because currents pass through the modal body. The modal value of CMA here referred to as "modal value" comprises several distinct features, including (i) the proprietary values (EV) and (ii) the angles (CA). For narrowband/wideband characteristics, energy behavior linked electrically/magnetically and radiative potential for far propagation [18] are prevalent.

2 MIMO Wireless Communication System

To implement multiple input multiple output (MIMO), use a complicated matrix that changes depending on the scenario you are looking at [19]. In each of the situations above, the Shannon extended capacity formula is used to determine the amount of capacity gained by the MIMO channel. Simulation results derived from the use of multiple antenna elements on the receiver and transmitter, their distance from one another, and the degree of correlation observed are proportional to the number of antenna elements and their distance from one another, as well as the number of simulations, run [20].

2.1 Shannon's Capacity Formula

The estimate of the theoretical maximal channel transmission is based on the Shannon capability formula, which estimates the channel the theoretical bandwidth, signal intensity transmitted, and single side noise spectrum and assumes that the track is white Gaussian (i.e., no explicit consideration is given to fading and interference effects) [21].

2.2 Extended Capacity Formula

Consecutive outputs: This kind of antenna is called a multiple input multiple output antenna (MIMO) because it enables both the receiver and transmitter to have multiple antennas. Using Shannon's capacity formula, we can determine the capacity of this antenna [22].

3 Microstrip Antenna Technology

For RF engineers, the most challenging challenge is to understand how complex systems work from the transmitter to the receiver. It is not enough to investigate a communications link's components one by one [23]. A vital element in the MIMO system is the antenna. MIMO systems make use of the channel's multipath characteristics. The antennas are chosen in line with the propagation channel to implement the best antenna design for the propagation medium. Correlations between channel coefficients are affected by antenna parameters. Mutual coupling effects may occur when antennas are co-located in a MIMO array. Think about how all of these issues affect the antenna array for MIMO systems when you are constructing it [24]. Due to this, it is an important challenge for MIMO system integration to accommodate many antennas in small portable devices. Because MIMO technology depends on practical antenna designs, functional antenna designs must be developed. Microstrip geometry was first used for high-performance airplanes, spacecraft, satellites, and missiles in the 1950s due to size, affordability, performance, ease of installation, and aerodynamic profile [25]. Figure 2 shows microstrip antenna.

It describes that the metal patch comprises parasitic and feed patches. Presently, commercial and government applications must comply with similar regulations, such as cellular radio and wireless communications. Almost simultaneously with its introduction in 1952, radiators suitable for microstrip transmission lines were developed by Grieg and Englemann. This is the first instance where a microstrip-like antenna was coupled with a microstrip transmission line to support communication. The idea of a microstrip antenna was patented by Gutton and Baissinot in 1955 [27]. Figure 3

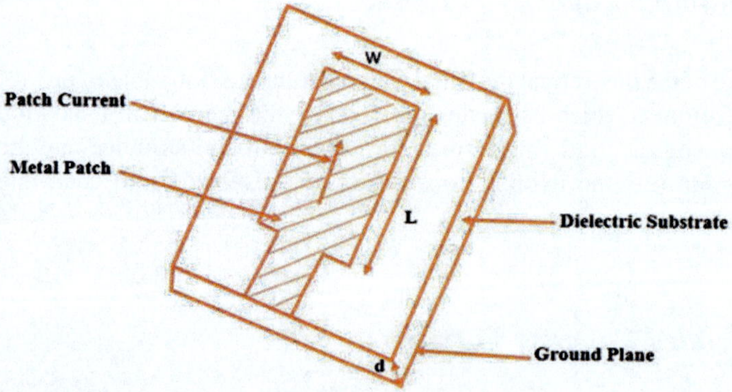

Fig. 2 Overview of MSA [26]

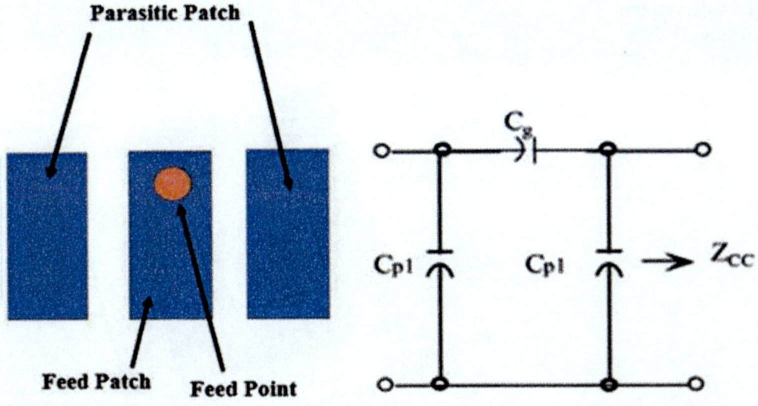

Fig. 3 MSA circuit theory [28]

shows the circuit theory of MSA. Parasite MSAs may be constructed using their analogous circuit diagram. These antenna types are shown as gap capacities (C_{p1} and C_{p2}) with a patch. C_c coupling is also established, and C_{P1}, $C_{P2,}$ and C_C values can be calculated.

These antennas have many contributing factors:

- They feature a low profile, making them less noticeable and easier to adapt to irregular surfaces.
- Their production is easy and inexpensive because they are printed circuit board designs.
- When they are connected to a hard surface, they are structurally strong.
- Depending on the simulation model and patch shape, the resulting variety of patterns and polarization may be very wide.
- Additional fabrication procedures may not be required if other microwave devices capable of being realized in microstrip are coupled with a microstrip antenna.

An early form of the microstrip antenna had a bandwidth of only a few megahertz, giving rise to several problems such as spurious feed radiation, a poor polarization purity, and limited power capacity [29]. This means that most research has been done to resolve the problems that arise from system needs, and therefore, this significant part of the research has been to meet higher system requirements. Additionally, these developments have included the development of novel microstrip antenna designs and transparent and flexible analysis of the inherent limitations of microstrip antennas. Figure 4 shows the design and performance issue of microstrip antenna.

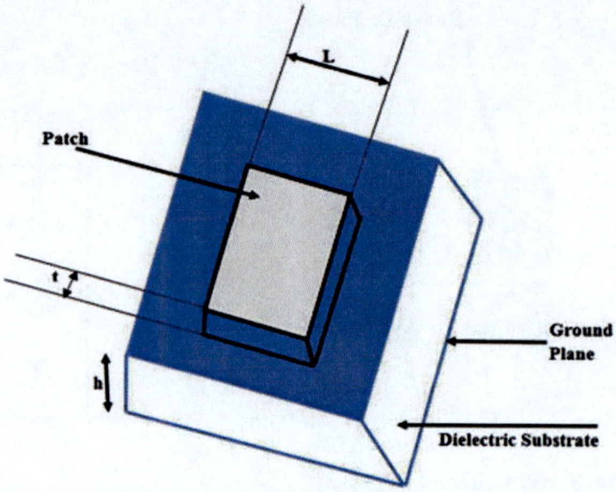

Fig. 4 MSA design and performance issue [30]

4 MIMO Antenna

In designing MIMO antenna designs, the broadband triangular patch is used. Once the simulations have been created, many construct and test prototypes. When designing MIMO antennas, two measures must be taken into consideration. Antenna performance parameters and available channel capacity are similar to conventional antennas. MIMO communications have two main disadvantages: mutual coupling and antenna size [31].

5 MIMO Antenna Advantages

- Multiplexing spatial
- Fading effects minimization
- Improved channel and spectral efficiency
- Cell coverage expansion and increase in average cell output
- Reliability and decreased vulnerability to unlicensed users
- Fast acquisition of data
- Need fewer antenna elements
- Good coherence for target response between signals may be imagined quickly changing target [32]

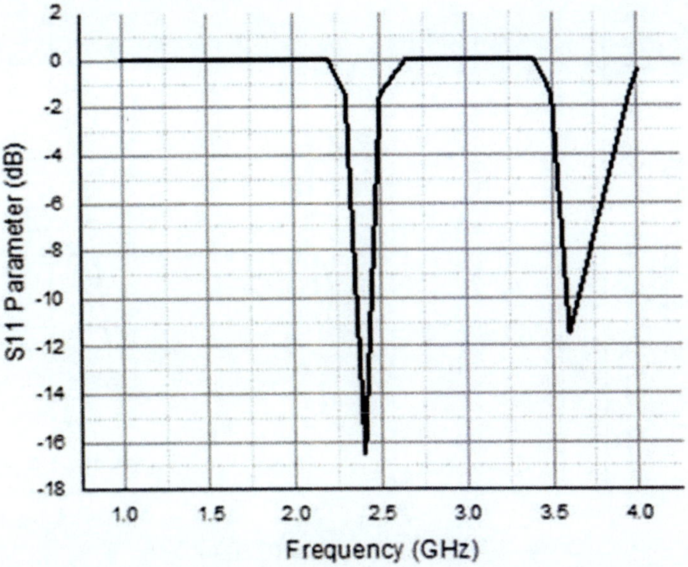

Fig. 5 Plotted frequency response of the 2.4 GHz MIMO antenna S_{11} parameter

6 Simulated Results and Discussion

The word "antenna techniques" refers to the testing process by antennas to verify that they satisfy or define the criteria. All antenna S-parameters, return loss, and ECC are described in this section.

Figures 5, 6, 7, and 8 will illustrate the observed and simulated frequency response of the 2.4 GHz MIMO, while the return loss will be shown in Figs. 9 and 10 ECC. S-parameter measurements are taken to determine the loss of return and isolation of the prototype generated. The S-parameter measures the antenna power because of its 2.4 GHz MIMO antennas. The lower return loss indicates higher efficiency, and the resonant frequency of the prototype is computed at the location of the lowest return loss.

7 Conclusion

The antennas presented are lightweight and small and may thus be modified for different purposes, such as aeronautical communication. MIMO is a technology that fulfills today's high data rate, enhanced security, and many other improvements in the field of antennas. Most of the newest MIMO-integrated wireless systems and much more research are underway.

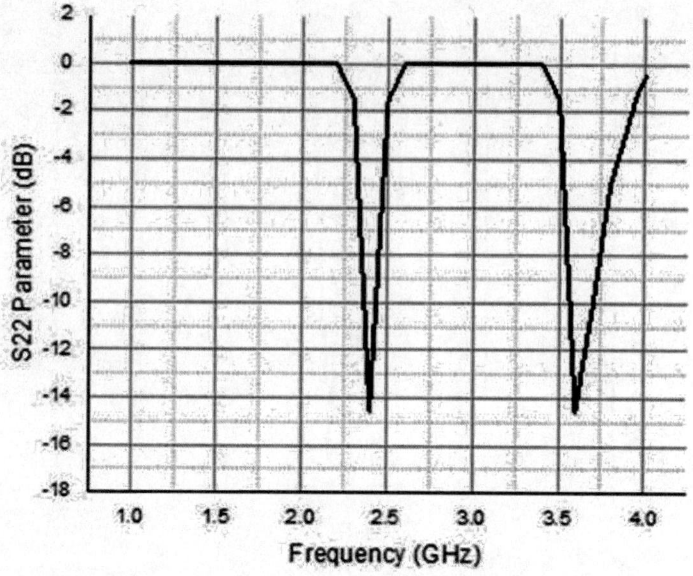

Fig. 6 Plotted frequency response of the 2.4 GHz MIMO antenna S_{22} parameter

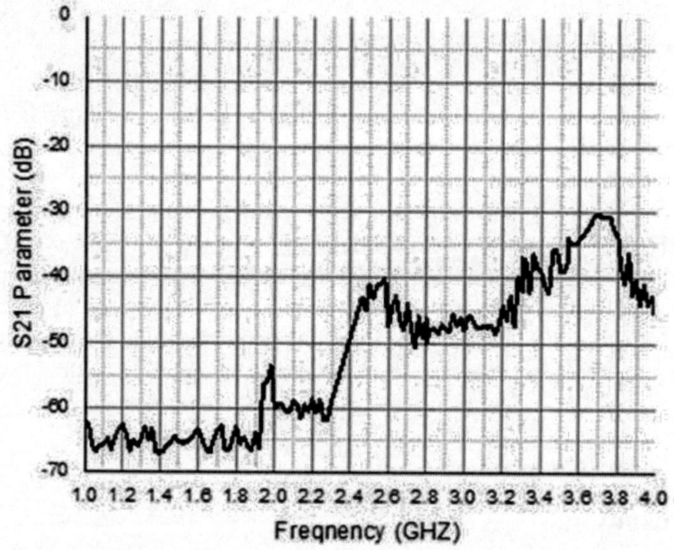

Fig. 7 Plotted frequency response of the 2.4 GHz MIMO antenna S_{21} parameter

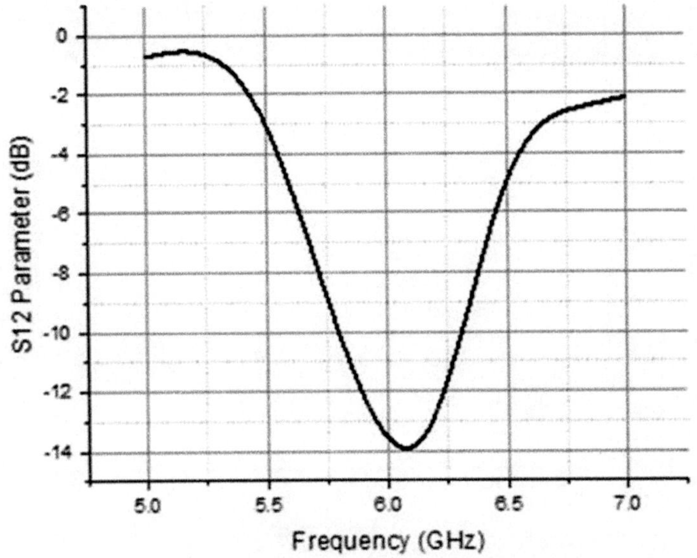

Fig. 8 Plotted frequency response of the 2.4 GHz MIMO antenna S_{12} parameter

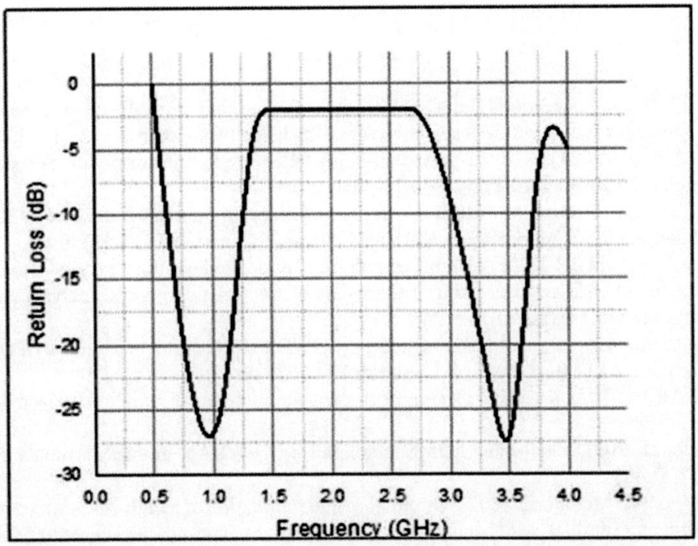

Fig. 9 Plotted frequency response of the 2.4 GHz MIMO antenna return loss parameter

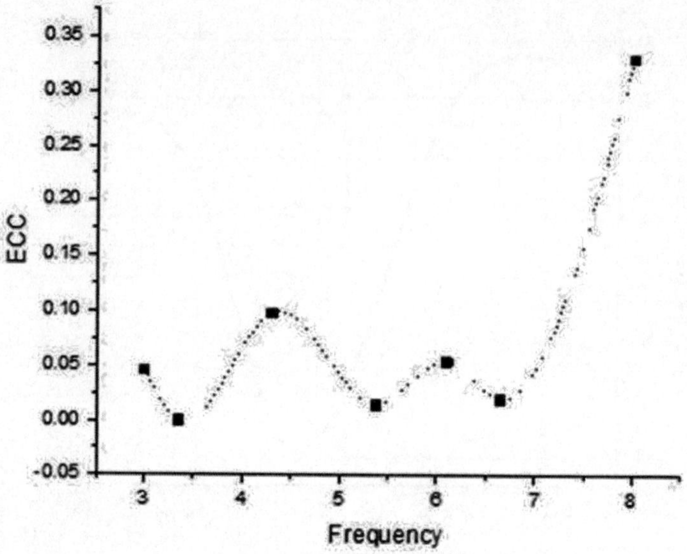

Fig. 10 Plotted frequency response of the 2.4 GHz MIMO antenna envelope correlation coefficient

References

1. Kulkarni J, Desai A, Desmond Sim C-Y (2021) Wideband four-port MIMO antenna array with high isolation for future wireless systems. AEU-Int J Electron Commun 128:153507
2. Huang J, et al (2021) A quad-port dual-band MIMO antenna array for 5G smartphone applications. Electronics 10(5):542
3. Pant A, Singh M, Parihar MS (2021) A frequency reconfigurable/switchable MIMO antenna for LTE and early 5G applications. AEU-Int J Electron Commun 131:153638
4. Mohanty A, Behera BR (2021) Insights on radiation modes and pattern diversity of two-element UWB fractal MIMO antenna using a theory of characteristics modes analysis. AEU-Int J Electron Commun 135(2021):153726
5. El-Hameed A, Anwer S, et al (2021) Quad-port UWB MIMO antenna based on LPF with vast rejection band. AEU-Int J Electron Commun 134:153712
6. Mutiara AB, Refianti R (2012) Design of microstrip antenna for wireless communication at 2.4 GHz
7. Huang J et al (2021) Dual-band MIMO antenna for 5G/WLAN mobile terminals. Micromachines 12(5):489
8. Bahmanzadeh F, Mohajeri F (2021) Simulation and fabrication of a high-isolation very compact MIMO antenna for ultra-wideband applications with dual band-notched characteristics. AEU-Int J Electron Commun 128:153505
9. Rajeshkumar V, Rajkumar R (2021) SRR loaded compact tri-band MIMO antenna for WLAN/WiMAX applications. Prog Electromagn Res 95:43–53
10. Khan AA et al (2021) Quad-port miniaturized MIMO antenna for UWB 11 GHz and 13 GHz frequency bands. AEU-Int J Electron Commun 131:153618
11. Peng H, et al (2021) Design of a MIMO antenna with high gain and enhanced isolation for WLAN applications. Electronics 10(14):1659
12. Ren Z, Zhao A, Wu S (2019) MIMO antenna with compact decoupled antenna pairs for 5G mobile terminals. IEEE Antennas Wirel Propag Lett 18(7):1367–1371

13. Ghosh CK (2016) A compact 4-channel microstrip MIMO antenna with reduced mutual coupling. IEEE Trans Antennas Propag 70(7):873–879
14. Wen Y et al (2017) Bandwidth enhancement of low-profile microstrip antenna for MIMO applications. IEEE Trans Antennas Propag 66(3):1064–1075
15. Gao C et al. (2018) Conceptual design and implementation of a four-element MIMO antenna system packaged within a metallic handset. Microw Opt Technol Lett 60(2):436–444
16. Saad AAR, Mohamed HA (2019) Conceptual design of a compact four-element UWB MIMO slot antenna array. IET Microw, Antennas Propag 13(2):208–215
17. Parvathi KSL, Gupta SR, Bhavarthe PP (2020) A novel compact electromagnetic band gap structure to reduce the mutual coupling in multilayer MIMO antenna. Prog Electromagn Res M 94:167–177
18. Mohanty A, Behera BR (2021) Characteristics mode analysis: a review of its concepts, recent trends, state-of-the-art developments and its interpretation with a fractal UWB MIMO antenna. Prog Electromagn Res B 92:19–45
19. Kumar N, Khanna R (2021) A two-element MIMO antenna for sub-6 GHz and mmWave 5G systems using characteristics mode analysis. Microw Opt Technol Lett 63(2):587–595
20. Sahu NK, Das G, Gangwar RK (2021) Circularly polarized offset-fed DRA elements & their application in compact MIMO antenna. Eng Sci Technol, Int J
21. Vasu BK, Anuradha B (2021) Design of UWB MIMO antenna to reduce the mutual coupling using defected ground structure. Wirel Pers Commun 118(4):3469–3484
22. Song R et al (2021) A graphene-assembled film-based MIMO antenna array with high isolation for 5G wireless communication. Appl Sci 11(5):2382
23. Fu Z, Shen W (2021) Eight-element self-decoupled MIMO antenna design for 5G smartphones. Int J RF Microwave Comput Aided Eng 31(3):e22523
24. Chattha HT, et al (2021) Compact multiport MIMO antenna system for 5G IoT and cellular handheld applications. IEEE Antennas Wirel Propag Lett
25. Khan MK, Feng Q, Zheng Z (2021) Experimental investigation and design of UWB MIMO antenna with enhanced isolation. Prog Electromagn Res C 107:287–297
26. Haupt RL (2010) Antenna arrays: a computational approach. Wiley
27. Mohanty A, Behera BR (2021) Design of a 3-port compact MIMO antenna based on characteristics model analysis approach. Prog Electromagn Res C 111:173–189
28. Singh A, et al (2020) A review: circuit theory of microstrip antennas for dual-, multi-, and ultra-widebands. Modul Electron Telecommun
29. Sanmugasundaram R, Natarajan S, Rajkumar R (2021) A compact MIMO antenna with electromagnetic bandgap structure for isolation enhancement. Prog Electromagn Res C 107:233–244
30. Bankey V, Anvesh Kumar N (2015) Design and performance issues of microstrip antennas. Int J Sci Eng Res 6(3):1572–1580
31. Andrade-González EA et al (2021) UWB four ports MIMO antenna based on inscribed Fibonacci circles. J Electromagn Waves Appl 35(9):1202–1220
32. Ren Z, Zhao A (2019) Dual-band MIMO antenna with compact self-decoupled antenna pairs for 5G mobile applications. IEEE Access 7:82288–82296

Design Stream

Design and Analysis of Single Screw Extruder for Hybrid Manufacturing Process

Saurabh Kausadikar, Mithilesh Kumar Tiwari, K. Ponappa, and Puneet Tandon

Abstract Metal additive manufacturing is a technology that has been developed for future industrial manufacturing. In the current scenario, metal additive manufacturing confronts several challenges, such as geometrical constraints or requiring post-processing on manufactured parts. Because of the constraints of processes, additive manufacturing is only used for prototyping in today's world. Individually, additive manufacturing and incremental sheet forming processes are inefficient for producing complicated geometries and have higher production time. To alleviate the limitations, hybrid manufacturing is a solution that combines two separate manufacturing technologies, i.e., combination of additive and incremental sheet forming which is capable of the development of new process routes that allows the production of more complex and higher quality parts in a more flexible manner. Hybrid manufacturing setup suitable for carrying out the method for building components, where the part is fabricated by extrusion-based additive manufacturing then subsequently locally formed by incremental sheet forming. This article's major contribution is design optimization, thermal analysis, and flow analysis of a single screw extruder for the development of a hybrid manufacturing setup. The extruder is designed with an attachable forming tip that serves the purpose of printing and then forming the material to obtain the finished component.

Keywords Screw-based extrusion · Design of single screw · Hybrid manufacturing

---

S. Kausadikar (✉) · K. Ponappa
Smart Manufacturing Lab, Mechanical Engineering Discipline, PDPM Indian Institute of Information Technology, Design and Manufacturing, Jabalpur 482005, Madhya Pradesh, India
e-mail: 20mmec07@iiitdmj.ac.in

M. K. Tiwari · P. Tandon
deLOGIC Lab, Mechanical Engineering Discipline, PDPM Indian Institute of Information Technology, Design and Manufacturing, Jabalpur 482005, Madhya Pradesh, India

N. M. Sivaram et al. (eds.), *Advances in Manufacturing, Automation, Design and Energy Technologies*, Lecture Notes in Mechanical Engineering,
https://doi.org/10.1007/978-981-99-1288-9_41

1 Introduction

Hybrid manufacturing (HM) is a combination of advanced and traditional production processes which integrates two manufacturing methods. The purpose of hybrid manufacturing is to recognize and integrate the benefits of both additive manufacturing (3D printing) and conventional subtractive/forming machining to produce a unified production process or single configuration that successfully combines both at the same time. By combining additive manufacturing with subtractive, forming, and other traditional processes, the long production time of additive manufacturing may be reduced. In addition, two or more types of additive manufacturing systems can be coupled to create multi-material systems for high-performance applications [1]. Last but not least, component production (through additive manufacturing) and post-processing can be integrated to bring up new research options for improving process capabilities, minimizing weaknesses, and expanding application areas [2].

Manufacturing costs and process time in additive manufacturing and forging could be reduced by mating both processes to new process chains. By combining additive manufacturing and metal forming in a single setup using a different power source, two possible process sequences can be put into practice, according to a report by the European Commission [3]. It is proposed to combine additive manufacturing with a subsequent forming operation as a process combination route. The process route has the potential to boost the additive manufacturing process route's efficiency by up to 360% [4]. Fischer G. et al. develop a new setup by a combination of additive and forming processes, and this invention aims to incorporate a forming process directly into the additive manufacturing single setup. The advantage of this setup is to smooth the surface roughness determined by additive manufacturing, but the limitation is power source which is different for both operations [5]. Holker et al. filed a patent for the method and device for the combined production of components by means of incremental sheet forming and additive methods in a one clamping setup. This invention is suitable for the method for producing components by means of ISF and additive methods; this configuration employs a variety of power sources to carry out operations [6]. As a result, an adequate hybrid manufacturing model for the manufacture of components using incremental sheet forming and additive methods in a single setup has yet to be developed. More precisely, the system in which additive manufacturing and incremental sheet forming will consume the same power supply has yet to be developed.

It is necessary to concentrate on the hybridization of additive manufacturing and incremental sheet forming in a single arrangement utilizing the same power source. There are still numerous drawbacks to additive manufacturing: costly initial investment and necessary maintenance expertise, deposited layers may cause weakening of components if not properly calibrated, far too slow for mass manufacture, and the equipment requires highly skilled maintenance personnel. On the other hand, the incremental forming process, confronts substantial obstacles such as the process duration being too long, geometrical inaccuracy, a higher amount of force is necessary, etc. As a result, a great potential innovation lies in the gap between the two: a vast range of feasible designs that combine complicated geometry with high precision manufacturing. An innovative approach of a hybrid manufacturing system needs to connect the gap by finding an appropriate solution to use additive manufacturing and incremental sheet forming process together on a single setup or in a single integrated design approach. The hybrid incremental forming additive manufacturing process enables the combination of incremental sheet forming and additive manufacturing. This paper major contribution is to design the single screw and optimization using parameters, and numerical analysis of a single screw extruder is employed in the hybrid manufacturing setup.

2 Single Screw

To design a single screw for the extrusion-based fused deposition modeling, there are several parameters of the single screw that must be taken into the consideration also the mathematical description of the screw geometry must be known to model an extrusion process. A single screw's design is mostly determined by its purpose; in this work, it is employed in the extruder to provide material for printing. A screw-based extruder is a type of extruder in which the material to be processed is forced into a barrel by the rotation of the screw. This system is made up of a rotor which is rotated by a drive system; for simultaneous rotation, the motor and rotor are linked together. The outside casing of the system is known as a barrel, and it stops material from spilling outside of the extruder. The cavity or channel depth defines the space between the barrel's inner and outer periphery and the screw.

The single screw must fulfill two essential conditions in order to work properly: (1) It must be tough enough to bear both load and torque, as well as extreme temperatures. (2) The movement of the screw, friction with the cylinder, and the heating system will all contribute to the high temperatures. This challenge is being addressed by selecting the best material for the single screw and barrel to withstand the high temperature and torque. A single screw is separated into three primary sections: the feeding zone, the transition zone, and the metering zone [7]. These zones are being provided to get accurate material extrusion through the nozzle shown in Fig. 1. Detail descriptions of these 3 zones are given below:

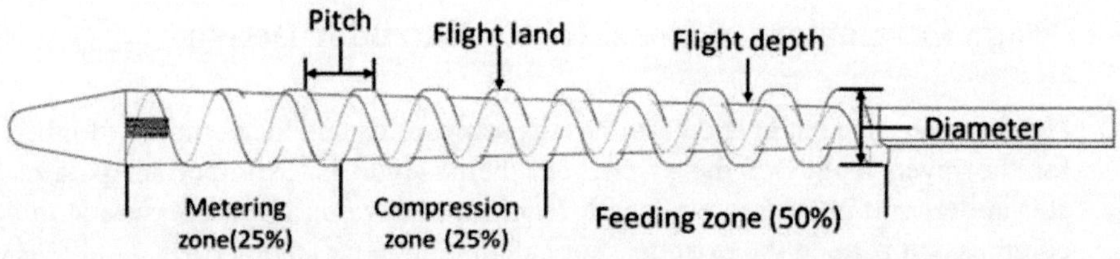

Fig. 1 Single screw terminology

(a) **Feed zone**—The feed zone aims to transfer a large volume of material to the transition zone whilst feeding through the hopper material is in a solid condition which is carried by thread depth of the single screw. The flight depth of the single screw in the feeding zone is varying from start to end. Some researchers fix the depth of the feeding zone at three to five times the depth of the metering zone [8, 9]. For this design work, the length of the feeding zone is half the length of the screw. A balancing of forces controls the extruder screw feeding material in the feeding area. As a thumb rule, the feed section of the screw should not be deeper than,

$$\text{Feed } d_{max} = 0.2 \times \text{ screw diameter [10]}$$

$$\text{Where } d_{max} = \text{Maximum flight depth}$$

(b) **Transition zone**—When the melt pool has formed to the point of piercing the solid substrate and generating a melt pool, the feed material should go through a phase shift in the transition zone. This melt pool coexists with the ever-shrinking breadth of the solid bed [11]. The transition zone is 25% of the whole screw length, and with varying diameter for melting material into the compression zone, this varying depth helps to compress the material during supplying heat.
(c) **Metering zone**—The solid bed is entirely melted in the metering zone, and there is just melt conveying. When the pressure gradient is positive, the drag flow is controlled, and when it is negative, the flow is facilitated. Whilst pressure develops, throughput remains constant along the channel [11]. The size of the metering zone should be 25% of the entire screw length.

Single screw's designed based on the material to be extruded as well as the amount of pressure that the screw must create to withstand the high forces encountered during printing and forming [12]. Because the main objective is to print the metal through the extruder and subsequently do forming on it, it is important to choose an extruded material with high strength and slip, as well as a high melting temperature. As reviewed many parameters of different materials for selecting the material for the additive manufacturing and incremental sheet forming coalescence extruder, selecting the Inconel 625 or high-speed steel as the material for the extruder.

3 Design Parameters to Consider for Extruder Design

(a) **Helix angle**—Helix angle determines the screw pitch and the number of flights for the given length of the screw. The helix angle is an important parameter in determining the screw length required to develop given pressure at the compression zone of the extruder. For calculating helix angle [10],

$$\tan \varnothing = \frac{P}{\pi D}$$

where $\varnothing$ = Helix angle; P = Pitch; D = Screw diameter.

(b) **Hopper angle**—To limit the possibility of bridging, the angle of the hopper should be chosen such that it is larger than the angle of repose. Wall friction will rise as the hopper angle increases.

(c) **Compression ratio**—The compression ratio is a crucial factor to consider whilst designing screws. It is stated as the ratio of the feed section channel depths to the metering section channel depths. The compression ratio is usually used in place of depth compression ratio, and it is defined as follows [10].

$$\text{Compression ratio} = \frac{\text{Feedzone flight depth}}{\text{metering zone flight depth}}$$

For specialized screws, such as discussed in further section extruder design, a volumetric compression ratio is a more suitable parameter than depth compression ratio, as the pitch of the single screw varies all the way from feed zone to the compression and metering zone.

(d) **Length-over-diameter (L/D) ratio**—For typical extruder, L/D ratios are 24:1 to 30:1, but there are some special applications where the extruder is built as short as possible having a 10:1 L/D ratio. The number of screw zones considered whilst designing a single screw estimate the length of the screw, which is determined by the application and materials being processed. Conventional screws with three functional zones (feeding, compression, and metering zones) have a standard L/D ratio of 24:1 [13].

(e) **Flight Clearance**—Flight clearance is defined as the distance between the screw's outer diameter and the barrel wall. It is commonly taken as 0.05% of the screw radius. Inadequate flight clearance will have a detrimental impact on extruder output efficiency; excessive wear on screw flights can be caused by less clearance, whilst large clearance will reduce the screw's melting efficiency [10].

Figure 1 shows the detailed design of the single screw design for the extruder which will use extruder assembly.

Figure 2 shows the design of the single screw extruder which is increasing compression ratio at the tip of the extruder and has been varying the depth on the screw from one end to another end, and the pitch of the screw is constant over the length. Forming tip which is detachable at the end of the screw where the diameter is 16 mm for the purpose of forming. The material will be inserted from the above part of a hopper, and a rotating force will be provided by the incremental sheet forming machine to a single screw extruder which will transport and drive the material toward the downward direction. During the transportation of material through extruder, the material will be converted into the solidus phase, whilst printing the hopper will be at its initial position, afterward the solenoid will lift the hopper toward the upward direction, and the forming tip will emerge outside of the hopper which is ready to do forming of

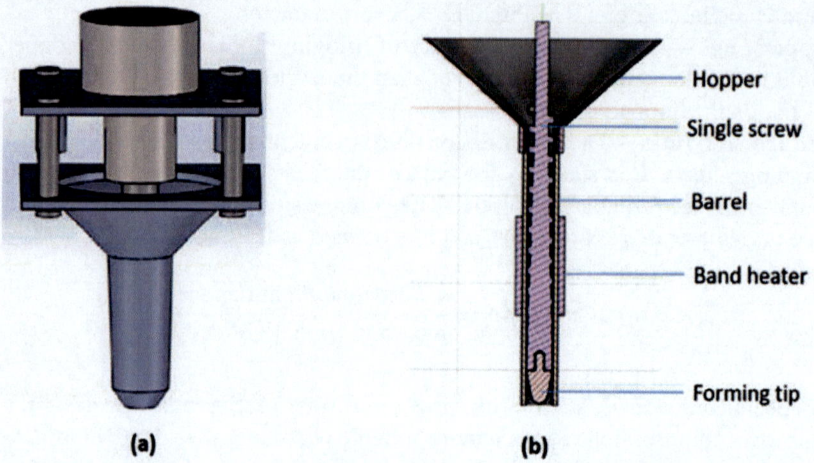

Fig. 2 **a** Proposed setup design for AM and ISF, **b** section view of single screw and forming tip with barrel and hopper

the printed part. This will be repeatedly done to print and form the material into the desired shape.

4 Result and Discussion

4.1 Thermal Analysis

The model environment is created in Autodesk Fusion 360, and the required geometry is created. The material used during the numerical analysis is Inconel 625 which is a metal that possesses high strength properties and resistance to elevated temperature. Properties of Inconel 625 are shown in Table 1.

From simulation results, it is visible that when heat is provided on band heaters, the single screw with forming tip, barrel and hopper heating up to some extents,

Table 1 Material properties of Inconel 625

Thermal conductivity (W/m K)	9.8
Melting point range (°C)	1,290–1,350
Density (g/cm^3)	8.44
Tensile strength, ultimate (MPa)	880
Thermal expansion coefficient (1/K)	1.28×10^{-5}
Elastic modulus (GPa)	205
Specific heat (J/g °C)	0.410

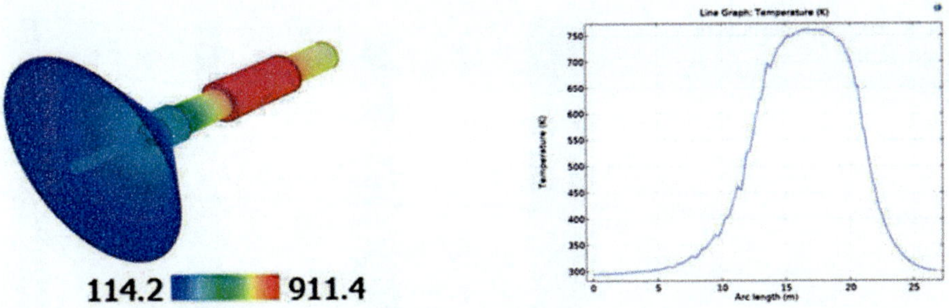

Fig. 3 Heat transfer analysis

this achieves temperature on the other parts will not affect the part of the extruder assembly. Figure 3 shows the thermal simulation results in Fusion 360. Heat is applied through the band heaters at achieves a temperature of 900 °C due to which maximum temperature is obtained at the contact surface between band heater and barrel, i.e., 911.4 °C, and minimum temperature is obtained at the hopper section around 114.2 °C. From this conclude, the extruder assembly will be safe at this much temperature from this point of view we have to feed material that goes in a semisolid state in temperature (above the solidus line which is passing through the extruder).

4.2 *Volumetric Flow Analysis*

The single screw is designed by taking extruder parameters into consideration. The first and most basic design is shown in Fig. 4, and it is more widely used and has an inner diameter that decreases from start to end, making compression zone and flow from a single screw impossible. To overcome the difficulties for positive flow, the single screw design in this study is optimized using the parameter volumetric flow and by analyzing whether flow via the single screw is feasible or not using the drag flow and pressure flow coefficient. Compression of material whilst flowing in between the barrel and single screw depends on the pitch and flight depth provided over the screw. Figure 4 Different designs of single screws in that B, C, and D are the optimized design of the single screw based on the volumetric flow through the extruder.

To make the safe design of the screw, it is important to lower the value of pressure flow coefficient than drag flow coefficient; otherwise, it is not possible any positive flow.

$$\text{Volumetric flow (Q)} = (\text{Dragflow} - \text{Pressureflow} - \text{Leakflow})$$

$$Q = \left[\frac{\alpha k}{k + \gamma(+\beta)}\right] \times \eta \ [14]$$

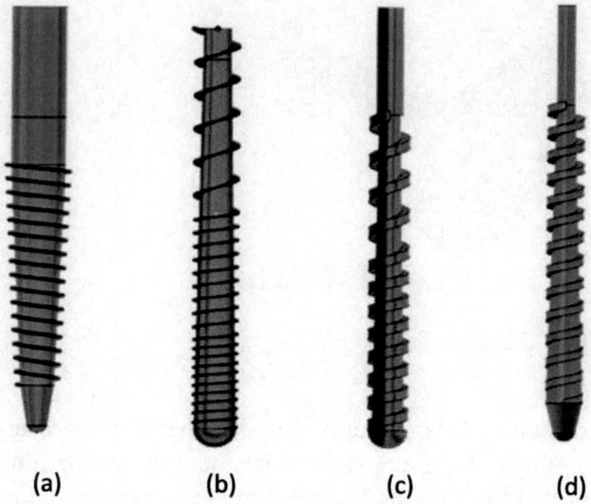

Fig. 4 Different designs of single screw

where α = Drag flow coefficient
δ = Filet clearance
β = Pressure flow coefficient
D = Screw diameter
k = Head geometrical constant
t = Pitch
γ = Filtration flow coefficient
h = Initial channel depth
η = Spindle speed
e = Range width (0.12 * D)

$$\alpha = \pi \cdot \text{mDh}\left[\frac{\text{t}}{\text{m}} - \text{e}\right]\frac{\cos 2\theta}{2}$$

$$\text{Filtration flow coefficient } (\gamma) = \frac{\pi^2 \text{D}^2 \delta^2 \tan\varnothing}{10\text{eL}}$$

The above equations define the relationship between pressure and drag flow which gives an idea about flow is possible or not.

From Table 2, it can be seen that design A has negative volumetric flow; it means that the pressure flow coefficient is greater than the drag flow coefficient; hence, flow is not possible moreover design C and D have more volumetric flow than the other 2 design; and also drag flow coefficient is more than pressure flow coefficient. Design D is more suitable for the application of extrusion-based additive manufacturing.

Table 2 Comparison between designs of single screw

	Pitch (mm)	Pitch diameter (mm)	Initial depth (mm)	Helix angle (°)	Screw length (mm)	Flight thickness (mm)	Relation between drag flow (α) and pressure flow coefficient (β)	Volumetric flow (mm^3/s)
Design A	4.38	Varying	1.49	8.07	150	0.5	β > α	−7358.34
Design B	Varying	13	2.5	12.58	150	0.41	β < α	22,160.32
Design C	Varying	15	3.3	16.78	150	4	β < α	50,638.64
Design D	11.54	16	4	17.21	150	4	β < α	71,409.19

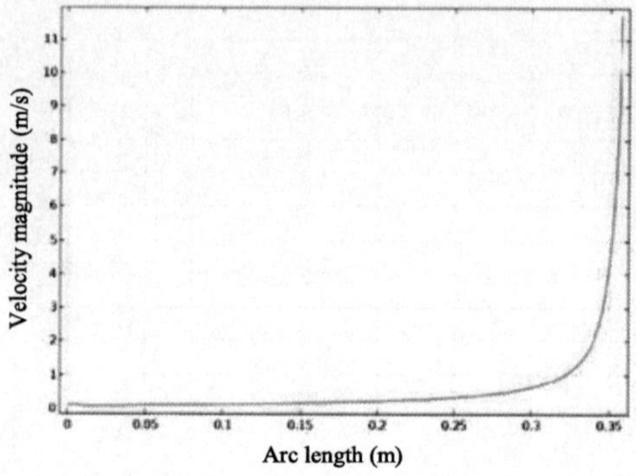

Fig. 5 Arc length (m) versus velocity magnitude (m/s)

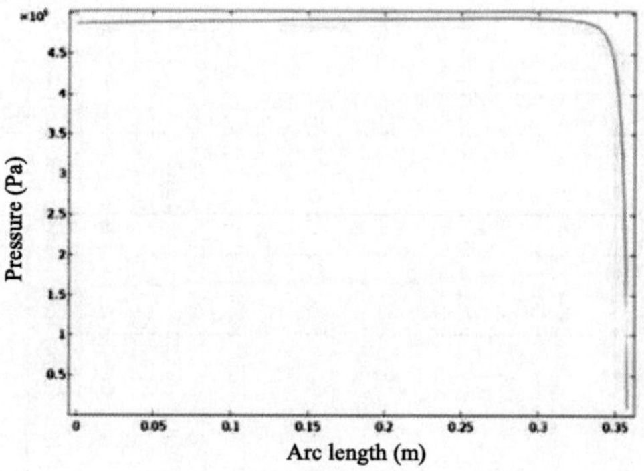

Fig. 6 Arc length (m) versus pressure (Pa)

4.3 Material Flow Analysis

Flow analysis is required to understand the behavior of the material inside the extruder. For the flow analysis, a path inside the extruder is required, which is obtained by performing a Boolean operation. The material flow is simulated using COMSOL multi-physics in the laminar flow module with a slip boundary. The computed flow velocity and pressure inside the extruder are shown in Figs. 5 and 6. The graph clearly shows that the pressure and velocity of the material flowing within the extruder at the compression and metering zones are rising.

5 Conclusion

The unique hybrid manufacturing setup is designed with the critical parameters of extrusion-based additive manufacturing in consideration for printing the material. Additive manufacturing and incremental sheet forming are proposed as one of the processes for hybrid manufacturing in which material is printed through the additive manufacturing process and then forming are done simultaneously through the incremental sheet forming. To get a significance of the thermal conditions, material flow and volumetric flow through single screw extruder simulation analysis were performed over the designed model, and the results were successful in terms of temperature rise, mathematical analysis of volumetric flow, and internal motion of the single screw extruder. The thermal simulation of the hopper shows that the temperature rise inside the hopper can reach up to 200 °C, whilst the temperature rise in the compression and metering zones can reach 600–700 °C, which can melt the inner material utilized in the study. The design of a single screw is optimized to obtain better volumetric flow and compression of the material inside the extruder.

References

1. Singh S, Ramakrishna S, Singh R (2017) Material issues in additive manufacturing: a review. J Manuf Process 25:185–200
2. Zhu Z et al (2013) A review of hybrid manufacturing processes—state of the art and future perspectives. Int J Comput Integr Manuf 26(7):596–615
3. Bambach M et al (2020) Hybrid manufacturing of components from Ti-6Al-4V by metal forming and wire-arc additive manufacturing. J Mater Process Technol 282
4. Rosenthal S et al (2021) Speeding up additive manufacturing by means of forming for sheet components with core structures. Int J Precis Eng Manuf Green Technol
5. Fischer G, Röhrich T, Bambach M (2019) Method and apparatus for combining additive manufacture and shaping. B.T.U.C.G. GmbH
6. Hölker R et al (2016) Method and device for the combined production of components by means of incremental sheet forming and additive methods in one clamping setup. T.U. Dortmund
7. Mount EM (2017) 12—Extrusion processes. In: Kutz M (ed) Applied plastics engineering handbook, 2nd edn. William Andrew Publishing, pp 217–264
8. Eslami H (2015) Understanding screw design for film extrusion process. Macro Engineering and Technology
9. Manohar B, Periasamy C (2016) Computational fluid dynamics simulation of single screw extruders in cable industries. IJRET: Int J Res Eng Technol 5(3):5
10. Campbell GA, Spalding MA (2021) Analyzing and troubleshooting single-screw extruders. In: Campbell GA, Spalding MA (eds) Analyzing and troubleshooting single-screw extruders, 2nd edn. Hanser, pp I–XIX
11. Béreaux Y, Charmeau J-Y, Moguedet MI (2007) A simple model of throughput calculation for single screw. In: AIP conference proceedings, pp 963–968
12. Abdel-Ghany WE, Ebeid SJ, Fikry I (2015) Mechanical design aspects of single screw extruders using finite element analysis. Int J Eng Tech Res (IJETR) 3(7):5
13. Shrivastava A (2018) 5—Plastics processing. In: Shrivastava A (ed) Introduction to plastics engineering. William Andrew Publishing, pp 143–177
14. Wankhade MH, Bahaley SG (2018) Design and development of plastic filament extruder for 3D printing. IRA-Int J Technol Eng 10(3). ISSN 2455-4480

Power Transmission in Electrodynamic Suspension (EDS) Type MAGLEVs Through 2G High-Temperature Superconducting Cable

Arijeet Roy Chowdhury and Vineet Sahoo

Abstract Contrary to conventional method of power transmission, HTS cables open a new front in the technology of MAGLEVs. The EDS type L0 series MAGLEVs are operational in Japan which can achieve the highest speed of 603 km/h and a maximum levitation of 3.9 in. from the ground. In this paper, the focus is to use the HTS cable in such MAGLEVs. HTS cable is proposed with liquid nitrogen (LN2) cooling which is in-built to power the on-board cryogenic systems and requisite electric components of the MAGLEV train. The width of the cable hereby proposed is determined considering a single bogie/compartment bearing in mind the optimum power requirement along with a buffer. This work is intended to provide a new reference for superconducting MAGLEV development.

Keywords HTS · Superconductivity · MAGLEV

1 Introduction

The increase in urbanization has led to an increased transport demand and an inherent urge to travel to destination in the least possible time-frame. Historically, train and airplanes are known to be the fastest mode of public transport. With the advent of modern age, any mode of transport that contributes less carbon to the environment or in other words produces a transport with less carbon footprint is also the need of the day. In this context, the high speed on wheel trains holds a lot of importance. Such trains in the last decade are mainly developed by three countries: Tr-07 of Germany, Transrapid SMG by China and L0 series by Japan. This L0 series of Japan is also known as the MAGLEVs which is entirely operated by the phenomenon of magnetism. It has achieved the highest speed of 603 km/h and is to be operational in Japan (Tokyo to Nagoya) by the year 2027.

A. R. Chowdhury (✉) · V. Sahoo
Department of Mechanical Engineering, NIT Jamshedpur, Jamshedpur 831014, Jharkhand, India
e-mail: 2020pgmees15@nitjsr.ac.in

N. M. Sivaram et al. (eds.), *Advances in Manufacturing, Automation, Design and Energy Technologies*, Lecture Notes in Mechanical Engineering,
https://doi.org/10.1007/978-981-99-1288-9_42

For wheel-on rail's speed up to 300 Km/h, catenary and pantograph-based system are used to transfer the required power. But, for speed more than 300 Km/h mechanical contact-based power transfer becomes impractical. So in MAGLEVs, transfer of electricity from groundside is crucial for powering its levitation, propulsion coils, and other on-board equipment [1].

2 Electrodynamic Suspension Type MAGLEV

This system uses magnetic repulsive force for accomplishing levitation. The levitation is achieved with the help of simple figurette (8-shaped) shaped coils which are not even powered. Many such coils are arranged in the guideway. To understand the levitation technique, we should understand the nature of a pair of superconducting magnets. The resultant magnetic field produced by this pair of SC magnets is very similar to a long permanent magnet. So for simplification of analysis, we hypothetically replace this pair with a bar magnet. If a permanent bar magnet moves parallel to these coils the varying magnetic flux with induce an emf on both the loops according to Faraday's law. The net emf is opposite in direction which means net emf induced on this coil due to the bar magnet movement is zero, and no current would flow through the loop. In short, the bar magnet moving through the center of the loop will not have any effect on the loop. Now if the magnet is slightly offset the loop, here the bottom loop faces magnetic flux of higher strength which means emf induced on the bottom loop will be higher than on the top. This higher strength also means that a net current will flow through the loop, this current flow produces a south pole on the top loop and a north pole on the bottom loop. If the force interaction between the magnetic poles is analyzed a resultant upward force is exerted on the SC magnet [1] (Fig. 1).

3 Conventional Method of Power Transmission and Its Problems

Power transmission in the existing L0 series of the Japanese MAGLEVs is achieved using inductive power collection method where the coils are spread along the tracks, and with the use of flux pumps, field strength is regulated. This regulation of field strength might turn unreliable when the train covers a considerable distance. Keeping this in mind, we propose an on-board HTS power cable for power transmission in MAGLEVs. The HTS when maintained below its critical temperature acquires the ability to carry current in kilo-ampere (kA) range. The cooling of HTS cable can be achieved with the help of on-board liquid nitrogen (LN2), which is already in use for maintaining temperature of the insulation shield of the super conductors (SC).

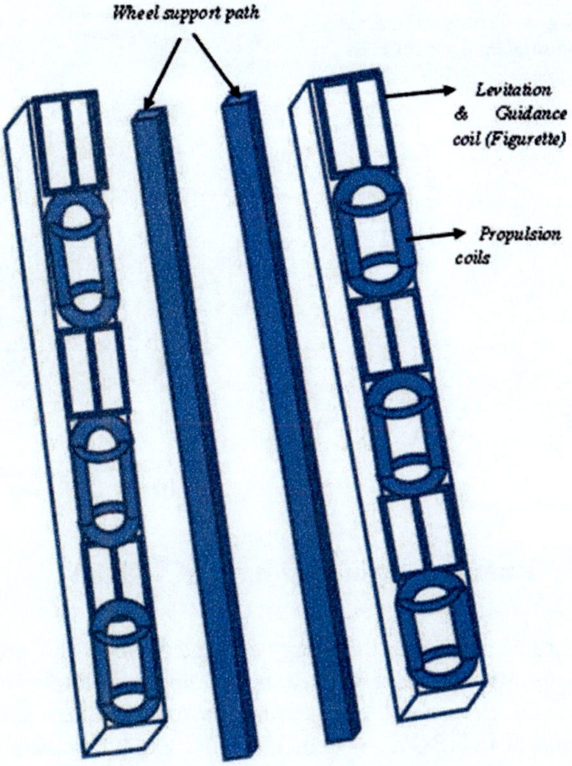

Fig. 1 Top view of EDS type MAGLEV with normal and figurette shaped magnets

4 HTS Cable

High-temperature superconducting (HTS) material is composed of stacks of various layers including a superconducting layer composed of rare earth material (e.g. Yttrium) through which the high intensity of current flows (see Fig. 2) The HTS material has greater advantages compared to a low-temperature superconducting (LTS) materials on higher critical current density and critical magnetic field. The LTS material is operated in liquid helium maintaining temperature of 4.2 K, while HTS materials are safely operated in cheaper liquid nitrogen at 77 K. The second generation (2G) of HTS material (e.g., YBC3O7-d) proposed here has enhanced property compared to its first generation (1G) version (e.g., $Bi_2Sr_2Ca_2Cu_3O_x$) including higher in-field current carrying ability and greater mechanical property [2].

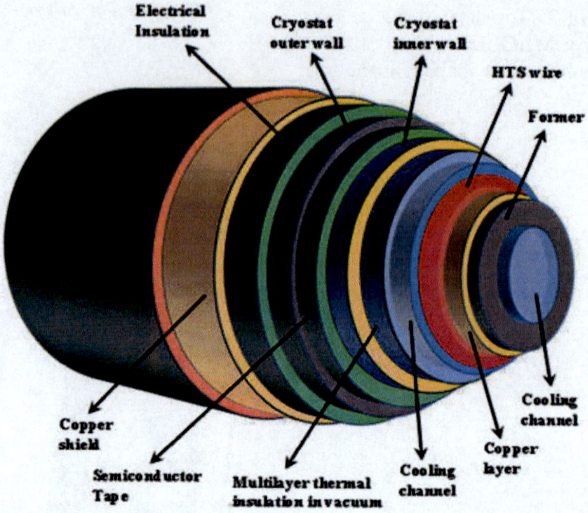

Fig. 2 Cross-sectional view of different layers of HTS cable

5 Energy Consumption in MAGLEV

For understanding, the energy demand of MAGLEV conventional method of energy consumption values per seat per kilometer is employed. The L0 series of Japanese MAGLEV usually has a total bogey/compartment strength of 12. This includes a total of two BOW sections (BS) and ten middle cars (MW). It covers a length of almost 300 m and width of 2.9 m. The BOW section has a seat capacity of 24, while the middle car (MW) has a maximum seat capacity of 68. So, calculating this the maximum seat capacity would be calculated as

$$(\text{Seat in BOW section} * 2) + (\text{Seat in middle car} * 10) \quad (1)$$

Energy consumption in MAGLEV can be calculated by two methods.

(1) Energy consumed in Watt-hour/effective area/kilometer (Wh/m^2/km)
(2) Energy consumed in Watt-hour/seat/kilometer (Wh/Ps/km) [3]

In our following proposal, **we have calculated the energy consumption of the middle car with an assumed distance traveled 20 km**, this being the approximate distance between two stations. So, all the distance calculation henceforth will be achieved assuming the above criteria (Table 1).

Table 1 Dimension and seating capacity of a MAGLEV

	Effective area (m^2)	Width (m)	Length (m)	Seating capacity
Middle car	56	2.9	24.3	58–68
Bow section	50	2.9	28	24

The MAGLEV achieves different phenomenon under different speed. At speed up to 300 km/h, the MAGLEV is wheel bound. But for speed above, this levitation occurs in MAGLEV. At speed of 500 km/h, the MAGLEV is known to consume power in the range of 112 Wh/m^2/km [energy consumption]. This high value of power is because of high aerodynamic resistance and high driving resistance of the vehicle [3].

So on calculating, the power consumption of the middle car of a MAGLEV would be

(Power by effective area)

$$\text{Power} * \text{Effective Area} * \text{Distance in km} = 125.44\ \text{KWh} \tag{2}$$

Again, the Japanese L0 series while cruising at 500 km/h through tunnels MAGLEV is observed to consume power at 99 Wh/seat/km (Fig. 3).

So on calculating, the power consumption of the middle car of the MAGLEV would be

(Power by seat capacity)

$$\text{Power} * \text{Seat Capacity} * \text{Distance in km} = 128.7\ \text{KWh} \tag{3}$$

The voltage fed from the transmission line is in the range of 3–20 kV (AC), at 50 Hz frequency (see Fig. 4).

The output transformer steps it down to the suitable range before feeding in the linear motor. The current in the HTS cable will also be in the kilo-ampere range.

On this context, we propose the use of HTS cable like the one developed under the AmpaCity Project in Essen, Germany, is operational for a distance of 1 km between substations, and is able to operate on a 10 kV system voltage and 2.3 kilo-amp of operating current. The high voltage and current carrying ability of this cable

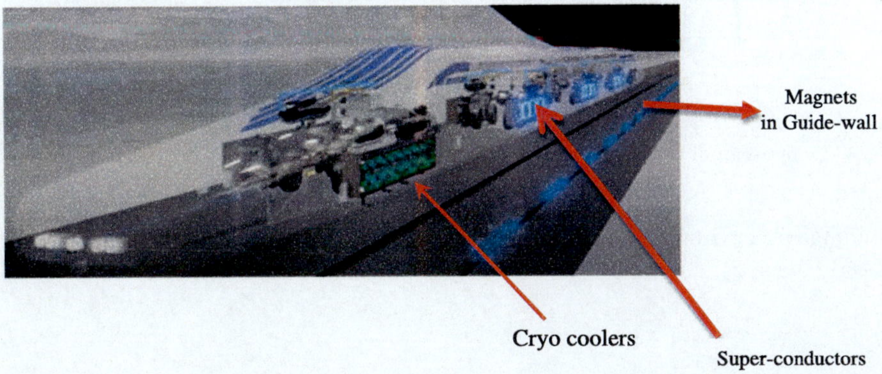

Fig. 3 X-ray view of EDS type MAGLEV [6]

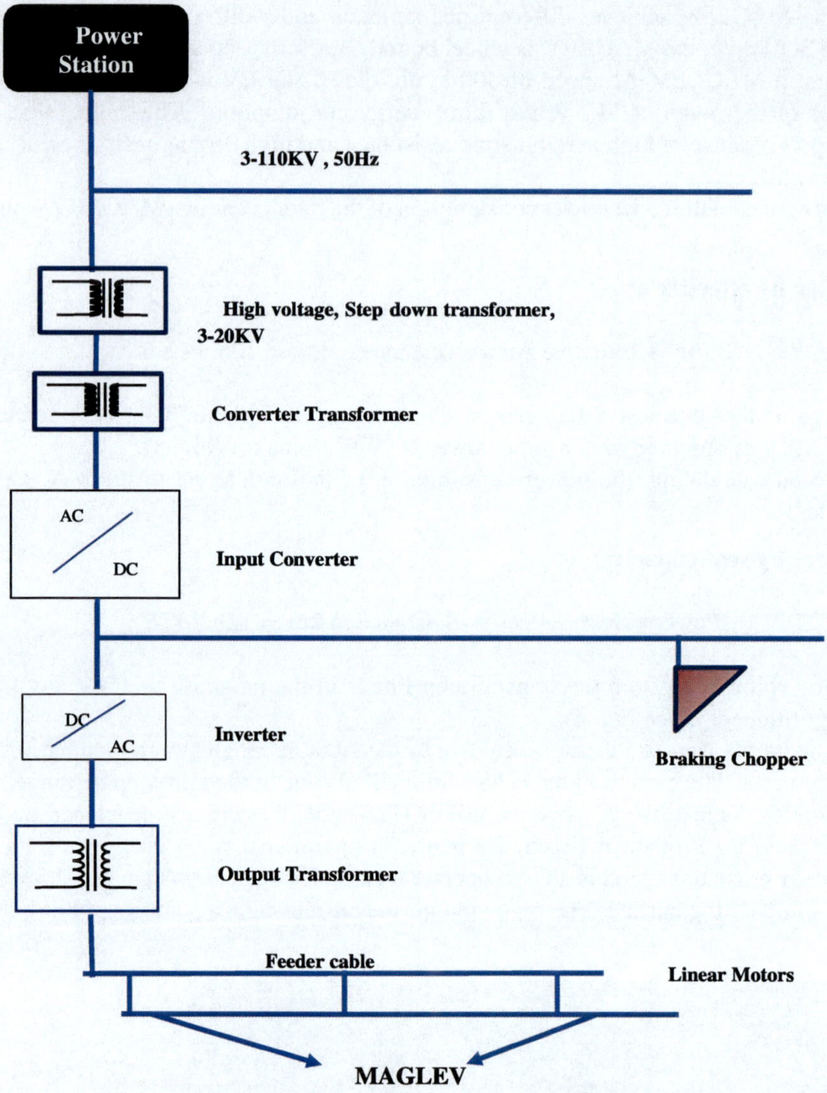

Fig. 4 Components of power supply from power station to MAGLEV [3]

at cryogenic environment are essential for the smooth power transmission in the MAGLEV [4, 5].

6 Conclusion

HTS as a medium of power transmission is growing popular day by day due to various advantages it has to offer. The thickness of such cable is in the range of millimeter making it easy to install and operationalize. Power transmission in MAGLEV using HTS can be very efficient and can open new gates of research due to its inherent advantage of on-board cryogenic system and high-current carrying ability.

References

1. Prasad N, Jain S, Gupta S (2019) Electrical components of maglev systems: emerging trends. Urban Rail Transit 67–79. https://link.springer.com/article/10.1007%2Fs40864-019-0104-1
2. Dong F, Huang Z, Hao L, Xu X, Jin Z, Shao N (2019) An on-board 2G HTS magnets system with cooling-power-free and persistent-current operation for ultrahigh speed superconducting maglevs. Sci Rep (nature.com)
3. Fritz E, Klühspies J, Kircher R, Witt M, Blow L (2018) Energy consumption of track-based high-speed trains: maglev systems in comparison with wheel-rail systems. Transp Syst Technol 134–155
4. Sudheer T, MuralidharBathula VAS, Choudhury UK, Nageshwar Rao B, Usoskin A, Rao VV (2018) Development of a lab-scale high-Tc superconducting power cable. Indian J Cryog 43
5. Stemmle M, Merschel F, Noe M, Hobl A (2014) AmpaCity—advanced superconducting medium voltage system for urban area power supply. In: 2014 IEEE PES T&D conference. IEEE
6. Central Japan Railway Homepage. https://global.jr-central.co.jp/en/. Last accessed 29 November 2021

Analysis of Energy Loss Through a Flow Divider Valve in Different Applications Using Various Hydraulic Drive Systems

Dharmendra Kumar and Anil C. Mahato

Abstract Flow divider valve (FDV) is a flow control valve used in power hydraulic system to split the pump flow in dual paths: primary path and auxiliary path. The flow rate ratio in each divided flow path is either predetermined or can be adjustable as per the loading conditions. Each flow path supplies flow to operate different loads to perform different work simultaneously. The FDV makes the system compact and efficient by reducing various components present in the system. This valve can be used in two different applications: a system that performing dual functions against dual loads simultaneously, e.g., *steering system in automobile* and another is in *wind turbine*, to reduce the power fluctuation in the absence of generator. In present study, two different hydraulic systems (wind turbine system and steering system) are identified to analyze the power loss through the FDV. The simulation results indicate that the power loss through the FDV is lower when it is used in wind turbine application for reducing the power fluctuation.

Keywords Flow divider valve · Power hydraulic system · Wind turbine · Steering system

1 Introduction

Flow divider valve (FDV) is a flow control valve that makes the hydraulic system compact and efficient as it reduces the multiple components of a power system by performing dual functions against dual loads simultaneously. Also, this FDV can be used in wind turbine hydraulic power transmission system to supply a stable power against a stable demand whenever the input wind speed is of unstable nature. By the help of a hydraulic power transmission with FDV and an accumulator, the electrical storage system of the wind energy generation system can be eliminated.

D. Kumar (✉) · A. C. Mahato
Department of Mechanical Engineering, Birla Institute of Technology, Mesra, Ranchi 835215, India
e-mail: dharmendrakmr81@gmail.com

N. M. Sivaram et al. (eds.), *Advances in Manufacturing, Automation, Design and Energy Technologies*, Lecture Notes in Mechanical Engineering,
https://doi.org/10.1007/978-981-99-1288-9_43

In this scheme, the energy is stored into an accumulator instead of the electrical storage system. More detail of the stated scheme is available in [1]. Usually, FDVs are classified as gear or rotary type and spool (slide) type. Accuracy and efficiency of rotary type FDV are better than spool type FDV. However, spool type FDVs are more flexible in manufacture and can be easily integrated with other component [2]. Besides, another special type FDV, i.e., split flow divider valve (SFDV) is discussed in [2]. The SFDV do not need any additional controlling device to control the split flow of the system. The stated feature of the valve reduces its maintenance cost and increases its flow adaption [3]. Some FDVs are auto regulated, i.e., it can control the flows according to the loading conditions effectively [4]. In [5], authors developed an algorithm, i.e., response surface methodology with steepest-decent method to simplify the design of the FDV and applied on a two way flow divider spool type valve to analyze the performance of it. This methodology helps to reduce the error percentage and repeatability with the experimental data. Another works [6, 7], authors develop a bond graph model of a hydraulic system with FDV. The stated system is based on performing dual functions against dual loads. The function of the FDV is experimentally validated whenever it is used to obtain dual functions. Also, authors addressed the influence of the damping value on the system responses. Recently, Coskun et al. [8] modeled FDV using one-dimensional solution technique and studied its behavior by applying various input. The system responses are also validated experimentally. This experimentally study establishes the function of the FDV and its application in dual functional schemes. Besides, FDV can also be used in wind power application to reduce the power fluctuation. In [1], authors replace the mechanical power transmission by a hydraulic power transmission to transmit the power from turbine rotor to the generator. The additional components of the hydraulic power transmission system are FDV and an accumulator. The FDV splits the pump flow in two ways, whereas the accumulator stores the excess flow into it whenever wind speed is higher than the cut-off speed or vice versa. The authors simulated the complete wind turbine scheme by considering a fluctuating wind speed. The system responses establish that the output power from the wind turbine is almost stable.

This article provides a single platform to analyze the power loss through the FDV using two different hydraulic schemes such as a hydraulic system that perform dual functions against dual loads simultaneously (steering mechanism) and the hydraulic system that used in wind turbine to obtain stable power against an unstable input wind speed.

2 Physical Model of Flow Divider Valve (FDV)

The physical model of the flow divider valve is shown in Fig. 1. The FDV consists of three ports, i.e., one for input and another two for outputs. The output flow is controlled by the spindle adjustment. The spindle is connected to the orifice of the primary port through spring. During the rotation of the spindle, it provides a spring force to the orifice plate of the primary port as a result primary flow is controlled.

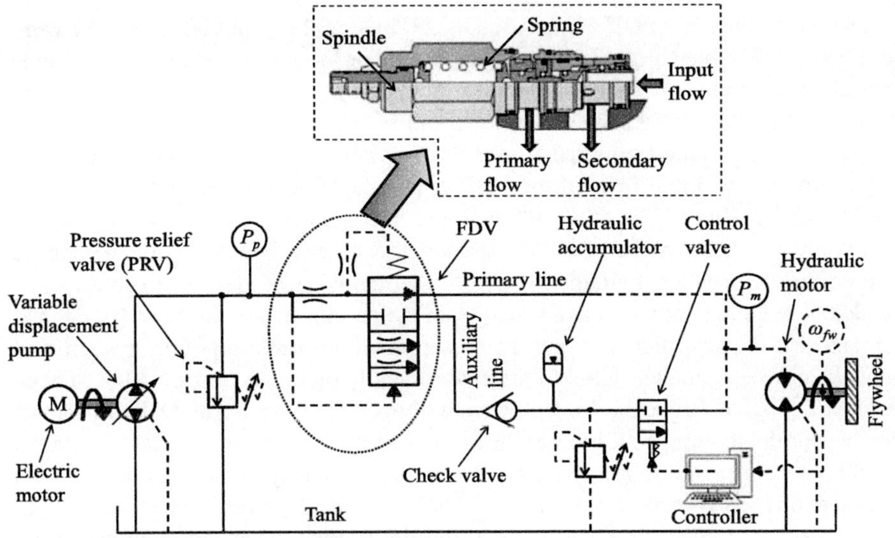

Fig. 1 Schematic diagram to obtain stable output from wind turbine using FDV

However, the orifice of the auxiliary port is not controllable; hence, the excess flow is passed through it.

3 Dual Application of FDV in Wind Turbine and Steering System in Automobile

The applications of the FDV in wind turbine and the steering system are shown in Fig. 1 and Fig. 2, respectively.

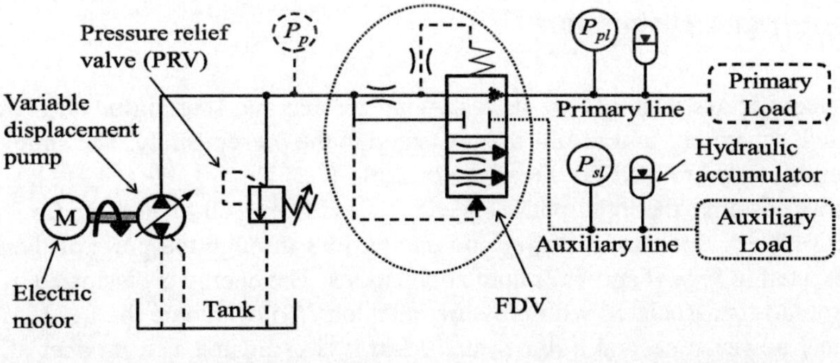

Fig. 2 Schematic diagram to perform dual function in steering mechanism using FDV

In wind turbine, the FDV can be used to obtain a stable output whenever the wind supply is of unstable nature. In this scheme, an open-loop power hydraulic system is considered as a power transmission unit to transmit the power from turbine rotor to the generator. The stated scheme is shown in Fig. 1. The hydraulic pump is rotated with variable speed and supplied a variable flow to the FDV as it is connected with turbine rotor. The input flow of the FDV is passed through both port of the FDV: primary and auxiliary. The primary line flow is the controlled flow as per the load of the system whereas auxiliary line flow is the excess flow of the system, and it is stored into storage device (accumulator). The stored energy is released whenever the system needed. The controller receives a signal from hydro-motor, that the flow through the primary line is not sufficient as per the present loading condition of the system, and hence, controller sends a signal to the electronically operated control valve to open or close the valve. Thus, the control valve will open to compensate the hydro-motor flow as per the loading condition of the system, and hence, it will always maintain a fixed operating speed of the hydro-motor. As the electric generator is coupled with the shaft of the hydro-motor, it always develops a fixed power whatever the input wind speed is and hence reduces the possibilities of the load shedding to the power supply point.

The application prospect of the FDV in power assist steering mechanism is shown in Fig. 2. The configuration of the figure is subdivided into three divisions: the power-assisted steering system with the steering control valve and the working cylinder, the load with the two load cylinders and the required circuit configuration, and the circuit for producing two pressure sources with the FDV. The FDV receives a variable flow from the hydraulic pump and splits the flow in two different paths: primary path and auxiliary path. The primary path line is connected with working cylinder servo through steering control valve. Steering control valve controls the operation of the working cylinder servo. The operational sequence of the other two cylinders: cylinder A and cylinder B that are connected with front wheels are controlled by the working cylinder servo operation.

4 Result and Discussion

Both hydraulic systems that are used in wind turbine and steering mechanism are simulated, separately, and results are presented in the present study. The simulation parameters related to the FDV are given in Table 1.

In both systems, the input pump speed is varied from 160 to 30 rad/s (as shown in Fig. 3). The responses of power and energy loss through the ports of the FDV are presented in Figs. 4 and 5 for both applications. The energy is dissipated as heat energy, and it is associated with pressure head loss. To overcome the head loss, an additional power is needed to the system, and it is estimated as a product of flow rate (Q) and pressure drop (ΔP) across FDV ports. Moreover, an integration of the power loss (P_l) provides the energy loss (E_l) of the system. The power and energy loss are expressed as

Table 1 Simulation parameters

Parameters	Value
Area of the FDV spindle (A_{sp})	5×10^{-5} m^2
Area toward primary port of the FDV (A_{pr})	2×10^{-5} m^2
Area toward auxiliary port of the FDV (A_{br})	5×10^{-5} m^2
Mass of the spring on FDV (m_s)	2×10^{-3} kg
Stiffness of the spring on FDV (K_{sp})	1×10^5 N/m
Viscous resistance on FDV spindle (R_{spdl})	2×10^3 N s/m
Coefficient of discharge through primary port of FDV (C_{d_pfl})	0.6
Maximum opening area of the primary port of FDV ($A_{i\,\max\_pfl}$)	3×10^{-6} m^2
Diameter of the indenter (d_i)	3×10^{-3} m
Density of the working fluid (ρ)	870 kg/m^3
Desired spindle displacement (x_i)	3.5×10^{-3} m
Coefficient of discharge through the auxiliary port of the FDV (C_{d_sfl})	0.6
Maximum flow area through the auxiliary port of the FDV ($A_{i\,\max\_sfl}$)	2×10^{-6} m^2

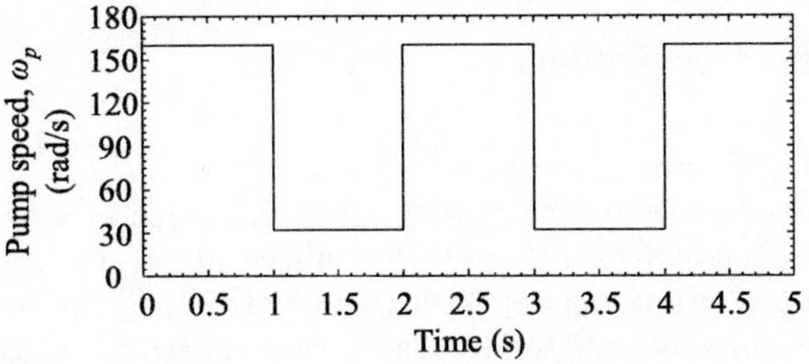

Fig. 3 Pump speed

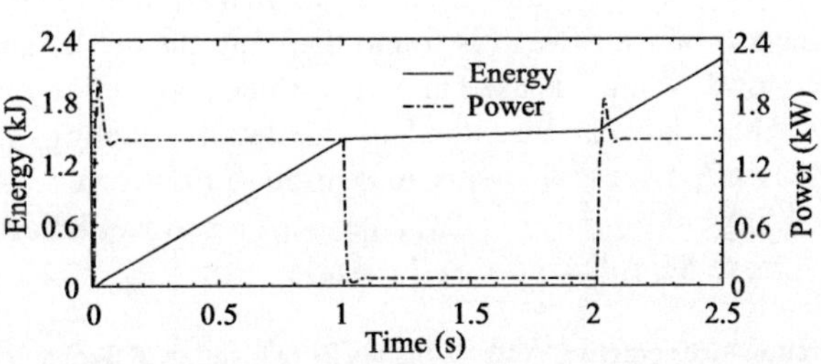

Fig. 4 Energy and power loss in hydraulic drive system with FDV used in *steering mechanism*

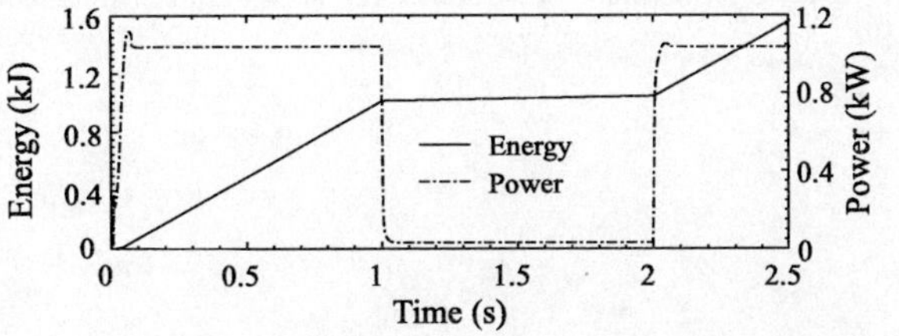

Fig. 5 Energy and power loss in hydraulic drive system with FDV used in wind turbine

$$P_l = (\Delta P.Q)_{\text{Primary port of FDV}} + (\Delta P.Q)_{\text{Auxiliary port of FDV}} \tag{1}$$

$$E_l = \int P_l dt \tag{2}$$

The simulation responses of the hydraulic system with FDV that is used for steering mechanism are presented in Fig. 4. The maximum power loss obtained is almost 1.8 kW at 2 s, and the energy loss is almost 2.3 kJ in given one cycle. However, the energy loss may be increased with the increase of cycle time period. Also, it is observed that the energy loss through the port of the FDV is more strips at the higher speed of the pump. Similarly, the power and energy loss through the port of FDV in the hydraulic system that is used in wind turbine are presented in Fig. 5. The power loss is highest at 2 s, and it is almost 1.1 kW. Also, the energy loss is almost 1.6 kJ in one cycle.

5 Conclusions

In this article, a simulation work is carried out to analyze the power and energy loss through the FDV in two applications such as in *steering mechanism* and in *wind turbine* application. The FDV is utilized in these two applications to perform different functions. In steering mechanism, FDV helps to operate two loads simultaneously, whereas in wind turbine, same valve is used to reduce the power fluctuation. Two hydraulic systems are simulated to analyze the power and the energy loss through the port of the FDV. It is found that both the power and the energy loss of the FDV are lower when it is used in wind turbine application to reduce the power fluctuation.

This study can be extended with detail modeling of both systems. Also, the variation of power and energy loss through FDV can be analyzed by changing various loading conditions. Finally, an experimental validation of simulation responses may increase the robustness of the work.

Acknowledgements Authors thanks Birla Institute of Technology Mesra for providing the necessary research infrastructure to carry out the research work. This study is funded by Birla Institute of Technology, Mesra. The reference no. is GO/Estb/SMS/2020-21/2262.

References

1. Mahato AC, Ghoshal SK, Samantaray AK (2018) Reduction of wind turbine power system by using priority flow divider valve in a hydraulic power transmission. Mech Mach Theory 128:234–253
2. Cheng M (2009) Modeling and analysis of a flow divider valve' ASME 21 forum on fluid machinery conference proceeding, V, pp 91–98
3. Metwally M, Abou El Azm AS, Sherif El (2012) Dynamic and steady-state analysis of split flow divider valve. In: Proceeding of the 15th international AMME conference, vol 29, p 31
4. Fedoroff M, Burton RT, Schoenau GJ, Zhang Y (1992) Dynamic and steady-state analysis of an auto-regulator in a flow divider and/or combiner valve. J Dyn Syst Meas Contr 114(2):306–314
5. Vacca A, Cerutti M (2007) Analysis and optimization of a two-way using response surface methodology. Int J Fluid Power 8(3):43–57
6. Kumar S, Dasgupta K, Ghoshal SK, Das J (2018) Dynamic analysis of a hydro-motor drive system using priority valve. Proc Inst Mech Eng J Process Mech Eng 233(3):508–525
7. Kumar S, Dasgupta K, Ghoshal SK (2019) Fault diagnosis and prognosis of a hydro-motor drive system using priority valve. J Braz Soc Mech Sci Eng 41(2):77
8. Coskun G, Kolcuoglu T, Dogramaci T, Turkmen AC, Celik C, Soyhan HS (2017) Analysis of a priority flow control valve with hydraulic system simulation model. J Braz Soc Mech Sci Eng 39(5):1597–1605

Fabrication of Mechanical Circuit Breaker Device for Overhead Transmission Lines

M. Purusothaman, Jayaparakash Venugopal, G. Senthil Kumar, M. Thirumurthy, and J. Naresh

Abstract Overhead power lines have risk of breakage which can cause severe damages to life and property. The electric shock hazard may lead to human or animal deaths and fire accidents, thus creating unsafe environment. So new techniques could be designed and developed to come over such risks and improve safety feature of the system. Mostly, power lines are cut down during rain, and contact of power line with the stagnated water in roads leads to unpredictable damage to lives. If a transmission wire cut down due to heavy rain or inclement weather, the cut wire automatically falls to road or submerged into stagnated water in roads certainly during rainy season, the electric power spreads through the water which makes a miserable accidents and deaths to humans and animals by the unbearable electric shocks when they walk through them. Till now, there is not an effective safety setup to handle overhead power line. This is an idea to handle power line failure accidents. A new spring mechanical device for safe power transmission is designed, developed, and implemented in this research which enables the transmission line to disconnect the power transfer for the rest of the current post towers.

Keywords Overhead line circuit breaker · Mechanical fuse · Overhead spring transmission

1 Introduction

In recent times, there are many overhead power lines downed incidents which has been reported, which costs many lives and leads to fire accidents if power is passing in it. If anything accidently comes into contact with downed wire, it will cause electrocution or accidents, it worsens if it falls on stagnant water on road. Even when it falls on vehicles, it causes severe damage to lives and objects [1, 2]. A

M. Purusothaman (✉) · J. Venugopal · G. Senthil Kumar · M. Thirumurthy · J. Naresh
School of Mechanical Engineering, Sathyabama Institute of Science and Technology, Chennai, India
e-mail: purusothmani@gmail.com

N. M. Sivaram et al. (eds.), *Advances in Manufacturing, Automation, Design and Energy Technologies*, Lecture Notes in Mechanical Engineering,
https://doi.org/10.1007/978-981-99-1288-9_44

new mechanical setup is implemented in the power lines which interrupts power transmission in case of cut down of power lines. In case, the power line cuts the mechanism which is actuated and interrupts the power transmission to the damaged line. Thus, many lives and property are saved [3–6]. In general, the power line is connected as a single wire for a specific length, and if there is any damage in the power line, the entire wire is to be replaced. In the following idea, each post will have a separate wire which is easy for changing and more reliable. Thus, the chances of failure in the wire will be less; load and stress acting the wire will be lesser than normal method of laying the wire [7].

The aim of that paper is that overhead power lines have chances to fall down which can cause severe damages to human and other living organism. Till now, there is not an effective safety setup to handle overhead power line. A new mechanical setup is implemented in the power lines which interrupts power transmission in case of cut down of power lines [8]. An engineer skillfully designs and builds a project which brings in a good change [9]. Before selecting this idea, various issues in day-to-day life were analyzed. Worldwide overhead power lines are used, and many accidents have been reported on downed overhead power line [10]. Till now, there is no safety device or mechanism to handle such situation. A new mechanism is to be implemented on the overhead power line post to deal with downed wire, to save many lives, and to avoid accidents [11–14].

2 Materials and Method

A spring is an elastic object that stores mechanical energy. Springs are typically made of spring steel. There are many spring designs. In everyday use, the term often refers to coil springs. When a conventional spring, without stiffness variability features, is compressed or stretched from its resting position, it exerts an opposing force approximately proportional to its change in length. There are many types of springs available like compression spring, tension spring or expansion spring, helical spring, coil spring, variable spring, etc. In this project compression, spring is used to accomplish the action required as shown in Fig. 1.

Threaded rod also called as stud is originally a long rod which is threaded by both the ends. The thread might extend throughout the entire rod's length. It is designed to be used whilst in tension. Based on the shape, stud bolts are classified into three basic types: "fully threaded stud bolts", "tap end stud bolts", and "double end stud bolts". Every one of those has variety of applications. In mechanical fuse, fully threaded stud bolt is preferred for the adjusting facility as shown in Fig. 2.

The overhead power lines are structures used for electric power transmission and distribute to transmit electrical current to various channels. Overhead power lines are generally the lowest-cost method of power transmission for large quantities of electric energy. The 7 core bare wire is selected as of traditional overhead power line. Aluminium cables are used because they are cheaper and half the weight to compare with a copper resistance cable (even though larger diameter because of lower specific

Fig. 1 Compression spring

Fig. 2 Threaded rod

conductivity. Copper was regularly used before for power transmission purpose, and it is still used rarely based on application, certainly at low voltages for grounding. Since large conductors may tend to lose less energy because of its lower electrical resistance, they are reasonably high expensive than smaller conductors as shown in Fig. 3.

Conductive bar is metal strip to transfer electrical energy. In mechanical fuse project, a highly conductive metallic plate is joined with the stud to conduct electricity. The conductive bar acts as the circuit breaker. The conductive bar is drilled at centre and welded with a nut in centre. The conductive bar is fastened to the threaded rod which makes the position of conductive bar in the threaded rod adjustable. This help in adjusting the conductive bar or for compensation when the wire sags due to loading and expansion as shown in Fig. 4.

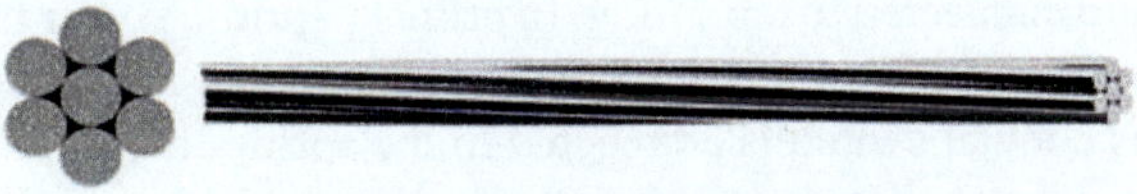

Fig. 3 Bare AAAC wire

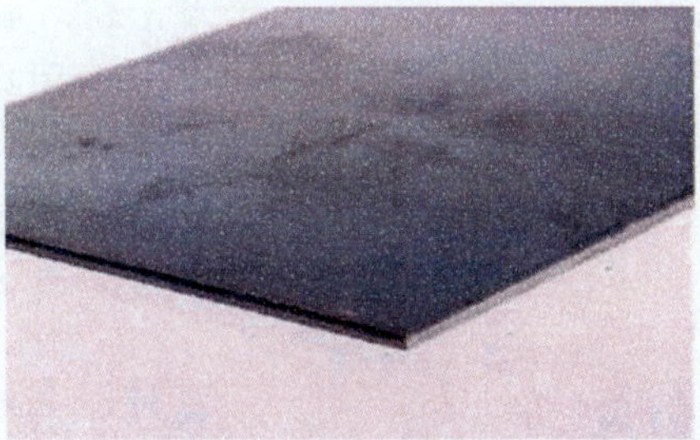

Fig. 4 Conductive bar

Fig. 5 Wooden plywood board

It is a wood often used in sheet form as a lightweight or shatter-resistant. Plywood is a strong, tough, and lightweight material. Plywood board is used for base and side wall member in the experimental setup as shown in Fig. 5.

3 Experimental Setup

The design of mechanical fuse is developed in Autodesk Fusion 360. Prototype is designed first, and it can be scaled to actual setup based on the load acting by wire on the spring.

Figure 6 shows the design of the model created in 360. The spring action is the basic mechanism used in the project. The compression spring is which compresses when the load is applied, and it expands to original condition when the load is removed. There are two spring used in the project that is primary spring and secondary spring. A rod is joined with the spring by means of a circular plate welded to springs. When the spring is compressed the rod moves. A conductive bar is used as contact switch and connected to the spring. In primary spring, power line is connected with the aid of loop, and in secondary spring, conductive bar is connected. A conductive bar is used as contact switch is connected to the spring through a rod. The compression spring is which compresses when the load is applied, and it expands to original condition when the load is removed. There are two spring used in the project that is primary spng and secondary spring. A rod is joined with the spring by means of a circular plate welded to springs. When the spring is compressed the rod moves. The rod and spring are joined with the vertical plate in the setup. In primary spring, power line is connected with the aid of loop, and in secondary spring, conductive bar is connected. A conductive bar is used as contact switch is connected to the spring through a rod.

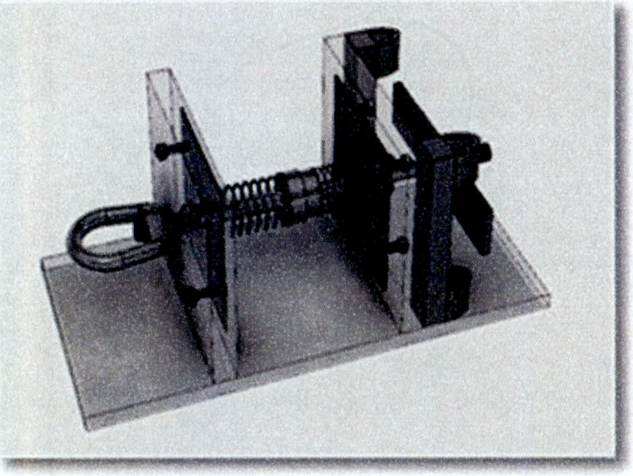

Fig. 6 Fabrication of mechanical circuit breaker

4 Result and Discussion

The basic function of this mechanical fuse is to act as a circuit breaker. Power is to be stopped passing to the downed wire to avoid electrocution and accidents. In this setup, principle of working is spring action. When the wire connected to the primary spring cuts, the spring expands due to the stiffness of the spring, i.e. lack of load (tension) by wire as shown in Fig. 7. Thus, primary spring compresses the secondary spring which has less stiffness than primary spring. The secondary spring moves conducting bar outward so the connection between two strips will be lost; thus, the power transmission to the follow-up power line will be interrupted. This will prevent power passing in the downed wire as shown in Fig. 8. Again, the power line is connected with the primary spring in the post with specific tension that overcomes the stiffness of the spring. For the repairing work to be done, there is no need of switching off the whole transmission line. Only the conducting bar or strip is disconnected, the work is carried on. After repairing the power line, the power will be transmitted normally.

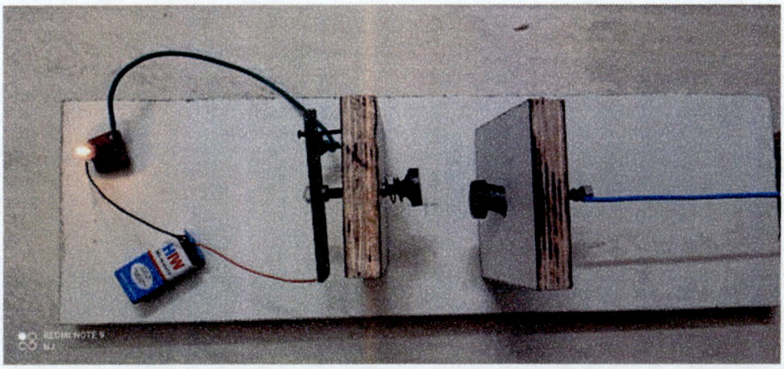

Fig. 7 Top view normal condition

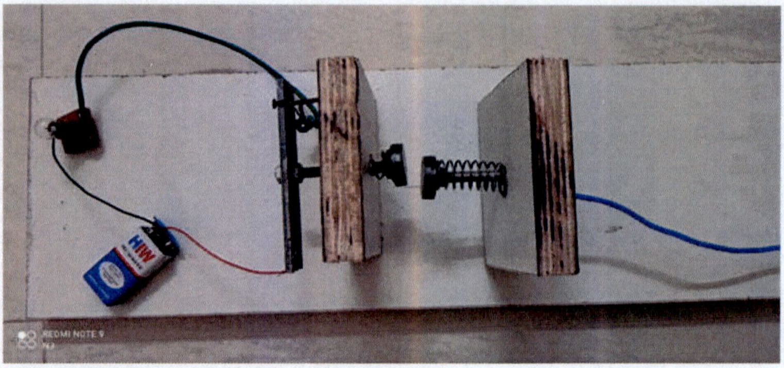

Fig. 8 Top view cable fall condition

The wooden board is cut to the dimension required. The wooden side wall is joined with the wooden base. The screw for the wire strip is fastened to the side wall. Holes are drilled in wooden board to place the threaded bolt. The threaded bolt and spring are joined to the circular metal plate. This is similar to both the primary and secondary spring. This setup is joined to the square metal plate and joined to the wooden board by bolts and nuts. The outgoing wire is joined to the loop and the feed wire from the previous transmission line strip 1 by traditional winding method as used in the overhead power line. Analysis of the setup is done to get the constraints and the maximum deflection, load. The prototype of the mechanical fuse is shown below. The project will be scaled to the original model depending upon the load acted by the original used power line.

The working of the prototype ensures the proper working of the mechanism when electricity is passing through the setup. The mechanical fuse project replaces the conventional line insulators used in the overhead power transmission post.

Applications:

- The mechanical fuse (this project) should be installed at the overhead power line post.
- This replaces the conventional post line insulators.
- Can also be used for insulated wires.
- Can also be used as circuit breakers in various fields.

Advantages:

- Saves many lives and avoids accidents
- Improves the safety factor of overhead power lines
- If the wire cuts, the accidents are prevents, thus saving living organism and properties
- Cost-efficient method available to avoid accidents
- More reliable over other method like electrical or electronics
- Less maintenance

5 Conclusion

Human lives are more valuable than money. The responsibility of every engineer is to design technologies with sustainable development in mind. Whilst rocket science is developing in a high rate, low end technologies which will improvise the lives of common people at a low cost should be developed and implemented world-wide. This not only upgrades the safety of power lines but creates a safer environment for the people around the region. As underground power cables are expensive, government uses overhead power lines. So this low-cost setup can be implemented for betterment of the system. Additional features like lot can be implemented to identify the spot of wire failure. Electronics and sensor can be used to indicate the failure of wire in the specified post which is easy for taking action. This can be implemented where immediate repair and fixing is needed, like military zones and other sensitive areas. The safety system for overhead power line can also be done by using electrical and electronic systems. RF components can be connected to each post and in contact with another RF component in the previous post. If the wire breaks, the power is lost to the RF component, and the contact with RF component is lost. If contact is lost any RF component, it should actuate the circuit breaker. Thus, the power transmitted to downed wire is stopped. Enhanced models could be made to personalize preferences according to the needs and applications.

Disclosure Statement No conflict of interests.

References

1. White L Jr (1966) Medieval technology and social change. Oxford Univ. Press, New York. ISBN 0-19-500266-0
2. Gönen T (2014) Electrical power transmission system engineering: analysis and design, 3rd edn. CRC Press. ISBN 9781482232233
3. Emerging trends in power systems, vol 1. Allied Publishers. ISBN 81-7023-417-4
4. Polymethylmethacrylate (PMMA, Wooden) archived 2015-04-02 at the Wayback Machine. Makeitfrom.com. Accessed 23 March 2015
5. Speck JA (1997) Mechanical fastening, joining, and assembly. CRC Press. ISBN 978-0-8247-9835-2
6. Mavrigian M (2008) High performance fasteners & plumbing. Penguin Group. ISBN 978-1-55788-523-4
7. Waygood A (2013) An introduction to electrical science. Routledge. ISBN 1-135-07113-6
8. Fink DG, Wayne Beaty H (ed) (1978) Standard handbook for electrical engineers, 11th edn. McGraw-Hill. ISBN 0-07-020974-X
9. Betz HD, Schumann U, Laroche P (2009) Lightning: principles, instruments and applications. Springer, pp 202–203. ISBN 978-1-4020-9078-3
10. Hughes TP (1993) Networks of power: electrification in Western Society, 1880–1930. Johns Hopkins University Press, Baltimore, pp 119–122. ISBN 0-8018-4614-5
11. Carlson WB (2013) Inventor of the electrical age. Princeton University Press, Tesla. ISBN 1-4008-4655-2
12. Jonnes J (2004) Empires of light: Edison, Tesla, Westinghouse, and the race to electrify the world. Random House Trade Paperbacks. ISBN 978-0-375-75884-3

13. Grigsby LL et al (2001) The electric power engineering handbook. CRC Press. ISBN 0-8493-8578-4
14. Purusothaman M, Valarmathi TN, Dada Mohammad SK (2016) Computational fluid dynamic analysis of enhancing passenger cabin comfort using PCM. IOP Conf Ser: Mater Sci Eng 149:012197. https://doi.org/10.1088/1757-899X/149/1/012197

Simulation of Hybrid CVT Motorcycle Using MATLAB

Jayaprakash Venugopal, M Purusothaman, K. Kulothungan, D. Lakshmipathy, A. Raja Rajeswari, and S. Pasupathy

Abstract In recent years, all over countries have targeted the reduction of on road emissions and pollution produced by vehicles. In this paper, design conventional continuously variable transmission (CVT) motorcycle to hybrid CVT motorcycle is performed. For transportation domain, hybrid vehicle concept is one of the solvable systems which were proposed and analyzing the result for environment effect. The consumption of usage fuel of fossil fuels at the same time maintaining vehicle performance. At the primary stage, motor unit and CVT engine connected together toward the measurement and precision of the hybrid CVT motorcycle concept by using MATLAB/Simulink. So, the simulation outputs were displayed and compared to the CVT motorcycle efficiency. The results obtained from MATLAB/Simulink. The MATLAB/Simulink motorcycle model delivers a simulation stage effect of flexible, prototype, and modular and can be changed to develop dissimilar types of motorcycle concepts and different cycle.

Keywords CVT motorcycle · Hybrid · MATLAB/Simulink · Drive cycle

1 Introduction

As the global economy and greenhouse effect direction area clean energy resources in the effect of climate severity, global warming, etc.… the running world is kept on researching and analyzing toward alternative resources of energy management/power like solar, wind, tidal, and geothermal. Subsequently vehicle industries are playing a major role of air contamination, soil pollution takes more significant of government action, and leading automotive corporations are cooperating to establish a best

J. Venugopal · M. Purusothaman (✉) · K. Kulothungan · D. Lakshmipathy · A. Raja Rajeswari · S. Pasupathy
School of Mechanical Engineering, Sathyabama Institute of Science and Technology, Chennai, India
e-mail: purusothmani@gmail.com

N. M. Sivaram et al. (eds.), *Advances in Manufacturing, Automation, Design and Energy Technologies*, Lecture Notes in Mechanical Engineering,
https://doi.org/10.1007/978-981-99-1288-9_45

industry revolution that will effect useful to reduction of world emissions generating by vehicle. Industrialized countries are studying the various types of fossil fuel energy conversion methods, and it is more over studying the power sources which are present world economy situation researched by many creators. Therefore, hybrid technology is one of the alternative concepts to research for upcoming years [1]. And the concept results are more than expected to become more expandable in current global warming situation. More importantly, it is eco-responsive [2]. Human made are best solutions so far is hybrid electric vehicles (HEVs) for best find of the century technology wise. The fuel consumption of motorcycles is less than the light commercial vehicle and passenger cars. This fuel economy is measured outstanding to the motorcycles mass. The toxic elements such as carbon monoxide (CO) and hydrocarbons (HC) produced by motorcycles generating more emission and affect living things [3]. Due that reason, automobile domain industry focusing how to effectively reduce pollution emissions from motorcycles to improve the quality of the carbon free environment has become a major issue that can no longer be ignored. So, now onwards take immediate action [4].

2 Requirement Specification

It describes the client data, group of members, and spectators to see the result and working function. It will provide the real-time data and different load for every cycle pattern [5]. Technology used: MATLAB/Simulink, detailed hybrid CVT model shown in Fig. 1.

2.1 System Layout

To expansion of challenging model, market-top organization adopts concept-based design by methodically using prototype using the whole process as shown in Fig. 2. A virtual combined model to simulate and test your established system early and often validate your idea design with physical models and fast prototyping. Instrument manufacture-quality C, C++, CUDA, PLC, Verilog, and VHDL code and deploy your fixed system into straight [6].

2.2 Motor Unit Model

The motor unit model is generally used in mechatronics engineering to express the values of like voltage, current, and power as shown in Fig. 3. It is used for modifiers and AC machines for control system analysis.

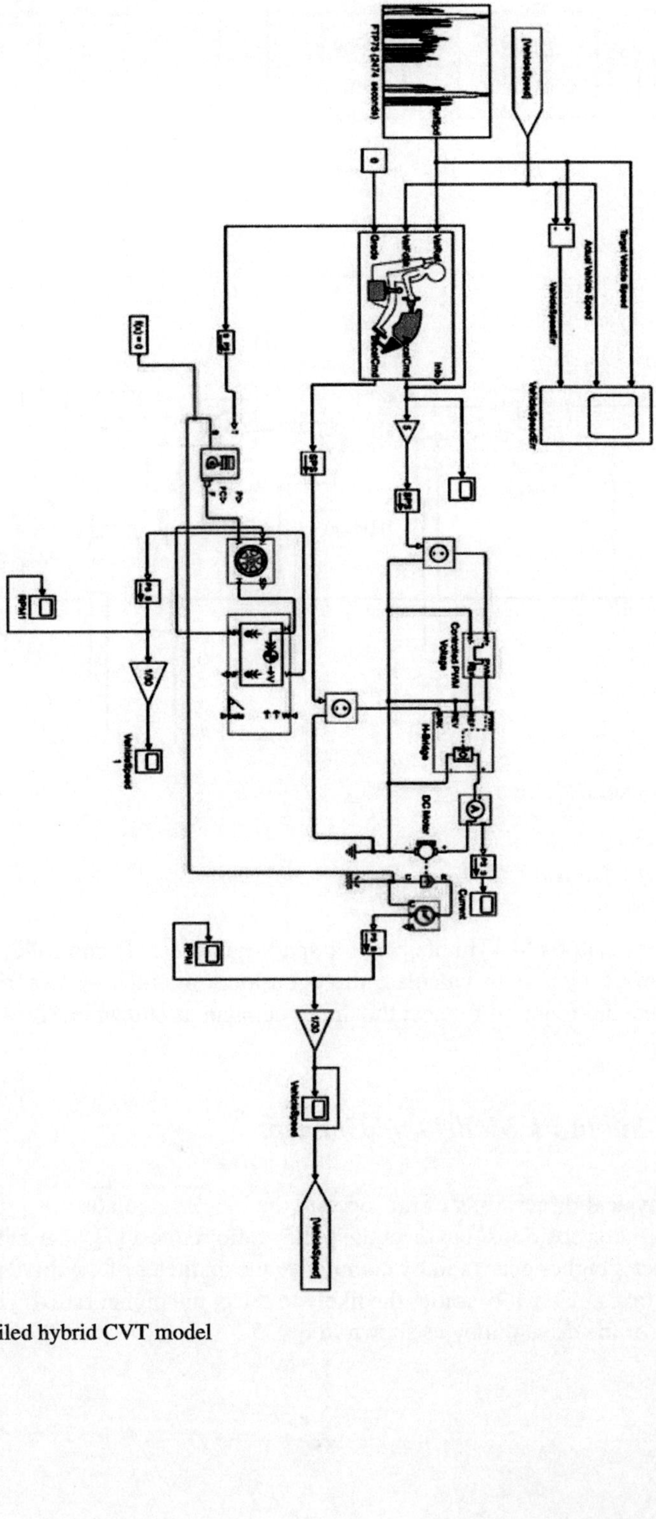

Fig. 1 Detailed hybrid CVT model

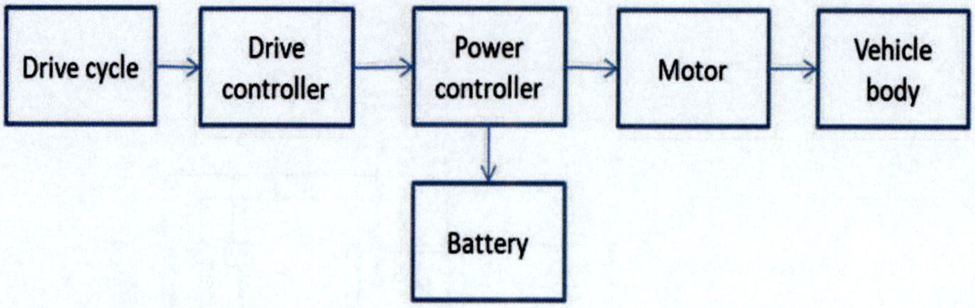

Fig. 2 System layout

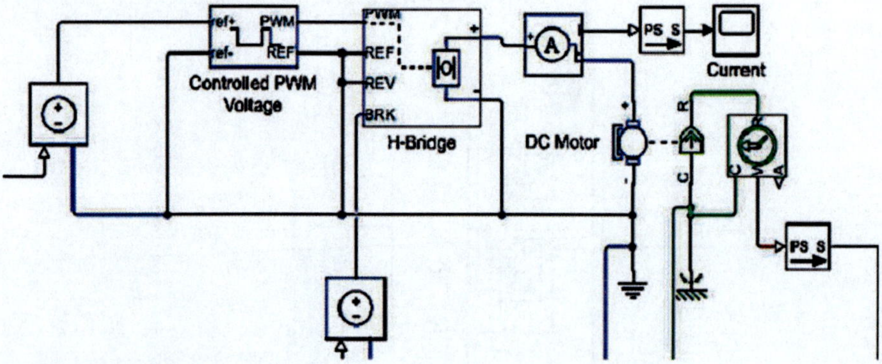

Fig. 3 Motor unit (MATLAB)

2.3 Engine Unit Model

The engine regulator block implements a spark-ignition (SI) controller that uses the driver torque request to calculate the open-loop air, fuel, and spark actuator commands that are required to meet the driver demand as shown in Fig. 4 [6].

2.4 CVT Simulink Model and Concept

Using the physical dimensions of the system, the block calculates the primary and secondary variator positions that meet the pulley ratio request [7]. The driven belt is likely to select a higher gear ratio by cumulative the diameter of the drive pulley. As acceleration takes place, it becomes the likely to select the higher ratio by increasing the diameter of the drive pulley as shown in Fig. 5.

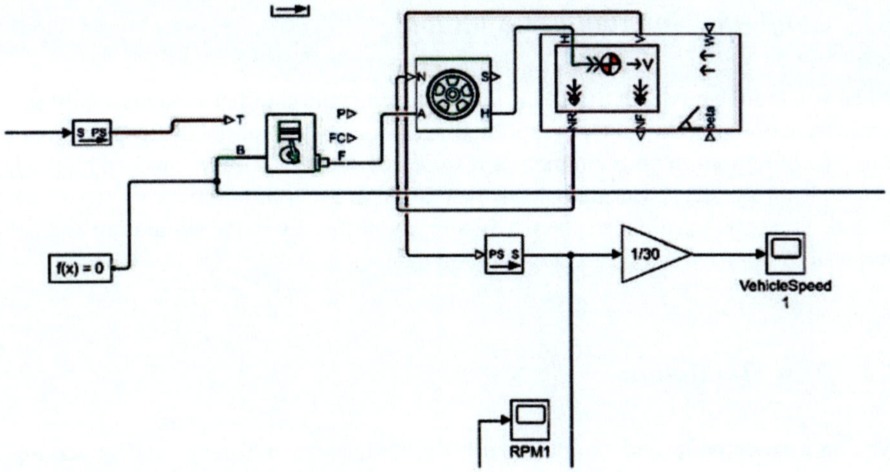

Fig. 4 Engine CVT Simulink unit (MATLAB)

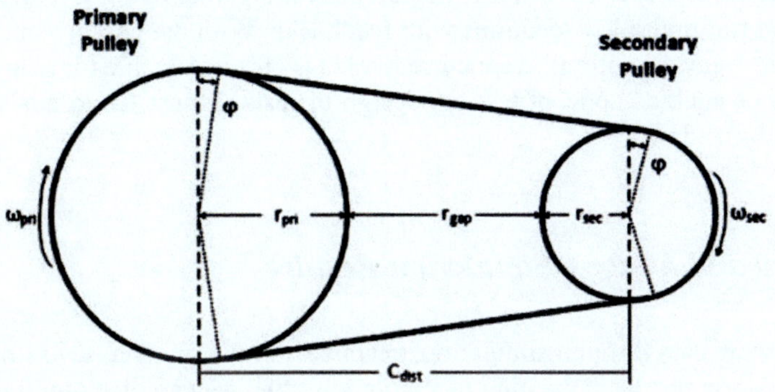

Fig. 5 Engine CVT Simulink unit

3 Prepare Model Inputs and Outputs

The test results are providing through the signals and data for simulation test results. Then MATLAB data Simulink data indefinitely save the time and execution data and log signals for researching purpose [8]. Create signal inputs for execution

a. Generate input records then apply capacity into a model as simulation input
b. Mass signal data for execution
c. Mass signal data for execution a perfect
d. Save run-time statistics from imitation

3.1 Configure Simulation Condition

When you build a model in Simulink, we can arrange the simulation to run easily and precisely without making any operational variations to the model design. The main stage in configuration your simulation is to select a solver [9]. By standard method, Simulink auto selects a variable-step solver. Sometimes, a simulation can slow down or stall. The solver profiler to classify blocks and difficulty in the simulation and get proposal to improve the solver presentation data.

3.2 Run Simulation

We can connectively and programmatically start, stop, and pause self-governing executions from the Simulink software. We can view our executions outcomes live and cooperate with the executions in many ways, with varying analysis and layout limitations [10]. Consuming whichever the more simulations board in the Simulink editor or the several executions API, we can deliver a gathering of several inputs to a model and run multiple executions with feedbacks. With the passim function, we can simulate many executions in concurrently [11]. This is valuable for many similar circumstances such as concept testing, design of tests, Monte Carlo analysis, and concept optimization.

3.3 View and Analyze Simulation Results

Generally, view data through simulation, examine simulation data, save simulation, and investigation results happening in simulation time, we can tune the limit values and view simulation outcomes using one or more tools [12].

a. Scope blocks—rapidly connects and appears data for prototyping and debugging the structure error.
b. Console blocks—View and monitor immediately provides signal values on dashboard display monitors, such as gages and lamps.
c. Simulation data checker—View many signals at the same time on imaginings such as spark lines, time plots, XY plots, and plans.
d. Calculating block—Attach and view signals implementation done by the system and can log to the workspace, to a file, or to the workstation and a file.
e. For whole information about inspecting simulation data, refer the below imagine simulation data.

Fig. 6 CUEDC petrol cycle run and simulation result (MATLAB)

4 Result and Discussion CUEDC Petrol Cycle

A system of composite urban emissions drive cycle (CUEDC) was advanced for light duty petrol vehicles. This UEDC cycle has been extended to include diesel vehicles. The petrol CUEDC was established to be characteristic of "factual world" problems.

4.1 Vehicle RPM Data

RPM stands for revolutions per minute, and it is used as a portion of how fast any motorcycle is functioning at a given time. In motorcycle, RPM measures how many times the internal combustion engine's crankshaft makes one full rotation every minute, and sideways with it, how many times each piston goes up and down in its cylinder as shown in Fig. 6 [13].

4.2 FTP75 Drive Cycle

FTP75 is a required dynamometer measurement on tailpipe emissions of a car that characterizes the driving city conditions. This was developed from the urban dynamometer driving schedule [14]. The UC test has a 3-bag construction comparable to the FTP, with bags on behalf of a cold-start passing phase, a constant stage, and a hot-start brief stage, but is a more destructive driving cycle in the segment. It has a more than expected higher regular speed, higher quickening rates, fewer stops

Fig. 7 FTP75 drive petrol cycle run and simulation result

per KM, and less idle time. The UC test also has a ten-minute hot soak as shown in Fig. 7.

4.3 WLTP CLASS 3 Drive Cycle

The world Forum for Organization of Vehicles Guidelines (WP.29) of the United Nations Economic Commission for Europe (UNECE) documented a not recognized group under its functioning party on pollution and energy (GRPE) to make a road map for the expansion of a worldwide harmonized light vehicles test as shown in Fig. 8 [15].

5 Conclusion

The MATLAB/SIMULINK SIMULATION OF HYBRID CVT MOTORCYCLE structure created based on motorcycle concept with all the necessities components. The power resources substances and the motorcycle controllers created in MATLAB/Simulink, whereas the motorcycle design and internal components were created in MATLAB.

MATLAB/Simulink for different pattern cycle,

a. CUEDC Petrol CYCLE
b. FTP75 Drive Cycle
c. WLTP CLASS 3 Drive Cycle

Fig. 8 WLTP Class 3 petrol cycle run and simulation result

Finally, our developed system layout ran successfully, and results are expected which is very satisfactory in terms showing how engine model and motor unit can be tested. Our proposed model eliminates the emission in order to provide by the current design.

References

1. Purusothaman M, Sunil Kumar M, Praveen kumar V, Suraj Kumar, Senthamizh Selvan S (2019) Design and thermal validation of four wheeler disc brake using different material. Int J Innov Technol Explor Eng 8(8):1739–1764
2. Purusothaman M, Mohan Krishna J, Jitendra Siva Prasad T, Manoj Kumar K (2019) Design and development of Tractor Tubular Tow Pin using Ahss. Int J Recent Technol Eng 8(1):2656–2660
3. Purusothaman M, Yogesh M, Vengatesan E, Vignesh B, Ramkumar A (2019) Design modification and improvement on automobile suspension system. Int J Innov Technol Explor Eng 8(11):277–281
4. Abramek KF (2018) Selected issues related to diagnostic testing of hybrid vehicles
5. Yousefi-Koma A, Asaei B (2010) Design, simulation, and implementation of an electric hybrid motorcycle
6. Aswin A (2018) A survey on power levels of battery charging and infrastructure For plug-in electric and hybrid vehicles
7. Arunraja M (2018) A literature review on hybrid electric vehicles
8. Balasubramani N, Hari Prasath S, Jagadeesh Kumar A (2018) Fabrication and performance analysis of hybrid two-wheeler
9. Bhanu Kiran K (2015) Design and performance of an effective mild-hybrid motorcycle prototype
10. Chen B-C, Huang C-N, Wu Y-Y, Huang Y-D (2014) Modeling and control of hybrid electric motorcycle with direct-driven wheel motor
11. Fan BS-M (2007) Modeling and simulation of a hybrid electric vehicle using MATLAB/Simulink and ADAMS

12. Feng C, Xia L (2021) Discussion on the technology of high voltage cable for hybrid electric vehicle
13. Frățilă G, Badea G (2020) Fuel consumption and emission improvements obtained using a mild hybrid electrical system on a vehicle equipped with a manual gearbox
14. Hazarathaiah, Pappuri (2017) Design and fabrication of hybrid electric bike
15. Kiran K (2014) Design and fabrication of hybrid two-wheeler

Design and Fabrication of Oil Skimmer with Metal Scrap Collector

M. Anish, M. Purusothaman, T. Viswanathan, and S. Sathish

Abstract The main aim of the paper is to collect metal chips on the shop floor and extract oily effluent and chemical compounds from industrial wastewater. Oil scavenging oil skimmers, regardless of design, use relative density, surface tension, and a flowing media to extract flying oil from a fluid's surface. Floating oil and grease have a stronger grip on skimming medium than water, while water has little love for it. In this method, a skimming route shaped like a belt, disk, drum, or other item can sweep through a fluid surface and pick it up flying oil and grease from a small volume of water. Wiper blades or pinch rollers are used to remove the greasy substance from the medium. Oil skimmers are simple devices for extracting oil, grease, and other hydrocarbon from water and cooling systems. Metal chips and scraps are produced as a by-product of the production process. Chips and scraps are frequently produced as a by-product of manufacturing operations such as cutting, turning, knurling, or threading. Both safety and efficiency may be jeopardized by these chips. In addition to the feature, we have also recovered metal scraps from coolants with this equipment by the magnetic force of attraction with help of magnetic plate.

Keywords Oil skimmer · Metal scrap collector · Magnetic plate

1 Introduction

Two definitions for oil are liquid that is viscously generated from petroleum, specifically for use as a lubricant or fuel, and any variety of viscous liquids that are soluble in organic solvents but insoluble in water and acquired by animals and favored over plants. Mineral oils (also known as crude oils) and organic oils (usually known as fats), which include essential oils, are divided into two categories depending on their chemical composition [1–4].

M. Anish · M. Purusothaman (✉) · T. Viswanathan · S. Sathish
School of Mechanical Engineering, Sathyabama Institute of Science and Technology, Chennai, India
e-mail: purusothmani@gmail.com

N. M. Sivaram et al. (eds.), *Advances in Manufacturing, Automation, Design and Energy Technologies*, Lecture Notes in Mechanical Engineering,
https://doi.org/10.1007/978-981-99-1288-9_46

Oil is a valuable crude that is used in a variety of everyday human activities. Because most oils are poisonous, they are extremely deadly to humans when they come into close touch with them. Throughout the last few decades, the world has seen numerous oil spill disasters and the resulting environmental devastation. Several countries have adopted rigorous safety laws for the disposal of wastewater containing oils, particularly from the petrochemical and manufacturing sectors, requiring that these companies build oil skimmers to extract the oils from the waste. The removal of oil from process fluids increases the life of the fluid and lowers the cost of machine fluid replenishing and improves the quality of wastewater disposal. In the current day, pollution is a major source of concern. Oil and oil spills are the primary cause of water contamination around the world. As a result, our goal is to reduce pollution by inventing technology that separates oil from water. As a result, adequate oil collection, disposal, and storage are required [5–8].

1.1 Metal Scrap

Ferrous and non-ferrous metals are the two types of scrap material that can be recovered. This category includes stainless steel as well as other metals that contain iron or iron alloys. Non-ferrous scrap metal includes copper, aluminum, nickel, zinc, tin, lead, titanium, and alloys like brass. Metal chips are typically produced by machining industries in ferrous wastes. As a result of the production process, metal chips are produced as a by-product. Cutting, turning, knurling, and threading all produce chips as a by-product. Both safety and efficiency could be jeopardized if these chips are used. Metal chips that have been encrusted with oil or lubricant can cause major damage to the machine and its final output. These chips not only contribute to a filthy and untidy workplace, but they can also constitute a major threat to co-workers. A comprehensive cleaning solution should be provided to the metal shop to ensure the safety of both the product and the worker. This page details the development of the oil skimmer with metal scrap collector, including a functioning model [3, 9, 10].

2 Materials and Methods

This project's purpose is to gather metal chips from contaminated water and recover oily effluent and chemical compounds from industrial wastewater. Oil skimmers use surface tension, surface gravity, specific gravity, using a moving media, irrespective of shape, to remove floating oil from a fluid's interface [11]. Floating oil and grease have a firmer grasp on the skimming medium than water, which has little affection for it. A skimming path, such as a belt, disk, drum, or other devices, can move across fluid and gather up flying grease from a little amount of water in this way [12]. To remove the greasy substance from the medium, pinch rollers or wiper blades are utilized. Oil skimmers are simple machines that gather oil, sludge, and other pollutants from liquid and coolants [13]. As a result of the production process, metal chips are produced as a by-product. Cutting, turning, knurling, and threading all produce chips as a by-product. Both safety and efficiency could be jeopardized by these chips. In addition to the feature, we have used this equipment to recover metal scraps from coolants using magnetic force and a magnetic plate.

2.1 *Model Components Mapping*

Three-dimensional model design with components' mapping is shown in Fig. 1, and actual fabricated prototype model with the equipment is shown in Fig. 2.

3 Working Principle

The polymer belt is adjusted for firm roll-on and is mounted to the roller with bearing support. A DC motor is coupled to the roller shaft, which is positioned at the top of the frame and is supported. Both rollers are connected to the belt arrangement by

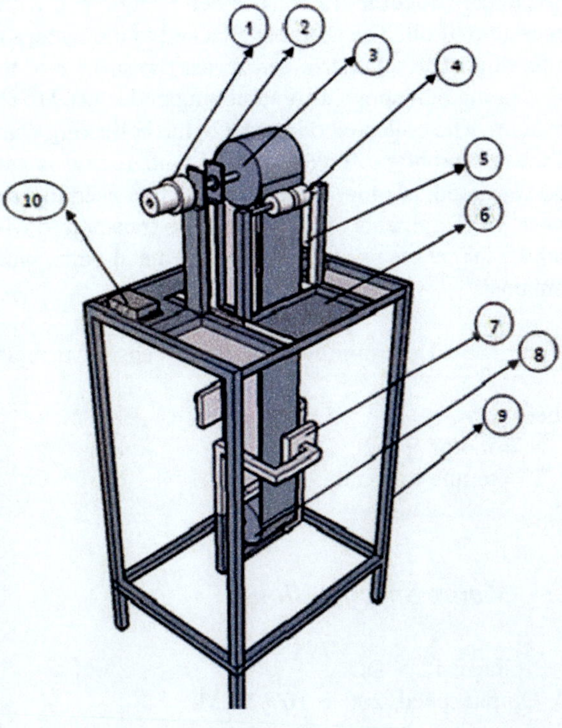

Fig. 1 Model components' mapping

Fig. 2 Fabrication of oil skimmer

rotating it. The opposite end is held in place by a DC motor. MS square tubes are used to construct the frame. The belt is equipped with wiping blades that aid in the separation of oil. The motor is attached to the battery and rotates with the belt at the same time. The oil distributed across the surface of the water will be sucked up by the rotating belt above it. A scrubbing pad is used to remove the oil and separate the particles. The collected oil is sent to the collecting chamber via a port. With the help of the wiping blades linked to the frame, the oil is gathered from the water surface and separated. Magnet installation is often used to extract metal fragments from the water. The fragments of metal parts are transported with the oil on the skimmer belt, and the magnet is used to retrieve the metal parts, which are collected in a separate container.

$$\text{Determining torque and tension in belt} = \text{NTP } 60/2,$$

where

N = motor speed.

T = torque transmitted by motor.

3.1 Motor Specification

1. Voltage: 12 V DC.
2. Output speed: $200 \pm 10\%$ RPM.

3. No-load output current: = <50 mA.
4. Rotation output: CW/CCW.
5. Noise: no gear noise.
6. Output shaft of the axial clearance: <0.1–0.3 mm.

Horizontal clearance requirement <0.05 Electrical Spec.

1. No-load speed: 5700 RPM.
2. No-load current: <30 mA.
3. Rotation: CW 4. Motor#: 370.

4 Results and Discussion

During testing, we have added one liter of oil and half kg of scrap (metal chips) into wastewater and poured them into a container. And then, insert our equipment into wastewater carrying container for performing the operation. We started the equipment with the power supply, and we took the reading of oil and scrap collection on oil skimming equipment with each minute and have taken ten readings with the one-minute gap.

Figure 3 shows the result of oil vs. scrap recovery rates of oil skimmer with scrap collecting mechanism. The maximum time this motor can run with full power is: 2.4 kWh/3 kW = 0.8 h = 48 min. A combination of transformers, diodes, and transistors is used to generate a 12 V DC output from a 120 V AC or 240 V AC input with this DC motor. With this equipment, we can extract 400 ml of oil contamination and 160 g of metal wastes every 10 min. The recovery rate of oil is 40% and scrap is 32% for every 10 min in the total quantity of contamination and scrap wastes.

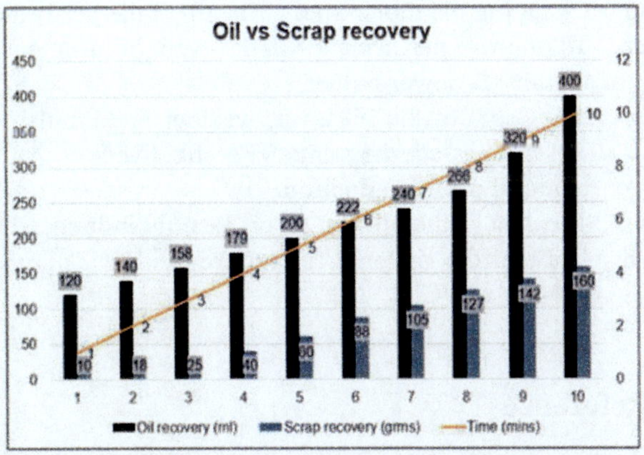

Fig. 3 Oil versus scrap recovery

5 Conclusions

The most effective strategy to deal with an oil spill is to prevent it from happening in the first place. The conventional approach for removing spilled oil and metal scraps is to do so by hand. As a result, they are both dangerous to people's health and time-consuming. A single blankets' belt economic oil skimmer with metal scrap collector construction work was finished after being concerned about the connected problem of spilled oil and scraps in the expensive belt oil skimmer. Due to a scarcity of oil skimmer belt replacement materials, an artificial tray model was successfully implemented to recover spill oil utilising a locally procured new or used blanket that had previously been used to warm. And, the magnetic plates with open and close types are used to collect and dispose of the metal scraps as well. The research clearly demonstrates that blankets can clean up oil spills in a wonderful way and that these powers must be made use of. In steady and moderately moving water, the suggested oil skimmer belt can skim most of the oil up to a 2 cm layer.

The use of a blanket belt instead of a traditional belt reduces maintenance costs and increases availability. The project's design and construction are straightforward, making it simple to adapt it to the various conditions that may arise. The width and quantity of the belts can be simply redesigned to acquire more absorption based on the volume of the spilled oil. As a result, the blanket belt oil skimmer is both cost-effective and practical for implementation. Several factors, however, have an impact on skimming performance. As a result, as shown below, the preceding advice is made for future development.

i. The belt utilized in the project is 4.6 inches wide and 38 inches long. For high recovery, belt dimensions should vary, and multi-belt projects may be considered.
ii. Because an oil skimmer is typically portable, the model must be light in weight.
iii. Changing the motor speed can affect the performance.
iv. To improve performance in the event of turbulence, a casing should be applied around the lower pulley.
v. The depth of the oil layer, weather and emulsifier of oil, and the amount of waste all affect the skimmer's effectiveness. As a result, actual experience is required prior to adoption.
vi. The cost of the blanket belt is heavily influenced by its durability.
vii. Only ferrous material can recover with this magnetic absorption type.

References

1. Annunciado TR, Sydenstricker THD, Amico SC (2005) Experimental investigation of various vegetable fibers as sorbent materials for oil spills
2. Alexander C, 17 December 2014. National Geographic Report

3. Ansell DV, Dicks B, Guenette CC, Moller TH, Santne RS, White IC (2001) Proceedings from '01: international oil spill conference. The International Tanker Owners Pollution Federation Limited, March 20, Tampa, Florida, USA
4. Agnew RW (1972) A free-floating endless belt oil skimmer, EPA–R2–72–006
5. ITOPF (2012) Use of skimmers in oil pollution response, Technical Information Paper No. 5, London, UK
6. Encyclopedia Britannica News (2010) Deepwater Horizon oil spill of 2010 (oil spill, Gulf of Mexico)
7. Fingas M (2004) Weather windows for oil spill countermeasures for Prince William Sound Regional Citizens' Advisory Council (PWSRCAC), Anchorage, Alaska, Environmental Technology Centre, Environment Canada
8. Government of Dubai, Technical Guideline Number–I (2011) Oil spill preparedness and response. Marine Environment & Wildlife Section, Environment Department
9. Gupta US, Jain S, Jain H, Rathi SS (2015) Experimental performance & evaluation of eco oil skimmer. Int J Res Eng Technol Manage 2(5):1–10
10. ITOPF (2011) Clean–up of oil from shorelines. Technical Information Paper No. 7, London, UK
11. Purusothaman M, Sunil Kumar M, Praveen Kumar V, Suraj Kumar, Senthamizh Selvan S, Design and thermal validation of four wheeler disc brake using different material. Int J Innov Technol Exploring Eng (IJITEE) 8(8):1739–1764. ISSN: 2278–3075
12. Purusothaman M, Mohan Krishna J, Jitendra Siva Prasad T, Manoj Kumar K (2019) Design and development of tractor tubular tow pin using Ahss. Int J Recent Technol Eng 8(1):2656–2660
13. Purusothaman M, Yogesh M, Vengatesan E, Vignesh B, Ramkumar A (2019) Design modification and improvement on automobile suspension system. Int J Innov Technol Exploring Eng (IJITEE) 8(11):277–281. ISSN: 2278-3075

Fault Bearing Detection from Vibrational Signal Data by Using Machine Learning Algorithms

Srinivasa Advaith Thutupalli, Grandhi Sri Sai Charith, Doppalapudi Manohar, Choppala Sarath, Saleel Ismail, and T. Jagadeesha

Abstract Rotating machines are of crucial importance in the modern industrial world where continuous and mass production is needed. Bearings are an essential part of the functioning of these machines. Hence, the conditional monitoring of the bearings in the machines is critical for easy maintenance and good production in the long run for the manufacturing firms. Advanced machine learning models can predict the failure of these bearings in advance and classify their faults efficiently and accurately using the vibrational signal data and can drastically reduce maintenance costs and maintain optimal production output. The vibrational data of the machine component is collected and featurized and available as standard datasets. Two machine learning models using advanced algorithms logistic regression and support vector machines (SVM) are developed and trained using these datasets. The performances of the models are compared. An initial accuracy of 94% is achieved for the SVM model. The model is optimized by tuning the hyperparameters and an optimized accuracy of 96% is achieved. The logistic regression classifies at an accuracy of 94%. The best model, which is the optimized SVM model, is adopted.

Keywords Logistic regression · Support vector machines · Machine learning · Bearing

1 Introduction

The advent of the modern industrial revolution and a fast-paced technology advancement in the last two decades paved way for a huge increase in the production capabilities of the modern industrial, especially manufacturing systems. This is made possible by the introduction of new and advanced machinery in the industrial shop

S. A. Thutupalli · G. S. S. Charith · D. Manohar · C. Sarath · S. Ismail · T. Jagadeesha (✉)
Department of Mechanical Engineering, National Institute of Technology Calicut, Calicut 673601, Kerala, India
e-mail: jagdishsg@nitc.ac.in

N. M. Sivaram et al. (eds.), *Advances in Manufacturing, Automation, Design and Energy Technologies*, Lecture Notes in Mechanical Engineering,
https://doi.org/10.1007/978-981-99-1288-9_47

floor which reduced the production lead time significantly and increased the production hundred folds. As a result, to meet the production demand in the modern world, these machines are expected to run for long hours continuously and so the machinery health has become one of the crucial factors that affect production. Conditional monitoring is the process which helps these systems to evaluate and continuously monitor the health of all the components of the machinery and thereby predict the upcoming failures in them and properly identify and classify which types of faults may occur.

So, various conditional monitoring systems which include advanced computing techniques have gained popularity in the recent years in the industry. Conditional monitoring is often achieved by the help of attaching sensors to various components in the machines and studying and analyzing the machine vibration signals through the sensors to detect the health of these machine components, identify the type of faults they belong to [1]. Failure in bearings can severely affect the lifetime of the machine and also causes obstruction in the mechanism and prevents the machine functioning. To monitor the health of the bearings, various sensors are mounted on the shafts and the vibration signals are recorded and analyzed.

Data mining is a process of extracting and discovering patterns in large data sets [2]. Data mining is used on the vibrational datasets obtained from the sensors in order to find patterns in the data and understand which dataset implies which type of fault. Machine learning is the most effective way to achieve data mining [1]. Today, various machine learning models are widely used in the modern industry for effective and cost-efficient conditional monitoring of bearings [3]. A robust machine learning model can help us accurately predict the failure in the bearings beforehand and can classify the type of fault just from the vibration signals of the machines [4].

This paper aims at developing a robust and highly accurate machine learning model by developing different machine learning models for the same datasets and optimizing them and finally taking the model with the best performance. The dataset used to train our models is Case Western Reserve University (CWRU) dataset. This is a standard dataset which is widely used by many researchers worldwide to standardize their works and is a trademark for vibration datasets corresponding to various types of faults in roller bearings [1]. The dataset is prepared by extracting many important features from the raw vibrational signals like mean, median, rms, etc. We developed two different machine learning models using different algorithms, namely multinomial logistic regression and support vector machine (SVM) which are among the best performing machine learning algorithms [5], and these algorithms have been proven to give accuracies as high as 97%. The performance of the models is compared based on various performance metrics, and then they are optimized. SVM is optimized by tuning the hyperparameters to achieve maximum accuracy. Finally, the model with the best performance is adopted.

2 Objectives

The primary objectives of this paper are to anticipate and classify faults in the bearings by an upgraded machine learning algorithm and analyze the accuracy of both models.

Standard datasets are utilized for building this model that were used in publishing many research and conference papers [1].

Following objectives will be targeted in the proposed paper:

- Collection, understanding and feature extraction from the CWRU dataset
- Build machine learning algorithms on the featurized dataset
- Training and testing the model
- Optimizing the model
- Obtaining a model with relatively high accuracy
- Documentation and publications.

3 Methodology

3.1 Proposed System Architecture

This work proposes a machine learning-based keen fault diagnosis framework for working on raw vibration signals (after data pre-processing). Our methodology consolidates the best machine learning models with an optimized approach to accomplish better results on the benchmark bearing fault dataset from Case Western Reserve University (CWRU).

3.2 Pre Processing of Dataset

To obtain the raw data, a test is carried out in an experimental setup consisting of a 2 horse power motor connected to shaft where the encoder and a moment transducer are mounted on the shaft in the middle and are coupled with dynamometer and an electronically controlled system. The details of bearings on each end of the apparatus are mentioned in Table 1.

Faults have been introduced in the bearings by Electrical Discharge Machining (EDM). The diameters of the faults have been mentioned.

Table 1 Specifications of the equipment used

S. No.	Location	Name	Specifications
1	Drive end	6205-2RS JEM	Radial ball bearing (Single row deep groove) Inner diameter—25 mm Outer diameter—52 mm 15 mm width
2	Fan end	6203-2RS JEM	

Table 2 Classification of faults

Type of fault	Size of defect (inch)	Number of samples	Label names
Normal	–	280	Normal
Ball	0.007	280	BL_7_1
Ball	0.014	280	BL_14_1
Ball	0.021	280	BL_21_1
Inner race	0.007	280	IR_7_1
Inner race	0.014	280	IR_14_1
Inner race	0.021	280	IR_21_1
Outer race	0.007	280	OR_7_1
Outer race	0.014	280	OR_14_1
Outer race	0.021	280	OR_21_1

Fault diameters: 7−28 mils (ranging from 0.007 to 0.021 in.).
Motor load: 0−3 horsepower
(With speeds approximately 1720−1797 rpm)

The data was collected at two sampling frequencies—12,000 samples per second and 48,000 samples per second [1]. In this work, only the data collected at 48 kHz frequency is considered. It implies that for the lowest speed of 1720 rpm, about 1670 data points are collected for each revolution of the shaft. So, all the data corresponding to each type of fault for each load is divided into samples of 1670 data points and about 280 samples corresponding to each fault type are obtained.

All categories were given class labels accordingly as shown in Table 2. Different conditions will be analyzed to check the accuracy of the method proposed to address various fault patterns.

3.3 *Features Extraction for the Model*

It is an essential part in model building. This vitally affects the accuracy of algorithms' performance and assists the algorithms that are used here to track the defect patterns resulting in increase of the accuracy of the model [5]. The raw vibrational signal recorded by the sensors is not suitable for training and testing of ML models [3]. For training and testing the model, data featurization is vital, so to avail this data, extraction of various features is carried out in python using Pandas library. All the features extracted are mentioned in Table 3.

When the rolling element comes across the faulty region, the amplitude of signal changes, resulting in a disturbance in the entire signal pattern, indicating the detection of a fault. For faulty members, the kurtosis of the sample generally increases and skewness vary across positive and negative region.

Table 3 Features extracted

No.	Feature	Formula	Description for statistical features
1	Mean	$\text{mean} = \frac{1}{n}\sum_{i=1}^{n} x_i$	It is the signal's average value, and it is termed as mean
2	Absolute mean	$\text{Abs mean} = \frac{1}{n}\sum_{i=1}^{n} \lvert x_i \rvert$	It is the absolute value of signal's average value
3	Maximum value	$\text{maximum} = (x_i)$	For the given vibration signal, it indicates the point of maximum amplitude of the signal
4	Minimum value	$\text{minimum} = (x_i)$	For the given vibration signal, it indicates the point of minimum amplitude of the signal
5	Peak to peak	Peak to peak = maximum − minimum	It denotes the difference between maximum signal point and minimum signal point
6	Absolute maximum	$\text{Abs max} = (\lvert x_i \rvert)$	Peak amplitude of the waveform
7	RMS	$\text{RMS} = \sqrt{\frac{1}{n}\sum_{t=1}^{n} x_i^2}$	It is the vibration signal's power content and energy content in it which is formulated by the formula mentioned
8	Variance	$\text{Var} = \frac{1}{n}\sum_{t=1}^{n} (x_i - x)^2$	
9	Clearance factor	$\text{CIF} = \frac{\text{Absolute max}}{\left(\frac{1}{N}\sum_{i=1}^{n}\sqrt{\lvert x_i \rvert}\right)^2}$	It is the signal points variance calculated by the formula mentioned
10	Kurtosis	$\text{Kurt} = \frac{\sum_i^n (x_i - x)^4}{n * \text{var}^2} - 3$	
11	Skewness	$\frac{\frac{1}{n}\sum_i^n (x_i - x)^3}{\left(\sqrt{\frac{1}{n}\sum_i^n (x_i - x)^2}\right)^3}$	It is a measure of signal's flatness relative to the normal distribution. Generally, high for the signal from faulty bearing as the signal is more spiky and low for the normal bearing because of the flat nature of the signal
12	Impulse factor	$\text{IF} = \frac{\text{Absolute max}}{\frac{1}{n}\sum_{i=1}^{n} \lvert x_i \rvert}$	It is the measure of absence of symmetry with respect to it respective mean distribution
13	Crest factor	$\text{CF} = \frac{\text{Absolute max}}{\text{RMS}}$	
14	Shape factor	$\text{SF} = \frac{\text{RMS}}{\frac{1}{n}\sum_{i=1}^{n} \lvert x_i \rvert}$	It is formulated by the peak amplitude of the waveform divided by the RMS value
15	Standard deviation	$\sqrt{\frac{n\sum x^2 - (\sum x)^2}{n(n-1)}}$	

For further analysis and classification purposes, the dimensionless parameters like impulse factor, crest factor and shape factor are determined. Shape factor varies as the shape changes since they are interlinked. Table 3 shows all the extracted parameters, significance and their mathematical formulas.

3.4 Training and Testing the Model and Its Optimization

After the signal feature extraction, the models built using standard machine learning algorithms give high accuracy. Following classification algorithms will be adapted:

- Support vector machine (SVM)
- Multinomial logistic regression (LR)

Logistic Regression

Multiple logistic regression is used to find relation between multiple input variables (features) and the target variable (bearing fault type). More accurate for multiple classifications and large data than that of linear regression. Works by fitting the best logarithmic curve, as shown below, between the output variable and multiple input variables.

$$\ln\left[\frac{y}{1-y}\right] = a + b_1x_1 + b_2x_2 + b_3x_3 + \ldots$$

Initially, the featurized data is split into training and testing data so as to avoid overfitting of the model to the current dataset. Then, multivariate logistic regression is first used to build the model and it is trained and tested against the dataset. The algorithm works by fitting the best logarithmic curve or boundary to the given data to classify the different classes accurately. The model accuracy is noted and the confusion matrix heatmap is plotted using the Seaborn library.

Support Vector Machine (SVM)

This algorithm is used for binary classification usually. It works by making a hyperplane boundary between the classes in the testing data by maximizing the margins to the nearest datapoints of each class [6]. The accuracy depends on parameters like gamma and regularization (C). In this case where we require multiple classification, the problem can be solved by SVM model by breaking down the test data to multiple binary classification problems. Initially, the model is built using the default values for the parameters. Further, the model is optimized by finding optimal values for the above parameters (gamma and 'C'), which is known as hyperparameter tuning, using GridSearchCV library, and then, the final optimized model is built, and accuracy is noted. The confusion matrix heatmap is plotted using Saeborn library.

Then after applying the machine learning algorithms like SVM and multivariable logistic regression, the model with the highest accuracy is chosen. After the grid of parameters is scrutinized to find the best one, the model with higher accuracy is optimized with the tenfold cross-validation and the best model is taken for the final consideration. In the process of performing this optimization, grid search optimization analysis is used. Then optimal C and gamma values were calculated. After the values are obtained, the model is built with the obtained values. For each model, respective accuracies are obtained. After building and verifying the classification reports, the best model is adopted.

4 Results

Here, the machine learning model is developed with support vector machine algorithm (SVM). Then it is optimized with a tenfold cross-validation for a more accurate outcome and compared with the non-optimized model and other machine learning model that uses the multinomial logistic regression algorithm. The confusion matrices of test sets are given, and the respective F-1 scores of each model were obtained and are clearly stated in Table 4. The labels in the confusion matrix are according to the class labels in Table 2. The initial classification report of the SVM model is mentioned in Table 4. An accuracy of 95% obtained. As SVM is a solid and robust classification algorithm, it can be optimized to improve accuracy [7].

The initial accuracy of the SVM model is 95% which could be still increased by tuning the hyperparameters. The optimal values of the parameters obtained are given in the Table 5.

An increase in accuracy is observed after tuning the hyperparameters which could be inferred from this 2D confusion matrix mentioned in Fig. 1, that our model

Table 4 Classification report of initial SVM model

	Precision	Recall	F1 score	Support
IR 7 1	0.99	1.00	0.99	75
IR-14-1	1.00	1.00	1.00	75
IR-21-1	0.95	1.00	0.97	75
BL 7 1	0.99	0.96	0.97	75
BL-14-1	0.90	0.76	0.83	75
BL-21-1	0.97	0.85	0.91	75
OR 7 1	1.00	0.99	0.99	75
OR-14-1	0.80	0.92	0.86	75
OR-21-1	1.00	0.99	0.99	75
Normal	0.88	0.99	0.93	75
Accuracy			0.95	750
Macro average	0.95	0.95	0.95	750
Weighted average	0.95	0.95	0.95	750

Table 5 Classification report comparison of both the models

Hyper parameters	Values
C	10
Gamma	0.1

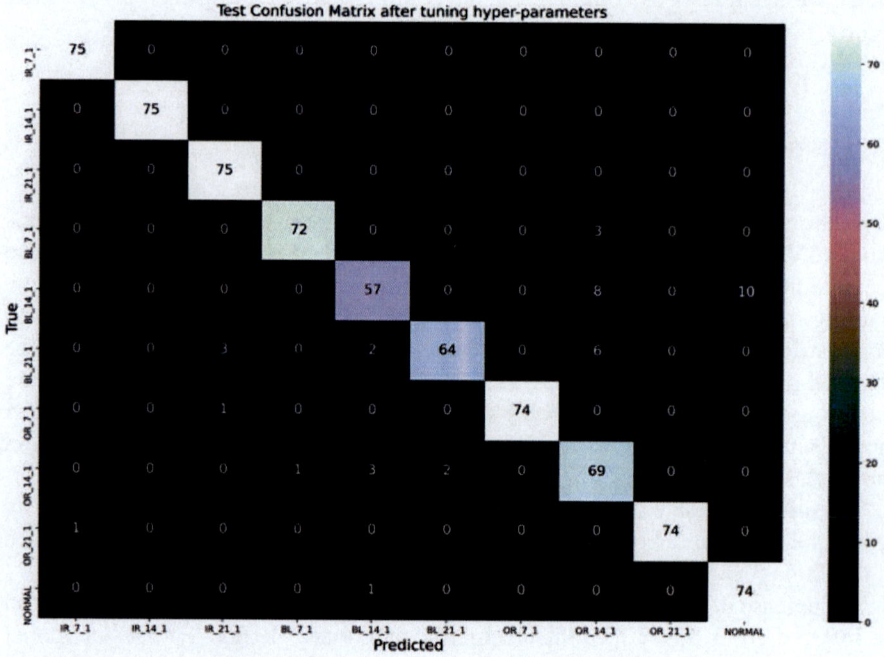

Fig. 1 Test confusion matrix of an optimized SVM algorithm

misclassified Ball fault of 0.014-in. defect (BL_14_1), Ball fault of 0.021-in. defect (BL_21_1), Outer race fault of 0.021-in. defect (OR_21_1). Model's classification accuracy is majorly affected in misclassifying Ball fault of 0.021 in. and Outer race fault of the 0.021-in. defect. A misclassification of nearly 10% majorly affected the potential of the classifier. However, the model's overall accuracy is 97% which is reasonably good and better than the initial SVM before tuning the hyperparameters. The detailed classification report for each class label is mentioned and compared in Table 6.

The confusion matrix of the logistic regression model is mentioned in the Fig. 2. From this 2D confusion matrix, it can be inferred that model misclassified the Ball fault of 0.014-in. defect (BL_14_1), Ball fault of 0.021-in. defect (BL_21_1) and Outer race fault of 0.014-in. defect (OR_14_1). However, the model's overall accuracy is 94% which is less than the initial SVM model. The detailed classification report for the logistic regression algorithm is mentioned in Table 6.

Table 6 Classification report comparison of both the models

Label	Precision		Recall		F1 score	
	LR	SVM	LR	SVM	LR	SVM
IR 7 1	0.97	1.00	1.00	1.00	0.99	1.00
IR-14-1	1.00	1.00	1.00	1.00	1.00	1.00
IR-21-1	0.96	0.97	1.00	1.00	0.98	0.99
BL 7 1	0.95	0.99	0.99	0.96	0.97	0.97
BL-14-1	0.91	0.92	0.81	0.93	0.86	0.93
BL-21-1	0.93	0.93	0.87	0.89	0.90	0.91
OR 7 1	1.00	1.00	1.00	1.00	1.00	1.00
OR-14-1	0.85	0.88	0.84	0.87	0.85	0.87
OR-21-1	1.00	1.00	0.97	1.00	0.99	1.00
Normal	0.90	0.96	1.00	1.00	0.95	0.98
Accuracy					0.94	0.97
Macro average	0.94	0.96	0.94	0.96	0.94	0.97
Weighted average	0.94	0.96	0.94	0.96	0.94	0.97

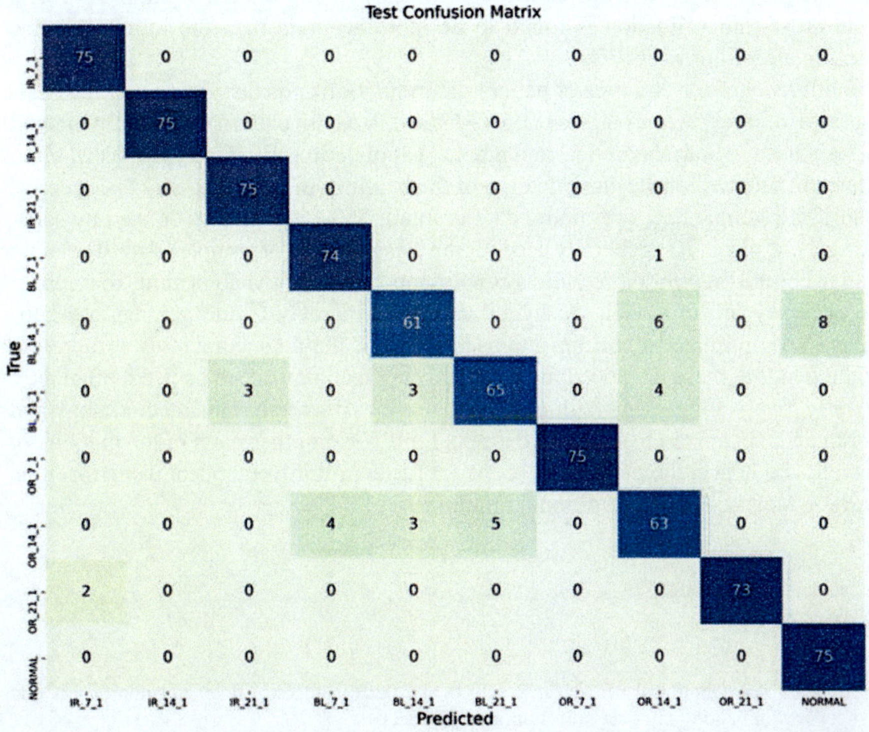

Fig. 2 Test confusion matrix of logistic regression model

On comparing the above classification reports, the best model is obtained by optimizing the SVM algorithm. The accuracy of the best model is 97% though the model misclassifies certain instances. For the optimized SVM model, the recall and precision values are ideal. The F1 score of the SVM model is nearing 1 in maximum instances indicating a greater potential of the classifier. From this, it can be inferred that this approach can be a practical and can be used in industry.

5 Conclusion

In this work, diagnosis of the faults in bearings of roll type has been demonstrated. This diagnosis is based on data obtained from vibration signals. Previous works in this domain have been studied and conclusions were made on their accuracies for predicting the faults in bearings. Their performance is benchmarked using signal data from the standard CWRU dataset. It is concluded that in order to have a better classification, enough features were considered from the signal data. It is found that when a high number of features are considered, the algorithm can make more accurate predictions. The optimized algorithm by tuning the hyperparameters with a ten cross-fold validation is found to be more accurate than the initial machine learning algorithm with SVM.

Fault recognition is a type of pattern determination and classification of the bearings into their respective classes [8]. SVM is a very useful pattern recognition method [6]. It is used in both theoretical and practical applications [7]. One vs all SVM algorithm can be used for the classification of the bearings of roll type. SVM is effective even if the sample data or training data is small. SVM is typically defined by separating hyperplane. It is generally used for the classification of complex data. The main motive behind this work is to build a concise model with SVM algorithm. To improve the accuracy of the model, the hyperparameters, namely C and gamma, are optimized. The optimized algorithm attains the state-of-the-art accuracies. It is therefore concluded that these advanced machine learning techniques can be used, efficiently and effectively, for bearing fault prediction and classification. This methodology and work can be extended to different datasets for different other parts of the machinery as well, and hence, there is a huge scope for further improvements of using machine learning techniques for conditional monitoring.

References

1. Farimani AB, Ghule L, Li J, Magar R, Li J et al (2020) FaultNet: a deep convolutional neural network for bearing fault classification. IEEE Struct 9
2. Wang B, Zhang S, Zhang S, Habetler TG et al (2020) Deep learning algorithms for bearing fault diagnostics—a comprehensive review. Proc IEEE 8
3. Kankar PK, Sharma SC, Harsha SP et al (2011) Fault diagnosis of ball bearings using machine learning methods

4. Prosvirin A, Piltan F, Kim J et al (2019) Rolling-element bearing fault diagnosis using advanced machine learning-based observer
5. Yu D, Cheng J, Yang Y et al (2006) A fault diagnosis approach for roller bearing based on IMF envelope spectrum and SVM
6. Chu F, Hao R, Peng Z, Feng Z et al (2012) Application of support vector machine based on pattern spectrum entropy in fault diagnosis of rolling element bearings 22
7. Widodo A, Yang B-S et al (2007) Support vector machine in machine condition monitoring and fault diagnosis
8. Ao HL, Cheng J, Li K, Truong TK et al (2014) A roller bearing fault diagnosis method based on LCD energy entropy and ACROA-SVM

Modelling and Simulation of Magneto-rheological Fluid in a Damper Using COMSOL

B. Ganga Nair, Nathan Job Antony, Sebin Sabu Mathew, and T. Jagadeesha

Abstract This paper deals with the modelling and simulation of magneto-rheological fluid in a damper using COMSOL software. MR dampers provide an edge over air or hydraulic dampers due to their variable viscous damping properties, which can be controlled with a magnetic field. This continuous controllability allows the MR damper to be optimized to give the best damping coefficient. In order to understand the functioning of an MR damper, we will have to simulate the flow of the MR fluid and understand how the rheological properties vary. In this study, a model of the MR damper will be first developed and the flow of MR fluid in the MR damper will be simulated using COMSOL. This study will aid us in designing an MR damper which will then be used in a dynamic vibration absorber.

Keywords Magneto-rheological fluid · MR damper · Vibration absorber

1 Introduction

Magneto-rheological fluids (MR fluids) are smart fluids whose properties such as viscosity, and thus shear stress can be varied with the magnetic field acting on them [1]. They are usually synthesized by adding ferromagnetic particles to a carrier fluid, with additives like surfactants to improve properties. When a magnetic field is applied, they aggregate and align along the magnetic field (changing from a Newtonian to a non-Newtonian fluid). This helps in changing the viscosity and shear stress of the fluid. The quality of the MR fluid for our application depends on its density, sedimentation rate and response to the applied magnetic field.

The utilization of MR fluids in dampers has many advantages: low power consumption, low response time, simple construction and very few moving parts.

B. G. Nair · N. J. Antony · S. S. Mathew · T. Jagadeesha (✉)
Department of Mechanical Engineering, National Institute of Technology Calicut, Calicut 673601, Kerala, India
e-mail: jagdishsg@nitc.ac.in

N. M. Sivaram et al. (eds.), *Advances in Manufacturing, Automation, Design and Energy Technologies*, Lecture Notes in Mechanical Engineering,
https://doi.org/10.1007/978-981-99-1288-9_48

In this study, an MR fluid will be modelled. Its flow in a damper can then be simulated and studied to understand how the rheological properties as well as parameters like pressure, velocity of fluid, etc., vary.

2 Modelling Using COMSOL

2.1 *General Design and Modelling of Damper*

To understand the behaviour of MR fluids in a damper with a varying magnetic field, a simple geometry was created to represent the setup. An axisymmetric 2D model was created to easily represent all the required features. The specifications are as given below.

The model consists of a damper, piston with a coil. The piston, named 1 and 2 in Fig. 1, consists of a coil (named 2 in Fig. 1) and a core (named 1 in Fig. 1). Copper was assigned as the material for the coil and the core. MR fluid is assigned to 3 in Fig. 1. Measurements are as given in Table 1.

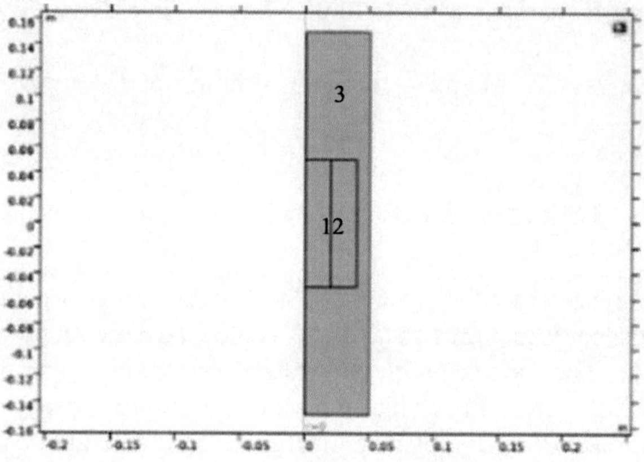

Fig. 1 Model of damper in COMSOL

Table 1 Design specifications of MR damper

Parameter	Value
Damper height	0.3 m
Damper radius	0.05 m
Height of piston	0.06 m
Radius of outer end of piston	0.02 m
Piston radius	0.045 m

2.2 *Modelling of MR Fluid*

A suitable MR fluid for the application has to be chosen before simulation. The alternatives considered for this decision were MRF 140-CG, MRF 132DG, MRF 122EG, MRF 126LF and MRF 140BC. MRF140BC was not recommended for damper applications and so was ruled out. MRF126LF was ruled out as it had lower settling resistance, which is an important property required for the damper application in the long run. Among the other options, the fluid that can achieve larger yield stress values at lower power was chosen, which was MRF 140-CG. It also has high plastic viscosity when no current is applied, allowing it to still possess good damping properties. The values for the parameters of LORD MRF 140-CG were obtained from its datasheet, and the required ones are given in Table 2 [2].

As MR fluid does not show a constant permeability, its B-H curve was utilized as obtained from the datasheet as in Fig. 2. The data was provided as an interpolation function in COMSOL [2].

Fluid Properties

The most important property of MR fluid is the variation of rheological properties with magnetic field. To model this, two different approaches were adopted. Both studies were done assuming laminar flow. The calculation for minimum velocity for

Table 2 Parameters from datasheet for LORD MRF 140-CG

Parameter	Value
Density	3540 kg/m^3
Relative permittivity	5
Electric conductivity	0.00000002 S/m

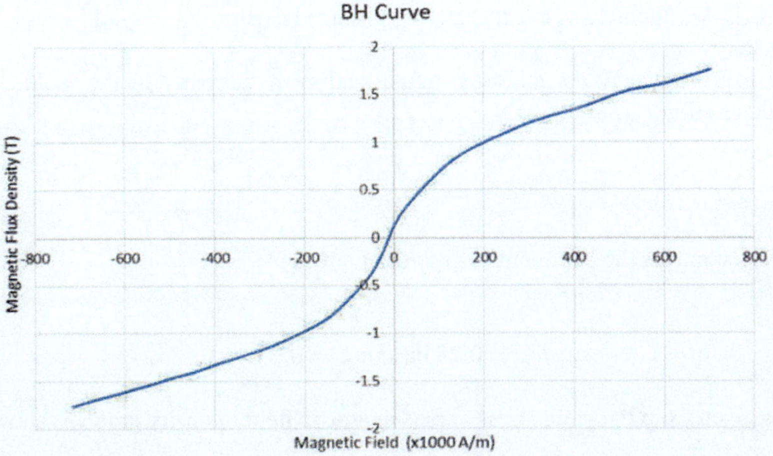

Fig. 2 B-H curve interpolated in COMSOL

laminar flow is as below. The maximum diameter of the damper, minimum viscosity and minimum Reynolds number of 2000 were considered for the calculation.

$$\text{Velocity } v = \frac{\mu \times \text{Re}}{\rho \times d} = \frac{0.28 \times 2000}{3450 \times 0.1} = 1.5819\,\text{m/s}$$

The maximum velocity observed in our study was 0.15 m/s. Therefore, laminar model is used.

Variation of Dynamic Viscosity with Magnetic Field

A simple laminar flow was chosen as the physics model. Dynamic viscosity versus magnetic flux density graph was put forward in [3]. This graph was used to specify the dynamic viscosity of the MR fluid with respect to magnetic flux density. The graph as a COMSOL interpolation function is given below.

Bingham-Papanastasiou Model

In the second method, COMSOL's built-in non-Newtonian fluid model was used. Inelastic non-Newtonian constitutive relation was chosen under fluid properties of the laminar flow study. Further, the Bingham-Papanastasiou inelastic model was chosen to represent the MR fluid. This model is proven to be able to closely approximate the ideal discontinuous model of MR fluids and describes fluid features accurately without adding unnecessary complications [4].

Here, viscoplastic fluid behaviour is characterized using yield stress, which is readily available from the LORD MRF 140-CG datasheet as a function of magnetic flux density. The constitutive relation is as follows:

$$\mu_{\text{app}} = \mu_p + \frac{\tau_y}{\dot{\gamma}}, |K| > \tau_y,$$

where μ_{app} = apparent viscosity, μ_p = plastic viscosity, τ_y = yield stress, $\dot{\gamma}$ = shear rate.

The following relation achieves regularization of viscosity in the yielded and unyielded region

$$\mu_{\text{app}} = \mu_p + \frac{\tau_y}{\dot{\gamma}}\left[1 - e^{-m_p \dot{\gamma}}\right],$$

where m_p controls the exponential growth of stress [4].

Here,

$$\mu_p = 0.28\,\text{Pa s};\ m_p = 0.25\,\text{s}$$

τ_y is given from the yield stress versus magnetic field intensity graph as shown in Fig. 3

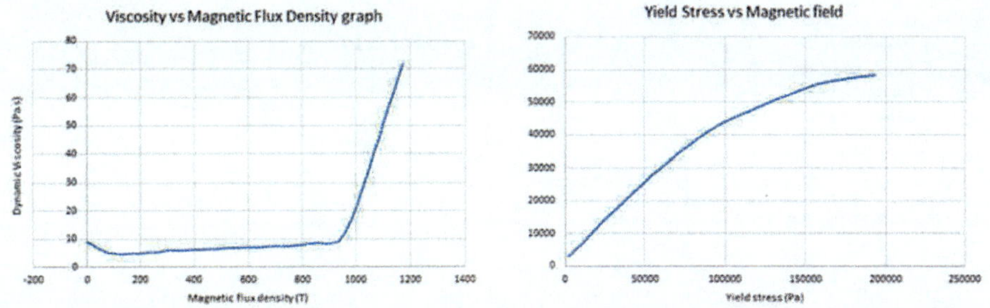

Fig. 3 **a** Dynamic viscosity versus B plot and **b** yield stress versus magnetic field intensity plot for MRF 140-CG plotted in COMSOL

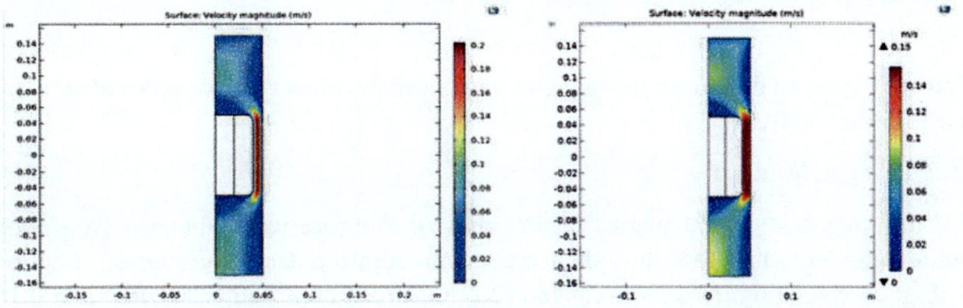

Fig. 4 **a** Dynamic viscosity with magnetic field model and **b** Bingham-Papanastasiou model

The Bingham-Papanastasiou model was chosen for further analysis. MR fluid shows non-Newtonian characteristics in the presence of a magnetic field, and this is taken into account by the inelastic non-Newtonian constitutive relations in COMSOL. The velocity profiles obtained through both methods are shown in Fig. 4. A faster change in velocity (smaller boundary layer) is observed in the Bingham-Papanastasiou model. The viscosity versus magnetic flux density graph obtained is calculated by comparing outflow time through a capillary tube to a reference fluid (glycerin). Factors like surface tension were not considered, which might have caused an apparent increase in viscosity value.

2.3 *Method of Study and Physics Used*

To obtain the required results, two physics models were required in COMSOL, magnetic fields (m_f) and flow physics [5]. For this model, laminar flow physics (spf3) was chosen and the mentioned non-Newtonian properties were introduced.

A grid independence test was conducted to get an optimized mesh setting. A free triangular mesh was utilized for the same. The results are shown in Fig. 5. As the mesh size decreases (towards left), the variation in output is observed to be negligible.

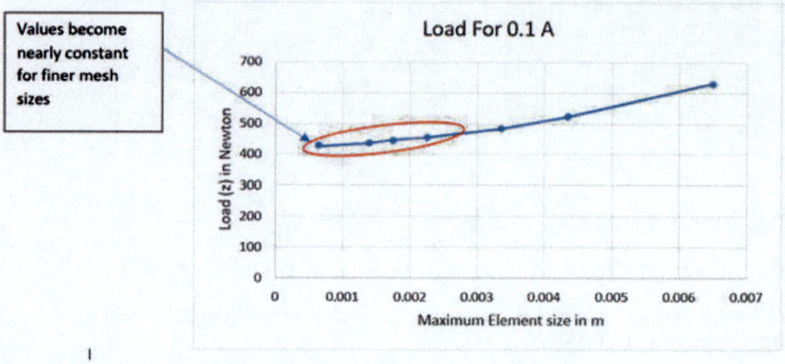

Fig. 5 Graph for the maximum element size versus load

The built-in extra fine mesh setting with a maximum element size of 0.00065 m was used for the study.

Magnetic Analysis

For the magnetic study, magnetic properties of the core were obtained from the built-in model, while MR fluid data was given from the dataset (Table 2). Region 2 (Fig. 1) was assigned as a homogenized multi-turn coil, with 700 turns and 0.1 A as the excitation current. The coil wire parameters were set as default, i.e. wire conductivity $= 6 \times 10^7 \frac{\text{S}}{\text{m}}$ and wire cross-sectional area $= 1 \times 10^{-6}\ \text{m}^2$.

Fluid Flow Study

The aim is to simulate a simplified model of MR fluid flowing over the region of influence of a magnetic coil in the piston. So, a stationary model was chosen. The uppermost boundary is chosen as an inlet with a fluid velocity of 5 cm/s. The lower boundary is an outlet with a gauge pressure of zero.

Fluid–Structure Integration

A significant result to be obtained from the model was the stresses and force on the piston. In order to obtain this, a fluid–structure interaction multiphysics study was also introduced. This simulates the piston as an elastic body and obtains the forces and stresses on it.

Initial Conditions

$$\text{For}\ t = 0,\ x \geq 0,\ y \geq 0\text{:}\ \begin{aligned} \text{Magnetic vector potential} &= 0\,\text{Wb/m} \\ \text{Velocity field}\ (u_r, u_z) &= 0\,\text{m/s} \\ \text{Pressure}\ (P) &= 0\,\text{Pa} \end{aligned}$$

Boundary Conditions

For $t \geq 0,\ x = 0$:		Axial symmetry
For $t \geq 0,\ x = 5\,\text{cm}$:		Wall boundary condition(for fluid flow)
		(No slip condition)
		Magnetic insulation
For $t \geq 0$,	Piston outer walls:	Wall boundary condition (for fluid flow)
		(No slip condition)
For $t \geq 0,\ y = 15\,\text{cm}$:		Magnetic insulation
		Inlet − velocity $(u_y) = -0.05\ \text{m/s}$
For $t \geq 0,\ y = -15\,\text{cm}$:		Magnetic Insulation
		Outlet − pressure $(P) = 1\ \text{atm}$

3 Results and Discussion

Before the simulation of the above model, a simpler model of fluid flow in a straight pipe was simulated. MR fluid (with properties as given in Sect. 2.2) was modelled to flow down a long pipe with magnetic field applied only at Sect. 2 (as shown in Fig. 6). A uniform flow was observed in Sect. 2. This shows that magnetic field changes the rheological properties of the fluid and thus can help control damping properties.

The results obtained from the model described in the previous sections is discussed in the following sections.

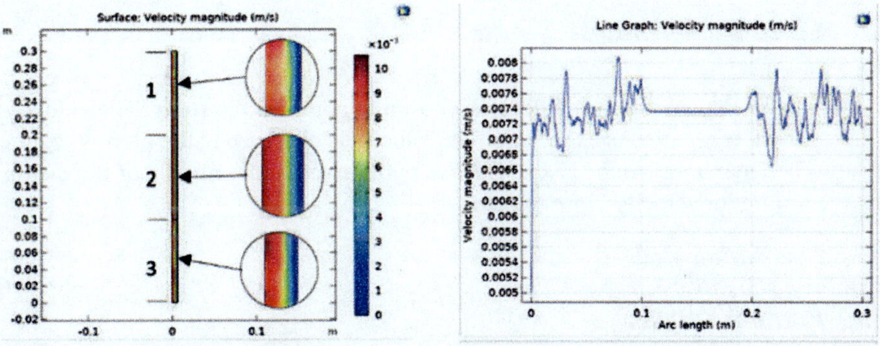

Fig. 6 Velocity profile with and without a magnetic field

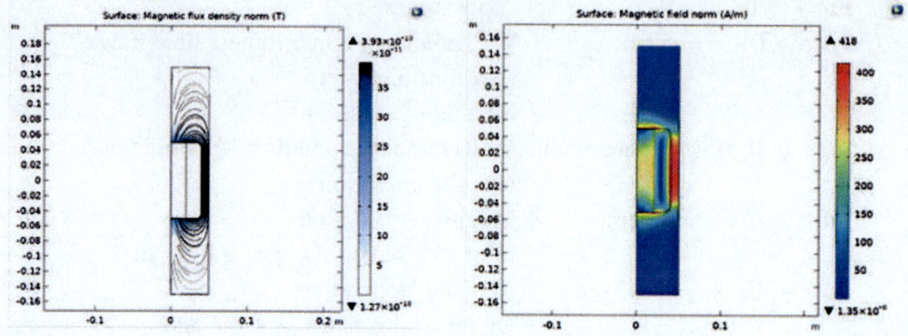

Fig. 7 **a** Magnetic flux density (B) and **b** magnetic field intensity (H)

3.1 Magnetic Analysis

Figure 7 shows a plot of magnetic flux density (B) and field intensity (H) in the damper model. The high H value in the annular gap gives higher resistance to the flow in this region. This allows us to control the damping force and associated properties.

A plot of the magnetic flux density shows the sudden drop of B in the MR fluid region. As shown in Fig. 7a, the maximum value is in the range of 10^{-10}. This might affect MR fluid's controllability and might call for an improvement in coil design and selection of MR fluids.

3.2 Fluid Flow Analysis

The Bingham-Papanastasiou model was chosen to continue the study. The velocity field is shown (Fig. 8a). As expected, high velocity is observed at the narrow region of flow. The maximum velocity of 0.15 m/s is observed at the middle of the piston and the minimum velocity close to zero.

3.3 Force Analysis

The effectiveness of a damper is obtained from the force applied on the piston. The load magnitude on the piston surface is plotted as a line plot to understand the load and its variation. The highest value of load magnitude is 1.56×10^4 N/m^2.

The total force was calculated as the integral of the z component of the load on the fluid–solid boundary. This was calculated for different values of current passing through the coil to understand the variation. As observed from Table 3, the load increases with an increase in current.

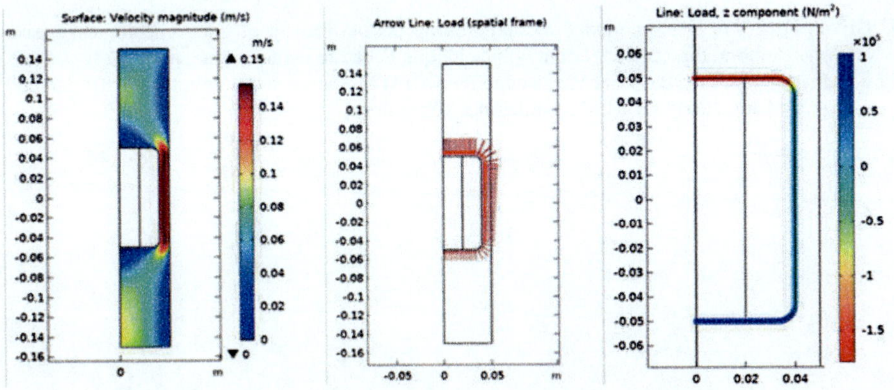

Fig. 8 **a** Velocity of flow and force on piston, **b** arrow line diagram and **c** line load diagram

Table 3 Variation of dependent variables with current in coil

Current (A)	Load (N)	Velocity (m/s)	Highest magnetic flux density (T)
0	–405.61	0.15	0
0.05	–414.78	0.15	2.57×10^{-4}
0.1	–423.65	0.15	5.14×10^{-4}

4 Conclusion

In this study, an attempt has been made to understand the dynamic behaviour of MR fluids in dampers. A detailed methodology on how to model MR fluids in a multi-physics software like COMSOL was discussed. It consisted of two methods, one using a simple laminar flow model and the other using the Bingham-Papanastasiou inelastic model. This modelled MR fluid was then simulated to understand varying parameters. Magnetic and fluid flow analyses were conducted in the model using COMSOL, and values of parameters like magnetic flux density, magnetic field intensity, velocity of flow and force on piston were obtained. The studies conducted will aid us in designing an MR damper suitable for application in a dynamic vibration absorber.

References

1. Genc S, Derin B (2014) Synthesis and rheology of ferrofluids—a review. Curr Opin Chem Eng 3:118–124
2. Lord Corporation: MRF-140CG Technical Data. www.lord.com. Last accessed on 27/22/2021
3. Roszkowski A, Bogdan M, Skoczynski W, Marek B (2008) Testing viscosity of MR fluid in magnetic field. Measure Sci Rev 8(3):58–60

4. Grivon D (2017) Design, modelling and sensing possibilities of magneto-rheological based devices. Doctoral dissertation, École Polytechnique Fédérale De Lausanne, pp 3–10
5. Khedkar YM, Joshi SB, Pawar PM, Ronge BP (2014) Modeling and simulation of MR damper using COMSOL: 2014 COMSOL conference, Bangalore

A Computational Study on B-Splines-Based Design Parameterization Strategy for Compressor Annulus for Throughflow Analysis

Mayuresh Damle and D. Arumuga Perumal

Abstract In this paper, a B-spline-based design parameterization strategy used for preliminary axial compressor design stage is presented. This strategy is critically important for multi-objective optimization problems with contrasting objectives for discovering better compressor configurations. It helps in reducing computation time drastically along with computation costs. In this case, B-splines are used for smooth parameterization of compressor annulus hub and tip lines. The target is to minimize discrepancies between original annulus contour and fitted data. Annulus geometry is modified by perturbing fitted annulus splines at required curve segments while maintaining necessary continuity conditions. Effect of geometry modifications on different flow parameters impacting compressor performance is also studied with the help of throughflow analysis based on streamline curvature method. Conducted investigations compare aerodynamic characteristics of fitted and modified annulus design with respect to the reference design. Additionally, by use of normalization approaches, the present method can be useful to compressor configurations with different length, number of stages and annulus contour.

Keywords B-splines · Parametric representation · Compressor · Annulus · Parametric continuity

1 Introduction

Over last few decades, due to globalization, air travel has become a major means of transportation. This increase in passenger kilometers is in conflict with concerns regarding reducing energy consumption and preserving existing resources. Along with increasing demands, there is need for addressing issues related with various significant elements including exhaust emission reduction and noise, safer air travel

M. Damle · D. A. Perumal (✉)
Department of Mechanical Engineering, National Institute of Technology Karnataka, Surathkal, Mangalore 575025, India
e-mail: perumal@nitk.edu.in

N. M. Sivaram et al. (eds.), *Advances in Manufacturing, Automation, Design and Energy Technologies*, Lecture Notes in Mechanical Engineering,
https://doi.org/10.1007/978-981-99-1288-9_49

and travel delays [1]. These arising concerns can be tackled by improving design technology for aero-engine production in terms of overall efficiency and minimum emissions.

Industrial aero-engine design comprises the iterative procedure of varying input parameters to an algorithm and analyzing the results with reference to the set of design objectives. This requires thousands of computers runs several weeks to months, when it performed manually. Modern technologies like expert systems, heuristics and machine learning allow computers to spend knowledge from design engineer along with the knowledge obtained after searching the design space using heuristics [2].

The compressor is the most vital and inspiring components within an aero engine, and it is responsible for 40–50% of weight, 50–60% of engine length and 35–40% of the manufacturing prices [3]. Throughflow analysis conventionally represents an important tool for turbomachinery design [4]. Cumpsty [5] investigated the implementation of throughflow analysis for study of primary flow in compressor annulus. Chen et al. [6] investigated rapid three-dimensional blade generation and blade parameterization of compressor rotor to improve adiabatic efficiency by 1.73%. Farin [7] presented the theory of Bézier curves and coon patches, later combined with B-spline methods. Bézier curves are commonly used by researchers because of the simplicity in computation of Bernstein polynomials. But it was found that, the use of Bézier curves with more than 10 control points is not recommended because they tend to oscillate with increasing number of control points.

Thus, for developing complicated parameterization strategies, B-splines can be used. Thus, parameterization and perturbation of strategy based on B-splines is devised for smooth modification of compressor annulus and its effect on aerodynamic properties is discussed in this paper.

2 Design Parameterization

As already mentioned, meanline evaluation process computes different aerodynamic and thermodynamic parameters such as stage pressure ratio and flow angles at the meanline of an annulus. In the next step of throughflow evaluation, annulus flow information is computed in radial direction of meridional plane using streamline curvature method. Essential parameters such as inlet Mach number, stage loading coefficient, diffusion factor and stage pressure rise exit whirl angles are obtained at the end of this stage which help making crucial design changes fairly early in design process. As these flow parameters have high impact on engine operation, they must be selected carefully.

Annulus geometry is described by (xH, rH) for inner hub and (xC, rC) for outer casing line. There are multiple ways by which parameterization can be done. In a first approach, either hub or casing line is parameterized and another spline defining radial thickness distribution $\Delta r\,(x) = rC(x) - rH(x)$ is parameterized. This approach requires a constraint so that radial thickness does not become negative. Still this approach is relatively easier to implement. In the second approach, a mid-height line rM (x) = (rC(x) + rH(x))/2 and a corresponding thickness distribution are parameterized. In the third approach, hub line and casing line are parameterized independently.

For current investigation, B-splines with $n + 1$ control points are used to parameterize the annulus geometry in two-dimensional space. B-splines are composed of piecewise polynomial functions defined within a particular curve segment [8]. Even though piecewise basis function computation is comparatively complicated than Bernstein polynomial calculation for Bézier curves, B-splines allow selection of number of control parameters irrespective of degree of spline and provide local spline control which helps in easier modification of geometry by perturbation of splines. Annulus lines parameterized by using B-splines are described as

$$\begin{pmatrix} x(t) \\ y(t) \end{pmatrix} = \sum_{i=0}^{n} N_{i,p}(t) P_i, \tag{1}$$

where B-spline representation in terms of x and y coordinates is computed using basis polynomials N_i, $p(t)$ with degree of spline p, number of control point $(n + 1)$ where control points P_i control the spline structure along with knot parameter $t \in [0, 1]$. It is important to note that in case of basis splines the $n + 1$ control points and the $m + 1$ polynomial degree are independent of each other [9].

In this case, universal method is used for parameter generation as it performs fairly well without complicating knot vector computation. Equidistant arrangement of knot might be acceptable, because in this case the data size (i.e., data points count) is not small, and data points are distributed uniformly across an annulus length [10]. Also, normalization of coordinates while calculations can avoid formation of loops during parameterization. Suppose the domain [0, 1] and $m + 1$ uniformly spaced knots are required. First and last knots must be 0 and 1 in order to pass the curve through first and last data points. Therefore, $t_0 = 0$ and $t_m = 1$. This can help in maintaining continuity at the end points of the curve which is similar to the working of Bézier curve. This can be achieved by increasing the multiplicity of first and last knot to $p + 1$ where p is degree of spline. Thus, we have

$$\begin{aligned} & t_0 = t_1 = \cdots = t_p = 0 \\ & t_{p+i} = \frac{i}{n-p+1} \quad \text{for } i = 1, 2, \ldots, n-p \\ & t_{m-p} = t_{m-p+1} = \cdots = t_m = 1. \end{aligned} \tag{2}$$

Therefore, a cubic spline would have first and last knot as 0 and 1 of multiplicity 4 and uniformly distributed interior knots. Then, knot vector for a clamped cubic B-spline with control points say 8 becomes {0, 0, 0, 0, 1/5, 2/5, 3/5, 4/5, 1, 1, 1, 1}.

Figure 1a illustrates the parameterization of annulus geometry by use of B-splines and subsequent parameter reduction. In order to determine position of control points for B-spline, combination of root-search methods and optimization strategies such as pattern-search method is used. Let $Y(t)$ be radial coordinates of the fitted curve and Y_i as the radial coordinate of contour points

$$S = \sum_{i=0}^{m} (Y(t_i) - Y_i)^2. \tag{3}$$

Pattern search method is used to search for the minimum value of S over a search domain. Because of their simplicity, pattern search methods are uncomplicated to use as primary optimization approach. Apart from their simplicity, these methods are adaptable and reliable [11].

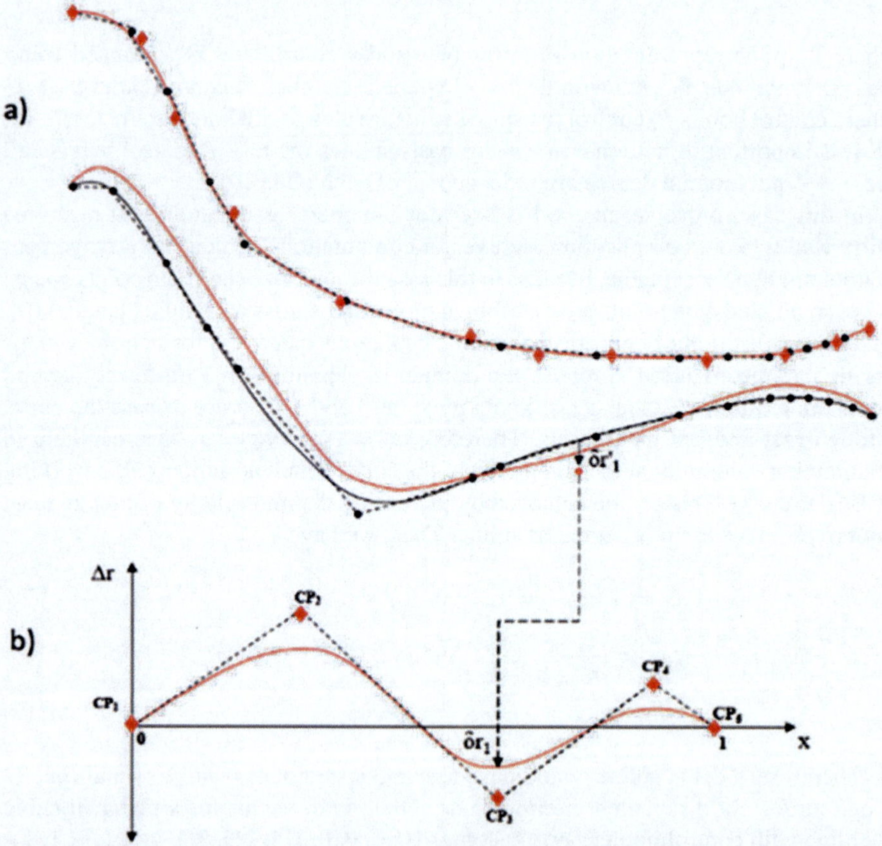

Fig. 1 **a** Parameterization and perturbation of annulus point cloud and **b** perturbation spline defined by 5 control points

Root search methods are used to determine the corresponding parameter value for axial coordinates of each data point on a B-spline. Once parameter value is obtained, radial coordinate can be obtained by using de Boor's algorithm and can be put in Eq. (3) for curve fitting [12]. The perturbation of fitted spline can be done by defining an additional perturbation spline. As perturbation spline has normalized dimensions, it is independent of the dimensions of an original annulus geometry. In case of fitted B-spline, number of control points depend upon required quality of fitting. Perturbation spline defines radial deviation Δr which is further to the place of fitted spline control points. It straight deviations the shape of annulus geometry. Thus, by moving control points of the perturbation spline, the annulus geometry can be modified. Figure 1b shows a perturbation spline with 5 control points which modifies fitted hub spline with perturbed spline (red hub spline in Fig. 1a). By modifying control point positions, variation of spline shape can be obtained which modifies original annulus points as B-spline curve, and the annulus points are correlated.

Annulus perturbation can be done over whole length of annulus or it can be done for certain length of annulus. On the other hand, local perturbation or the perturbation between any two required points on spline can be complex as it may raise continuity issues at endpoints of perturbation domain. Let curve segment 'A' has to be perturbed without modifying other two curve segments. One way to achieve this is by implementing subdivision algorithm so that fitted spline is divided into three separate B-spline segments. Therefore, another approach is followed in which required order of continuity is established first at the endpoints of perturbation domain by using fundamental B-spline properties of knot insertion and local modification of control points. After this step, perturbation spline is used to modify the perturbation domain without affecting other curve segments. In order to maintain C1 continuity at a point on the cubic spline, two neighboring control points should be equidistant from the point (which is also made control point by knot insertion), i.e., $P_1^* = P_d + (P_d - P_d - 1)$ as shown in Fig. 2.

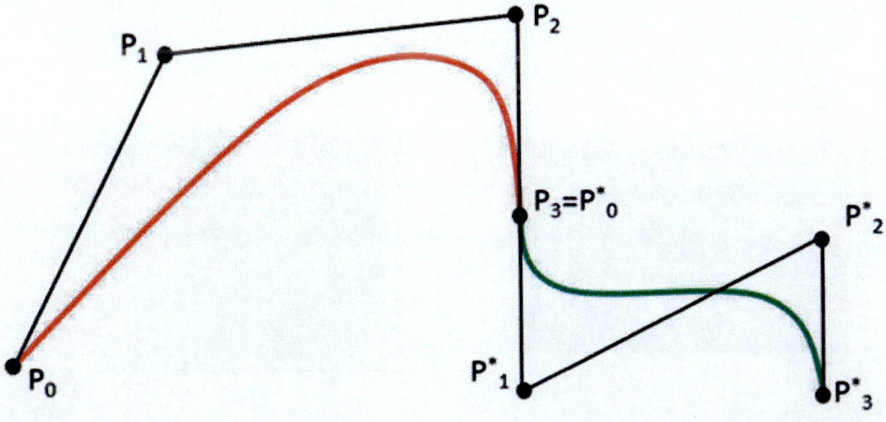

Fig. 2 C^1 continuity

In order to obtain smoother splines after modification, parametric continuity of second order, i.e., C^2 continuity, can be defined at end points of perturbation. C^2 continuity means first and second derivative do not change magnitude and direction when crossing one curve to the other. For maintaining curvature continuity or C^2 continuity at a point (which is also made control point by knot insertion) on a cubic B-spline, neighboring control points has to be maintained in such a way that $P_2^* = P_{d-2} + 4 \times (P_d - P_{d-1})$.

3 Results and Discussion

In order to test functionality of this strategy, the proposed annulus parameterization and perturbation process is integrated in throughflow design environment. Fitted annulus geometry by parameterization is considered as a reference design. In this paper, annulus geometry with perturbed hub spline over whole length as an initial design is compared with globally (over whole length) perturbed tip spline geometry along with reference design.

For better understanding of geometry modification on aerodynamic parameters, different flow parameters are analyzed using throughflow analysis. Figures 3a, 4a, 5a and 6a show the rotor relative inlet Mach number, stator inlet Mach number, rotor de Haller number (DH) and stage loading (Ψ) distribution for axially and radially normalized annulus geometry for initial (above), reference (middle) and modified (below) compressor geometry configurations. Figures 3b, 4b, 5b and 6b on the right-side show difference between initial (upper) and modified (lower) designs with relation to the reference design, respectively. The present results prove the performance of annulus parameterization and perturbation process.

The current geometry modifications give insight of mass distribution, as it is a significant criterion for complete design of an aero engine.

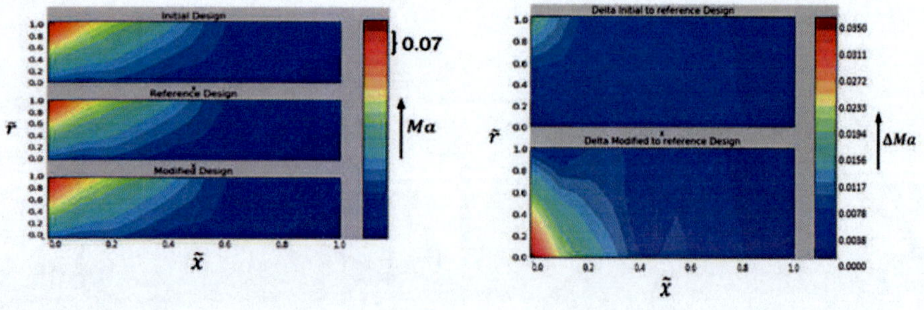

Fig. 3 Throughflow plots for annulus perturbation of a reference to a new compressor design in **a** rotor relative inlet, **b** abs. deltas for rotor relative inlet Mach number

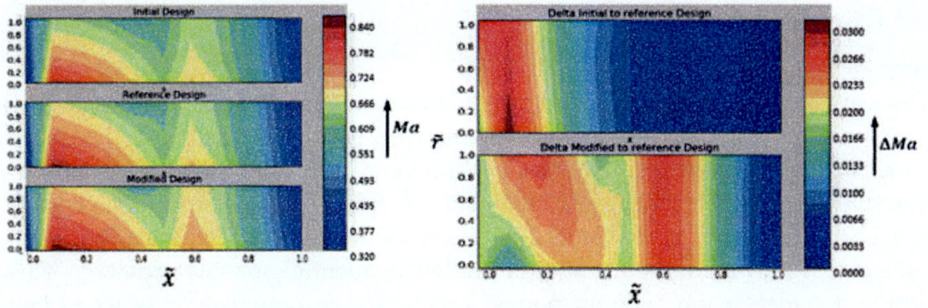

Fig. 4 Throughflow plots for annulus perturbation of a reference to a new compressor design shown in **a** stator inlet and **b** absolute deviation for stator inlet Mach number

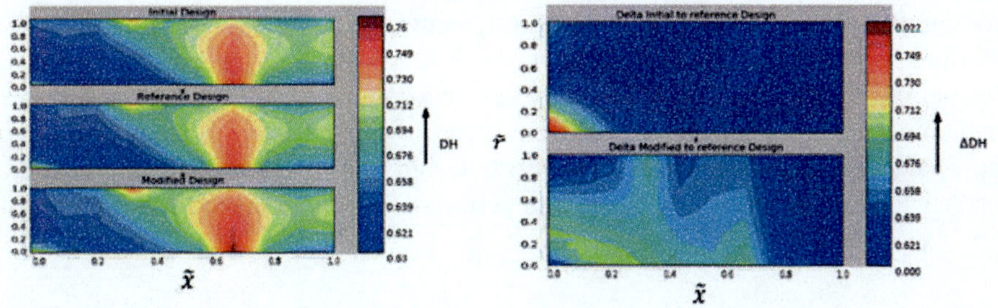

Fig. 5 Throughflow contour plots for annulus perturbation of a reference design to a new compressor design shown in **a** rotor de Haller number and **b** absolute deviation for rotor de Haller number

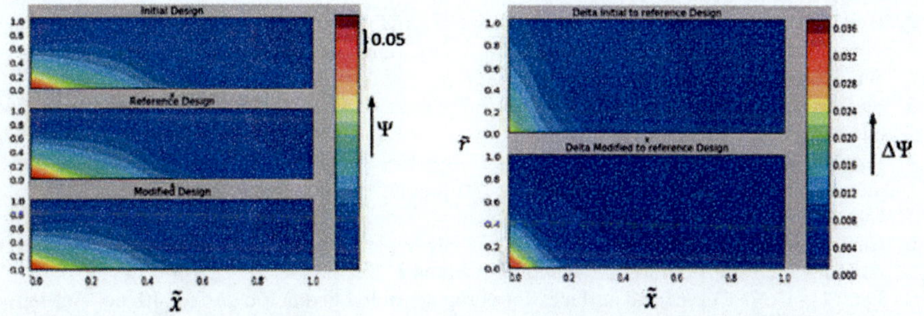

Fig. 6 Throughflow contour plots for annulus perturbation of a reference design to a new compressor design shown in **a** rotor stage loading and **b** absolute deviation for rotor stage loading

4 Conclusion

The present paper puts forward a design parameterization approach based on B-splines for axial compressor annulus geometry. Smooth annulus geometries can be obtained along with reduction of design parameters which helps in further optimization studies. Alternate strategy of modification of fitted splines by defining separate perturbation spline is used. Control points of this spline act as new design parameters and as they are normalized, they are independent of annulus geometry. These design parameters (i.e., control points) can be fed to optimizer in order to automate the process instead of manually implemented conventional methods. Also, this parameterization strategy can be extended to different compressor design parameters such as rotor/stator exit whirl angles and stage pressure ratio. A general approach for comparison is used to check the aerodynamic matching between reference and modified compressor for relevant flow parameters such as stage loading, diffusion factor or Mach number. This can finally lead to a complete improvement of the aerodynamic features according to the reference aerodynamic behavior. These changes in flow parameters can be fed back to previous design steps to make early modifications in design which can save overall design period for new compressor configurations.

References

1. ACARE, European Aeronautics: A Vision for 2020. Advisory Council for Aeronautics Research in Europe (2001)
2. Holt G, Bassler S (1991) Preliminary design of axial compressors using artificial intelligence and numerical optimization techniques. In: International gas turbine and aero engine congress and exposition Orlando FL
3. Steffens K, Schaffler A (2000) Engine compressors-key technology for success in aviation engines. DGLR Annual Meeting
4. Pacciani R, Marconcini M, Arnone A (2017) A CFD-based throughflow method with three-dimensional flow features modelling. Int J Turbomach Propul Power 2(4):11
5. Cumpsty NA (2004) Compressor aerodynamics. Krieger, Malabar
6. Chen N, Zhang H, Xu Y, Huang W (2007) Blade parameterization and aerodynamic design optimization for a 3D transonic compressor rotor. J Thermal Sci 16(2):105–114
7. Farin G (1997) Curves and surfaces for computer aided geometric design, 4th ed. Academic Press, Inc.
8. Schönberg IJ (1946) Contributions to the problem of approximation of equidistant data by analytic functions. Q Appl Math 4(1):45–99
9. Piegl L, Tiller W (1997) The NURBS book, 2nd edn. Springer, Berlin
10. Likhachev DV (2017) Selecting the right number of knots for B-spline parameterization of the dielectric functions in spectroscopic ellipsometry data analysis. Thin Solid Films 636:519–526
11. Lewis RM, Torczon V, Trosset MW, William C (2000) Direct search methods: then and now. J Comput Appl Math 124:191–207
12. Lu L, Zhang L, Gu Y, Zhao J (2018) Fast parametric curve interpolation with minimal feedrate fluctuation by cubic B-spline. Proc Inst Mech Eng Part B J Eng Manuf 232(9):1642–1652

Magnetic Analysis of Magnetorheological Brake with Multiple Conductor Paths Using COMSOL

Peri Krishna Karthik, Kotipalli Hemanth Harsha, Natla Vishnu Vardhan Reddy, and T. Jagadeesha

Abstract Magnetorheological fluids are a class of smart fluids that have the potential for many automobiles, aerospace, human prosthetics, military and defence applications due to their property of changing viscosity under the application of an external magnetic field. Extensive research is taking place on magnetorheological brakes. Conventional magnetorheological brakes use the shear mode operation of magnetorheological fluid to produce the required braking torque when the fluid is excited by a magnetic field created by a current-carrying conductor. A comparative study was done to analyse the effect of having multiple current-carrying conductors on the magnetic field induced. A conventional disc-type magnetorheological brake was considered for the simulation. Simulations were carried out with a varying number of current-carrying conductors, and the corresponding magnetic field intensity in the magnetorheological fluid and the torque produced were compared. The simulations were carried out using COMSOL Multiphysics software. This work would give an insight into optimizing the conventional design of the magnetorheological brake.

Keywords Magnetorheological fluid · Magnetorheological brake · COMSOL Multiphysics

1 Introduction

Magnetorheological (MR) fluid is a non-Newtonian fluid, which has the property of changing its viscosity when there is a change the magnitude of magnetic field intensity applied on it. The viscosity of this fluid increases as there is a increase in the magnetic field intensity. This property makes this liquid very useful for many purposes. These MR fluids are widely used in prosthetics, vehicle suspensions these days.

P. K. Karthik · K. H. Harsha · N. V. V. Reddy · T. Jagadeesha (✉)
Department of Mechanical Engineering, National Institute of Technology Calicut, Calicut 673601, Kerala, India
e-mail: jagdishsg@nitc.ac.in

N. M. Sivaram et al. (eds.), *Advances in Manufacturing, Automation, Design and Energy Technologies*, Lecture Notes in Mechanical Engineering,
https://doi.org/10.1007/978-981-99-1288-9_50

There are three main constituents in the MR fluid. They are magnetic particles, base fluid and surfactants or additives. The magnetic particles used are the pure carbonyl powder particles which are the reason for the change in the viscosity of the fluid when there is a change in the magnetic field. These magnetic particles are suspended in a carrier fluid, which is a non-magnetic fluid, having preferably low viscosity, low density and low permeability. And to prevent the sedimentation and coagulation of the particles, surfactants and additives are also added to the liquid.

Moreover, there are three different modes of applications in MR fluids. They are namely, flow mode, shear mode and squeeze mode. But, most of the MR brake applications use shear mode. A typical MR brake has a disc or drum connected to the shaft and surrounded by a film of MR fluid. And, an electric coil is wound around this setup. Electricity is passed through the windings. The required magnetic field is generated under the action of which chain-like structures are formed inside the MR fluid, creating mechanical resistance in opposition to the shearing surfaces. So, when the brakes need to be applied, the coils are energized, and the magnetic field intensity in the MR fluid increases. Then the viscosity of this MR fluid also increases rapidly. This increases the yield strength of the fluid. When the disc tries to rotate in the MR fluid, it opposes the motion by applying shear stress to the disc surface. So, this shear stress gives braking torque required to stop the disc. The torque generated is dependent and can also be instantaneously and continuously regulated via the magnetic field applied. The impact of changing the number of current-carrying conductors on the magnetic field created in the MR fluid is investigated in this work.

2 Literature Review

For a mid-sized passenger vehicle, Nguyen and Choi [1] offered an optimal design of an MR brake as an alternative to a normal hydraulic brake. It states an idea of making the whole disc with silicon steel as an alternative for employing a non-magnetic hub amid the disc's teeth. This change was significant with respect to the manufacturing efficiency and accuracy of the brake. Nguyen and Choi [2] proposes a T shaped magnetorheological brake that can substitute the usual drum brake in a mid-sized bike. They have done two optimizations, the first being maximizing the braking torque and the second being minimizing the weight. It was concluded that the MR fluid MRF-132DG is suitable for obtaining the optimal braking results. Nguyen et al. [3] proposed a disc-type magnetorheological brake for an average sized bike. It is an optimized design that focuses on reducing the weight of the brake. It was claimed that raising the current to a greater value (3A here) causes the braking torque to achieve saturation. Thus, the current applied should be limited to that value. Sohn et al. [4] is an experimental evaluation of the performances of the designed MR brake. The model is optimized and fabricated, stating the methodology to be followed. This helps to understand the real-time performance of the brake and the deviations that can be expected from the simulated results. Guoliang et al. [5] propositioned a double-disc magnetorheological brake with a higher torque to volume ratio than many other MR brake models. It has also shown that changes in the rotational speeds have almost zero effect on the braking performance at fixed applied current. This characteristic reduces the design complication of the control strategy of the MR brake and broadens its application fields. Hu et al. [6] proposes a rotary MR brake with several fluid flow passages. It demonstrates that placing a non-magnetic ring in the middle of the magnetic conduction sleeves improves torque performance by directing the magnetic flux to the exterior fluid flow channel and increasing the effective damping gaps from two to four sections. In order to stimulate more shearing areas within a confined volume, Huanhuan Qin et al. [7] presented a new multi drum design. This paper also taught about the Ferro fluidic sealing approach, which employs the fluid itself as a sealing medium.

3 Designing and Modelling

3.1 Geometric Modelling

The basic design of a conventional disc-type magnetorheological brake used in the simulation analysis is shown in Fig. 1. The outer casing and the rotor disc, which is the core, is constructed of highly magnetic materials. The magnetorheological fluid is filled in the cavity amid the housing and the rotor disc. The copper coil is wound around the fluid gap and is covered by a bobbin made of material with very low permeability.

To ease the computational load, the 2D axisymmetric model, Fig. 2, is developed in COMSOL. Three different geometries were modelled with varying numbers of current coil paths, as shown in Fig. 3.

The coil case is divided equally into two and three parts for the two and three paths model, respectively.

3.2 Magnetic Analysis in COMSOL

COMSOL Multiphysics software was chosen for the simulation study because of its seamless integration of multiple physics modules and ease of Multiphysics which would be helpful in further studies. After the geometrical construction, the materials are assigned to the domains as mentioned previously. The material properties are also defined. For the MR fluid and magnetic steel core, the B-H curves are defined in COMSOL material data, and for the non-magnetic bobbin material, permeability value is assigned.

The magnetic fields physics is setup. Three separate Ampere's laws are defined each for MR fluid, rotor disc and housing and the bobbin, respectively. For the MR fluid and magnetic core, the magnetization model uses the B-H curves, Fig. 4,

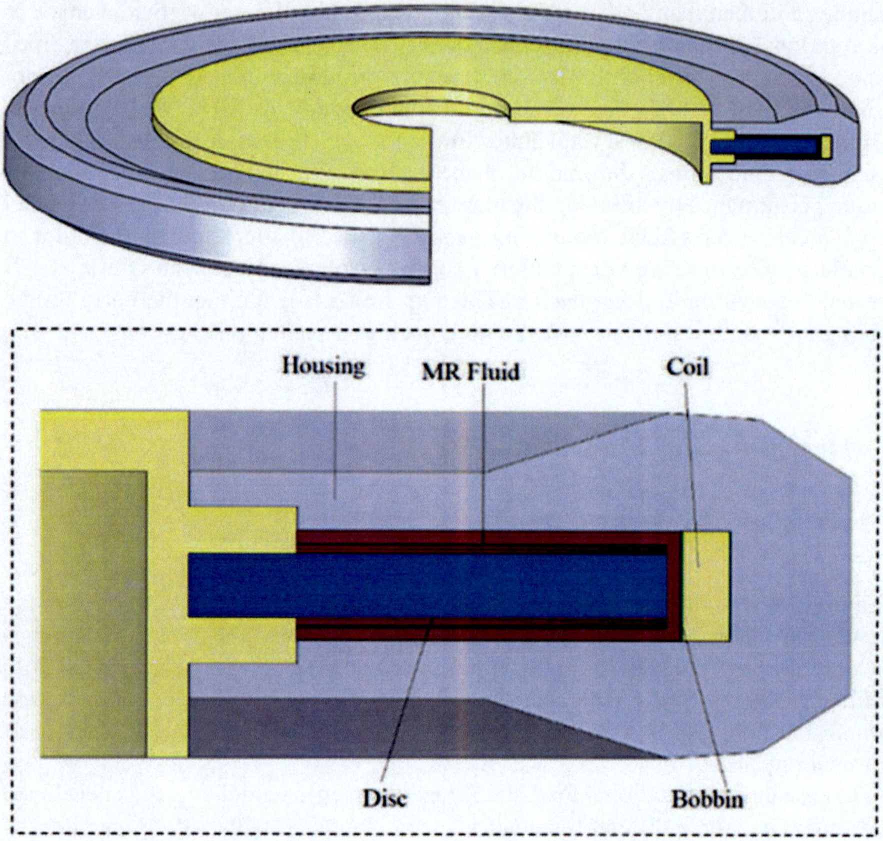

Fig. 1 Schematic design of MR brake

Fig. 2 2D-axisymmetric model of MR brake in COMSOL with single conductor path

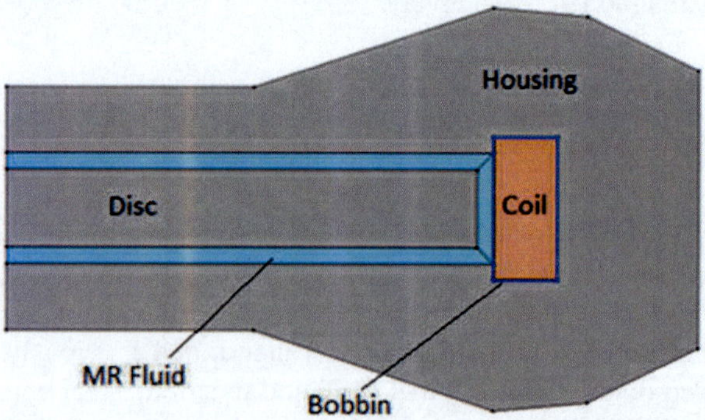

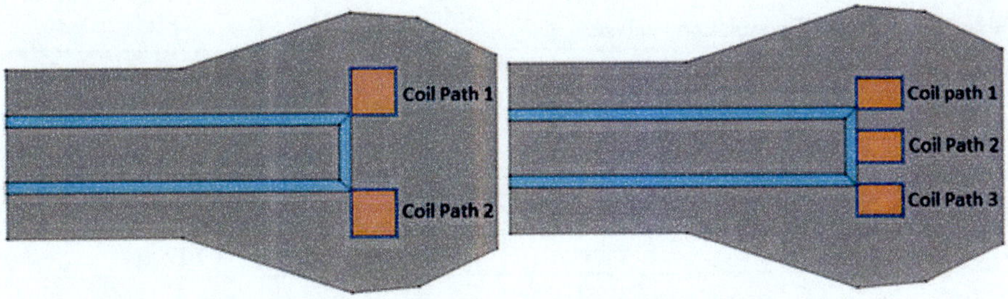

Fig. 3 Axisymmetric models of two (left) and three (right) conducting paths

for the materials which are defined in the material library. For bobbin material, the magnetization model uses the relative permeability model assigned. Since the bobbin is very thin, it is modelled using the Thin Low Permeability Gap option in magnetic physics with a thickness of 0.5 mm.

In the case of the single-coil setup, the coil has 136 turns of AWG 24 wire, and the current passed is 1A. For the two-coil setup, the number of turns is divided equally, and the same is the case with three coil setups. The current in each of the setups through each coil remains constant, i.e. 1A, to maintain equal power consumption in all three configurations.

The meshing parameters were automatically chosen by the software based on the physics. The physics induced sequence has triangular elements over the entire domain with a maximum and minimum element size of 1.55 mm and 5.25 μm, respectively.

A stationary setup was setup, which follows the Newtonian nonlinear method for solving, and the termination technique is based on iterations or tolerance with the maximum number of iterations as 100.

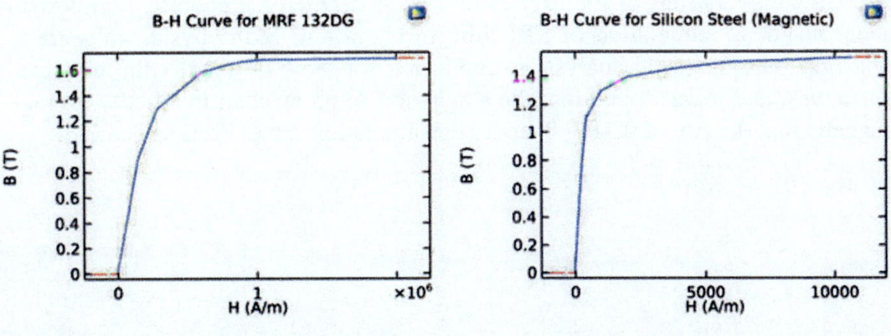

Fig. 4 B-H curves for MRF 132DG (left) and Magnetic Silicon Steel (right)

Table 1 Results from simulation analysis

Configuration	Average magnetic flux density in MR region (T)	Current through each of the coil (A)	Power for each coil (W)	Total electrical power consumed by the setup (W)
Single coil setup	0.3167	1	9.12	9.12
Two coil setup	0.1603	1	4.56	9.12
Three coil setup	0.1154	1	3.04	9.12

4 Results and Discussion

With the setup as mentioned earlier, the simulations were carried out for three different setups. A domain probe was also setup to find the average magnetic flux density in the magnetorheological fluid region.

From the simulation analysis, a bar graph showing the average magnetic flux density in each setup, Fig. 6, was plotted, and it can be found that the average magnetic flux density decreases as the number of conducting paths increases. Since all other geometrical parameters remain constant, the braking torque generated, which depends on the magnetic flux density, also decreases. The results obtained are consolidated in Table 1. From the magnetic flux density plots, Fig. 5, it can also be seen that there are no or very few magnetic lines passing through the gap in between the coil arrangement in two and three coil setups.

5 Conclusion

In this study, an attempt was made to understand the behaviour of MR brake under different configurations of the excitation coil. A detailed methodology on how to model magnetic stimulation of MR fluid in COMSOL Multiphysics software is discussed. The magnetic analysis is conducted for three different coil configuration setups, and it was found that the single-coil setup resulted in a better average magnetic flux density of 0.3167 T under the simulation setup discussed earlier.

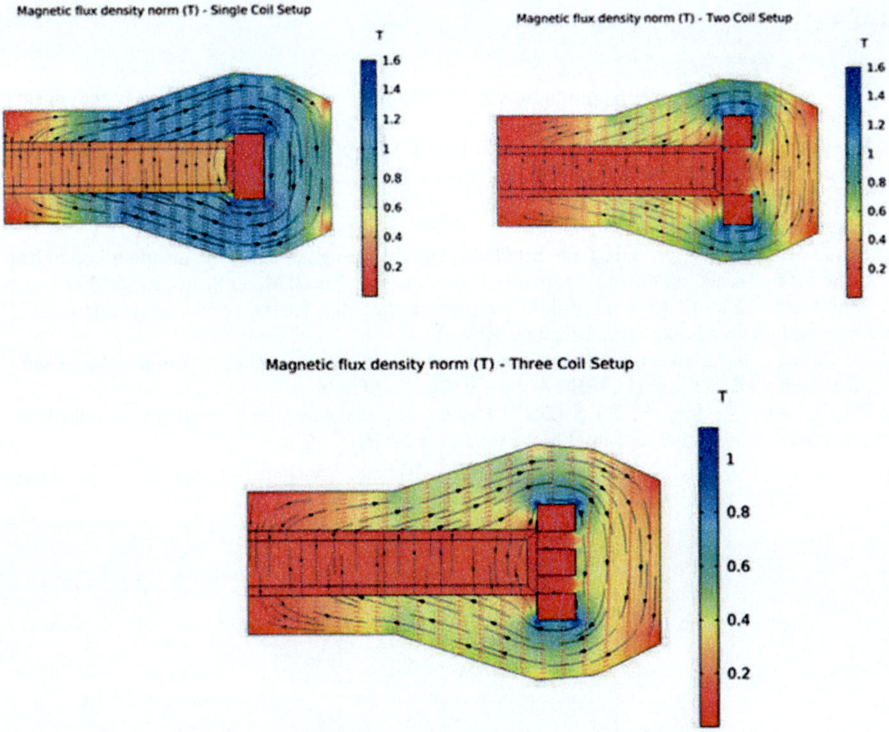

Fig. 5 Magnetic flux density plots for single, two, three coil setup MR brake

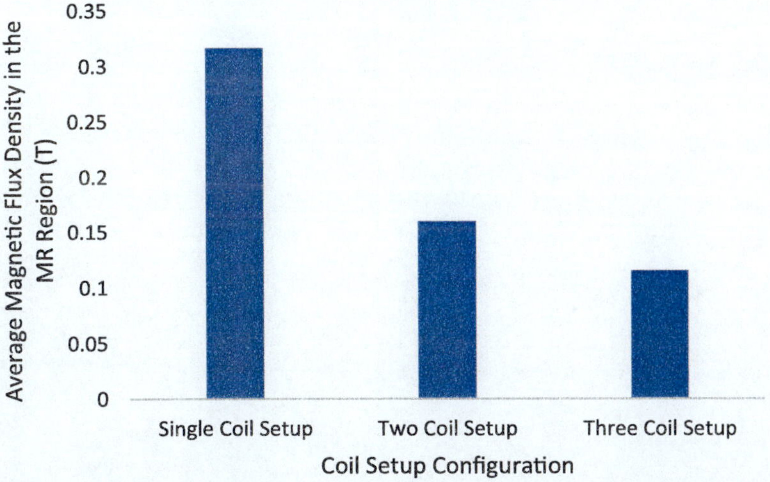

Fig. 6 Variation of average magnetic field w.r.t coil setup

References

1. Nguyen QH, Choi SB (2010) Optimal design of an automotive magnetorheological brake considering geometric dimensions and zero-field friction heat. Smart Mater Struct 19:115024–1–24
2. Nguyen QH, Choi SB (2012) Optimal design of a novel hybrid MR brake for motorcycles considering axial and radial magnetic flux. Smart Mater Struct 21:055003
3. Nguyen QH, Jeon J, Choi SB (2011) Optimal design of a disc-type MR brake for middle-sized motorcycle. Proc SPIE 7977:797717
4. Sohn JW, Jeon J, Nguyen QH, Choi SB (2015) Optimal design of disc-type magnetorheological brake for mid-sized motorcycle: experimental evaluation. Smart Mater Struct 24:085009
5. Guoliang H, Lifan W, Linsen L (2021) Torque characteristics analysis of a magnetorheological brake with double brake disc. Actuators 10(2):23
6. Hu G, Wu L, Li L, Yu L (2020) Performance analysis of rotary magnetorheological brake with multiple fluid flow channels. IEEE Access 8:173323–173335
7. Qin H, Song A, Zeng X, Hu S (2018) Design and evaluation of a small-scale multi-drum magnetorheological brake. J Intel Mat Syst Struct 29:2607–2618

Finite Element Analysis of Silicon Heat Sink with CNT Nanofluids for Microelectronics Applications

M. Appadurai, E. Fantin Irudaya Raj, S. Darwin, and I. Jenish

Abstract The recent development in power electronics and minimizing the size of electronic components requires an effective cooling system for the system's existence. Conventional electronic cooling methods are not preferable for high-temperature applications since it does not maintain the electronic components in low temperature at all boundary conditions. Nanofluids are used as the base fluid in modern devices since it has higher thermal conductivity. The carbon nanotube nanoparticles are the recent innovation in the materials that have better performance even at higher temperatures. This paper presents the numerical analysis of silicon heat sinks used in high-temperature applications. The water is used as the coolant in the heat sink. The effect of integration of wavy and rectangular fin performance on different velocity flow rates are studied. The numerical analysis is done for various Reynolds numbers varying from 100 to 350. The wavy heat sink provides better performance on various Reynolds numbers. The performance of carbon nanotube nanofluids as wavy heat sink coolant is further studied. The analysis proved that the wavy fin integrated heat sink with 0.03 volume fraction nanofluids has the highest Nusselt number on all the boundary conditions.

M. Appadurai (✉)
Department of Mechanical Engineering, Dr. Sivanthi Aditanar College of Engineering, Tiruchendur, India
e-mail: appadurai86@gmail.com

E. Fantin Irudaya Raj
Department of Electrical and Electronics Engineering Engineering, Dr. Sivanthi Aditanar College of Engineering, Tiruchendur, India
e-mail: fantinirudhayaraj@drsacoe.com

S. Darwin
Department of Electronics and Communication Engineering Engineering, Dr. Sivanthi Aditanar College of Engineering, Tiruchendur, India
e-mail: darwin@drsacoe.com

I. Jenish
Department of Applied Mechanics, Seenu Atoll School, Addu City, Maldives
e-mail: jenish@satollschool.edu.mv

N. M. Sivaram et al. (eds.), *Advances in Manufacturing, Automation, Design and Energy Technologies*, Lecture Notes in Mechanical Engineering,
https://doi.org/10.1007/978-981-99-1288-9_51

Keywords Silicon heat sink · ANSYS fluent · Electronic cooling · Nusselt number · Fin · Nanofluid

1 Introduction

The revolution in power electronics increased the performance of electronic devices used in various applications. The current electronic devices are compact in size, reliable and low price. The miniature size of the electronic devices required effective cooling methods to prevent the devices from high temperature. Several innovations in the electronic cooling methods help to maintain the electronic device at low temperatures.

Ebu [1] did the finite element analysis of a heat sink integrated with several types of fins for microelectronic applications. The thermal efficiency of the difference in fine shape such as rectangular, circular, elliptical, triangular and hexagonal fin is calculated. The two-layered heat sink design is optimized by Vafai [2]. They concluded that the temperature rise at the high-temperature applications is rapidly reduced and the pressure drop is minimum in this microchannel heat sink. Tateishi [3] fabricated the 3D printed heat sink for electronics application. The newly fabricated 3D printed heat sink manufacturing cost is higher when compared to the conventional manufacturing process, but the thermal resistance is reduced to 15% when compared to the same.

The heavy compact electric motors [4] used in a hybrid [5, 6] and electrical vehicles required proper fin design to dissipate the waste heat [7] to the surroundings. Several authors optimized the geometrical parameters by the neural network [8], deep [9, 10] or machine learning methods. The machine learning algorithm [11] gave a feasible solution among several physical variables. Chen [12] did the numerical simulation on the CPU heat sink. They optimized the various geometrical parameters by the genetic algorithm. The optimization algorithm sets the objective as the possible minimum entropy generation.

Koga [13] did the topological optimization in the heat sink physical parameters. The width, length and breadth of the heat sink channel are optimized by sequential linear programming. The feasible optimized model is identified for heat sink manufacturing. Appadurai [14] did the numerical analysis [15] on induction motor [16] external axial fins used in hydraulic applications. Multi-objective optimization is done to find the optimal number of fins, fin length and height. The finite element analysis [17] provided monetary benefits [18] by reducing the real-time experimentation methods [19]. Levin [20] proposed the heat sink is integrated with fin and phase change material. The numerical simulation is made to find the optimal parameters on integration with phase change material and fins.

2 Numerical Analysis of Heat Sink

The 3D model of the heat sink is modelled in ANSYS workbench and its dimensions are 20 mm × 20 mm. The heat sink is divided into 40 rectangular ducts. The wall thickness between the ducts is 0.4 mm. Figure 1 illustrates the 3D model of the microchannel silicon heat sink. The heat sink is square in shape and constantly absorbs the heat from the electronic devices. The nanofluids that pass through the heat sink continuously absorb the heat from the sink. The heat sink has 40 rectangular ducts which are the same in dimensions. Due to the shape similarity, one duct is modelled and symmetrical temperature distribution in the flow direction, only one duct is considered for numerical analysis. The fin arrangements inside the duct are in the form of the wavy fin (WF), rectangular (RF) and no fin (NF) condition is considered in this analysis. The volume fraction of nanofluid is varied from 0 to 3%.

The heat sink is meshed into small size finite nodes. The grid independence study is made to find the correct finite number of nodes for numerical analysis. The 86,457 nodes are found to be the optimum number of nodes from model discretization. The mesh element is fully hexahedral in the solid region as well as in the fluid domain. The mesh element is uniformly swept along the flow direction. The discretization study is done at the 250 Reynolds number.

This numerical simulation of a microchannel heat sink is a conjugate heat transfer type problem. The SIMPLE type of solution method is given in the pressure velocity coupling. The analysis is done in the ANSYS Fluent. The momentum and energy equations are resolved in a second-order discretization method. The velocity inlet boundary condition is given in the water inlet and the water inlet temperature is considered as 300 K. The pressure outlet condition of value zero-gauge pressure is given in the water outlet. A constant heat flux of 120 W/cm^2 is applied on the bottom of the heat sink and the no-slip condition is applied on the inner wall. All wall thickness is the same and is made of silicon material. The remaining walls are

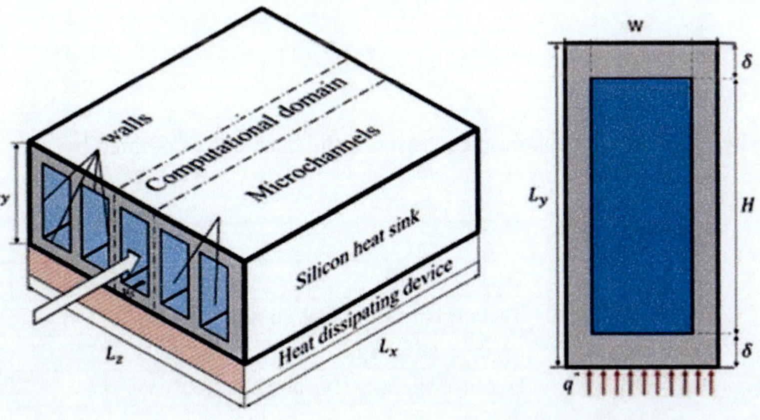

Fig. 1 Heat sink

considered as adiabatic conditions. The residual value of 10–6 is deemed for the momentum, energy and continuity equations.

3 Themo-Physical Properties of Nanofluid

Table 1 shows the thermo-physical properties of materials. The ratio of the density of nanofluid to that of base fluid is given by

$$\frac{\rho_{nf}}{\rho} = 1 + k_\rho \varphi \text{ where } k_\rho = \left(\frac{\rho_s}{\rho} - 1\right) \tag{1}$$

The thermal conductivity of nanofluid increases with increasing particle volume concentration. The ratio of thermal conductivity of nanofluid to that of base fluid is given by

$$\frac{k_{nf}}{k} = 1 + k_k \varphi \text{ where } k_k = 3 \tag{2}$$

The specific heat of the nanofluid is calculated as

$$\frac{C_{p,nf}}{C_p} = \frac{1 + k_c \varphi}{1 + k_\rho \varphi} \text{ where } k_c = \left(\frac{\rho_s C_{p,s}}{\rho C_p} - 1\right) \tag{3}$$

The effective viscosity of a fluid of lower volume concentration is calculated as

$$\frac{\mu_{nf}}{\mu} = 1 + k_\mu \varphi \text{ where } k_\mu = 2.5 \tag{4}$$

The Reynolds number is calculated by the equation

$$Re = \frac{\rho u D}{\mu} \tag{5}$$

The hydraulic diameter of the each slot of the heat sink is defined by

Table 1 Thermo-physical properties

	Water	Silicon	CNT
Density (kg/m^3)	997	2230	1350
Thermal conductivity (W/m K)	0.613	148	3000
Specific heat (J/kg K)	4179	710	3910
Dynamic viscosity (N s/m^2)	0.000855	–	–

$$D_h = \frac{4A}{P} = \frac{4ab}{2(a+b)} \tag{6}$$

The average heat transfer co-efficient for the convective heat transfer between the heat sink and coolant is calculated by

$$h = \frac{q}{T_w - T_m} \tag{7}$$

The mean temperature of the working fluid is calculated by

$$T_m = \frac{T_i - T_o}{2} \tag{8}$$

The average Nusselt number is given by

$$Nu = \frac{hD}{k} \tag{9}$$

The thermal resistance is given by

$$R = \frac{T_w - T_i}{q} \tag{10}$$

4 Results and Discussion

While flowing through the heat sink channel, the temperature of the water is gradually increased along the flow direction as identified from the results. The highest temperature is shown in the outlet point. The outlet temperature of no fin condition is minimum since it has a higher thermal resistance. The other case such as wavy fin and fin has lower thermal resistance due to the turbulence of the fluid due to fins. Figure 2 represents the velocity magnitude of all fin-type heat sinks at the middle plane in the x-direction. The velocity distribution of no fin condition reached the top section of the heat sink since the flow sensed the least resistance on its path. The wavy fin heat sink has uneven flow in its channel due to the presence of fins. The path blockage effect is induced and increases the pressure drop [21–23].

Figure 3 shows the temperature distribution of all fin-type heat sinks at the middle plane in the x-direction. The results presented that the wavy fin-type heat sink has better results. From the streamline contour, it is identified that the no fin boundary condition has minimum flow resistance and so the pressure drop is minimum [9, 24–26]. Figure 4 denotes the Nusselt number with respect to Reynolds number for different fin configurations. As the Reynolds number is increased the Nusselt number is enhanced is clearly correlated from the illustrated graph. The Nusselt number is

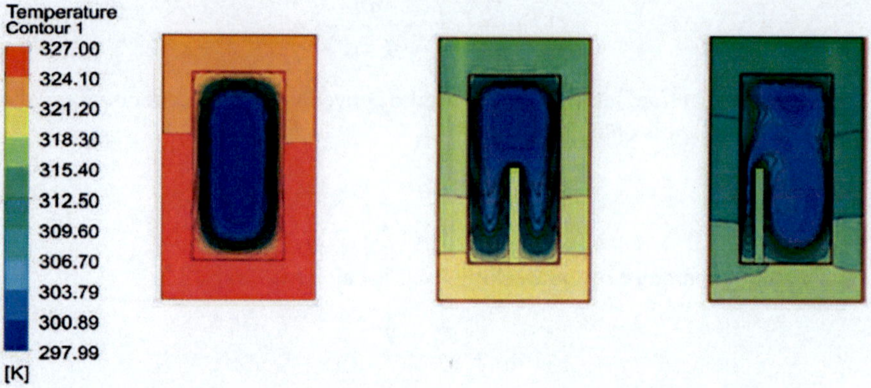

Fig. 2 Temperature distribution of all fin-type heat sink at a middle plane in x-direction

the convective heat transfer factor which determined the useful heat transfer rate [27–29].

Now the wavy fine is analysed with different volume concentrations CNT nanofluids differ from 0 to 0.03 volume fraction. The numerical simulation is run to attain the lowest base face wall average temperature. The analysis is done for the volume fraction 0, 0.01, 0.02, 0.03 volume fraction. Figure 5 shows the Nusselt number for various Reynolds numbers. Increasing the volume fraction of CNT nanofluids increases the Nusselt number proportionally. Also Fig. 5 compares the temperature difference of wall and inlet fluid to the corresponding Nusselt number. The 0.03 volume fraction CNT nanofluids give better results on higher Reynolds numbers.

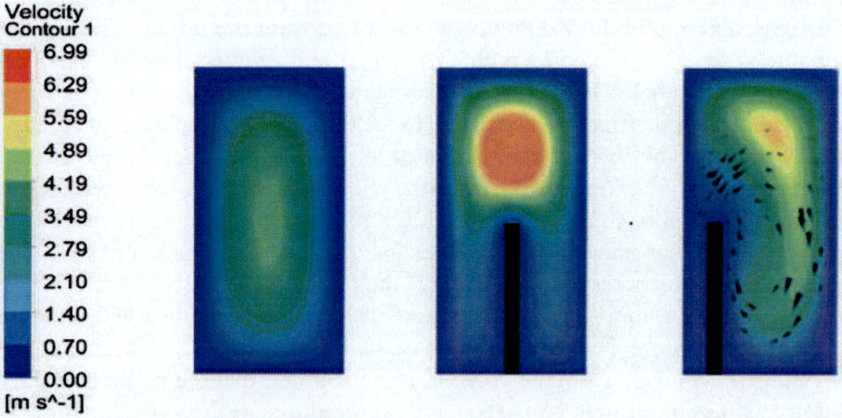

Fig. 3 Velocity magnitude of all fin-type heat sink at a middle plane in x-direction

Fig. 4 Nusselt number with respect to reynolds number

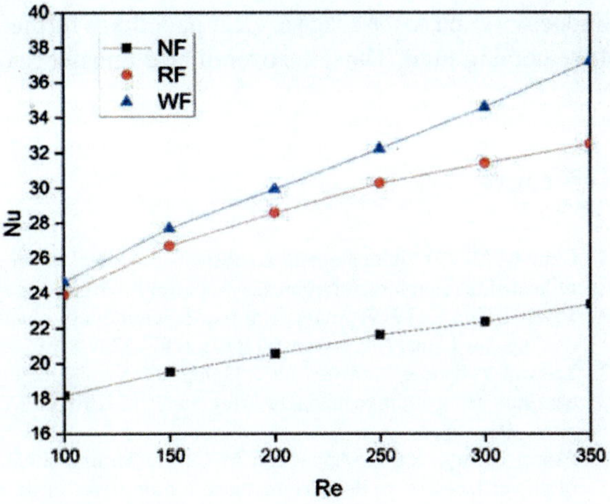

Fig. 5 Nusselt number of nanofluids with respect to Reynolds number

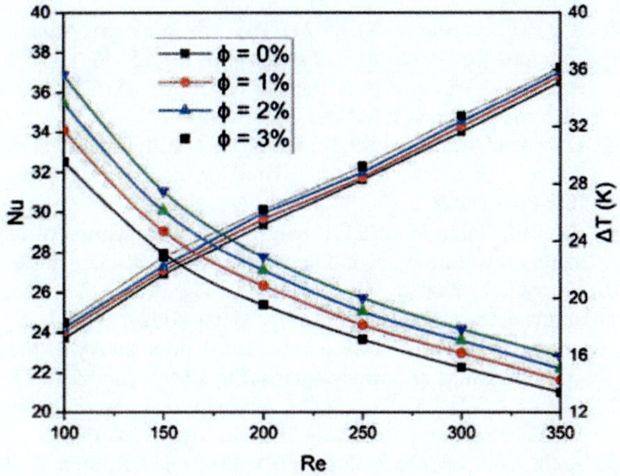

5 Conclusion

The numerical simulation of two fin models such as wavy and rectangular fin integrated with the conventional silicon heat sink is done in this paper. The effect of CNT nanotubes on heat sink performance is studied. This study gives a new view of the heat sink design process. From the numerical study, it is identified that the wavy fin gives the best performance. The minimum average heat sink base face is attained in the way fin-type heat sink. And this wavy fin is further studied for various volume fraction CNT nanofluids at the Reynolds number 350 value. The 0.03 volume fraction CNT nanofluid gives good results and the temperature is greatly reduced and it is preferred for electronic cooling applications. Since the wavy fin has a higher solid–fluid interface and flows turbulence. The CNT nano particle has higher thermal conductivity and so the usage of CNT nanofluids further enhances the Nusselt number of the cooling fluid. Thus, the overall heat transfer rate is enhanced.

References

1. Ekpu M (2019) Finite element analysis of the effect of fin geometry on thermal performance of heat sinks in microelectronics. J Appl Sci Environ Manag 23(11):2059–2063
2. Vafai K, Zhu L (1999) Analysis of two-layered micro-channel heat sink concept in electronic cooling. Int J Heat Mass Transf 42(12):2287–2297
3. Tateishi Y, Parque V, Miyashita T, Gohara H, Kato R, Ikeda Y (2017) Design optimization of heat sink using additive manufacturing. In: 2017 IEEE CPMT Symposium Japan (ICSJ). IEEE, pp 91–94
4. Fantin Irudaya Raj E, Appadurai M (2021) Minimization of torque ripple and incremental of power factor in switched reluctance motor drive. In: Recent trends in communication and intelligent systems: proceedings of ICRTCIS 2020. Springer Singapore, pp 125–133
5. Sijini AC, Fantin E, Ranjit LP (2016) Switched reluctance motor for hybrid electric vehicle. Middle-East J Sci Res 24(3):734–739
6. Raj EFI, Appadurai M (2021) The hybrid electric vehicle (HEV)—an overview. Emerging solutions for e-mobility and smart grids, pp 25–36
7. Raj EFI (2016) Available transfer capability (ATC) under deregulated environment. J Power Electron Power Syst 6(2):85–88
8. Deivakani M, Kumar SS, Kumar NU, Raj EFI, Ramakrishna V (2021) VLSI implementation of discrete cosine transform approximation recursive algorithm. J Phys Conf Ser 1817(1): 012017. IOP Publishing
9. Raj EFI, Balaji M (2021) Analysis and classification of faults in switched reluctance motors using deep learning neural networks. Arab J Sci Eng 46(2):1313–1332
10. Gampala V, Kumar MS, Sushama C, Raj EFI (2020) Deep learning based image processing approaches for image deblurring. Mater Today: Proc
11. Jain D, Sangale DMD, Raj E (2020) A pilot survey of machine learning techniques in smart grid operations of power systems. Eur J Mol Clin Med 7(7):203–210
12. Chen CT, Wu CK, Hwang C (2008) Optimal design and control of CPU heat sink processes. IEEE Trans Compon Packag Technol 31(1):184–195
13. Koga AA, Lopes ECC, Nova HFV, De Lima CR, Silva ECN (2013) Development of heat sink device by using topology optimization. Int J Heat Mass Transf 64:759–772
14. Appadurai M, Raj EFI, Venkadeshwaran K (2021) Finite element design and thermal analysis of an induction motor used for a hydraulic pumping system. Mater Today: Proc 45:7100–7106
15. Jenish I, Appadurai M, Fantin Irudaya Raj E (2021) CFD Analysis of modified rushton turbine impeller. Int J Sci Manag Stud IJSMS 4:8–13
16. Raj EFI, Kamaraj V (2013) Neural network based control for switched reluctance motor drive. In: 2013 IEEE international conference on emerging trends in computing, communication and nanotechnology (ICECCN). IEEE, pp 678–682
17. Appadurai M, Raj EFI (2021) Finite element analysis of composite wind turbine blades. In: 2021 7th international conference on electrical energy systems (ICEES). IEEE, pp 585–589
18. Appadurai M, Raj E (2021) Epoxy/silicon carbide (sic) nanocomposites based small scale wind turbines for urban applications. Int J Energy Environ Eng 1–16
19. Appadurai M, Fantin Irudaya Raj E (2022) Finite element analysis of lightweight robot fingers actuated by pneumatic pressure. In: Recent advances in manufacturing, automation, design and energy technologies. Springer, Singapore, pp 379–385
20. Levin PP, Shitzer A, Hetsroni G (2013) Numerical optimization of a PCM-based heat sink with internal fins. Int J Heat Mass Transf 61:638–645

21. Appadurai M, Velmurugan V (2017) Experimental analysis of stepped basin pyramid solar still integrated with mini solar pond. Desalin Water Treat 84:1–7
22. Appadurai M, Velmurugan V (2015) Performance analysis of fin type solar still integrated with fin type mini solar pond. Sustain Energy Technol Assess 9:30–36
23. Mangalaraj A, Vellaipandian V (2018) Performance analysis of modified solar still with forced water circulation. Thermal Sci 22(6 Part B):2955–2964
24. Appadurai M, Prasath AB, Thanu MC, Richard S (2015) Determination of dual phase forced convective heat transfer of nano fluids by means of CFD. Int J Appl Eng Res 10(12):32627–32635
25. Appadurai M, Raj E, Jenish I (2021) Application of aluminium oxide–water nanofluids to augment the performance of shallow pond: a numerical study. Process Integr Optim Sustain 1–12
26. Jenish I, Sahayaraj AF, Suresh V, Appadurai M, Irudaya Raj EF, Nasif O, Alfarraj S, Kumaravel AK (2022) Analysis of the hybrid of mudar/snake grass fiber-reinforced epoxy with nano-silica filler composite for structural application. Adv Mater Sci Eng
27. Jenish I, Felix Sahayaraj A, Appadurai M, Fantin Irudaya Raj E, Suresh P, Raja T, Salmen SH, Alfarraj S, Manikandan V (2021) Fabrication and experimental analysis of treated snake grass fiber reinforced with polyester composite. Adv Mater Sci Eng
28. Ravindran P, Mohamed MJS, Gnanaraj SJP, Appadurai M (2021) Effect of graphite addition on dry sliding wear behavior of self-lubricating Al-SiC-Gr hybrid composites by PM process. Mater Today: Proc
29. Raj E, Appadurai M, Athiappan K (2021) Precision farming in modern agriculture. In: Smart agriculture automation using advanced technologies. Springer, Singapore, pp 61–87

SAR Exposure Assessment in Human Head Tissue Model at GSM Frequency

S. Jemima Priyadarshini, J. Immanuel Nargunathan, S. Anitha Christy, and A. Kanimozhi

Abstract The specific absorption rate (SAR) is a measuring value for investigating human exposure caused by nonionizing radiation. The output radiation from the mobile phone sources is categorized as class 2B a possible carcinogen by World Health Organization. There is a need for the evaluation of the biological impact caused due to radiation on human beings. This paper suggests a new experimental method of investigation for measuring SAR in human head tissue models such as the cerebellum, cerebrospinal fluid, gray and white matter with average brain and skull. The inverted F antenna is used as a source of radiation at a GSM frequency of 900 MHz band. The dielectric equivalent properties of human tissues are obtained from databases for the design of linear tissue structures. The simulation environment is set using Altairs FEKO for SAR computation and analysis. The gain characteristics and S_{11} parameters are also studied for more understanding of radiation source behavior.

Keywords Specific Absorption Rate · Inverted F antenna · Dielectrics · Head

1 Introduction

There is a day-to-day exposure surrounded by every human being due to some constant sources of radiation. This paper emphases on investigations on head tissues properties and the effects of radiation exposure from devices such as mobile phones.

S. J. Priyadarshini (✉)
Immanuel Hospital, Chennai, Tamil Nadu, India
e-mail: mailjemi@gmail.com

J. I. Nargunathan
Sri Balaji Medical College and Hospital, Chennai, Tamil Nadu, India

S. A. Christy
Govt. Stanley Medical College, Chennai, Tamil Nadu, India

A. Kanimozhi
VHS-M.A. Chidambaram College of Nursing, Chennai, Tamil Nadu, India

N. M. Sivaram et al. (eds.), *Advances in Manufacturing, Automation, Design and Energy Technologies*, Lecture Notes in Mechanical Engineering,
https://doi.org/10.1007/978-981-99-1288-9_52

Radiofrequency exposure to the human head can lead to acute effects induced in brain functions [1]. The long-term and heavy users are at risk of development of brain tumors and there is a need for more investigation for oncology studies [2]. The SAR parameter is an essential value for understanding and the estimation of absorption on a mass of exposed tissue. The SAR can be either estimated with the equivalent exposure mass of 1 g for a smaller exposure area or 10 g for a larger exposure area [3]. The Inverted F antenna is frequently used in communication devices for its expediency and less weight. The dipole antenna or monopole has much experimented however, they are not widely used these days due to its size and shielding issues [4]. This paper details head tissues including brain layers with skull and its surrounding tissues. An experimental study on each tissue model is performed with the designed IFA and tissue properties such as permittivity, conductivity is applied for the medium at 900 MHz frequency [5].

1.1 *Specific Absorption Rate*

The SAR estimates energy absorption of an exposed human medium. When the exposed medium is subjected to radiofrequency, there is an occurrence of absorption due to its electromagnetic field. The SAR depends on power absorption and tissue characteristics [6]. The SAR is measured as model that can be determined by the exhibited near-field E (V/m) is given by the following Eq. (1).

$$\mathrm{SAR} = \frac{\sigma|E|}{\rho}(\mathrm{W/Kg}) \quad (1)$$

1.2 *Inverted F Antenna*

Inverted F antenna was proposed for missiles but the variant is widely used for communication purposes. It is a modified monopole antenna consist of conductive rod on the ground plane region. The length of the antenna is shorted with an intermediate point for feed instead of the base. The traditional IFA design is shown in Fig. 1. The input impedance depends on the feed point and the distance between the ground plane that is similar to short-circuit stub [7]. FEKO IFA design is displayed in Fig. 2.

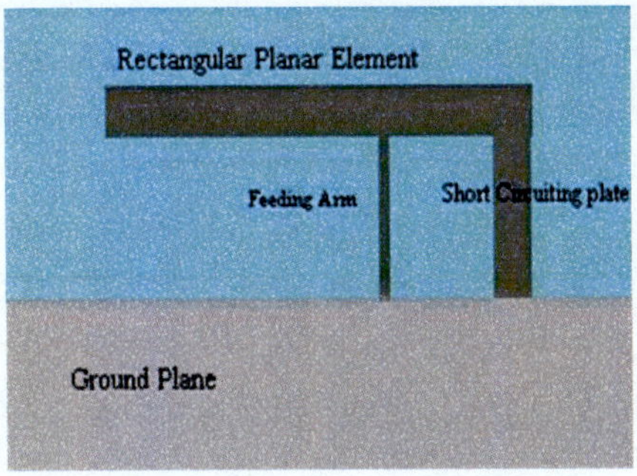

Fig. 1 Inverted F Antenna

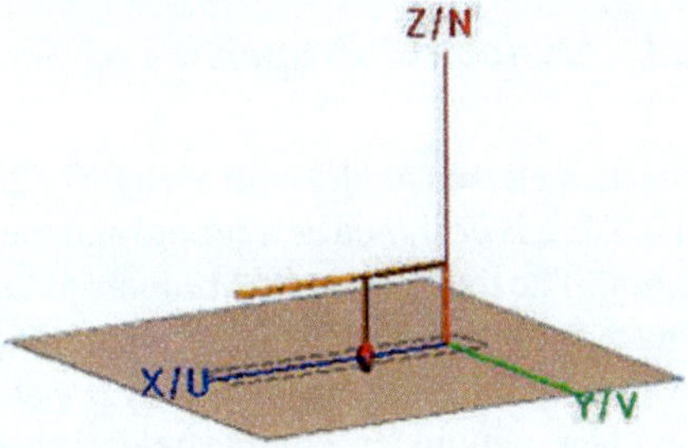

Fig. 2 IFA model designed using FEKO

2 Numerical Modeling and Experimental Setup

2.1 Modeling Method

There is a ban on any experiments conducted with human under radiation exposure. This investigation is done on numerical human models and their supporting calculations with Method of Moments (MoM) [8]. MoM method is the most implemented method for its simplicity and accuracy in computation. In this linear tissue model and IFA antenna, MoM is a suitable method for implementation using the EM platform.

2.2 Geometry and Dimensions of IFA

IFA appears to be a wire antenna however it is classified as an aperture antenna with the feed placed on the upper arm. The vertical polarization with a donut shape is the output due to the short-circuited stub. The IFA mandates the ground plane to the length of its upper arm and $\lambda/4$ in height. The height of the upper arm should be with appropriate value else there is a decrease in bandwidth. IFA has a unique advantage of limited backward radiation thus minimizing SAR values. The following IFA design values of the upper arm, planar arm, and feed offset are implemented for

Table 1 Inverted F antenna

IFA parameters	Values (mm)
The planar element (length and width)	23.7 and 2.5
The folded upper arm (width and length)	3 and 5.3
The feed arm (width and length)	0.85 and 5.3
The offset feed position	6.97

the designed IFA using CADFEKO for GSM 900 MHz frequency. The dimensions are listed in Table 1.

2.3 *Dielectric Properties of Tissues*

The linear tissue model with a length of 200 mm and width of 100 mm with a thickness of 0.3 mm is designed as a cuboid and the dielectric medium is applied to the designed region. The tissue dielectric properties are derivative from the preexisting biomedical data over a frequency [9]. The conductivity σ, relative permittivity ε_r, and mass density ρ of the tissue over the frequency are acquired from the database and are applied to the linear cuboid structure. The designed tissue medium is placed above IFA and is set for EM interaction as shown in Fig. 3. The design properties are listed in Table 2.

Simulation setup. The simulation is instigated in FEKO with applying mathematical model of the Method of Moments [10]. The tissue equivalent has placed a distance of 0.1 mm from the radiation IFA for exposure assessment. The IFA is placed at free space condition and the exposure area field is created for SAR computation.

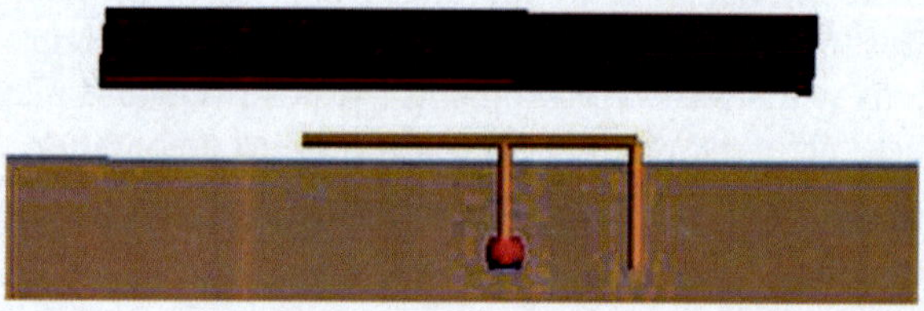

Fig. 3 IFA with linear tissue equivalent

Table 2 Dielectric properties of head tissues

Tissue	ε_r	σ
Cerebellum	49.4431	1.26265
Cerebra Spinal fluid	68.6383	2.41258
Gray matter	52.7247	0.94219
White matter	38.8863	0.59082
Avg. brain	45.8055	0.7665
Avg. skull	16.6208	0.24164

The EM flow occurs between the head medium and the radiating IFA. The 1 g SAR estimation and computation is done based on Fresnel field.

3 Results and Validation

3.1 *Operation Characteristics*

S_{11} or reflection coefficient (return loss) is a crucial constraint for determining the operational performance of an antenna. It is a ratio of its input power to the radiated power over a frequency represent in dB value. If the value tends to be zero dB, it implies no power is radiated by the operating antenna. For any simulation with antenna S parameters is tested whether the setup is efficient in terms of operation over a particular frequency. In this experiment, the S_{11} is calculated for the head tissues. The return loss is calculated for all the tissues and tested and output values are shown in Table 3. Their return loss characteristics of the brain are shown in Fig. 4.

The S_{11} coefficient of IFA for all the head tissues are simulated and the obtained results are shown in Table 3. The values in the table indicate that for all tissues the IFA operates optimally and it can generate an adequate field characteristic for SAR computation [11].

Table 3 Reflection coefficient

Tissue	*s*11
Cerebellum	49.4431
Cerebra Spinal fluid	68.6383
Gray matter	52.7247
White matter	38.8863
Avg. brain	45.8055
Avg. skull	16.6208

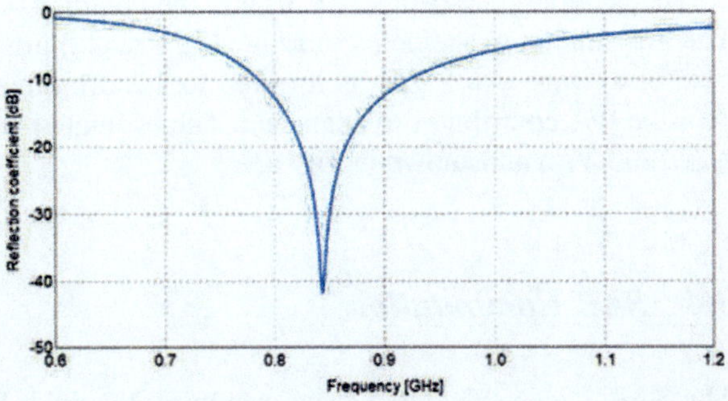

Fig. 4 Return loss of avg. brain (dB)

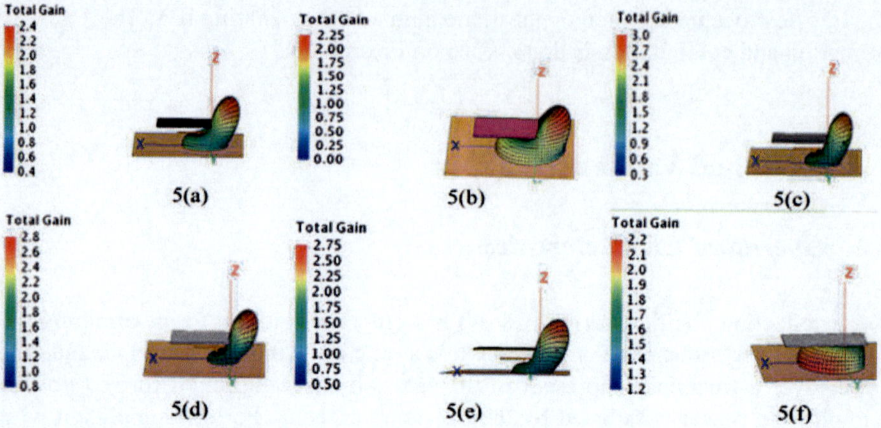

Fig. 5 3D Far-field characteristics of **a** Cerebellum, **b** CSF, **c** gray matter, **d** white matter, **e** avg. brain and **f** avg. skull

3.2 Far-Field in 3D

The field behavior of head tissues is investigated gain values. The IFA exhibits radiation around itself to a particular distance. The extreme distance where its field is divulged is its radiation behavioral pattern. The far-field pattern of head tissues with a corresponding total gain in dB is given in Fig. 5.

3.3 Gain and Near-Field

The antenna gain measures the ability of the radiating structure or device to convert the given input power to the electromagnetic waves for propagation. The gain of IFA with each dielectric tissue equivalent is measured, and tissue interactions with IFA are studied.

The general gain of IFA is 3 dB and the obtained values are compared in Fig. 6. The gray matter tissue helps in achieving a maximum gain of 3 dB, whereas in the case of average skull, gain is lowered to 2.2 dB value. The scattering around and near the IFA contributes to near-field. The estimated near-field value is a maximum of 1.125 kV/m and shown in Fig. 7.

3.4 SAR Computation

The SAR 1 g is calculated from averaging the peak SAR value and the acquired values are added until a mass of 1 g is obtained. Similarly, SAR around each point

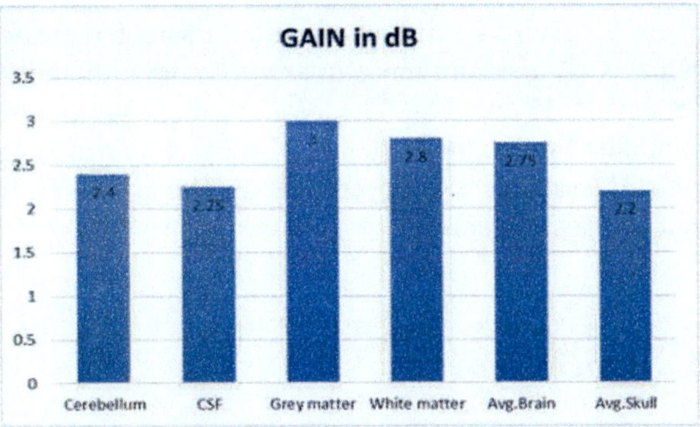

Fig. 6 Gain of the head tissues (dB)

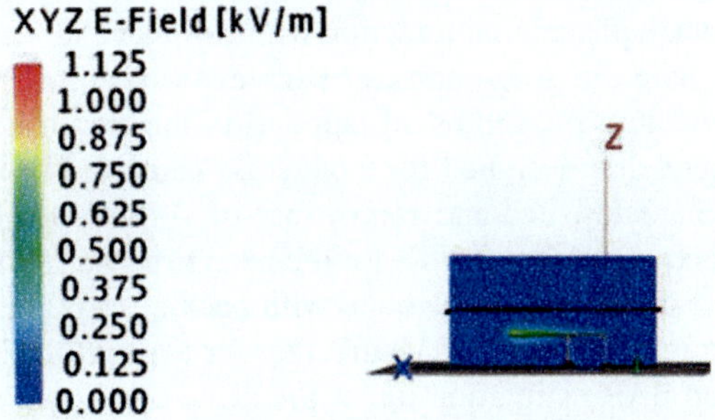

Fig. 7 Near-field distribution

in volume is calculated and averaged to 10 g. For both cases, a cubic structure of mass is required [12]. In this case, a linear structure of tissue equivalents is used for comparison and peak SAR value is calculated. The dielectric values are attained from a standard database for research for evaluation. The tissues and their behavior are analyzed in this SAR computation displayed in Fig. 8.

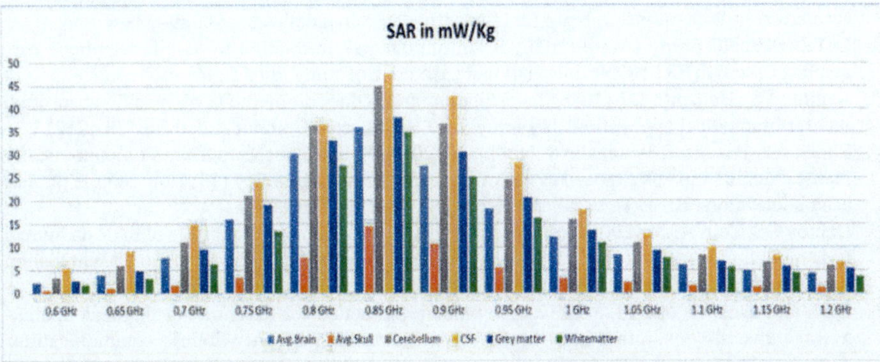

Fig. 8 SAR comparison of head tissues (mW/Kg)

Each tissue has different medium characteristics contributing to different SAR values. CSF medium tends to absorb more radiation 47 mW/Kg followed by cerebellum of 45 mw/Kg. The Avg. skull displays resistance to absorption with a minimum SAR value of 10 mW/Kg. The gray matter absorbs more than white matter but still less than CSF or cerebellum. The avg. brain tissue has SAR values close to that of gray matter.

4 Conclusion

This paper intends to experiment and study SAR impact on six different human head tissues for IFA at 900 MHz frequency. This work further concludes that the tissues when placed in interaction with IFA can alter the gain and S_{11} characteristics of IFA. There are sole characteristics of tissues observed and the difference in dielectric medium properties are applied to the experimental study. The homogenous linear model is designed for evaluation using the MoM method in FEKO. The operation characteristics and return loss of -15 to 30 dB are achieved. The gain is varied between 2.2 and 3 dB for the head tissues. There is a maximum SAR value obtained near 850–900 MHz band with peak absorption values of 47 mW/Kg for CSF and a minimum value of 10 mW/Kg for avg. skull. Each head tissue has a unique impact on SAR estimation under the same condition. The future work can be extended to cubic structure for 1 g and 10 g comparison with suggested reduction methods [13]. The analysis can also be extended for the various tissues and tissue properties can be investigated comprehensively for the biomedical area of research.

References

1. Zhang J, Sumich A, Wang G (2017) Acute effects of radiofrequency electromagnetic field emitted by mobile phone on brain function. Bioelectromagnetics 38(5):329–338
2. INTERPHONE Study Group (2010) Brain tumour risk in relation to mobile telephone use: results of the INTERPHONE international case-control study. Int J Epidemiol 39(3):675–694
3. Jemima PS, Hemanth DJ (2018) Investigation and reduction methods of specific absorption rate for biomedical applications: a survey. Int J RF Microw Comput Aided Eng 28:e21211
4. Simba AY, Hikage T, Watanabe S, Nojima T (2009) Specific absorption rates of anatomically realistic human models exposed to radiofrequency electromagnetic fields from mobile phones used in elevators. IEEE Trans Microwave Theory Tech 7(5):1250–1259
5. Dautov OS, Zein AEM (2005) Application of FEKO program to the analysis of SAR on human head modeling at 900 and 1800 MHz from a handset antenna. In: IEEE international symposium on electromagnetic compatibility and electromagnetic ecology, pp 274–277
6. IEEE Standard (2003) 1528 IEEE recommended practice for determining the peak spatial-average specific absorption rate (SAR) in the human head from wireless communications devices: measurement techniques
7. Ronold King CH, Denton D (1960) Transmission-line missile antennas. IRE Trans Antennas Propag 8(1):88–90

8. Caorsi S, Gragnani GL, Pastorino M (1993) Use of redundant testing functions in moment-method solutions for block models. IEEE Trans Microwave Theory Tech 41(2):305–310
9. Gabriel C (1995) The dielectric properties of biological materials. In: Klauenberg BJ, Grandolfo M, Erwin DN (eds) Radiofrequency radiation standards. NATO ASI series (series a: life sciences, vol 274)
10. https://www.altair.com/feko
11. Jensen MA, Rahmat-Samii Y (2009) EM interaction of handset antennas and a human in personal communications. In: Proceedings, vol 83. IEEE, pp 7–17
12. Kawai H, Ito K (2004) Simple evaluation method of estimating local average SAR. IEEE Trans Microwave Theory Tech 52(8):2021–2029
13. Stephen JP, Hemanth DJ (2019) An investigation on specific absorption rate reduction materials with human tissue cube for biomedical applications. Int J RF Microw Comput Aided Eng. 29:21960–21972

Bending and Contact Stress Analysis of Helical Gear: A Comparative Study

R. Mohanraj, S. Elangovan, S. Pratheesh Kumar, S. Arun Srivatsan, N. Kulasekaran, B. Lakshmana Kumar, and R. Natarajan

Abstract Gears are most commonly used and critical component in power transmission systems. Helical gears have gained popularity as a result of their lower dynamic load and noise level during its working, as well as the requirement for lighter and smaller vehicle transmissions. This paper includes a comparative research of helical gear design and performance characteristics, which was carried out using Finite Element Analysis (FEA) and theoretical methods. Theoretical analysis of bending stress in helical gears was evaluated using Python Tkinter Graphical User Interface (GUI) and Microsoft excel solver, based on American gear Manufacturing Association (AGMA) specifications and the Lewis equation. Similarly, using AGMA standards, FEA of contact stress in helical gears have also been carried out in this work. The stress distribution on helical gears is evaluated using a comparative evaluation with four distinct face widths and helix angles. Finite element analysis is used to investigate the effects of bending stress and contact stress on helical gear systems, and the results are compared with theoretical results.

Keywords Helical gears · Finite element analysis · Theoretical analysis · Contact stress · Bending stress

1 Introduction

Power transmission plays a major role in our everyday life and has high priority. The quantity of power lost in the process determines a machine's efficiency. Gears are well-known for being one of the most efficient ways of transmitting power between shafts. From cars to clocks, gears are used in a variety of situations, including shipbuilding, aero planes, and railways.

R. Mohanraj (✉) · S. Elangovan · S. Pratheesh Kumar · S. Arun Srivatsan · N. Kulasekaran · B. Lakshmana Kumar · R. Natarajan
Department of Production Engineering, PSG College of Technology, Coimbatore, Tamil Nadu 641004, India
e-mail: mohanraj839@gmail.com

N. M. Sivaram et al. (eds.), *Advances in Manufacturing, Automation, Design and Energy Technologies*, Lecture Notes in Mechanical Engineering,
https://doi.org/10.1007/978-981-99-1288-9_53

Helical gears are one type of cylindrical gear, with teeth cut at a constant angle to the gear's axis, known as the helix angle. Helical gears are known for power transmission between two shafts parallel to the axis of rotation. They are smooth and silent in operation due to less vibration, have high load carrying capacity, and can operate at high speeds. Helical gears may have a longer life for similar loading circumstances than spur gears, have a greater contact ratio, and can transmit force of greater value.

This paper presents about behavioral study of tooth bending stresses by the use of two distinct standards, namely the Lewis Equation and the AGMA theory and contact stress is also studied using the AGMA Theory. The comparison study is performed by varying face width (75 mm, 65 mm, 55 mm, and 45 mm) and helix angles (15°, 20°, 25°, and 30°) to find the distribution of stress in helical gears and stress analysis was performed using the FEA (ANSYS) approach. The FEA results were compared and validated with theoretical results and conclusions were drawn.

2 Literature Review

Sarkar et al. [1] investigated three-dimensional Computer-Aided Design (CAD) model in FEA for the purpose of keeping track of the stress that has been created during meshing by tooth flank and tooth filet. AGMA bending and contact stresses equations were used to obtain analytical results. The maximum bending stresses were obtained for small face width and larger helix angle gears. Investigation of helical gears by measuring contact stresses for various pressure angles, helix angles, and face widths was performed by Anusha et al. [2] and Singh [3]. Contact stress of mating gear was found using contact stress equation which was used as the theoretical values. For variable face width, the stress values derived using the AGMA equation was higher than FEA values, and the contact stress between the pinion and gear decreases as the face width of the gear increases.

Devaraj et al. [4] attempted to make a comparison on contact stresses for various helical gear pairs using different standards. Comparison was made between FEA stress values and analytical values and results indicate that the contact stress decreases with the increase in helix angle because of increase in the contact area length for a maximum difference of 0.86% between theoretical and FEA results. Patil et al. [5] studied contact stresses for three-dimensional helical gear sets using FEA. Contact stresses for different helix angles and friction coefficients were calculated and suitable inference was drawn based on difference in results. Gidado et al. [6] used bending stress calculations to study fatigue fracture in gear teeth. The results show that there is a minor variation with a difference of 4.70%. As the face width increases, there is a corresponding decrease in the tooth bending stresses values of helical gears for both AGMA theory and numerical simulation.

Vishwakarma et al. [7] made an attempt to study the three-dimensional CAD model in FEA for the purpose of determining the stress that has been created during meshing by tooth flank and tooth filet. Analytical bending stress values were calculated using Lewis and AGMA bending equations. Hertz and AGMA contact equations were used for contact stresses. As the face width varied, their effects were studied and shows that larger the face width, smaller the stress. Khunti et al. [8] attempted to study the contact stresses which were found out from Hertz equation standard and FEA. The material selected for this analysis was EN8. The values obtained from the analytical calculation using the Hertz equation is 715.69 MPa and the value obtained from FEA is 714.77 MPa. This shows that there is very minor variation between FEA and theoretical results and is found to be within limits.

Ramakrishna et al. [9] used ANSYS modal analysis and fatigue analysis was carried out on herringbone gears of alloy steel and carbon fiber-reinforced composite. It was observed that carbon fiber is performing almost 70% of the performance of 6150 alloy steel. Jyothirmai et al. [10] made a study of comparison between the performance of various helical gears using the numerical approach with AGMA standards and FEA for listed specifications. The maximum error in the FEA results for various performance metrics (tooth bending stress, contact stress, and surface fatigue stress) was less than 2% when compared with the obtained theoretical results. So far, the work has aimed to improve the gear's efficiency. The majority of literatures have concentrated at the performance of gear systems based on material characterization in operating settings. The estimation of acceptable tooth bending stress distribution on helical gears has received less attention. Bending stresses and contact stresses were studied in this research for better knowledge.

3 Methodology

In this work, initially bending stresses and contact stresses on helical gears were studied and finally FEA and theoretical results was compared and validated.

4 Gear Design

The helical gear pair has to be designed for transmitting 20 kW power, at a speed of 1440 rpm with a gear ratio of 1, that is, both the gear and pinion will be rotating at the same speed and the gears have to be designed for a helix and pressure angle of 15° and 20°, respectively, as given in Table 1.

The parameters of the helical gear are calculated using the standard formulae [11] and listed in Table 2. Since gear ratio (i) is 1, both the pinion and gear will be identical. Pinion will have right-hand helix and gear will have left hand helix.

Table 1 Gear design parameters

Parameters	Symbols	Specifications
Gear material	–	C45
Power source	–	Uniform
Load type	–	Continuous
Factor of safety	–	1.1
Poisson ratio	ν	0.3

Table 2 Helical gear design parameters

Parameters	Symbols	Unit	Pinion and gear
Normal module	m_n	mm	5
Module	m	mm	5.18
Pitch circle diameter	d	mm	129.04
Number of teeth	Z	–	25
Face width	b	mm	75
Center distance	a	mm	129.04
Pitch line velocity	V	m/sec	9.76

5 Bending Stress Analysis of Gear Tooth

5.1 Analysis of Bending Stress Using Lewis Equation

The Lewis equation is a well-known method for calculating bending stresses in gear teeth. By modeling a gear tooth as a simple cantilever beam with the load concentrated at the tip of the teeth, the refined form of Lewis equation is used to calculate tooth bending stress which models a gear tooth as a simple cantilever beam as shown in Eq. (1).

$$\sigma_b = \frac{F_t \cos(\beta)}{m.b.y} \tag{1}$$

where β-helix angle (degree), *y*-form factor.

The tangential force (F_t) is expressed using the following Eq. (2),

$$F_t = \frac{M_t.2}{d_1} \tag{2}$$

5.2 Analysis of Bending Stress Using AGMA Equation

The AGMA bending stress equation is a detailed refinement of the Lewis stress equation. The AGMA bending stress is determined by using Eq. (3).

$$\sigma_b = \frac{F_t}{bm_n J} K_v K_o (0.93 K_m) \quad (3)$$

where σ_b = bending stress (MPa), F_t = normal tangential load, b = face width, m_n = normal module, J = geometry factor, K_v = dynamic factor, K_o = overload factor, K_m = load distribution factor. The dynamic factor (K_v) is determined by Eq. (4),

$$K_v = \sqrt{\frac{78 + \sqrt{200.\mathrm{V}}}{78}} \quad (4)$$

where $K_o = 1$ for uniform power source and $K_m = 1.3$ for face width up to 150 mm and precision gears.

6 Contact Stress Analysis of Gear Tooth

6.1 Analysis of Contact Stress Using AGMA Equation

In order to determine the total performance of gear pairs in terms of tooth energy and stresses generated at low speeds, AGMA theory uses the idea of calculating contact stresses and fatigue stresses. The following Eq. (5) is used by AGMA theory to describe contact stress.

$$\sigma_h = C_p \sqrt{\frac{F_t}{b.d_1.I} \cdot \frac{\mathrm{Cos}\beta}{0.95 * \mathrm{CR}} K_v . K_o .(0.93 K_m)} \quad (5)$$

where σ_h-AGMA contact stresses (MPa), C_p-elastic coefficient factor, d_1-pitch circle diameter of the pinion (mm), I-geometry factor, CR-contact ratio. The elastic coefficient factor C_p, geometry factor (I) and contact ratio (CR) are defined by the using the following Eqs. (6), (7), and (8):

$$C_p = 0.564 \sqrt{\frac{1}{\left(\frac{1-\mu_1^2}{E_1^2} + \frac{1-\mu_2^2}{E_2^2}\right)}} \quad (6)$$

C_p = 191 MPa (For Pinion and Gear as Steel Material) [11]

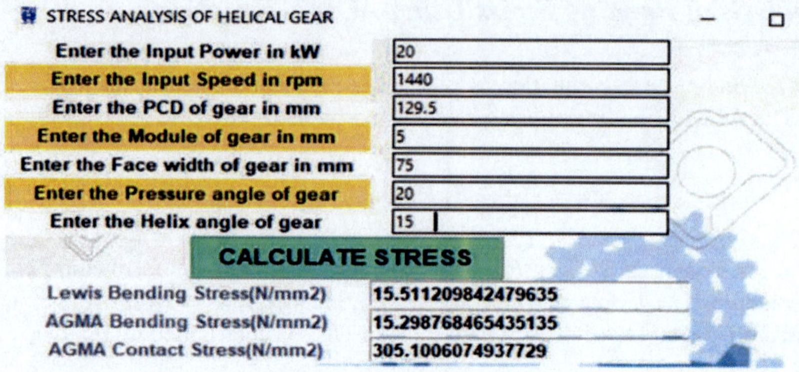

Fig. 1 Python program results

$$I = \frac{\text{Sin}\varnothing.\text{Cos}\varnothing}{2}.\frac{i}{i+1} \tag{7}$$

$$\text{CR} = \frac{\sqrt{(r_1+a)^2 - r_{b1}^2} + \sqrt{(r_2+a)^2 - r_{b2}^2} - ((r_1+r_2).\text{sin}\varnothing)}{\pi.m.\text{cos}\varnothing} \tag{8}$$

6.2 Analysis of Contact Stress Using Python Tkinter GUI

Python Tkinter graphical user interface was used for calculation of both the stresses theoretically. Lewis bending and AGMA standard equations were used to write the python program. For different face width and helix angles as input parameters, the corresponding bending and contact stresses were obtained. The Python program output is shown in Fig. 1.

7 Stress Analysis of Helical Gears Using FEA

FEA was performed to find numerical values for bending and contact stress in the designed helical gears. Bending stress analysis was done in ANSYS for individual right-hand helical gear (75 mm face width and 15° helix angle). C45 material was adopted for the simulation and property of the material such as elastic modulus of 200 GPa, ultimate tensile strength of 565 MPa, density of 7800 kg/m^3, and Poisson ratio of 0.3 was provided as input to the simulation. The boundary conditions of the gears are shown in Figs. 2 and 3. Adaptive convergence was used for the model and the maximum refinement loops and refinement depth were limited to 3. The convergence type was set to maximum with allowable change of 5%. The FEA model was solved

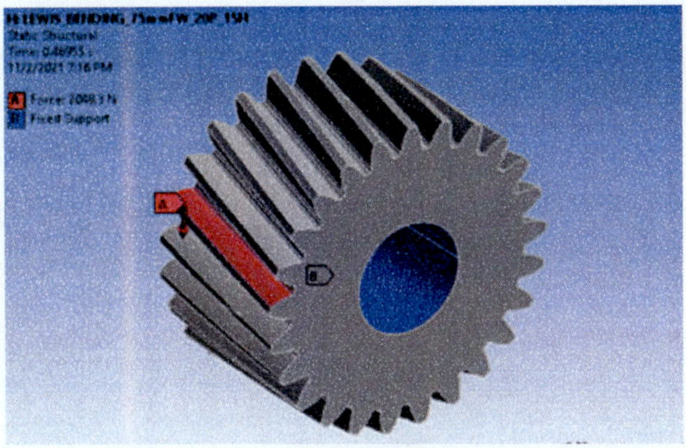

Fig. 2 Boundary condition in right-hand helical gear

Fig. 3 Boundary condition in helical gear pair

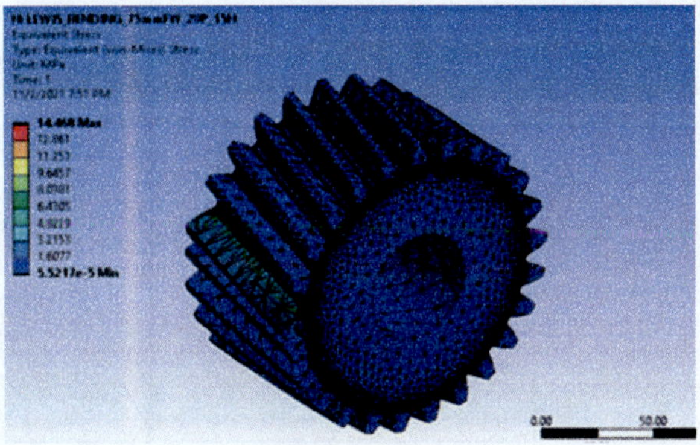

Fig. 4 Bending stress for right-hand helical gear

for the equivalent bending stress and the result is shown in Fig. 4. The obtained stress value from FEA for the face width of 75 mm is 14.468 MPa. Similarly, the stresses were also calculated using FEA for the other three right-hand helical gears of different face widths (65 mm, 55 mm, and 45 mm), keeping helix angle constant (15°).

Fig. 5 Contact stress for helical gear pair

For contact stress analysis in helical gear pairs, the boundary conditions such as frictionless support and load is provided. The FEA model was solved for the equivalent contact stress and the result is shown in Fig. 5. The obtained stress value from ANSYS for the helix angle of 15° is 303.13 MPa Similarly, the stress were also calculated using ANSYS for the other three helical gear pairs of different face helix angles (20°, 25°, and 30°), keeping the face width constant (75 mm).

8 Results and Discussion

Comparison of theoretical bending stress with FEA results for right-hand helical gear is shown in Table 3. Figure 6 shows that the face width and the bending stress are inversely proportional to each other where the bending stress is lower for a larger face width and this is due to the larger contact area during the mating of two gears. Comparison of theoretical contact stress with FEA results for helical gear pair as shown in Table 4. Figure 7 shows that contact stresses decreases for increasing helix angle due to increase in the contact area.

Table 3 Comparison of theoretical bending stress with FEA results for right-hand helical gear

Face width (b) (mm)	Lewis bending stress (MPa)	AGMA bending stress (MPa)	FEA bending stress (MPa)
45	25.862	25.510	25.531
55	21.160	20.872	20.884
65	17.904	17.660	16.611
75	15.517	15.305	14.468

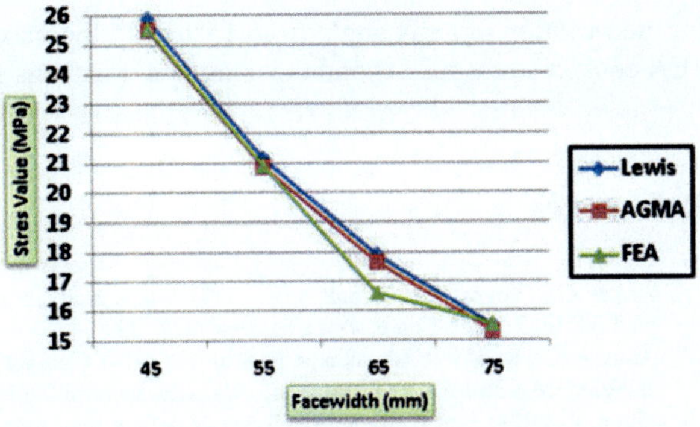

Fig. 6 Theoretical bending stress for various face widths

Table 4 Comparison of theoretical contact stress with FEA results for helical gear pair

Helix angle (β) (degree)	AGMA contact stress (MPa)	FEA contact stress (MPa)
15	305.281	303.13
20	298.989	297.96
25	286.766	284.82
30	272.302	275.7

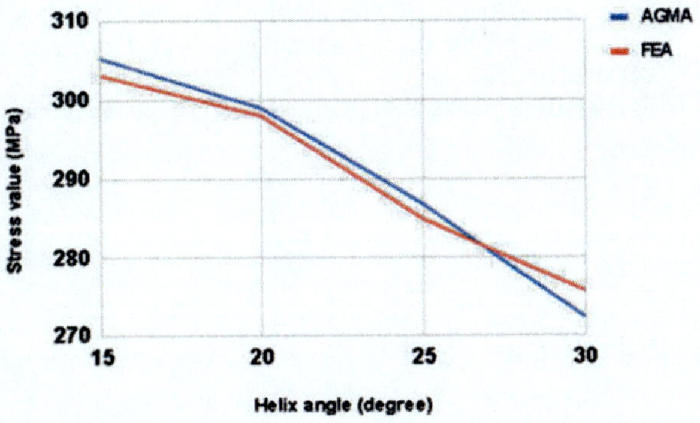

Fig. 7 Theoretical contact stress for various helix angles

9 Conclusion

Helical gears were designed and the results were compared and validated for bending and contact stresses. The results obtained from both theoretical and numerical simulations were in close agreement to one another. The bending stress is least for larger face width and this is due to the larger contact area during the mating of two gears. Contact stresses also decrease for increasing helix angle due to increase in the contact area. The maximum deviation between the FEA bending stress results and the theoretical results is 7.8%, for the case of face width variation from 45 to 75 mm, while for the variation of helix angle from 15°to 30°, the maximum deviation between the FEA contact stress results and the analytical results is 1.29%.

References

1. Sarkar GT, Yogesh LY, Dipak VB (2013) Stress analysis of helical gears by finite element method. Int J Mech Eng Robot Res 02(4):322–329
2. Anusha SS, Reddy S, Bhaskar P, Manoj M (2014) Contact stress analysis of helical gear by using AGMA and ANSYS. Int J Sci Eng Adv Technol 02(12):1012–1016
3. Singh V (2018) Finite element analysis of helical gear pair for bending and contact stresses. Int J Comput Eng Res Trends 05(5):136–140
4. Devaraj R (2015) Contact stress analysis of a helical gear. Appl Mech Mater 05(02):1070–1075
5. Patil SS, Karuppananan S, Atanasovskab I, Wahab AA (2014) Contact stress analysis of helical gear pairs, including frictional coefficients. Int J Mech Sci 85(2):205–211
6. Gidado AY, Muhammad I, Umar AA (2014) Design, modeling and analysis of helical gear according to bending strength using AGMA and ANSYS. Int J Eng Trends Technol 8(3):1–5
7. Vishwakarma B, Kumar JU (2014) Finite element analysis of helical gear using three-dimensional cad model. Int J Eng Sci Res Technol 03(4):1952–1958
8. Khunti MK, Rathod PP (2015) Contact stress analysis of helical gear of PTO drive using finite element analysis. Int J Adv Eng Res Develop 02(5):943–946
9. Ramakrishna D, Arul S (2019) Material selection for herringbone gear through modeling and finite element analysis. J Emerg Technol Innov Res 06(4):177–186
10. Jyothirmai S, Ramesh R, Swarnalatha T, Renuka D (2014) A finite element approach to bending, contact and fatigue stress distribution in helical gear systems. Int Conf Mater Process Char 06(1):907–918
11. Shigley JE, Mischke CR (1996) Standard handbook of machine design. Mc-Graw-Hill, USA

Theoretical and Numerical Analysis of Bending Stress on Spur Gears

R. Mohanraj, S. Elangovan, S. Pratheesh Kumar, K. Venkateshan, R. Manojh Kumar, and N. Sathish Kumar

Abstract Gears are used for the transmission of power due to their high efficiency. Spur gears are generally utilized in parallel power transmission. The appropriate operation of a spur gear drive design is dependent on a number of aspects that must be thoroughly examined and comprehended. This is done to obtain an acceptable relationship between design factors and operational factors for effective optimization. Bending stress developed in the tooth surface of gear is one of the important factors that lead to the failure of gears. In this paper, comparative studies were done to find the tooth bending stress of spur gears using theoretical and finite element analysis (FEA) methods. The theoretical method was performed with Lewis bending equation, and a Python program was developed to validate the theoretical results. The gear was modeled using computer-aided design (CAD) software. The effect of bending stresses on the modeled spur gears was analyzed using FEA methods. The comparative study was performed for various face widths of the pinion of spur gears, and the resulting bending stresses obtained from both the simulation and theoretical methods were compared and evaluated. The results show that the bending stress of spur gears decreases with an increase in its face width.

Keywords Spur gears · Theoretical model · Finite element analysis · Bending stress

1 Introduction

A gear is a revolving cylindrical wheel into which the teeth are cut. These teeth engage with the teeth of another gear to transmit torque or power from one shaft to the other. During motion, both gears rotate in opposite directions or in the same

R. Mohanraj (✉) · S. Elangovan · S. Pratheesh Kumar · K. Venkateshan · R. Manojh Kumar · N. Sathish Kumar
Department of Production Engineering, PSG College of Technology, Coimbatore, Tamil Nadu 641004, India
e-mail: mohanraj839@gmail.com

N. M. Sivaram et al. (eds.), *Advances in Manufacturing, Automation, Design and Energy Technologies*, Lecture Notes in Mechanical Engineering,
https://doi.org/10.1007/978-981-99-1288-9_54

way. The type of gearing system, such as external or internal gearing, determines the motion's direction. Through their gear ratio, gears usually cause a change in torque, resulting in a mechanical advantage. It is one of the most efficient ways to convey power and rotating motion, with or without speed or direction changes as needed. Gears are broadly classified on the basis of axis of the shaft, type of gearing, and amount of power to be transmitted. As per this criteria, number of gears are there such as spur, helical, bevel, double helical, hypoid, rack, and pinion. These gears are well-equipped in modern engineering machinery to make their operation as simple as possible.

Spur gears find their application in various industries such as vehicles, aircraft, manufacturing, power transmission machinery, lifting and hoisting machinery, and many more. Out of this vast category of gears, spur gear can be categorized as the most basic and widely used type of gear. A cylinder or disc is the most common shape. The teeth are projected radially. The leading edges of these straight-cut gears are oriented to the rotational axes in parallel direction. They can mesh properly only when the gears are mounted on parallel axes. The calculation of the torque ratio is performed through calculating the load exerted by one gear's tooth on another gear's tooth. The efficiency of the spur gear is approximately from 94 to 98% [1]. Spur gears can be used in a wide range of rapid and high-burden applications, in wide ranges of gear train, and over a wide range of velocity ratios, thanks to their varied preferences. As a result, spur gears are used in a variety of mechanical applications, such as tickers and electric screwdrivers.

2 Literature Review

Lisle et al. [2] have investigated the use of finite element analysis software for evaluating the root bending stresses. The bending stresses of spur gears were determined using theoretical (ISO and AGMA) and simulation methods. The results were compared to experimental strain gauge values for validation in the study.

Sivakumar and Michael [3] have investigated the stress analysis of spur gears with cast iron material as the traditional model. The material possessed high modulus properties and the FEA was done. The deformations and the stress for each material were determined. On comparison of the results of the composite materials with the traditional material, the deformations and stresses of composite materials were lower.

Ram and Abishek [4] have investigated the stress analysis of spur gears. FEA was done for 3D model of spur gears, and comparison was made with the results obtained from theoretical method. In order to reduce the fatigue failure, ideal size and stress relieving locational features of spur gears were discussed. The results showed that there was a reduction in bending stress by 21% and life was enhanced.

Patil [5] has investigated the bending stress of spur gear which was designed based on beam strength criteria. Bending stress of gears was calculated theoretically. The gear was modeled in CATIA V5, and static structural analysis was carried out with the given boundary conditions in Ansys. The MATLAB code was utilized to find the bending stresses of spur gears and validate the results obtained by the analytical methods. Based on the results obtained by both methods, the design was found to be safe as the stress values were within the safe limits.

Prabhakaran et al. [6] have investigated the stresses developed in tooth during the meshing of spur gears with material steel 15Ni2Cr1Mo28. Ansys simulation was done to find the bending stresses in spur gear. Lewis bending strength equation was utilized to calculate the bending stresses of spur gears. The stress values of spur gear with steel 15Ni2Cr1Mo28 material indicated that a reduction in the bending stresses of the gears with the higher values of module and the bending stress was reduced.

Vigneshwaran et al. [7] have studied the stresses on a spur gear developed in a stone crusher machine and analyzed it using Lewis bending strength equation theoretically. The root bending stresses of spur gears with different modules were analyzed using theoretical and analytical methods. When plain carbon steel was replaced by medium carbon steel, the life of gear and the efficiency was increased due to the reduction of bending stress.

Santosh et al. [8] have investigated the bending stresses of a misaligned spur gear pair. The analysis was evaluated with and without friction by employing corresponding analytical methods and finite element method (FEM). The maximum bending stresses increased by 18.9%, respectively, for axial misalignment, similarly for angular misalignment, the maximum bending stresses increased by 62.4%, respectively.

The root surface of the spur gear teeth is exposed to excessive bending stresses [9]. This causes breakage of spur gears. Only by reducing bending loads or changing the shape or characteristics, gear tooth failure can be prevented. Gear analysis is considered as an important challenge in the realm of design and production of gears [10]. The work done so far has attempted to increase the gear efficiency. The performance of gear systems has been investigated in the majority of the literatures based on material characterization in operating conditions. In spur gears, the estimate of permissible tooth bending stress distribution has received less attention. So in this paper, bending stresses of spur gear were analyzed using theoretical and numerical methods for application in gear manufacturing industry.

3 Research Methodology

The methodology adopted in this study starts with the basic understanding of the spur gears. The parameters which affect the process are selected. The results obtained from theoretical methods are validated with Python program and compared with the simulation results.

Table 1 Gear parameters and factors

Parameters	Symbols	Specifications
Transmission power	P	2.2 kW
Speed of the gear	N	750 rpm
Face width	b	57.5, 60, 62.5, 65 mm
Pressure angle	ϕ	20°
Number of teeth on pinion	Z_p	36

4 Gear Parameters

The various spur gear parameters and factors utilized in this study are provided in the Table 1.

5 Design of Spur Gear

The parameters of the spur gears are calculated using the standard formulae [10] and listed in Table 2.

Table 2 Spur gear design values

Parameters	Symbols	Unit	Pinion	Gear
Module	m	mm	6	6
Pitch circle diameter	d'	mm	216	108
Number of teeth	Z	–	36	18
Circular pitch	p	mm	18.84	18.84
Diametral pitch	P	mm	0.167	0.167
Addendum	h_a	mm	6	6
Dedendum	h_f	mm	7.5	7.5
Tooth thickness	s	mm	9.04	9.04
Tip circle diameter	d_a	mm	228	120
Root circle diameter	d_t	mm	201	93
Base circle diameter	d_b	mm	202.97	101.48
Center distance	a	mm	324	324

6 Bending Stress Analysis of Gear Tooth

6.1 Bending Stress Analysis of Spur Gears Using Lewis Bending Strength Model

Wilfred Lewis published a method for calculating the bending stress acting at the tooth surface in 1893. He designed a gear tooth as a simple cantilever beam that takes the whole force at its tip. Lewis modeled the gear tooth as a cantilever beam with a static normal force F at the tip. The following are the assumptions used throughout the derivation:

1. In a static situation, the whole force is delivered to a single tooth's tip.
2. The radial component is negligible.
3. The force is uniformly distributed across the gear's full face width of the gear.
4. Forces due to tooth sliding friction are negligible.
5. Stress concentration in the tooth filet is negligible.

The Eq. (1) describes the Lewis bending equation [6] used to calculate the bending stress of the gears.

$$\sigma_b = \frac{F_t}{b.m.y} \tag{1}$$

where σ_b—tooth bending stress in pascal (Pa), F_t—tangential force in Newton (N), b—face width of gear in millimeter (mm), m—module of the gear in millimeter (mm), and y—Lewis form factor, which is a constant used in the bending stress equation. The form factor is selected based on the number of teeth (Z) and pressure angle (φ) of the gear.

$$M_t = \frac{\mathrm{P.60}}{2.\pi.\mathrm{N}} \tag{2}$$

$$F_t = \frac{2.M_t}{d'} \tag{3}$$

The power to be transmitted and the speed of the gear were substituted in the Eq. (2) to obtain the moment (M_t). The moment and pitch circle diameter of the gear substituted in Eq. (3). The speed of the gear should be substituted in revolutions per minute (rpm), and the power should be substituted in watts (W) to obtain the tangential force in newton (N). The tangential force can also be calculated using pitch line velocity (v). The bending stress of spur gears with different face widths is calculated.

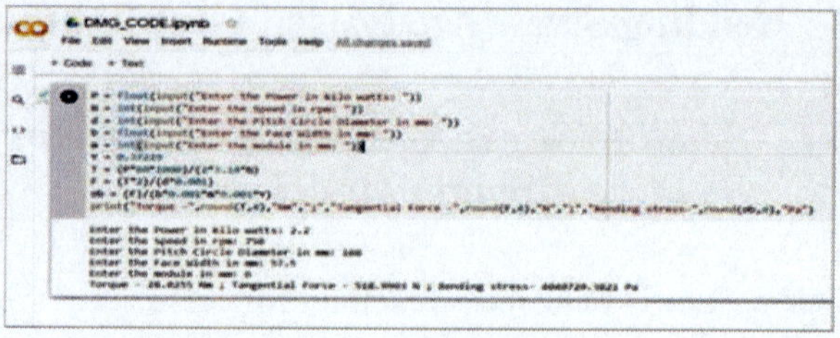

Fig. 1 Python program

6.2 *Bending Stress Analysis of Spur Gears Using Python Programming*

Python is one of the most widely used and rapidly expanding programming languages. In numerical analysis, the problems take more time to obtain the final solution, and Python helps to reduce the solving time and also helps to obtain the results quickly.

Google Colab is a tool used for writing Python programs easily and quickly. It helps to create and run arbitrary Python code directly from browser. It is especially useful for solving problems related to machine learning, data analysis, and numerical analysis.

The Python program was developed using Google Colab tool. Figure 1 displays the Python program developed to calculate the tooth bending stresses. The tooth bending stresses of spur gears with different face widths are calculated using Python program. The program takes power, speed, pitch circle diameter, face width, and module as the input parameters. The program calculates moment, tangential force, and bending stress. Lewis bending equation was used in the Python program to determine the moment, tangential force, and bending stress.

6.3 *Bending Stress Analysis of Spur Gear Using Finite Element Analysis*

The spur gears were modeled using Solidworks software with respect to the design values. Ansys simulation was performed to evaluate the bending stresses of spur gears. The major steps involved in the analysis are (i) selection of required material in engineering data source, (ii) importing the IGES model, (iii) assigning material and contacts for the spur gear model, (iv) generating mesh for the spur gear model, (v) provision of frictionless supports, displacements to the spur gear model, (vi) applying the required force of 518 N on the teeth of the model, and (vii) generating bending stress from the model. Aluminum alloy is taken as the material for the model. Material property of aluminum alloy such as density of 2.77 e−06 kg/mm^3, Young's modulus of 71,000 MPa, and Poisson's ratio of 0.33 was used for analysis. The mesh type used was quadrilateral. The fixed support and load applied were provided as

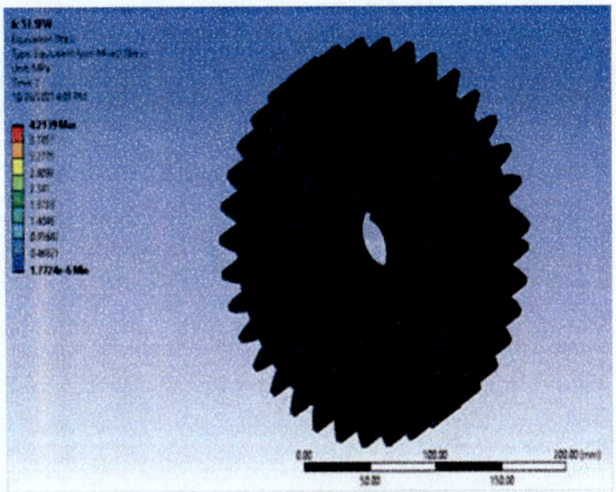

Fig. 2 Bending stress of spur gear with of 57.5 mm face width

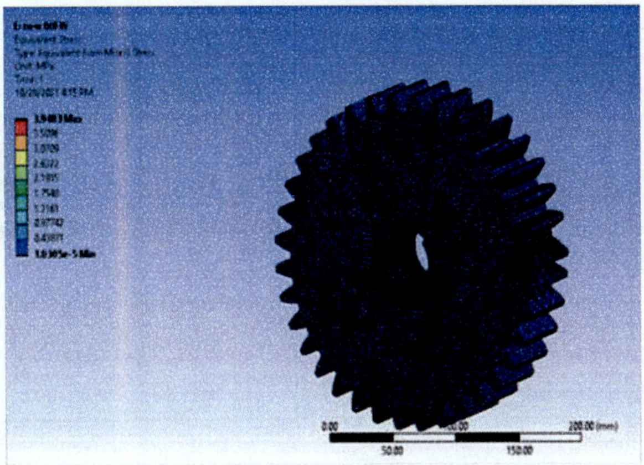

Fig. 3 Bending stress of spur gear with of 60 mm face width

input. The bending stress of spur gear for different face width such as 57.5 mm, 60 mm, 62.5 mm, and 65 mm is shown in Figs. 2, 3, 4, and 5, respectively.

7 Results and Discussion

The bending stress of the pinion of spur gears was obtained using theoretical and numerical methods. Table 3 shows the results of bending stress obtained using various methods and error percentage between the theoretical and numerical results. The impact of face width change on bending stresses obtained using Lewis bending equation is shown in Fig. 6. The impact of face width change on bending stresses obtained using Python programming is shown in Fig. 7. The impact of face width change on bending stresses obtained using Ansys is shown in Fig. 8. During the FEA analysis, the gear movement along the shaft is limited in its displacement by the

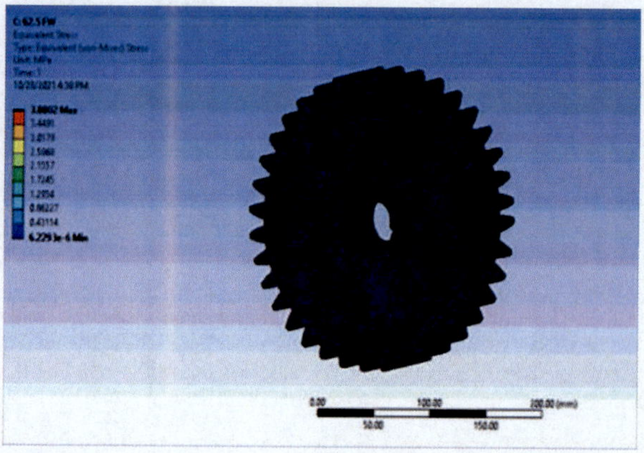

Fig. 4 Bending stress of spur gear with of 62.5 mm face width

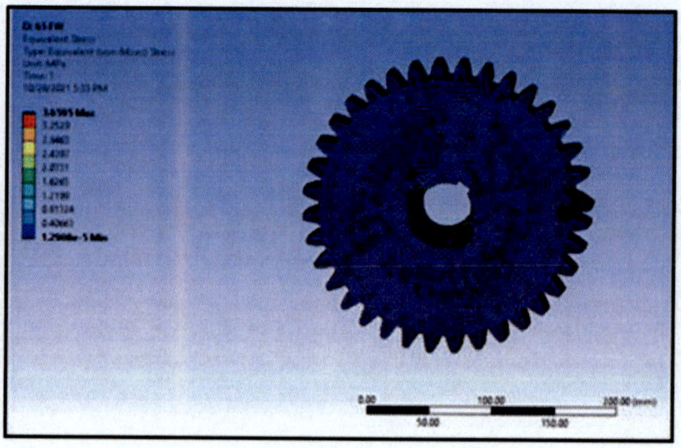

Fig. 5 Bending stress of spur gear with of 65 mm face width

Table 3 Bending stress results

Face width (mm)	Bending stress using theoretical methods (MPa)	Bending stress using Python programming (MPa)	Bending stress using Ansys results (MPa)	Error (%) between theoretical and numerical results
57.5	4.038	4.0407	4.2139	4.35
60	3.8703	3.8723	3.9483	2.01
62.5	3.7155	3.7174	3.8848	4.55
65	3.572	3.5744	3.6595	2.44

application of displacement constrain to the spur gear model generated using CAD software.

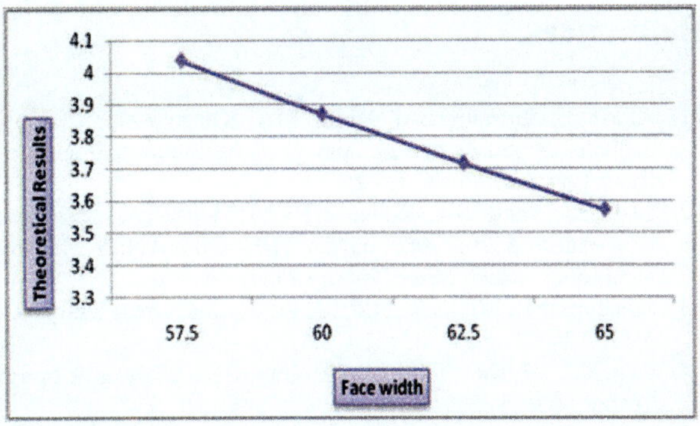

Fig. 6 Results obtained using theoretical methods

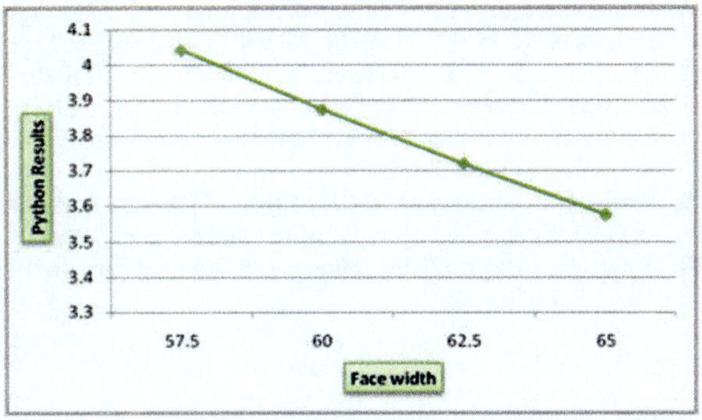

Fig. 7 Results obtained using Python program

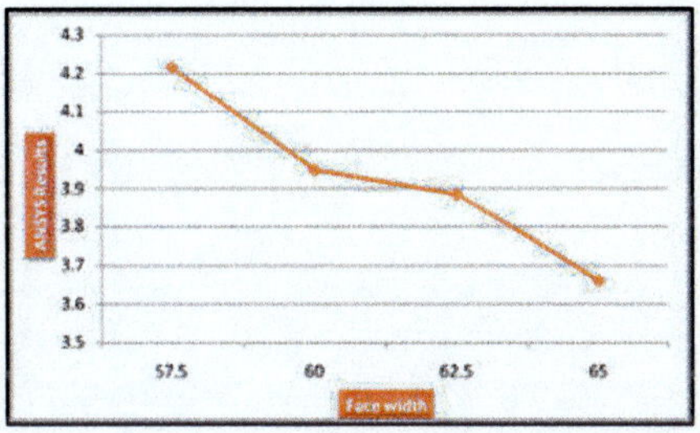

Fig. 8 Results obtained using FEA

8 Conclusion

Tooth bending stresses of the spur gears were performed with four distinct face widths. The developed FEA model and Python program for bending stresses were compared with the results obtained from the theoretical model. The results were found to be in good agreement with each other. Lower values of stresses were obtained in spur gears having higher face widths. At lower values of speed and load, the effectiveness of the spur gears was determined to be higher. It is achieved due to the reduction in the stresses developed at the tooth surfaces of the gear.

References

1. Akash J, Chitranjan A, Saloda MA, Khidiya MS (2020) Contact stress and bending stress analysis of pair of mating spur gears by theoretical and finite element analysis approach. Int Res J Eng Technol 07:12–25
2. Lisle TJ, Shaw BA, Robert CF (2017) External spur gear root bending stress: a comparison of ISO 6336:2006, AGMA 2101–D04, ANSYS finite element analysis and strain gauge techniques. Mech Mach Theory 111:1–9
3. Sivakumar B, Michael IJ (2018) Design and stress analysis of spur gear. Int Res J Eng Technol 5(5):112–122
4. Ram KR, Abishek T (2014) Bending stress analysis & optimization of spur gear. Int Res J Eng Technol 03(05):322–331
5. Patil A (2017) Bending stress analysis of spur gear. Int J Res Appl Sci Eng Technol 05(06):101–112
6. Prabhakaran S, Balaji DS, Praveen Kumar R (2017) Bending stress analysis of spur gear for material steel 15Ni2Cr1Mo28. ARPN J Eng Appl Sci 12(19):32–41
7. Vigneshwaran K, Shanmugam D, Balasuthagar C (2020) Experimental and analytical stress analysis of spur gear. IOP Conf Ser Mater Sci Eng 912(1):64–72
8. Santosh P, Saravanan K (2018) Frictional stress analysis of spur gear with misalignments. J Mech Eng Sci 12(02):72–84
9. Krishanu G, Sushovan C (2015) Effect of pressure angle of spur gears on bending and contact stresses: a comparative study using finite element software. Int J Adv Res Sci Eng 4(01)
10. Tufan GY, Fatih K (2018) Influence of root geometry on bending stress for involute spur gears. Mater Sci 05(6):12–24

Frequency Calculation of Shear Deformable Beams by Isogeometric Approach

Gourav Prasad Sinha and Bipin Kumar

Abstract Natural frequency of shear deformable beam on which free vibration is applied is considered in the research of the paper. Timoshenko beam is considered to be thick which counts the angular deformation and inertia due to rotation. Isogeometric approach is involved to calculate non-dimensional frequency of Timoshenko beam. Vibration analysis is done under the effect of different boundary conditions. Only two elements are considered in this paper. Importance of isogeometric approach can be understood by the results obtained as it was very much accurate with degree of BFs employed. It was found that the error occurred during the calculation is different for different boundary conditions. The percentage of error was within (3–6)%. As it can be observed that the percentage of error is very much within the range of accuracy even when the degree of shape function is less as compared to other method. Isogeometric analysis or spline-based analysis is found to be very effective numerical computation method.

Keywords IGA · Timoshenko beam · Knot vectors · Vibration · Dimensionless parameter

1 Introduction

Analysis of vibration of beam is considered to be firstly done by Bernoulli in his experiments. Bernoulli found that the curvature of beam largely depends upon the bending moment. The theory provided by Bernoulli further enhanced by the addition of remarks by Leonhard Euler which is combinedly named as Euler–Bernoulli beam theory. It was observed that result obtained by Euler–Bernoulli beam theory overestimates the results. It happened because of certain factors are not picked up

G. P. Sinha (✉) · B. Kumar
Department of Mechanical Engineering, NIT Jamshedpur, Jamshedpur, India
e-mail: jsrgourav@gmail.com

B. Kumar
e-mail: bipinkumar.me@nitjsr.ac.in

N. M. Sivaram et al. (eds.), *Advances in Manufacturing, Automation, Design and Energy Technologies*, Lecture Notes in Mechanical Engineering,
https://doi.org/10.1007/978-981-99-1288-9_55

in Euler–Bernoulli beam theory. Later on, Lord Rayleigh added an effect of rotary inertia in it and tried to correct the results up to some extent. Still there was an option of correction regarding shear generated in the beam which is taken care by Timoshenko. It was Timoshenko who introduced a theory on beam which included the efficacy of rotary inertia and the angular deformation [1].

Timoshenko beam theory has been examined by different models, methods at various points of time by researchers. Kapur [2] created a finite element model of Timoshenko beam element which includes rotational inertia and angular deformation into account. Davis et al. [3] in their paper modelled a Timoshenko beam element depended upon total differential equation in equilibrium of static type. The efficacy of shear coefficient on natural frequency is considered. Abbas and Thomas [4] in their paper went for finite element model of Timoshenko beam with nodal DOF. Matrices of stiffness and mass were derived out of strain and kinetic energies. Shahstry and Rao [5] in their paper found that beam with short length was subjected to effect of rotational inertia and angular deformation. They have done calculation on different boundary condition. Laura et al. [6] in their research employed Differential Quadrature Method (DQM) for vibration analysis of Timoshenko beam. The method mentioned in paper is executed by analyst by selecting finite number of points over which approximate solution of differential system is calculated through polynomial collocation approach. Lee et al. [7] derived variational formulation of shear deformable beam, i.e. Timoshenko beam by separating of deformation mode into shear deflection and bending deflection. Houmat [8] in his paper presented a four-noded finite element for Timoshenko beam added with variables DOF. The sine terms of variable numbers added with cubic polynomial were used by author to describe rotation and transverse displacement of element cross section. The compatibility between the elements was achieved by paralleling generalized coordinates at the end nodes of element. Leung et al. [9] in their paper recommended that instead of going for single polynomial, one must employ product of Fourier series and polynomial in P-element shape function. Malekzadeh et al. [10] in their paper applied the Differential Quadrature Element Method (DQEM) based on (FSDT) lower order shear deformation theory on non-uniform beam located on elastic foundation. Hamilton's principle was employed for deriving governing equation, conditions of natural compatibility between the elements. Lee et al. [11] in their research have done study on Timoshenko beams and Mindlin plates. The analysis done by authors were depended on Chebyshev psuedospectral process. Ferreira [12] in his paper discussed about the implementation of (RBF) Radial Basis Function for the vibration analysis of angular deformable beam and Mindlin plates. Eigen problem was defined by lower order shear deformation theory or in other words FSDT which was calculated by different algorithms applied by authors. Authors found that RBF becomes more powerful in case of the worse conditioned problems. Torgut et al. [13], in their paper, formulated a finite element model of Timoshenko beam considering the effect of rotational inertia and angular deformation. Rotation and deflection of the beam's cross section were expressed through polynomials. Eigen values of Timoshenko beams were calculated in different boundary conditions with different L/h ratios. Ferreira et al. [14] in their paper have done the analysis of vibration of angular deformable beam and theoretical Mindlin plates through employing RBF with addition of psuedospectral method. Civalek et al. [15] in their paper presented the analysis of natural vibration of Timoshenko beam by Discrete Singular Convolution (DSC) method. Motion equation of Timoshenko beam was solved by DSC method as numerical solution. Hsu [16] in his paper mentioned about another system for analysing natural vibration of Timoshenko beams. Hierarchical approximation and Partition of Unity Method (PUM) were two different approaches which were employed by author to enrich the mathematical space made out of existing finite element method. Osadebe et al. [17] in their paper presented the combination finite element method (FEM) and energy method to analyse the natural vibration of Timoshenko beam. In process, Timoshenko beam was virtually split into two beams, shear layer beam and Euler beam.

Hughes et al. [18, 27] created a new concept for the solution of problem based on governing equation. It was said to be more advanced than finite element method in terms of accuracy, time consumption and degree of basis functions. Reali [19] in his paper calculated natural frequency of structures through isogeometric approach. When results obtained through isogeometric approach is compared with the results of classical FEM, it was found to be in sound accord. Cotrell et al. [20] in their paper have done research on analysis of vibration of different structure through isogeometric approach. A theory of K-refinement was investigated for analysis, and it was shown that it enhances the accuracy of results. It was observed that for implementing boundary conditions in bar and Euler beam problems was not simple because NURBS basis function in general is not an interpolation function. Penalty method and Lagrangian multiplier method have to be implemented in order to apply boundary condition. Wang et al. [21] in their research paper deduced an enhanced formulation for isogeometric analysis. Authors employed a transformation method which directs the control variable to collocated nodal value at essential boundary conditions. Results show that it provide greater accuracy and have good agreement with the exact results. Lee et al. [22] in their research have done the natural vibration analysis of Timoshenko beam through isogeometric approach. They have considered rotary inertia effect as well as shear deformation effect along with shear locking factor. Authors applied h-p-k refinement for greater accuracy with exact results. Jockovic et al. [23] also done their analysis of natural vibration of beams. They considered rotation free thin (Euler–Bernoulli) beam and Timoshenko beam with different BC's and variable L/h ratios. Results obtained through isogeometric approach is than collate with exact analytical solutions and classical finite element (FEM) solutions. Mondal et al. [24] in their research have presented a novel method to imply Neumann boundary condition to one-dimensional beam problem. They applied penalty approach method to the cantilever beam for the deflection of beam. The results obtained through this approach were in good agreement with exact results.

2 Basics of B-Splines

2.1 *Knot Vectors*

A knot vector ξ is known to be group of real number, in increasing order and never decreasing in nature which is used for constituting coordinates in parametric dimension.

$$\xi = [\xi 1, \xi 2, \xi 3, \ldots \xi n + p + 1] \tag{1}$$

In the above written equation, n stands for number of basis function and p stands for order of concerned basis function of B-spline. Knot vectors are classified as uniform knot vector and non-uniform knot vector. Uniform knot vectors are those whose knots are uniformly spaced, whereas non-uniform knot vectors are those whose knots are not uniformly spaced. Another classification of knot vector is open and closed knot vector. If last and first knot vectors found to be in multiplication of $p + 1$ time, then it is said to be open knot vectors otherwise close. Throughout in this paper, we will use open knot vectors because open knot vectors are naturally interpolatory at the end points of parametric range $[\xi_1, \xi_{n+p+1}]$. Open knot vectors are non-interpolatory at inner point in general.

2.2 *Basis Functions*

Definition of B-spline function for $p = 0$ is given by

$$\begin{aligned} N_{i,0}(\xi) = & 1, \text{ if } \xi_i < \xi < \xi_{i+1} \\ & 0, \text{ else} \end{aligned} \tag{2}$$

For higher orders

$$N_{i,p}(\xi) = \frac{(\xi - \xi_i)Ni, p - 1(\xi)}{\xi i + p - \xi i} + \frac{\xi i + p + 1 - (\xi)\ Ni + 1, p - 1(\xi)}{\xi i + p + 1 - \xi i + 1} \tag{3}$$

As it is well known that continuity of B-spline function is dependent on the multiplicity of knots. A basis function is said to be having C^{p-1} continuity if they are not repeated. If any knot is repeated k times, then the continuity will be given as C^{p-k} [25].

2.3 B-spline Curve

The curve of B-spline generated is piecewise polynomial in nature of order p obtained through merger between basis function and control points.

Given "n" number of basis functions, $N(i,p)$, where $i = 1, 2,\ldots, n$, and correlated control points $Bi \in$ Rd, $i = 1, 2,\ldots, n$, a B-spline curve of polynomial and piecewise in nature is given by

$$C(\xi) = \sum_{i=1}^{n} Ni,\ p(\xi)Bi, \tag{4}$$

where p and n stand for order of engaged basis function and number of control points. i stands for the corresponding control points [26].

3 Isogeometric Formulation

In this paper, linear elastic and isotropic beam element is considered for analysis. As it is already mentioned that we are analysing on the basis of theory of Timoshenko beam while considering shear deformation of transverse nature. The deformation in beam can be represented as shown in Fig. 1.

For calculation of natural frequency of free vibration in Timoshenko beam, a dynamic equilibrium equation has been derived on the principal of virtual work.

$$\int_0^L EI\frac{\partial\theta}{\partial x}\delta\frac{\partial\theta}{\partial x} + \int_0^L KGA\left(\frac{\partial w}{\partial x} - \theta\right)\delta\left(\frac{\partial w}{\partial x} - \theta\right)dx = \int_0^L \delta w\rho A\ddot{w}dx + \int_0^L \delta\theta\rho I\ddot{\theta}dx \tag{5}$$

In above mentioned equation, following terms stands for.

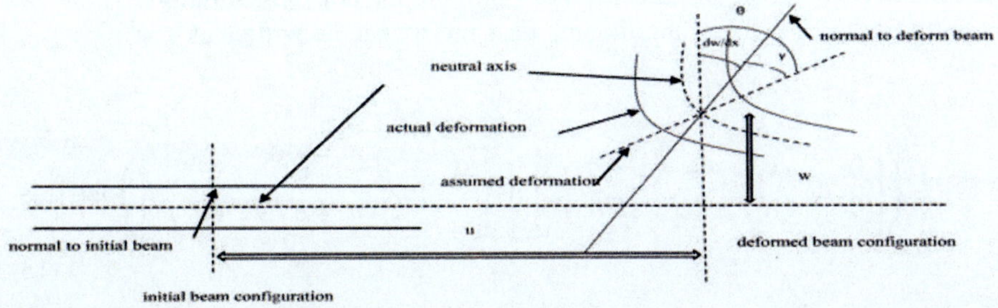

Fig. 1 Deformation of cross section of thick beam [22]

w = displacement, θ = angular displacement or rotation, E = modulus of elasticity, I = moment of inertia, k = shear correction factor, G = modulus of rigidity, $\ddot{w}$ = acceleration, $\ddot{\theta}$= angular acceleration, ρ = density δ = virtual terms of equations and L stands for length of the beam.

Above-mentioned virtual statement of problem is transformed into group of algebraic equations with the involvement of B-spline basis functions. They are represented as

$$w = Nw, \quad \ddot{w} = N\ddot{w}, \quad \theta = N\theta, \quad \ddot{\theta} = N\ddot{\theta} \tag{6}$$

Virtual terms related to displacement and acceleration are given as

$$\delta w = N\delta w, \quad \delta\theta = N\delta\theta, \quad \delta w = N\delta w, \quad \delta\theta = N\delta\theta \tag{7}$$

where N is the basis functions generated from Eqs. (2) and (3). Substituting the value of Eqs. (6) and (7) into Eq. (5), we get simplified equation as

$$\delta \text{uT}[\text{Ku} - \text{M}u] = 0 \tag{8}$$

$$\text{Ku} - M\ddot{u} = 0 \tag{9}$$

$$u = \varphi_k e^{i\omega k} \tag{10}$$

$$[\text{K}-\omega_{\text{k}}^2\text{M}]\varphi_{\text{k}} = 0 \tag{11}$$

ωk stands for natural frequency of natural vibration of kth mode, whereas φk is amplitude of motion at control points.

$$\left|K-\omega_k^2 M\right| = 0 \tag{12}$$

As it can be observed from above written equation that it is of determinant type. We have to solve an eigen value problem and find the non-zero solutions.

From Eq. (12), stiffness matrix and mass matrix can be written as

$$\begin{aligned} \text{K} &= K_{ab}^s + K_{ab}^b \\ &= \int_e^{e+1} \left(\begin{bmatrix} 0 & 0 \\ 0\ \frac{1}{j}\nabla\xi Na,\ pEI\nabla\xi Nb,\ p \end{bmatrix} + \begin{bmatrix} 0 & 0 \\ 0\ \frac{1}{j}\nabla\xi Na,\ pEI\nabla\xi Nb,\ p \end{bmatrix} \right) \text{det(j)d}\xi \end{aligned} \tag{13}$$

$$M = \int_e^{e+1} (Na,\ p\rho ANb,\ p + Na,\ p\rho INb,\ p)\text{det}(j)d\xi \tag{14}$$

In Eqs. (13) and (14), *Na,p* and *Nb,p* are the basis functions generated with the help of Eqs. (2 and 3), e and $e + 1$ stands for the elements considered while calculating the values mentioned in matrices. J represents the Jacobian between ξ and x.

3.1 Calculation of Basis Functions:

Let say knot vectors are given as.

ξ1	ξ2	ξ3	ξ4	ξ5	ξ6	ξ7
0	0	0	0.5	1	1	1

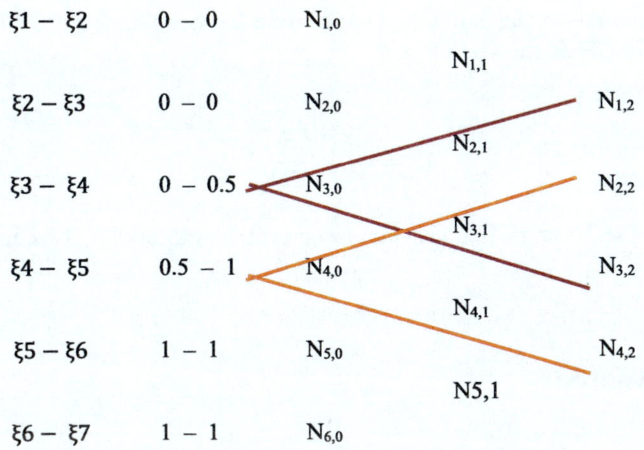

Here are the calculated values of basis functions of order 1 & 2
$N1,0 = 0$ $N2,0 = 0$ $N3,0 = 1$ $N4,0 = 1$ $N5,0 = 0$ $N6,0 = 0$.
$N1,1 = 0$ $N2,1 = 1-2\xi\ (N3,0)$ $N3,1 = 2\xi\ (N3,0) + 2(1-\xi)\ (N4,0)$ $N4,1 = 2(\xi-0.5)\ (N4,0)$ $N5,1 = 0$.

4 Numerical Results

Putting the values of basis functions as given in Table 1 in Eqs. (13 and 14) prepares stiffness and mass matrix. As we can observe from the equation that it will create 2 × 2 matrix for each segment of each element. Upon assembling the individual matrices along to prepare a stiffness matrix for a single element result in 6 × 6 matrix. After obtaining stiffness matrices for element 1 "(0 – 0.5)" and element 2 "(0.5 – 1)", we have to assemble them into a single matrix as same as in finite element analysis. The

Table 1 B-spline basis function

Basis function	Element 1	Element 2
$N_{1,2}$	$(2\xi-1)^2$	–
$N_{2,2}$	$2\xi(2-3\xi)$	$2(\xi-1)^2$
$N_{3,2}$	$2\xi^2$	$(2-2\xi)(3\xi-1)$
$N_{4,2}$	–	$(2\xi-1)^2$

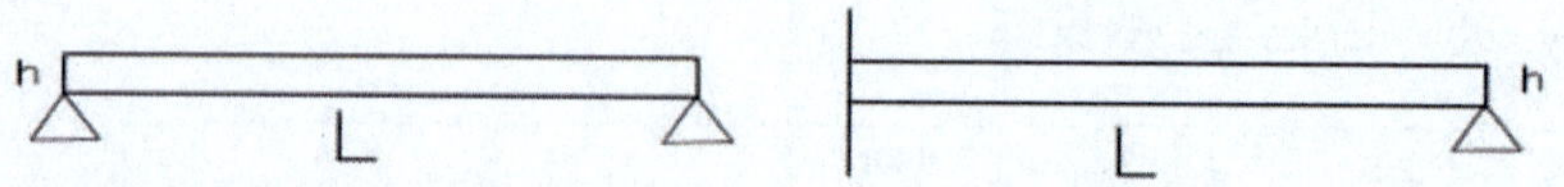

Fig. 2 Boundary condition

resulting matrix will be in the size of 8 × 8 matrix for stiffness matrix. $L/h = 100$ has been considered for the solution. The solution has been done for the boundary condition HH & CH as shown in Fig. 2.

$$\lambda^2 = \omega L^2 \sqrt{\frac{m}{EI}} \tag{15}$$

In Eq. 15, the letter m denotes mass per unit length and E, I as mentioned previously.

5 Result Analysis

Calculation of every boundary condition has been done. Four different boundary conditions have been applied to isogeometric problem. We will analyse each and every boundary conditions. Transverse beam which is shear deformable nature is been subjected to natural vibration under different BCs.

As we can observe from the results and error percentage mentioned in Table 2 that results under boundary conditions have some error on it. Error of the obtained results is shown in Fig. 3. From there we can observe that the calculated results are in close approximation of the referred results.

Table 2 Comparison of results obtained through Isogeometric analysis

B C	Hinged–hinged	Clamped-hinged	Clamped–clamped	Clamped-free
λ (calculated)	8.7514	12.8064	3.1610	19.2517
λ [17]	9.1259	13.6141	3.3312	19.9256
Error in %	4.10	5.93	5.10	3.38

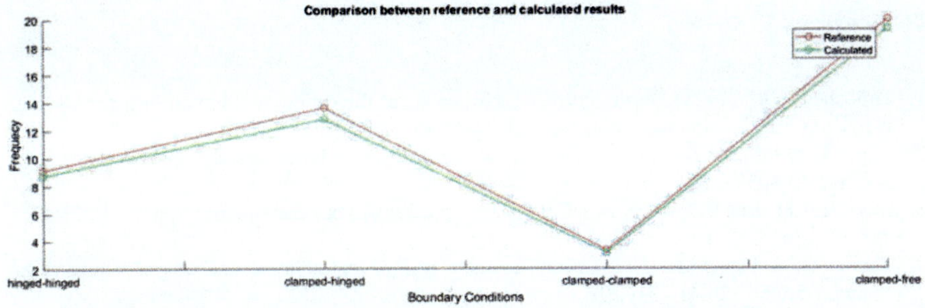

Fig. 3 Graphical representation of results

6 Conclusion

All the calculation mentioned in this paper is done manually without any aid of programming in any software. The result as it is mentioned in Table 2 is in satisfactory agreement with the previous calculated results. From the methods I applied and result we have got I made some point wise conclusions. They are as follows:

Use of isogeometric approach gives the freedom of selecting the order of basis function to be employed. Results we got from FEA by employing higher order basis functions can be easily calculated by employing lower order basis function in isogeometric approach, thus making it simpler with great accuracy.

The level of accuracy in results obtained through isogeometric approach is much higher than the parallel FE analysis.

On the basis of outcome acquired through isogeometric approach, it can easily be concluded that isogeometric approach is a powerful tool to derive the natural frequency of the vibration indulged in beam. Isogeometric approach becomes an automatic choice when the shape of beams are not the conventional ones.

References

1. Timoshenko SP (1921) On the correction for shear of the differential equation for transverse vibrations of prismatic bars. Phil Mag 41:744–746
2. Kapur K (1966) Vibrations of a Timoshenko beam using a finite element approach. J Acoust Soc Am 40:1058–1063
3. Davis R, Henshell RD, Warburton GB (1972) Timoshenko beam element. J Sound Vib 22:475–487
4. Thomas J, Abbas BAH (1975) Finite element model for dynamic analysis of Timoshenko beams. J Sound Vib 41:291–299
5. Shastry BP, Rao GV (1985) Free vibration of shorts beam. J Sound Vib 100:305–308
6. Laura PAA, Gutierrez RH (1993) Analysis of vibrating Timoshenko beams using the method of differential quadrature. Shock Vib 1:89–93
7. Lee SS, Koo JS, Choi JM (1994) Variational formulation for Timoshenko beam element by separation of deformation mode. Commun Numer Methods Eng 10:599–610
8. Houmat A (1995) Vibrations of Timoshenko beams by variable order finite elements. J Sound Vib 187:841–849
9. Leung AYT, Chan JKW (1998) Fourier p-element for the analysis of beams and plates. J Sound Vib 212(1):179–185
10. Malekzadeh P, Karami G, Farid M (2003) DQEM for free vibration analysis of Timoshenko beams on elastic foundations. Int J Comput Mech 31:219–228
11. Lee J, Schultz WW (2004) Eigenvalue analysis of Timoshenko beams and axisymmetric Mindlin plates by the pseudospectral method. J Sound Vib 269:609–621
12. Ferreira AJM (2005) Free vibration analysis of Timoshenko beams and Mindlin plates by radial basis functions. Int J Comput Methods 2:15–31
13. KocatÜrk T, Simsek M (2005) Free vibration analysis of Timoshenko beams under various boundary conditions. Sigma J Eng Nat Sci 2005:30–44
14. Ferreira AJM, Fasshauer GE (2006) Computation of natural frequencies of shear deformable beams and plates by an RBF-pseudospectral method. Comput Methods Appl Mech Eng 196:134–146
15. Civalek O, Kiracioglu O (2010) Free vibration analysis of Timoshenko beams by DSC method. Int J Numer Meth Biomed Engg 26:1890–1898
16. Hsu YS (2016) Enriched finite element methods for Timoshenko beam free vibration analysis. Appl Math Model 1–22
17. Osadebe NN, Agunwamba JC, Onyia ME, Rowland Lato EO (2018) Free vibration analysis of Timoshenko beam using energy separation principle. Int J Sci Eng Res 9(2)
18. Hughes TJR, Cottrell JA, Bazilevs Y (2005) Isogeometric analysis: CAD, finite elements, NURBS, exact geometry and mesh refinement. Comput Methods Appl Mech Eng 194:4135–4195
19. Reali A (2006) An isogeometric analysis approach for the study of structural vibrations. J Earthquake Eng 10:1–30
20. Cottrell J, Reali R, Bazilevs Y, Hughes TJR (2006) Isogeometric analysis of structural vibrations. Comput Methods Appl Mech Eng 195:5257–5296
21. Wang D, Xuan J (2010) An improved NURBS-based isogeometric analysis with enhanced treatment of essential boundary conditions. Comput Methods Appl Mech Eng 199:2425–2436
22. Lee SJ, Park KS (2013) Vibrations of Timoshenko beams with isogeometric approach. Appl Math Model 37: 9174–9190
23. Jočković M, Nefovska Danilović M (2016) Free vibration analysis of beam element using isogeometric analysis. In: Conference proceeding 4th international conference contemporary achievements in civil engineering 2016, pp 269–278
24. Mondal SK, Gondegaon S, Voruganti HK (2017) A novel method to apply Neumann boundary conditions in the Isogeometric Analysis (IGA) of beam with 1-D formulation. World J Eng 14(6):538–544

25. De Boor C (1978) A practical guide to splines. Springer
26. Rogers D (2001) An introduction to NURBS: with historical perspective. Academic Press, London, UK
27. Cottrell JA, Bazilevs Y, Hughes TJR (2009) Isogeometric analysis: towards integration of CAD and FEA. Wiley

Deflection of Thin Plate Through Isogeometric Analysis

Gourav Prasad Sinha, Bipin Kumar, and K. Priya Ajit

Abstract The aim of the paper is to obtain the deflection of plates when plates are subjected to different types of boundary conditions. The approach used for the deriving the results is isogeometric approach (IGA). IGA is the latest approach in the field of numerical computation. Basis function capable of replicating geometries is used for approximating solution field. Plates considered for research are thin and isotropic in nature. Kirchhoff thin plate theory is applied to obtain the deflection of plates. Shape of the plate employed is square. Two different boundary conditions are applied over the plates (CCCC and SSSS). Isogeometric analysis was carried out through MATLAB. Results obtained through program were in good agreement with the previous results.

Keywords IGA · Kirchhoff plates · Knot vectors · Deflection

1 Introduction

Kirchhoff plate theory which is also known as classical plate theory is one of the earliest theories given on plates. Classical plate theory is applicable to thin plates as it ignores the effect of shear deformation. This theory still holds an importance because of its simplicity and provides a base for further higher shear deformation theory. Again, there are number of methods through which theories of plates were examined. Plates are subjected to different types of loading conditions with different boundary condition. Deflection is the one of the main parameters to be observed

G. P. Sinha (✉) · B. Kumar
Department of Mechanical Engineering, NIT Jamshedpur, Jamshedpur, India
e-mail: jsrgourav@gmail.com

B. Kumar
e-mail: bipinkumar.me@nitjsr.ac.in

K. Priya Ajit
Department of Mining and Machinery Engineering IIT (ISM), Dhanbad, India
e-mail: ajitkp@iitism.ac.in

N. M. Sivaram et al. (eds.), *Advances in Manufacturing, Automation, Design and Energy Technologies*, Lecture Notes in Mechanical Engineering,
https://doi.org/10.1007/978-981-99-1288-9_56

under loading condition. Many attempts have been made in past to examine the deflection of plates with different loading condition and boundary condition.

The detailed analysis of large flat plates is first observed by Von Karman. He provided large plate large plate theory based on Von Karman Equation [1]. Marguerre [2] in his paper explained the nature of imperfect plates subjected to transverse and in-plane loads. Timoshenko et al. [3] have penned down their theories of plates and shell and drawn attentions on the limitation of Von Karman's equations. Allami et al. [4] done calculations on basis of finite difference method for more than 100 cases of elastic plates with different boundary conditions and loading conditions. Bhattacharya [5] in his paper developed a new finite difference method to analysis the deflection of plates under dynamic and static loading condition. Liew et al. [6] in their paper done the static analysis of plates which are rectangular in shape and having three dimensions with their own derivation of differential quadrature and harmonic differential quadrature method. Defu et al. [7] in their research paper provided a mathematical approach for calculating deflection of large plates. Solution for different boundary conditions can be easily calculated through this method. Bakker et al. [8] have done their research on method of approximate analysis for deflection of large magnitude on thin plates under the action of normal loading condition, and boundary condition was all simply supported.

Hugh et al. [9, 20] proposed a new numerical computation method to be known as isogeometric approach. They observed that out of total time consumption in analysis of any problem, 80% of time is consumed in only designing or meshing and remaining 20% for analysis work. In finite element analysis, first approximation of unknown solution field is done and then same approximation is used for designing, whereas in Isogeometric approach method has been reversed. In isogeometric approach, method is to select the basis functions which are capable of represent design as exact as possible, and then same basis function is used to approximate the unknown solution field. Cotrell et al. [10] have done intense research on the refinement, continuity, and smoothness of basis function. The same basis function is to be employed for approximating the design and unknown solution field. Deng et al. [11] in their research have found some error while implementing basis function prepared through NURBS. To reduce the error, T-spline basis functions are formulated. Schillinger et al [12] in their paper raised an issue related to implementation of T-spline, as T-spline proofs to favorable for local refinement but at the same time insertion of knots are complex. Wall et al. [13] focused their research on shape optimization through IGA. They observed the analytical responsiveness of discretization of B-spline by applying the gradient-based algorithm of optimization. Hassani et al. [14] in their paper studied about the shape optimization of 2D and 3D problems with the help of isogeometric analysis. Kiendl et al. [15] in their paper applied isogeometric approach to have the analysis of Kirchhoff-Love shell structures and found that smoothness of non-rational uniform B-spline has allowed the researchers to form rotation free formulation of shell and plate elements. Vanam et al. [16] have done analysis of plates deflection through FEA and that results will be taken for validation. Mondal et al. in their paper researched on the method to apply boundary conditions on the Timoshenko beams [17].

As it can be observed through literature review of isogeometric approach, that IGA is bridging the gap between the design and post processing in the most efficient way. Though it is in still developing mode, but its application has made a deep impact in many aspects such as medicals and various field of engineering.

2 Basics of B-Splines

2.1 Knot Vectors

A knot vector ζ is known to be group of real number, in increasing order and never decreasing in nature which is used for constituting co-ordinates in parametric dimension.

$$\zeta = [\zeta_1, \zeta_2, \zeta_3, \ldots, \zeta_{n+p+1}] \tag{1}$$

In the above written equation, n stands for number of basis function and p stands for order of concerned basis function of B-spline. Knot vectors are classified as uniform knot vector and non-uniform knot vector. Uniform knot vectors are those whose knots are uniformly spaced, whereas non-uniform knot vectors are those whose knots are not uniformly spaced. Another classification of knot vector is open and closed knot vector. If last and first knot vectors found to be in multiplication of p+1 time, then it is said to be open knot vectors otherwise close. Throughout in this paper, we will use open knot vectors because open knot vectors are naturally interpolatory at the end points of parametric range $\zeta = [\zeta_1, \zeta_{n+p+1}]$. Open knot vectors are non interpolatory at inner point in general.

2.2 Basis Functions

Definition of B-spline function for $p = 0$ is given by

$$\begin{aligned} N_{i,0}(\zeta) = 1,&\ \text{if}\ \zeta_i <; \zeta < \zeta_{i+1} \\ 0,&\ \text{else} \end{aligned} \tag{2}$$

For higher orders

$$N_{i,p}(\zeta) = \frac{(\zeta - \zeta i)\, N_i,\ p-1(\zeta)}{(\zeta i + p - \zeta i)} + \frac{(\zeta i + p + 1 - \zeta)\, N_i + 1,\ p - 1(\zeta)}{\zeta i + p + 1 - \zeta i + 1} \tag{3}$$

As it is well known that continuity of B-spline function is dependent on the multiplicity of knots. A basis function is said to be having C^{p-1} continuity if they are not repeated. If any knot is repeated k times, then the continuity will be given as C^{p-k} [18].

2.3 B-spline curve

The curve of B-spline generated is piecewise polynomial in nature of order p obtained through merger between basis function and control points.

Given "*n*" number of basis functions, $N_{(i,p)}$, where $i = 1, 2, \ldots, n$, and correlated control points $Bi \in Rd, i = 1, 2, \ldots, n$, a B-spline curve of polynomial and piecewise in nature is given by

$$C(\xi) = \sum_{i=1}^{n} N_i p(\zeta)\mathrm{B}i \tag{4}$$

where *p* and *n* stand for order of engaged basis function and number of control points. *i* stands for the corresponding control points [19].

3 Isogeometric Formulation

In this paper, static analysis of an isotropic plate, square in shape, is to be done. The plate considered for studies is thin. As plate is thin, effect of shear deformation can be neglected; therefore, we can apply the theory of classical plate or Kirchhoff plate theory. Shape of the plate is shown in Fig. 1.

The basic governing equation of classical plate theory is given by

$$\frac{\partial^4 \mathrm{w}}{\partial \mathrm{x}^4} + 2\frac{\partial^4 \mathrm{w}}{\partial^2 \mathrm{x} \partial^2 \mathrm{y}} + \frac{\partial^4 \mathrm{w}}{\partial \mathrm{y}^4} + \frac{q}{D} = 0 \tag{5}$$

where $w =$ displacement,

$q =$ normal load distribution per unit area,

$D =$ flexural rigidity of plate material.

Fig. 1 Plate of equal side

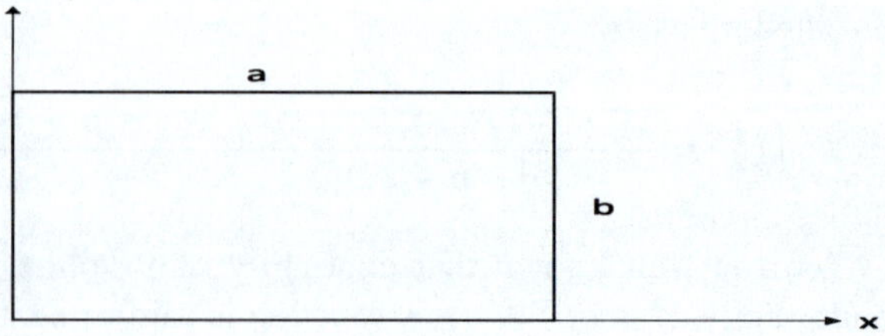

The weak form of governing equation is given as

$$\int_{\Omega} \epsilon(u) : D : \epsilon(\delta u) = \int_{\Omega} b.\delta u d\Omega \tag{6}$$

where b is body force and others are same.

By applying the Bubnov–Galerkin method where the same shape function (ϕ) is applied to define both u and δu.

$$u(x) = \sum_{I}^{n} \phi_I(\xi) u_I \qquad \delta u(x) = \sum_{I}^{n} \phi_I(\xi) \delta u_I \tag{7}$$

Substitution of Eq. (7) into Eq. (6) provides the following relation:

$$K.u = f \tag{8}$$

$$K_{IJ} = \int_{\Omega} B_I^T D B_j d\Omega \qquad f_I = \int_{\Omega} \phi_I(\xi) \delta u_I \tag{9}$$

4 Calculation of Basis Functions

Let say knot vectors are given as

ζ_1	ζ_2	ζ_3	ζ_4	ζ_5	ζ_6	ζ_7	ζ_8	ζ_9	ζ_{10}	ζ_{11}	ζ_{12}	ζ_{13}
0	0	0	0.125	0.25	0.375	0.5	0.625	0.75	0.875	1	1	1

$\zeta_1 - \zeta_2$	$0 - 0$	$N_{1,0}$		
			$N_{1,1}$	
$\zeta_2 - \zeta_3$	$0 - 0$	$N_{2,0}$		$N_{1,2}$
			$N_{2,1}$	
$\zeta_3 - \zeta_4$	$0 - 0.125$	$N_{3,0}$		$N_{2,2}$
			$N_{3,1}$	
$\zeta_4 - \zeta_5$	$0.125 - 0.25$	$N_{4,0}$		$N_{3,2}$
			$N_{4,1}$	
$\zeta_5 - \zeta_6$	$0.25 - 0.375$	$N_{5,0}$		$N_{4,2}$
	$0.375 - 0.5$		$N_{5,1}$	
$\zeta_6 - \zeta_7$		$N_{6,0}$		$N_{5,2}$
			$N_{6,1}$	

(continued)

(continued)

$\zeta_1 - \zeta_2$	$0 - 0$	$N_{1,0}$		
$\zeta\_7 - \zeta\_8$	$0.5 - 0.625$	$N_{7,0}$		$N_{6,2}$
	$0.625 - 0.75$		$N_{8,1}$	
$\zeta\_8 - \zeta\_9$		$N_{8,0}$		$N_{7,2}$
	$0.75 - 0.875$		$N_{8,1}$	
$\zeta\_9 - \zeta\_10$		$N_{9,0}$		$N_{8,2}$
			$N_{9,1}$	
$\zeta 10 - \zeta 11$		$N_{10,0}$		$N_{9,2}$
	$0.875 - 1$		$N_{10,1}$	
$\zeta 11 - \zeta 12$	$1 - 1$	$N_{11,0}$		$N_{10,2}$
			$N_{11,1}$	
$\zeta 12 - \zeta 13$	$1 - 1$	$N_{12,0}$		

We have calculated the basis function for set of 13 knot vectors. The knot vectors employed are having $P + 1$ multiplicity. These knot vectors are open knot vector. The selected order for this study is 2. So, in total we have final ten basis functions as it can be observed in Fig. 2. Same knot vectors are applied for both x and y direction.

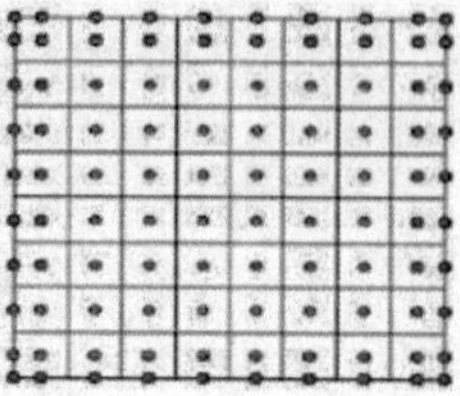

Fig. 2 Node representation of isogeometric formulation

5 Numerical Results

The numerical values considered for the problem are given as follows:

$E = 70 * 10^9$ Gpa, $\mu = 0.33$, $a = b = 1$, UDL $= 500$ N/m^2 in the downward direction.

The number of degrees of freedom is considered to be one over every control points. As it can be observed that the total number of final basis function of order 2 is ten so problem requires 10 control points in each direction. So, the total degree of freedom is hundred. The prepared stiffness matrix is of 100 * 100 dimension, whereas the matrix of force vector is of 100 * 1. Maximum deflection is recorded for different boundary condition. The boundary condition applied is CCCC and SSSS.

6 Result Analysis

The expressions are derived for exact calculation of maximum deflection through Kirchhoff plate theory. Isogeometric approach has been considered for the numerical computation of the required results. Second order of NURBS function is used for knot vectors of both directions.

The calculation has been done on the different thickness. The range of thickness covered in the calculation is from 0.01 to 0.20. The maximum deflection varies from $9.86 * 10^{-5}$ to $1.69 * 10^{-8}$. All the dimension of length parameters is in meter. As shown in Fig. 3 and Table 1 comparison between the calculated results and exact values are shown.

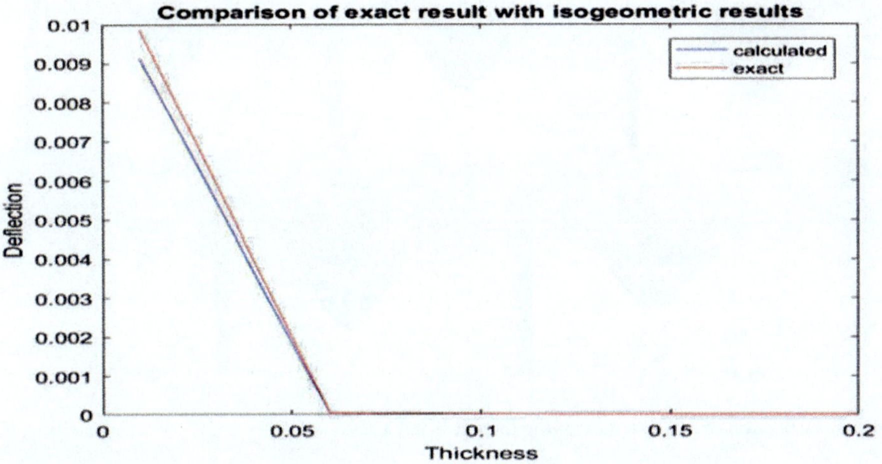

Fig. 3 Graphical representation of results

Table 1 Comparison of exact and calculated results

Thickness	Maximum deflection calculated (CCCC)	Exact result [16]	Maximum deflection calculated (SSSS)	Exact result [16]
0.01	$9.1276 * 10^{-3}$	$9.86 * 10^{-3}$	$3.0883 * 10^{-5}$	$3.15 * 10^{-5}$
0.06	$4.2258 * 10^{-5}$	$4.56 * 10^{-5}$	$1.4298 * 10^{-6}$	$1.43 * 10^{-6}$
0.12	$5.2822 * 10^{-6}$	$5.559 * 10^{-6}$	$1.7827 * 10^{-6}$	$1.79 * 10^{-6}$
0.18	$1.5651 * 10^{-6}$	$1.66 * 10^{-6}$	$5.2954 * 10^{-8}$	$5.3 * 10^{-8}$
0.20	$1.141 * 10^{-6}$	$1.521 * 10^{-6}$	$3.8604 * 10^{-8}$	$3.845 * 10^{-8}$

7 Conclusion

It can be observed through Figs. 4 and 5 that plate of given dimensions and material properties will deflect according to thickness provided. The more thickness you increase, the magnitude of deflection decreases.

Through results, it can be observed that the calculated results are in very close agreement with the previous results. Previous results are calculated on the FEA software, whereas this calculation has been done on second order function, still isogeometric analysis provides accurate results. This concludes that isogeometric

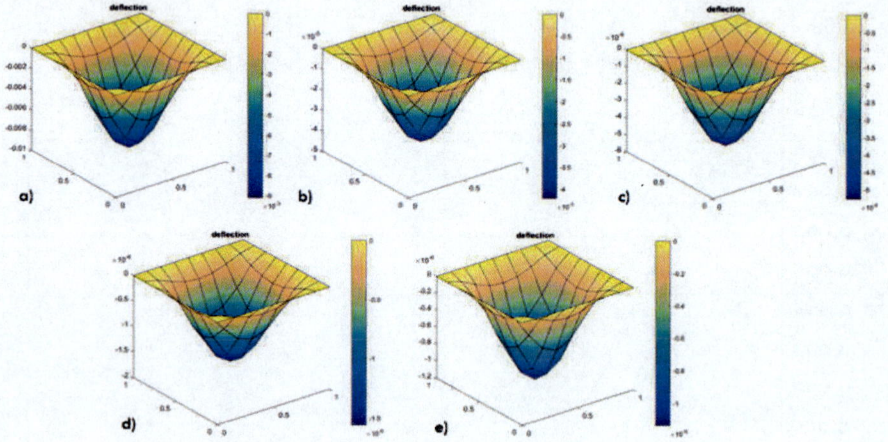

Fig. 4 Deflection of plates with varying thickness in CCCC

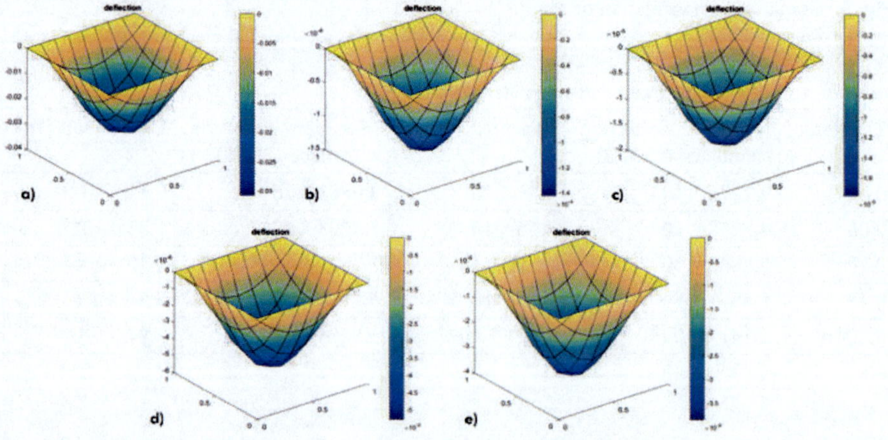

Fig. 5 Deflection of plates with varying thickness in SSSS

analysis if far ahead in terms of numerical computation. Through isogeometric analysis, problem related of complex shape geometries can be solved with ease. Very less amount of computation effort is required when it comes to isogeometric analysis.

Isogeometric approach works on a basis function which were formulated to design geometry and same basis function is used for approximating unknown field. With Lagrange basis function used in FEA, there is C^0 continuity between the elements, whereas the continuity in IGA depends on number of knot multiplicity and higher order NURBS basis function. All these factors make it more computational efficient with respect to conventional computational method.

References

1. Von Karman T (1910) Festigheitsprobleme im Maschinenbau. Encyclopaedie der Mathematischen Wissenschafte 3:349
2. Marguerre K (1938) Zur Theorie der gekrümmten Platte grosser formänderung. In: Proceedings of the fifth international congress for applied mechanics, Cambridge, UK
3. Timoshenko SP, Woinowsky-Krieger S (1959) Theory of plates and shells. McGraw-Hill Book Company Maidenhead, England
4. Aalami B, Williams DG (1975) Thin plate design for transverse loading. Crosby Lockwood Staples, London
5. Bhattacharya MC (1986) Static and dynamic deflections of plates of arbitrary geometry by a new finite difference approach. J Sound Vib 107(3):507–521
6. Liew KM, Teo TM, Han JB (1999) Three-dimensional static solutions of rectangular plates by variant differential quadrature method. Int J Mech Sci 43:1611–1628
7. Defu W, Sheikh E (2005) Large deflection mathematical analysis of rectangular plates. J Eng Mech 131(8):809–821
8. Bakker MCM, Rosmanit M, Hofmeyer H (2008) approximate large deflection analysis of simply supported rectangular plates under transverse loading using plate post-buckling solutions. Struct Des 46(11):1224–1235
9. Hughes TJR, Cottrell JA, Bazilevs Y (2005) Isogeometric analysis: CAD, finite elements, NURBS, exact geometry and mesh refinement. Comput Methods Appl Mech Eng 194:4135–4195
10. Cottrell JA, Hughes TJR, Reali A (2007) Studies of refinement and continuity in isogeometric structural analysis. Comput Methods Appl Mech Eng 196(41):4160–4183
11. Deng J, Chen F, Li X, Hu C, Tong W, Yang Z, Feng Y (2008) Polynomial splines over hierarchical T-meshes. Graph Models 70(4):76–86
12. Schillinger D, Dede L, Scott MA, Evans JA, Borden MJ, Rank T, Hughes TJ (2012) An isogeometric design-through-analysis methodology based on adaptive hierarchical refinement of NURBS, immersed boundary methods, and T-spline CAD surfaces. Comput Methods Appl Mech Eng 249:116–150
13. Wall WA, Frenzel MA, Cyron C (2008) Isogeometric structural shape optimization. Comput Methods Appl Mech Eng 197(33):2976–2988
14. Hassani B, Tavakkoli SM, Moghadam NZ (2011) Application of isogeometric analysis in structural shape optimization. Sci Iranica 18(4):846–852
15. Kiendl J, Bletzinger KU, Linhard J, Wüchner R (2009) Isogeometric shell analysis with Kirchhoff–Love elements. Comput Methods Appl Mech Eng 198(49):3902–3914
16. Vanam BCL, Rajyalakshmi M, Inala R (2012) Static analysis of an isotropic rectangular plate using FEM. J Mech Eng Res 4(4):148–162

17. Mondal SK, Gondegaon S, Voruganti HK (2017) A novel method to apply Neumann boundary conditions in the Isogeometric Analysis (IGA) of beam with 1-D formulation. World J Eng 14(6):538–544
18. De Boor C (1978) A practical guide to splines. Springer
19. Rogers D (2001) An introduction to NURBS: with historical perspective. Academic Press, London, UK
20. Cottrell JA, Bazilevs Y, Hughes TJR (2009) Isogeometric analysis: towards integration of CAD and FEA, Wiley

Specific Energy Absorption and Bending Resistance of Hybrid Bumper Beam

Gaurav Tiwari and Amita Shinde

Abstract In this study, a numerical investigation was carried out on an innovative double hat shape beam that involves DP800 steel–aluminum 6061-T6 hybrid material to evaluate crashworthiness characteristics such as specific energy absorption (SEA) and bending resistance for automobiles application. The length of the double hat shape beam was 500 mm and thickness as 1.5 mm. For both the hat, the width was taken as 80 mm, and the flange width was 20 mm. The bending response was evaluated with the help of a three-point bending test by using commercial code ANSYS. Three different models were made by changing height ratios (1:3, 1:1, 3:1) of both the beams to analyze the effect on crashworthiness parameters. The results were validated through experimental results available in literature and found close to each other.

Keywords Double hat shape beam · Hybrid material · Bending resistance

1 Introduction

Nowadays, with the rapid growth in the market, higher safety requirements were observed in automobile and aviation industries which led to increase in use of thin-walled structures applications for energy absorbing devices. Bumper is the first element which comes in contact during collision. In this regard, many literature addresses the bending response of beams with varying their material and shapes.

Huang et al. [1] performed both experimental and numerical studies on the bending behavior of a thin-walled beam under quasi-static loading. It was concluded that the deformation pattern of thin-walled structures or tubes changes with the change in the load position. Sadowski et al. [2] investigated elastic response and degradation of thin-walled specimen up to the last stage of elastic response. Three-point bending

G. Tiwari (✉) · A. Shinde
Department of Mechanical Engineering, Visvesvaraya National Institute of Technology, Nagpur 440010, India
e-mail: gauraviitdelhi@gmail.com

N. M. Sivaram et al. (eds.), *Advances in Manufacturing, Automation, Design and Energy Technologies*, Lecture Notes in Mechanical Engineering,
https://doi.org/10.1007/978-981-99-1288-9_57

tests were performed, and it was concluded that when aluminum undergoes local buckling, then damaging observed in adhesive. Bilston et al. [3] studied the different cross-section shapes of aluminum tubes in a three-point bending test, and it was found that tubes with square cross-sections gave the highest specific energy absorption. Ayhan et al. [4] investigated geometric sensitivities and nonlinear bending behavior for tubular beams, and it was revealed that the smaller diameter of the punch shows higher displacement than the larger punch diameter. Bai et al. [5] investigated a thin-walled dual rectangle tube using a three-point bending test to analyze the bending collapse of the tube, and the conclusion was made that derived expression correctly predicts three-point bending characteristics of thin thin-walled dual rectangular tubes. Zhao and Zhai [6] studied experimentally and numerically the bending strength of aluminum alloy having a circular cross-section, and it was observed that local buckling occurred in the specimen having a high slenderness ratio. Qi et al. [7] studied lateral impact behavior of empty and foam-filled double hat beams. Peak forces, energy absorption as well as bending strength decreased as the pusher moved from the middle position to the supports. Xiao et al. [8] studied the bending response of functionally graded and empty foam-filled bumper beams. Specific energy absorption was more improved in functionally graded foam-filled bumper beams. Belingardi and Scattina [9] investigated the bending behavior of steel and hybrid thin-walled box shape beams. It was concluded that a hybrid specimen can easily build with the help of adhesive and it reduces weight up to 28%. From the literature study, it was found that bumper beam made from different design and materials had been commonly used for study under quasi-static loading, whereas response of hybrid beams and influence of pusher has studied limited. In current study, the bending resistance and specific energy absorption of the hybrid bumper beam (steel DP800-aluminum 6061-T6) have been explored using a three-point bending test.

2 Material Properties and Numerical Modeling

Aluminum 6061-T6 and DP800 steel were selected for the research work. Aluminum 6061-T6 gives good strength to weight ratio as well as it is heat treatable. T6 indicates a degree of hardness. True stress–strain curves were used to define the plasticity properties of the material, see Table 1.

Table 1 Mechanical properties of the material

Material	Density (gm/cm^3)	Young's modulus (GPa)	Yield stress (Mpa)	Poisson's ratio
Aluminum 6061-T6	2.7	68.1	162	0.3
DP800 steel	7.85	195	495	0.33

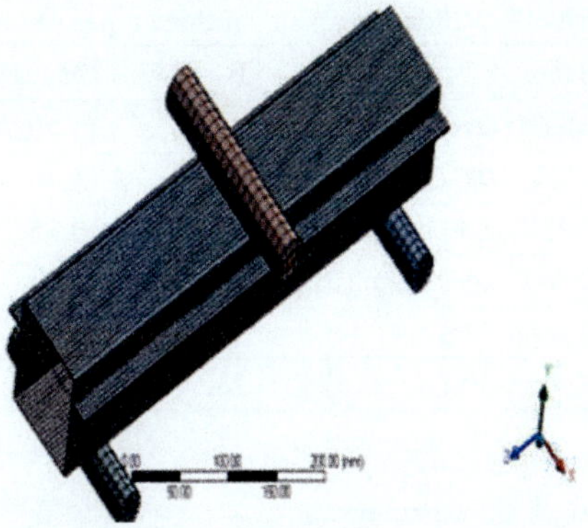

Fig. 1 Meshing model of double hat shape beam

The length of the double hat shape beam was 500 mm and thickness as 1.5 mm. For both the hat, the width was taken as 80 mm, and the flange width was 20 mm. For all the models, total height of beam was taken as 80 mm. The finite element modeling was performed in Ansys software. The total four contacts were found in the assembly. The upper hat beam and lower hat beam were connected to each other by keeping the bonded (tied surface to surface contact) contact between them. Automatic surface to surface contact was detected between pusher and upper hat and between lower hat and supports. Dynamic frictional coefficient of 0.3 was given to the contacts. The simulation was performed without considering material failure, and beam solution type was taken as bending. The total number of nodes and elements was 14,736 and 13,896. The element size of 4 mm was used, whereas the double hat shape beam was made as a deformable body. Mesh model was developed as shown in Fig. 1.

3 Result and Conclusion

3.1 Validation of Numerical Model

FE model of all three types of beam was validated by using Chang qi model as per literature [7]. Once the FE model was validated for its bending behavior and strength-based behavior, the same procedure could be applied for the different conditions to perform parametric studies for various configurations. The validation was carried out for a three-point bending test for both the beam where the impact velocity of the pusher was taken as 10 m/s. Table 2 shows the validation results for both the beam. From the results, it was cleared that the numerical model shows a very close result with the experimental model. This indicates the FE model could perform a low-velocity impact test for several configurations.

Table 2 Validation results of three types of beam

Beam	F_{ip} (KN)	M_b (KN m)	F_m (KN)	SEA (KJ/Kg)	Mass (Kg)
Al6061-T6 beam (Nu)	14.9	1.29	12.9	1.59	0.81
Al6061-T6 (Ex)	11.61	1.14	11.37	1.40	0.81
DP800 steel beam (Nu)	42.14	3.35	33.5	1.42	2.36
DP800 steel beam (Ex)	48.96	3.04	30.44	1.30	2.36
Hybrid (Nu)	12.56	2.83	28.35	1.66	1.42
Hybrid (Ex)	13.6	2.69	26.83	1.55	1.42

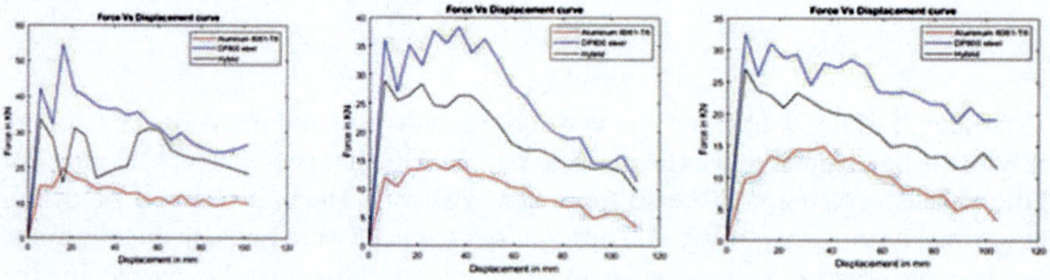

Fig. 2 Force–displacement response for models 1, 2, and 3 under low-velocity impact

3.2 *Effect of Different Geometries on Crashworthiness*

Numerical simulations were carried out on a 1.5 mm thick double hat shape beam having a length of 500 mm, see Fig. 2. In model 1, the height ratio was taken as 1:3 (upper hat having the height of 20 mm and lower hat having a height of 60 mm respectively.). Similarly, in model 2, the height ratio was 1:1 (upper and lower hat having a height of 40 mm each), and in model 3, height ratio was 3:1 (upper hat having a height of 60 mm and lower hat height as 20 mm), see Table 3. Two homogeneous and one hybrid beam were evaluated to compare the results. Two homogeneous beams were made from DP800 steel and aluminum 6061-T6 respectively. The beams were pushed from center. A velocity of 10 m/s was applied for 0.01 s. From the results, it was concluded that hybrid beam gives higher SEA than homogeneous beam.

3.3 *Comparison of Model 1, Model 2, and Model 3*

After performing simulation for all three models, a comparison of each model was done on basis of different parameters as given in Table 3. The mass ratio for model 1 was 1.93, and for model 2 and model 3, it was 2.9 and 4.36, respectively. The mass of model 1 was taken as 1.42 kg; for model 2, it was taken as 1.58 kg, and for model 3, it was 1.73 kg. We mainly focused on specific energy absorption and bending resistance. Model 1 showed higher specific energy absorption around 1.26 times

Table 3 Comparison of model 1, model 2, and model 3

Comparison	Model 1	Model 2	Model 3
Section	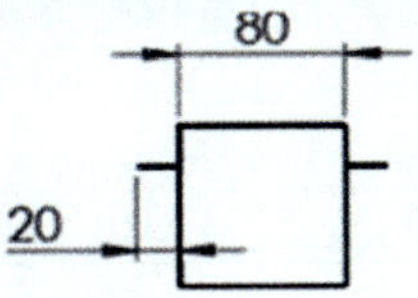	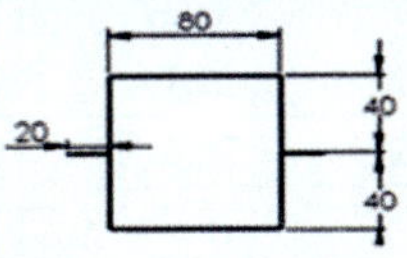	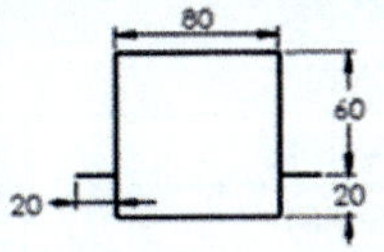
Height ratio	1:3	1:1	3:1
Mass ratio	1.93	2.9	4.36
Total mass (Kg)	1.42	1.58	1.73
Specific energy absorption (KJ/Kg)	1.66	1.31	1.06
Bending moment (KN m)	2.33	2.03	1.69

higher than model 2 and 1.56 times higher than model 3. Also, the bending moment observed in the case of the model was 87.12% that of model 1, and in model 3, it was only 72.53% of model 1. Higher energy absorption results in a higher bending moment. From this result, it was found that an increase in the mass of the double hat shape hybrid beam results in decreasing specific energy absorption and bending resistance.

3.4 *Different Positions of Pusher*

To understand the real-life application scenario of crash events and to explore the bending behavior of hybrid double hat shape beam, the low-velocity impact was performed on hybrid beam with different positions of the pusher as shown in Fig. 3.

Four different positions were examined as shown in Fig. 3. It was observed that when pushers started shifting from mid-span toward support, specific energy absorption and bending moment started continuously decreasing, see Table 4. At the last position (133 mm), specific energy absorption was found only 83.13% of SEA observed at mid-span. A bending moment of 2.3 KN m was observed in mid-span, and at the last position, it was found 1.99 KN m.

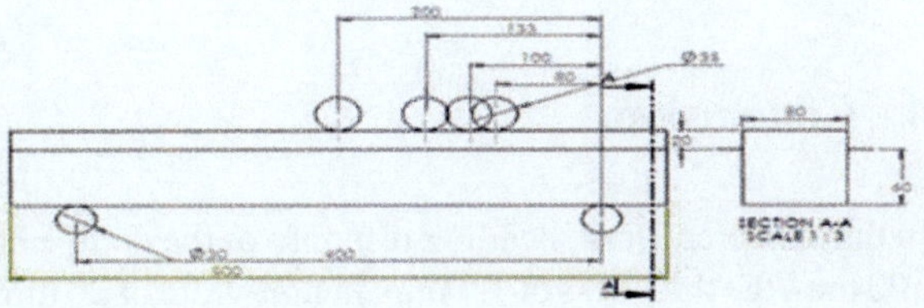

Fig. 3 Different positions of pusher

Table 4 Crashworthiness indices for different pusher position

Position of pusher	EA (KJ)	F_M (Kg)	M_b (KN m)	SEA (KJ/Kg)	Mass (Kg)
l/2 (200)	2.35	23.37	2.35	1.66	0.42
l/3 (133)	2.11	21.1	2.11	1.48	0.42
l/4 (100)	2.05	20.5	2.05	1.42	0.42
l/5 (80)	1.99	19.9	1.99	1.40	0.42

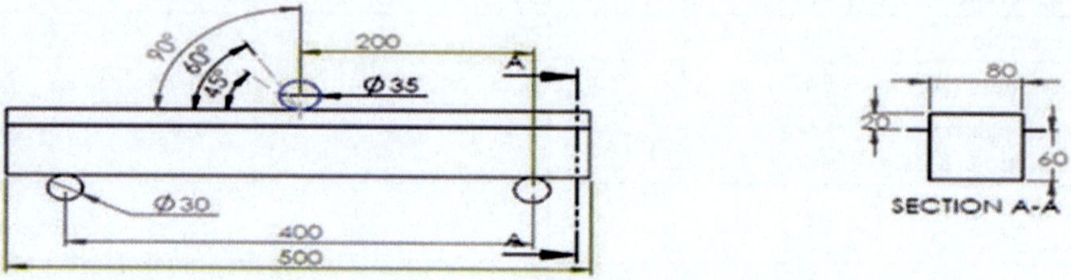

Fig. 4 Impact of cylindrical pusher with different angles

Table 5 Simulation result for different angles impact of cylindrical pusher

Angles	F_{ip} (KN)	Mb (KN m)	F_m (KN)	SEA (KJ/Kg)	Mass (Kg)
90°	33.63	2.35	23.5	1.66	1.42
60°	29.47	2.2	22.2	1.56	1.42
45°	26.54	1.35	13.5	0.95	1.42

3.5 Impact of Cylindrical Pusher at Different Angles

Impact at different angles with the cylindrical pusher at mid-span on a hybrid double hat shape beam was analyzed using three-point bend test such type of different situation may be observed in crash accidents. Y-axis of pusher made an angle of 90°, 60°, 45° with the z-axis of double hat shape beam, see Fig. 4.

From the simulation, it was found that SEA was 1.66 kJ/Kg when the cylindrical pusher was impacting with of 90°. SEA of the hybrid beam with a pusher angle of 90° and 45° was found 93.97 and 57.22% of the beam with a pusher angle of 90°. In the case of bending moment, it was found 2.35, 2.2, and 1.35 KN m for a beam with the pusher angle of 90°, 60°, 45°, respectively, see Table 5.

4 Conclusion

In this research work, bending response of the double hat shape beam made of length 500 mm, the thickness of 1.5 mm, flange width of 20 mm, and overall height of 80 mm were evaluated under dynamic three-point bending test with the velocity of 10 m/s.

Numerical simulations were configured using the explicit module of commercially available software Ansys.

Major conclusions obtained from this research work are as follows:

- The specific energy absorption of model 1, model 2, model 3 was found to be 1.66, 1.31, and 1.06 kJ/Kg, respectively. Also bending resistance of model 1, model 2, and model 3 was found as 2.23, 2.03, and 1.69 KN m, respectively. It was found that model 1 having a height ratio of 1:3 shows the highest specific energy absorption such as 1.26 times higher than model 1 and 1.56 times higher than model 2.
- The effect of pusher positions on crashworthiness parameter was also evaluated by taking four different pusher positions, l/2, l/3, l/4 and l/5 (l = 400 mm). SEA and bending resistance were observed higher when cylindrical pusher at mid-position.

References

1. Huang Z, Zhang X, Fu X (2020) On the bending force response of thin-walled beams under transverse loading. Thin-Walled Struct 154:106807
2. Sadowski T, Golewski P, Pietras D, Nowicki M (2019) Gradual degradation of a thin-walled aluminum adhesive joint with omega cross section under bending. Int J Adhes Adhes 89:72–81
3. Bilston D, Ruan D, Candido A, Durandet Y (2019) Parametric study of the cross-section shape of aluminium tubes in dynamicthree-point bending. Thin-Walled Struct 136:315–322
4. Ayhan A, Eks-I S, Genel K (2012) Simulation of nonlinear bending behavior and geometric sensitivities for tubular beams with fixed supports. Thin-Walled Struct 51:1–9
5. Bai J, Meng G, Wu H, Zuo W (2019) Bending collapse of dual rectangle thin-walled tubes for conceptual design. Thin-Walled Struct 135:185–195
6. Zhao Y, Zhai X (2020) Bending strength and design methods of the 6082–T6 aluminum alloy beams with circular hollow sections. Structures 26:870–887
7. Qi C, Yang S, Sun Y, Hu H, Cao G, Wang D (2016) On design of hybrid material double-hat thin-walled beams under lateral impact. Int J Mech Sci 168:MS3414
8. Xiao Z, Li Q, Sun G, Fang J (2015) Crashworthiness design for functionally graded foam-filled bumper beam. Adv Eng 85:81–95
9. Belingardi G, Scattina A (2013) Experimental investigation On the bending behaviour of hybrid and steel thin walled boxbeams—the role of adhesive joints. Int J Adhes Adhes 40:31–37

Modeling of Fine Korai Fiber Extraction Machine to Reduce user's Fatigue

M. Vignesh Kumar, R. Harish, and Surya Bharath

Abstract The paper emulates the modeling and analysis of korai fiber extraction machine to extract high-quality fine korai fibers. Manual extraction of korai fiber is tedious and time-consuming. The labor expenses and efficiency are the key issues with the conventional method of extraction. Therefore, the enhanced output can be achieved only through a machine, making it a viable one for mass production. The existing machine doesn't envisage in the fine korai fiber extraction and also with an average production of poor quality fibers. The paper proposes a model of machine to be designed for fine korai fiber extraction machine. The machine is designed to follow a simple mechanism by meeting the ergonomic comfort of the users as well as enabling fine korai fiber extraction.

Keywords Fine korai fiber · Uniform fiber extraction · Efficiency

1 Introduction and Literature Review

Korai grasses are generally split very fine and woven with cotton warp on a ground loom. It is generally soaked in water before weaving, as it may break on rigid uses [1]. The issue of manual splitting and retting of the korai grass is focused as the ultimate aim of the project. The machine proposed will mechanize the process of splitting and retting, thus reducing the fatigue caused due to intense work [2]. The total population falling prey to the issue is along the belts of Pattamadai and Karur chiefly at the banks of their respective rivers Thamirabarani and Kaveri. Also the

M. Vignesh Kumar
Assistant Professor, Department of Mechanical Engineering, KCG College of Technology, Chennai, India

R. Harish (✉)
UG Scholar, Department of Mechanical Engineering, KCG College of Technology, Chennai, India
e-mail: harishraj0126@gmail.com

S. Bharath
UG Scholar, School of Fashion Technology, KCG College of Technology, Chennai, India

N. M. Sivaram et al. (eds.), *Advances in Manufacturing, Automation, Design and Energy Technologies*, Lecture Notes in Mechanical Engineering,
https://doi.org/10.1007/978-981-99-1288-9_58

artisans of the other parts of Tamil Nadu and Kerala also practice the korai grass mat weaving [3]. The korai mats are processed manually for three days involving intense man power. The artisans and the weavers practicing korai grass weaving are highly affected by the issue, as the labor intense manual processing deteriorate the health of the workers [4]. The reduction in the korai mat weavers and awareness to protect the traditional handicraft have given limelight on the issue. The GI tag conferred on the korai mats is also part of the reason for us getting aware on the problem [5]. This traditional craft of Tamil Nadu will survive if the processing technique is eased using mechanization and the latest technology. The issue has not yet been addressed at present. The retting and splitting process of the korai grass is still done manually involving three days of eight hours physically tiring work [6]. The fiber extraction machines generally made using aluminum alloys [7, 8]. The performance of korai grass in the small scale and handicraft industries has great impact on the overall growth of an economy of a country in terms of small-scale industrial units, variety of production, employment, standard of living, and exports to foreign countries [9]. Considering all these, the main focus of the project is to reduce the fatigue and labor intense pre-processing work involved in korai grass. The splitting of the korai grass takes hours to complete, which the machine proposed will complete with less manual force involved and high efficiency [10]. This will pave way to produce huge unified quality of the korai grass for mat weaving. The solution would be highly beneficial for the artisans and weavers of the korai grass mat. Apart from the artisans, the exporters and large-scale korai grass mat manufactures would also be the active customers for the product developed [11]. The artisans and laborers are in need of mechanization for the retting and splitting process. As the manual process causes various health issues such as back pain, eye irritation, numbness is legs, muscle cramps. These health related problems can be avoided when the machine we fabricated is utilized. The artisans would be relieved completely of these issues making them wanting to buy the machine, as a necessary one (Table 1).

2 Problem Statement

See Table 1.

3 Extraction of Fiber

See Fig. 6.

Table 1 Problems identified from the field visit

Si. No.	Field Visit Images	Problem Identified
1	 Fig. 1. Manual arranging of fibres	Manual arranging of the korai stem before the removal of the pith from the underside of the korai stem. The process depicted in Fig.1, is tedious and time consuming.
2	 Fig. 2. Bundles of coarser korai fibres	The bundles of coarser variety of korai stems, stored in a storage unit. These korai stems have to be converted into finer fibers. These coarser variety of fibers, as depicted in Fig. 2, has to be split into finer variety.
3	 Fig. 3. Fresh korai stem	The raw korai stems, before the drying process depicted in Fig. 3.The dry korai stems, has to be dipped in water for three long days before the pith removal process.
4	 Fig. 4. Manual removal of pith	Manual removal of pith from the underside of the korai stems. The worker holds two to three korai stems, together and carries out the process as depicted in Fig. 4.
5	 Fig. 5. Bundles of dried korai fibres	Bundles of dried korai fibres hanged in air, to prevent rotting of the fibres in a storage unit. The Fig. 5, depicts the bundles of coarser korai fibres.

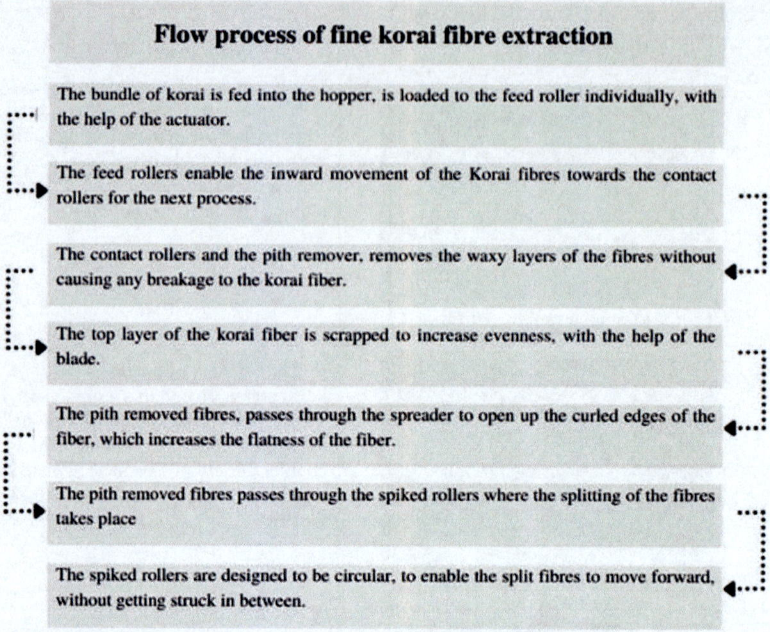

Fig. 6 Flow process of working

4 Modeling of Parts

4.1 Hopper

The hopper functions to feed the bundle of korai stem into the machine, with a capacity of 2–3 Kgs at a time depicted in Fig. 7. The hopper is designed with a broad opening at the top and a tapered bottom. As the bottom is narrower, it enables the hopper to feed individual korai stem into the feed roller, thus allowing only one korai stem at a time in the process of fine korai extraction.

Fig. 7 Hopper

Fig. 8 Feed roller

4.2 Feed Rollers

The feed roller is used to insert korai stem into the machine, which is placed between hopper and the spreader depicted in Fig. 8. Insertion of the stem fully into machine is carried out by the feed roller, ensuring the safety of the operator. The feed roller is designed to be 50 mm in diameter and 400 mm in length.

4.3 Spreader

The spreader roller functions to completely open up the korai stem, which is placed between feed roller and contact roller, as depicted in Fig. 9. The opening of the curled or uneven korai stem is essential to evenly remove the pith. The spreader roller is designed to be 20 mm in diameter and 400 mm in length.

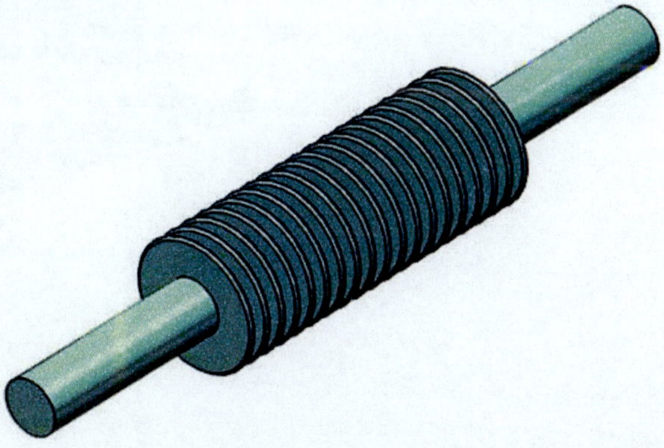

Fig. 9 Spreader

Fig. 10 Contact roller

4.4 Contact Rollers

The spreader roller functions to push the korai stem toward the blade, which is placed between the spreader and blade, as depicted in Fig. 10. The contact roller helps reducing the distance between the blade and contact roller, thus enabling the clean removal of the complete pith content from the underside of the korai stem. The contact roller supports the stem to prevent breaking of the stem, during the pith removal process. The contact roller is designed to be 50 mm in diameter and 400 mm in length.

4.5 Spiked Rollers

The spiked roller functions to split the stem into finer fiber segments of equal width, which is placed between the spreader and delivery chamber, as depicted in Fig. 11. The initial korai stem inserted into the machine is of broad width, which needs to be reduced to obtain finer korai fibers at the end of the delivery chamber. The spiked roller is designed to be 35 mm in diameter ad 300 mm in length.

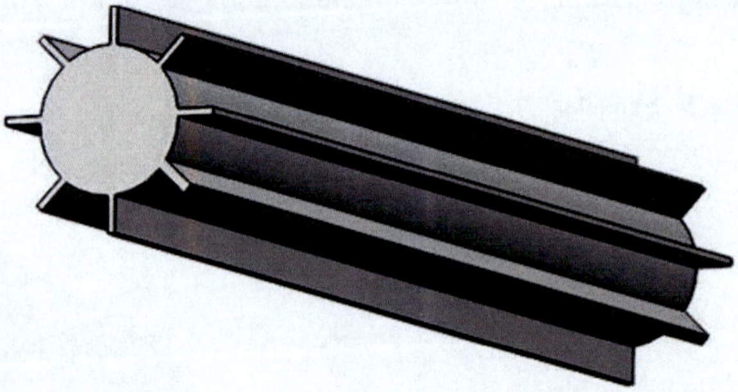

Fig. 11 Spiked roller

Fig. 12 Pith remover

4.6 Pith Remover

The pith remover functions to remove the bulk amount of pith present underside of the korai stem, as depicted in Fig. 12. It is designed to semi-circular in shape, with a slot at the center and sharp edges at the top. The contact roller pushes the korai stem toward the pith remover, and the additional pressure posed by the roller helps in even removal of the pith from the stem without causing any breakage of the korai stems. The pith remover is designed to be 15 mm depth slot and 300 mm in length.

4.7 Electronic Actuator

The actuator functions to push individual korai stem from the bulk of korai stems fed into the hopper, as depicted in Fig. 13. This allows the entry of only one stem into the feed roller.

5 Assembly of Parts

The complete assembly of the machine is depicted in Fig. 14, and the splitting unit is shown in Fig. 15. This model proposes to extract fine korai fibers without human intervention with pith removed from the fibers falling away from the motor. The hopper and actuators are key drivers in the initiation of the process of fine korai fiber

Fig. 13 Electronic actuator

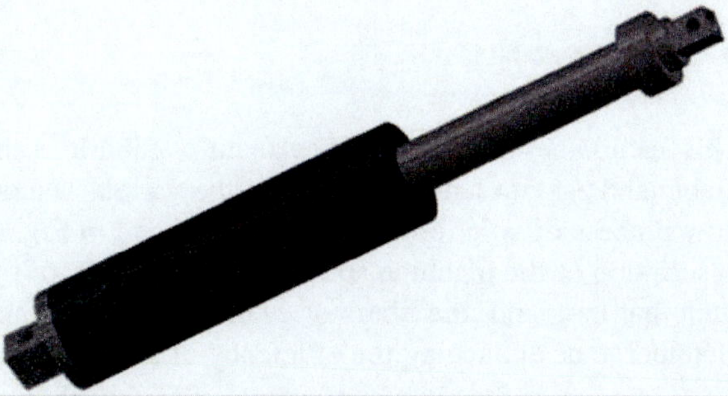

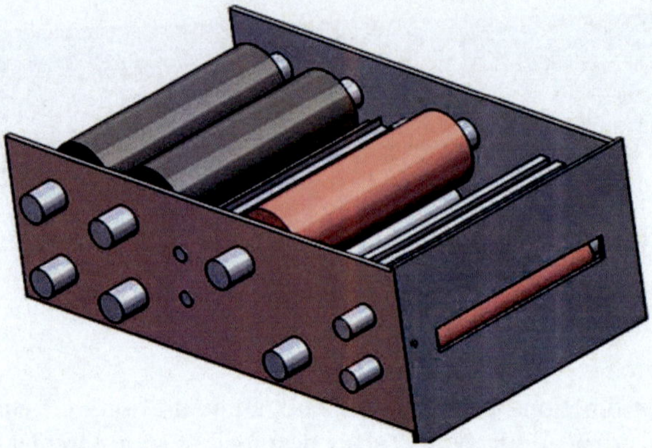

Fig. 14 Splitting unit of the machine

Fig. 15 Complete assembly of the machine

extraction, as it allows the entry of individual stems at a time. The three rollers, namely feed rollers, spiked rollers, and contact rollers enable the removal of pith evenly from the fibers and extraction of fine korai fibers, with the help of pith remover.

6 Conclusion

This section summarizes the key features inbuilt in the machine, which makes it a viable and easy to handle. Based on the various issues as inference in Table. 1, the flow process of solution proposed is depicted in Fig. 6. Table 2 gives the complete description of the machine specification; the spec of the machine is set out in a way such that long and fine fibers of even length are obtained at the end of the delivery chamber, thus enhancing the efficiency of the weavers.

Table 2 Machine specification

Capacity	3–5 kg
Material	Aluminum and mild steel
Size	Splitting unit (20 × 30 × 30 cm) Hopper (150 × 50 × 25 cm)
Weight	20–25 kg
Motor	0.25 single phase induction motor
Power supply	220–240 V

6.1 Key Features

- Compact, in-line design, high quality, widely available hardware
- Minimum manual monitoring and physical work required.
- Protective case shield, during the splitting of the korai grass to prevent the worker.
- Broad bed to accommodate the fibers, easy setup guides, easy operation and care.
- The comb pin blade helps in accurate splitting of the fibers.
- High-quality fiber splitting takes place at the end of the process.

References

1. Ganthi S (2009) Indigenous knowledge on natural dyeing of Korai grass mat in Pattamadai, Tirunelveli district, Tamil Nadu. Nat Prod Radiance 8(5):542–545
2. Census of India 1961, Part VII-A-IV, Handicrafts and artisans of madras state, Fine Mats of Pattamadai. Accessed on 10 Nov 2021
3. https://en.wikipedia.org/wiki/Kaveri. Accessed on 10 Nov 2021
4. Economics of Korai mat weaving in Thanjavur District, A thesis submitted by V. Ethirajan, The University of Cochin for the degree of doctor of Philosophy in Economics, April 1982. Accessed on 10 Nov 2021
5. Government of India Geographical Indications Journal No. 47 october 30, 2012/Kartika 08, Saka 1934. Accessed on 10 Nov 2021
6. Venkateswaran A (2017) A study on problems and prospects of Pattamadai handloom mat weaving industry in Tirunelveli district of Tamilnadu. Int J Interdisc Res Arts Hum (IJIRAH) 2(1). ISSN (Online): 2456 - 3145 (www.dvpublication.com)
7. Rajeshkannan S, Vignesh Kumar M (2021) Optimization and characterization of AA5083 and AA7075 dissimilar aluminium alloy joints produced by friction stir welding. Int J Veh Struct Syst. https://doi.org/10.4273/ijvss.13.3.19
8. Vignesh Kumar M, Veerappan A, Padmanaban G et al (2021) Microstructural evolution of a cast ZK60 Mg Alloy/SiCp surface composites induced by friction stir process. Phys Metals Metallogr 122:403–415. https://doi.org/10.1134/S0031918X21040049
9. Muthiah S, Venkateswaran A (2014) Economic importance of mat weaving industry in Pattamadai of Tirunelveli district: an overview. Shanlax Int J Commer 2(2). ISSN: 2320-4168
10. Archunan K (2017) A study on the problems of manufacturing of korai mats in tirunelveli and Thoothukudi districts. Shanlax Int J Commer 5(2). ISSN: 2320-4168
11. Archunan K (2017) A study on the marketing of korai mats in tirunelveli and Thoothukudi districts. Shanlax Int J Commer 4(4). ISSN: 2321-4643

EMI-Based Fatigue Life Assessment of Friction Stir Welded AA5083-O and AA6063-T6 Aluminum Alloy

Reetesh Kumar Shukla and K. N. Pandey

Abstract In the present work, electro-mechanical impedance (EMI)-based fatigue life assessment of friction stir welded (FSW) dissimilar joint of AA5083-O and AA6063-T6 at different loading conditions is presented. The dissimilar FSW joints of aluminum alloys AA 6061and AA 5083 strength were found to be almost same of the parent materials. It is necessary to monitor the fatigue crack growth for life assessment, safety, and reliability of the machine components and structures during their useful life. The present experimental study uses piezoelectric transducer (PZT) pasted on the specimen to get the EMI signature of the FSW welded samples at different stages of the damage. The technique was applied to develop a relationship between fatigue crack propagation and real part of the admittance, which is called conductance. Crack growth rate was also obtained by conventional crack growth method, and a correlation was developed between the conventional and EMI-based method of crack growth measurement. Because of higher sensitivity for damage, the real part of admittance (conductance) signature is preferred as a damage identifier compared to susceptance signature. During fatigue damage, the leftward horizontal shift of the resonance peak was used as damage identifier. The frequency range selected for experiments was set as 100–180 kHz to minimize bonding and temperature effect. The results show a good relationship between conventional and EMI-based damage identification of the FSW welded samples under fatigue loading conditions.

Keywords EMI · Friction stir welding · Fatigue crack growth · Crack propagation

R. K. Shukla (✉) · K. N. Pandey
Department of Mechanical Engineering, Motilal Nehru National Institute of Technology Allahabad, Prayagraj 211004, India
e-mail: reeteshshukla@gmail.com

K. N. Pandey
e-mail: knpandey@mnnit.ac.in

N. M. Sivaram et al. (eds.), *Advances in Manufacturing, Automation, Design and Energy Technologies*, Lecture Notes in Mechanical Engineering,
https://doi.org/10.1007/978-981-99-1288-9_59

1 Introduction

In present scenario, most of the components and structures are fabricated by welded joints, and therefore, quality of weld is the most critical area for performance perspective of the components and structures. In welded structures, fatigue cracking, i.e., the most common failure mechanism occurs without warning and often leads to serious consequences. Reduced fatigue strength in welded joints is mainly due to weld shape and joint geometry, stress concentration due to weld imperfections and welding residual stresses. Therefore, under normal working conditions, monitoring of components and structures made by welded joints is very important. Most of the conventional monitoring methods are based on inspection of cracks and its propagation rate during routine inspection period. In the present work, fatigue crack growth behavior of base materials (AA5083-O and AA6063-T6) and friction stir welded joints of dissimilar AA5083-O/AA6063-T6 aluminum alloys is presented.

In 1991, the Welding Institute of UK developed the mechanism of friction stir welding [1, 2]. Material twisting creates and reorders heat [3], delivering the temperature field in the joint. When the path of tool rotation is similar as the path of welding, it is termed as the advancing face, and the opposite face is said to be the retreating face. As the tool moves ahead with explicit revolution [4] and travel speed, the pin is inclined at a particular point, which manages the weld appearance, and it becomes thin [5, 6]. The FSW development is shown in Fig. 1.

The FSW tool encompasses three sections: shank, probe, and shoulder [7, 8]. Since the pin length is variable, this apparatus could be utilized for various thicknesses of sheets. FSW can be utilized to make the lap and the butt joins [9]. For fatigue crack growth (FCG), American Society for Testing and Materials (ASTM) standard E647 [10] provides the accepted guideline. Precise measurements of fatigue crack

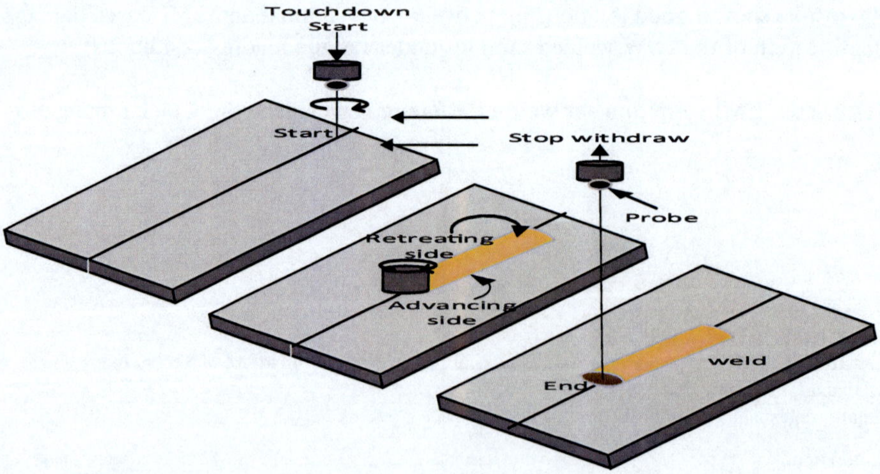

Fig. 1 FSW process

extension are critical for the determination of reliable crack growth rates. ASTM E647 requires a minimum resolution of 0.1 mm in crack length measurement. Frequency of the test is kept constant in order to make crack length measurements. Frequency effects are usually not observed in metals in typical test conditions (1–100 Hz).

2 Material and Experimental Setup

2.1 Test Setup and Procedure for Friction Stir Welding

Welding was done by joining two 250 mm × 50 mm × 6 mm plates of dissimilar AA5083-O/AA6063-T6 aluminum alloys to make a welded joint of 250 mm long and 100 mm wide. Cylindrical tools with tapered pin were used for FSW and were made of H13 tool steel. Profile and dimensions of the tools are shown in Fig. 2. The composition of materials is given in Tables 1 and 2.

The primary test parameters and their levels are given in Table 3.

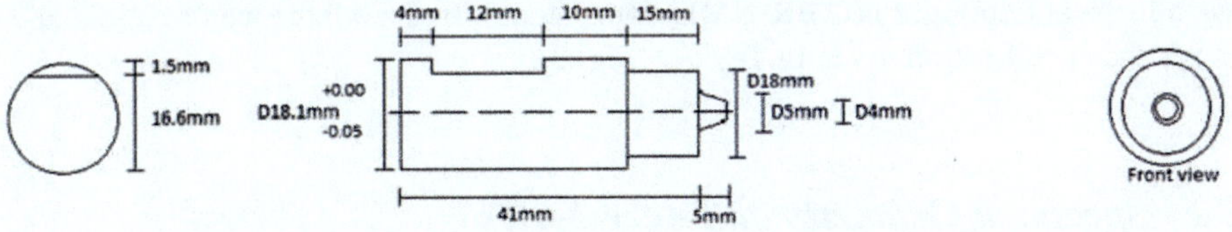

Fig. 2 Tool geometry for friction stir welding

Fig. 3 Vertical milling machine (make: Bharat Fritz Werner Ltd., Bengaluru, India) at CSIR-NML, Jamshedpur

Table 1 Chemical composition of aluminum alloys (%wt)

Al alloy	Si	Fe	Cu	Mn	Mg	Ti	Cr	Zn	Al
AA5083	0.190	0.230	0.02	0.6	4.53	0.030	0.08	0.15	94.350
AA6063	0.420	0.160	0.001	0.001	0.53	0.013	0.00	0.00	98.875

Table 2 Mechanical properties of aluminum alloys

Aluminum alloy	Ultimate tensile strength (MPa)	Yield strength (MPa)	% of elongation
AA5083-O	290	181	24.4
AA6063-T6	220	185	12.6

Table 3 Process parameters and their unit and level

Symbol	Parameters	Unit	Levels		
			1	2	3
N	Tool rotational speed (TRS)	rpm	1000	1200	1400
S	Welding speed (WS)	mm/min	25	30	35
D	Shoulder diameter	mm	15	18	21
D	Pin diameter	mm	4.5	5	5.5

FSW was carried out at Council of Scientific and Industrial Research-National Metallurgical Laboratory (CSIR-NML) Jamshedpur, India. All the welds were made as per the specification given in Table 5 (Fig. 3).

2.2 *Specimen Geometry and Accessories*

Compact tension (CT) specimen for fatigue crack growth test was prepared as per ASTM (E647). The dimensions of the CT specimen were 47, 46, and 6.1 mm in the direction of length, breadth, and thickness, respectively, as shown in Fig. 4a. The CT specimen was made from the friction stir welded plates in keeping the weld in lateral and horizontal position as shown in Fig. 4b, c. The CT specimen with weld in longitudinal direction and horizontal direction will be called CT-L and CT-H, respectively. The specimen has been subjected to the constant amplitude fatigue crack propagation (VACFP) tests. Straight notch, which is shown in Fig. 4b, was made on middle position of the specimen. The PZT patch of size 10 mm × 10 mm × 0.2 mm, procured from Central Electronics Limited, Industrial Area, Sahibabad, Uttar Pradesh, (India) was surface-bonded on CT specimen using high strength Araldite epoxy.

2.3 *Test Setup and Procedure for Fatigue Testing*

From the welded specimen, the CT specimen was made, and cyclic load was applied on 25 kN servo-hydraulic UTM (make: BISS TWI USA) at MNNIT Allahabad India

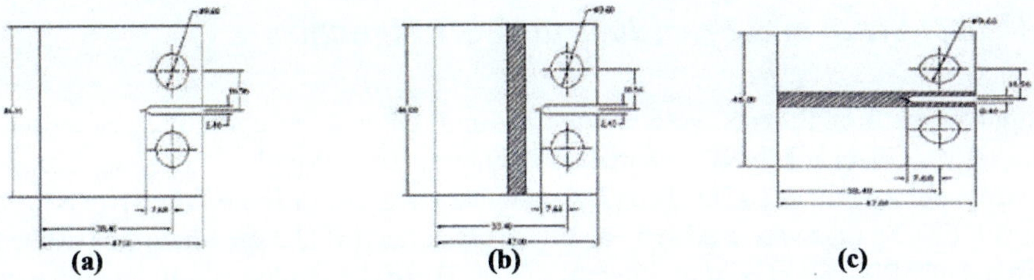

Fig. 4 **a** CT specimen geometry **b** CT specimen with weld in lateral direction, CT-L **c** CT specimen with weld in longitudinal direction, CT-H

as shown in Fig. 5. Crack growth was measured with the help of a COD gauge. The EMI signatures were recorded before and after loading the specimen on LCR meter of HIOKI IM 3536, 2015 make as shown in Fig. 6.

Fig. 5 Figure 25 kN servo-hydraulic UTM (make: BISS TWI USA) at MNNIT, Allahabad

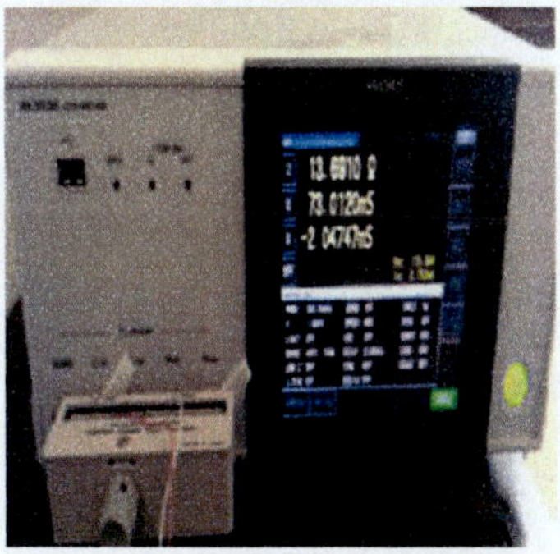

Fig. 6 Hioki IM3536 LCR Meter, For Electro-(Mechanical impedance signature Machine used—HIOKI IM 3536 LCR meter, Control—Sweep Measurement, Frequency range 50–200 kHz, End Condition—Free end

2.4 Test Setup and Procedure for EMI Signature

Fatigue crack growth tests were conducted on a 25kN servo-hydraulic test system (BISS-TWI USA) at MNNIT Allahabad Prayagraj (In-dia) under load control mode and frequency equal to 10 Hz. The maximum load is given in Table 1, and stress ratio was 0.1. COD gauge was used to measure the crack length. The specimen was excited at frequency 20–180 kHz at an interval of 100 Hz by LCR meter, and conductance signatures were taken. The LCR meter was kept on sweep measurement mode. The signatures were obtained at different crack lengths by stopping the test. Specimen at initial crack length was considered as the initial or healthy-state signature. The process was repeated till fracture of the specimen. Results related to this analysis are given in Table 4.

Table 4 Test results for dissimilar FSW welded samples under the different loading conditions

Sr. no	Maximum load (kN)	Crack length *a* (mm)	No. of cycles *N*	Resonance frequency *F*	Life used (%)	Resonant frequency shift, ΔRF
1	3	12	980	1,40,240	4	0
		14	8070	1,40,040	29	200
		16	13,112	1,39,740	46	500
		18	17,880	1,39,240	70	1000
		21	22,180	1,38,640	88	1600
		24	24,890	1,37,840	98	2400
2	4	12	810	1,40,500	4	0
		14	5700	1,40,200	26	300
		16	9600	1,39,800	45	700
		18	12,710	1,39,200	68	1300
		21	15,920	1,38,500	86	2000
		24	17,870	1,37,600	97	2900
3	5	12	927	1,40,300	2	0
		14	8001	1,40,100	26	200
		16	12,900	1,39,700	47	600
		18	17,070	1,39,200	65	1100
		21	19,826	1,38,500	82	1800
		24	21,750	1,37,600	95	2700

2.5 Results and Discussions

Study of FSW and effects of its process parameters on joining of other dissimilar materials combinations are challenging to other welding processes using EMI. Effect of post-weld heat treatment conditions and corrosion environment on mechanical and microstructural properties of welded joints may be studied using EMI. First, conventional method of measurement of crack by COD gauge was employed, and graphs between crack length versus number of cycles and crack growth rate versus range of stress intensity factor were plotted. The effects of cyclic loading conditions and FSW process parameters on fatigue crack growth rate of welded joints were also studied.

From Fig. 7a, it can be concluded that after applying the 3 kN loading on friction stir welded CT specimen, the crack length is growing very slowly up to 13,000 numbers of cycles, but after 130,000 number of cycles, the crack is very severe till the fracture. When load increases to 4 and 5 kN on the same specimen, the severity of crack starts at 5700 cycles (Fig. 7b) and 4900 cycles (Fig. 7c), respectively. Figure 8 presents a plot between shifts in resonant frequency versus percent of elapsed life. From comparison of Figs. 7 and 8, it is clear that shift in resonant frequency presents the quantification of the crack growth or percent of elapsed life of the specimen. Figure 8a shows the percentage life decrease in comparison with resonance frequency occurs at 90% life used of specimen for 3 kN, whereas for 4 and 5 kN, it occurs at 65% (Fig. 8b) and 43% (Fig. 8c), respectively.

Figures 9 and 10 are the plot between rate of crack growth (da/dN) versus shift in resonant frequency (ΔRF) and rate of crack growth (da/dN) versus range of stress intensity factor (ΔK). For all constant amplitude tests with maximum stress amplitude of 3kN, 4kN, and 5 kN, the trends of Figs. 9 and 10 are same. In conventional fatigue crack growth test, range of stress intensity factor is taken as damage quantifier. Therefore, after comparison of Figs. 9 and 10, shift in resonant frequency, ΔRF, can also be considered as damage quantifier under fatigue loading conditions.

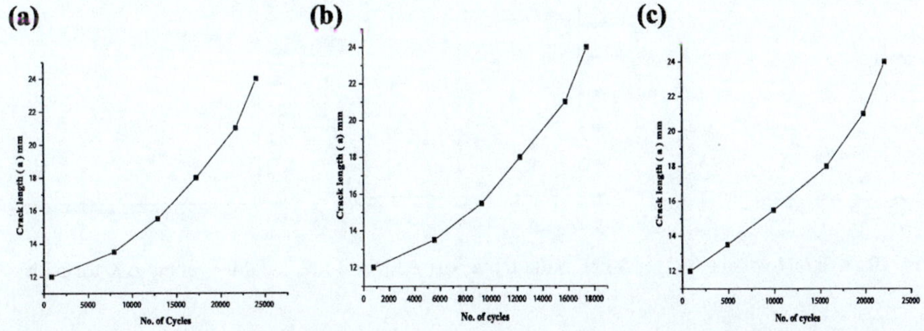

Fig. 7 **a** Crack length versus number of cycles for 3 kN, **b** Crack length versus number of cycles for 4 kN, **c** Crack length versus number of cycles for 5 kN

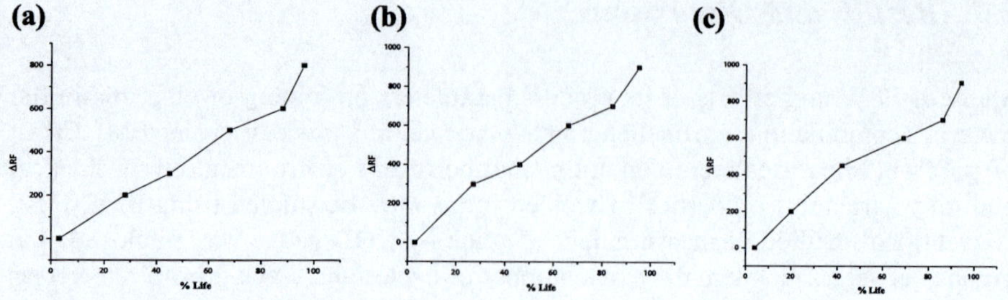

Fig. 8 **a** ΔRF versus percentage life used for 3 kN. **b** ΔRF versus percentage life used for 4 kN, **c** ΔRF versus percentage life used for 5 kN

The variation of crack with the number of cycles was also observed in Fig. 10. From this figure, it can be concluded that on increasing the number of cycles for 3, 4, and 5 kN, the crack size increases by 26.8 mm (Fig. 10a), 27 mm (Fig. 10b), and 30 mm (Fig. 10c), respectively.

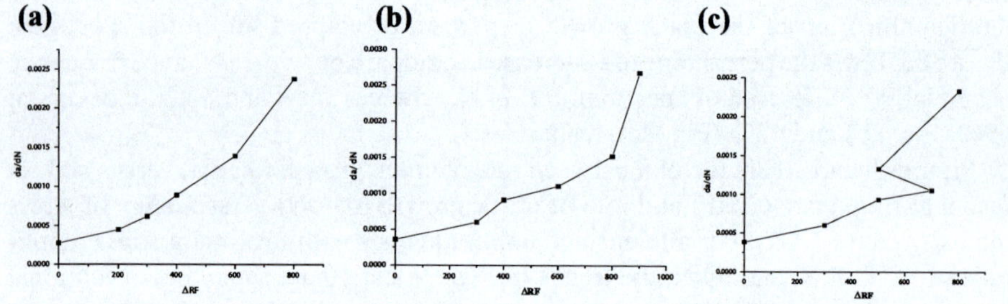

Fig. 9 **a** da/dN versus ΔRF for 3 kN, **b** da/dN versus ΔRF for 4 kN, **c** da/dN versus ΔRF for 5 kN

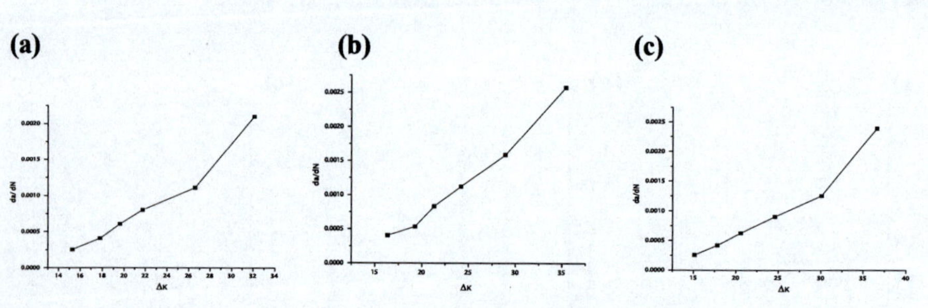

Fig. 10 **a** da/dN versus ΔK for 3 kN, **b** da/dN versus ΔK for 4 kN, **c** da/dN versus ΔK for 5 kN

3 Conclusions

The following conclusions are derived from the application of EMI technique on FSW welded CT specimen:

a. The gradual increment in load increases the severity of damage to crack length, but initially, there is a rapid change in crack length which becomes less severe with the increase of load.
b. The percentage life decrease in specimen with the change in resonance frequency is severe at high load and less at low load.
c. The EMI-based shift in resonance frequency, ΔRF, can be taken as a damage quantifier for FSW welded joints similar to the range of stress intensity factor.
d. The EMI-based method provides a rapid and cheap method for quantifying damage in a FSW welded components or structures.

References

1. Mishra RS, Ma ZY (2005) Friction stir welding and processing. Mater Sci Eng R 50(1–2):1–78
2. Thomas WM, Johnson KL, Wiesner CS (2003) Friction stir welding-recent developments in tool and process technologies. Adv Eng Mater 5(7):485–490
3. Dawes C, Thomas W (1995) Friction stir joining of aluminium alloys. TWI Bull 6:124–127
4. Muruganandam D, Balasubramaniyan C, Gokulachander B (2015) Review paper on friction stir welding of aluminium and magnesium alloys. Ind J Sci Technol 8(35):1–6
5. Peel MJ et al (2006) Dissimilar friction stir welds in AA5083-AA6082. Part I: process parameter effects on thermal history and weld properties. Metall Mater Trans A 37(7):2183–2193
6. London B, Mahoney M, Bingel W, Calabrese M, Waldron D (2001) Experimental methods for determining material flow in friction stir welds. In: Proceedings of the third international symposium in friction stir welding, Kobe, Japan
7. Peel MJ, Steuwer A, Withers PJ (2006) Dissimilar friction stir welds in AA5083-AA6082. Part II: process parameter effects on microstructure. Metall and Mater Trans A 37(7):2195–2206
8. Bhattacharya TK, Das H, Pal TK (2015) Influence of welding parameters on material flow, mechanical property and intermetallic characterization of friction stir welded AA6063 to HCP copper dissimilar butt joint without offset. Trans Nonferrous Metals Soc China 25(9):2833–2846
9. Jamshidi Aval H, Serajzadeh S, Kokabi AH (2012) Experimental and theoretical evaluations of thermal histories and residual stresses in dissimilar friction stir welding of AA5086-AA6061. Int J Adv Manuf Technol 61(1–4):149–160
10. Gupta SK, Pandey KN, Kumar R Comparison of ANN and regression modeling for predicting the responses of friction stir welded dissimilar AA5083-AA6063 aluminum alloys joint. Appl Mech Mater 813–814:415–419

Low-Cost Emergency Ventilator

R. Mathiyazhagan, P. R. Prakatish, P. Muthamizharasu, and G. Sathish Kumar

Abstract Ventilator and oxygen shortage have become a problem of immediate concern due to the COVID-19 pandemic. Numerous people worldwide fight for survival due to the lack of getting ventilator assistance at the right time. Hence, there is an inevitable need for rapidly deployable, emergency use lifesaving ventilating devices. In a way to provide an effective solution to address these problems, we have designed and developed a prototype of a low-cost portable ventilating device. During the course of this work, various ventilating parameters were compared and the most important parameters which includes respiratory rate, tidal volume, minute ventilation, and positive end-expiratory pressure (PEEP), for the development of this device. The software simulation has been performed using Siemens NX software for designing the CAD, performing the motion simulation and static force analysis. Working prototype has been developed on the basis of determined values and parameters. Algorithm has been developed to operate the prototype. The results showed us that this emergency use ventilator could be useful to non-invasive type ventilation and could also conservative oxygen delivery to the patients. The test was performed using a spirometer and not using artificial lungs. Future iterations of the device will include development of the device following the standards of Central Drugs Standard Control Organization with, peak inspiratory pressure monitoring sensor, pressure relief valve, and PEEP capabilities.

Keywords Ventilator · COVID-19 · Low-cost · Tidal volume

1 Introduction

A ventilator is a device that is designed to mechanically move air to breathe in and out of the lungs to provide a respiration mechanism for a patient who is physically unable to breathe satisfactorily. The ventilators used in hospitals are very expensive

R. Mathiyazhagan (✉) · P. R. Prakatish · P. Muthamizharasu · G. Sathish Kumar
Madras Institute of Technology Campus, Anna University, Chennai 600044, India
e-mail: mathi_raj06@yahoo.co.in

N. M. Sivaram et al. (eds.), *Advances in Manufacturing, Automation, Design and Energy Technologies*, Lecture Notes in Mechanical Engineering,
https://doi.org/10.1007/978-981-99-1288-9_60

and highly sophisticated and require well-trained staff for their operation, as a result in developing countries like India where the medical resources are concentrated in major urban centers; the poor people and the population living in rural and outlying areas have no or very little access to mechanical ventilators; hence, the need for an low-cost portable ventilators is therefore paramount. To ventilate a patient, it is necessary to set the ventilator parameter according to the lung compliance of the patient [1]. To bring an economical and effective solution for this problem, we have ideated to develop a DIY emergency ventilator; we used the MADVent developed by Aditya Vasan et al. [2] and E-Vent being developed by MIT, USA, as reference models for this work.

Though the above emergency ventilators are currently in development, these equipment tend to be more costly in terms of Indian National Rupees and require synchronization of two circuits. In addition, the above models are awaiting approval from the Food and Drug Administration, USA. Hence, a need for similar but economical and user-friendly emergency ventilators which can control vital ventilating parameters of the patients is inevitable [3].

The ongoing COVID-19 pandemic has created a huge shortage of ventilators, and the access to ventilators by the poor and rural segment population has almost nearly been impossible. So, there is a need to eliminate this disparity and at the same time help the people overcome this pandemic. This work aims to provide an alternative solution by developing a low-cost portable ventilating device [4]. The feasibility of design and prototyping of a low-cost portable mechanical ventilator by using cam-actuation for use in mass casualty cases and resource-poor environments was described by Abdul Mohsen Al Husseini et al. The ventilator delivers breaths by compressing a conventional bag valve mask with a pivoting cam arm, eliminating the need for a human operator for the bag valve mask [5]. The basics of respiratory mechanics and various important ventilator-derived parameters that need to be monitored and taken care of while intubating a patient was described by Pedro Leme Silva et al. This paper explains the benefits and harms of mechanical ventilation which is dependent not only on the operator's input setting of the machine, but also on their interpretation of the output parameters [6].

For patients with significantly healthier lung, a maximum tidal volume of 10 ml/kg could be used for mechanical ventilation. Studies in the recent times have shown decreased mortality with the use of lower tidal volume in patients with acute lung disease. Therefore, patients with an acute lung disease should be ventilated with tidal volumes of 6–8 ml/kg. This was inferred from the paper by Salyer et al. [7]. Bunting Leonard et al. describe about the adjustable positive end-expiratory pressure valves. The inline positive end-expiratory pressure valves enables to decrease peak inspiratory pressure; this helps us to increase positive end-expiratory pressure, PEEP valve also ensures unidirectional flow of the gas flow [8]. Bucher et al. depicted how bag valve mask ventilation can be aided using PEEP valve which is attached the pressure range is set between 5 and 15 cm H_2O to improve oxygenation prior to intubating the patients. It instructs not to exceed a PEEP of 20 cm H_2O on a bag valve mask [9].

The research works of Pedro Leme Silva et al. depicted that the respiratory rate (RR), inspiration to expiration ratio, minute ventilation, tidal volumes, and positive-end-expiratory pressure (PEEP) are the most important ventilator parameters to be taken into consideration. From the work of Bunting Leonard et al., we concluded that adjustable PEEP valves can be used to regulate the positive end-expiratory pressure during ventilating a patient.

The following values were obtained from the Literature review:

i. The respiratory rate ranges from 12 to 16 breaths per minute.
ii. The tidal volume values lie between 6 and 8 ml per kg body weight.
iii. The PEEP value should be in the range of 3 to 5 cm of H_2O.

The intention of this project is to provide an effective and economical solution to the above problems and to design the equipment using CAD software and develop a working prototype of the model.

2 Materials and Methodology

Methodology refers to the specific steps, tools, and procedures used to collect and analyze the data. The process involved in completion of the project is listed in Fig. 1.

2.1 *Design Calculations*

When a number of gears are mounted, their teeth get engaged, and this mechanical process forms a gear train. The gear train model and its assembly with BVM are shown in Fig. 2a, b. Gear teeth are designed in such a way that the pitch circles of the meshing gears do roll on so that there would be no collision between the gears and ensure smooth transmission between one and another gears.

Gear train formula is given by Eq. 1 [10],

$$\frac{N_i}{N_j} = \frac{D_j}{D_i} = \frac{T_j}{T_i} \tag{1}$$

where

N_i, N_j speed of Driving and Driven gears, respectively.
D_i, D_j diameter of Driving and Driven Gears, respectively.
T_i, T_j Number of Teeth of Driving and Driven Gears, respectively.

Minute ventilation calculation was done by using the formula shown in Eq. 2 [11].

$$\text{Minute Ventilation} = \text{Mass} \times \text{Respiratory Rate} \times \text{Tidal Volume} \tag{2}$$

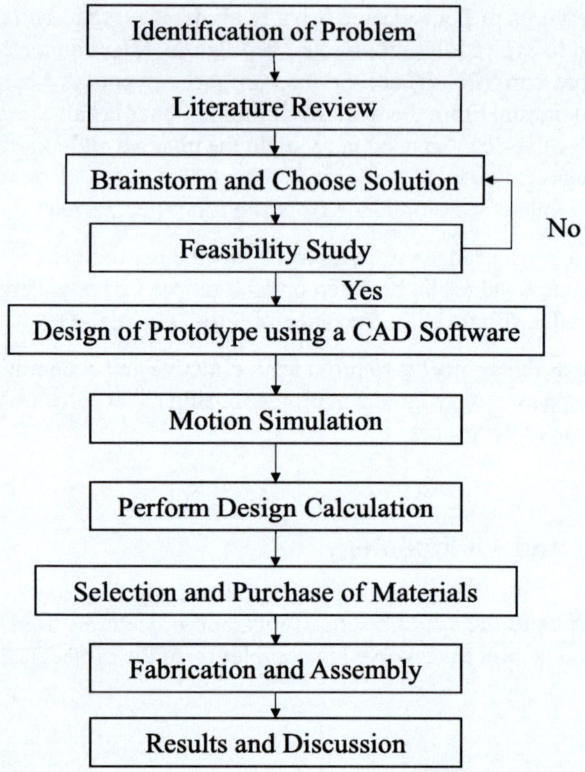

Fig. 1 Methodology

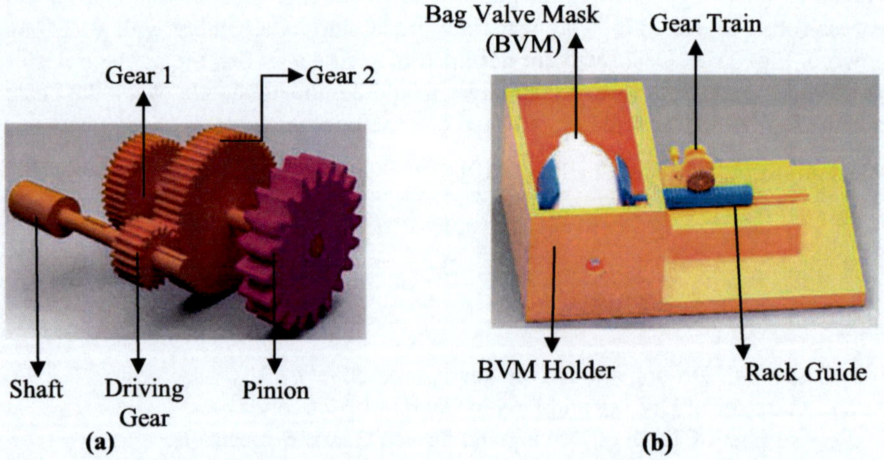

Fig. 2 **a** Gear train, **b** 3D model

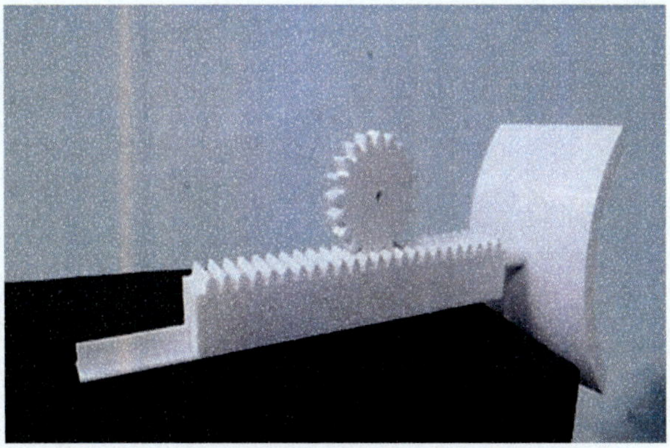

Fig. 3 Rack and pinion

Distance moved by the rack was calculated by using the formula shown in Eq. 3 [12, 13].

$$\text{Distance moved by rack, } D_R = \frac{2 \times \pi \times R_p \times t \times N_p}{60} \tag{3}$$

2.2 *Compression Mechanism*

The compression mechanism should be choosing to mimic the effect of a hand used to compress the bag valve mask; linear actuation mechanisms like rack and pinion as shown in Fig. 3 were developed for this. Hence for simpler operational construct, we chose the rack and pinion mechanism connected to a gear train in order to achieve variable drive speed according to the requirement.

2.3 *Assembly*

All the components are assembled to form the complete device as shown in Fig. 4. The driving shaft of the motor fitted with a gear which meshes with the gear of the driven shaft. This gear train is used to vary the speed of the pinion according to our requirements. The pinion is mounted on the rack and the rack on the rack guide. Thus, after assembly, the device is operated for testing.

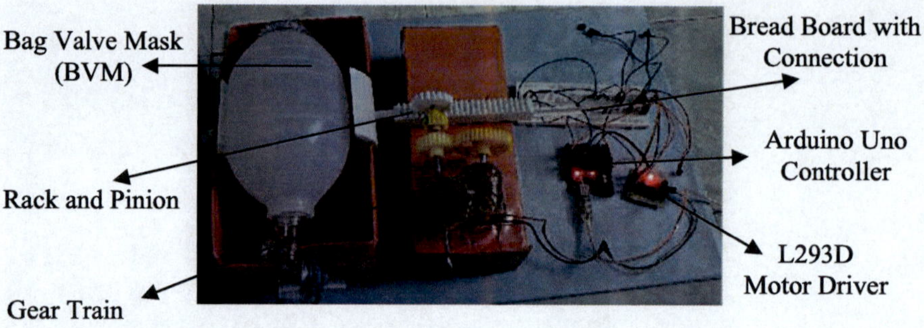

Fig. 4 Final assembly

3 Result and Conclusion

The designed and fabricated ventilator is tested to know its strength and capabilities. The critical components are shown in the block diagram given in Fig. 5 to understand the process flow. The Arduino board sends signals to the L293D motor driver based on the push button activation, and the motor rotation conditions are defined through programming such that the retraction stroke time is twice the compression stroke to achieve Inspiration: Expiration ratio of 1:2 [14]. This motor rotation connected to the gear train enables to and fro movement of the rack, thereby imitating the inspiration and expiration of a human.

3.1 Spirometer Results

The instrument that is used to measure the volume of air inspired and expires by lungs is called spirometer. The spirometer test indicates that this ventilator can be used to deliver minute ventilation from 4.38 to 11.9 L/min as given in Table 1, which is a very satisfactory range since most of the minute ventilation falls between the ranges of 4 and 8 L/min. The motor rotation speed and duration were identified by trial-and-error method such that it could accommodate all the practical minute ventilation values possible.

We conducted the spirometer experiment to determine the volume of air delivered by the ventilator for various motor timings. The volume of air delivered for various stroke lengths of the rack is shown in Fig. 6.

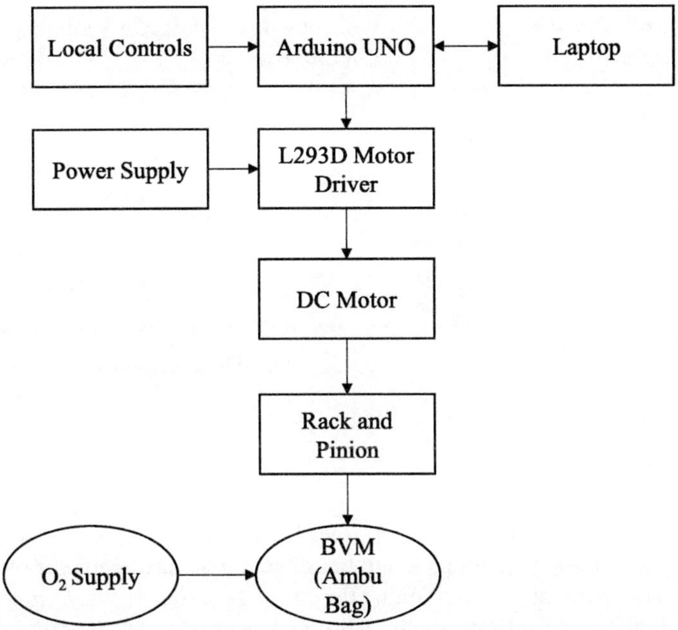

Fig. 5 Block diagram

Table 1 Minute ventilation delivery

Gear PCD (mm)	Motor rotation time (s)	RPM	Possible rack distance movement (mm)	Volume of air displaced per stroke (L)	No. of strokes per minute possible	Minute ventilation (L/min)
45	1	12	22	0.219	20	4.38
45	1.5	12	33	0.423	13	5.499
45	2	12	44	0.677	10	6.77
30	1.5	18	99	0.795	13	10.335
30	2	18	132	1.19	10	11.9

3.2 Results

Thus, the bag valve mask ventilator has been designed in the NX software and then required components are purchased, and the final assembly of the Ventilator has been done. The prototype has been tested in the spirometer, and the outputs obtained satisfy that this ventilator can supply the required amount of air to the patients weighing between 40 and 100 kg. In this ventilator, the ventilation is volume controlled. By varying the speed of the pinion, the volume of air delivered is controlled.

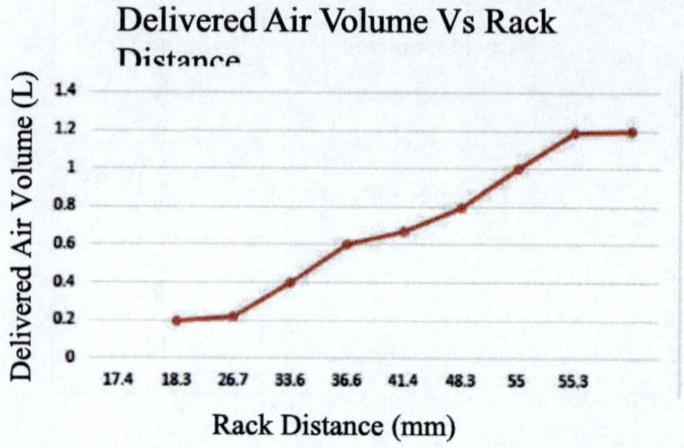

Fig. 6 Delivered air volume versus rack distance

3.3 *Discussion*

The surge in COVID-19 cases in recent times has raised the concern about people's accessibility to ventilators, though a number of solutions like ventilator by splitting it to two of more patients, reconditioning the expired ventilators were proposed. The posed solutions lacked serious applicability and seemed to be dangerous as well. Hence, an effective makeshift solution was the need of the hour. Various researchers around the world worked to find better solution to address this problems. The challenges of feasibility and affordability to large sector of people, low-cost emergency ventilator comes out to be more effective alternate solution.

3.4 *Conclusion*

The increased need for ventilators pushed many renowned ventilator manufacturers by reapportioning resources to increase production. More of the solutions have also been made open source. A large number of initiatives focusing on producing low-cost emergency ventilators suitable for fast implementation to meet momentary need gaps. This project has detailed the construction of a functional, low-cost, and open-source mechanical ventilator aiming at providing a makeshift solution to overcome the tough situation.

3.5 *Future Developments*

Future versions will be planned to include changes prompted by the results of current prototype testing. Currently, this ventilator is working in an open loop. In the future development, a flow sensor can be added to the circuit, and the output can be connected to the input through a feedback thereby the outputs can be further improved through a closed loop monitoring [15]. Also, when there are some critical values, alarms can be integrated and visualized by clinicians using an alarm screen or sound speakers [16].

References

1. Mora Carpio AL, Mora JI (2020) Ventilator Management. In: StatPearls [Internet]. Treasure Island (FL): StatPearls Publishing
2. Vasan A et al (2020) MADVent: a low-cost ventilator for patients with COVID-19.' Med Dev Sens e10106. https://doi.org/10.1002/mds3.10106
3. Ranney ML, Griffeth V, Jha AK (2020) Critical supply shortages—the need for ventilators and personal protective equipment during the Covid-19 pandemic. N Engl J Med 382:e41
4. Acho L, Vargas AN, Pujol-Vázquez G (2020) Low-cost, open-source mechanical ventilator with pulmonary monitoring for COVID-19 patients. Actuators 9(3):84
5. Al Husseini AM, Lee HJ, Negrete J, Powelson S, Servi AT, Slocum AH, Saukkonen J (2010) Design and prototyping of a low-cost portable mechanical ventilator. Trans ASME J Med Dev 4:027514
6. Silva PL, Rocco PRM (2018) The basics of respiratory mechanics: ventilator-derived parameters. Ann Transl Med 6(19):376. https://doi.org/10.21037/atm.2018.06.06
7. Salyer SW (2009) Essential emergency medicine for the healthcare practitioner, 1st edn. Saunders Publication
8. Bunting L et al (2020) A novel inline positive end expiratory pressure valve design for differential multi-ventilation. Am J Emerg Med 38(10):2045–2048. https://doi.org/10.1016/j.ajem.2020.06.089
9. Bucher JT, Vashisht R, Ladd M, Cooper JS (2021) Bag mask ventilation. In: StatPearls [Internet]. Treasure Island (FL). StatPearls Publishing. https://www.ncbi.nlm.nih.gov/books/NBK441924/
10. Bansal RK (2012) Strength of materials, Revised Edition. Laxmi Publications (pvt) Ltd
11. Kerechanin CW, Cytcgusm PN, Vincent JA, Smith DG, Wenstrand DS (2004) Development of field portable ventilator systems for domestic and military emergency medical response. John Hopkins Apl Tech Dig 25(3)
12. Khurmi RS, Gupta JK (2005) A textbook of machine design, .14th edn. Eurasia Publishing House (P.V.T) Ltd
13. Bhandari VB Design of machine elements, 3rdedn. The McGraw Hill Companies
14. Boidin MP (1982) A portable positive end expiratory pressure valve for 0–20 cm H_2O. Acta Anaesthesiol Belg 33(1):69–74
15. Hallett S, Toro F, Ashurst JV (2020) Physiology, tidal volume. In: StatPearls [Internet]. Treasure Island (FL): StatPearls
16. Mora Carpio AL, Mora JI (2020) Positive end-expiratory pressure. In: StatPearls [Internet]. Treasure Island (FL). StatPearls Publishing

Comparison of Fault Detection Data from Defective Ball Bearings Using Artificial Neural Networks

V. M. Akhil, S. L. Aravind, and Ravikiran Nayak

Abstract This paper tries to integrate artificial neural networks with the data sets acquired from the experimental set-up and compares which classifier is the best one that gives more accurate result. The data sets acquired from an experimental test rig which consists of defective ball bearing is analysed with the help of different classifier algorithms of artificial neural network. It was found that the 10% random set of data have given more accuracy than all other data sets subjected study in this proposed work. This method can be effectively used not only for data sets obtained from vibration analysis but also can be extended to all datatypes acquired from various experimental set-ups.

Keywords Bearing defects · Vibration analysis · Artificial neural networks · Algorithms · Classifiers

1 Introduction

Ball bearings are used in almost all the industries for the smooth functioning of the machineries [1–3]. It is very difficult to find a machinery which is working without a ball bearing. As a result of this, ample amount of ball bearings are operating in different machineries all over the globe. Since the main purpose of using ball bearings is to reduce the load on the rotating shafts, these ball bearings are subjected to enormous loads during its lifetime. This will lead to a lot of defects in the bearings in the form of inner ring defects, outer ring defects, cage defects, etc. [4]. As a result of this, vibrations will be induced in the machinery. In most of the cases, vibrations are undesirable which will lead to unnecessary sound pollution and later failure of the machinery. To prevent such kind of things, the different bearing defects must be

V. M. Akhil (✉)
Amrita Vishwa Vidyapeetham, Amritapuri, Kerala, India
e-mail: akhilvmtvm@gmail.com

S. L. Aravind · R. Nayak
The National Institute of Engineering, Mysuru, India

N. M. Sivaram et al. (eds.), *Advances in Manufacturing, Automation, Design and Energy Technologies*, Lecture Notes in Mechanical Engineering,
https://doi.org/10.1007/978-981-99-1288-9_61

detected early to prevent the failure and thereby shutting down of the unit which will lead to loss of time and money.

Lots of methods are used to find out the bearing defects. The most accurate and common method used is by obtaining vibration signals from accelerometers, and later by the interpretation of peaks obtained, we can easily find out the type of defect which have occurred to the bearing. The augmentation of artificial neural network with these data sets has proved to be a very much useful technique to compare the effectiveness of this type of vibration data acquisition from these ball bearings [5].

In this research work, the vibration data from a test rig is obtained experimentally using accelerometers, and later, this data is put into different artificial neural network logarithms, namely classifiers. Therefore, the prime objective of this work is to compare the accuracy of the different classifiers used in artificial neural network with different data sets obtained from the experimental data.

2 Methodology

This work analyses different classification algorithms with different percentage of bearing data set. The data was taken from the NSF I/UCR Centre for Intelligent Maintenance Systems [1]. The test set-up consists of four bearings (Rexnord ZA-2115 double row bearings) that were installed on a shaft and maintain a constant speed of 2000 rpm using an AC motor. A spring-loaded device applies a radial load of 6000 pounds to the shaft and bearing (Fig. 1).

Accelerometers (PCB 353B33 High Sensitivity Quartz ICP) were installed on the bearing housing for taking acceleration data. All failures occurred after exceeding designed lifetime of the bearing which is more than 100 million revolutions.

Four bearing data with different failure (Two no failure, Inner ring Failure, Outer ring Failure) is selected for analysis. For classification, different classification algorithms were selected with different data percentage. Case 1 analysed with 100%

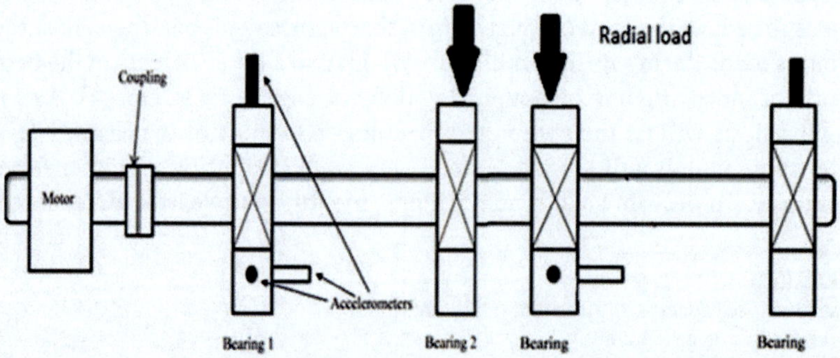

Fig. 1 Bearings test rig and sensor location

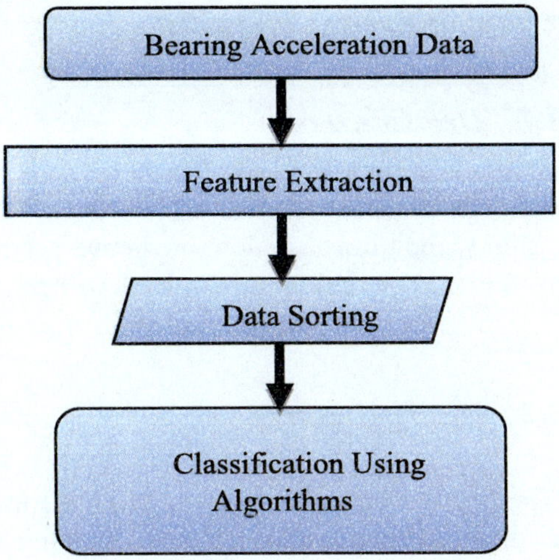

Fig. 2 Methodology

bearing data; case2 analysed with 50% bearing data selected randomly; case 3 analysed with 10% bearing data selected randomly. All cases run with five trails. The flow diagram for methodology is shown in Fig. 2.

2.1 *Time-Domain Features*

Kurtosis

The flatness or spikiness of a signal is depicted by the kurtosis plot. Due to the spiky character of the signal, its value displays a rising trend from normal bearing condition to poor bearing condition.

$$Kurtosis = \left\{ \frac{n(n+1)}{(n-1)(n-21)(n-3)} \sum \left(\frac{x_i - \overline{x}}{S} \right)^4 \right\} - \frac{3(n-1)^2}{(n-2)(n-3)}$$

Peak magnitude to RMS ratio (PMRR)

Calculate the ratio of the object's biggest absolute value to its root-mean-square (RMS) value.

Skewness

Skewness is a distortion or asymmetry in a collection of data that deviates from the symmetry bell-shaped curve, or normal probability plot.

3 Classification Algorithms

3.1 Decision Tree

Although decision trees are easy to learn, fit and forecast quickly, and need little memory, they can have low predictive accuracy. Develop simpler trees to avoid overfitting. You can control the depth with the maximum number of splits parameter.

3.2 Discriminant

Discriminant analysis is a popular first classification algorithm to use since it is quick, accurate, and simple to understand. Discriminant analysis works well with large data sets. Discriminant analysis is based on the assumption that different classes create data using distinct Gaussian distributions. In order to train a classifier, the fitting function estimates the parameters of a Gaussian distribution for each class.

3.3 Svm

An SVM classifies data by determining the optimum hyperplane that divides data sets from a class from some of those from another. The optimum hyperplane for an SVM is one with the greatest difference between the two classes. The maximum width of the slab parallel to the hyperplane with no inside data points is referred to as the margin. The support vectors are the data points closest to the separating hyperplane; these points are on the slab's border.

3.4 Knn

Classifying points based on their closeness to coordinates in a training sample could be a simple and effective method of doing so.

3.5 Ensemble

Ensemble classifiers combine the findings of numerous poor learners into a single high-quality ensemble model. The qualities are determined by the algorithm used.

3.6 Neural Network

Despite the fact that neural network models have a high predicted accuracy and may be used for multiclass classification, they are difficult to understand. The neural network's size and number of completely linked layers boost model adaptability..

4 Result and Discussion

All the selected features are showing the better classification results. The skewness feature is analysed as an objective function for vibrational bearing data analysis. Figure 3 is showing the variation in skewness during running time for bearing 1 to bearing 4. Form this, it is clear to understand the skewness variation during failure and it is showing a maximum of 14.8 for inner ring failure.

Crest factor or peak magnitude to RMS ratio may imply incipient defects of the bearings. Figure 4 shows crest factor for bearing during running time for bearing 1 to bearing 4. From this, it is clear to understand that crest factor variation during failure and it is showing a maximum of 22 for inner ring failure. Kurtosis is another feature that mostly used for rolling element bearing fault signal analysis. Figure 5 shows the kurtosis feature variation during bearing running. Here, also the maximum variation is shown by the inner ring fault bearings.

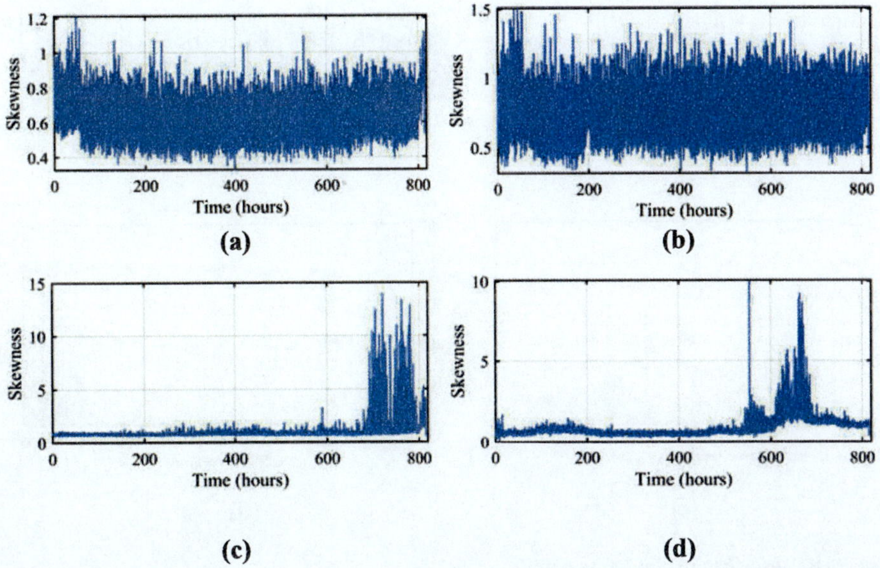

Fig. 3 Skewness feature plot for bearings (a—Bearing1, b—Bearing 2, c—Bearing 3, d—Bearing 4)

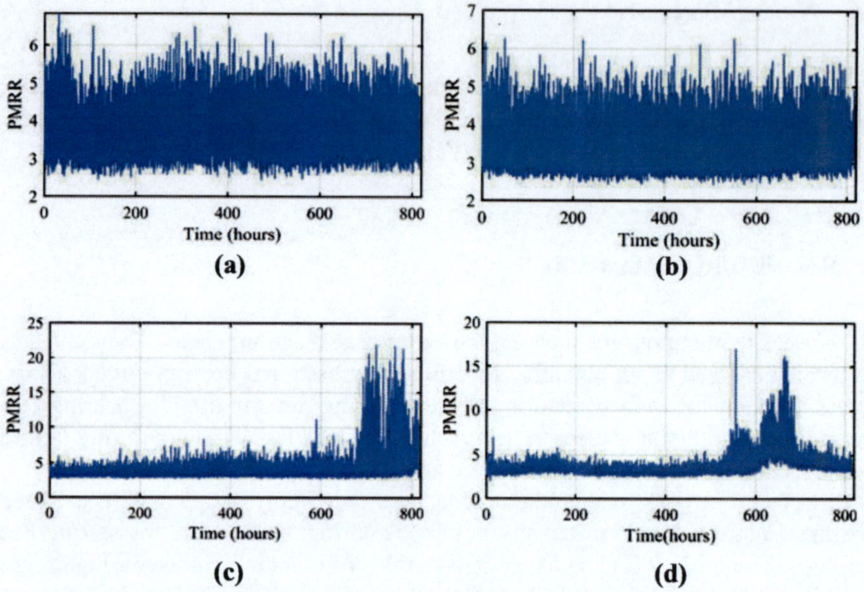

Fig. 4 Peak magnitude to RMS ratio feature plot for bearings

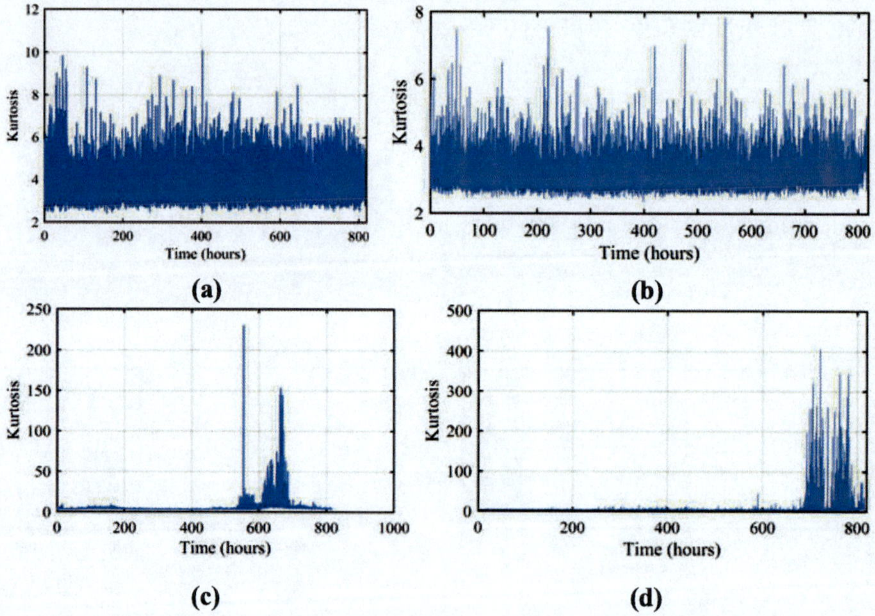

Fig. 5 Kurtosis feature plot for bearings

Table 1 Classification accuracy variation under different percentage of data set

Classification Algorithm	Accuracy		
	100% Data	50% Data	10% Data
Fine tree	69.1	58.3	55.6
Quadratic Discriminant	52.8	49.7	46.3
Linear SVM	53.9	50.1	48.6
Fine Gaussian SVM	75.8	74.1	69.3
Medium Gaussian SVM	74.2	**75.9**	70.7
Coarse Gaussian SVM	66.0	69.6	64.9
Fine KNN	66.6	54.7	52.0
Medium KNN	73.5	71.8	68.4
Coarse KNN	72.5	69.1	66.8
Cubic KNN	73.5	74.2	69.9
Ensemble Bagged tree	66.5	71.6	64.2
Medium NN	75.8	72.3	69.1
Wide NN	**76.2**	74.9	**71.0**
Trilayered NN	76.0	75.4	69.7

This work analysed different classification algorithms with different percentage of data set. When training is done with 100% data set, wide neural network (76.2%) showed highest level of accuracy (Table 1). Quadratic Discriminant algorithm gives the poor result (52.8%) for classification. Half of data set is given for next iteration of algorithms and Medium Gaussian SVM shown the highest accuracy of 75.9%, whereas Linear SVM exhibits the least percentage (50.1%). While doing the experiment with 10% of data set, total accuracy of classification is directly reduced. But here again wide neural network shows the highest accuracy of 71%. Form this, there is reduction of 90% of data set even through only 5.2% reduction in classification accuracy. Confusion matrix for KNN medium and wide NN is shown in Fig. 6. Scatter plot for wide NN model prediction correctly classified Kurtosis with skewness is shown in Fig. 7. A receiver operating characteristic curve, or ROC curve (Fig. 8), is a graphical representation of a binary classifier system's diagnostic performance when its discrimination threshold is modified.

5 Conclusion

In this work, four bearing data with different failure (Two no failure, Inner ring Failure, Outer ring Failure) is selected for classification with different classification algorithms along with different data set percentage. Three time domain features like kurtosis, skewness, Peak to RMS were selected for classification. When training is done with 100% data set, wide neural network showed highest level of accuracy.

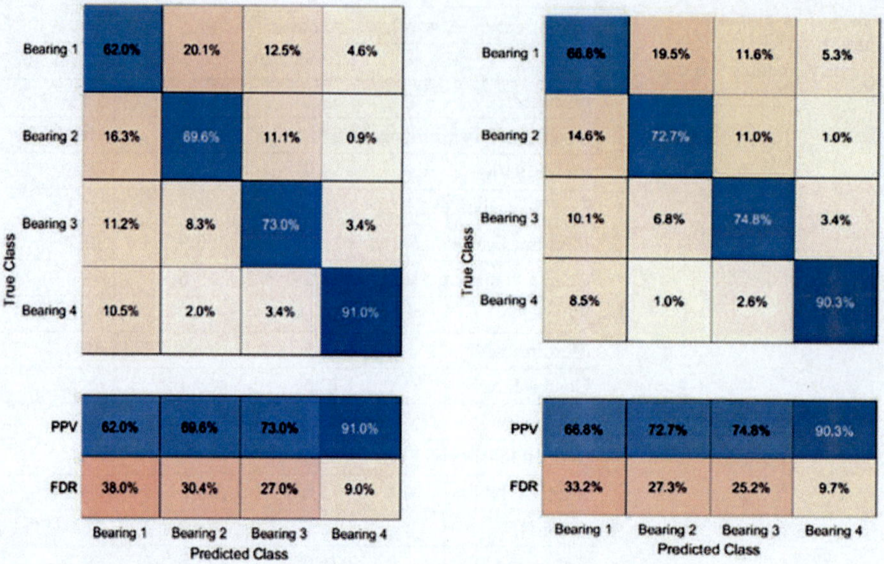

Fig. 6 Confusion matrix for KNN medium and wide NN

Fig. 7 Scatter plot for wide NN model prediction correctly classified Kurtosis with skewness

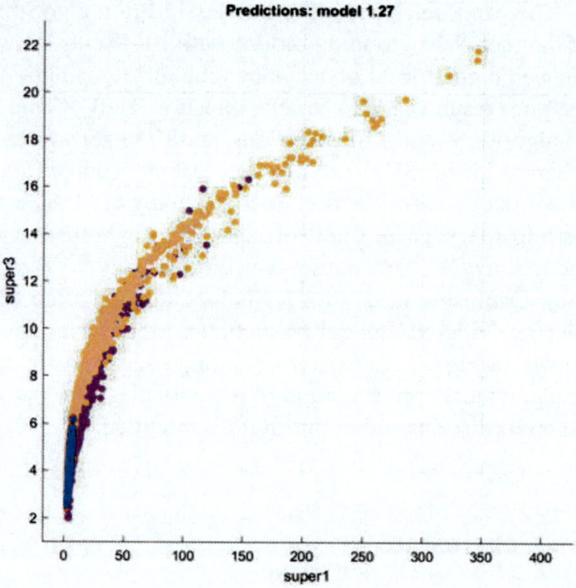

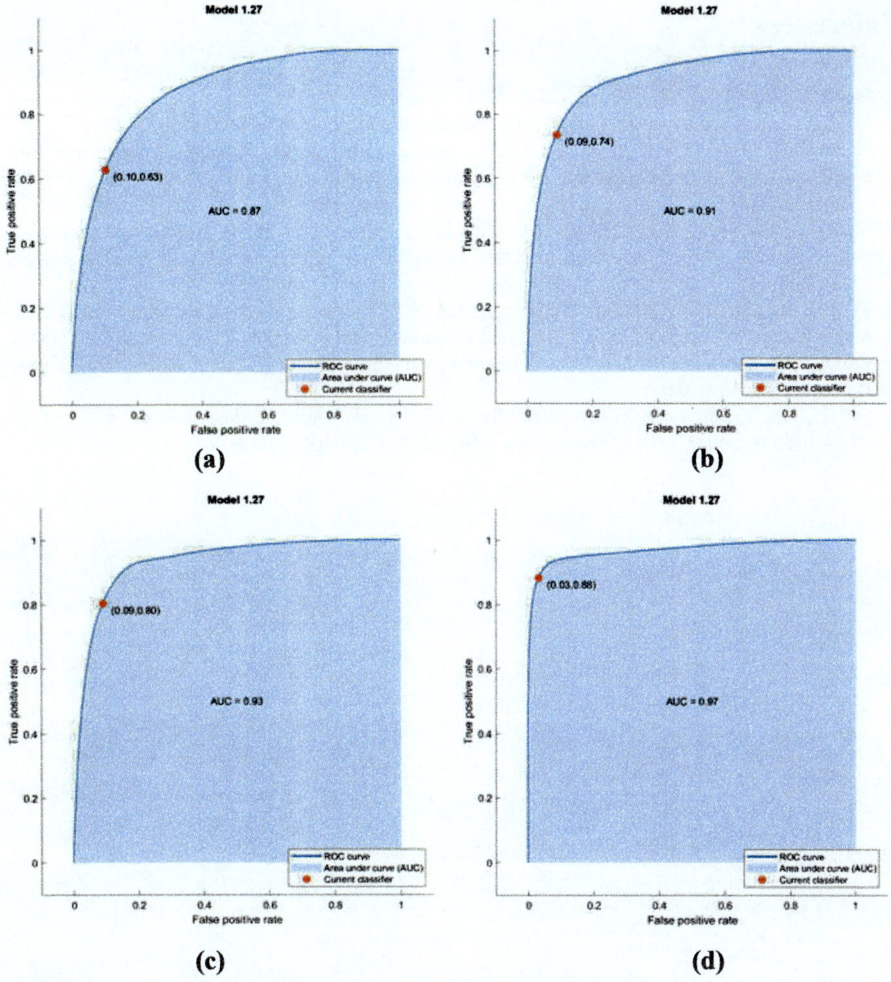

Fig. 8 ROC plot for bearing

The 50% of data set is given for next iteration of algorithms and Medium Gaussian SVM shows the highest accuracy of 75.9%, whereas Linear SVM exhibits the least percentage. While doing with 10% of data set, wide neural network shows the highest accuracy of 71%. There is just a slight decrease in classification accuracy even if the data set is smaller.

References

1. Fan Y, Zhang C, Xue Y, Wang J, Gu F (2020) A bearing fault diagnosis using a support vector machine optimised by the self-regulating particle swarm. Shock and Vibration, 2020
2. Hwang DH, Youn YW, Sun JH, Choi KH, Lee JH, Kim YH (2015) Support vector machine based bearing fault diagnosis for induction motors using vibration signals. J Electr Eng Technol 10(4):1558–1565
3. Zhang J, Yi S, Liang GUO, Hongli GAO, Xin HONG, Hongliang SONG (2020) A new bearing fault diagnosis method based on modified convolutional neural networks. Chin J Aeronaut 33(2):439–447
4. Tyagi S, Panigrahi SK (2017) A DWT and SVM based method for rolling element bearing fault diagnosis and its comparison with Artificial Neural Networks. J Appl Comput Mech 3(1):80–91
5. Randall RB, Antoni J (2011) Rolling element bearing diagnostics—A tutorial. Mech Syst Signal Process 25(2):485–520
6. Qiu H, Lee J, Lin J (2006) Wavelet Filter-based weak signature detection method and its application on roller bearing prognostics. J Sound Vib 289:1066–1090

Optimization of Processing Parameter for Optimal Performance of Dyneema HB50 Composite

Arunesh Kumar Srivastava and K. N. Pandey

Abstract The present work is an experimental study to evaluate the performance of a new composite material Dyneema HB50, one of the types of Ultra-High-Molecular-Weight Polyethylene (UHMWPE)-based composite material, for ballistic applications, viz. bullet proof vests and helmets. This material can also be use in various other purposes where tensile strength is main concern with temperature range like aircrafts, submarines, space technology. A three factor, three-level Taguchi experimental design was employed for maximizing ultimate tensile strength (UTS) of Dyneema HB50 composite based on different experimental data obtained in L9 orthogonal array. Three independent variables pressure ranging from 100 to 180 Kg/cm^2. The temperature and time ranges from 125 to 135 °C and 10 to 20 min as $\times$ 1, $\times$ 2 and $\times$ 3, coded at three consecutive levels (1, 2 and 3) and then to evaluate the responses. The regression equation is obtained. The significance of the independent variables and their interactions were obtained by variance analysis (ANOVA) with a 95% confidence limit (= 0.05). The optimized process parameters pressure, temperature, and time were 100 Kg/cm^2, 152 °C, and 10 min, respectively.

Keywords UHMWPE · Composite material · Dyneema HB50 · Analysis of variance (ANOVA)

1 Introduction

Composites can be counted as one of the most advanced and adaptable man made engineering material. Research and development can be one of the reasons behind the creation of these interesting and fantastic materials. Usually, process of creation of composites required filler of two or more components or fiber reinforcing and a compactable matrix [1].

A. K. Srivastava (✉) · K. N. Pandey
Mechanical Engineering Department, Motilal Nehru National Institute of Technology, Prayagraj-211004, Uttar Pradesh, India
e-mail: arunesh_srivastava15@yahoo.com

N. M. Sivaram et al. (eds.), *Advances in Manufacturing, Automation, Design and Energy Technologies*, Lecture Notes in Mechanical Engineering,
https://doi.org/10.1007/978-981-99-1288-9_62

The matrix material may be metal, ceramic, or polymer composite. The matrix provides texture, surface texture, environmental tolerance, and overall durability to the composite, but the structural load is largely swallowed by fibrous reinforcement that provides strength and macroscopic rigidity [2]. The versatile property of UHMWPE has been studied by many researchers [3]. reveal the microstructure of 0/90° laminate of Dyneema fiber laminate, and the relationship between tensile properties of the laminate, yarn, and fiber for Dyneema SK76 was studied over a wide range of strain rate. Researchers [4] observed the performance of UHMWPE fiber at high temperatures. They run various tests to calculate the tenacity and modulus of Dyneema SK76 fiber as function of the temperature and strain rate. Various experimental tests were performed on composite materials for ballistic application [5]. The creep characteristics of UHMWPE at elevated temperatures would seem that there is an underlying problem of localized excessive heat build-up on the bearing surface of the polyethylene acetabular cups. For determining, the extent of influence ANOVA is a standard statistical technology. To estimate the confidence level of the influence and to calculate the errors in the experiment [6, 7], these values are calculated with the help of F and P values of the factors. The ratio between sums of the square of each factor's average deviations to that of the experimental error is equivalent to F [7, 8]. Studied the influence of elastic constant on the model of Dyneema yarn with a single layer Dyneema woven fabric with ballistic impact by applying the orthogonal experiment method. For analyzing, the six factors orthogonal table L25 was used with each having five levels, and result shows that ballistic modeling was formalized with the experimental results, viz. absorption of energy, time of failure, and the number of failed yarn. As per the orthogonal analysis, G13 is the most influencing for simulated result in which confidence level is more that 95%, and v is the least significant. The orthogonal study helps in identification of the practical result and the elastic constant that led to a significant agreement between the finite element method and experimental outcome. A special statistical method is developed by Taguchi and Konishi [9]. The prior purpose to develop this process was to improve the quality of the manufacturing but later on ita are of application got broaden like Biotechnology [10], etc. The application of design developed by Taguchi also includes a practitioner's handbook [11]. This development in methodology has brought design of experiments to a totally new world of manufacturing from the area of statistician. This development has also made researchers work easier because it supports few design of experiments and provides clear picture of economical cost effectiveness of quality engineering in manufacturing world [11, 12].

In agriculture, Design of Experiments (DOE) method was started by Sir R. A. Fisher with a motive to improve the crop production trails and treatments [13].

Fisher first created layout of all involved factors in the study and then developed full factorial design with all varied factors simultaneously. The experimental design study has given the effective way to study the interaction of variables instead of trial and error.

DOE is a statistical way to study the any system or process and take a decision on input variables to the output, either on their own influences, the interactions between the inputs, or not at all. Scientists who studied the optimization of injection molding conditions through customary DOE methods were separated in full factorial design and fractional factorial design [14]. The main purpose of factorial design is to recognize all the permutation combination in a given set of factors. In fact, all experimental experiments normally required significant number of factors and full factorial design to perform the experiments [15]. Thus, fractional factorial design provides maximum information in limited number of experiments in all permutation and combination.

On the other hand, fractional factorial design is a well-known complex design. There is no general guideline which is available for this application or analysis for the result obtained by this experiment [16]. To solve all these problems, a new method is developed by Dr. Genichi Taguchi which is known as Taguchi method and utilizes a modified and standardized form of DOE in the late 1940s [17, 18]. Nowadays, Taguchi method is applied to different areas like manufacturing system [19], mechanical component design [20], and process optimization [21, 22], and in last 20 years, it has drawn attention various researches and literature.

Taguchi method is so popular because of its practical design in designing high quality systems which provides much-reduced variance for experiments with best possible conditions for process control parameters.

2 Material and Method

The current work is a pilot study to test the performance of the new composite Dyneema HB50, which is one of the Ultra-High-Molecular-Weight Polyethylene (UHMWPE)-based composite materials, for ballistic use, i.e., bullet proof vests and helmets. This material can also be use in various other purposes where tensile strength is main concern with temperature range like aircrafts, submarines, space technology. Firstly, Dyneema HB 50 sheets were converted into a panel using various pressure, temperatures, and time cycles. These sheets were then tested for tensile properties using standard tensile test.

Classical experimental design methods can be defined as a complex process which are difficult to use. A number of experiments need to be conducted as process parameter increase to 86. For solving this issue as special design of orthogonal arrays, the Taguchi method is applied to study the whole parameter space with a very small number of experiments. Pressure, temperature, and time are considered as controlling factors for super plastic formation. For each parameter, there are three levels low, medium, and high, and for denoting these, level 1, 2, and 3 numbers are used, respectively. As per the Taguchi method three parameters with 3 levels has to be employed for L_9 orthogonal array. Superplastic formation parameters are given in Table 1, and their levels are used for experiment.

Orthogonal arrays are important design of experiment which required only a small number of experiments to find out the major factor effect on result. Prior to the selection of the orthogonal array, small number of experiment are supposed to be run using Eq. (1)

Table 1 Process parameter and level

Process Parameter	Levels		
	Level 1	Level 2	Level 3
Pressure (Kg/cm^2)	100	140	180
Temperature (°C)	125	130	135
Time (minutes)	10	15	20

$$N = 1 + NV(L - 1) \quad (1)$$

where N = Number of experiment to be conducted, NV = number of parameter, L = Number of levels. After conducting tests, S/N ratios for testing samples are calculated on the basis of the Eq. (2) which is used for the cases of larger-the-better type of quality characteristic/objective function.

$$S/N = -10 * log(\Sigma(1/Y^2)/n \quad (2)$$

where n denotes observations number (test runs); y_i is the observed data for i^{th} test run. The SNR calculation was combined with the ANOVA and utilized for the experimental results and selection of optimal level of ultimate tensile strength (UTS) testing conditions. Such optimal conditions were confirmed with the help of confirmation experiment. Analysis of effect of all testing parameter was carried out with the help of statistical tool box MINITAB 16.0. Table 2 shows the orthogonal arrays based on L9 and using Eq. (2). L9 Experiment is performed on 100 kN BISS make universal testing machine (UTM) at Advanced Center for Material Science (ACMS), IIT, Kanpur, India, as shown in Figs. 1, 2 shows tensile test specimens.

Table 2 Experimental design using orthogonal arrays (OAs)

Run no	Level of variables and experimental values				
	Pressure (Kg/cm^2)	Temperature (°C)	Time (Minute)	S/N Ratio	Ultimate Tensile Strength (Kg/cm^2)
1	100	125	10	32.94	44.4
2	100	130	15	31.73	38.63
3	100	135	20	30.8	34.7
4	140	125	15	30.56	33.74
5	140	130	20	30.27	32.65
6	140	135	10	28.98	28.13
7	180	125	20	28.91	27.9
8	180	130	10	26.48	21.09
9	180	135	15	26.23	20.49

Fig. 1 Universal testing machine

Fig. 2 Tensile test specimen based on L_9 OAs

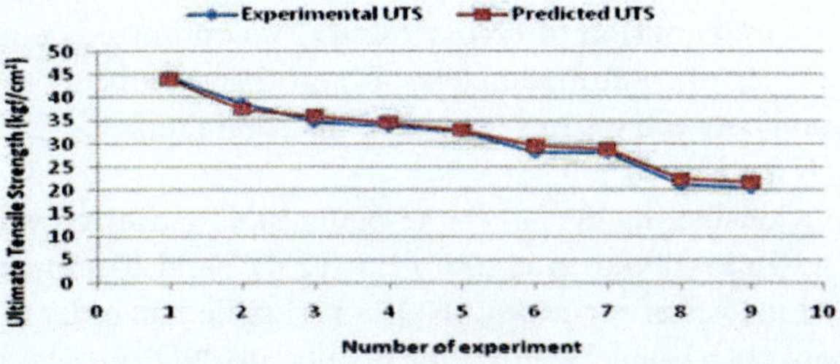

Fig. 3 Comparison curve of experimented, predicted values

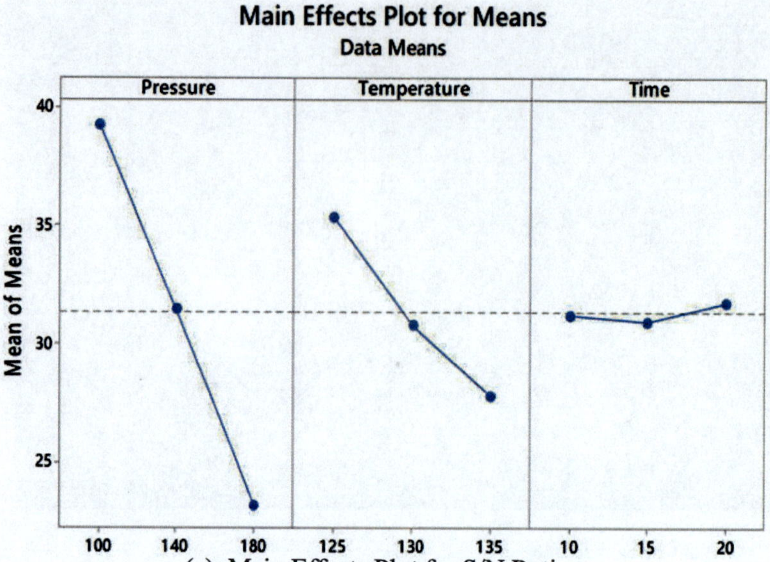

(a). Main Effects Plot for S/N Ratios

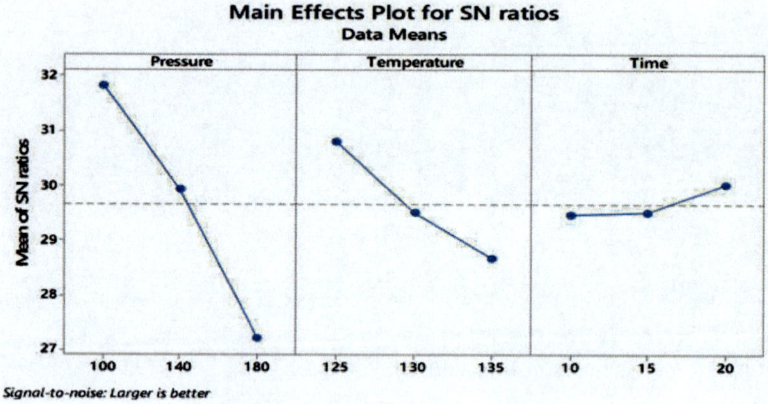

(b). Main Effects Plot for Means

Fig. 4 **a** Main effects plot for S/N ratios **b** Main effects plot for means

Confirmation of Experiment: The optimized control factors are validated with the help of confirmation tests. Figure 3 shows the more or less similar nature of variation between the experimented data and predicted data for ultimate tensile strength (UTS) (Fig. 4).

Considering the larger-the-better S/N, optimized values of pressure, temperature, and time is obtained as $100 MPa$, $125^{o}C$, and 20 min, respectively, clear from Fig. 2a, b. Analysis of variance (ANOVA) is applied, in order to analyze the influence of each parameter. Table 3 shows the results of ANOVA obtained with the help of statistical tool MINITAB 16.

Table 3 Analysis of variance (ANOVA)

Source	DF	Adj SS	Adj MS	F- value	P-value
Regression	3	473.844	157.948	60.78	0
Pressure	1	386.709	386.709	148.8	0
Temperature	1	86.648	86.648	33.34	0.002
Time	1	0.4888	0.4888	0.19	0.683
Error	5	12.994	2.599		
Total	8	486.838			

The mathematical modeling is developed using response surface methodology (RSM) and the developed mathematical regression Eq. (3)

$$UTS = 157.4 - 0.2007Pr - 0.760Temp + 0.057Time \quad (3)$$

The calculated F-values for the regression models of ultimate tensile strength (UTS) are 60.78 which falls in the range of acceptance. Because of noise there are very few chances for F-value to reaching this range. For the regression models, the associated p-values are linear, and square terms are lower than 0.01($\alpha = 0.01$ or 99% confidence level) which shows the statically significance of the model.

The coefficients of determination R^2, adjusted R^2, and predicted R^2 values, in Table 4, of the developed models are being calculated to find the accuracy of the data.

The values of R^2, adjusted R^2, and predicted R^2 are found as 0.9733, 0.9573, and 0.8951 for tensile strength, respectively, which shows that the prediction of data by developed models for each and every response are well fitted. Associated data of Table 4 in Tables 5, 6 shows the pressure is most contributed input parameters for affecting the ultimate tensile strength (UTS).

Table 4 Model summary

S	(R^2)	(R^2_{adj})	$(R^2_{prediction})$
1.61209	97.33%	95.73%	89.51%

Table 5 Coefficient terminology of model

Term	Coeff	SE Coeff	T-value	P-value	VIF
Constant	157.4	17.4	9.05	0	
Pressure	–0.2007	0.0165	–12.2	0	1
Temperature	–0.76	0.132	–5.77	0.002	1
Time	0.057	0.132	0.43	0.683	1

Table 6 Variation difference in model

	SS	Percentage variation
Pressure	386.71	79.43
Temp	86.648	17.79
Time	0.488	0.1
Total	486.84	

Table 6 also shows the significant variation, 79.43%, of experimented pressure values on UTS from mathematical model data. The percentage variation in pressure of model data from calculated also approves the pressure is most contributed parameters.

3 Results and Discussion

For Optimal Performance of Dyneema HB-50 Composite the optimize fabrication process parameter is done by Taguchi design of experiment. Pressure, temperature, and time were taken as the process parameters. Range and levels of process parameters were selected on the basis of preliminary experiments and selected for the cases of larger-the-better type of quality characteristic/objective function. F- and p-value test was done to find out the capability of the well-developed mathematical models. For the regression models, calculated F-values of ultimate tensile strength (UTS) are 60.78 which must be in the range of acceptance. The coefficients of determination R^2, adjusted R^2, and predicted R^2values, are found as 0.9733, 0.9573 and 0.8951 for tensile strength, respectively, which shows that each responses are well fitted with data predicted by developed models. Associated data shows the pressure is most contributed input parameters for affecting the ultimate tensile strength (UTS) and also shows the significant variation, 79.43%, of experimented pressure values on UTS from mathematical model data. The percentage variation in pressure of model data from calculated also approves the pressure is most contributed parameters. The optimized process parameters pressure, temperature, and time were 100 Kg/cm^2, 125 °C, and 10 min, respectively, using the Taguchi L_9 orthogonal array.

4 Conclusion

On the basis of three factors, three-level Taguchi experimental design implementation the maximum tensile strength (UTS) of the Dyneema HB50 composite was obtained in L9 orthogonal array. Three independent variables pressure ranging from 100 to 180 Kg/cm^2, temperature and time ranges from 125 to 135 °C and 10 to 20 min respectively as $\times$ 1, $\times$ 2 and $\times$ 3, coded at three consecutive levels (1, 2 and 3) and then responses is evaluated. The regression equation is obtained. The significance of the independent variables and their interactions was obtained by variance analysis (ANOVA) with a 95% confidence limit =0.05 The significance of independent variables and their interactions were obtained by means of the analysis of variance (ANOVA) with 95% confidence limits ($\alpha = 0.05$). The optimized process parameters pressure, temperature, and time were 100 Kg/cm^2, 125 °C, and 10 min, respectively.

References

1. Ferguson, Clive (2008) Historical introduction to the development of material science and engineering as a teaching discipline. The Higher Education Academy, UK Centre for Materials Education, Liverpool
2. Roylance, David (2000) Introduction to composite materials. Department of material science and engineering, Massachusetts Institute of Technology, Cambridge
3. Russell BP et al (2013) The high strain rate response of Ultra High Molecular-weight Polyethylene: From fibre to laminate. Int J Impact Eng 60:1–9
4. Meulman JH, et al. (2010) Ballistic performance of Dyneema at elevated temperatures, extreme for body armor. In: PASS Conference papers
5. Czechowski, Leszek, Jacek Jankowski, and Tomasz Kubiak (2012) Experimental tests of a property of composite material assigned for ballistic products. Fibres & Textiles in Eastern Europe
6. Wu, Xuan, Dennis YC Leung (2011) Optimization of biodiesel production from camelina oil using orthogonal experiment. Appl Energy 88.11, 3615–3624
7. Roy, Ranjit K (2001) Design of experiments using the Taguchi approach: 16 steps to product and process improvement. John Wiley & Sons
8. Yuan Z, Chen X, Zeng H, Wang K, Qiu J (2018) Identification of the elastic constant values for numerical simulation of high velocity impact on dyneema woven fabrics using orthogonal experiments. Compos Struct 204:178–191
9. Taguchi G, Konishi S (1987) Taguchi Methods, orthogonal arrays and linear graphs, tools for quality American supplier institute. Am Supplier Inst 8–35
10. Rao, Ravella Sreenivas, et al. (2008) The Taguchi methodology as a statistical tool for biotechnological applications: a critical appraisal. Biotechnol J: Healthc Nutr Technol 3.4 510–523
11. Taguchi G (1990) Introduction to quality engineering, Tokyo. Asian Productivity. Organization 4(2):10–15
12. Ross PJ (1988) Taguchi techniques for quality engineering
13. Dowlatshahi, Shad (2004) An application of design of experiments for optimization of plastic injection molding processes. J Manuf Technol Manag
14. Park, Keun, Jong-Ho Ahn (2004) Design of experiment considering two-way interactions and its application to injection molding processes with numerical analysis. J Mater Process Technol 146.2 221–227
15. Farkas, Károly et al. (2007) NWC: node weight computation in MANETs. In: 2007 16th International Conference on Computer Communications and Networks. IEEE
16. Rao, Ravella Sreenivas, et al. (2008) The Taguchi methodology as a statistical tool for biotechnological applications: a critical appraisal. Biotechnol J: Healthc Nutr Technol 3.4 510–523
17. Singh, Hari, Pradeep Kumar (2005) Optimizing cutting force for turned parts by Taguchi's parameter design approach
18. Kamaruddin S, Zahid A Khan, Foong SH (2010) Application of Taguchi method in the optimization of injection moulding parameters for manufacturing products from plastic blend. Int J Eng Technol 2.6 574

19. Mahfouz, Amr, Said Ali Hassan, Amr Arisha (2010) Practical simulation application: Evaluation of process control parameters in Twisted-Pair Cables manufacturing system. Simul Model Pract Theory 18.5 471–482
20. Shim, Hee-Jin, Jung-Kyu Kim (2009) Cause of failure and optimization of a V-belt pulley considering fatigue life uncertainty in automotive applications. Eng Fail Anal 16.6 1955–1963
21. Mohan NS, Ramachandra A, Kulkarni SM (2005) Influence of process parameters on cutting force and torque during drilling of glass–fiber polyester reinforced composites. Compos Struct 71(3–4):407–413
22. Ariffin, MK A Mohd, et al. (2009) An optimise drilling process for an aircraft composite structure using design of experiments. Sci Res Essays 4.10 1109–1116

Design, Thermo-Mechanical Analysis and Optimization of an IC Engine Piston with Aluminium and AlSi4032

Nishant Patel and Ashok Atulkar

Abstract This work is to present an analysis of thermo-mechanical behaviour of the automotive IC engine piston using FEA in static structural and steady-state thermal analysis for an engine system. In this work, thermo-mechanical analysis is performed on two materials: aluminium and its alloy AlSi4032 for thermal, structural and optimization of IC engine piston, which is based on a problem using ANSYS-18. A piston work is to transfer the force produced by gas during expansion strokes known as power stroke in which the piston has to suffer highly thermal and mechanical stress. The first analysis is performed on the aluminium with the inclusion of convection, temperature and pressure elements. Characteristics for piston models such as stress concentration von Mises stress, equivalent strain and heat flux are predicted. Next, thermo-mechanical analysis is performed on the alloy material of the piston in the same manner and required results are predicted. The results based on the piston's maximum stress, deformation, strain, heat flux and critical region using FEM. AlSi4032 can be hot forged with a close die and wieldable with inert metal gas welding. Responsive for heat treatment generally annealing is preferred for AlSi4032. There are many materials available in industries with different proportions of alloying elements but silicon in aluminium is most widely used. For modelling and analysis, the software tool namely CREO and ANSYS Workbench have been used. AlSi4032 performs well in every criterion.

Keywords IC engine piston · Aluminium alloy · Thermo-mechanical analysis · Optimization

---

N. Patel (✉) · A. Atulkar
Department of Mechanical Engineering Shri Govindram Seksaria Institute of Technology and Science, Indore, India
e-mail: nishantpatel738@gmail.com

N. M. Sivaram et al. (eds.), *Advances in Manufacturing, Automation, Design and Energy Technologies*, Lecture Notes in Mechanical Engineering,
https://doi.org/10.1007/978-981-99-1288-9_63

1 Introduction

This research work has been completed in the design, thermal and optimization of the piston of an engine with thermo-mechanical analysis in ANSYS. Because of the growing use of vehicles, automobile industries are in high demand these days. The rising demand is attributable to engine components, enhanced performance and lower cost. To reduce the time it takes to launch new goods, R&D and testing engineers should build key components as quickly as feasible. In the current scenario of automobile industries, aluminium alloy is the most preferred material for pistons manufacturing. Higher UTS is characterized in reduction of ductility in Al-based alloys as silicon content rises, to gain a better understanding of the mechanisms that cause hardening in Al-based piston alloys [1]. With increasing Si concentration, the ductility of Al-based cast alloys decreases so in commercial applications, content of silicon is limited to roughly 19% by mass [2]. An essential approach for enhancing the characteristics of Al alloys is structure modification. Heat treatment after casting has long been a standard method for AlSi alloys. The addition of Na, P or Sr can enable the eutectic silicon phase to change from coarse flakes to tiny fibres, improving mechanical characteristics, particularly ductility [3]. As the temperature of alloy material increases and reaches up to 400 °c the strength of the material reaches approximately half of its initial value. Repeated cycles with the impact of thermal and mechanical stresses have a magnificent impact on the function of the piston. There are two main ways to improve the strength of piston material: (i) using alloy material with improved properties with different amounts of alloying element and (ii) by using a coated layer over the piston surface. This work shows the location of maximum stress and stress concentration which can cause fatigue failure as cyclic rotation and linear movement are involved in piston-cylinder arrangement fatigue failure becomes important criteria of concern. On the other hand, pistons experience an excess thermal condition in case of insufficient in the clearance area and sometimes due to improper scavenging. The thermal load is the most critical concern since the piston has heating surface and inadequate heat transfer [4]. The piston is subjected to periodic heat load under high-temperature conditions, which can cause surface and structural damage under irregular temperature or burning conditions [5, 6]. Used a commercial algorithm to model the thermal characteristics of functional graded coatings on AlSi and steel piston materials in his thermal analysis of piston. Based on the elastic–plastic finite element computation of an aluminium alloy piston, [7] investigated and find a method for reliable life of an aluminium alloy piston at elevated temperature. Yan Feng et al. [8] calculated the coupled stress of the piston after estimating the piston temperature field and conducting multi-body dynamics analysis in a diesel engine [9–11]. Calculated the effects of coating thickness on temperature and thermal stress distributions and also used existing experimental results to simulate the temperature distribution in a CCD-coated piston used in an HCCI engine. Using three different combustion boundary conditions, forged aluminium and forged magnesium alloys are the materials widely used in internal combustion engines. Under 1.1Mpa pressure load, the piston head is between 340 °c and 370 °c for alloys of aluminium and magnesium [12]. Al alloy is a well-blended mixture of silicon in aluminium manufactured by powder metallurgy helps to increase mechanical properties, as well as some copper and magnesium additions that aid to increase mechanical properties and heat treatment response [13–16]. worked on calculation of stress generation and heat loads at different pressure levels for the piston. In his two studies, [17, 18] discovered that AlSi has a better casting characteristic and a higher strength [19]. Performed a comparative analysis of different materials for piston and give their justification and find out that aluminium-based alloy has more dimensional stability compare to magnesium-based alloy and can with stand at high pressure for maximum duration for a torque based engine [20]. Presents Homogenous charge compression ignition (HCCI) engines that combine the benefits of SI and CI engines to reduce fuel consumption and emissions.

The purpose of this research is to investigate the various types of pistons and their designs in terms of enhancing the performance of gasoline-powered HCCI engines [21]. Use FEM and mechanical property testing to calculate the stress due to pressure and temperature for five piston materials.The simulation parameters are operational gas pressure, temperature and piston material characteristics. The suitable material is selected based on the results of structural and thermal analysis [22]. Offer a numerical technique for calculating thermal stress generated only by unequal temperature distribution utilizing thermo-mechanical decoupled FEM (Finite Element Method). The primary focus of this work is on the thermal behaviour of functionally graded materials produced by utilizing the commercial code ANSYS on aluminium alloy piston surfaces [23]. Present Aluminium Silicon Carbide (AlSiC), an aluminium matrix composite, is employed as a substitute to aluminium in this project. AlSiC outperforms aluminium in terms of abrasion resistance, creep resistance, dimensional stability, stiffness-to-weight and strength-to-weight ratios and high-temperature performance. Making a piston out of AlSiC is also less difficult than making one out of aluminium [24]. Give a piston was numerically evaluated using the FEM, its thermo-mechanical capacity under a predetermined thermal and structural stress. To improve engine efficiency, the weight of the piston has been maintained to a minimum by optimizing various parameters [25]. Describe the stress distribution on the four-stroke engine piston is analysed using FEA in this paper. Couple field analysis is used to compute stresses and deflections caused by heat loads and gas pressure [26]. Present the stress and displacement distribution due to flue gas pressure and temperature are calculated individually and integrated into this article using the finite element method. FEA is carried out using CAD and CREO software. The results are compared to those acquired using the analytical approach, and conclusions are formed [27]. Performed an analysis to calculate the influence of piston stress on piston crown. There results revealed that the best piston design was 3 mm piston crown based on the stress values obtained for different piston diameters. The ideal piston design lowered overall stress and mass while increasing the temperature generated in the piston [28]. Present a model and study of piston rings that takes into account both structural and thermal studies to calculate the maximum stress that the rings can endure [29]. Work on thermal analysis and optimization of piston using aluminium alloy with a two-layer coating of TBC followed by zirconia find out that maximum stress value decreases from 85 to 55 MPa and also deformation too [30]. Work on piston design and analysis for A2618, Al-GHS and Ti-6Al-4 V material and find out Al-GHS can prefer in three piston materials as a factor of safety is maximum, deformation and mass is least [32]. Work on fatigue occurrence at elevated temperature and oxidation characteristics of forged steel piston materials and find out 42CrMo4 material piston has oxidation and carbon deposition occurrence is weaker than that 38MnVS6 piston for maximum optimization value [32]. Investigate ablation for a diesel engine piston with AlSi alloy, alumina fibre composite and aluminium borate whisker composites and find out aluminium borate whisker composites has very good anti ablative property as ablation which is due to flow of gas and high temperature [33]. Perform design optimization and thermo-mechanical analysis of a diesel engine piston, they reduce the mass of piston by 4.7% with a decrease in maximum coupling stress by 1.3% which enhances the structural strength of piston.

FEM method is used in this analysis and model is created in CREO which then implemented in ANSYS where the static structural and thermo-mechanical analysis performed.

1.1 Design Considerations in Designing of Piston

- The piston can withstand high inertia and pressure forces.
- It should light in weight to reduce inertia.
- Heat of the combustion should be dispersed quickly to the cylinder walls.
- It should be less noisy at high speed.
- It must form an effective seal against the leakage of gas and oil.
- It must have sufficient bearing area.
- It should have enough strength to withstand thermo-mechanical distortion.

1.2 Designing Procedure of Piston [34]

Assuming piston diameter 100 mm and other dimensions are calculated according these empirical relations given in design data book (see Table 1) and Table 2 shows the properties of materials.

- Thickness of piston head (t_H)

$$t_H = \sqrt{\frac{3pD^2}{16\sigma_t}} \tag{1}$$

- Heat flow through piston head (H)

Table 1 Dimensions of piston model

S.no	Description	Notation	Unit	Value used
1	Length of piston	L	mm	150
2	Outside diameter of piston	D	mm	100
3	Thickness of piston head	tH	mm	15
4	Thickness of radial ribs	t_R	mm	6
5	Radial thickness of the ring	t_1	mm	4
6	Axial thickness of the ring	t_2	mm	3
7	Width of the top land	b_1	mm	16
8	Width of other ring lands	b_2	mm	2.5
9	Maximum thickness of barrel	t_3	mm	11.32
10	Length of skirt	l	mm	70

Table 2 Properties of aluminium and AlSi032

	Aluminium	AlSi4032
Young's modulus	70 Gpa	79 Gpa
Poisson's ratio	0.35	0.33
Density	2700 kg/m^3	2680 kg/m^3
Thermal conductivity	237 W/m-k	155 W/m-k

$$IP = \frac{P_m LAN}{60x2} \tag{2}$$

$$BP = IPx\eta_m \tag{3}$$

$$H = C \times HCV \times m \times BP \tag{4}$$

- Radial thickness of ring (t_1)

$$t_1 = \sqrt{\frac{3p_w}{\sigma_t}} \tag{5}$$

- Axial thickness of the ring(t_2)

$$t_2 = 0.7\ to\ t_1 = 1 \tag{6}$$

- Length of piston skirt

$$Ls = 0.6D \ to \ 0.8D \tag{7}$$

- Total length of piston

$$L = D \ to \ 1.5D \tag{8}$$

2 Meshed Model

Elements are known as the smallest division of a cross section of a model. Elements utilized for a coincided model are tetrahedral 3D components. The piston model divided in such a way that the components came to 162,329 elements with an aggregate of 261,533 nodes (see Fig. 1).

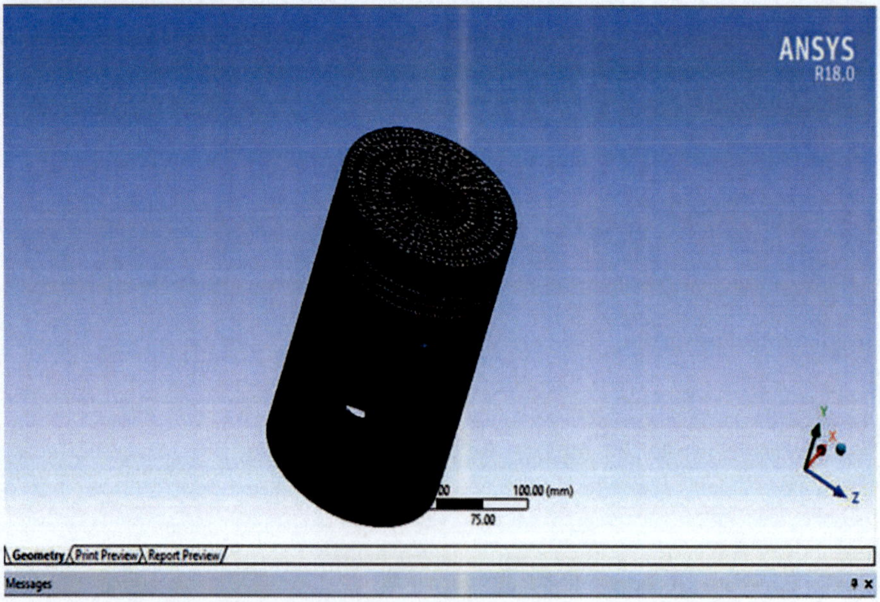

Fig. 1 Meshed model of piston in Ansys

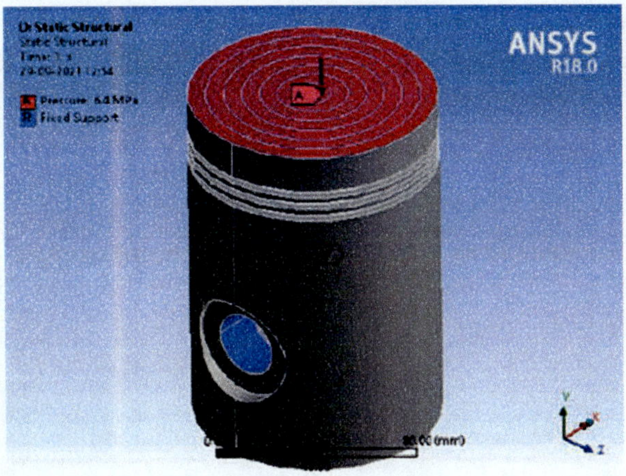

Fig. 2 Boundary condition

2.1 Boundary Condition and Loading

For an on-going task, we see that various parameters are need as forced in a piston system to accomplish an assignment. A loadings and boundary condition of a limited component model was such that corresponding conditions were forced as appeared in Fig. 2.

- Fixed support at the piston pin location.
- Pressure act on top of piston $p = 6.4\ N/mm^2$.

2.2 Thermal Boundary Conditions

Thermal Analysis is done on piston for which boundary conditions are taken from design data book [34] range is given for pressure and surface temperature so for our suitable calculation values are as follow:

- Temperature in the surface of piston head ranges from 200–300 °C.
- Convection is 16 W/m^2 k at 220 °C.
- To calculate heat transfer in a piston model, a few thermal limit conditions and initial condition should be characterized. As shown in Fig. 3, the piston head temperature rises quickly, especially around the centre and in corners, and for the exposed region of the piston, heat is exchanged with the environment via convection.

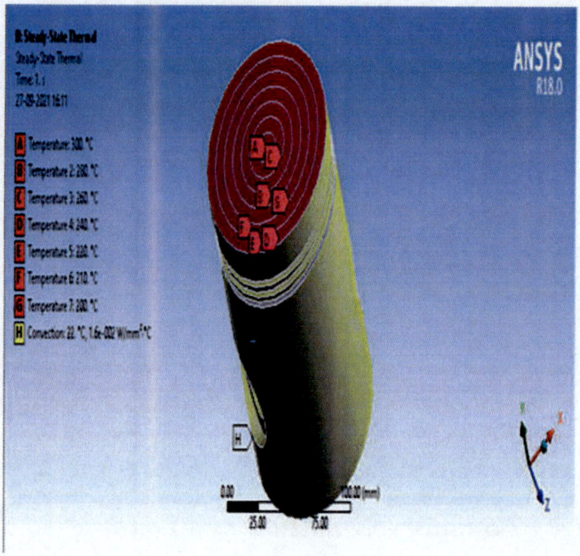

Fig. 3 Thermal boundary condition

3 Results

3.1 Thermo-Mechanical Deformation

Figure 4 presents the piston deformations against Aluminium and AlSi. It is noted from the figures that large deformations occur or found at the top surface of piston in both materials. It is noted that Aluminium alloy (AlSi) deforms less compare to aluminium. The highest deformation is 66 μm for aluminium.

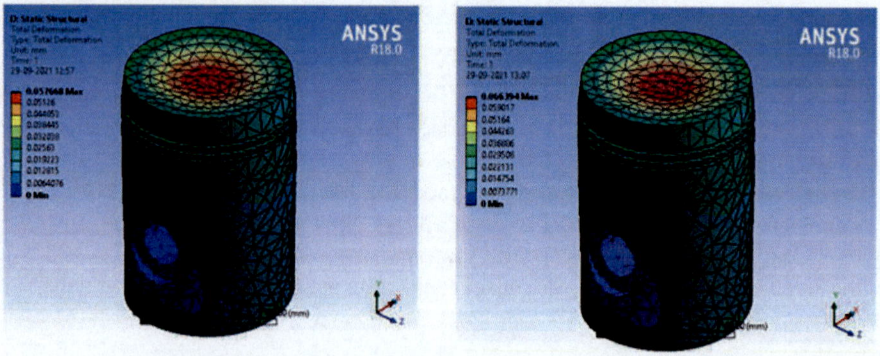

(a) (b)

Fig. 4 Total deformation **a** AlSi4032 **b** Aluminium

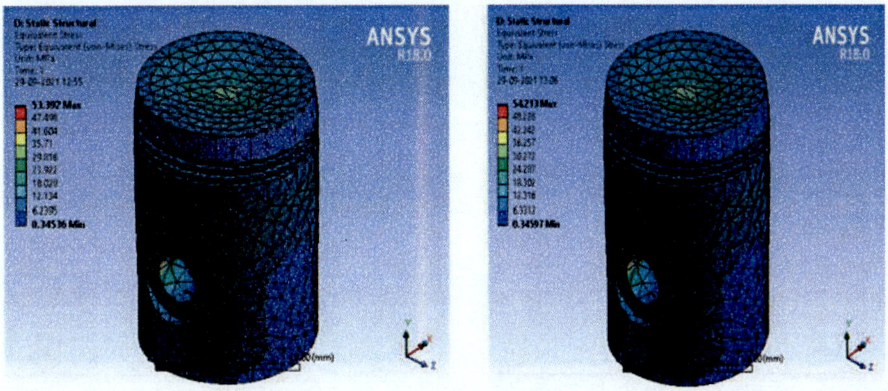

(a) (b)

Fig. 5 Equivalent (von Mises) stress **a** AlSi **b** Aluminium

3.2 *Thermo-Mechanical Stress*

According to this analysis, the distribution of thermo-mechanical stress is less for AlSi compares to aluminium as shown in Fig. 5. The maximum stress is more in aluminium with 54Mpa and appears at the pin-hole region. It can be concluded that this doesn't cause structural damage or plastic deformation since the stress is far less than the yield limit of piston material. AlSi performs best in terms of stress generation.

3.3 *Thermo-Mechanical Strain*

Figure 6 shows the equivalent strain of the Aluminium and AlSi model. It can be seen that the strain is minimum in AlSi material piston then compare to aluminium based on thermo-mechanical coupling.

The maximum strain is 0.00078 mm/mm in aluminium and it occurs at the pin-hole region.

3.4 *Thermo-Mechanical Heat Flux*

Figure 7 shows the variation of heat flux in the piston model. It can be seen that maximum heat flux generated at the top land corners of the piston where most of the high temperature-related activities occurs. The maximum intensity of heat flow occurs for the aluminium than AlSi with a value around 4.54 W/mm^2.

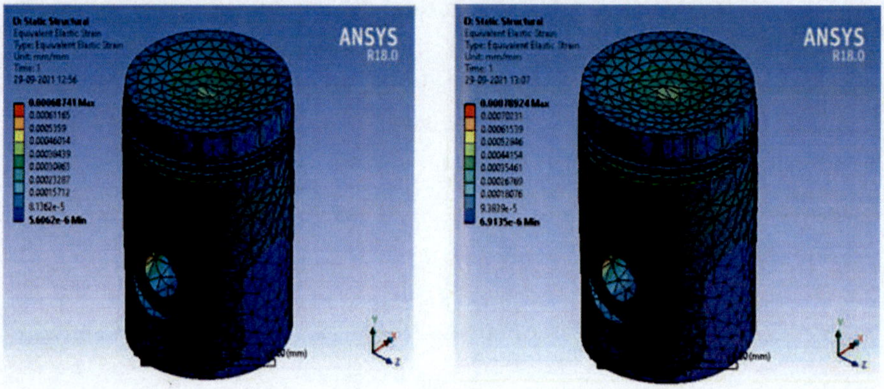

(a) (b)

Fig. 6 Thermo-mechanical strain **a** AlSi **b** Aluminium

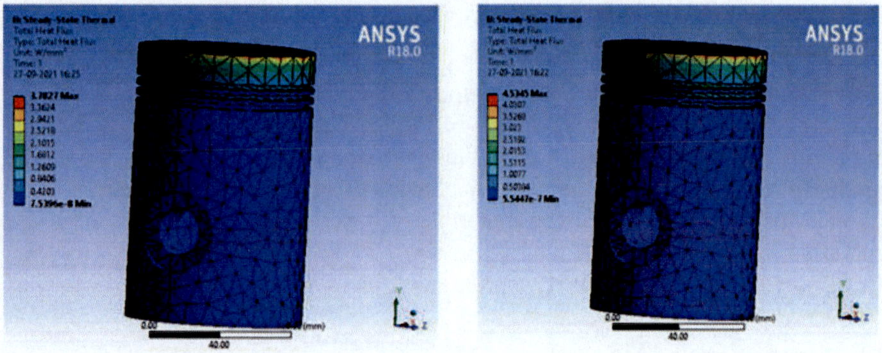

(a) (b)

Fig. 7 Heat flux **a** AlSi **b** Aluminium

4 Conclusions

The research tried to investigate the design, structural, thermal and optimization of the piston using the ANSYS workbench for static structural and stable thermal analysis. To improve the efficiency and offer better stability to the piston, while designing, deformation, von Mises stress, equivalent strain and heat flux distribution of piston model were carried out. Using thermo-mechanical analysis, the piston model was compared using two materials, namely aluminium and AlSi4032. The following findings were made (from Table 3):

- The pin-hole region of the piston experiences more deformation. In terms of overall deformation, AlSi4032 works admirably.

Table 3 Results based on thermo-mechanical analysis

	AlSi 4032		Aluminium	
	Maximum	Minimum	Maximum	Minimum
Total deformation, μm	0.057668	0	0.066394	0
Von Mises stress, MPa	53.39	0.3456	54.213	0.34597
Equivalent strain, mm/mm	0.00068741	5.6062 e^{-6}	0.00078924	6.9135 e^{-6}
Heat flux, W/mm^2	3.7827	7.539 e^{-8}	4.5345	5.5447 e^{-7}

- AlSi performs less distribution of thermal stress compared to aluminium based on thermo-mechanical coupling. The pin-hole region generates the majority of the stresses.
- More uniform equivalent strain distribution is noticed for both the materials of piston yet AlSi has the advantage of minimum generation compared to aluminium.
- Von Mises stress and total deformation of piston model increase in a noticeable way in thermo-mechanical analysis.
- AlSi4032 has a better temperature distribution effect than aluminium.

References

1. Zeren M (2005) Effect of copper and silicon content on mechanical properties in Al–Cu–Si–Mg alloys. J Mater Proc Technol 169:292–298
2. Tomida S, Nakata K, Shibata S, Zenkouji I, Saji S (2003) Improvement in wear resistance of hyper-eutectic Al–Si cast alloy by laser surface remelting. Surf Coat Technol
3. Hengcheng L, Yu S, Guoxiong S (2003) Restraining effect of strontium on crystallization of Mg2Si phase during solidification in Al–Si–Mg casting alloys and mechanisms. Mater Sci Eng. A
4. Hu Zhen, Qiao Xinqi, Wang Qudong, Ye Bing (2014) Piston temperature field analysis based on ANSYS. Small Intern Combust Engine Mot
5. Silva FS (2006) Fatigue on engine pistons–A compendium of case studies. Eng Fail Anal, 13(3)
6. Buyukkaya E, (2008) Thermal analysis of functionally graded coating AlSi alloy and steel pistons. Surf CoatS Technol, 202(16)
7. Zhang Weizheng, Wei Chunyuan, Su Zhiguo, Liu Jian, Xiang Jianhua (2003) Prediction of fatigue life of aluminum alloy piston [J]. China Mechanical Engineering
8. Yang Y, Ni PY, Dai F, Jin HC (2016) Mechanical and thermal load analysis in 186F diesel engine piston. Mech Res & Appl, p 05
9. Cerit M, Coban M (2014) Temperature and thermal stress analyses of a ceramic-coated aluminum alloy piston used in a diesel engine. Int J Therm Sci, 77
10. Cerit M, Soyhan HS (2013) Thermal analysis of a combustion chamber surrounded by deposits in an HCCI engine. Appl Therm Eng, 50(1)
11. Cerit M (2011) Thermo mechanical analysis of a partially ceramic coated piston used in an SI engine. Surface and coatings Technology, 205(11)
12. Vaishali RN, Khamankar SD (2015) Stress analysis of piston using pressure load and thermal load. Int J Mech Eng, 3(8)
13. Singh L, Rawat, SS, Hasan T, Kumar U (2015) Finite element analysis of piston in ANSYS. International Journal of Modern Trends in Engineering and Research (IJMTER) Volume, 2

14. Zamani M. (2017) Al-Si Cast alloys-microstructure and mechanical properties at ambient and elevated temperatures (Doctoral dissertation, Jönköping University, School of Engineering)
15. Schenkel JL, Briggs, Stratton Corp (2002) Anodized magnesium or magnesium alloy piston and method for manufacturing the same. U.S. Patent 6,495,267
16. Singh AR, Sharma PK (2014) Design, analysis and optimization of three aluminium piston alloys using FEA. PARAMETERS 2618:A4032
17. Zamani M, Seifeddine S, Jarfors AE (2014) Effects of microstructure and defects on tensile and fracture behaviour of an Hpdc component: potential properties and actual outcome of En AC-44300 Alloy. In Light Metals 2014
18. Zamani M, Seifeddine S, Azizderouei M (2016) The role of Sr on microstructure formation and mechanical properties of Al-Si-Cu-Mg cast alloy. In Light Metals 2013
19. Singh U, Lingwal J, Rathore A, Sharma S Kaushik V (2020) Comparative analysis of different materials for piston and justification by simulation. In: Materials Today: Proceedings, 25
20. Aljaberi HA, Hairuddin AA, Aziz NA (2017) The use of different types of piston in an HCCI engine: A review. Int J Automot Mech Eng, 14
21. Kumar AN, Adithyan V, Hemachandran P, Kumar PR, Manikandan J (2019) Review on design and analysis of IC engine piston with five different material, Int J Manag, Eng Technol, 9(1)
22. Havale SA, Wankhade S (2017) Design, thermal analysis and optimization of a piston using ANSYS. Int Res J Eng Technol, 4(12).
23. John A, Mathew JT, Malhotra V, Dixit N (2015) Design and analysis of piston by SiC composite material. Int J Innov Res Sci & Technol, 1(12)
24. Sinha D, Sarkar S, Mandal SC (2017) Thermo mechanical analysis of a piston with different thermal barrier coating configuration. Int. J. Eng. Trends Technol, 48(6)
25. Kumar KS (2016) Design and analysis of IC engine piston and piston-ring on composite material using CREO and ANSYS software. J Eng Sci, 1(1)
26. Cioată VG, Kiss I, Alexa V, Rațiu SA (2017) Mechanical and thermal analysis of the internal combustion engine piston using Ansys. In: IOP Conference Series: Materials Science and Engineering (Vol. 163, No. 1, p 012043)
27. Badri Y, Shamseldin A, Renno J, Sassi S, (2021) Thermal structural optimization of IC engine piston. Int J Mech Eng Robot Res 10(1)
28. Kiran S, Palanivendhan M, Kumar M (2020 December) Modeling and analysis of Piston and Piston rings of SI engine for different materials using FEM. In: IOP Conference Series: Materials Science and Engineering (Vol. 993, No. 1, p 012155)
29. Reddy SS, Kumar DBSP (2013) Thermal analysis and optimization of IC engine piston using finite element method. Int J Innov Res Sci, Eng Technol, 2(12)
30. Dhamecha Y, Saptarshi V, Parikh S, Parnami T, (2020) Design and analysis of piston using different materials
31. Niu D, Zhang J, Xiong P, Hao G, Liu S, Guo W (2019) High temperature fatigue and oxidation characteristics of forged steel piston materials. Eng Fail Anal, 97
32. Qin Z, Jia C, Zhang W, Wang L (2019) Investigations on ablation for highly-intensified diesel engine piston material. Case Stud Therm Eng 13:100371
33. Zhaoju Q, Yingsong L, Zhenzhong Y, Junfa D, Lijun W (2019) Diesel engine piston thermo-mechanical coupling simulation and multidisciplinary design optimization. Case Stud Therm Eng 15:100527
34. Bhandari VB (2010) Design of machine elements. Tata McGraw-Hill Education

Design of Spherical Hopping Terrain Surveillance Robot

K. Karthik, K. Kishorekumar, K. Raja, R. Saravanan, and R. Annamalai

Abstract This paper deals with the design of rolling, steering and jumping mechanisms of a spherical surveillance robot that can be operated in hilly and rough terrain regions. The driving mechanism of the robot is designed such that the robot resembles the movement of a spherical ball. The steering mechanism is designed to enable the robot to achieve on spot turning. The principle behind the design of steering is the conservation of angular momentum. The jumping mechanism adds a special advantage to the locomotion of ground robots, in terms of the ability to avoid obstacles. The parameters considered for the selection of optimum mechanism design for the robot are also presented in this paper.

Keywords Rolling mechanism · Steering mechanism · Jumping mechanism · Conservation of angular momentum

1 Introduction

Spherical robots are one of the fascinating areas to study and have many unique features that are of greater interest. They can be designed to encounter and operate

K. Karthik · K. Kishorekumar (✉) · K. Raja · R. Saravanan · R. Annamalai
Department of Mechanical Engineering, Coimbatore Institute of Technology, Civil Aerodrome Post, Coimbatore, Tamilnadu 641014, India
e-mail: 1802089m@cit.edu.in

K. Karthik
e-mail: karthik.k@cit.edu.in

K. Raja
e-mail: 1802107m@cit.edu.in

R. Saravanan
e-mail: 1802065m@cit.edu.in

R. Annamalai
e-mail: 1802113m@cit.edu.in

N. M. Sivaram et al. (eds.), *Advances in Manufacturing, Automation, Design and Energy Technologies*, Lecture Notes in Mechanical Engineering,
https://doi.org/10.1007/978-981-99-1288-9_64

more efficiently in any harsh environment. Among the land locomotive modes such as walking, crawling, jumping and rolling, jumping mode has a special advantage that locomotive time is less in jumping as it can easily avoid obstacles and move to different heights. The spherical hopping terrain surveillance robot (SPHOT) is assisted with rolling and jumping locomotive modes. SPHOT is designed to be operated for applications such as earthquake rescue missions, military reconnaissance, space explorations and some secondary applications such as wildlife videography and photography, aquatic life and marine exploration, mining inspection and radioactive exploration.

In earthquake rescue missions, during rescue, rescuers' rubble movement over the damaged buildings may kill survivors trapped under them. A large amount of time is taken to locate the people trapped under buildings which results in an increase in recusal time. Thus, SPHOT can be used to locate survivors trapped, thereby reducing the rescue time and saving the lives of the survivors.

Nowadays, military surveillance is expected at a great standard. SPHOT can be used in hostage rescue missions for path determining and surveillance. Drones are now used for surveillance. But the path for the soldiers to enter buildings and the traps set on their way can be effectively detected only with the surface level bot.

SPHOT is equipped with rolling, steering and jumping mechanisms for enhancing robot locomotion. The optimum mechanisms are designed and assembled in CAD Software. Proof of concepts was done for both rolling and steering mechanisms, and they were observed. Based on the inferences made from the proof of concept and the literature study, the design of each mechanism was optimized for a number of iterations to enhance the functionality of SPHOT.

2 Design of Rolling Mechanism

The objective of the rolling mechanism is to analyse and study the principle of existing driving mechanisms of spherical robots and design an optimum mechanism such that no vibration and instability occurs in the motion of the robot which puts light on the novelty of the design. To create an innovative and best outcome model, the inferences from the study on existing systems were keenly noted. The prime objectives are:

- To ensure that not much energy loss occurs due to internal friction.
- To make sure the centre of mass (COM) of the robot does not shift often as it leads to instability of the robot.
- To develop a mechanism arrangement that can also be used as provided for internal parts used in the robot.
- The further objective is to design a simple mechanism without higher maintenance.

Table 1 Comparison of mechanical driving principles in spherical robots

S. No	Driving mechanisms	Disadvantages
1	Hamster Ball	Internal friction loss and Internal slipping occurs during rotation
2	Pendulum Driven	Shifting of the pendulum causes shifting in COM which results in the wobble of the robot
3	Universal Wheel	Like Hamster ball, there is high energy loss due to surface friction
4	Shell Transformation—Pressurized bladders	Usage of bladders, valves and actuators adds more weight to the robot and makes the design more complex

2.1 Concept Generation

There are a greater number of spherical robots which use various mechanical driving principles. On comparing and focusing on the advantages and disadvantages of the driving principles, an efficient driving mechanism is chosen and implemented in SPHOT.

Out of the referred driving principles and as well the other driving mechanisms available, the motor and arm-driven rolling mechanism is considered for its better efficiency and it satisfies the design objectives.

2.2 Motor and Arm-Driven Rolling Mechanism

Detailed design and working explanation of the motor and arm-driven rolling mechanism is depicted.

2.3 Components of Rolling Mechanism

The motor-driven mechanism is comprised of the following components (Fig. 1).

1. Outer shell.
2. DC motor.
3. Driving arm—Motor side.
4. Driving arm—Bearing side.
5. Main axle.
6. Bearing and shaft.

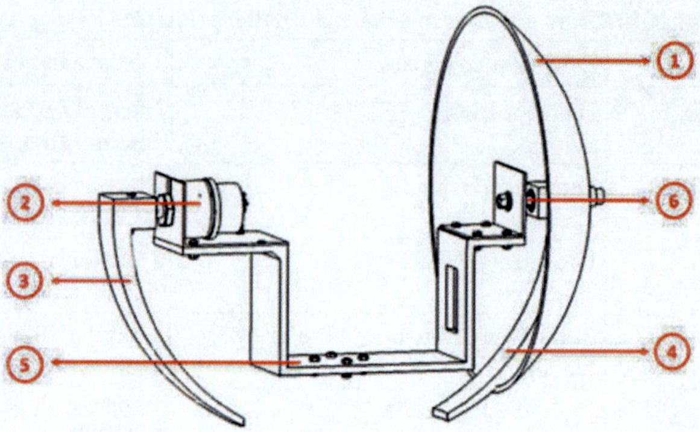

Fig. 1 Assembled view of the rolling mechanism

2.4 Working of the Mechanism

The following specification particularly describes the mechanism and the way it is to be performed.

The rolling motion is achieved by rotating the driving arms (3) & (4) attached to the spherical shell (1) of SPHOT. One driving arm is attached to a DC motor (3) which provides the necessary torque to the robot. When SPHOT must move forward or backward, the DC motor is powered in the direction. This DC motor is a high torque motor with enough RPM to drive the entire robot.

This DC motor is the only power source that is responsible for the driving motion of the robot.

The other driving arm (4) is connected to the main axle (5) by means of a bearing and shaft arrangement (6). The reason to use a bearing and shaft arrangement (Fig. 3) on the other side of the main axle is to reduce the weight of the robot. One DC motor was found to be sufficient to drive the entire robot. The bearing arrangement just ensures the rotation of the bearing end driving arm does not affect the main axle by causing any kind of twist or rotation.

As this mechanism is used for only the forward and backward motion of SPHOT, the steering of SPHOT is made by using a separate mechanism that does not cause much vibration and wobble as compared to the previously discussed driving principles.

3 Design of Steering Mechanism

In SPHOT, the steering mechanism is so unique that it is rarely found in some of the spherical surveillance robots. The principle behind the flywheel is the conservation of angular momentum (COAM). A flywheel is rotated at a high RPM in the direction to be steered so that the spherical bot will be able to achieve a turn due to the transfer of angular momentum from the flywheel to the outer spherical shell.

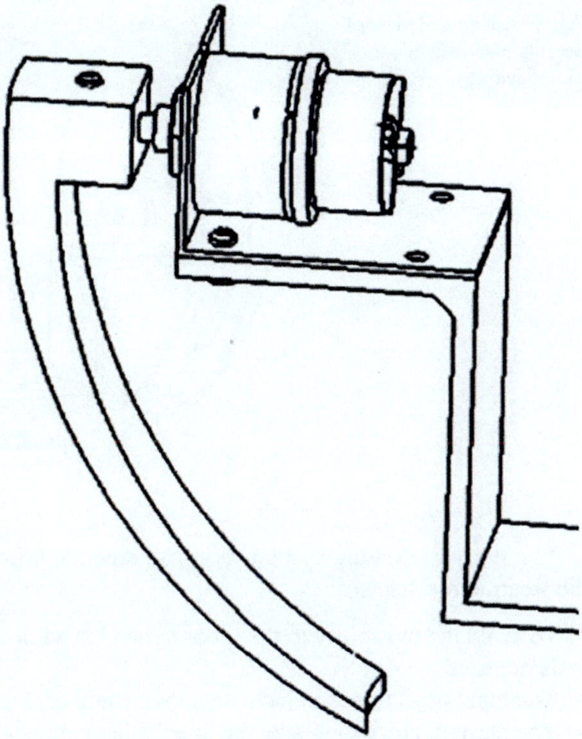

Fig. 2 Motor shaft arrangement with driving arm

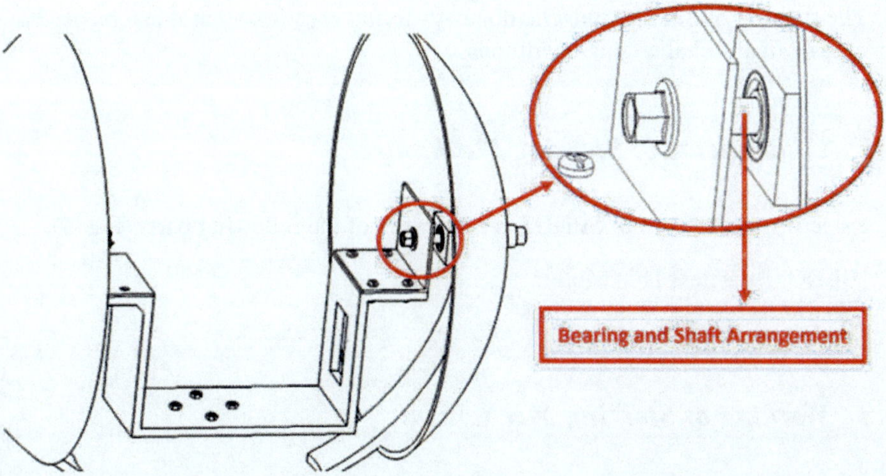

Fig. 3 Bearing shaft arrangement with driving arm

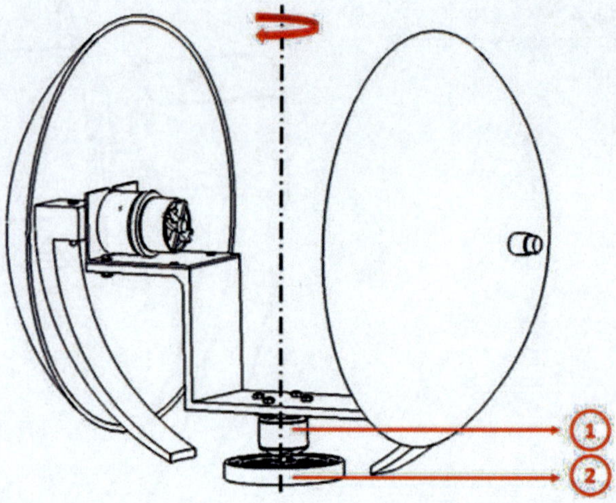

Fig. 4 Assembled view of steering mechanism components

The study of existing systems is considered the prime objective for the design of the steering mechanism.

- To avoid the instability in the robot dynamics while driving, the shifting of COM is not used.
- The main objective is to have on spot turning of the robot using COAM.
- The design aims to make the mechanism much simpler so that no specific maintenance is required for the robot.
- The steering of SPHOT must be done at varying speeds so that it can be used to operate in any challenging conditions.

3.1 Components of Steering Mechanism

The steering mechanism of SPHOT is comprised of the following parts (Fig. 4).

1. BLDC (Brushless DC) Motor.
2. Flywheel.

3.2 Working of Steering Mechanism

The following particularly explains the working of the steering mechanism. As stated in the design of the rolling mechanism, the forward and backward motion is alone achieved. Thus, for steering an additional mechanism must be accompanied.

When the robot must be steered at rest or when in motion, the flywheel (1) is rotated at a high RPM in the direction it must be steered with the help of a BLDC motor (2). When flywheel acquires a certain RPM, the angular velocity of the flywheel (1)

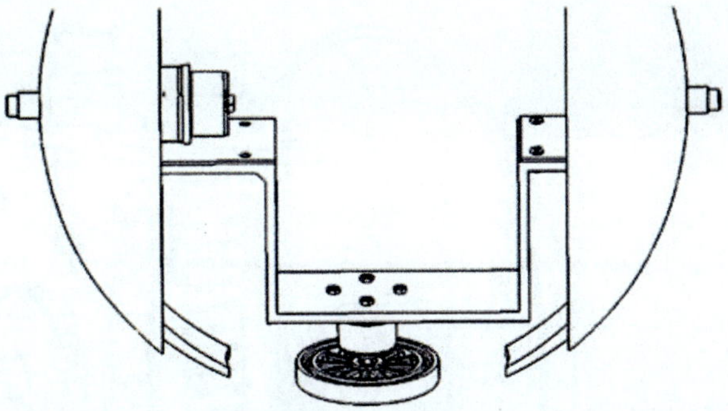

Fig. 5 Steering Mechanism closer view

increases and the output torque also increases. Due to COAM, the angular momentum from the flywheel (1) is transferred to the outer shell of the robot which results in the steering of the robot.

The robot moves in forward/backward motion and until there is a necessity for the robot to make a turn, the BLDC motor (2) is not powered on. The acceleration of the turn can be controlled by spinning the flywheel faster or slower. This can also attain on spot turning which provides some degree of holonomy that it can be used at high speeds and on rough terrain.

4 Design of Jumping Mechanism

Jumping is used in nature by many small animals to locomote in cluttered environments or in rough terrain. It offers small systems the benefit of overcoming relatively large obstacles at a low energetic cost.

To design the robot, which is capable of performing jumps, the following objectives are considered:

- The mechanism should be able to produce a possible higher jump.
- The jumping mechanism should be designed in an optimum size, such that it fits inside the shell of SPHOT.
- The ability to change the direction of jumping should be implemented in the design.

4.1 Components of Jumping Mechanism

The jumping mechanism of SPHOT consists of the following components (Fig. 6).

1. Top plate.
2. Helical Compression Spring.
3. Gear Arrangement.

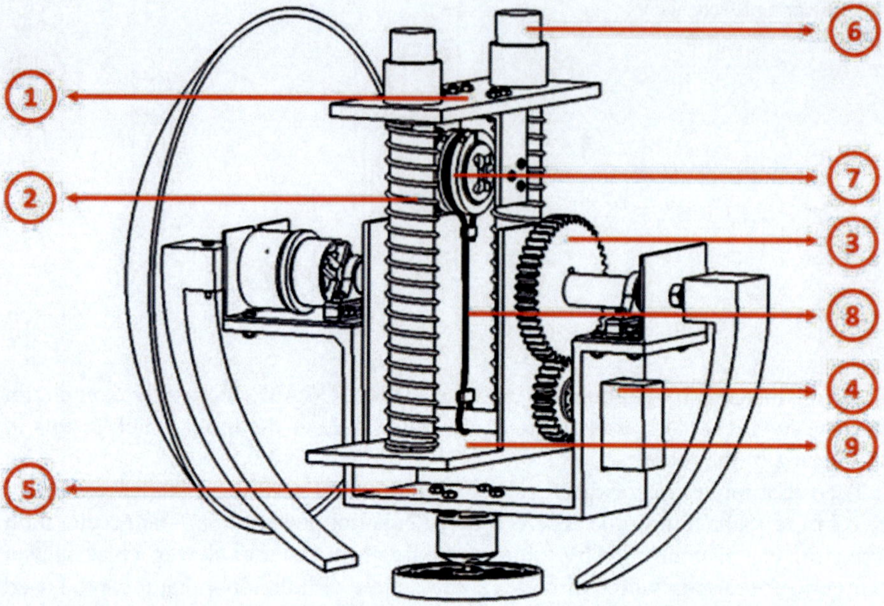

Fig. 6 Components of jumping mechanism

4. Servo Motor.
5. Main Axle.
6. Spring Guider.
7. Pulley.
8. Wire.
9. Bottom plate.

4.2 Working of Jumping Mechanism

The jumping mechanism is designed such that the direction of jumping can be changed whenever required. Thus, the entire jumping mechanism which is mounted over the main axle (5) and is provided with a gear arrangement (3) powered by a servo motor (4) to change the direction of the jumping mechanism. The servo motor (4) rotates the gear arrangement (3) within the range of + 45 to -45 degrees to tilt the entire jumping assembly.

When the high torque DC motor mounted with the pulley (7) is powered, the pulley (7) wounds the wire (8) over it. Thus, the length of the wire (8) decreases. The wire (8) pulls the top plate (1) toward the bottom plate (9) and compresses the helical springs (2) as well. By this process, a lot of compression energy is stored in the helical springs (2). When the required compression level is achieved, the power supply to the DC motor is cut off. Since the helical springs (2) have stored a large amount of energy in them, it forces the top plate (1) to retract upwards. The wire (8) also gets unwounded from the pulley (7). Since the top plate (1) and bottom plate (9) are connected by means of a wire (8), while the top plate (1) protrudes it also pulls the bottom plate (9) which is mounted over the main axle (5). Thus, the entire main axle (5) along with the spherical shell is pulled upwards by the top plate (1) resulting in the jumping of SPHOT.

5 Conclusions

We proposed the design of rolling, steering and jumping mechanisms that can be implemented in any kind of spherical robot in order to enhance the locomotion of the robots. SPHOT uses a driving mechanism that improves the stability in the motion of the robot by means of driving arms, a steering mechanism that has an efficient and versatile mechanism of locomotion to achieve better steer of the robot in rough and terrain regions and a jumping mechanism which adds a special ability to overcome obstacles in hard rescue and surveillance regions.

SPHOT will be a better helping assistant for Border Security Forces, Soldiers during military surveillance and reconnaissance and National Disaster Response Force (NDRF), firefighters during earthquake rescue and fire and rescue missions.

In addition, we aim to develop a prototype by fabricating the designed mechanisms and testing the robot to react with real-world problems.

References

1. Karthik K, Kishorekumar K, Raja K, Saravanan R and Annamalai R: Spherical Hopping Robot for Reconnaissance and Rescue Purpose (Application No.: 202141025222, Official journal of the patent office, June 2021)
2. Richard Chase, Abhilash Pandya (2012) A review of active mechanical driving principles of spherical robots
3. Aarne Halme, Torsten Schonberg, Yan Wang (1996) motion control of a spherical mobile robot
4. Landa K, Pilat AK (2015) Design and start-up of spherical robot with internal pendulum
5. Akira Kamei, Masaya Jinno, Naoya Chishima, Kenji Sato, Koki Kikuchi (2014) Soft landing condition for stair-climbing robot with hopping mechanism

Design of Special Purpose Hydraulic Forging Press for 15000 Tons Capacity

M. S. Tufail, R. B. Chadge, K. S. Ansari, and D. B. Meshram

Abstract Forging presses with hydraulic mechanisms are used in many industrial applications. These applications include forging pressure with flat dies used for hot working to break down ingot and shape them into roll, pressure vessel, forged bar, rods, plates, etc. Forging presses with closed dies, used to process preheated billet into various shapes. In India, forging presses are available only up to 12,000 tons, above this available capacity always need to design a customized hydraulic system. This paper aims to design the hydraulic circuit and pumping requirement, i.e., amount of fluid handled by the pumps and the size of the accumulator for a 15,000-ton capacity. The paper highlights the complete working process, components of the forging press and the design procedure calculations. The sophistication of control desired will determine the signal sources to create the cycle pattern of power transmission system. All these information about the requirement of different component will provide the basis for individual design of the same.

Keywords Open circuit hydraulic pump · Forging press · Accumulator

1 Introduction

Forging is a process used to form metal into large quantities of identical parts. It is often used in the manufacturing process of automobiles and to improve the properties of the metal being used [1]. The products made by forging may be large or very small and can be made of steel or any other metal [5]. Forging changes the shape

M. S. Tufail (✉) · R. B. Chadge
Mechanical Engineering Department, Nagpur, Y.C.C.E, India
e-mail: mstufail@rediffmail.com

K. S. Ansari
Civil Engineering Department, Nagpur, Y.C.C.E, India

D. B. Meshram
Centre for Skilling and Technical Support, CIPET, Korba, India

N. M. Sivaram et al. (eds.), *Advances in Manufacturing, Automation, Design and Energy Technologies*, Lecture Notes in Mechanical Engineering,
https://doi.org/10.1007/978-981-99-1288-9_65

and size, but not the volume, of a part. Hydraulic presses, in which high-pressure fluid produced by hydraulic pumps drives a ram, are about 100 times slower than hammers [9]. They are used for large or complex die forgings. Press forging is more economical than hammer forging and closer tolerances can be obtained. A greater proportion of the work done is transmitted to the work piece, differing from that of the hammer forging operation, where much of the work is absorbed by the machine and foundation [2]. The basic difference between hydraulic press and other method is that pressure is applied in squeezing manner rather than by impact [4]. The paper presents a development of hydraulic system with their component. Hydraulic circuit is the main structure of whole system is developed by considering the requirement of the system. The selection of hydraulic components such as pump, pressure control valves and direction control valves in term of specification as per the requirements are given. These requirements are obtain from the hydraulic design of the press which contains the cylinder dimension, pumping requirement and accumulator size calculations.

2 Open Die Forging Press

The dies used in open die forging are either flat or rounded. Large forgings can be made by applying force to different sections of the material in a series of steps. Closed die forging uses both hydraulic presses and forging equipment. In terms of tolerances, die forging is the best process, and it also produces a finished item that is entirely filled up and has the least amount of flashing [7]. The operator's expertise determines the final shape and improvement in metallurgical characteristics. Closed die forgings can hold tighter dimensional tolerances than open die forgings, and the operator requires less experience [11]. A the open die forging press as shown in Fig. 1, the application is characterized by huge flow demands and forces and also requires a dynamic performance in the critical decompression phase which is only achievable in open loop control. The process of open die forging is used for large steel bars. In order to achieve the required forging dimension, typically an iterative approximation algorithm is use (Figs. 2, 3).

The actuation principle of the top die is depicted in the schematic of the actuation principle. Pumps supply flow into the upper chamber of the differential cylinder. Only the flow out of the cylinder is controlled via the forging valve.

Drawbacks of forging press for heavy application are.

- Proportional increase in shock, pressure and flow control valves.
- Step changes in flow and sudden ram stops and starts were accommodated with the large pressure spikes or droops.
- Enormous energy consumption occurred—electric power
- Consumption increases more than 50%.
- High level of noise.

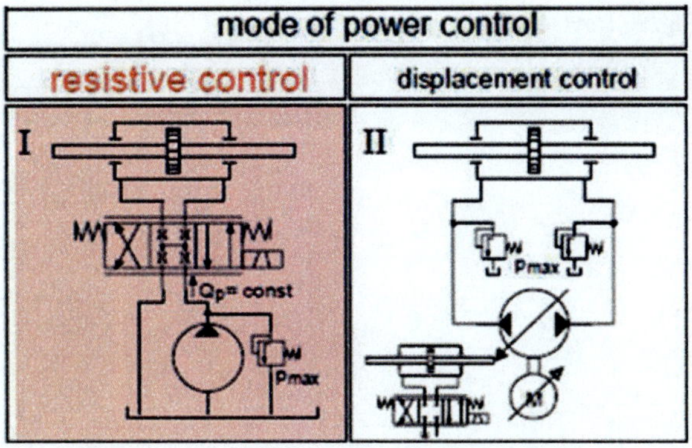

Fig. 1 Open die forging press power control

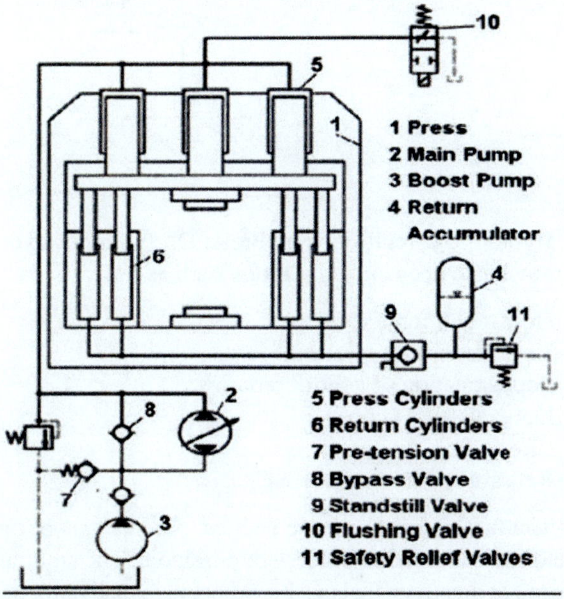

Fig. 2 low capacity circuit

- Selection of the hydraulic components, design of the system and the cooling type are not appropriate. Lack of cushion equipment, out-of-date the connection type of the system and the unreasonable pipelines mounting and fixation, lead to great vibration and big noise. The pump and the system glows greatly, which leads to the failure of the seal and leakage, furthermore big power was wasted.
- The collocation of the system is not well-designed, which lead to incommodity of service for the hydraulic components. The absence of oil level control and Alarming equipment also caused the heavy loss of hydraulic oil.

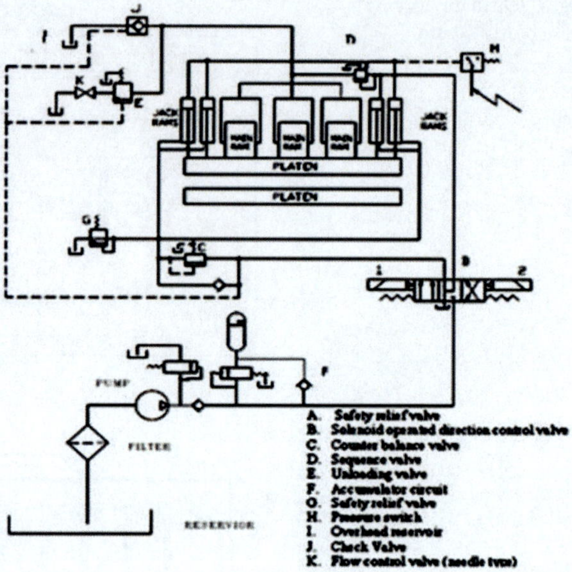

Fig. 3 Developed hydraulic circuit for press

Hydraulic Circuit For The Press: Designing of hydraulic circuit is predicated on certain basic need and conditions such as,

- How much force is needed.
- How fast circuit must run.
- Sophistication of control requires
- Duty cycle
- Life cycle
- Reusability of components.

The first concern is force needed. An analysis of operation to be performed will yield some indication of force requirement. The amount of information available and degree of accuracy that can be expected in analysing this information will determine the other circuit's requirement [6]. After force requirement have been established, cylinder and motor size can be determined. The determination of component size may hinge heavily on the second factor—actuating speed [13]. The simple ratio of pressure to area for force determination must be related to gallon per minute fluid flow that will determine the speed of movement of ram in cylinder. The sophistication of control desired will determine the signal sources to create the cycle pattern of power transmission system. The duty cycle may require high performance for limited time, as machine tool and construction machine circuit.

At the end of production cycle, the machine can be altered for new functional operation at much lower cost than would be possible with all new components.

Figure shows the circuit diagram for higher capacity press. The circuit consist of pump, reservoir, pressure control valves, direction control valve, flow control valves and hydraulic actuators. The pressurized fluid from pump flows to the main and jack rams through direction control valve B which is solenoid hydraulically actuated valve operated by the pressure switch H. This pressurized fluid passes to the main ram after the jack rams counter balance valve C through sequence valve D which is set at particular value. The circuit doesn't provide the pump compensation; water hammer and accidental intensification may cause damage, etc., to the press components due to the large tonnage of press.

To compensate above limitation in the circuit, we added the accumulator circuit, safety valve G and check valve J with overhead reservoir I so that circuit can work smoothly without with the required pressure to provide required capacity for pressing. Figure shows the circuit diagram and working of hydraulic press with their components. Counter balance valve receives the signal from the main inlet line. This signal is normally taken from point within counterbalance valve structure. Valve C is externally drained. It can be internally drained if there is to be no resistance on the cylinder line when the flow is diverted to the tank. But it is safest to provide an external drain. Valve D is sequence valve. The signal from ahead of valve and is picked up within valve envelope. The drain in this is independently connected to tank because there will be pressure exerted on the discharge of the valve. An integral check valve may or may not be used in a sequence valve applied as indicated on the circuit. If valve E were not included in the circuit, the amount of fluid passing back through the check valve incorporated in valve D as an alternate circuit design might be used to assist in decompression of the press. Then the check valve would pass a certain part of the flow back through the four way valve to the tank. Valve E is used in the circuit as unloading valve. It will be externally piloted and externally drained if there is resistance in the return to the tank port.

The needle valve K on the downstream side as shown, the spring pocket must be externally drained so that back pressure does not provide an additive pressure to the spring. The needle valve used in conjunction with the unloading valve, establishes the amount of flow necessary to decompress the press in proper time interval. The unloading valve receives the signal and starts and stops the fluid flow. Pilot check valve J is working as unloading valve during this portion of cycle. A safety valve G is used at the rod end of the jack rams to prevent accidental intensification of pressure. The accumulator is used to hold the pressure when the desired maximum pressure is attained. Valve A is safety valve it is adjusted to pressure higher than that of established by pumps for providing compensation when pump mechanism fails.

Valve B is the direction control valve. This valve is solenoid controlled, hydraulically actuated unit. When side 1 is actuated, it directs the fluid to the head end of the small pilot cylinder or jack rams. Valve D will prevent the fluid flow to the main ram until certain minimum pressure reached. As the fluid entering to the jack rams causes the platen to move downward, the main cylinder acts as a pump and the vacuum is thus created in the main cylinder is nullified by atmospheric pressure, causing fluid flow through valve. There is a reservoir above the main ram called as overhead reservoir, and valve J is in the bottom of the tank so that gravity also assist the fluid flow. When platen meets and resistance causes a rise in fluid pressure, the flow from the pump is diverted through valve D at its set value, and the full pressure will be effective on the both jack ram and main ram. Full tonnage will then provide between the platens. Pressure switch H creates a signal at pressure value somewhat less than the maximum value created by the pump control. The signal from the pressure switch H will start a timer that times out at pre-determined period. This denergizes side 1 of valve B and activates the side 2. Fluid is then directed by the side 2 to the rod end of the jack rams through the check valve around valve C. This check valve may be built integral with valve C.

The pressure opens the valve E, relatively small two way valve that allows all stresses in the press is to be relived and the pressure drop to a low value. After this decompression, the pilot pressure opens the pilot check valve J and permits fluid flow to exhaust from the main ram. At this time, the platen can rise and open the press, ready for unloading and reloading. When the reaches the upper level of travel, it contacts a limit switch, or some other sensing devices that cause the solenoid on the direction control valve to denergize. Denergizing the solenoid causes the platen to stops its upward travel. The press may not produce at its maximum rate because of the time necessary to move the ram. If the ram needs to be raised only halfway up to clear the work piece, the moving time can be halved. The press can then function at more economical rate.

When changing dies or clearing jams ups, automatic control may have to be blocked in order to raise the ram to its full travel. When the press opens the predetermined amount and the direction control has been returned to neutral position, the pump will be short stroked, holding pressure on other devices. The flow from the pump may also be used for other functional operation such as stock feeding or ejecting.

3 Design of hydraulic forging press for 15,000 tonnage capacity:

Given data:

- Press rating T = 15,000 tons
- Pullback rating Tr = 1500 tons
- Total stroke St = 125 in (3.175 m)
- Maximum pressing stroke = 40 in (1.016 m)
- Fast advance speed Va = 90 in (2.286 m)
- Fast return speed Vr = 90 in (2.286 m)
- Pressing speed Vp = 40 in/min (1.016 m/min)
- Fast advance = 15 in (0.381 m)
- Pressing = 25 in (0.635 m)
- Return = 25 in (0.635 m)
- Cycle time Tc = 12 s

- Pressing time Tp = 4 s
- Selected oil pressure in the system P = 5000 lb/in2 (34,473.786 kN/m2)

- Step I:

– Determining the required effective pressing area of main plunger of the press.

$$A = \frac{T * 2000}{P}$$

$$= \frac{15000 * 2000}{5000} = 6000\text{in}^2 = 3.8712\text{m}^2$$

- Step II:

– select no.of cylinder = 3

  – each cylinder of capacity = 5000 tons

- Step III:

  – sub divide A in accordance with selected rating and determining the individual diameters:

$$A1 = A2 = A3 = 1/3 * 6000 = 200in^2(1.2903\text{m}^2)$$

$$\text{D1} = \text{D2} = \text{D3} = \surd(2000/0.786) = 50.44 \text{ in}$$
$$\sim 51 \text{ in } (1.29 \text{ m})$$

– Adjusted area = 6130 in2 = 39.55 m

- Step IV:

– geometric volume corresponding to stroke

$$\text{Vt} = \text{st} * \text{A}$$
$$= 125 * 6130 = 766250 \text{ in } 3$$
$$\equiv (12.55 \text{ m}^3)$$

$$-\text{for cogging Vpc} = 4 * 6130$$
$$= 24520 \text{ in}^3$$

$$\left(158.2\text{m}^3\right)$$

- Step V:

- assuming hoop stress along the inner surface of cylinder
- barrel Sh =30000 lb/in^2
- cylinder expansion de will be

$$\begin{aligned}\text{de} &= (\text{Sh}/\text{E}) * \text{D}\\ &= \frac{30000 * 51}{30 * 10^3}\\ &= 0.051 \text{ in } \left(1.29 * 100^3 \text{ m}\right)\end{aligned}$$

- Step VI:

- Compression of liquid takes place along the entire length of cylinder barrel

$$\begin{aligned}&\text{assuming length of barrel} = \text{S} + \text{S}/2\\ &= 125 + 125/2\\ &= 187.5 \text{ in}(4.76\text{m})\end{aligned}$$

Volume of compressed liquid Vec:

$$\begin{aligned}&\text{Vec1} = \text{Vec2} = \text{Vec3}\\ &= \frac{187.5(\Pi/4)(2\text{D} * 30000 * \text{D})}{30 * 103}\\ &= 291\text{in3}\left(0.0048\text{m}^3\right)\\ &\text{Vec} = \text{Vec1} * 3 = 573\text{in}^3\left(0.088\text{m}^3\right)\end{aligned}$$

Volume required for return stroke.
Vr= 6743 in^3

- Total additional volume

= 6743 + 573
= 7316 in^3 /stroke

- Total amount of liquid required / cogging stroke V

= 24520 + 7316
= 31836 in^3 (0.5217 m^3)

- Determining the pumping requirement for pump

- without accumulator

$$\text{Gpm} = \text{V} * 60$$
$$231 * \text{Tp}$$
$$= 2067\left(7.82545\text{m}^3/\text{min}\right)$$

- With an accumulator

$$\text{Gpm(acc)}$$
$$= \text{Vtc} * 60231 * \text{Tp}$$
$$= 835\left(3.16\text{m}^3/\text{min}\right)$$
$$= 3461\text{lit/min}$$

- Determining the size of accumulator to store the accumulator volume Vs of liquid

$$\text{V s} = \text{g p m(a c c)*T c} - \text{T p} - \text{V r}$$
$$60$$
$$= 82.4\ \text{gal}$$
$$= 82.4/7.48$$
$$= 10.98\ \text{ft}^3\left(0.3109\ \text{m}^3\right)$$

- Air volume required

$$\text{Vair} = \underline{\text{Vs}}$$
$$1.11(1/n) - 1$$
$$= 10 * 82.14$$
$$= 821.4\ \text{gal}$$
$$= 109.81\ \text{ft}^3$$

- Total accumulator volume

$$= \text{Vs} + \text{Vair}$$
$$= 109.81 + 10.98$$
$$= 120.8\text{ft}^3\left(3.42\text{m}^3\right)$$

4 Conclusion

Design of hydraulic circuit is the most important part of development of the hydraulic system. The simple ratio of pressure to area for force determination must be related to gallon per minute fluid flow that will determine the speed of movement of ram in cylinder. The sophistication of control desired will determine the signal sources to create the cycle pattern of power transmission system. The design part of press provides the pumping requirement for the system on the basis of which decision on the number of pumps with capacity, piping requirement, number of different types of valves with requirement of the same and size of accumulator and reservoir can be determined. All these information about the requirement of different component will provide the basis for individual design of the same.

References

1. Min-Qiang Dai, Sheng-Dun Zhao, Fan Xu And Xiao- Mei Yuan (2011) The 20mn energy savings fast forging press hydraulic system design, engineering plasticity and its applications, pp 368–372
2. Yu HY, Tang JL, Hu J (2012) Research on hydraulic system of speed forging hydraulic machine. Appl Mech Mater 192:123–127
3. Ai C, Liu YJ, Song Y, Kong XD Research on flow pressure compound position control of open circuit hydraulic pump-controlled forging press system. C
4. Ferriera JA, Sun P, Gracio JJ, Design and control of hydraulic press. In: IEEE Conference 2006, Germany
5. Pippenger, John H (Author), Hicks, Tyler Gregory Industrial Hydraulics 3rd edition ,2001 McGraw Hill publication
6. Bhagwati Prasad Gupta “Industrial Hydraulic Systems and Circuits” feed forward publication
7. Industrial fluid power basics, Bud Trinkel 1980 3-Ring Binder
8. Industrial hydraulics manual vickers Corp. 1993
9. Industrial Hydraulic Technology, Schrader, Lawrence F., Parker Hannifin, Corp
10. Valve Handbook, Skousen, Philip L.
11. Vickers industrial hydraulics manual, Vickers training center
12. Piping Design Hand Book,G.K. Sahu-2002 Edition
13. Hydraulic pump hand book, L.K. Karassik-97.
14. Valve Selection hand book-Zapper R.W.3rd Edition 1995. Bombay
15. Machine tool design, Donaldson
16. Production technology by P.N.Rao, Dhanpat Rao publication

A Review on the Auxiliary Drones Used as Safety System for Passenger Aircraft

Ujjal Kalita and Krishnanand K. Anandakrishnan

Abstract The issues of safety have always been a deep concern in the field of aviation industry, as it deals with millions of lives each day. Our intention for contributing some innovation in the field of safety made us think about what is lacking in the same. We have gone through so many safety measures that is taken by the civil aviation authority, but we couldn't find much lacking in the field of safety as safety is the first priority in the aviation filed. While going through the different safety system it was observed that a major flaw is found in the recovery of missing aircrafts. Sometimes, we could recover the lost aircraft and could not able to know the position from which it was missing. Different researchers have worked on using different types of drones or UAV's to detect those lost aircraft or knowing the position of aircraft. In this work, a literature review is conducted on implementation of auxiliary drone or UAV in the aircraft for detecting those aircraft or deploying the drone from aircraft at times of emergency. The position of drones on the aircraft and its deployment were studied. From the review, we get a clear idea on the position and implementation of auxiliary drone on the aircraft used for safety of the aircraft.

Keywords Auxiliary drone · Safety system · Aircraft · Position and deployment

1 Introduction

A drone is a type of unmanned aerial vehicle. It does not carry a human pilot, crew or passengers. Drones area unit a part of a pilotless craft system that embrace the drone itself, a ground-based controller, and a system of communications between the two. The flight of UAVs might operate beneath remote by a personality's operator as remotely piloted craft, or with varied degrees of autonomy, like autopilot help, up to completely autonomous craft that don't have any provision for human intervention. Drones are used for different purpose. It can be used for different purposes such as

U. Kalita (✉) · K. K. Anandakrishnan
School of Mechanical Engineering, Lovely Professional University, Phagwara, Punjab, India
e-mail: ukalita3487@gmail.com

N. M. Sivaram et al. (eds.), *Advances in Manufacturing, Automation, Design and Energy Technologies*, Lecture Notes in Mechanical Engineering,
https://doi.org/10.1007/978-981-99-1288-9_66

Cargo Delivery, Combat, Irrigation, Research and Development and many more but none of the drone can give information about an airplane got missing or an aircraft who gets into an accident. Even though many researches are going for aircraft safety and stability but this is a whole new and a unique idea which will take the whole aviation to a different level [1].

Air Transport is considered to be the safest form of traveling but even after highly developed technologies and strict safety measures we lost many passenger aircraft, some of which can't retract the location of the aircraft as the connection between them is lost, these kind of scenes are getting usual these days and the cause is pretty unknown to us. This project will enhance the safety measure for flight from getting lost from the network. So, here even if the aircraft is lost or it met with any accident the drone will be deployed from the aircraft and then it will use its own network in finding the location of the aircraft. This system will be helpful by giving quick response to the rescue team. This drone safety system works like a flare system in ship, which deploys different drones at a time from the aircraft during any problem in the flight, this sudden deployment of drones carries information of the aircraft which will help in accessing the data in a very fast and reliable way without any damage, it will take information from the pilot regarding the location and happening of the flight and then the system will help in finding the location of the aircraft. It will find the aircraft by using a different network other than the ATC which will enhance the safety of flight.

In today's world, security is a top priority and this system will give a quick response which will make the aircraft more secure and safe.

The Drone components are:

1. Body of the Drone.
2. Power supply system.
3. Sensors.
4. Actuators.
5. Software like Raspberry Pi.
6. Communication.

Adobe Photoshop may be a computer code application for graphic design, image editing and digital art. This software could be implemented for editing pictures, background file formatting and many real life painting could be done for actual conditions.

2 Literature Review

Literature review reveals all the ideas of the auxiliary drone safety system to design and fabrication. All the ideas regarding this project has been gathered with the help of various research papers. The purpose of literature reviews is to study the theory compare and analyze people's methods of safety with their past studies. This method can be applied as a guidance for this project. Table 1 shows the literature review of few of the works performed by various researchers in tabulated form.

Further, Swaminathan in his work has performed about varied bird strike tests on airplanes. For this, he made a mimicked point collision between pair of 1-pound DJI Phantom, taken two Quad-copter and a Mooney M20 plane. The drone is bored into the loads on plane like a bird's feather and when required to deploy, the feather is broken [15].

Yong a representative for DJI said the show presented a drone that the company no longer assemblies. They created a video which shows an infrequent worst case perfect hit. He also believed the company deals drones that contain built-in geofencing and elevation limitations [16, 17].

Federal Aviation Administration (FAA) in their research center made comparable findings, saying drones are more stiff materials which allow them to cause bigger damage than birds. That examination team estimated potential impacts of two quad-copters and two fixed-wing drones assembled on a single-aisle commercial transport jet as well as on business jet [18].

ICAO and other organization said that today's fast rising drone market and sensor technologies are frequently the unheralded undisclosed sauce from inside. In this article, we discourse approximately some of the categories of sensor technologies that control today's drones [19, 20].

1. Accelerometers.
2. Inertial Measurement Units.
3. Tilt Sensors.
4. Current Sensors.
5. Magnetic Sensors.
6. Engine Intake Flow Sensors.

Ansys Blog in his work defined that how to retain Airspace Safe through absolutely Autonomous Drones and craft, these often the encounter we have a tendency to square measure presently functioning on. It need new rubrics, air/road control systems, predefined ways and increased technology to sense, react, and avoid complications. Sensors should be ready to clearly intellect their environments while not presence fooled by fog or striking conditions, like glare or shadows. These sensors should be cheap and light-weight to make the most of the UAV's carrying capability or mobility [21, 22].

3 Research Gap

Even though many researches are going on in the field of Aircraft Safety and Reliability, we haven't installed any kind of enhanced security features to find the lost aircrafts, the current systems are installed in the aircrafts which can get damaged and fails to provide the information that might be needed during an emergency situation. Sudden deployment of drones which carries the information from the aircraft will

Table 1 Literature review of work performed by different researchers

Authors	Year	Component of study	Methods/outcome
Azar et al.	2021	UAVs, navigation, control, machine learning, deep reinforcement learning (DRL)	• DRL use becomes fruitful as a result of it ought to be pushed to change several different tasks for UAVs. One in every of such employments is that the landing situation for the vehicle. Normally, it depends on powdered cameras, vary sensors, GPS, and additional to land exactly within the finish position, escaping any failure. With deep reinforcement learning, a rising plan discusses the combination of it and solely a downward-looking camera to land exactly • The DRL may sustenance 2 main stages of marking the tip position to be noticed and also the vehicle's vertical downward-sloping. Those networks' coaching are going to be most useful for UAVs and dead in a quicker manner [1]
Sriram et al.	2019	Autonomous drone, wireless communication, image processing, line of control, UAV	• In their work a pick hawk flight controller is used for controlling the operation of an autonomous unmanned aerial vehicle (UAV). Their research mainly focus on UAV • Quad-Copter is also used in their work which is capable of steering autonomously without implementing any time period • This could normally use by police for investigation, machinery maintenance as well as for defense purpose to locate many unknown places during war time in war zones. They mainly design the quad-copter for it [2]

(continued)

Table 1 (continued)

Authors	Year	Component of study	Methods/outcome
Gregersen	2021	Aviation disaster	• The flight 370 disappearance, jointly called MH370 disappearance, disappearance of a Asian country Airlines traveler jet on 8th March 2014, throughout a flight from Malaysian capital to national capital • The Boeing 777 disappearance with twelve crew members and 227 passengers on board junction rectifier to an exploration effort stretching from the Indian ocean to Central Asia • Flight 370 disappearance become one altogether history's most illustrious missing craft [3]
Schaul	2014	67 years of aircraft missing	• In the last seven decades, nearly ninety business airliners have gone missing, while not one piece of portion ever placed • A lot of folks were lost on Asian country Airlines Flight 370 in March of this year than on any flight before • AirAsia confirmed that portion from Sunday's AirAsia Flight 8501 has been found, thus it's been far away from this list [4]
Majhiri et al.	2018	Reason for airlines disaster	• In his work the Statistical analysis was performed and it indicates that hull lose accidents are caused worldwide by flight crew errors and human errors. 56% of worldwide accidents is because of crew errors and 70% human errors • These data shows that every accidents have some type of human error connected to their causes [5]

(continued)

Table 1 (continued)

Authors	Year	Component of study	Methods/outcome
Lee et al.	2021	Main components are Artificial intelligence and artificial neural network, multilayer network, IOT machine, UAV's	• In their work said that they need to stipulate the thought of autonomy for drones, with a scan to recognize altogether all the autonomous navigation continuously at different levels • Against these level the rising navigation choices in analysis. We've an inclination to use the six levels of autonomy. SAE had implemented these levels for ground vehicles • The thought of autonomy for drones and cars is comparable, implying a removal of driver among the navigation of path finding and obstacles • This, planning to fully freelance independent navigation in spite of limitation as a result of surface positive movement. By examining the SAE levels of cars autonomy, we've an inclination to notice but each level is applicable to drone [6, 7]
Kumar et al.	2021	To develop an UAV equipped with new technologies various military civil applications	• On their paper aims to develop an UAV with advanced technologies various civil military applications. It's associate automatic system • The increasing capabilities and shrinking size of electronics devices in latest years has detached the doors to a lot of auto pilot and moved for a lot of latest UAVs applications [8] • The UAV market grown by 2020, as civil, business, and military applications still develop. Changes in the traffic management embrace the creation of associate info UAV to be defined as the associate craft that is management system to exchange info among suppliers and traffic Management users • The impact of every state of affairs on the police investigation and air traffic is summarized, and then the associated problems known. The paper concludes by telling the requirement for a longer term UAV roadmap • This paper main aims to supply an easy associated cheap answer of an independent aerial surveyor which may do aerial police investigation, acknowledge and trace numerous objects, in a position in creating easy 3d map [9]

(continued)

Table 1 (continued)

Authors	Year	Component of study	Methods/outcome
Moshe et al.	2021	Geofence	• In their work said that the approaching generation of LBSs are going to be considerably determined by background chase and geofencing applications • A geofence could be a little geo-graphic space that's outlined to get a location event as shortly as a user enter or leave this geofence Associate in Nursingd to method this event within the context of an LBS. Background chase incessantly monitors the whereabouts of a user and is so the key necessity for sleuthing location events • Each functions may be considered vital enablers for Associate in Nursing improvement and broad institution of data connection in mobile environments. when an summary of the emergence of LBSs within the recent years, this text provides Associate in Nursing introduction into back-ground chase and geofencing and demonstrates their operating for location-based recommender systems [10, 11]
Zaheer et al.	2021	Geomapping	• In their work they conducted an enquiry victimization Inspire a pair of quadcopter drones with RGB cameras, developing 3D models victimization scenario mapping, and photogrammetric uses geographic data systems • The drone used an exceedingly wider vary of areas with adequate power support. For image stability drone is additionally supported by a high-quality dreadlocks camera, therefore it's appropriate to be used in mapping activities [12, 13]

(continued)

Table 1 (continued)

Authors	Year	Component of study	Methods/outcome
Patrik et al.	2021	Navigation systems of autonomous drone	• In their work they presented their analysis using a Global Navigation Satellite system (GNSS) which was implemented on drone for operating it. GNSS and compass were used for navigation of drone • The objective of their analysis is to transport medical assistances for patients and implementation in agriculture in country, as a section of the huge mission 5.0 and connected with immense data. In effort technique, drone ought to be able to reach goal position and discover object and come back carefully exploitation GPS • We've a bent to project a navigation formula for drone in addition because the usage of heading material to support drone in doing independent navigation. At various intervals the experiment that we've a bent to did, the common of purpose deviance of landing location between the specified landing position and conjointly the actual landing position at intervals the flight tests of flying from begin to goal is one.1125 m and for the tests that use the formula that uses heading, the purpose deviance has average of 2.39 m. Navigation exploitation heading formula is not tons of definitive than the navigation formula with compass and GNSS at a navigation distance of 1 m. Chien conjointly performed work on almost like it [14]

help in accessing the data without any damage in fast and reliable way. This feature will surely be an added advantage on safety that will help the whole aviation industry. The civil aviation industry gives safety as the first priority for the manufacturing and operation of aircrafts, even after prioritizing safety more than economy the industry is facing cases that the aircrafts are completely lost without having any hints for the recovery. This reduced the credence of the common passengers on the assurance of safety that is provided by the aviation industry and the airlines.

4 Conclusion

The commencement of this project will be new technology used in the aviation industry. This technique is referred from the flare system used in the ships. By the implementation of auxiliary drone safety system in the aircraft, it will lead to an aviation industry with no missing aircrafts. The idea behind this project make a true meaning for the safety as the first priority in the industry of aviation. Safety is indispensable among all other vital needs in the industry and that's what the civil aviation authority depicts for the operation of aircrafts and its airworthiness. This project will make the aircrafts for a complete airworthy and safe operations.

References

1. Azar AT, Koubaa A, Ali Mohamed N, Ibrahim HA, Ibrahim ZF, Kazim M, Ammar A, Benjdira B, Khamis AM, Hameed IA, Casalino G (2021) Drone deep reinforcement learning: a review. Electron 10:1–30
2. Sriram PR, Ramani SK, Shrivatsav RV, Mankiandan MM, Ayyappaa N (2019) Autonomous drone for defence machinery maintenance and surveillance. In: Proceedings of the 3rd world conference on smart trends in systems security and sustainability, WorldS4 2019, pp 288–292
3. Gregersen E (2014) Malaysia Airlines flight 370 disappearance aviation disaster
4. Schaul K (2014) 67 years of missing flights. The Washington post
5. Mahjri A, Dhraief A, Belghith S, Gannouni I, Mabrouki M, AlAjlan (2018) Collision risk assessment in flying Ad Hoc aerial wireless networks. J Netw Comput Appl 124:1–13
6. Lee T, McKeever S, Courtney J (2021) Flying free: a research overview of deep learning in drone navigation autonomy. Drones 5:1–18
7. Sharma T, Acharya B, Daga H, Sharma A, Srinivasan K, Bajpai V, Chang J-R (2020) Int J Soc Human Comput 3:245–269
8. Mellinger D, Shomin M, Kumar V (2010) Control of quadrotors for robust perching and landing. In: International powered lift conference, pp 119–126
9. Pounds P, Mahony R (2009) Design principles of large quadrotors for practical applications. In: Proceedings—IEEE international conference on robotics and automation, pp 3265–3270
10. Ben-Moshe B (2021) Power line charging mechanism for drones. Drones 5:1–14
11. Talebi H, Milanfar P (2016) Fast multilayer Laplacian enhancement. IEEE Trans Comput Imaging 2:496–509
12. Zaheer Z, Usmani A, Khan E, Qadeer MA (2016) Aerial surveillance system using UAV. In: IFIP international conference on wireless and optical communication networks, WOCN

13. Do HT, Truong LH, Nguyen MT, Chien C-F, Tran HT, Hua HT, Nguyen CV, Nguyen HTT, Nguyen NTT (2021) Energy-efficient unmanned aerial vehicle (UAV) surveillance utilizing artificial intelligence (AI). Wirel Commun Mob Comput 2021:1–11
14. Bouabdallah S (2007) Design and control of quadrotors with application to autonomous flying. École Polytech. Fédérale Lausanne, À La Fac. Des Sci. Tech. L'Ingénieur, vol 3727, p 61
15. Swaminathan V (1987) Microwave-powered unmanned drone: a poorman's satellite. IETE J Educ 28:22–28
16. Yong SP, Yeong YC (2018) Human object detection in forest with deep learning based on Drone's vision. In: 2018 4th international conference on computer and information sciences: revolutionising digital landscape for sustainable smart society, ICCOINS 2018—Proceedings, pp 1–5
17. Jeon I, Ham S, Cheon J, Klimkowska AM, Kim H, Choi K, Lee I (2019) A real-time drone mapping platform for marine surveillance. In: The international archives of photogrammetry, remote sensing and spatial information sciences—ISPRS Archives, vol 42, pp 385–391
18. Han S, Shen W, Liu Z (2016) Deep drone: object detection and tracking for smart drones on embedded system, pp 1–8
19. ICAO (2001) Cover sheet to amendment 11, aircraft accident and incident investigation, Annex 13, Conv. Int. Civ. Aviat., pp 1–42
20. Royal air force flying accidents, flight magazine, London: Reed Business Information. XX(41):898. Retrieved July 22 (2011)
21. Ansys Blog (2021) The challenges to developing fully autonomous drone technology
22. Wikarek J, Sitek P, Zawarczynski (2019) An integer programming model for the capacitated vehicle routing problem with drones. In: Lecture notes in computer science 11683 LNAI, pp 511–520

Energy Stream

Experimental Study on Water Quality Enhancement: Micro-nano-bubbles Technology

J. Dilip Singh and G. Senthilkumar

Abstract The present research article enumerates the potential features of micro-nano-bubbles (MNBs) for the treatment of surface water without affecting the environmental circumferences. The formation of free radicals increased the surface area density with reduction in radius of MNBs. The sustainability for prolonged duration, optimum interface potential, high oxygen transfer efficiency, and ease of production methods increased the research in MNBs in recent years. The surface water is treated for reduction in chemical oxygen demand, biochemical oxygen demand, total dissolved solids and total suspended solids with this unique MNB technology, and the results are highlighted. The 11% drop in total dissolved solids of surface water and 45% increase in dissolved oxygen in domestic water are some of the worthy deliverables of this research. Also the promotion of dissolved oxygen in the selected water samples from tap, pond, domestic waste, and sea water after treatment with MNBs is discussed.

Keywords Dissolved oxygen · Micro-nano-bubbles · Sonication · Surface water

1 Introduction

The sustainability of all the organisms like plants, human, and animals is the function of quality of water [1]. The MNBs were tiny gas bubbles in liquid medium with particle size ranges between 200 nm and 10 microns [2]. The application of MNB includes treatment of waste water [3], drug delivery [4], ozone treatment [5], marine operation, H2 storage, seed germination activity [6], anti-microbial activity, aquaculture [7], sterilization of oysters, heat transfer [8], engine performance [9], and vessel cleaning though the research is in infant stage. The buoyancy force of MNBs was found to be less than the large bubbles due to their particle size led them to sustain for even months [10]. The MNB is a spherical air cavity in a liquid medium

J. Dilip Singh · G. Senthilkumar (✉)
Sathyabama Institute of Science and Technology, Chennai, India
e-mail: Tosenthilgs79@gmail.com

N. M. Sivaram et al. (eds.), *Advances in Manufacturing, Automation, Design and Energy Technologies*, Lecture Notes in Mechanical Engineering,
https://doi.org/10.1007/978-981-99-1288-9_67

and the surface area density A/V = 3/r, (where A = Surface area of MNBs, V = Volume of MNBs, r = radius of MNBs of spherical shape) the density increases with reduction in size of MNBs. As the smaller bubbles subject to large surface tension, leading for shrink and increase in more internal pressure inside the bubble cavity. If attempts were made to shrink the MNB to such low size, resulted in pressure applied to infinite causing collapsing of MNBs [13]. This results in formation of OH-free radicals. The salient feature of OH-ions is that it's oxidizing behavior and removal of pollutants such as phenol which is very hard to be removed by any other procedures. The zeta potential influences the stability of MNBs directly [11]. The MNBs zeta potential (particle's electrical potential at the slip plane) ranges between −34 and −45 mV for oxygen, −29 and −35 mV for nitrogen, −20 and −17 mV for air, −20 and −27 mV for CO2 and −11 and −22 mV for Xenon. The zeta potential directly influences the stability of MNBs. The ability of dissolution of MNBs was extremely more than that of normal large bubbles [12]. The air NBs were spherical in shape when they exist in water medium [13, 14]. The dissolved oxygen (DO) % in water was increased with the presence of MNBs in water [15, 16]. The MNBs were long lasting without reduction in size for months. The presence of MNBs also depended on pH of water and other salt contents if any. For a 100 nm bubble size, the normal Laplace pressure was around 30 N/m^2. The water with MNBs was found to be charged electrically and its zeta potential values were negative in all the O_2, CO_2, Xe, and N_2 samples. Nano-bubbles water curcumin extract might be one of the competent substitute nutrition used with exercises. The vacuum degasification method was employed for the production of bulk nano-bubbles in ultrapure water with the aid of a syringe. The molecular dynamics simulation was employed to investigate the Brownian movement of nano-bubbles in deionized water. The ozone microbubbles were capable of removing pollutants which were organic in nature. The gas–oil engine with air nano-bubbles contributed for energy savings. The physiochemical properties of water were influenced by the presence of nano-bubbles. The coalescence of nano-bubbles generated was the function of temperature variations as predicted by response surface methodology.

2 Micro-nano-bubbles Generation Procedure

The hydrodynamic cavitations, constant temperature stirring, electrolysis, ultrasonic cavitations, oscillation of fluid, solvent–water mixing, and plunger mechanism were some of the unique techniques practically in existence for the generation of MNBs based on the characteristics of liquid medium and application criterions. In our experimental procedure, we used an ultrasonic sonicator combined with an air compressor for the creation of tangential force in water to be tested.

The sample was first vibrated for 15 min at isothermal conditions and the process was repeated 3–4 times for the observation of MNBs as shown in Fig. 1. The stability, zeta potential, and the size distribution of the MNBs were ensured with the addition of oleic acid as a surfactant with a concentration of 0.05% by volume basis, based

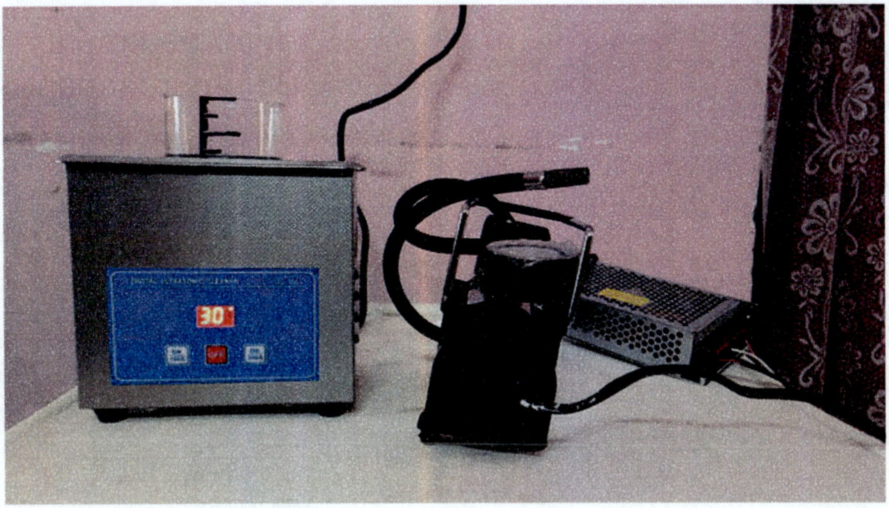

Fig. 1 Experimental photograph

on the previous studies. The size of MNBs generated in our work was between 100 and 400 nm. This surfactant was chosen owing to the better impact on humans. The 300 mL of water to be tested was filled in a 500 mL beaker, kept in the sonicator and subjected to vibration for nearly 1 h at a constant temperature of 30 °C. An air compressor of capacity 0–5 bar was incorporated for injection of air with shearing action in the water at a constant pressure of 1.5 bar so that MNBs was produced. The macro bubbles generated during the process moves up due to the buoyancy and collapses near the top surface of the beaker.

3 Result and Discussion

The chemical oxygen demand (COD) specifies the oxidation behavior of organic substances present in water which in turn is a measure of normal water and waste water quality. The chemical oxygen demand and biochemical oxygen demand (BOD) are the measure of level of pollutants present in water. From the previous research studies, it was confirmed that many commercially available COD analyzers incorporated in the measurement of COD. COD is the deciding factor for plant efficiency in wide range of waste water treatment plants. One of the contradictory facts was that higher the COD resulted in lower DO in any water samples. The percentage decrease in COD was 11, 121, 400, and 430% in tap water, pond water, domestic waste water, and sea water, respectively, as reported in the Fig. 2. Under any testing conditions, the quantity of compounds oxidized biologically is inferior to the quantity of compounds oxidized chemically.

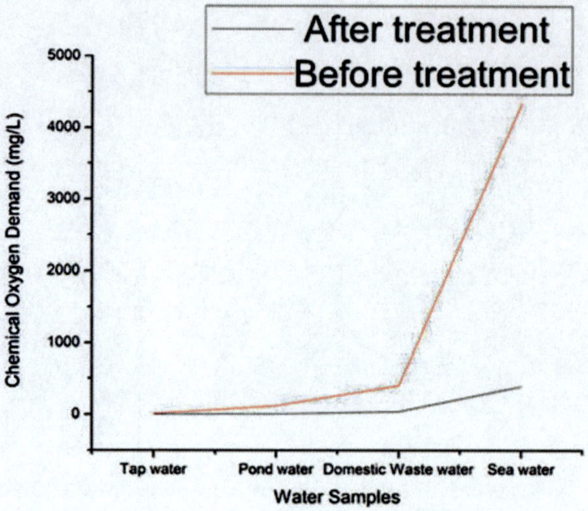

Fig. 2 COD variation after treatment with MNBs

The increase in aerobic microorganisms due to the presence of biological organic pollutants in water require higher amount of oxygen to degrade this pollutants resulted in higher biochemical oxygen demand (BOD) in any tested water samples. The percentage decrease in BOD for tap water, pond water, waste water, and sea water was about 3, 18, 105, and 2500% after the formation of MNBs in above-mentioned samples, respectively, as depicted in Fig. 3. The waste water and sea water were found to be with higher BOD due to the existence of organic matters in biological nature.

From the ambient, the oxygen diffused into the water based on the intensity of heat that exists in tested water samples at controlled thermal environments to ensure

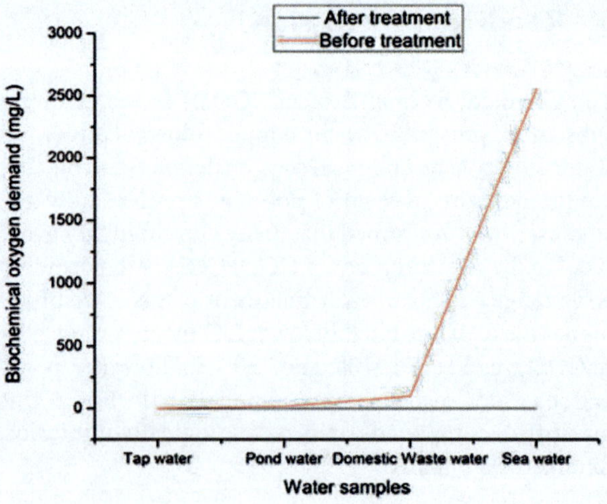

Fig. 3 BOD variation after treatment with MNBs

the proper degree of saturation of dissolved oxygen. The dissolved oxygen presence is indirect proportion to the temperature in any liquids due to the activities of inter molecular force. The other contributors for reduced DO level in water are biological oxygen demand, fertilizers, and the bacterial growth. The presence of MNBs in surface water contributed for the enhancement of biological activity. The minimum DO requirement for living sea fishes is in the order of 5 mg/L. The enrichment of quality of sea water is possible by the generation in MNBs in sea water due to the increase in O_2. The volume fraction of dissolved oxygen (DO) was measured using a high accurate DO meter (0–100 mg/L range). From the figure, it was evident that 50% increase in DO could be achieved in tap water after the generation of free radicals of MNBs. In case of stagnant pond water, it was about 66% due to the MNBs. In domestic waste water and sea water, this increase in trend was measured to be 300 and 44%, respectively, as given in Fig. 4. The interesting factor of the experimental results was that stable zeta potential values for the entire DO measured outputs.

The particles (size not less than 2 μm) which were not dissolved in liquid medium due to their weight namely total suspended solids (TSS) attributed for significant research in many fields of science and engineering. Except algae and bacteria, most of the TSS was inorganic in nature. In country like United States, it was called as conventional pollutant in water. One of the simplest techniques for the measurement of TSS is filtering through a permeable membrane. The porosity, mesh size, and surface roughness of filter are the other contributing factors for proper determination of TSS in any liquid sample. 30.77, 78.9, 300, and 1500% reduction in TSS was attained after the generation of MNBs in tap water, pond water, domestic waste water, and sea water, respectively, as shown in Fig. 5.

Total dissolved solids (TDS) comprise both the organic and inorganic components present in liquids. The waste water treatment plants major cost incurred in the reduction of TDS due to their varying sizes and behaviors. The 83 and 84.26%

Fig. 4 DO variation after treatment with MNBs

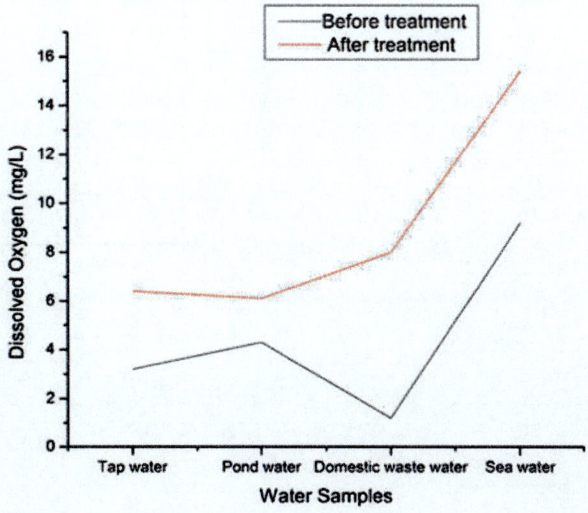

Fig. 5 TSS variation after treatment with MNBs

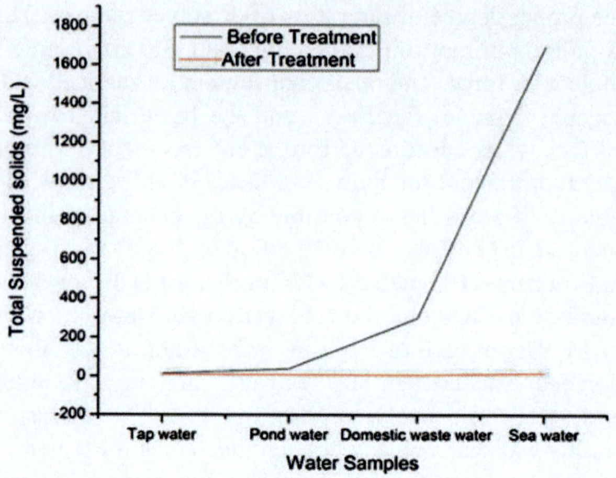

drop in TDS were measured in tap water and pond water, respectively, before and after treatment with MNBs. In domestic waste water and sea water, the percentage reduction was measured to be 89.6%, 97.7% as revealed in Fig. 6. One of the major drawbacks before treatment was that corrosion increase and decrease in conductivity of the water samples. The various test results were carried out at mettex lab, Chennai, with different type of analyzers.

Fig. 6 TDS variation after treatment with MNBs

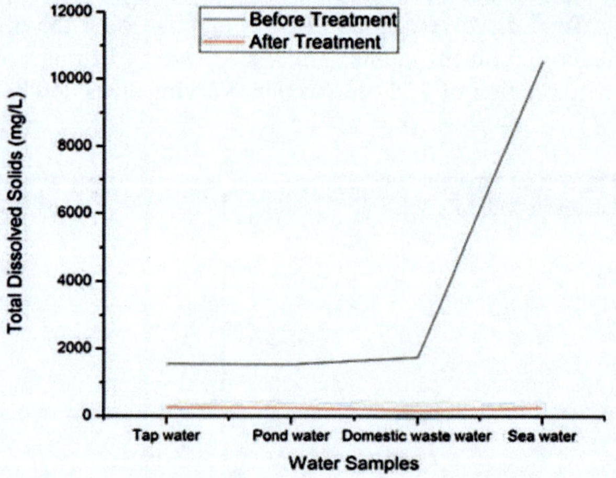

4 Conclusion

The MNBs generation with the unique hydrodynamic cavitations technique was adopted in all the water samples such as tap, pond, domestic waste, and sea water. It was evident that the dissolved oxygen enrichment was increased to a great extent in all the cases. The TSS and TDS level dropped down to the safe limit. The COD was reduced nearly by 400% in sea water and least in tap water by 11% due to the presence of more chemical substances that exist in sea water. The organic pollutants which are biological in nature reduced by 2500% BOD in sea water with this MNBs technique. Without affecting the zeta potential values, the DO % is increased in all four tested water samples. The conventional pollutant called TSS was reduced in the range of 30–300% in as tap, pond, and domestic waste and sea water samples.

Disclosure statement No conflict of interests.

References

1. Tekile A, Kim I, Lee J-Y (2016) Extent and persistence of dissolved oxygen enhancement using nanobubbles. Environ Eng Res 21(4):427–435
2. Gurung A, Dahl O, Jansson K (2016) The fundamental phenomena of nanobubbles and their behaviour in wastewater treatment technologies. Geosys Eng 19(3):133–142
3. Khalifa AE, Lawal DU (2016) Application of response surface and Taguchi optimization techniques to air gap membrane distillation for water desalination—a comparative study. Desalin Water Treat 57(59):28513–28530
4. Vergara D, Bellomo C, Zhang X, Vergaro V, Tinelli A, Lorusso V, Rinaldi R, Lvov YM, Leporatti S, Maffia M (2012) Lapatinib/Paclitaxel polyelectrolyte nanocapsules for overcoming multidrug resistance in ovarian cancer Nanomedicine. Nanomed Nanotechnol Biol Med 8(6):891–899
5. Gobinath S, Senthilkumar G, Beemkumar N (2018) Air nanobubble-enhanced combustion study using mustard biodiesel in a common rail direct injection engine. Energy Sources Part A Recov Util Environ Effects 41(15):1809–1816
6. Nasr P, Sewilam H (2016) Investigating fertilizer drawn forward osmosis process for groundwater desalination for irrigation in Egypt. Desalin Water Treat 57(56):26932–26942
7. Rameshkumar C, LakshmiSankar S, Senthilkumar G (2019) Characterisation of seed germination using nanaobubbled water. Int J Ambient Energy
8. Rameshkumar C, Senthilkumar G, Subalakshmi R, Gogoi R (2019) Generation and characterization of nanobubbles by ionization method for wastewater treatment. Desalin Water Treat 164:98–101
9. Rameshkumar C, Senthilkumar G, Subalakshmi R (2021) Purification of tap water to drinking water: nanobubbles technology. Desalin Water Treat 233:11–18
10. Kim S, Kim H, Han M, Kim T (2019) Generation of sub-micron (nano) bubbles and characterization of their fundamental properties. Environ Eng Res 24(3):382–388
11. Senthilkumar G, Lakshmi Sankar S (2018) Bubble dynamic behavior of air water two phase flow mixing in hydraulic structures. Int J Ambient Energy 42(5):505–507
12. Senthilkumar G, Purusothaman M, Rameshkumar C (2020) Generation and characterization of nanobubbles for heat transfer applications. Mater Today Proc 43(6):3391–3393

13. Senthilkumar G, Rakesh AM, Aravind Sri Krishna BRS (2020) Experimental investigation on the effect of nanobubbled water in evaporative cooling and oxygen enrichment in air cooler. Mater Today Proc 44:3633–3626
14. Senthilkumar G, Rameshkumar C, Nikhil MNVS, Navin Ram Kumar J (2018) An investigation of nanobubbles in aqueous solutions for various applications. Appl Nanosci 8(6):1557–1567
15. Lvov YM, Pattekari P, Zhang X, Torchilin V (2011) Converting poorly soluble materials into stable aqueous nanocolloids. Langmuir 27(3):1212–1217
16. Zheng Z, Zhang X, Carbo D, Clark C, Nathan C-A, Lvov Y (2010) Sonication-assisted synthesis of polyelectrolyte-coated curcumin, nanoparticles. Langmuir 26(11):7679–7681

A Novel Design of Internal Heat Exchangers in Metal Hydride System for Hydrogen Storage

Swaraj D. Lewis and Purushothama Chippar

Abstract Performance of the metal hydride system depends on effective heat transfer systems. A numerical model of the $LaNi_5$ metal hydride reactor is developed having two different concepts such as compartmentation and embossment. Compartmentation has an advantage when the metal bed is having larger volume. Splitting the volume into number of parts showed better performance than the a single large volume. Around 33% improvement is observed when the volume is split and stacked. Embossment can give the benefit of heat transfer by both fluid and heat plate. Various design of embossed heat exchangers such as pin, parallel, vertical serpentine, and horizontal serpentine is modeled and compared with the helical coil heat exchanger. Vertical serpentine embossed plate heat exchanger gave optimum performance than the other configurations. More uniformity is observed when using an embossed heat exchanger due to the availability of wider area for heat extraction. This study shows the design optimization and performance analysis of the two different design concepts which can be employed in the metal hydride systems.

Keywords Hydrogen · Metal-hydride · Heat exchangers

1 Introduction

Extensive use of nonrenewable energy worldwide such as fossil fuel resulted in scarcity of this energy which is a major source of fuel for automobiles, industries, and household. Currently, 85% energy requirement comes from fossil fuels, and small amount of energy from renewable energy sources [1]. Emission from fossil fuel is the main cause of environmental pollution leading to global warming. Hydrogen

S. D. Lewis (✉) · P. Chippar
St Joseph Engineering College (Affiliated to Visvesvaraya Technological University, Belagavi), Vamanjoor, Mangaluru 575 028, India
e-mail: swarajl@sjec.ac.in

P. Chippar
e-mail: purushothama@sjec.ac.in

N. M. Sivaram et al. (eds.), *Advances in Manufacturing, Automation, Design and Energy Technologies*, Lecture Notes in Mechanical Engineering,
https://doi.org/10.1007/978-981-99-1288-9_68

can serve as an alternate energy because of its high energy output more than the fossil fuel [2, 3] and also, it is pollution free making it environmentally friendly. Hydrogen can be stored by different methods such as high-pressure compression, cryogenic cooling, and solid state [4, 5]. Solid state storage is much safer when compared with other methods due to lower pressure and temperature. In solid state storage, hydrogen gets absorbed on the metal surface forming a Metal Hydride (MH). This reaction being an exothermic process, the heat generated can slow down the absorption process. Hence, removal of heat is very much important. Since the metal hydride has low thermal conductivity [6, 7], heat exchanger is needed to speed up the removal of heat from the system. Researchers have developed various types of heat exchanger which helps in dissipation of heat from the reactor bed such as providing an external cooling bath [8, 9], fins on the outer wall surface [10]. To remove the heat generated from the center of the reactor, internal heat exchanger such as placing Cooling tubes [11–13], Helical coil [14–17], and tube with fins [18, 19] were used.

Thermal conductivity being poorer in metal hydride powders, adding the Copper flakes, and metal foam inside the reactor [19–23] improved the heat transfer ability of metal bed leading to better performance. Phase change material (PCM) technology was used as an alternative to heat exchanger. Here, heat generated in the MH bed is extracted and stored in PCM for later usage. Stored heat is extracted during the desorption process which eliminated the need for an external HTF supply. PCM melts when it stores heat energy and solidifies when the heat energy is released. PCM is positioned in the external wall surface [24–27] extracts heat from the bed surface and stores it. To remove heat internally, PCM are in placed in encasement [28] or in sandwiched between MH bed layers [29]. Further metal foam and nano material can be incorporated to raise the effective thermal conductivity of PCM system [30, 31].

In this paper, we discuss novel heat exchangers developed, to extract the heat generated during the absorption process and to reduce the time required to store hydrogen. Two different types of MH-heat exchangers are developed. The design 1 has compartmentation inside the reactor and the design 2 has an embossed plate heat exchanger. Performance and optimization of two designs are discussed.

2 Numerical Model and Validation

The simulation is done using ANSYS-Fluent 17.2 software. $LaNi_5$ materials is taken as a metal powder for both the cases. Simulation is done at 10 bar inlet hydrogen pressure at 293.15 K temperature. The numerical model is governed by the following conservation equation [15, 31–33].

2.1 Mass Conservation Equation [15, 32]

For hydrogen gas

$$\frac{\partial \varepsilon \rho^{g}}{\partial t} + \nabla \cdot \left(\rho^{g} \vec{u}\right) = -\dot{m} \tag{1}$$

For metal powder

$$\frac{(1-\epsilon)\partial \rho^{s}}{\partial t} = \dot{m} \tag{2}$$

$$\dot{m} = C_{a} exp\left(-\frac{E_{a}}{RT}\right) ln\left(\frac{P^{g}}{P_{eq}}\right)\left(\rho_{sat}^{S} - \rho^{S}\right) \tag{3}$$

where, ρ^{g} and ρ^{s} are the density of gas and metal powder, ϵ = porosity, Source term $\dot{m}$ depends on Equilibrium Pressure P_{eq}.

Momentum conservation equation [32, 33]

$$\frac{1}{\varepsilon}\frac{\partial \rho^{g} \vec{u}}{\partial t} + \frac{1}{\varepsilon^{2}} \nabla \cdot \left(\rho^{g} \vec{u}\, \vec{u}\right) = -\nabla p + \nabla \cdot \tau + S_{\mathrm{u}} \tag{4}$$

where, $S_{u} = -\frac{\mu}{K}\vec{u}$ is the momentum source term.

Energy conservation equation [34]

Assuming the local thermal equilibrium exists, for non-porous zone, energy equation is given as

$$\frac{\partial\left(\varepsilon \rho^{g} E^{g} + (1-\varepsilon)\rho^{s} E^{s}\right)}{\partial \mathrm{t}} + \nabla \cdot \left(\vec{u}\left(\rho^{g} E^{g} + P\right)\right) = -\nabla\left[k_{\mathrm{eff}} \nabla T + \tau . \vec{u}\right] - \dot{m}\left[\Delta H + T\left(C_{\mathrm{P}}^{\mathrm{g}} - C_{\mathrm{P}}^{\mathrm{s}}\right)\right] \tag{5}$$

The following model is validated [32] using data of an experimental study conducted by Jemni et al. [35]. A theoretical model similar to the experimental model is developed and compared. Results are in better concord with the current model. The properties of $LaNi_5$ used for the mathematical model are mentioned in the Table 1 [32, 33].

Table 1 $LaNi_5$ properties used for simulations

Parameters/description	Value
Activation energy, E_a	21,179.6 J mol^{-1} K^{-1}
Absorption rate, C_a	59.187 s^{-1}
Specific heat, $C_{p,M}$	419 J^{-1} mol K^{-1}
Thermal conductivity of metal, K_m	0.167 Wm^{-1} K^{-1}
Porosity, ϵ	0.63

3 Implementation of Numerical Model to Various Cases

3.1 *Case1: Stackable Reactor with Compartmentation*

To store more quantity of the hydrogen, the simple method is to increase the volume of the cylinder which accommodates more quantity of metal powder for absorption. But increasing the size of reactor increases the distance between the core and wall surface. The heat generated at the core surface takes more time to reach the wall surface resulting in slower performance. A novel concept of stackable MH bed reactor with $LaNi_5$ is developed which is similar to a fuel cell system. Here, the Multiple MH reactor unit can be stacked together to form a larger storage volume.

Further, the individual component having single MH bed container is divided into compartmentation using a thin wall. The thin wall acts as a heat fin which helps in the transfer of heat from the core to the wall surface. Also, the compartment divides the volume into further smaller segment which reduced the heat stress of bulk material and provide uniformity in heat distribution. Figure 1 shows the various compartmentation model and a single stackable reactor. A grid independent test is done to study the ideal mesh size for accuracy and to save computational time.

Figure 2a shows the performance of the metal hydride reactor for various configurations of compartmentation. The single reactor is further split into 3 and 6 smaller volumes shown in Fig. 1.

From the Fig. 2a, it is observed that providing more compartments helps in better removal of heat which lowers the bed temperature at a faster rate. This results in faster hydrogen absorption. The case 1 took around 618 s for 90% hydrogen storage. Adding compartmentation in case 2 and case 3 took around 492 and 415 s which gave an

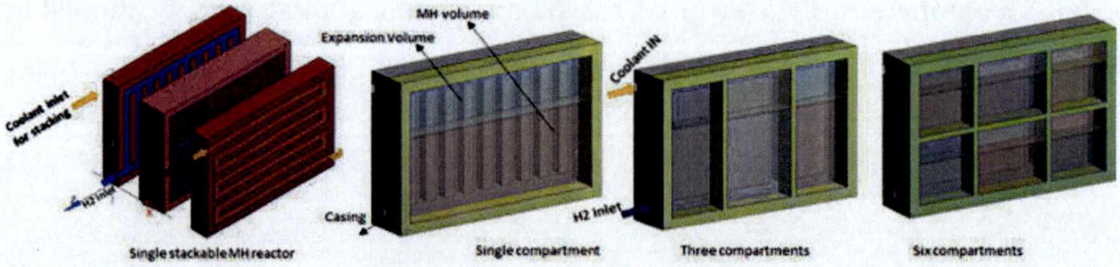

Fig. 1 Stackable MH bed reactor and various cases of compartmentation

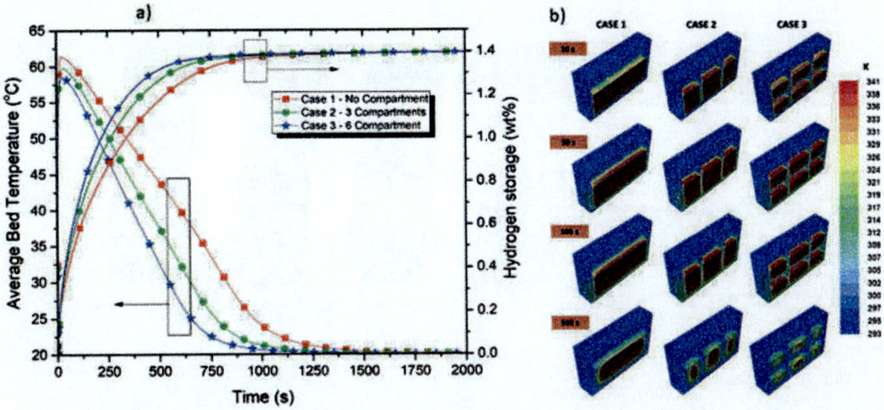

Fig. 2 **a** Average MH bed temperature and hydrogen storage evolution for all cases of compartmentation. **b** Temperature contour plot for case 1, case 2, and case 3

improvement of 20 and 33% when compared to case 1. Figure 2b shows the temperature plot for different cases. From contour plot, at 500 s, it is noticed that the high temperature intensity in the core is reduced significantly when compartmentation is introduced.

Adding a compartment will reduce the volume of metal powder. This loss can be overcome by faster storage performance. Compartment can minimize the heat accumulation at the center of the reactor.

3.2 *Case 2: Embossed Plate Heat Exchanger*

A numerical model with $LaNi_5$ metal hydride, having a plate embossed with a coolant channel is modeled. The coolant flow through the embossed path and transfer the heat from the bed. Further, the metal plate acts as a heat fin and it helps in removal of heat from the reactor bed to the coolant channel. Various configuration of embossment such as parallel type, pin type, vertical serpentine type, and horizontal serpentine type is modeled shown in Fig. 3. Grid independence test is also done for each model to obtain the optimum mesh size.

Figure 4 shows the reactor bed temperature variation, hydrogen storage performance for the various cases. It is observed that the serpentine channel provides optimal performance due to more torturous path when compared to other cases. Further vertical serpentine showed the best performance when compared to horizontal serpentine case. Time for 90% hydrogen storage is 2236, 2092, 1863, and 1897 s for case of parallel type, pin type, vertical serpentine type, and horizontal serpentine type channel. More uniformity in heat transfer is observed in embossed plate heat exchanger (EPHX) model when differentiated with the same volume of helical coil heat exchanger (HCHX) model which is observed in Fig. 5.

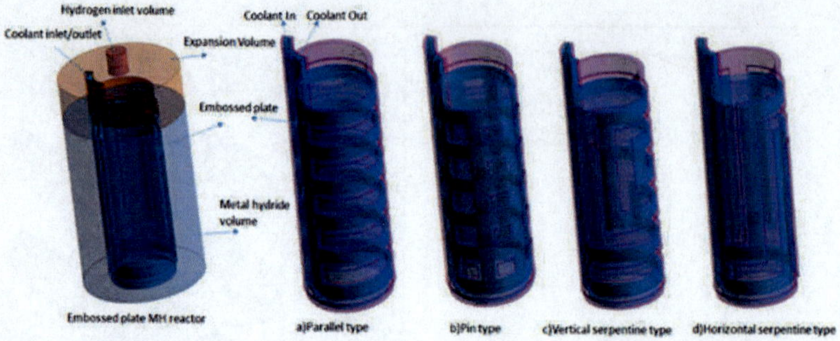

Fig. 3 Various embossed plate heat exchanger models **a** Parallel **b** Pin **c** Vertical serpentine, and **d** Horizontal serpentine

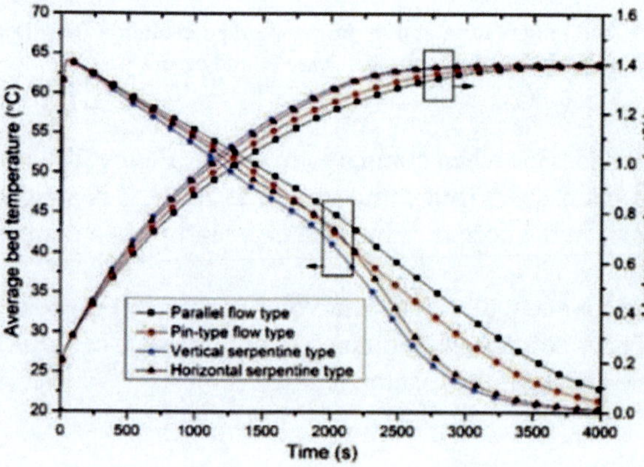

Fig. 4 Reactor bed temperature and hydrogen storage evolution comparison for various cases

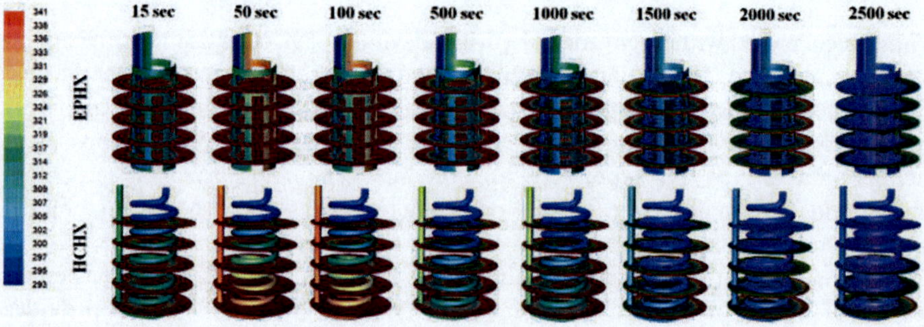

Fig. 5 Contour plot showing the temperature distribution in embossed plate heat exchanger model (EPHX) and helical coil heat exchanger model (HCHX)

At 1000 s, uniformity in the temperature distribution observed in EPHX when compared to HCHX. At 2500 s, HCHX has a higher temperature at the bottom whereas the entire bed is found to be at equilibrium in cases of EPHX. More uniformity is observed when using embossed heat exchanger due to availability of wider area for heat extraction.

4 Conclusion

Numerical modeled is developed and applied to two different configurations. A novel concept of compartmentation and embossment is shown for improvement in performance of the hydrogen storage. Following results is observed.

1. Compartmentation shows the importance of splitting the large volume of the reactor into smaller subparts which can be stacked together to achieve better performance than the single reactor.
2. Adding compartmentation in case 2 and case 3 took around 492 and 415 s which gave an improvement of 20 and 33% when compared to case 1.
3. Embossed plate heat exchanger imparts the application of embossment to supply coolant fluid. Using embossment techniques, various configuration can be designed.
4. Vertical serpentine configuration gave better performance than other configurations. Time for 90% hydrogen storage is 2236, 2092, 1863, and 1897 s for case of parallel type, pin type, vertical serpentine type, and horizontal serpentine type channel.
5. Uniformity throughout the MH bed can be obtained due to embossed plate acting as heat fin and distributing the heat from the reactor to the fluid.

Acknowledgements The research work was funded by the (VGST), Government of Karnataka.

References

1. Rivard E, Trudeau M, Zaghib K (2019) Hydrogen storage for mobility: a review. Materials (Basel) 12(12):1973
2. Dawood F, Anda M, Shafiullah GM (2020) Hydrogen production for energy: an overview. Int J Hydrog Energy 45(7):3847–3869
3. Wallace JS, Ward CA (1983) Hydrogen as a fuel. Int J Hydrog Energy 8(4):255–268
4. Zhang J, Fisher TS, Ramachandran PV, Gore JP, Mudawar I (2005) A review of heat transfer issues in hydrogen storage technologies. J Heat Transf 127(12):1391–1399
5. Abe JO, Popoola API, Ajenifuja E, Popoola OM (2019) Hydrogen energy, economy and storage: review and recommendation. Int J Hydrog Energy 44(29):15072–15086
6. Sun DW, Deng SJ (1990) A theoretical model predicting the effective thermal conductivity in powdered metal hydride beds. Int J Hydrog Energy 15(5):331–336

7. Suissa E, Jacob I, Hadari Z (1984) Experimental measurement and general conclusions on the effective thermal conductivity of powdered metal hydride. J Less-Common Metal 104(2):287–295
8. Patil SD, Ram Gopal M (2013) Analysis of a metal hydride reactor for hydrogen storage. Int J Hydrog Energy 38(2):942–951
9. Jemni A, Nasrallah SB, Lamloumi J (1995) Study of two-dimensional heat and mass transfer during absorption in metal-hydrogen reactor. Int J Hydrog Energy 20(1):43–52
10. Kaplan Y (2009) Effect of design parameters on enhancement of hydrogen charging in metal hydride reactors. Int J Hydrog Energy 34(5):2288–2294
11. Visari M, Mudawar I (2012) Coiled-tube heat exchanger for high-pressure metal hydride hydrogen storage systems—Part 1. Experimental study. Int J Heat Mass Transf 55(5–6):1782–1795
12. Chung CA, Chen YZ, Chen YP, Chang MS (2015) CFD investigation on performance enhancement of metal hydride hydrogen storage vessels using heat pipes. Appl Therm Eng 91:434–496
13. Bhouri M, Bu I, Linder M (2015) Numerical investigation of hydrogen charging performance for a combination reactor with embedded metal hydride and coolant tubes. Int J Hydrog Energy 40(20):6626–6638
14. Wu Z, Yang F, Zhu L, Feng P, Zhang Z, Wang Y (2016) Improvement in hydrogen desorption performances of magnesium based metal hydride reactor by incorporating helical coil heat exchanger. Int J Hydrog Energy 41(36):16108–16121
15. Lewis SD, Chippar P (2021) A numerical study of metal hydride reactor embedded with helical coil heat exchanger. Gruyter Ser Appl Math Eng Inf Sci 7:201–221
16. Dhaou H, Souahlia A, Mellouli S, Askri F, Jemni A, Nasrallah SB (2010) Experimental study of a metal hydride vessel based on a finned spiral heat exchanger. Int J Hydrog Energy 35(4):1674–1680
17. Eisapour AH, Naghizadeh A, Eisapour M, Talebizadehsardari P (2021) Optimal design of a metal hydride hydrogen storage bed using a helical coil heat exchanger along with a central return tube during the absorption process. Int J Hydrog Energy 40(27):14478–14493
18. Raju M, Kumar S (2012) Optimization of heat exchanger designs in metal hydride based hydrogen storage systems. Int J Hydrog Energy 37(3):2767–2778
19. Singh A, Maiya MP, Srinivasa MS (2017) Experiments on solid state hydrogen storage device with a finned tube heat exchanger. Int J Hydrog Energy 42(22):15226–15235
20. Ferekh S, Gwak G, Kyoung S, Kang H, Chang M, Yun S, Oh Y, Kim W, Kim D, Hong T, Ju H (2015) Numerical comparison of heat-fin- and metal-foam-based hydrogen storage beds during hydrogen charging process. Int J Hydrog Energy 40(42):14540–14550
21. Mellouli S, Dhaou H, Askri F, Jemni A, Nasrallah SB (2009) Hydrogen storage in metal hydride tanks equipped with metal foam heat exchanger. Int J Hydrog Energy 34(23):9393–9401
22. Ferekh S, Kyoung S, Ju H (2016) Hydrogen discharging and thermal behaviors of ZrCo based hydrogen storage beds equipped with heat-fins or metal-foam. Fusion Eng Des 109–111(Part A):884–889
23. Laurencelle F, Goyette J (2007) Simulation of heat transfer in a metal hydride reactor with aluminium foam. Int J Hydrog Energy 32(14):2957–2964
24. El Mghari H, Huot J, Xiao J (2019) Analysis of hydrogen storage performance of metal hydride reactor with phase change materials. Int J Hydrog Energy 44(54):28893–28908
25. Darzi AAR, Afrouzi HH, Moshfegh A, Farhadi M (2016) Absorption and desorption of hydrogen in long metal hydride tank equipped with phase change material jacket. Int J Hydrog Energy 41(22):9596–9610
26. Maad HB, Askri F, Nasrallah SB (2016) Heat and mass transfer in a metal hydrogen reactor equipped with a phase-change heat-exchanger. Int J Therm Sci 99:271–278
27. Mellouli S, Askri F, Abhilash E, Nasrallah SB (2017) Impact of using a heat transfer fluid pipe in a metal hydride-phase change material tank. Appl Therm Eng 113:554–565
28. Mellouli S, Abhilash E, Askri F, Nasrallah SB (2016) Integration of thermal energy storage unit in a metal hydride hydrogen storage tank. Appl Therm Eng 102:1185–1196

29. Alqahtani T, Mellouli S, Bamasag A, Askri F, Phelan PE (2020) Thermal performance analysis of a metal hydride reactor encircled by a phase change material sandwich bed. Int J Hydrog Energy 45(43):23076–23092
30. Elarem R, Alqahtani T, Mellouli S, Edacherian A, Askri F, Jemni A (2021) Numerical analysis of a built-in thermal storage system of metal hydride and nanoparticles enhanced phase change material and nanofluid. Int J Energy Res 45(4):5881–5893
31. Lewis SD, Chippar P (2020) Analysis of heat and mass transfer during charging and discharging in a metal hydride—phase change material reactor. J Energy Storage 33:102108
32. Chippar P, Lewism SD, Rai S, Sircar A (2018) Numerical investigation of hydrogen absorption in a stackable metal hydride reactor utilizing compartmentalization. Int J Hydrog Energy 43(16):8007–8017
33. Lewis SD, Chippar P (2020) Numerical investigation of hydrogen absorption in a metal hydride reactor with embedded embossed plate heat exchanger. Energy 194:116942
34. ANSYS Fluent theory guide, chapter 7.2.3
35. Jemni A, Nasrallah SB, Lamloumi J (1999) Experimental and theoretical study of a metal–hydrogen reactor. Int J Hydrog Energy 24(7):631–644

Simulation Analysis of Wind Turbine Generator System

Prakash Malaiyappan, A. Buckshumiyan, and K. R. Shanmuga Vadivu

Abstract The aim of this paper is to investigate the performance of induction generator and synchronous generator ratings in the wind turbine system. Both induction and synchronous generator rating are compared with help of MATLAB Simulink software. The best suitable for model and analyzing various ratings of induction and synchronous generators is Simulink library function. The result obtained from analysis is the induction generator rating (200 HP, 460 V, 50 Hz, 1765 RPM) which is gaining more attention than the synchronous generator rating. Induction generators are low cost of manufacturing, and maintenance is simple compared to synchronous generators.

Keywords Induction generator · Synchronous generator · MATLAB Simulink · Library function

1 Introduction

B. Chitti Babu and K. B. Mohanty projected the application of wind turbine-driven induction generator with the two directional power flow analysis. Advancement in induction machine with the pulse width modulation and dynamic vector approach are described in detail [1]. Y. D. Song et al. investigated automatic control which represents one of the most important factors responsible for the efficiency. The new

P. Malaiyappan (✉)
Mechanical Engineering, St. Peter's Institute of Higher, Education and Research, Chennai, Tamil Nadu, India
e-mail: prakashvictorz@gmail.com

A. Buckshumiyan
Mechanical Engineering, New Prince Shri Bhavani College of Engineering & Technology, Chennai, Tamil Nadu, India

K. R. S. Vadivu
Electrical and Electronics Engineering, B.S. Abdur Rahman Crescent Institute of Science and Technology, Chennai, Tamil Nadu, India

N. M. Sivaram et al. (eds.), *Advances in Manufacturing, Automation, Design and Energy Technologies*, Lecture Notes in Mechanical Engineering,
https://doi.org/10.1007/978-981-99-1288-9_69

method of nonlinear and adaptive control technology in induction generator used to improve the performance of the wind turbine [2]. Ramtharan et al. [3] proposed the options for low voltage ride through (LVRT) are discussed for a wind turbine with a synchronous generator and fully rated converter. A combination of torque control and braking resistor helps to ride through multiple successive grid faults without shutting down the wind turbine. M. N. Mansouri et al. proposed that feed control analysis consists of asynchronous generator control amplifier to avoid the speed the variation. The major failure system occurs only due to the fluctuation induced in voltage and vibration impact of the generator reified by the RL circuit induced in the line current of the energy motion transf. The optimum pitch angle of asynchronous generator generally $\phi \sim 6°$ for input voltage = 9 m/s, $\phi \sim 12°$ for V_{in} = 18 m/s and $\phi \sim 34°$ for input voltage = 28 m/s varied according to the geometrical climate of the country [4].

Anastasios Balouktsis et al. enlightened the degree of accuracy depends on constructional analyses as concerns suitability and employment of the human worker [5]. Z. Hameeda et al. progressives the various arrangements fault detection procedures and set of rules advanced to monitor the performance of overall system [6]. Dadone and Dambrosio [7] proposed the fuzzy logic control procedures to increase the efficiency of the wind turbine. The proposed results of various wind turbine generators are analyzed [8–15].

2 Simulink Model

2.1 Simulation of Induction Generator

2.1.1 Parameters of Induction Generator

See Table 1.

2.1.2 Simulink Library Browser of Induction Generator

For the first cycle of simulation, the held constant to the value specified by the initial magnitude of input parameter. Fundamental frequency used is 50 Hz. Asynchronous type mechanical input given torque with respect to stator or rotor reference frame. Figure 1 shows three phase convertor MOSEFT used in circuit to reduce the poor power factor, power fluctuation output. For the above Table 1 shows comparing with varies rating (HP), voltage (V), frequency (Hz), and speed (RPM) configuration of model.

Table 1 Represent the parameters of the induction generator

S. No.	Rating (HP)	Voltage (V)	Frequency (Hz)	Speed (RPM)
1	5	460	50	1740
2	10	460	50	1745
3	20	460	50	1745
4	50	460	50	1765
5	100	460	50	1765
6	150	460	50	1770
7	200	460	50	1765
8	5	575	50	1745
9	10	575	50	1745
10	20	575	50	1750

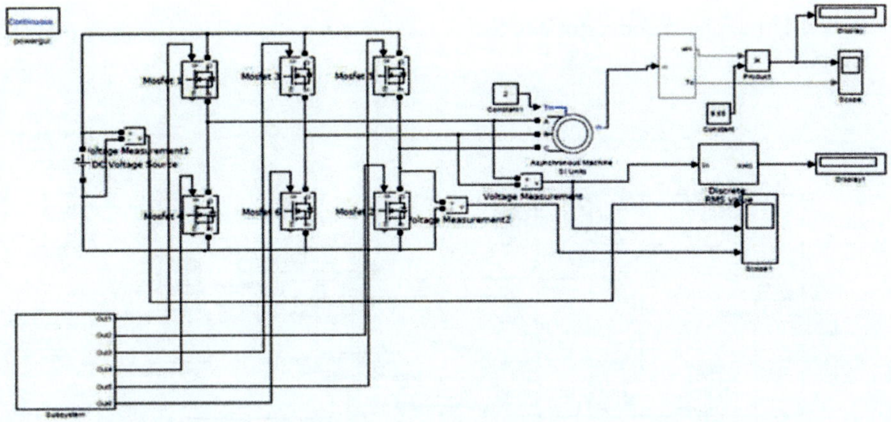

Fig. 1 Simulink model of induction generator

2.2 *Simulation of Synchronous Generator*

2.2.1 Parameters of Synchronous Generator

See Table 2.

2.2.2 Simulink Library Browser of Synchronous Generator

For the first cycle of simulation, the held constant to the value specified by the initial magnitude of input parameter. Fundamental frequency used is 50 Hz. Figure 2 shows asynchronous type mechanical input given torque with respect to stator or rotor reference frame. For the above Table 2 comparing with varies configuration

Table 2 Represent the parameter of the synchronous generator

S. No.	Rating (KVA)	Voltage (V)	Frequency (Hz)	Speed (RPM
1	8.1	400	50	1500
2	16	400	50	1500
3	31.3	400	50	1500
4	42.5	400	50	1500
5	60	400	50	1500
6	85	400	50	1500
7	250	400	50	1500
8	325	400	50	1500
9	670	400	50	1500
10	910	400	50	1500

of model. Three phase convertor MOSEFT used in circuit to reduce the poor power factor, power fluctuation output.

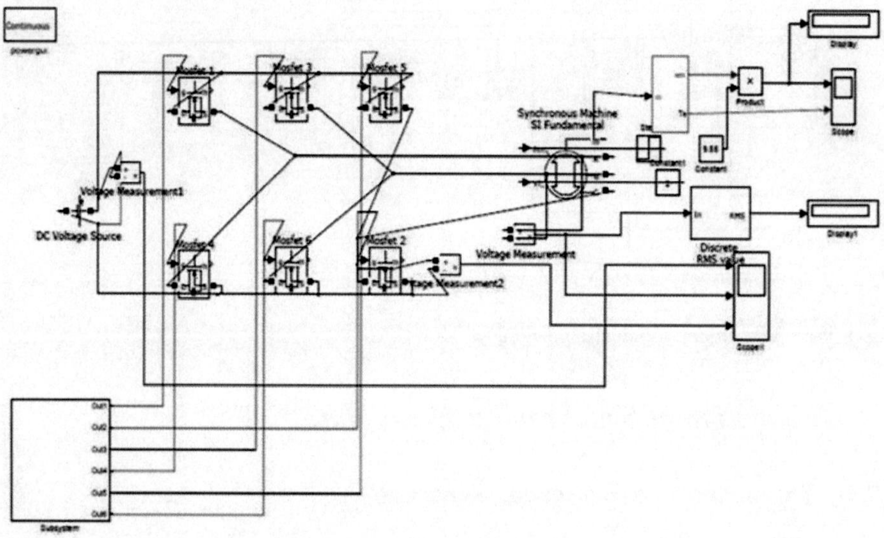

Fig. 2 Simulink model of synchronous generator

3 Results and Discussion

Figure 3 shows simulation of (a) input and (b) output voltage and time results of induction generator, respectively, of 5 HP, 460 V, 50 Hz, 1765 RPM is the best result with compared other result of Table 1.

Figure 4a shows simulation of torque (Nm), (b) speed (rpm) and time (s), respectively, torque required produce to the high power factor and low power fluctuation output, Fig. 4b shows simulation of speed (rpm) and time high wind energy conversion efficiency and low mechanical stress on turbine components.

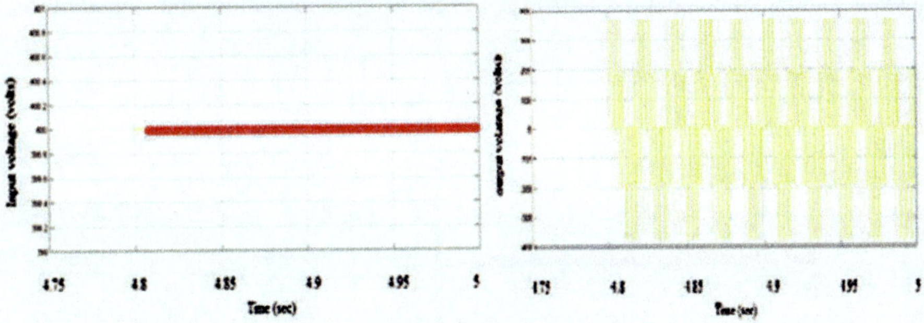

Fig. 3 **a** Input voltage (v) and time (s) **b** output voltage (v) and time (s)

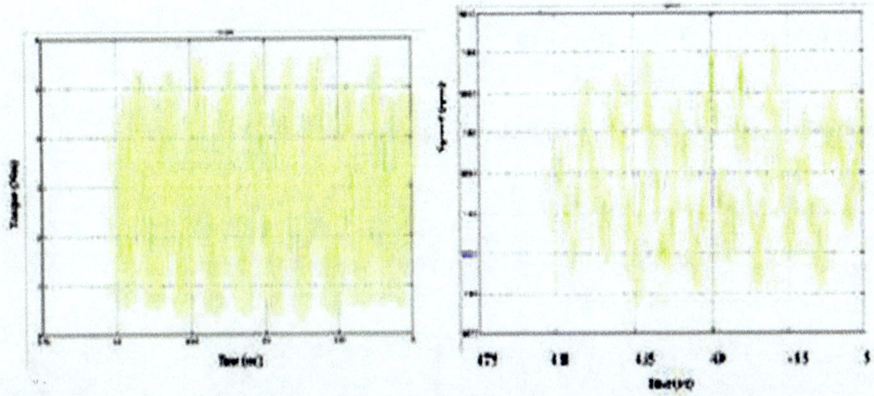

Fig. 4 **a** Torque (Nm) and time (s) **b** speed (rpm) and time (s). Time starts after 4.8 sec in x axis

3.1 Simulation Results of Synchronous Generator

Figure 5 shows simulation of (a) input and (b) output voltage and time results of synchronous generator, respectively, of 8.1 KVA, 400 V, 50 Hz, 1500 RPM is the best result with compared other result of Table 2.

Figure 6a shows simulation of torque (Nm), (b) speed (rpm) and time (s), respectively, torque required produce constant speed from no load to full load and high power factor power fluctuation output. Electro-magnetic power varies linearly with voltage.

Figure 6b shows simulation of speed (rpm) and time low wind energy conversion efficiency and high mechanical stress on turbine components. This type of generation

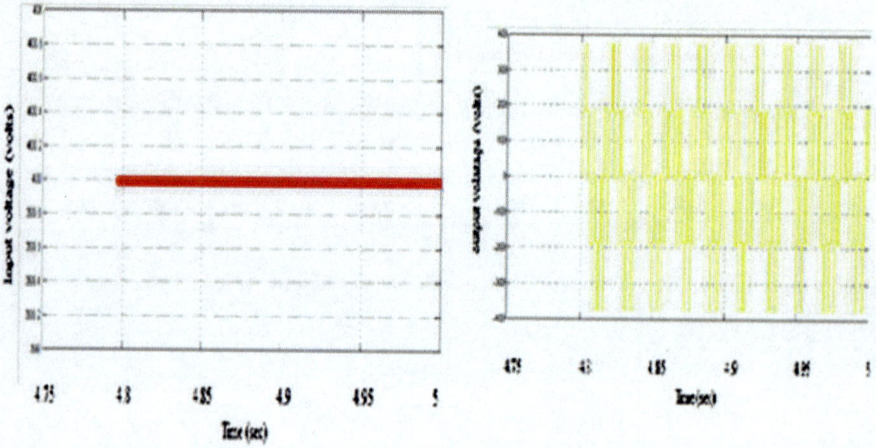

Fig. 5 **a** Input voltage (v) and time (s) **b** output voltage (v) and time (s)

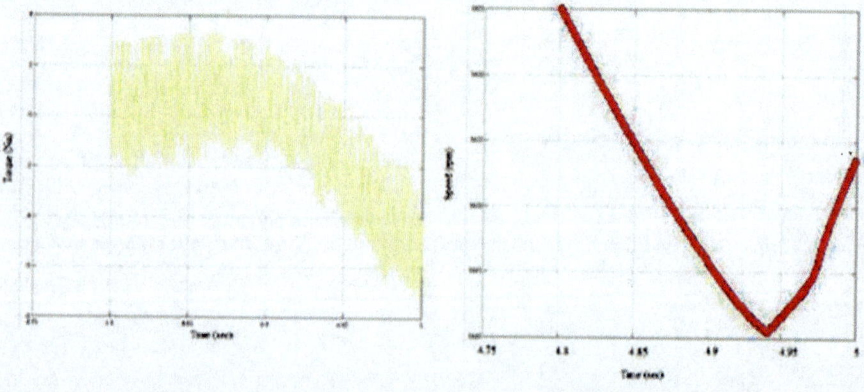

Fig. 6 **a** Torque (Nm) and time (s) **b** speed (rpm) and time (s). Time starts before 4.8 sec in x axis

cannot be used for variable speed job as there is no possibility of speed adjustment. This type of generation cannot be started under loaded condition. Its starting torque is zero.

3.2 Discussion

Input voltage given to the induction generation 400 V and corresponding output obtained is sinusoidal AC output. By the use of MOSFET converter reduce the power fluctuation and low mechanical stress. For the 5 HP, 460 V, 50 Hz simple in construction and low maintenance, extract sinusoidal AC output is produced with minimum loss. Mechanical variable torque given to the parameter corresponding variable speed is obtained from the torque. Variable speed is best suitable for the induction generator. Input voltage given to the synchronous generation 400 V and corresponding output obtained is sinusoidal AC output. By the use of MOSFET converter reduce the power fluctuation and low mechanical stress. For the 8.1 KVA, 400 V, 50 Hz simple in construction and low maintenance, extract sinusoidal AC output is produced with minimum loss. Mechanical variable torque given to the parameter corresponding synchronous speed is obtained from the torque. Synchronous speed is best suitable for the synchronous generator.

3.3 Conclusion

For the induction generator 5 HP, 460 V, 50 Hz, 1765 RPM is best suitable for the application compared with the other parameter of the generator simulation done on Table 1. Hence for required torque give high wind energy conversion efficiency, high power factor, low power fluctuation output, low mechanical stress on turbine components in asynchronous speed. The comparison between the two systems was 50 Hz only.

For the synchronous generator 8.1 KVA, 400 V, 50 Hz, 1500 RPM is best suitable for the application compared with the other parameter of the generator simulation done on Table 2. Hence for required torque give high wind energy conversion efficiency, high power factor, low power fluctuation output, low mechanical stress on turbine components in synchronous speed. From the simulation result, induction generators are simple and low cost in manufacturing and low maintenance. Synchronous generators give constant speed from no load to full load. Electromagnetic power varies linearly with voltage. Main drawback of the synchronous generator cannot be started under loaded condition. Hence, the starting torque is zero.

References

1. Chitti Babu B, Mohanty KB (2010) Doubly fed induction generator for wind energy conversion systems modelling & simulation. Int J Comput Electr Eng 2:1793–8163
2. Song YD, Dhinakaran B, Bao XY (2000) Variable speed control of wind turbines using nonlinear and adaptive algorithms. J Wind Eng Ind Aerodyn 85:293–308
3. Kristoffersen JR, Christiansen P (2003) Horns Rev offshore wind farm: its main controller and remote control system. Wind Eng 27(5):351–360
4. Mansouri MN, Mimouni MF, Benghanem B, Annabi M (2004) Simulation model for wind turbine with asynchronous generator interconnected to the electric network. Renew Energy 24:421–431
5. Anastasios B, Dimitrios C, Thodoris DK (2002) A nomogram method for estimating the energy produced by wind turbine generator. Sol Energy 72:251–259
6. Muyueen SM (2011) Variable speed wind turbine generator system with current controlled voltage. Energy Convers Manag 52:2688–2694
7. Dadone A, Dambrosio D (2003) Estimator based adaptive fuzzy logic control technique for a wind turbine generator system. Energy Convers Manag 44:135–153
8. El Mokadem M, Courtecuisse V, Sauidemont C, Robyns B, Deuse J (2009) Experimental study of variable speed wind generator contribution to primary frequency control. Renew Energy 34:833–844
9. No TS, Kim E, Moon JH, Kim SJ (2009) Modeling, control, simulation of dual rotor wind turbine generator system. Renew Energy 34:2123–2132
10. Iqbal AR, Mann T, Iqbal KIG (2008) Controller for a small induction-generator based wind-turbine. Appl Energy 85(4):218–227
11. Miao Z, Fan L (2008) The art of modeling and simulation of induction generator in wind generation application using high-order model. Simul Model Pract Theory 16:1239–1253
12. Duenas L, Basu B (2008) Unavailability of wind turbine due to wind induced accelerations. Eng Struct 30:885–893
13. Schatzer C, Muller W, Binder A (1999) Vector optimization of two-dimensional numerical field problems applied to the design of a wind turbine generator. Math Comput Simul 49:105–117
14. Senthil Kumar N, Gokulakrishnam J (2011) Impacts of facts controller on the stability of power systems connected with doubly fed induction generators. Electr Power Energy Syst 33:1172–1184
15. Pulgar H, Sauer PW (2011) Reduced order model of Type C wind generators. Electr Power Syst Res 81:840–845

Bio-plastic Preparation Using Potato, Corn and Rice-Based Starches

G. Senthilkumar, A. Mubeen Banu, and M. Prasanth

Abstract Plastics can be used effectively in many applications only when proper procedures to be addressed to decompose them effectively. Increasing environmental concerns strain on plastic waste and broad interest on fostering the eco-accommodating procedures. The starch-based polymers are the promising option in contrast to the petrol-based plastics. These items not just can possibly be biodegradable or potentially recyclable, yet completely practical across all aspects of their life cycle. To decrease the interest on single source starch, this current research is aimed to set up various starches and consolidating them for bio-plastic readiness. So forth the principle objective of undertaking is to plan biodegradable plastic by utilizing starches (potato, rice and corn–in mixed ratio) and utilizing the bio item for buyer use. The current works yield will be bio-plastics by utilizing corn, potato and rice starch. Then the results pertaining to thickness test, water solubility test, moisture content test and biodegradable also carried out to improve the properties of bio-plastics. The outcome showed starch-based bio-plastic for different applications.

Keywords Biodegradable plastics · Environmental issues · Starch

1 Introduction

From recent years, a ton of examination has been taken to track down the reasonable option however the right organization is as yet considered. Bio-plastics are inexhaustible and eco cordial. Many explores were finished by utilizing starch, cellulose like. The huge utilization and amassing of plastics have become one of the serious issues all through the world [1]. In light of these issues, there has been impressive interest in the turn of events, creation of biodegradable microbial bio-plastics [2]. Starch has a high substance of amylose, this adds to the film strength of the polymer and on the off chance that the starch contains spread amylopectin, this will

G. Senthilkumar (✉) · A. Mubeen Banu · M. Prasanth
Sathyabama Institute of Science and Technology, Chennai, India
e-mail: Tosenthilgs79@gmail.com

N. M. Sivaram et al. (eds.), *Advances in Manufacturing, Automation, Design and Energy Technologies*, Lecture Notes in Mechanical Engineering,
https://doi.org/10.1007/978-981-99-1288-9_70

prompt a film with low mechanical properties can be overhauled through plasticizers like glycerol and sorbitol [3]. Some constraint is seen about unique microbial strains when contrasted with the recombinant organisms for the amalgamation of novel polyesters. Bio-plastic materials, because of exceptional qualities and expansive biotechnological application, offer a very encouraging future [4, 5]. The removal issues related with traditional oil-based plastics comprise grave natural hazard across the globe considering the sums produced as waste consistently. Starch has been fused into regular plastics trying to give some degree of biodegradability on the subsequent composites. In a ternary (three-stage) composite framework, there are two prospects; these incorporate (1) both the filler and the minor parts being independently scattered in the consistent lattice; or (2) the minor part embodying the filler particles. The greater part of the work done is on oat and root/tuber crops [6, 7]. There is consequently the need to extend the extent of this examination by accepting different wellsprings of starch [8, 9]. It was likewise seen that the PHEE interlayer (covering) on the starch granules raised the de bonding stress span which thusly converted into better tractable and yield qualities for the starch-PE-PHEE mix. It is obvious that a considerable amount has been done in the improvement of starch-based items in the move toward biodegradable plastics [10].

2 Problem Identification

Plastic is highly resistant to degradation, which makes it popular material and used for various departments like packaging and engineering field [11]. Non-biodegradability of the plastic which creates many difficulties in environment both rural and urban countries. Usual problems like stifling of channels, stagnation of water and delivery poisonous gases due to open incineration. While disposing by landfill method will spoils the quality of ground water and soil also. The water and soil pollution plays the major role on environmental causes. The need to depend on reasonable, rely less upon fossil items and fossil fuel byproducts leads on the necessity of biomaterials and bioplastics as substitutes of oil-based plastics. Because of the environmental impacts on fossil plastics, the development of biodegradable plastics for both commercial and industrial applications is essential nowadays. This experimentation shows the possibilities on starch-based bio-plastic creation and performing water soluble test, moisture content test and biodegradability test with results. The main aim of this research was to use lower cost products such as potato, corn and rice for the production of bio-plastic, which can help to reducing environmental issues that are caused by oil-based bio-plastic. Bio-plastics are renewable and ecofriendly. This bio-plastic preparation process must be a easy and accessible form for all. The different physical, mechanical and flexibility properties are found in bio-plastics. Because of the low value, inexhaustibility and having good mechanical properties, it is utilized to deliver decomposable polymers. The common concern of the public regarding plastic waste products prompted us to undertake an overview of the biodegradation of plastic products [12].

3 Experimental Methodology

On bio-plastic preparation what we are going to discuss is shown in Fig. 1 process flow chart available below. We are starting in on to set up the biodegradable film, by extricating the example of starch from different ecological amicable merchandise of reproducible things like potatoes, rice and corn [13]. The accompanying bit by bit technique, clarifies the materials and strategies are associated with the planning of various starch by person. The required ingredients for preparing starch is available in Fig. 2 as well as the required items for preparing bio-plastic film.

From potatoes only potato starch is extracted. The root tubers of the potato plant had leucoplasts itself which have an ability to produce the grains of starch. For starch extraction first potatoes are crushed and the starch grains are released from the obliterated cells. The washed starch is then taken out and powder is allowed to dry. The schematic process flow chart to produce starch identify and select the great

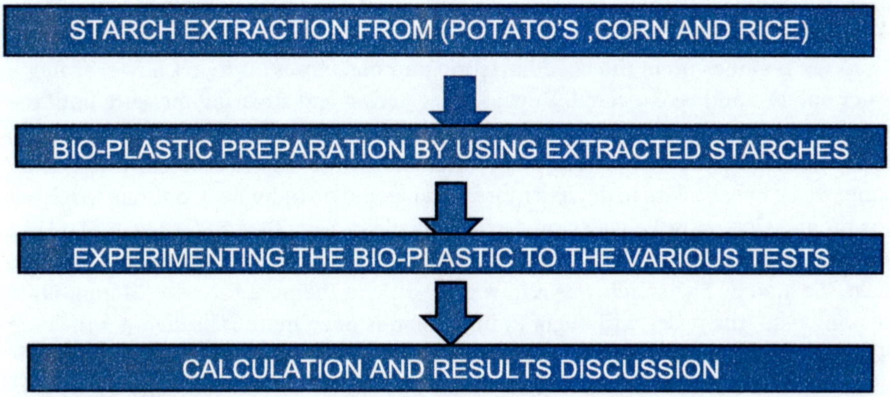

Fig. 1 Process flow chart

Fig. 2 Required items for preparing starch

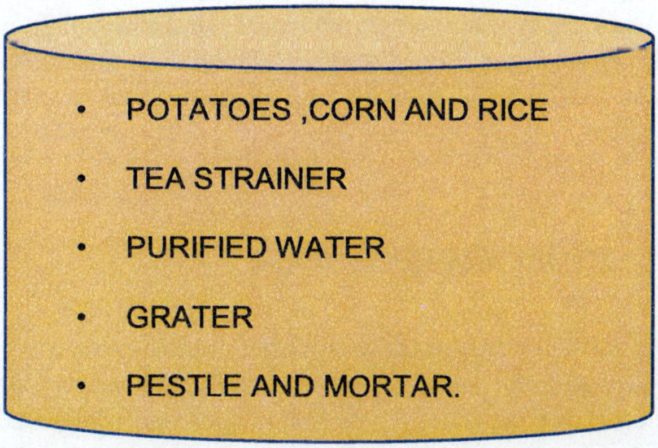

quality potatoes. Scrub the potatoes clean first utilizing warm water and an inferior brush. The principle reason for cleaning in technique for planning of potato starch is to eliminate silt from the external layer of potato and eliminate the epidermis of potato. Cleaning unrefined substance stage is the premise assurance of starch quality. The cleaner the starch, the better the starch quality. Scrap the perfect surface of potatoes with a blade and cut into pieces and spot them in a blend and mix for 1–2 min. Crushing serves to annihilate the design of the potato cell, with the goal that the small starch granules can be flawlessly isolated from potato. The pounding measure is rehashed three to multiple times until we get a homogenous granular particles however much as could reasonably be expected by break the cells of the material and delivery all the more free starch granules. Starch extraction, otherwise called mash partition, is a critical connection in technique for planning of potato starch, straightforwardly influencing starch extraction rate and starch quality. The squashed material is in the condition of fine fiber and isolated from the starch by utilizing water as a medium. Scoop the ground potatoes into a pot, and afterward cover them with enough amount of warm water. Place a sifter sort of cotton fabric or cheddar material onto another pot. Then, at that point, empty the ground splashed potatoes into it. Wrap the potatoes up in the cheddar fabric and make press it to get however much water out as could be expected. Repeat the watering and stressing measure until the water runs clear. Spot the potatoes back into the vacant pot. Cover them with all the more warm water, and afterward empty them back into the cheesecloth-lined sifter. Squeeze the cheesecloth to deplete more water as shown in figure. Continue to rehash this progression until the water runs generally clear. Save the stressed water and put the ground potatoes away. Furthermore, we just need the stressed potato water to make the starch. The starch was allowed to settle at the base for next 20 min. Over the long haul, the water will begin to turn clean is once more. The starch will settle at the lower part of the pot. The water might take on a brown or orange color. This is entirely fine. Pour out the abundance water. The starch will be generally adhered to the lower part of the pot, discard this abundance water by separating it from top. Add new water to the holder and stand by an additional 20 min or thereabouts. Empty some more water into the holder attempt to utilize the very sum that you spilled out. Allow the starch to settle to the base, an additional 20 min least. This new water will clean the muck. Pour the water off and save the starch. Utilize a similar method as you did the initial time. By and by, be mindful so as not to blend the starch into the water.

4 Result and Discussion

Initially, 100 mL water was poured in the beaker. Then weighing the starch powder was performed accurately. Additionally, 15 g starches powder was added for preparation of mixture. Then 5 g each starch powders (potato, corn and rice) was mixed properly for uniform porosity. Once again the weight of mixed starches was confirmed. Append the 15 mL glycerol. Add 0.1 mL of hydro-chloric acid. Figure 3 showed the

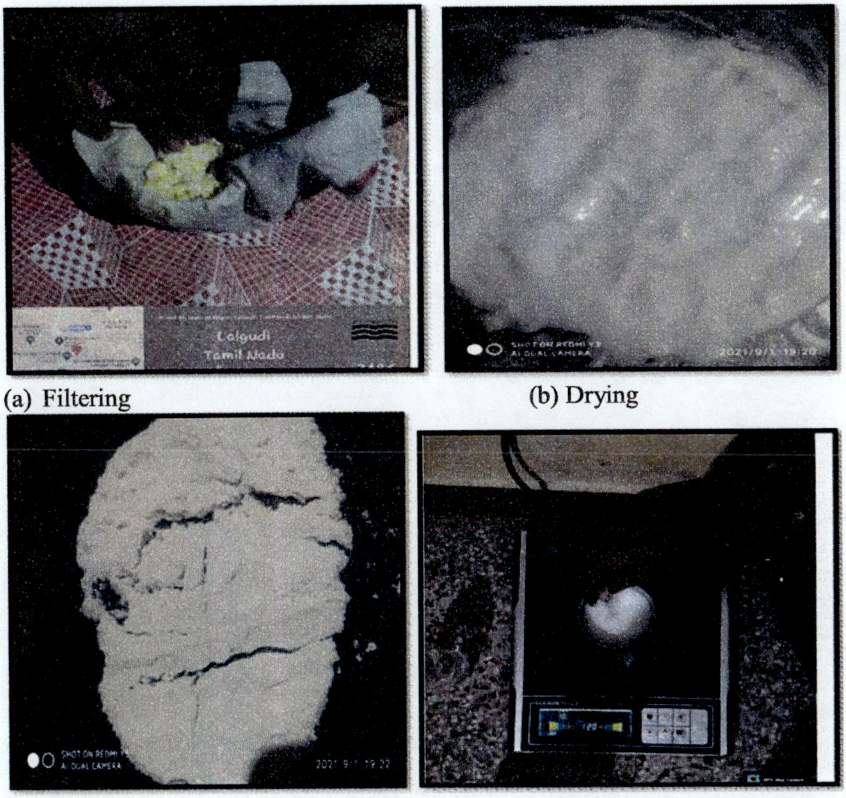

(a) Filtering (b) Drying

(C) Dried powder (D) Heating

Fig. 3 Bio-plastic preparation at different phases

photo gallery of addition process procedure of bio-plastic preparations. The sample solution was stirred by manually before heating. By using the burner, the mixture heated on 120 °C for 15 min followed by mechanical stirring. After 15 min of heating and stirring, the mixture was examined to pH value. The universal indicating paper was used to check the acidic value of prepared solution. While examining the solution with universal indicating paper, the indicating paper indicated green color. So, the solution was free from acidic value.

5 Conclusion

Biodegradability of this prepared bio-plastics is 18.42% in a 7 days. With this proportion, the achieved biodegradability was about 78.94% in a one month. According to the Government of India regulations, the plastic bag thickness should not be less than

50 microns. The average thickness of our prepared bio-plastics was 1.47 mm. Water solubility of the prepared bio-plastics was 61.29% which enhanced the biodegradable and dissolving property of the bio-plastics. Moisture content of the prepared bio-plastics was 26.48%. The count of citric acid and moisture content which advances the life of plastic film and progress in its mechanical properties. By adopting the methodology properly, it would be possible to prepare waste carrying bags, medical plastic papers and catering service wares using this bio-plastics with minimum expenditure.

Disclosure Statement No conflict of interests.

References

1. Bayer IS, Guzman-Puyol S, Heredia-Guerrero JA, Ceseracciu L, Pignatelli F, Ruffilli R, Cingolani R, Athanassiou (2014) Direct transformation of edible vegetable waste into bioplastics. Macromolecules 47:5135–5143
2. Emadian SM, Onay TT, Demirel (2017) Biodegradation of bioplastics in natural environments. Waste Manag 59:526–536
3. Jain R, Tiwari A (2005) Biosynthesis of planet friendly bioplastics using renewable carbon source. J Environ Health Sci Eng 13:11
4. Azahari N, Othman N, Ismail H (2011) Biodegradation study of polyvinyl alcohol—corn starch blend films in solid and solution media. J Phys Sci 22:15–31
5. Siakeng R, Jawaid M, Ariffin H, Sapuan S, Asim M, Saba N (2019) Natural fiber reinforced polylactic acid composites: a review. Polym Compos 40:446–463
6. Muscat D, Adhikari B, Adhikari R, Chaudhary D (2012) Comparative study of film forming performance of low and high amylose starches using glycerol and xylitol as plasticizers. J Food Eng 109:189–201
7. Johansson C, Bras J, Mondragon I, Nechita P, Plackett D, Simon P, Svetec DG, Virtanen S, Baschetti MG, Breen C (2012) Renewable fibers and bio-based materials for packaging applications—a review of recent developments. BioResources 7:2506–2552
8. Fakhouri FM, Costa D, Yamashita F, Martelli SM, Jesus RC, Alganer K, Collares-Queiroz FP, Innocentini-Mei LH (2013) Comparative study of processing method s for starch/gelatin films. Carbohydr Polym 95:681–689
9. Schirmer M, Höchstötter A, Jekle M, Arendt E, Becker T (2013) Physicochemical and morphological characterization of different starches with variable amylose/amylopectin ratio. Food Hydrocoll 32:52–63
10. Borges J, Romani V, Cortez-Vega W, Martins V (2015) Influence of different starch sources and plasticizers on properties of bio degradable films. Int Food Res J 22:2346–2351
11. Podshivalov A, Zakharova M, Glazacheva E, Uspenskaya M (2017) Gelatin/potato starch edible bio composite films: correlation between morphology and physical properties. Carbohydr Polym 157:1162–1172
12. Gómez-Heincke D, Martínez I, Stading M, Gallegos C, Partal P (2017) Improvement of mechanical and water absorption properties of plant protein based bioplastics. Food Hydrocoll 73:21–29
13. Kulshreshtha Y, Schlangen E, Jonkers H, Vardon P, Van Paassen L (2017) Corncrete: a corn starch based building material. Constr Build Mater 154:411–423

Cooling of Gearbox Oil Using Peltier Module

M. Purusothaman, A. Karthikeyan, V. Srinivasa Rao, and J. Vales Thomson

Abstract The industrial gearbox requires better lubrication to reduce the friction. Better lubrication demands better cooling of oil. Cooling is the heart of the lubrication system, if it fails then it results in worn of gears. Temperature is one of the parameters that largely decides the properties of the oil. The temperature of the oil increases after the lubrication of gears. To cool gear oil, large size of coolers is used or need to be replaced continuously. This project introduces a new, compact, and efficient way of cooling using Peltier's principle. This is accomplished by incorporating Peltier's module. It can be observed that there is a massive amount of research being done in this topic about thermoelectric materials, but in this case if use multiple connections of Peltier with correct specification then its efficiency is slightly increase due to its compatibility it consume smaller in space. In addition, the heat extracted from the oil is used for preheating. If Peltier is built large in size means, then it can used to large size industrial gearbox, and solar power is used to power the Peltier module.

Keywords Peltier · Cooling · Gearbox oil · Thermoelectric materials

1 Introduction

The design of this prototype is based on thermo-cooling because the target is to cool down industrial gearbox oil. We use Peltier modules in our project to cool the gearbox oil [1]. When using copper or aluminum heat sinks in a Peltier module then it will increase cost, but still it can overcome the disadvantages of existing large external coolers which consume more space and high construction cost and maintenance cost [2]. So in this project, heat sinks also need to be selected wisely to increase efficiency. And also correct specifications of Peltier modules also need to be selected, so we

M. Purusothaman (✉) · A. Karthikeyan · V. Srinivasa Rao · J. Vales Thomson
School of Mechanical Engineering, Sathyabama Institute of Science and Technology, Chennai, India
e-mail: purusothmani@gmail.com

N. M. Sivaram et al. (eds.), *Advances in Manufacturing, Automation, Design and Energy Technologies*, Lecture Notes in Mechanical Engineering,
https://doi.org/10.1007/978-981-99-1288-9_71

have to study about Peltier also. Due to the usage of Peltier in this project, we have to check the performance of Peltier [3–5].

1.1 Heating of Industrial Gearbox Oil

A number of factors can contribute to overheating in your industrial gearbox. Factors to consider include operating temperature, ambient temperature, viscosity index, load conditions, contamination levels, lubrication requirements, humidity, air velocity, and air temperature [6]. So cooling of gearbox is absolutely necessary and cooling the oil with traditional forced lubrication which causes more cost and space, etc. So in this project, we used Peltier module to cool the gear oil temperature [7].

1.2 Design of Gearbox Cooling Using Peltier Module

1.2.1 Thermoelectric Effect

The thermoelectric effect is the direct conversion of temperature contrasts to electric voltage and vice versa via a thermocouple. A thermoelectric device generates a voltage when the temperatures on each side are different [8]. The term "thermoelectric impact" refers to three independently recognised impacts: the Seebeck impact, the Peltier impact, and the Thomson impact.

1.2.2 Peltier Effect

Peltier plant there is a contradictory wonder to the Seebeck impact, in which atomic force could be burned through at one different pith intersection and delivered at the additional junction while an electric inflow spilled inside the congested circuit [9].

The circuit is intended to acquire a unlike setup that demonstrates the PE, a fantastic inverse of the SE as shown in Fig. 1. If a voltage (Ein) is functional to outstations T1 and T2, an electrical inflow (I) will flow through the circuit [10]. As a result of the inflow conduit, there will be an insignificant cooling sway (QC) at thermocouple intersection A and a warming effect (QH) at intersection B. Joule warming, with a degree of I2 x R, also occurs in the transmitters when the inflow conduit is considered. This Joule warming effect works in opposition to the PE resulting in a net decrease in the accessible cooling.

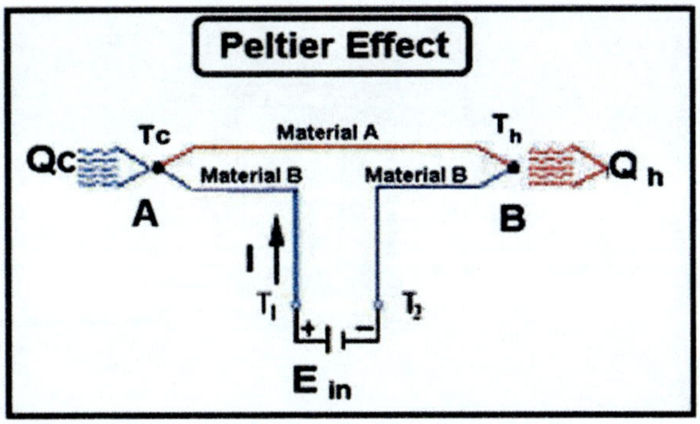

Fig. 1 Peltier effect

2 Working of Peltier

The Peltier impact happens at whatever point electrical flow courses through two different transmitters; contingent upon the heading of the flow stream, the junction of both channels will either hold or convey heat [11]. SCs are the substance of alternative for conveying the Peltier outgrowth in the domain of thermoelectric creation because they can be improved even more effectively for siphoning heat. In this plan, the other far off material is the bobby affiliation approaches to the power to have PE as shown in Fig. 2.

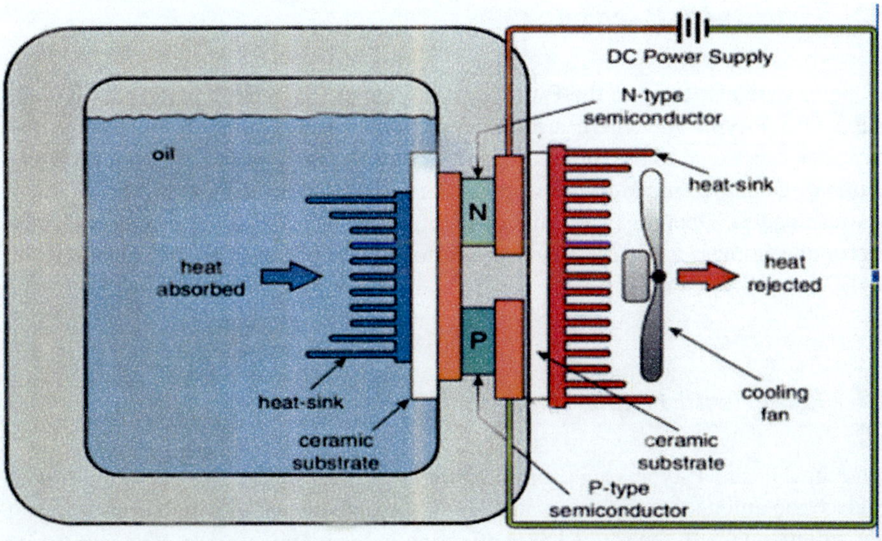

Fig. 2 Model setup of prototype

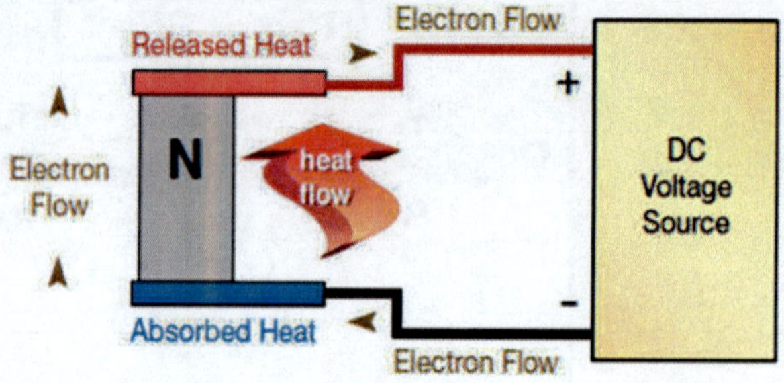

Fig. 3 Peltier with N-model SC

2.1 Peltier with N-model Semiconductor (SC)

The projectile is made of N-model SC material, so electrons with a -ve charge will charge the transporter used to make the primary piece of the PE. In the Fermi position of the N-model SC, a new electron has been introduced (progressed energy position). DC source-related electrons will rebuff the negative post and pull in the positive shaft of the power; in light of this enticing quality [9, 12]. As shown in Fig. 3, electrons at position shift toward the positive workstation by conveying hotness and creating openings in the Fermi position. Currently, electrons shift from the valence band to the Fermi position by holding power from the intersection due to an amicable supply of current [13].

2.2 Peltier with P-model SC

SC pellets are also used in the P-model as shown in Fig. 4 depicts the energy band graph of a P-type SC. There are openings at the Fermi point in this. When DC current is functional to the route, openings are drawn in toward the source's adverse terminal. Openings move to the negative terminal by delivering heat as a result of this fascination. Openings from the conduction band move to Fermi point through engrossing hotness as of the intersection due to the constant inventory of current [14].

2.3 Peltier with P- and N-model SC

Planning N- and P-type slugs that will be connected as couple and outlining an intersection among them with a plated tab, it's promising to construct a string circuit that will maintain all the roguishness affecting in the identical route. The free (base) finish of the P-model slug is associated with the positive voltage possibility, while the at no cost (base) finish of the N-model projectile is associated with the negative voltage possibility. It is possible to convey roguishness and hold from a different

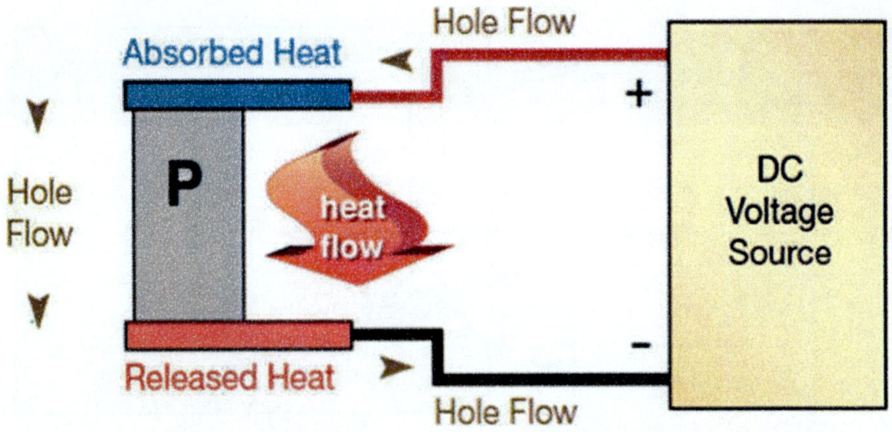

Fig. 4 Peltier with P-model SC

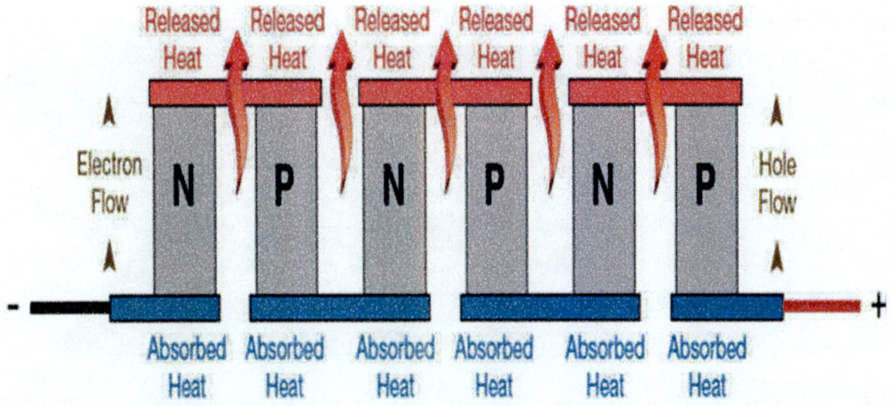

Fig. 5 Peltier by multiple pallets

side. The TE couples excellent packages as shown in Fig. 5. It is possible to assemble incalculable shots in blockish that shows to make sensible thermoelectric modules.

3 Experimental Setup

The schematic block diagram is shown in Fig. 6. As per diagram model, gearbox is produced from mild steel plates using cutting and welding. On one side of the box, a small square size hole is cut to insert an aluminum heat absorber. Then, 6 Amp Peltier is taken; cold side of the Peltier is attached to aluminum heat absorber with the help of thermal paste, and hot side of the Peltier is attached to the heat sink.

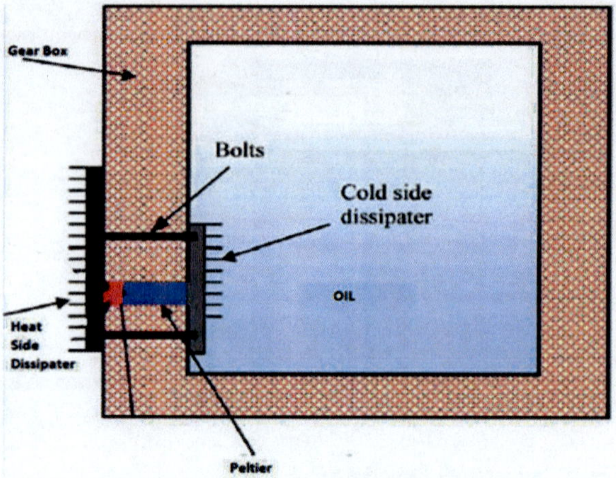

Fig. 6 Block diagram

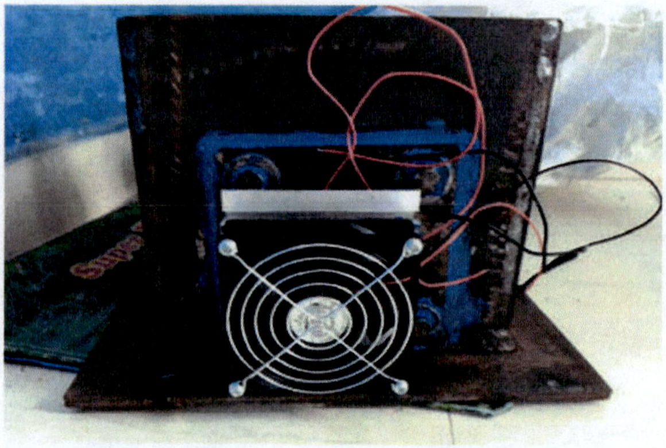

Fig. 7 Prototype setup

Electric DC fan is attached with heat sink to transfer heat to atmosphere [15]. Then, the aluminum absorber side placed inside the square hole to submerge in gear oil, and sealant is used to seal the passage to avoid oil leakage. 12 V 6 Amp DC battery is connected to this setup to give current to the Peltier module. As shown in Fig. 7, the prototype setup is made.

4 Result and Discussion

The Peltier element is used in the gearbox to cool the engine oil. The different specification Peltier elements were used to check the performance variation like temperature, maximum heat absorbed by the Peltier element. The 6 amps Peltier

Table 1 Peltier specification (6 Amps)

Hot Side Temperature (°C)	25 °C	50 °C
Qmax (Watts)	50	57
Delta Tam (°C)	66	75
Imax (Amps)	6.4	6.4
Vmax (Volts)	14.4	16.4
Module Resistance (Ohms)	1.98	2.3

Table 2 Peltier specification (15 Amp)

Hot Side Temperature (°C)	27 °C	50 °C
Qmax (Watts)	150	164.2
Delta Tmax (°C)	70	79
Imax (Amps)	15	15
Vmax (Volts)	16	17.2
Module Resistance (Ohms)	0.8	0.88

element and their performance variations are tabulated in Table 1, and likewise 15 amps Peltier element were tabulated in Table 2.

5 Conclusion

Peltier cooling isn't proficient relatively and because of its small size applications, it isn't broadly utilized. It discovered its application just in hardware cooling and so forth. However, colossal extent of exploration in this area regarding thermoelectric resources, but in this case if we use multiple connections of Peltier with correct specification, then its efficiency is slightly increase due to its compatibility it consume smaller in space. In addition, the heat extracted from the oil is used for preheating. If Peltier is built large in size means, then we can use it to large size industrial gearbox, and solar power is used to power the Peltier module.

References

1. Julian Goldsmid H (2014) Bismuth telluride and its alloys as materials for thermoelectric generation. Materials 7:2577–2592
2. Goldsmith HJ (2010) Introduction to thermoelectricity. Springer-Verlag, Berlin Heidelberg
3. Melcor homepage http://www.melcor.com. Accessed on October, 2015
4. Riffat S, Omer SA, Ma X (2001) A novel thermoelectric refrigeration system employing heat pipes and a phase change. Renew Energy 23:313–323
5. Chein R, Chen Y (2005) Performances of thermoelectric cooler integrated with microchannel heat sinks. Int J Refrig 28:828–839

6. Riffat S, Ma X (2004) Improving the coefficient of performance of thermoelectric cooling systems, international journal of energy research. Int J Energy Res 28:753–768
7. Hermes C, Barbosa J (2012) Thermodynamic comparison of Peltier, Stirling, and vapor compression portable coolers. Appl Energy 91:51–58
8. Vian J, Astrain D (2008) Development of a heat exchanger for the cold side of a thermoelectric module. Appl Therm Eng 28:1514–1521
9. Kaseb S, El-hairy G, Electronics Cooling, Mechanical Power Engineering Department, Faculty of Engineering, Cairo University, Egypt
10. Thermoelectric cooling FAQ, Tellurex Corporation
11. www.hyperphysics.phy-astr.gsu.edu/. Accessed on September, 2015
12. Marc H. (2011) Thermoelectric modules: principles and research. InterPACK, Portland
13. Enescu D, Virjoghe EO (2014) A review on thermoelectric cooling parameters and performance. Renew Sustain Energy Rev 38:903–916
14. Zhao D, Tan G (2014) A review of thermoelectric cooling: materials, modeling and applications. Appl Therm Eng. https://doi.org/10.1016/j.applthermaleng.2014.01.074
15. Mani P, Debnath A, Pradeep LA, Natesan VT, Ponnappan VS (2020) Experimental investigation of dual power generation using peltier cooler. AIP Conf Proc 2311, 090018:1–6

Assessment of Diverse Characteristics of Diesel Engine Fueled with Various Biofuels: A Review

Geddam Prasada Rao, L. S. V. Prasad, and V. Dhana Raju

Abstract Energy consumption has been rising globally as a result of numerous developments in many sectors. Various forms of fossil reserves are being utilized to fulfill the rising energy needs, resulting in fossil reserves depletion accompanied by ecological degradation. The growing demand must be met by alternative energy sources that can compensate for the depletion of fossil resources while also balancing the ecosystem. Biodiesel is one of them, which is a proven alternative to overcome the energy crisis. For biodiesel synthesis, edible vegetable oils are a potential feedstock, but their use might imperil the food supply and destabilize the environment. As a result, non-edible vegetable oils are being investigated as possible replacements for edible vegetable oils. This study presents a detailed analysis of many elements of biodiesel, its properties, and the performance and emission characteristics of various diesel–biodiesel blends in conventional diesel engines and variable compression ratio (VCR) engines as well.

Keywords Biodiesel · Performance · Combustion · Emission

1 Introduction

Moderate utilization of energy is regarded as an indicator of global economic development and a basic requirement for the infrastructure development of a country [1]. Simultaneous growth of population and industrialization leads to increased energy consumption. Petroleum fuels are critical for industrial expansion, agricultural development, building, transportation, and meeting a variety of other fundamental human

G. P. Rao (✉) · L. S. V. Prasad
Department of Mechanical Engineering, Andhra University College of Engineering (A) Andhra University, Visakhapatnam, A.P. 530003, India
e-mail: prasad.zee@gmail.com

V. D. Raju
Department of Mechanical Engineering, Lakireddy Bali Reddy College of Engineering, Mylavaram, A.P., India

N. M. Sivaram et al. (eds.), *Advances in Manufacturing, Automation, Design and Energy Technologies*, Lecture Notes in Mechanical Engineering,
https://doi.org/10.1007/978-981-99-1288-9_72

requirements. Out of these, road transport sector itself is consuming maximum share [2–4]. The transportation sector's energy requirements are met by gasoline and diesel which are petroleum-based fuels, both of which are expected to run out in the near future as these are non-renewable in nature [5]. Booming energy demand, increased crude oil prices, global climate change triggered by greenhouse gas emissions, ecological degradation, and the depletion of traditional fossil fuel sources all contribute to the quest for clean, ecologically friendly, and economically viable alternative energy sources to satisfy global energy demand. This draws the attention of several researchers to biodiesel as a possible replacement for diesel fuel due to its environmental friendliness, biodegradability, greater cetane number, and reduced sulfur level [6–8].

2 Objective of the Review

Following a thorough examination of the biodiesel literature, it was noted that no evaluation has examined the characteristics of edible and non-dibble oils performance- and emission-wise when utilized in unmodified diesel engines and variable compression ratio engines concurrently. It was the primary motivation for writing this review article. Four biodiesels were selected from a variety of feedstocks, which include soy bean oil, rapeseed oil from edible oils, and cotton seed oil, jatropha oil, from non-edible oils. The feedstock performance and emissions characteristics in conventional diesel engines as well as variable compression ratio engines with each biodiesel are discussed.

3 Performance and Emission Characteristics of Engines with Various Biodiesel Blends

The performance and emission characteristics of various biodiesel blends were compared to that of diesel by many researchers, out of them Soybean oil Methyl Easter (SME), Rapeseed oil Methyl Easter (RME), Cotton seed oil Methyl Easter (CME), and Jatropha oil Methyl Easter (JME) performance and emission characteristics were studied, and results are tabulated in Table 1.

4 Conclusions

Immense research was conducted by a number of researchers and varying in terms of performance and emission characteristics with diesel and biodiesel on unmodified diesel and variable compression ratio (VCR) engines have evaluated. Several critical

Table 1 Performance and emissions attributes with various biodiesel blends in conventional diesel engine and VCR engine [9–16]

S No	Biodiesel feedstock	Engine details	Test conditions	Performance characteristics		Emission characteristics		References
				BTE (%)	BSFC	NOx	CO	
1	Soybean	Single-Cylinder, 4-Stroke, C.R:17.5:1, R.P:8.1 kW, AC, Diesel Engine	Variable speed (max3000 rpm) and various loads (%). B10, B20, B50, B100	Less for BD (BD) blends. Decrease with rise of BD content	More for BD. Increasing with increase of BD content	More for BD blends. Increase with BD content	Less for BD. Decrease with rise of BD content	[9]
2	Soybean	Single-Cylinder, 4-Stroke, VCR:13–17, R.P:3.5 kW, WC, Diesel Engine	Speed (1500 rpm) and various loads, C.R:13,15,17 and B20, B40, B60	Increase with load and C.R. Decrease with rise of BD content	Decrease with load rise and C.R. Increase with rise of BD content	Increase with load and BD content	Increase with load. Decrease with rise of C.R and BD content	[10]
3	Rapeseed	4-Cylinder, 4-Stroke, C.R:17:1, R.P:46 kW, Diesel Engine	Variable speed (max2400 rpm) Rated load, various IPs, and B100	Less for BD, increase and then decrease with increase of IP	More for BD, decrease with increase of IP and at higher speed	More for BD, increase with increase of injection pressure	Less for BD, decrease with increase of injection pressure	[11]
4	Rapeseed	Single-Cylinder, 4- Stroke, VCR:12–18, R.P:3.5 kW, Diesel Engine	Speed (1500 rpm) and various loads, C.R:12, 15, 17.5 and B20	Less for BD. Increase with load and C.R	More for BD. Decrease with load rise and C.R	Increase with load and C.R	Less for BD. Increase with load	[12]
5	Cotton seed	Single-Cylinder, 4-Stroke, C.R:18:1, R.P:7.5 kW, Diesel Engine	Variable speed (max3600 rpm) Full load & B5, B20, B50, B75, B100	Less for BD, increase with rise of speed	More for BD, decrease with rise of speed	Increase up to 50% rpm and then decrease	Less for BD. Decrease with rise of speed	[13]

(continued)

Table 1 (continued)

S No	Biodiesel feedstock	Engine details	Test conditions	Performance characteristics		Emission characteristics		References
				BTE (%)	BSFC	NOx	CO	
6	Cotton seed	Single-Cylinder, 4-Stroke, VCR:5–22, R.P:3.7 kW, Diesel Engine	Speed (1600 rpm) and various loads, C.R:22:1 and B15,B30,B45,B60,B75,B90,B100	Less for BD. Increase with rise of bmep. Decrease with rise of BD content	More for BD. Decrease with rise of bmep. Rise with BD content	More for BD. Increase with rise of bmep and BD content	Increase with increase of bmep for all blends except B15 and B60	[14]
7	Jatropha	Single-Cylinder, 4-Stroke, C.R:17.5:1, R.P:7.4 kW, Diesel Engine	Speed (1500 rpm) and various loads and B5, B10,B20, B30, B100	Less for BD blends. Increase with rise of bmep. Decrease with rise of BD content	More for BD blends. Decrease with rise of bmep	More for BD blends. Increase with rise of bmep and BD content	Less for BD blends. Increase with increase of bmep	[15]
8	Jatropha	Single-Cylinder, 4-Stroke, VCR:12–18, R.P:3.5 kW, WC, Diesel Engine	Speed (1500 rpm) and various loads, C.R:16, 17, 18, IP:150, 200, 250 bar and B100	Less for BD. Increase with load and with rise of C.R and IP	More for BD. Decrease with rise of load and C.R	Less for BD. Increase with load and C.R	High for BD. Decrease with rise of C.R. Increase with rise of IP	[16]

* C.R—Compression Ratio, IP—Injection Pressure, R.P—Rated Power, WC—Water Cooled, BD—Biodiesel, and B10 indicates 10% biodiesel and 90% diesel

performance characteristics, such as Brake Thermal Efficiency (BTE), Brake Specific Fuel Consumption (BSFC), and emission characteristics, such as Oxides of Nitrogen (NOx) and Carbon monoxide (CO), have been studies in this review and mixed results and conclusions observed; hence, preponderance inferences have postulated here.

4.1 Inferences from the Review on BTE

- BTE is lower for biodiesels than diesel as well as BTE drops whenever the percentage of biofuels increases in the mixture due to the inferior combustion features of biodiesels owing to their higher density, viscosity, and low volatility which aid the creation of bigger droplets of fuel all through the combustion chamber.
- BTE increases as load increases; the reasons for this are
  - When load increased suction pressure increases, resulting in more efficient combustion.
  - Losses are less at higher loads due to excess of oxygen and high lubricity property.
  - Due to comparatively reduced heat loss at greater engine loads, the percentage increase in fuel required to operate the engine was less than the percentage increase in braking power; hence, BTE will be more for a given fuel input.
- BTE increases when the Compression Ratio (CR) increases in a VCR engine owing to increase in-cylinder temperature and pressure; the increased combustion temperature resulting in more efficient fuel combustion and conversion of biodiesel's chemical energy to mechanical energy.
- BTE increases as Injection Pressure (IP) is increased owing to smaller breakdown of fuel droplets results in increased surface area and proper mixing of the fuel with the air, leading in enhanced combustion. A combination of a higher CR and a higher IP will result in a greater BTE.
- BTE increases with increase of speed due to the increased fuel supply at higher speed.
- BTE increases with increase of bmep due to supply of large amount of fuel at higher bmep.

4.2 Inferences from the Review on BSFC

- BSFC is more for biodiesels than for diesel and grows in proportion to the quantity of biodiesel in the blend. This is because biodiesel, and its blends have a lesser calorific value, greater viscosity and density, and to generate equivalent power as diesel does, excess biodiesel is required.

- BSFC increases as load decreases because the fuel–air mixture is too lean at low loads, as well as BSFC decreases as load increases because the percentage increment consumption is smaller than the increment in braking power, and heat transfer losses are likewise less at greater engine loads.
- The BSFC value decreases as the Compression Ratio (CR) increases in a VCR engine, owing to the better combustion characteristics at higher CR.
- BSFC increase with increase of Injection Pressure (IP) due to supply of more fuel at high pressures.
- BSFC increases with speed due to the higher fraction of fuel that is not completely combustible.
- BSFC decreases when bmep increases, as the percentage of brake power becomes greater than the percentage of fuel.

4.3 *Inferences from the Review on NOx*

- NOx emissions are greater in biodiesels than in diesel and grow as the biodiesel content in the blend increases due to the higher exhaust gas temperature. Biodiesel components have a large molecular weight, which causes them to combust during late phase of combustion, leading to higher exhaust temperature. NOx formation will occur due to the combined effect of the oxygen present in biodiesel and the increased temperature of the exhaust.
- NOx emissions increase with increase of load as the load increase, more fuel will be consumed and the in-cylinder temperature will increase and causes the formation of NOx.
- NOx emissions increase as the Compression Ratio (CR) is increased in a VCR engine, owing to better combustion characteristics that result in a greater combustion temperature within the engine cylinder.
- NOx emissions increase as injection pressure (IP) is raised, as a result of the engine cylinder developing a high temperature as a result of better combustion.
- NOx emissions grow as speed increases up to middle range, then drop as the possible NOx production window narrows.
- NOx emissions rise when bmep increases because a greater quantity of fuel is burnt at a higher bmep, consequently, the peak cylinder temperature rises, resulting in the generation of NOx.

4.4 *Inferences from the Review on CO*

- CO emissions are lower in biodiesels than in diesel and drop with the biofuels amount in blends. CO which is a result of incomplete combustion is converted into CO_2 with fuel bound oxygen of biodiesel. But, due to the lean nature of the engine, decrease of CO emissions due to the fuel-bound oxygen in biodiesel was insignificant. As a result, there will be no linear decrease in CO content as the biodiesel proportion of blends grows.

- CO emissions increase as load increases; with greater loads, more CO is generated as a result of the burning of a richer air–fuel combination. High viscous biodiesel has poor atomization, producing rich air–fuel mixture combinations and therefore causes inferior combustion, resulting in increased CO emissions during combustion.
- CO emissions drop as compression ratio (CR) increases in a VCR engine, owing to enhanced combustion at greater CRs as a result of the oxygen molecules added to biodiesel blends and the elevated temperature in the cylinder associated with higher CRs.
- CO emissions drop when injection pressure (IP) rises because of improved fuel atomization, resulting in improved combustion at increased IP.
- CO emissions increase as engine speed increases; the absolute time available for complete combustion reduces due to reduced air intake, resulting in incomplete combustion and increased CO production.
- CO emission increases as bmep increases; as bmep increases, the fuel ratio increases, and the excess fuel–air ratio produces more smoke, which prevents CO from being oxidized to CO_2, resulting in increased CO emission.

References

1. Ahmad T et al (2019) Optimization of process variables for biodiesel production by transesterification of flaxseed oil and produced biodiesel characterizations. Renew Energy 139:1272–1280. https://doi.org/10.1016/j.renene.2019.03.036
2. Wan Ghazali WNM, Mamat R, Masjuki HH, Najafi G (2015) Effects of biodiesel from different feedstocks on engine performance and emissions: a review. Renew Sustain Energy Rev 51:585–602. https://doi.org/10.1016/j.rser.2015.06.031
3. Basha SA, Gopal KR, Jebaraj S (2009) A review on biodiesel production, combustion, emissions and performance. Renew Sustain Energy Rev 13(6–7):1628–1634. https://doi.org/10.1016/j.rser.2008.09.031
4. Atabani AE, Silitonga AS, Badruddin IA, Mahlia TMI, Masjuki HH, Mekhilef S (2012) A comprehensive review on biodiesel as an alternative energy resource and its characteristics. Renew Sustain Energy Rev 16(4):2070–2093. https://doi.org/10.1016/j.rser.2012.01.003
5. Mahmudul HM, Hagos FY, Mamat R, Adam AA, Ishak WFW, Alenezi R (2017) Production, characterization and performance of biodiesel as an alternative fuel in diesel engines—a review. Renew Sustain Energy Rev 72:497–509. https://doi.org/10.1016/j.rser.2017.01.001
6. Salvi BL, Panwar NL (2012) Biodiesel resources and production technologies—a review. Renew Sustain Energy Rev 16(6):3680–3689. https://doi.org/10.1016/j.rser.2012.03.050
7. Demirbas A (2009) Progress and recent trends in biodiesel fuels. Energy Convers Manag 50(1):14–34. https://doi.org/10.1016/j.enconman.2008.09.001
8. Balat M, Balat H (2008) A critical review of bio-diesel as a vehicular fuel. Energy Convers Manag 49(10):2727–2741. https://doi.org/10.1016/j.enconman.2008.03.016
9. Özener O, Yüksek L, Ergenç AT, Özkan M (2012) Effects of soybean biodiesel on a DI diesel engine performance, emission and combustion characteristics. Fuel 115:875–883. https://doi.org/10.1016/j.fuel.2012.10.081

10. Kulkarni HB, Kubade PR (2018) Performance optimization of VCR diesel engine using soybean oil-based biodiesel. Springer
11. Çelikten I, Koca A, Ali Arslan M (2010) Comparison of performance and emissions of diesel fuel, rapeseed and soybean oil methyl esters injected at different pressures. Renew Energy 35(4):814–820. https://doi.org/10.1016/j.renene.2009.08.032
12. Arumugam S, Sriram G, Ellappan R (2014) Bio-lubricant-biodiesel combination of rapeseed oil: an experimental investigation on engine oil tribology, performance, and emissions of variable compression engine. Energy 72:618–627. https://doi.org/10.1016/j.energy.2014.05.087
13. Aydin H, Bayindir H (2010) Performance and emission analysis of cottonseed oil methyl ester in a diesel engine. Renew Energy 35(3):588–592. https://doi.org/10.1016/j.renene.2009.08.009
14. Santhosh M, Padmanaban KP (2016) Experimental investigation on engine performances, combustion characteristics and emission of exhaust gases of VCR engine fuelled with cottonseed oil methyl ester blended with diesel. Int J Green Energy 13(14):1534–1545. https://doi.org/10.1080/15435075.2014.977438
15. Chauhan BS, Kumar N, Cho HM (2012) A study on the performance and emission of a diesel engine fueled with Jatropha biodiesel oil and its blends. Energy 37(1):616–622. https://doi.org/10.1016/j.energy.2011.10.043
16. Jindal S, Nandwana BP, Rathore NS, Vashistha V (2010) Experimental investigation of the effect of compression ratio and injection pressure in a direct injection diesel engine running on Jatropha methyl ester. Appl Therm Eng 30(5):442–448. https://doi.org/10.1016/j.applthermaleng.2009.10.004

A Mathematical Steady-State Energy Balance Model for Studying the Physiology of the Human Body

Shashikant Das, Deepak Kumar, and Sudhakar Subudhi

Abstract To improve the efficiency of work and quality of life of human beings, the study of thermal comfort is essential. Thermoregulation of the human body is examined analytically in this study. The human body's steady-state heat balance model is used to design a mathematical model. The convective, radiative, respiratory, and evaporative heat transfer are the mechanisms of heat transfer in this model. The behaviour of mean body, core, and required skin temperatures to attain thermal comfort are observed by increasing metabolic rate and keeping all other parameters constant. The behaviour of different modes of heat transfer is examined by increasing ambient temperature and keeping all other parameters constant. The convection, radiation and respiration heat losses to the environment decrease; however, evaporation heat loss increases due to the increase of environmental temperature. With the increase of ambient temperature by keeping all other parameters constant, the behaviour of mean body and core temperatures are examined.

Keywords Thermoregulation · Steady-state · Thermal comfort · Skin temperature · Core temperature

Nomenclature

T_a	Ambient temperature [°C]
T_b	Mean body temperature [°C]
M	Metabolic rate [W/m^2]
Q_{st}	Heat stored rate inside the body [W/m^2]
T_{cr}	Body core temperature [°C]

S. Das (✉) · S. Subudhi
Department of Mechanical and Industrial Engineering, IIT Roorkee, Roorkee 247667, India
e-mail: sdas@me.iitr.ac.in

D. Kumar
Department of Mechanical Engineering, National Institute of Technology, Uttarakhand 246174, India

N. M. Sivaram et al. (eds.), *Advances in Manufacturing, Automation, Design and Energy Technologies*, Lecture Notes in Mechanical Engineering,
https://doi.org/10.1007/978-981-99-1288-9_73

W	Work acquire [W/m^2]
T_{sk}	Skin temperature [°C]
Q_{rd}	Radiation heat transfer [W/m^2]
T_{cr}	Core temperature [°C]
Q_{cv}	Convection heat transfer [W/m^2]
T_{mrt}	Mean radiation temperature [°C]
Q_{evp}	Evaporation heat transfer [W/m^2]
ε	Body surface emission coefficient
σ	Stefan Boltzmann's constant
Q_{res}	Respiration heat exchange [W/m^2]
f_{cl}	Ratio of body area with fabric to without cloth
h_{cv}	Coefficient of convective heat transfer [W/m^{2}K]
I_{cl}	Insulating factor due to clothingclo
h_{rd}	Radiation heat transfer coefficient [W/m^{2}K]
m_{bl}	Per unit of skin area, the blood flow between the core and the skin
m_{rsw}	Sweat generation rate per unit skin area
$WSIG$	The body's warm signals
$CSIG$	The body's cold signals
LR	Lewis ratio
T_{cl}	Average temperature of outer garment [°C]
MRT	Mean radiant temperature [°C]
RH	Relative humidity %
w_{rsw}	Skin wittedness as a result of regulating sweating
α	Percentage of total body mass localized in the skin section

1 Introduction

Thermal comfort has become the area where a large number of researchers are contributing. As the environment changes every day and becomes polluted, global warming is another challenging threat to the individual. So the study of thermal comfort is required to solve these problems. The human body is exposed to a variety of environmental conditions, such as mining, military service, and industry. Experiments on physiological thermoregulation are difficult to conduct in these conditions. In many industries, human being safety is a major concern, and also, production efficiency depends on the working thermal environment. Under these conditions, the analytical study can be a handy tool that can predict the human body's physiological behaviour. It can also aid in the effective design of an indoor environment, which is a big challenge in cities.

Thermal comfort is a mental state in which the human body's thermal sensations are neutral to the thermal surroundings. It is significant to determine the state where an individual feels neutral to the thermal environment. The mathematical model can help to determine the thermally neutral state of the human body, where the body maintains a certain body and core temperature to achieve this state.

It is important to determine factors affecting thermal comfort. Since one single factor cannot define the whole comfort completely, many factors, such as resistance due to clothing, physiological, and environmental factors, affect it. Jang et al. [1] have investigated the factors affecting body metabolic rate: health, activity, acclimation, gender, and age. And they concluded among them activity is the most important factor affecting comfort.

There is a demand of development of mathematical models to predict precise physiological thermoregulation of the human body. Gagge et al. [2] developed a mathematical two-node model that divides the human body into two concentric cylinders: the outer cylinder is called the skin, while the inner cylinder is called the core. This model predicts core temperature, skin temperature, different modes of heat transfer, shivering heat production, sweat rate, and blood mass flow rate as a function of time. Fanger [3] has developed a thermal balancing model that operates in a steady-state where heat stored within the human body is negligible.

Foda et al. [4] have worked on predicting local skin temperature with better accuracy. Here, they have used Pierce two-node model and got an accuracy of ±0.1 °C for neutral conditions. They have also worked for warm and cold conditions and got an accuracy of ±0.3 °C. Tanabe et al. [5] have worked on a manikin and investigated different conditions of thermal comfort at constant ambient temperature. The investigations on different body parts and evaluated sensible heat transfer from these body parts have been performed. The evaluation of total heat transfers and thermal resistance of these body parts is determined.

In this study, the physiological thermoregulation of the human body is analysed by the steady-state heat balance model. Different heat transfer modes over the human body are examined at different temperatures. The core and mean body temperatures are explored as the ambient temperature rises. The physiological behaviour of the human body is examined by varying different metabolic rates.

2 Analytical Model

2.1 Model of Steady-State Energy Balance

The steady-state energy balance model is based on the assumption that energy stored is negligible within the body and presents its mathematical equation per unit area as [6, 7]

$$Q_{st} = (M - W) - \left(Q_{rd} + Q_{cv} + Q_{evp} + Q_{res}\right) = 0 \tag{1}$$

where M metabolic rate, Q_{st} denotes heat storage, Q_{cv} denotes convective heat exchange, Q_{rd} denotes radiation heat transfer, Q_{res} denotes respiration heat transfer, Q_{evp} denotes heat loss due to evaporation, W stands for work acquired, and in the case of light activity, work obtained is ignored [8]. The radiation heat transfer is the exchange of heat by radiation between the human body and its surroundings [4, 5], and it can be calculated using the formula.

$$Q_{rd} = f_{cl} h_{rd} (T_{cl} - T_{mrt}) \tag{2}$$

$$h_{rd} = 4\,\varepsilon\,\sigma\,(0.72)\,(273.15 + ((T_{cl} + T_{mrt})/2))^3 \tag{3}$$

where T_{cl} is the average outer garment temperature, σ represents Stefan Boltzmann's constant, ε is the body surface emission coefficient, and T_{mrt} denotes the mean radiation temperature of the surrounding environment and is expressed as

$$MRT = \left[\left(T_g + 273.15\right)^4 + \frac{v_a^{0.6} \times 1.1 \times 10^8}{\varepsilon D^{0.4}} \times \left(T_g - T_a\right) \right]^{0.25} - 273.1 \tag{4}$$

where D is globe diameter in metre, T_g is globe temperature in °C, ε refers to emissivity of black globe, V_a is velocity of air, and T_a is air temperature in °C.

The convective heat exchange from the human body's surface to the adjacent environment is expressed as [9]

$$Q_{cv} = f_{cl} h_{cv} (T_{cl} - T_a) \tag{5}$$

The convective heat transfer coefficient in W/m^2 °C is represented by h_{cv}, which can be derived as the greatest value of the three equations below [10, 11]:

$$h_{cv} = 3(P_B/760)^{0.53} \tag{6}$$

$$h_{cv} = 3(V P_B/760)^{0.53} \tag{7}$$

$$h_{cv} = 5.66(M/58.2 - 0.85)^{0.53} \tag{8}$$

where V denotes air velocity in m/s and P_B is barometric pressure in mmHg. f_{cl} is an abbreviation for clothed area factor, which is the ratio of total body area covered by fabric to total body area unclothed and is determined by [12].

$$f_{cl} = 1 + 0.3 I_{cl} \tag{9}$$

The average outer surface temperature of garments, denoted by T_{cl}, can be computed using [9].

$$\begin{aligned}T_{cl} = &- R_{cl}[(M - W) - 0.42\,((M - W) - 58.2) \\ &- 3.05\,(5.73 - 0.007\,(M - W) - P_a) \\ &- M(0.0173\,(5.87 - P_a) + 0.0014\,(34 - T_a))] \\ &+ 35.7 - 0.0275\,(M - W)\end{aligned} \tag{10}$$

The respiratory heat transfer is the summation of heat loss due to evaporation (latent) and sensible heat transfer, and it can be calculated using the formula [13].

$$Q_{res} = M(0.0173\,(5.87 - P_a) + 0.0014\,(34 - T_a)) \tag{11}$$

The evaporation heat loss is frequently expressed as the sum of naturally occurring water diffusion from the skin ($Q_{e,dif}$) and thermoregulation evaporation of sweat secretion ($Q_{e,rsw}$), and can be estimated by [14].

$$Q_{e,rsw} = m_{rsw} h_{fg} \tag{12}$$

$$Q_{e,dif} = 0.06\,Q_{e,max}(1 - w_{rsw}) \tag{13}$$

w_{rsw} represents skin wittedness because of thermoregulation mechanism of body, and it can be estimated by

$$w_{rsw} = \frac{Q_{e,rsw}}{Q_{e,max}} \tag{14}$$

$Q_{e,max}$ refers to maximum evaporative potential and is estimated by [14].

$$Q_{e,max} = \frac{P_{sk,s} - P_a}{\frac{R_{cl}}{i_{cl} LR} + \frac{1}{h_{cv} f_{cl} LR}} \tag{15}$$

where LR refers to Lewis ratio and is determined by [14].

$$LR = \frac{15.15(T_{sk} + 273.15)}{273.15} \tag{16}$$

m_{rsw} refers to rate of production of sweat per unit area of skin and is determined by [9].

$$m_{rsw} = 4.7 \times 10^{-5} WSIG_b \times exp\left(\frac{WSIG_{sk}}{10.7}\right) \tag{17}$$

$$WSIG_b = T_b - 36.49 \tag{18}$$

$$WSIG_{sk} = T_{sk} - 33.7 \tag{19}$$

where T_b refers to the average body temperature. The required comfort skin temperature is determined by [9].

$$T_{sk,req} = 35.7 - 0.0275 \times (M - W) \tag{20}$$

$$T_b = T_{cr}(1 - \alpha) + T_{sk} \times \alpha \tag{21}$$

where α denotes body mass fractions that are concentrated in the skin layers. *WSIG* and *CSIG and* are the thermoregulatory body's warm and cold signals. The flow of blood per unit skin area among the core and the skin is denoted by m_{bl} [13].

$$\alpha = 0.0418 + \frac{0.745}{3600 \times m_{bl} + 0.585} \tag{22}$$

$$m_{bl} = \frac{6.3 + 200WSIG_{cr}}{3600(1 + 0.5CSIG_{sk})} \tag{23}$$

$$WSIG_{cr} = T_{cr} - 36.8 \tag{24}$$

$$CSIG_{sk} = 33.7 - T_{sk} \tag{25}$$

Equation (1) can be rewritten by taking steady-state condition where there is no heat stored inside the body and considering all the above equations. The core temperature is obtained from Eq. 26 using MATLAB software, and further using core temperature, all modes of heat transfer are obtained.

$$\begin{aligned} M - W =& \left[\varepsilon\, f_{cl}\sigma\, 4\,(A_{rd}/A_{Du})\,(273.15 + ((T_{mrt} + T_{cl})/2))^3 (T_{cl} - T_{mrt})\right] \\ &+ [\, f_{cl} \times h_{cv} \times (T_{cl} - t_a)] \\ &+ M[0.0014(34 - T_a) + 0.0173\,(5.87 - P_a)] \\ &+ \left[4.7 \times 10^{-5}(T_b - 36.49) \times \exp\left(\frac{Tsk - 33.7}{10.7}\right) h_{fg} \right. \\ &\left. + 0.06 \times \frac{P_{sk,s} - P_a}{\frac{R_{cl}}{i_{cl}LR} + \frac{1}{h_{cv} f_{cl} LR}} (1 - w_{rsw}) \right] \end{aligned} \tag{26}$$

2.2 *Flow Chart of the Simulation*

The flow chart of the simulation is shown in Fig. 1. The input parameters that are required in this simulation are work obtained, relative humidity, velocity of air, mean radiant temperature, insulation for garments, air temperature metabolic rate, and core

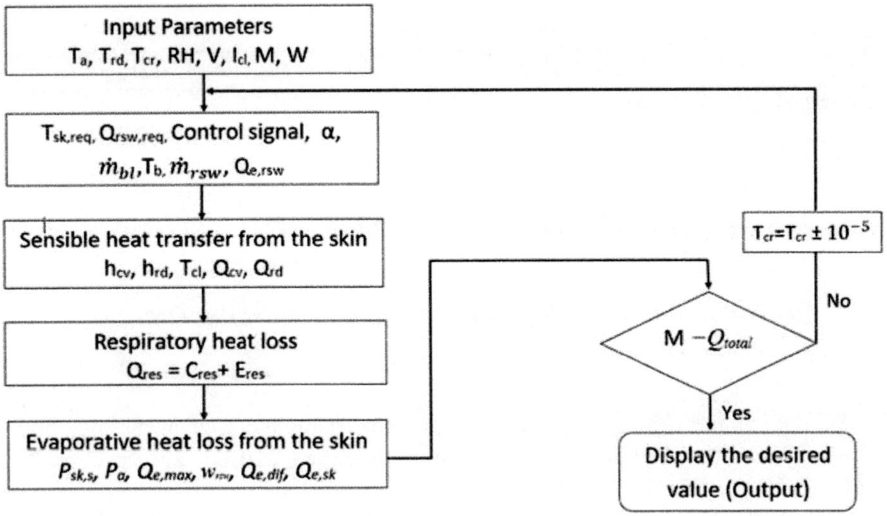

Fig. 1 Simulation of the present model

temperature. To run this simulation initial guess is required, core temperature is used here as an initial guess of the simulation. Further, all the heat transfer modes are determined; evaporation heat transfer from the skin, respiratory heat transfer from body, and sensible heat transfer from the skin. Further, total heat loss is equated with the metabolic rate for getting the desired output. This steady-state heat balance model is observed and found core temperature is the only unknown. Once this core temperature is determined, furthermore all modes of heat transfer and different body parameters can be determined. This simulation is used to determine this core temperature using initial guess.

2.3 *Model Validation*

Initially, this model is simulated and solved at the environmental condition of 23.9 °C and 40% RH. After that, it is compared with ISO-7730 result. For the validation of the present model with ISO-7730, the fixed input parameters used here are RH = 40%, M = 58.2 W/m^2, V = 0.05 m/s, I_{cl} = 1 Clo, $T_a = T_{rd}$ = 23.9. The current model findings are found to be in close accord with ISO-7730, as given in Table 1, hence can be used for further study.

Table 1 Validation of the present model with ISO-7730

	$Q_{cv} + Q_{rd}$	Q_{esk}	Q_{res}	Q_{total}
ISO-7730 [15]	38.05	12.64	5.46	56.14
Present model	36.15	15.39	6.55	58.09

3 Result and Discussion

3.1 Variations of Physiological Parameters with the Increase of Metabolic Rate

Table 2 depicts the changes in body physiological parameters when the metabolic rate rises from 60 to 120 W/m^2. The fixed input parameters used here are $T_a = T_{rd}$ = 24 °C, $RH = 40\%$, $V = 0.05$ m/s, $I_{cl} = 0.5$ *clo*. The core temperature rises from 36.79 to 37.39 °C, and the mean body temperature rises from 36.38 to 37.14 °C. The variances mentioned above are shown in the second and fourth columns of the table below. The third column of the table shows the variations of required skin temperature, which provides thermal comfort and decreases from 34.05 to 32.4 °C. It shows that for the compensation of the increase of core and mean body temperature, our required skin temperature should decrease in order to attain comfort.

The fifth column shows the variations of the mass flow rate of blood, and it increases from 1.8 to 21.2 g/(sm^2). The accumulation of heat inside our bodies grows as our metabolic rate rises. However, for proper operation of the human body, this heat must be released into the environment. Here, the thermoregulation comes into the picture. With the help of vasodilatation, blood vessel dilates, which result in an increase in blood flow rate, and accumulation of heat release from the core to the skin and from the skin to the environment. The sixth column shows the variations of heat release from the skin due to evaporation heat transfer, and it increases from 0.043 to 72.49 W/m^2. The sensible heat transfer is not enough to release this heat. The latent heat transfer mechanism is heat transfer due to evaporation; a lot of heat is dissipated via this process, and the body maintains comfort. The last column represents the

Table 2 Variations in physiological parameters when metabolic rate increases

M (W/m^2)	T_{cr} (°C)	T_{sk} (°C)	T_b (°C)	m_{bl} (g/sm^2)	Q_{esk} (W/m^2)	Q_{res} (W/m^2)
60	36.79	34.05	36.38	1.8	0.043	6.74
70	36.81	33.78	36.45	2.5	4.59	7.87
80	36.85	33.50	36.55	4.1	15.73	8.99
90	36.93	33.23	36.67	7.2	28.47	10.12
100	37.05	32.95	36.81	11.6	42.33	11.24
110	37.21	32.68	36.97	16.4	57.06	12.37
120	37.39	32.40	37.14	21.2	72.49	13.49

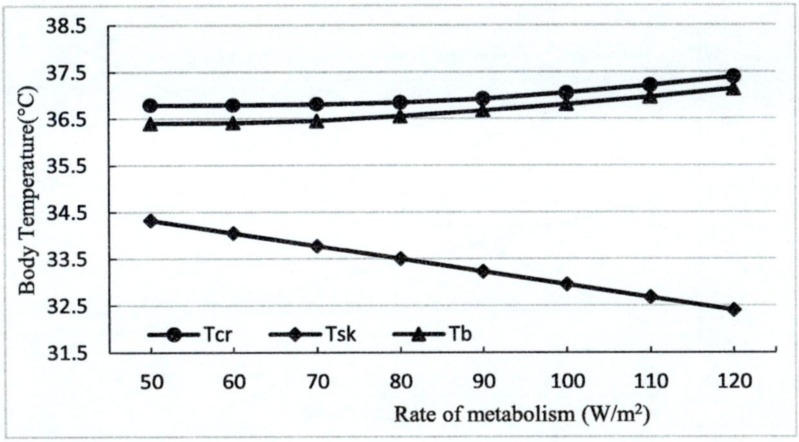

Fig. 2 Body temperature variation when exercise level increases ($V = 0.05$ m/s, $T_{rd} = T_a = 24$ °C, $RH = 40\%$, $I_{cl} = 0.5$ Clo)

heat transfer due to respiration, which does not vary significantly in this condition, ranging from 6.74 to 13.49 W/m^2.

3.2 Deviations of Body Temperature with the Rise of Metabolic Rate

Figure 2 shows skin and body temperature deviations with the increased metabolic rate, which provide comfort. The following parameters are fixed to obtain this deviation: $V = 0.05$ *m/s*, $T_{rd} = T_a = 24$ *°C*, $RH = 40\%$ *and* $I_{cl} = 0.5$ *Clo*. The mean body temperature remains constant when the activity level is less; however, when the activity level increases, body temperature also increases. Equation 20 is used to calculate the required comfort skin temperature in this graph. With the rise of activity level, the required skin temperature which delivers comfort found reduced. When the metabolic rate increases, the skin temperature is supposed to fall in order to attain thermal comfort. The graph also demonstrates that when metabolic rate increases, so does body core temperature. To compensate for this rise in core temperature, the needed skin temperature for comfort must fall.

3.3 Deviations of Heat Transfer with Air Temperature

Figure 3 depicts the variations in heat transmission modes as the ambient temperature rises from 23 °C to 29 °C. The following fixed parameters used in this calculation are $I_{cl} = 0.5$ clo, $V = 0.05$ m/s, $M = 80$ W/m^2, $RH = 40\%$. Among different modes

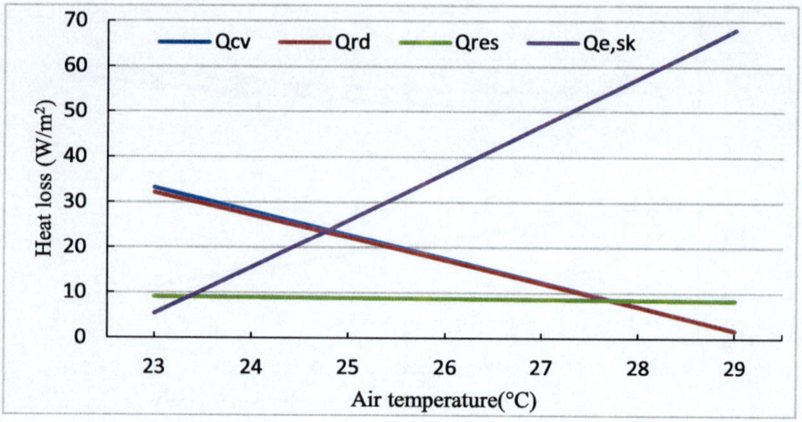

Fig. 3 Deviation of heat transfer of the body with increase of temperature ($I_{cl} = 0.5$ clo, $M = 80$ W/m^2, $RH = 40\%$, $V = 0.05$ m/s)

of heat transfer at 23 °C evaporative heat transfer found minimum and convective heat transfer found maximum. The difference between the core and ambient temperatures diminishes as the ambient temperature rises, resulting in sensible heat transfer decreases. This behaviour is also shown in the figure below, where convective and radiative heat transfer decrease with the increase in ambient temperature, and at 29 °C, it becomes a minimum value. The graph shows that the evaporative heat transfer is increasing with the increase in temperature. As with the increase of temperature, heat accommodation inside the body increases due to evaporation, a lot of heat can lose during this process, and body temperature can be maintained. In this range of temperature, the contribution of respiration heat transfer is very less compared to other modes. The pattern exhibited in the graph below shows a decline in the behaviour of respiration heat transfer.

The deviations of mean body and core body temperatures with the increase of temperature are shown in Fig. 4. The input conditions used to get this behaviour are $I_{cl} = 0.5$ clo, $M = 80$W/m^2, $V = 0.05$ m/s, $RH = 40\%$. The increase in air temperature results in an increase in core temperature, starting from 36.82 °C to 37.24 °C (see Fig. 4). A similar pattern is also found for the mean body temperature from 36.45 °C to 37.05 °C. The core temperature should not increase beyond a certain limit; it can cause serious health problems like heat stroke. Even in this situation, the body can be lethal, so certain prevention must be required before reaching this limit.

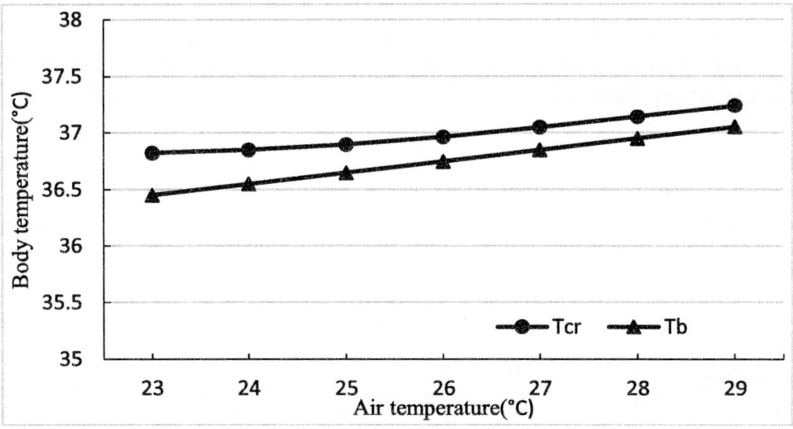

Fig. 4 Mean body and core temperature variations with air temperature ($I_{cl} = 0.5$ clo, $M = 80$ W/m^2, $V = 0.05$ m/s, $RH = 40\%$)

4 Conclusions

Human performance depends extensively on the environmental parameters. So to increase the performance of human beings, many researchers are contributing to this field. Mathematical models of physiological study can help to design the indoor environment for better efficiency. An analytical model is developed in which the behaviour of the human physiological body is investigated. The steady-state heat balance equation is used to develop this model, in which the body does not store any heat. This heat balance is consisting of heat transfer in the form of radiation, convection, respiration, and evaporation. The mathematical model is simulated to solve all physiological parameters by equating metabolic rate with total heat loss. An initial guess of core temperature is required to solve this simulation; once the only unknown core temperature is obtained, other physiological parameters is calculated using this simulation.

With the increase of activity level, the behaviours of skin, mean body, core temperature are investigated keeping other parameters constant. The core temperature of the body increases as the activity level rises. The required skin temperature, which provides comfort, must decrease so that the rise of core temperature can be compensated. This can be obtained by the increase of air velocity or decrease of air temperature. At a low activity level, the mean body temperature remains constant. However, it rises with an increase in activity level.

The effect on heat transfer of the body is investigated with the rise of the ambient temperature from 23 to 29 °C by fixing all other parameters constant. By increasing ambient temperature, the convection and radiation heat transfer decreases and it becomes low value at 29 °C. The evaporative heat loss increases with the rise in air temperature; however, the heat loss by respiration decreases.

For a certain ambient temperature, only body thermoregulation works. If the ambient temperature exceeds this limit, thermoregulation is not enough. The core temperature is the indicator if it exceeds this limit with the increase of ambient temperature and the core temperature increases in the simulation. So certain prevention is required when core temperature exceeds this limit and to avoid lethal condition.

Acknowledgements The authors acknowledge IIT Roorkee for providing the required facility for this project.

References

1. Jang TS, Iyoho AE (2009) Modeling individual variations in thermal stress response for humans in transient environments. ASHRAE Trans 115:552
2. Gagge AP, Fobelets AP, Berglund L (1986) A standard predictive Index of human reponse to thermal enviroment. Trans Soc Heating, Refrig Air-Cond Eng 92:709–731
3. Fanger PO, Toftum J (2002) Extension of the PMV model to non-air-conditioned buildings in warm climates. Energy Build 34:533–536
4. Foda E, Sirén K (2011) A new approach using the Pierce two-node model for different body parts. Int J Biometeorol 55:519–532
5. Tanabe S, Arens EA, Bauman F, et al (1994) Evaluating thermal environments by using a thermal manikin with controlled skin surface temperature. 39–40
6. Das S, Subudhi S (2021) A review on different methodologies to study thermal comfort. Int J Environ Sci Technol 1–17
7. Das S, Subudhi S (2022) Study of human body temperature and different modes of heat transfer using steady-state energy balance model. In: Advances in energy technology. Springer, pp 571–581
8. Standard A (2013) Standard 55-2013 Thermal environmental conditions for human occupancy. ASHRAE, Atlanta, GA 30329
9. Kilic M, Kaynakli O, Yamankaradeniz R (2006) Determination of required core temperature for thermal comfort with steady-state energy balance method. Int Commun Heat Mass Transf 33:199–210. https://doi.org/10.1016/j.icheatmasstransfer.2005.10.003
10. Fanger PO (1970) Thermal comfort. Analysis and applications in environmental engineering. Therm Comf Anal Appl Environ Eng
11. Doherty T, Arens EA (1988) Evaluation of the physiological bases of thermal comfort models. ASHRAE Trans 94
12. Li B, Yang Y, Yao R et al (2017) A simplified thermoregulation model of the human body in warm conditions. Appl Ergon 59:387–400
13. Prek M (2005) Thermodynamic analysis of human heat and mass transfer and their impact on thermal comfort. Int J Heat Mass Transf 48:731–739
14. Pala U, Oz HR (2015) An investigation of thermal comfort inside a bus during heating period within a climatic chamber. Appl Ergon 48:164–176
15. Standardization IO for (1994) ISO 7730: moderate thermal environments-determination of the PMV and PPD indices and specification of the conditions for thermal comfort. ISO

Investigation on Hydrogen Production Using Concentrated Solar Thermal (CST) Technology Through Thermochemical Water Splitting and Solid Oxide Electrolysis (SOEC)

Senthil Kumar and K. Ravi Kumar

Abstract In this article, an investigation on hydrogen production using concentrated solar thermal technology is presented. The thermal energy generated from the central tower is used in two different technologies, namely (a) thermochemical water splitting and (b) high temperature solid oxide electrolysers for continuous H_2 production. The 10 MW_{th} central tower plant is modeled and simulated using SolarPILOT software for the Jaisalmer location in India on an hourly basis. The results revealed that the average daily energy yield from the central tower is 45.43 MWh with the peak instantaneous power delivery of 6.3 MW_{th}. The data is further used as input to the hydrogen models for estimating the hourly hydrogen production rate. The obtained result denotes that the proposed model shall generate hydrogen at a rate of 31.46 kg/hr. and 25.2 kg/hr using thermochemical water splitting and high temperature solid oxide electrolysers, respectively.

Keywords Solar energy · Green hydrogen · Thermochemical water splitting · Solid oxide electrolyser

1 Introduction

The global demand for pure hydrogen is steadily increasing in recent days and the energy sector slowly entering the hydrogen economy from an oil-based economy [1]. The main attraction for hydrogen as a fuel is that it does not create carbon dioxide emission like fossil fuels [2], and it can also be generated from many diversified resources such as conventional (using natural gas, coals, etc.) and non-conventional (through solar, wind, etc.) resources. Globally, more than 95% of overall hydrogen generated nowadays are gray hydrogens (i.e., produced majorly using fossil fuels) [3], but the drawback of such H_2 production processes is that it contributes

S. Kumar · K. R. Kumar (✉)
Department of Energy Science and Engineering, Indian Institute of Technology Delhi, Hauz Khas, New Delhi—110 016, India
e-mail: krk@ces.iitd.ac.in

N. M. Sivaram et al. (eds.), *Advances in Manufacturing, Automation, Design and Energy Technologies*, Lecture Notes in Mechanical Engineering,
https://doi.org/10.1007/978-981-99-1288-9_74

to CO_2 emissions and requires hydrocarbon-based fuel resources or supply chain [4]. A very promising alternative is to use high temperature to split water through thermochemical water splitting [5] and high temperature solid oxide electrolysers [6].

The majority of the studies are limited in estimating the hydrogen generation under the constant solar radiation scenario. The effect of the transient nature of solar thermal technology in hydrogen generation is not widely covered in the literature. Further, a feasibility study of generating hydrogen particularly for the Indian weather scenario is not available in the literature. In view of the above, the study is carried out to analyze and estimate the hourly hydrogen generation throughout the year using central tower technology [7, 8].

The concept of using metal oxides as a substrate for thermochemical production of hydrogen was suggested by Kodama et al. [9]. Further, different studies have presented the use of different oxide materials such as Fe_2O_3 [10], ZnO [11], cerium oxide (Ce_2O_3), and doped ceria [12] over the span of four decades. Among the promising cycles, for this study, the cerium oxide–cerium sulfate water splitting cycle was focused on owing to its low thermal reduction temperature ~1200 °C and high thermal stability at the higher temperature.

Similarly, the required electrical power in high temperature electrolysers (HTE's) is reduced at higher temperatures as the unavoidable joule heat of an electrolysis cell is used in the water splitting process. Also, it reduces the electrodes over potential at elevated temperature supply [13]. The minimum electrical energy supply required for the electrolysis process is equal to the change in Gibbs free energy (ΔG), where $\Delta G = \Delta H — T\Delta S$. Thus, the heat energy from the central tower is used to estimate the hourly hydrogen production in both the above technologies at their respective operating conditions.

2 Model Development

Figures 1, 2 show the complete process flow of hydrogen generation through two different processes using heat energy from concentrated solar tower technologies. Model 1 is representing the H_2 generation through the thermochemical water splitting process, and Model 2 represents the H_2 generation through high temperature solid oxide electrolysis. Both the models are divided into two system boundaries, the first one describes, solar field and particle receiver network, the second one includes the hydrogen generation unit of the respective technology.

The subsystem of the models includes (i) a solid–solid recuperator in which the ceramic particle from the tower transfers heat energy to the metal oxides (CeO_2). Metal oxides are allowed to thermally reduce in the same reactor, (ii) the vacuum pumps, (iii) the steam generator, (iv) water splitting, (v) using condenser, (vi) filters, (iv) solid oxide electrolysers with controlled power supply source, (v) steam network, (vi) air supply unit for cooling the stacks/reactors, (vii) condensing unit to remove

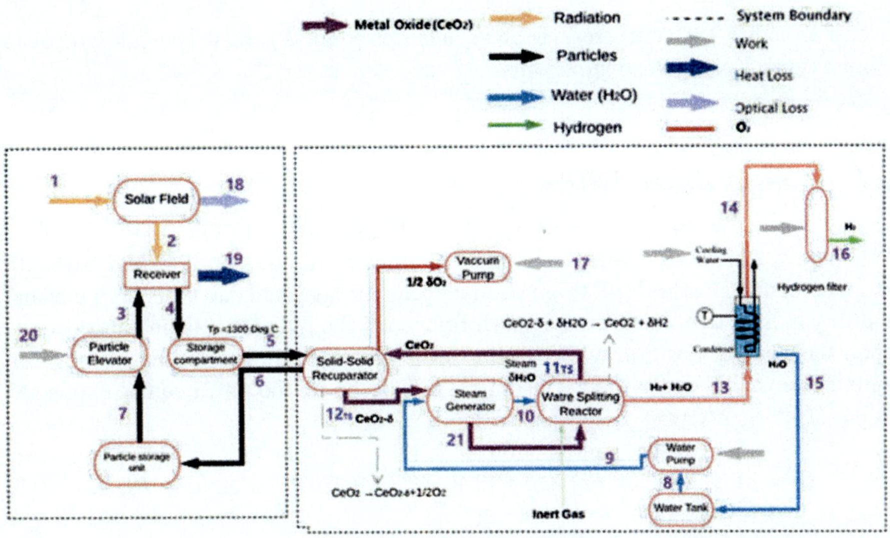

Fig. 1 Process flow diagram of hydrogen generation through thermochemical water splitting using central tower technology

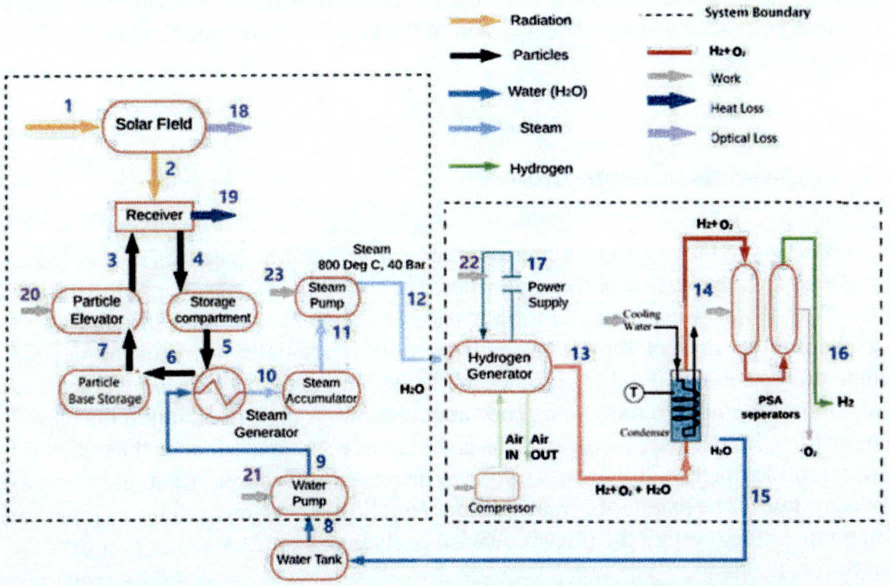

Fig. 2 Process flow diagram of steam supply for hydrogen generation through high temperature solid oxide electrolysis

excess moisture, and (viii) pressure swing adsorption (PSA) unit to separate hydrogen from oxygen hydrogen stream.

2.1 Central Tower System

The radially staggered, central tower plant is simulated for this study. A high-strength proppant named Carbo-HSP (high strength proppant) is used due to its high thermal stability at high temperatures. The performance of the individual heliostats depends upon their actual position with respect to the towers and their respective position with the sun at any given time. The optical loss of the central tower can be expressed in Eq. (1).

$$Q = \rho i + \gamma + \cos(\theta i) + Sp.Li + So.Li + Sh.Li + Attn + Bl.Li \quad (1)$$

where ρi is the reflectivity of the mirror, γ is the intercept factor, $Sp.L_i$ is spillage losses, $So.L_i$ is soiling losses, $Sh.\ L_i$ is shading loss, Attn is attenuation, and *Bl. Li* is blocking losses of the respective heliostat in the field.

The position of the individual heliostat is indicated in X and Y coordinates from the center point of the tower. The location of the tower is (0, 0), and the height of the tower is specified as h.

2.2 Solid–Solid Recuperator

In the solid–solid recuperator, the heat energy from ceramic particles is allowed to transfer to the metal oxide particles through a counter flow heat exchanging arrangement. The temperature over the tube walls is assumed as constant throughout its peripheral. The flow of the metal oxide particle is regulated to meet the desired temperature. The metal oxide particles are heated to the reduction temperature T_{TR} and allowed to undergo reduction. The vacuum pump is used to maintain the partial pressure lesser than that of oxygen. This gives the advantage to enhance the reduction process unlike the use of inert gas sweep to decrease partial pressure in the conventional systems. The extent of ceramic oxide reduction at a different range of pressure and temperature through the linear equation is given in Eq. (2) & (3) as proposed by Sorgulu et al. [14].

$$\log[\delta(P, T)] = a_1 + a_2 P + a_3 P^2 + a_4 P^3 + a_5 T + a_6 PT + a_7 P^2 T + a_8 PT^2 \quad (2)$$

and

$$\Delta_r H(\delta) = b_1 + b_2\delta + b_3\delta^2 + b_4\delta^3 \quad (3)$$

Table 1 Fitting parameters for CeO_2, ΔH, δ

Parameters	a1	a2	a3	a4	a5	a6	a7	a8
Values	−8.9549	1.8861	0.056207	0.000334	0.003717	−0.00244	−3.5E-05	6.9E-07
	b1 (kJ/mol)	**b2 (kJ/mol)**	**b3 (kJ/mol)**	**b4 (kJ/mol)**				
	−971	3106	−16,281	24,785				

where δ is the extent of CeO_2 reduction and $P = \log(\frac{PO_2}{PO})$; PO_2 is oxygen partial pressure and the PO is the standard atmospheric pressure at sea level. $\Delta_r H(\delta)$ indicates the enthalpy needed for the reaction of δ particles. It shall be noted that the decrease in PO2 increases δ, which indicates the extent of reduction per unit volume of metal oxide is increased; this reduces the huge sensible energy load needed in the thermal reduction chamber. Hence, maintaining the low pressure using a vacuum pump increases the overall conversion efficiency. The value of the fitting parameter is given in Table 1.

2.3 Water Splitting Reactor

The thermally reduced metal oxide is directly introduced with the steam for carrying out the water splitting reaction. The partial pressure of oxygen in the vapor phase is always maintained greater than the equilibrium vapor pressure of oxygen by an excess amount of steam supply, which provides enough thrust for water splitting reaction in the reactor. The mass and energy conservation equation for the W-S reaction is expressed in Eqs. (4) and (5).

$$m_{steam} \cdot y_{13} = m_{m.o.} x(\delta_{11} - \delta_{12}) \tag{4}$$

$$E_{11}(T_{11} x \delta_r) - E_{12}(T_{12} x \delta_{ox}) = E_{13}(T_{13} x y_{13}) - E_{10}(T_{10}) \tag{5}$$

2.4 Solid Oxide Electrolysers and its Sub-systems

The second system in Fig. 2 shows the broad schematic of the SOEC and its subsystems. The electrical energy requirement of the electrolysers stack is met by the constant power supply source. The thermal energy requirement is supplied in the form of steam extracted from concentrated solar technology. The general reaction equation for solid oxide electrolysis is given in Eq. (6).

$$H_2O(g) \rightarrow H_2 + -0.5O_2 \tag{6}$$

Firstly, the operating temperature and pressure of the electrolysers have been selected as 800 °C and 40 bar, respectively, for optimal output delivery by the SOEC stacks. The energy requirement for the steam is calculated using steam tables, and accordingly, the hourly steam generation from the central tower is calculated. The value of $\Delta H = 247.9 \frac{kJ}{Mol^{-1}}$.

3 Results and Discussion

The objective of this exercise is to design and estimate the energy generation from the 10 MW_{th} central tower system which is used to study the hydrogen generation using two different processes for its technical feasibility. The optimized physical dimensions and optical performances obtained from SolarPILOT software are given in Table 2 and Table 3, respectively. The simulated heliostat field layout with its coordinates is given in Fig. 3.

The hourly simulation for one complete year is carried out initially for the design DNI of 750W/Sq.m. Further, the hourly DNI from NSRDB is processed and used in the simulation. The difference in the power simulated at constant design point DNI, and the power estimated based on the actual radiation data is shown in Fig. 4. Further, the cosine efficiency of the central tower for four months has been plotted to understand the monthly variation. The better cosine efficiency is observed during the winter months (i.e., Oct–Mar) as compared to the summer months as shown in Fig. 5. This is due to the average angle of incidence b/w heliostat and fixed receiver being high during summer, especially the heliostats positioned far away from the tower. It is observed that the actual power generated by the tower field is strongly depends up on the DNI and the cosine loss. Therefore, there is an irregular variation of actual power generation which is observed with respect to time.

The model validation is given in Table 4. It is observed that the parameters such as annual attenuation factor, shading, and blocking factors, annual efficiency of heliostat are in good agreement. The parameter cosine factor is more sensitive to the location and the actual heliostat positions which are ~ 3.5% than the existing literature. In the present model, the parameters such as cosine factor, annual shading, and blocking

Table 2 Optimized solar field output

Optimized design	Values	Unit
Tower height	107.6	m
Heliostat aperture width	12	m
Heliostat aperture height	8.08	m
Total no of heliostats	291	No's
Receiver aperture area	9	Sq. m
Receiver width and height	3 × 3	m
Concentration ratio	3000	–

Table 3 Optical performance of the central tower

Particulars	Units	Value	Mean	Minimum	Maximum	Std. dev
Cloudiness efficiency	%	100	100	100	100	0
Shading efficiency	%	98.40	98.40	89.12	100	2.8236
Cosine efficiency	%	96.17	96.19	84.19	99.99	3.4144
Reflection efficiency	%	90.25	90.25	90.25	90.25	0
Blocking efficiency	%	97.16	97.18	49.65	100	7.0372
Attenuation efficiency	%	96.95	96.94	95.59	97.94	0.6871
Image intercept efficiency	%	70.19	69.88	38.04	85.54	10.6052
Absorption efficiency	%	94	–	–	–	–
Solar field optical efficiency	%	56.47		20.7	73.57	10.6523
Optical efficiency incl. receiver	%	53.08		19.45	69.16	10.0131
Annualized heliostat efficiency	%	0		18.58	70.98	12.5367
Incident flux	kW/m^2	1337.4		442.40	2095.40	429.363
Simulated heliostat area	m^2	28,227	–	–	–	–
Simulated heliostat count	–	291	–	–	–	–
Power incident on the field	kW	21,316	–	–	–	–
Power absorbed by the receiver	kW	11,314	–	–	–	–
Power absorbed by HTF	kW	9946.6	–	–	–	–

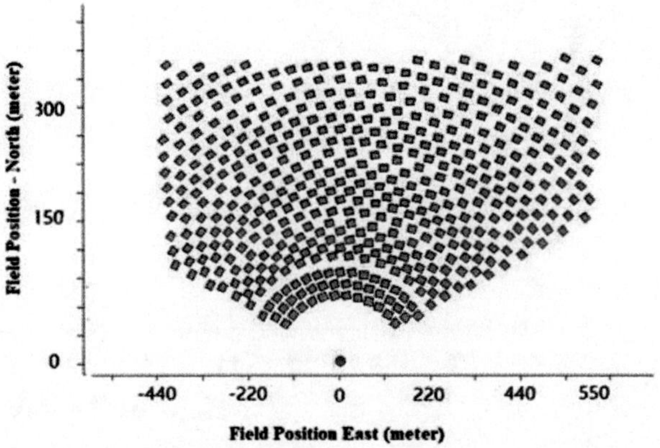

Fig. 3 Simulated heliostat plot

factor are better than the models presented in the existing literature due to the selection of the location and field layout selected during the plant design.

The value of δ_r is simulated for different thermal reduction temperatures (Tr) and partial pressure of the oxygen (PO_2) in the reactor. From Fig. 6, it shall be observed that the lesser the partial pressure of O_2 in the reduction process, increases the value of δ_r, i.e., the lower the pressure of the reactor, the more oxygen atoms are easily dissociated by the metal oxides at elevated temperature. The pressure is adjustable

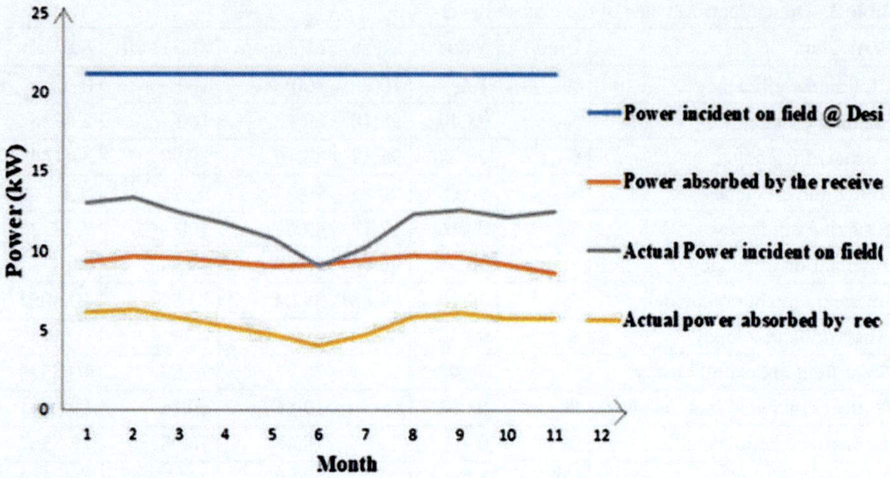

Fig. 4 Annual power absorbed at design v/s actual DNI

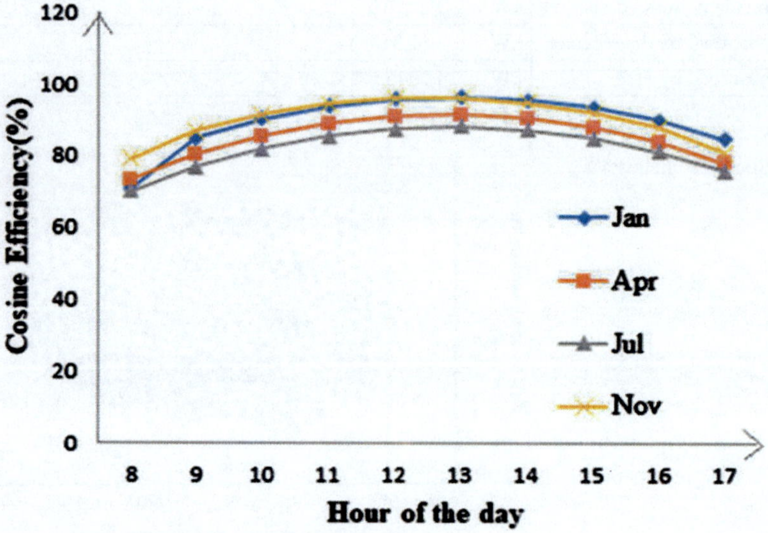

Fig. 5 Annual cosine efficiency of the proposed design

with the help of a vacuum pump. The value of δ_o is initially assumed to be 1 for avoiding the complexity of the problem.

The obtained results have been cross validated with the data published by [19] and [20] and found that the data is exactly fitting with the curve plotted by the present model. Similarly, the change in δ_r with respect to the reduction temperature (Tr) has been plotted in Fig. 7. The baseline values and the obtained result from the simulation are given in Tables 5, 6, 7, respectively.

Table 4 Validation of the present model with the existing data

Parameter	Noone et al. [15]	Besarati et al. [16]	Yao et al. [17]	Abaza et al. [18]	Present Model
Total heliostat area (M2)	439,000	72,075	436,000	222,600	28,227
Heliostat size	121.34	121.34	121.34	119.9	97
Tower height	115	115	93.2	120	108
Field type	Radial staggered	Radial staggered	Radial staggered	Surrounding and radial staggered	Radial staggered
Receiver type and dimension	Flat rectangular	Cavity	Flat rectangular	Circular, Dia -11 m	Flat rectangular
Annual cosine factor	82.83%	83.15%	84.53%	82.56%	86.45%
Annual attenuation factor	94.98%	94.98%	95.07%	94.99%	96.95%
Annual spillage factor	99.26%	99.31%	99.93%	99.32%	–
Annual shading and blocking factor	92.55%	91.61%	90.18%	91.94%	97.16%
Reflection efficiency	88%	88%	Unpublished	Unpublished	90.25%
Annual efficiency	63.40%	63.80%	Unpublished	Unpublished	70.98%

The thermal efficiency is achieved as 35.79%, and the solar to fuel efficiency is obtained as 27.8%. The obtained solar to fuel efficiency is lower compared to the efficiency of the Budama et al. (38.2%), Rahul et al. (45.1%) model, as they have simulated for constant design point DNI values.

Similarly, Table 7 shows the simulated yield from high temperature SOEC on monthly basis.

4 Conclusions

Demonstration of hydrogen production using concentration thermal technology for Indian weather conditions using thermochemical water splitting and solid oxide electrolysis is discussed. The hydrogen generation rate is obtained as 19.58 g/s for the constant heat energy delivery of 10 MW_{th} from the solar concentrator and assuming

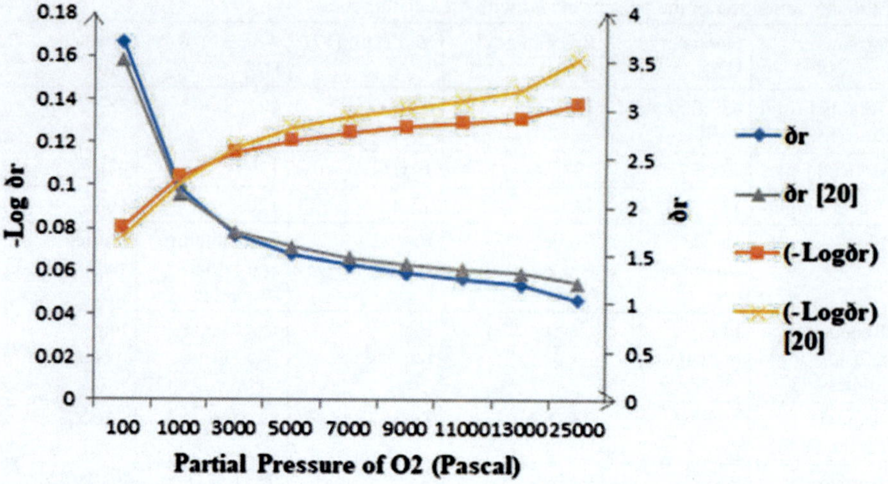

Fig. 6 Effect of PO_2, δ_r, $Log\delta_r$

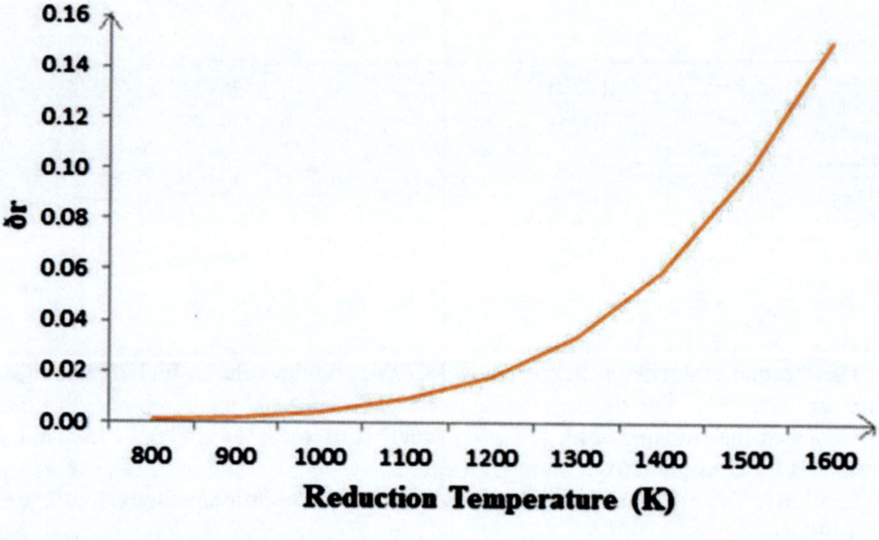

Fig. 7 Reduction temperature $T_{sV/s}\delta_r$

the constant temperature of reduction, and splitting is maintained. In this case, the extent of reduction and oxidation that takes place in their respective reactors is analyzed at a partial pressure of oxygen at 1000 Pa. The flow rate of particles was adjusted in such a way that the reduction temperature is maintained throughout the process. Under such circumstances, the near actual annual average hydrogen production rate is obtained as 8.74 g/s which is 55% lesser than the ideal scenario. Similarly, the near actual annual average hydrogen production rate from SOEC is obtained as

Table 5 Baseline parameters for thermochemical water splitting

Component	Parameters	Values	Units
Solid–solid recuperator/thermal reduction reactor	Partial pressure of the reactor	1000	Pascal
	The partial pressure of atm	101,312	Pascal
	δ_r	0.099	
	Recuperator effectiveness ε_r	0.9	
	Reduction temperature	1500	K
Water Splitting reactor	Water splitting temp	750	K
	Effectiveness of reactor	0.9	–
	δ_o	0.0099	–
	Total heat required for H2 generation, Q_{H_2}	1029.42	kJ/mol of H_2
Solar field	Heat input from solar after losses $Q_{th, Ideal}$	10,000	kW
General	Mass of CeO_2 needed for 1 mol of hydrogen generation	11.22	mol
	The average flow rate of the CeO_2 in the solid–solid recuperator	48.63	mol/s
	H_2 Production rate (ideal)	19.58	g/s
	Rate of steam delivered (Ideal)	32.64	g/s

Table 6 Hydrogen production rate through thermochemical water splitting

Average metal oxide flow rate	Mol/s	48.6
Average hydrogen production rate	kg/hr	31.5
Average steam supply rate	kg/hr	52.4

Table 7 Hydrogen production rate through SOEC

Average annual hydrogen production rate	kg/hr	25.5
Average electricity needed	kWh	727.9
Average steam supply rate	kg/hr	1961.7

7.08 g/s. Assuming the steam quality inlet to the system is always constant while the quantity of steam varies as per the intermittent thermal output received from the central tower. The reduction thermal efficiency is achieved as 35.79%, and the solar to fuel efficiency is obtained as 27.8%. It is also observed that the hydrogen generation rate in thermochemical water splitting is slightly higher compared to the SOEC. This is due to multiple factors such as the individual reactor's conversion efficiency, the temperature limit, operational conditions, etc. Thus the direct comparison of these two technologies based on the hydrogen production rate may not yield proper validation. This study shall be used as a reference for understanding the potential of large-scale green hydrogen production, especially for the Indian context.

References

1. Iea IEA (2019). The future of hydrogen Futur Hydrog. https://doi.org/10.1787/1e0514c4-en
2. BP (2014) Statistical review of world energy. Stat Rev World Energy 67:1–56
3. IRENA (2019) Hydrogen: a renewable energy perspective
4. Liu H, Liu S (2021) Life cycle energy consumption and GHG emissions of hydrogen production from underground coal gasification in comparison with surface coal gasification. Int J Hydrogen Energy 46:9630–9643. https://doi.org/10.1016/j.ijhydene.2020.12.096
5. Reeves D (1980) Energy requirements in rye production. Handb Energy Util Agric
6. Doenitz W, Schmidberger R, Steinheil E, Streicher R (1980) Hydrogen production by high temperature electrolysis of water vapour. Int J Hydrogen Energy 5:55–63. https://doi.org/10.1016/0360-3199(80)90114-7
7. Srilakshmi G, Ramaswamy MA, Thirumalai N (2019) Design of solar field and performance estimation of solar tower plants. J Chem Inf Model 53:1689–1699
8. Ho CK (2017) Advances in central receivers for concentrating solar applications. Sol Energy 152:38–56. https://doi.org/10.1016/j.solener.2017.03.048
9. Kodama T, Nakamura Y, Mizuno T (2006) A two-step thermochemical water splitting by iron-oxide on stabilized zirconia. J Sol Energy Eng Trans ASME 128:3–7. https://doi.org/10.1115/1.1878852
10. Statnatiou A, Loutzenhiser PG, Steinfeld A (2010) Solar syngas production via H2O/CO2-splitting thermochemical cycles with Zn/ZnO and FeO/Fe3O4 redox reactions. Chem Mater 22:851–859. https://doi.org/10.1021/cm9016529
11. Steinfeld A (2002) Solar hydrogen production via a two-step water-splitting thermochemical cycle based on Zn/ZnO redox reactions. Int J Hydrogen Energy 27:611–619. https://doi.org/10.1016/S0360-3199(01)00177-X
12. Bhosale RR, Takalkar G, Sutar P (2019) A decade of ceria based solar thermochemical H2O/CO2 splitting cycle. Int J Hydrogen Energy 44:34–60. https://doi.org/10.1016/j.ijhydene.2018.04.080
13. Laguna-Bercero MA (2012) Recent advances in high-temperature electrolysis using solid oxide fuel cells: A review. J Power Sources 203:4–16. https://doi.org/10.1016/j.jpowsour.2011.12.019
14. Sorgulu F, Dincer I (2019) Design and analysis of a solar tower power plant integrated with a thermal energy storage system for cogeneration. Int J Energy Res 43:6151–6160. https://doi.org/10.1002/er.4233
15. Noone CJ, Torrilhon M, Mitsos A (2012) Heliostat field optimization: A new computationally efficient model and biomimetic layout. Sol Energy 86:792–803. https://doi.org/10.1016/j.solener.2011.12.007
16. Besarati SM, Yogi GD (2014) A computationally efficient method for the design of the heliostat field for solar power tower plant. Renew Energy 69:226–232. https://doi.org/10.1016/j.renene.2014.03.043
17. Yao Y, Hu Y, Gao S (2015) Heliostat field layout methodology in central receiver systems based on efficiency-related distribution. Sol Energy 117:114–124. https://doi.org/10.1016/j.solener.2015.04.029
18. Abaza MA, El-Maghlany WM, Hassab M, Abulfotuh F (2020) 10 MW Concentrated Solar Power (CSP) plant operated by 100% solar energy: Sizing and techno-economic optimization. Alexandria Eng J 59:39–47. https://doi.org/10.1016/j.aej.2019.12.005
19. Ermanoski I, Siegel NP, Stechel EB (2013) A new reactor concept for efficient solar-thermochemical fuel production. J Sol Energy Eng Trans ASME 135:1–10. https://doi.org/10.1115/1.4023356
20. Charvin P, Abanades S, Flamant G, Lemort F (2007) Two-step water splitting thermochemical cycle based on iron oxide redox pair for solar hydrogen production. Energy 32:1124–1133. https://doi.org/10.1016/j.energy.2006.07.023

A Study on Rayleigh Plateau Instability in Slender Jets of Nuclear Materials

M. Chandra Kumar, A. Jasmin Sudha, V. Subramanian, R. Venkatesan, S. Athmalingam, and B. Venkatraman

Abstract Nuclear fuel pins arranged inside a hexagonal stainless steel wrapper form a fuel subassembly. The subassemblies are supported on the grid plate making the nuclear reactor core. In the unforeseen event of the core disruptive accident, these fuel pins melt through the grid plate and interact with the coolant sodium in the reactor main vessel in sodium-cooled Fast Reactor. This interaction of the fuel and coolant is commonly termed as Molten Fuel Coolant Interaction (MFCI) in nuclear industry. The molten material will fall in the form of jet or stream into the coolant under the action of gravity. The study of MFCI is important because of its potential to add more thermal load on the core catcher based on the jet breakup and debris settling pattern on the core catcher. The molten jet will be subjected to thermodynamic as well as hydrodynamic forces which will eventually lead to its breakup. Hydrodynamic forces such as surface tension and inertial forces play a prominent role in the breakup of the jet. Depending on the jet velocity, the jet breakup can happen in dripping or jetting mode. Rayleigh Plateau (RP) instability prevails in dripping regime of the jet. In this work, nature of flow of jet which breaks up due to RP instability is investigated numerically, applying stability analysis theory. Breakup length, breakup time and primary droplet diameter of uranium plutonium mixed oxide fuel (MOX) jet in liquid sodium at low jet velocities were studied. The trend of breakup characteristics is also evaluated for a few other nuclear reactor materials such as metallic fuel and stainless steel.

Keywords Jet breakup · Molten corium · Fuel coolant interaction · Rayleigh plateau instability

---

M. Chandra Kumar (✉) · A. Jasmin Sudha
Indira Gandhi Center for Atomic Research, A CI of Homi Bhabha National Institute, Kalpakkam 603102, Tamil Nadu, India
e-mail: chandrakumar@igcar.gov.in

A. Jasmin Sudha
e-mail: jasmin@igcar.gov.in

A. Jasmin Sudha · V. Subramanian · R. Venkatesan · S. Athmalingam · B. Venkatraman
Safety, Quality & Resource Management Group, IGCAR, Kalpakkam 603102, Tamil Nadu, India

N. M. Sivaram et al. (eds.), *Advances in Manufacturing, Automation, Design and Energy Technologies*, Lecture Notes in Mechanical Engineering,
https://doi.org/10.1007/978-981-99-1288-9_75

1 Introduction

Melting of core leading to core disruption in a SFR is one of the severe accident scenarios. Such an accident is considered in Design Extension Conditions (DECs) of the nuclear reactor [1]. The reactor is equipped with two diverse, fast acting and reliable shutdown systems which kick in as soon as one or more SCRAM parameters crosses the threshold value, bringing the reactor to safe shut down state. Moreover, there are many inherent design features and other safety provisions installed in the reactor core which make CDA highly improbable. The probability of such kind of core disruptive accidents (CDA) is as low as 10^{-6} per reactor year [2]. Nonetheless, CDAs cannot be ignored and are studied worldwide for safety purposes. In the core-melt scenario, the core-melt gets deposited in the grid plate. The core-melt then melts through the grid plate and falls in the form of multiple streams of jet into the lower plenum of reactor main vessel. The illustration of the phenomenon in the lower part of the reactor main vessel is shown in Fig. 1. This jet will undergo different types of hydrodynamic and thermodynamic phenomena leading to its breakup. The study of this breakup is particularly important in post-accident heat removal and in understanding the fragmentation of the molten fuel. Here in the present work, we consider only hydrodynamic part which leads to the breakup of the jet.

A falling stream of jet is inherently unstable by nature. Jet breakup has always been a topic of research since very long time. Jet instability finds its uses in many industrial applications such as spray painting, fuel injection, inkjet printing, micro-fluidic lab on chip, drawing of optical fibres, long-term stability of electronic devices and many more. There are different forces which come into play during the jet breakup such as inertial, gravity, surface tension and viscous forces. A flowing fluid doesn't have a shape of its own; the dominance of these forces mentioned above will determine the shape of the fluid. Similarly in our case, molten fuel in the form of jet falling into the liquid sodium will be subjected to different forces leading to changing of its shape.

Depending on the relative velocity of the jet, breakup pattern is divided into two modes. At very low velocities, it is known as dripping mode and at higher velocities

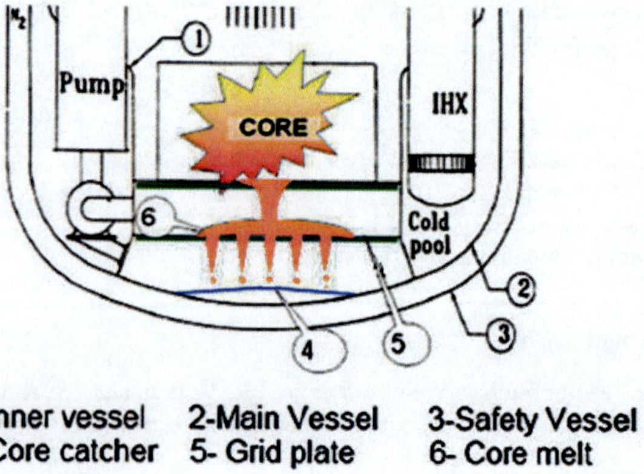

Fig. 1 Illustration of the degraded core

it is known as jetting mode. Generally, when jet weber number $We_j > 8$, jetting mode sets in. It is observed in dripping mode surface tension force plays dominant role than inertial forces and gravitational forces. The instability prevailing in this mode is known as Rayleigh Plateau (RP) instability. At higher velocity, i.e., in jetting mode inertial forces play a major role along with other forces. Major instabilities driving the breakup in jetting mode are Rayleigh Taylor instability and Kelvin Helmholtz instability. Rayleigh Taylor instability will happen at the leading edge of the jet whereas Kelvin Helmholtz instability is observed on the sides of the jet due to shear forces.

Here in this work, we are dealing with very low velocity of the jet stream falling under the influence of gravity. The jet will appear like thread like smooth cylindrical jet before the surface tension forces dominate and break the jet into droplets [3]. In 1992, Eggers and Dupont [4] developed a one dimensional equation of motion of velocity derived from Navier Stokes equation in order to study the nature of singularities developed in column of fluids. Singularity develops when the jet is pinching off. Bang et al. [5] experimentally studied the breakup of melt jet in water under non boiling conditions. The variables in this experiment were diameter of the jet and jet velocity, the debris collected was analysed and mass fractions were calculated. In 2004, Cheong et al. [6] analysed the influence of the gravity in the capillary jet. It was found from the numerical analysis that under the influence of gravity, the elongation of the disturbance takes place. It was concluded that for low velocity capillary jet, the formation of satellite drop is highly unlikely for high local wavenumber at the point of pinch off. Later, they validated their results with the experiments. Omocea et al. [7] studied the influence of viscosity, ratio of jet viscosity and surrounding medium viscosity on the breakup length of the jet experimentally. They studied the effect of jet material density and viscosity on the surrounding medium which was oil in their case; they increased the density of the jet by adding glycerine to the water. They reported that increasing the viscosity and the density of the jet leads to the decrease in the breakup length of the jet. In 2016, Rosello et al. [8] studied the effect of bond number in the formation of droplet in the dripping mode of jet. It was observed that with the increase in the bond number at low jet velocities, the stretching of the neck portion of the jet takes places which results in the increase in the satellite drop volume when compared with drop volume.

In the present study, the breakup of nuclear fuel materials such as MOX, metallic fuel and stainless steel due to RP instability is undertaken. Investigation of breakup length is crucial in the case of MFCI because it is important for jet to breakup before reaching the core catcher to avoid damage due to jet ablation. In the cases, where the breakup of the jet doesn't happen then the molten material will agglomerate at the core catcher which will drastically affect its coolability, increasing the thermal load on core catcher. If the breakup of the jet occurs much before the jet reaching the core catcher, then fragmentation of the molten droplet will take place and much of the heat will be dissipated to the coolant. So the positioning of the core catcher has to be such that there is sufficient sodium column to break up the jet before it reaches the core catcher.

2 Governing Equation

Boundary Conditions and Initialisation

Here, sinusoidal perturbation with time is applied on the velocity and is carried through the downstream [9].

$$u_0(0,t) = U_0\left[1 + \varepsilon sin\left(\frac{CU_0 t}{R_0}\right)\right] \tag{1}$$

where $h(z, 0) = R_0$, $U_0 = 0.5$ m/s, $\varepsilon = 0.05$ and $C =$ scaling constant.

To incorporate the effect of acceleration due to gravity, interface radius and initial velocity can be initialised as [5]

$$h(z,0) = R_0\left(1 + \frac{2g' z}{U_0{}^2}\right)^{-1/4} \tag{2}$$

$$u_0(z,0) = U_0\left(1 + \frac{2g' z}{U_0{}^2}\right)^{1/2} \tag{3}$$

Here in order to incorporate buoyancy of the surrounding media, acceleration due to gravity is modified, the modified acceleration due to gravity can be represented as

$$g' = g\left(1 - \frac{\rho_{\text{amb}}}{\rho_{\text{jet}}}\right) \tag{4}$$

In our case, MOX is chosen as jet and sodium is chosen as ambient medium.

When the length of the jet is much more than that of the diameter of the jet, slender jet approximations can be used. In our case, we have assumed slender jet approximation so the length of the jet is taken to be fifty times that of the diameter. Equation 5 gives expressions for slender jet approximation for velocity and pressure. In the slender jet approximations, length of the jet is much larger than its radius so the approximations for velocity, pressure and temperature can be expanded in the form of power series of r. Since we are using axis symmetric condition, the first derivative of the variables vanishes. So, variable like velocity and pressure can be approximated in the power series of r as

$$u_z(r,z,t) = u_0(z,t) + r^2\frac{\partial^2 u_z(z,t)}{\partial z^2} + \cdots$$

$$p(r,z,t) = p_0(z,t) + r^2\frac{\partial^2 p(z,t)}{\partial z^2} + \cdots \tag{5}$$

Here, the continuity equation and momentum equations are conserved, by applying the boundary condition and slender jet approximation, the final differential governing equation comes out to be [10]

$$\frac{\partial u_0}{\partial t} = -u_0 \frac{\partial u_0}{\partial z} - \frac{1}{\rho} \frac{\partial}{\partial z}\left(\sigma \frac{\partial^2 h}{\partial z^2}\right) + \frac{2}{\rho h} \frac{\partial \sigma}{\partial z} + \frac{3\mu}{\rho h^2} \frac{\partial}{\partial z}\left(h^2 \frac{\partial u_0}{\partial z}\right) + \frac{3}{\rho} \frac{\partial \mu}{\partial z} \frac{\partial u_0}{\partial z} + g \tag{6}$$

$$\frac{\partial h}{\partial t} = -\left(\frac{h}{2} \frac{\partial u_0}{\partial z} + u_0 \frac{\partial h}{\partial z}\right) \tag{7}$$

3 Computational Model

The code DROP which is an in-house developed code has been employed with suitable modification on acceleration due to gravity as given in Eq. (4) to study the effect of jet velocity on the breakup length and breakup time for the case of dripping flow. The validation of this code was done by Dipin et al. [9] with the experimental data of wood's metal jet breakup in water. In the present study, the effect of Weber number and material properties on breakup length and time has been evaluated addressing the RP instability.

In the first case study, breakup of MOX jet in sodium is analysed in dripping regime by varying Weber number from 2 to 7. Diameter is varied in order to change Weber number. For each diameter, the fastest growing disturbance is found out by varying scaling constant given in the Eq. (4). The minimum breakup time corresponding to the fastest growing disturbance is found out for each case. In the second case study, breakup length of different nuclear materials is assessed by keeping jet diameter of the jet as 1 mm and its velocity as 0.5 m/s. Since the mixed oxide fuel contains 80% uranium oxide and 20% plutonium oxide, its properties are approximated as that of uranium oxide properties. Similarly, properties of uranium are used for the metallic fuel containing about 80% uranium along with a small fraction of plutonium and zirconium.

4 Results and Discussion

A sinusoidal perturbation is given to the jet initial velocity as it enters into the coolant as defined in Eq. (1). As we can see from the graph, breakup time decreases from the initial value and then starts to increase after reaching a minimum value.

The corresponding value of wave number at this particular point indicates the most disturbing wave number of the flow. The wave number for $U_0 = 0.5$ m/s and diameter $= 2R_0 = 1$ mm comes out to be 2000. Weber number for this case is 4.7. For

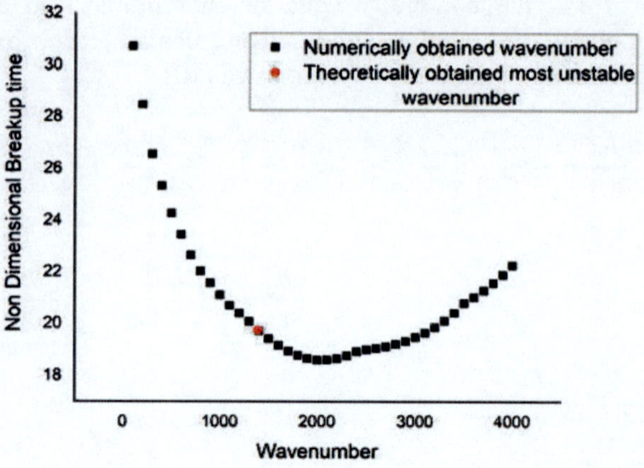

Fig. 2 Variation of breakup time with wave number for MOX with $U_0 = 0.5$ m/s and $D = 1$ mm

all of the case studies undertaken in this study, the reported values are obtained with the fastest growing disturbance corresponding to minimum breakup time. Rayleigh in 1879 derived the expression for most unstable wave number which is given by $kR_0 = 0.697$.

The wave number calculated from the above expression is for an ideal case considering only surface tension forces [10]. The value of most unstable wave number from above equation comes out to be 1390. The deviation of the wave number obtained numerically from that of the ideal case can be observed in the Fig. 2.

This is due to the fact that in present numerical model, more realistic approach is used considering gravitational forces and viscous forces along with surface tension forces whereas only surface tension forces were considered while deriving the analytical expression given by Rayleigh's theory.

4.1 *Case 1: Parametric Study on Weber Number in Dripping Mode*

Here, a parametric study is conducted by varying diameter of the jet, in order to investigate the changes in breakup time and breakup length. Table 1 lists the scaling constant (C) which corresponds to minimum breakup time. It is observed that breakup length increases with the increase in the Weber number in dripping mode. The breakup length and breakup time are reported in non-dimensional form in the table. The symbol t is the actual breakup time in seconds and L is actual breakup length in metres. R_0 is the unperturbed radius of the jet which is 0.5 mm.

In the jetting mode, which is not addressed here, the breakup length and breakup time are bound to decrease because of the shear forces coming into play. Hence, the breakup length and time obtained from dripping mode can be considered as the upper bound of the parameters.

Table 1 Jet breakup characteristics for MOX in liquid sodium

Weber number	C	Breakup time ($t^* = U_0.t/R_0$)	Breakup length ($l^* = L/R_0$)	Pinch off diameter ($D^* = D/D_0$)
2	0.9	14.8	13.9	1.91
3	0.95	16.5	16.4	1.87
4.7	1	18.6	20.1	1.85
6	1.1	19.9	22.9	1.76
7	1.2	20.7	24.9	1.72

Oren Breslouer [10] obtained an expression for the breakup length of the jet in dripping mode with the help of the dimensional analysis. It was found that the breakup length has functional dependency on the jet velocity, unperturbed radius of the jet and density and surface tension of the jet fluid. The final expression suggested for breakup length is given in Eq. (8).

$$l^*_{\text{Analytical}} = \left(\frac{L}{R}\right)_{\text{Analytical}} \sim (\text{We})^{0.5} \tag{8}$$

From the breakup length of MOX jet in sodium, evaluated from the present numerical analysis, the constant of proportionality is found to be 9.5. Therefore,

$$l^*_{\text{Analytical}} = 9.5(\text{We})^{0.5} \tag{9}$$

where l^* is the non-dimensional breakup length, L is the breakup length of the jet and R is unperturbed radius. The parity plot between $(L/R)_{\text{analytical}}$ and $(L/R)_{\text{numerical}}$ is given in Fig. 3.

4.2 *Case 2: Jet Breakup Study with Different Nuclear Materials*

This case study has been undertaken to study jet breakup parameters in metallic fuel and stainless steel also; these materials also find application in SFR. The variation of C, breakup time, breakup length and pinch off diameter is analysed for these materials keeping initial velocity and unperturbed radius of the jet as constants. Table 2 shows the variation of Weber number, C, breakup time, breakup length and pinch off diameter for different material. As it can be noted from the table, the breakup time for MOX is more than metallic fuel and stainless steel. Similar trend can be observed in breakup length also for the respective materials. So it can be inferred that higher the surface tension, lower will be the breakup time and breakup length of the jet. It is because higher the surface tension will lead to higher energy release on

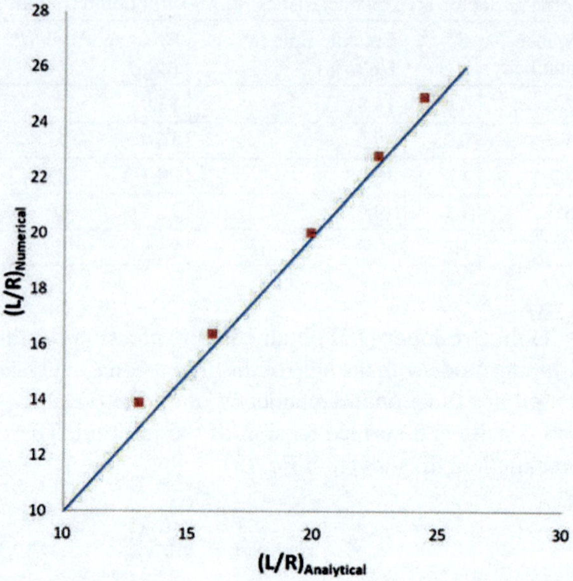

Fig. 3 Parity plot between numerically and analytically obtained breakup lengths

breakup. So the jet will breakup faster in order to reduce its surface area and lower its potential energy.

Figure 4 shows interface profile at the time of pinch off for stainless steel jet. The criterion for jet pinching is fixed such that the jet breaks off when jet radius diminishes to 5% of the unperturbed initial radius.

Table 2 Breakup characteristics for different materials ($D_0 = 1$ mm; $U_o = 0.5$ m/s)

Material	Weber number	C	Breakup time ($t^* = U_0t/R_0$)	Breakup length ($l^* = L/R_0$)	Pinch off diameter ($D^* = D/D_0$)
SS	1.1	1.05	10.7	14.2	0.9
Metallic fuel	2.4	1.05	17.6	18.8	1.8
MOX	4.7	1	18.6	20.0	1.85

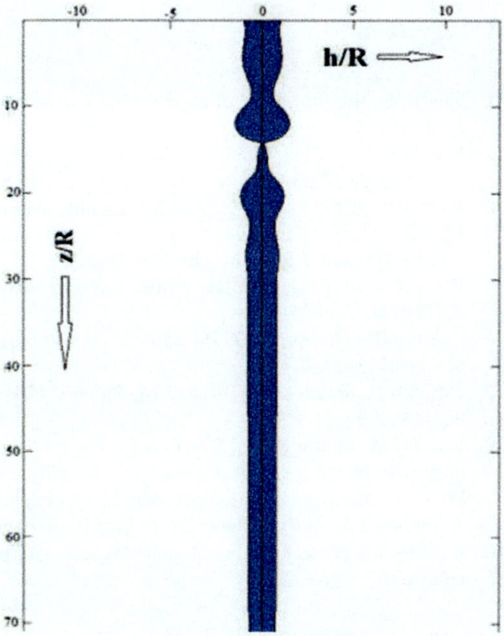

Fig. 4 SS jet interface at the time of jet pinching

5 Conclusion

The semi analytical analysis of Rayleigh Plateau instability developing in very low Weber number flow has been conducted in this study for a few materials which are of interest in the nuclear reactor industry. In the first case, the Weber number is varied and its influence on breakup length and breakup time has been studied with mixed oxide fuel jet dispersing in sodium. It is observed that the breakup length increases with the increase in the Weber number in the dripping mode of the jet whereas there is a slight decrease in the pinchoff droplet diameter. The primary droplet pinching off from the jet can be as big as twice the jet diameter, as inferred from the analysis. In the second case, the molten jet material is changed and its influence on breakup length and breakup time is investigated keeping jet velocity and the radius of the jet constant. It has been noted that the jet breakup length can lie between 10 and 20 times of the jet radius. The present study is done with dripping mode regime. The estimates of breakup length and drop size obtained in this study are the most conservative ones since shear forces, ambient liquid entrainment and thermal effects are neglected here. Further advanced models are being developed for jetting flows.

References

1. Chellapandi P, Rao PRV, Kumar P (2015) Fast reactor programme in India. Pramana J Phys 85:525–538
2. Kumar CS, Arul AJ, Singh OP, Rao KS (2005) Reliability analysis of shutdown system. Ann Nucl Energy 32:63–87
3. Rapp BE (2017) Plateau-Rayleigh instability. Microfluid Model Mech Math 467–477
4. Eggers J, Dupont TF (1994) Drop formation in a one-dimensional approximation of the Navier-Stokes equation. J Fluid Mech 262:205–221
5. Bang KH, Kim JM, Kim DH (2003) Experimental study of melt jet breakup in water. J Nucl Sci Technol 40:807–813
6. Cheong BS, Howes T (2004) Capillary jet instability under the influence of gravity. Chem Eng Sci 59:2145–2157
7. Omocea IL, Patrascu C, Turcanu M, Balan C (2016) Breakup of liquid jets. Energy Procedia 85:383–389
8. Rosello M, Maitrejean G, Roux DCD, Jay P (2016) Numerical investigation of the influence of gravity on the Rayleigh-Plateau jet instability. Fluid Dyn Res 48:1–13
9. Pillai DS, Narayanan P, Pushpavanam S, Sundarajan T, Jasmin Sudha A, Chellapandi P (2012) A nonlinear analysis of the effect of heat transfer on capillary jet instability. Phys Fluids 24
10. https://www.princeton.edu/~stonelab/Teaching/Oren%20Breslouer%20559%20Final%20Report.pdf

Numerical Investigation of a Channel During Loss of Coolant Accident

Bhuwanesh Kumar, Ravi Kumar, and Akhilesh Gupta

Abstract This paper numerically investigated the thermal behavior in the nuclear reactor channel during the Loss of Coolant Accident (LOCA) along with the failure of the Emergency Core Cooling System (ECCS). The temperature distribution has been thoroughly analyzed and its effect on the behavior of the reactor channel is simulated using steady-state thermal analysis in ANSYS Workbench 19.2. In steady-state thermal analysis, surface-to-surface correlation property, and perfect enclosure type property were used to determine the temperature distribution and heat fluxes in the channel. The temperature was found to range from 1186.8 to 1184.5 °C, 1129.3 to 965.83 °C, 932.63 to 757.88 °C, and 619.01 to 612.72 °C in surfaces of center clad, middle clads ring, outer clads ring, and pressure tube (PT), respectively. Heat loss from clad tubes to PT inner surface and calandria tube (CT) inner surface is approximately 1.84 and 1.89%, respectively.

Keywords Steady-state thermal · LOCA · PT · CT

1 Introduction

In 220 MWe, Indian Pressurized Heavy Water Reactor (IPHWR), heavy water is used as the primary heat transport system and circulating through 306 horizontal channels which are housed in calandria vessel and submerged in moderator (heavy water). In the reactor, heavy water flows in the core of the channels and picks up heat generated in the fuel bundles and transfer it to the secondary fluid (normal water) to generate the power. Each pressure tube contains 12 fuel bundles and each fuel bundle consists of 19 fuel rods [1]. The 19 fuel elements are arranged in a concentric ring having 1, 6, and 12 elements in different rings. Fuel clads caps are welded at both ends which acts as a barrier to avoid the release of radioactive emissions and also protects the

B. Kumar (✉) · R. Kumar · A. Gupta
Department of Mechanical and Industrial Engineering, Indian Institute of Technology Roorkee, Roorkee 247667, India
e-mail: 347bhuwanesh@gmail.com

N. M. Sivaram et al. (eds.), *Advances in Manufacturing, Automation, Design and Energy Technologies*, Lecture Notes in Mechanical Engineering,
https://doi.org/10.1007/978-981-99-1288-9_76

fuels from corrosion from the coolant. Each channel consists of a PT 82 mm inner and 90 mm outer diameter, respectively [2]. PT is concentrically placed in CT of 109 mm ID and 110 mm OD, respectively. The length of each channel is 6000 mm and the gap between PT and CT is 8.5 mm [3] and is filled with CO_2 which work as a thermal insulator [3, 4].

Under postulated accident scenarios in IPHWRs like Loss of Coolant Accident (LOCA) (e.g., break of primary system pipe) coexisting with Loss of Emergency Core Cooling System (LOECCS). In such a scenario, decay heat transfer by the radiation becomes the dominating mode of heat transfer over the conduction and convection mode of heat transfer from clad tubes to PT and PT to CT. Fuel bundles heat up as it continues to produce decay heat even after the reactor is tripped. The temperature of the PT increases, if the pressure inside the PT is less than 1 MPa, the PT sags, due to high temperature, fuel bundle, and its weight, and comes in contact with CT [3]. PT eccentricity effects on circumferential temperature distribution on channel components of IPHWR fuel bundle.

Governing Equations:

Radiation exchange in 220 MWe, nuclear power plant channel.

G_i Irradiation at zone i; W/m^2

J_i Radiosity at zone i; W/m^2

Radiosity expression for a gray opaque surface.

$$J_i = \epsilon_i E_{bi} + \rho G_i \tag{1}$$

q_i Net heat flux leaving at zone i; W/m^2

$$q_i = J_i - G_i \tag{2}$$

$\alpha + \rho + \tau = 1$ for a gray opaque body $\tau = 0$, $\alpha = \epsilon$ from Kirchhoff's law, $\rho = 1 - \alpha = 1 - \epsilon$ put in Eq. (1)

$$J_i = \epsilon_i E_{bi} + (1 - \epsilon_i) G_i \tag{3}$$

Radiation energy leaving the surface A_J that strikes surface-to-surface A_i,

$$J_j A_j F_{ji} = J_i A_i F_{ij}$$

Radiation from all the surface that strikes A_i (including itself)

$$= \sum_{J=1}^{N} J_j A_j F_{ji} = A_i \sum_{J=1}^{N} J_j F_{ij} \big(A_i F_{ij} = A_j F_{ji}\big).$$

Irradiation $G_i = \sum_{J=1}^{N} J_j F_{ij}$

Net radiation leaving surface i, $q_i = J_i - \sum_{J=1}^{N} J_j F_{i-j}$

$$E_{bi} = \frac{J_i}{\epsilon_i} - \frac{1-\epsilon_i}{\epsilon_i}\sum_{J=1}^{N} J_j F_{i-j} \tag{4}$$

$$\frac{J_i}{\epsilon_i} - \frac{1-\epsilon_i}{\epsilon_i}\sum_{\mathrm{J}=1}^{N} J_j F_{ij} = E_{bi} => [M][J] = [E]$$

where $[M] = \begin{bmatrix} m_{11} & \dots & m_{1N} \\ \vdots & \ddots & \vdots \\ m_{1N} & \cdots & m_{NN} \end{bmatrix}$.

$$m_{ij} = \frac{\delta_{ij} - (1-\epsilon_i)F_{ij}}{\epsilon_i} \tag{5}$$

$$m_{ij} = \delta_{ij} - F_{ij} \tag{6}$$

$$\delta_{ij} = 1\,(\text{if } i = j), = 0\,(\text{if } i \neq j)$$

Equation (5) is used if the average temperature of all the clad tubes and PT are known.

Equation (6) is used if the heat flux of the all clad tubes are known.

Radiosity vector $[\mathrm{J}] = \begin{bmatrix} J_1 \\ J_2 \\ \vdots \\ J_N \end{bmatrix}$, the surface input vector $[\mathrm{E}] = \begin{bmatrix} E_{b1} \\ E_{b2} \\ \vdots \\ E_{bN} \end{bmatrix} = \begin{bmatrix} \sigma T_1^4 \\ \sigma T_2^4 \\ \vdots \\ \sigma T_N^4 \end{bmatrix}$

$q_i = \frac{\epsilon_i(E_{bi}-J_i)}{1-\epsilon_i} = \frac{E_{bi}-J_i}{\frac{1-\epsilon_i}{\epsilon_i}} = \frac{E_{bi}-J_i}{\mathrm{R}}$ where $\frac{1-\epsilon_1}{\epsilon_1}$ is surface thermal resistance to radiation between the potential E_{bi} and J_i.

2 Modeling and Simulation

A 3D model of a channel is generated in ANSYS SpaceClaim. The dimensions of the channel model are given in Table 1.

In this analysis, thermal conductivity and emissivity of the channel were assumed to be constant as shown in Table 2.

220 MWe nuclear power plant, a bundle of 19-element is generating decay heat with a radial power factor 1: 1.1: 1.4. The applied heat flux in different clad tubes is shown in Table 3.

Table 1 Channel model dimensions

	OD (mm)	ID (mm)	Length (mm)
Clad	15.2	14.44	10
PT	82	90	10
CT	107	110	10

Table 2 Thermal properties of channel material

	Thermal conductivity (W/m K)	Emissivity [5]
Clad	16	0.8
PT	20	0.8
CT	16	0.8

Table 3 Heat flux applied to the concentric clad tubes' rings

	Heat flux (W/m^2)
Center clad	34,952
Middle clads	6407.8
Outer clads	4077.7

The meshing of the channel has been done in ANSYS Workbench as shown in Fig. 1. Mesh sensitivity test has been done.

The quality of the meshing is illustrated in Fig. 2. The skewness mesh metrics are used to check the skewness of all elements of the channel. The skewness of the element is found out to be approximately equal to 0.1.

Radiation is the major source of heat transfer in LOCA along with ECCS in nuclear reactor channels. In the setup model, within steady-state thermal analysis simulation of a channel "radiation" load is available to utilize the surface-to-surface radiation capabilities to include radiation exchange between surfaces of clad tubes, PT, and CT. Heat transfer between the surfaces by the radiation mode of heat transfer which is a highly non-linear mode of heat transfer is given in Eq. 7.

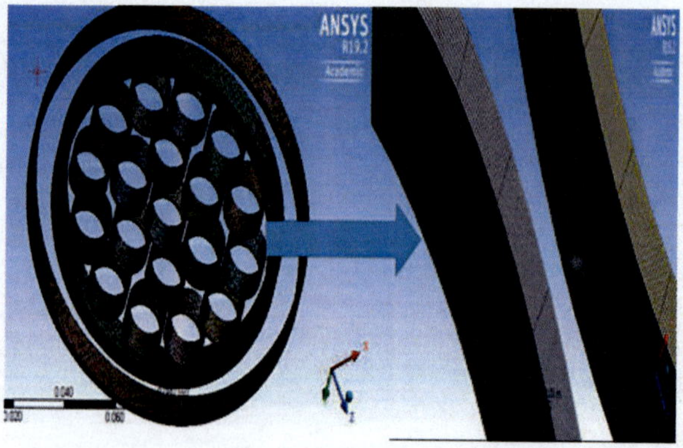

Fig. 1 Mesh configuration for steady-state thermal analysis

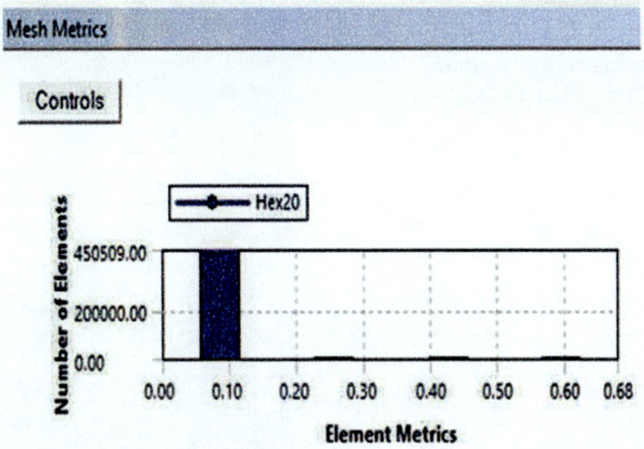

Fig. 2 Skewness mesh metrics

$$Q_{ij} = A_i \in_i F_{ij} \sigma (T_i^4 - T_j^4) \tag{7}$$

Surface-to-surface correlation was used to calculate the radiation heat transfer between the participating surfaces, while perfect enclosure type property was used to confine the heat transfer boundary within nuclear reactor channel tubes. The order of accuracy for the convergence criteria taken is 1 × e−06.

3 Results and Discussions

The circumferential temperature variations over clad tubes, PT, and CT have been studied through steady-state thermal simulation using ANSYS Workbench 19.2. The temperature distribution for the innermost clad tube is shown in Fig. 3. The temperature ranges from 1186.6 to 1184.5 °C on the surface of the center clad tube. The circumferential temperature distribution in middle clads and outer clads is shown in Figs. 4 and 5, respectively. The temperature difference of 163.47 and 156.75 °C is found along the circumference of middle pins and outer pins respectively. The circumferential temperature over PT and CT is shown in Figs. 6 and 7, respectively. The net heat reached the inner surface of the CT and then heat is transferred from CT to the moderator by the convection mode of heat transfer. In this model, the convective heat transfer coefficient is taken 500 W/m^2 K.

The energy balance is shown in Tables 4 and 5. The energy outgoing and energy received is shown by positive and negative signs, respectively.

Table 5 demonstrates that the energy provided to the clads tubes is identical to the energy found in the PT inner and CT inner surfaces with 1.84 and 1.89% inaccuracy, respectively.

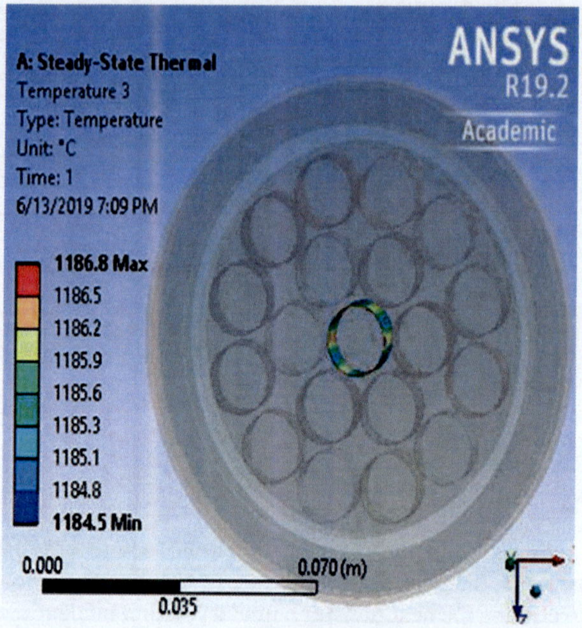

Fig. 3 Circumferential temperature variation over-center clads

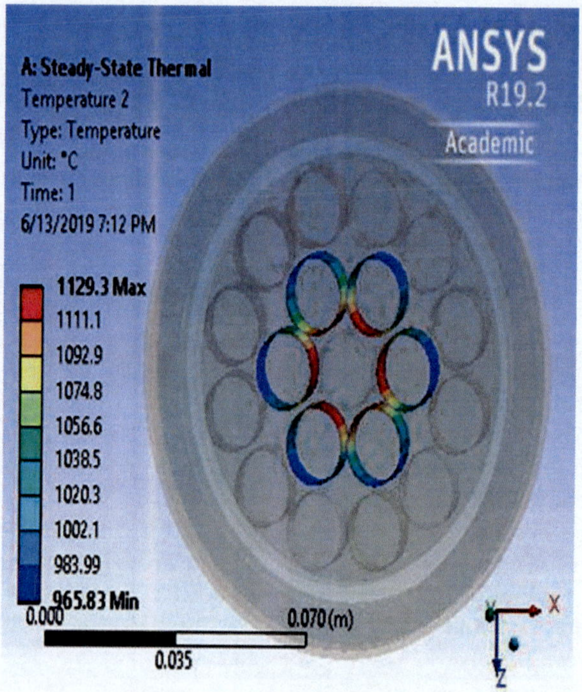

Fig. 4 Circumferential temperature variation over middle clads

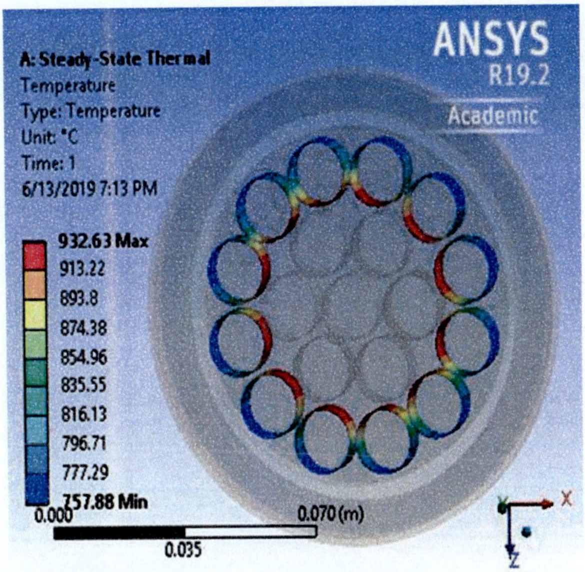

Fig. 5 Circumferential temperature variation over outer clads

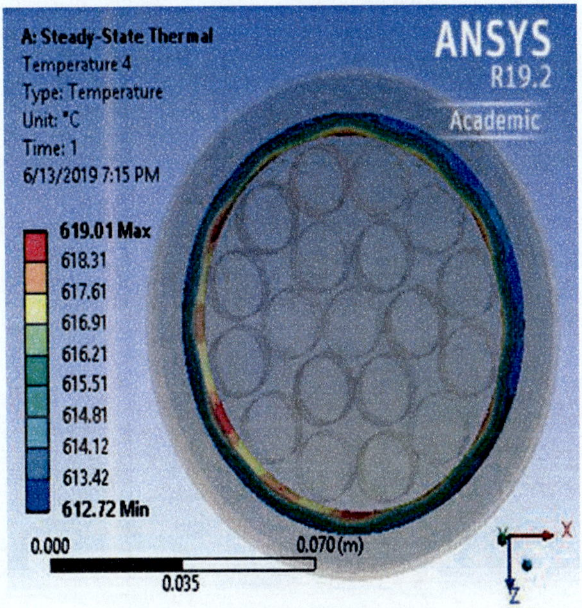

Fig. 6 Circumferential temperature variation over PT

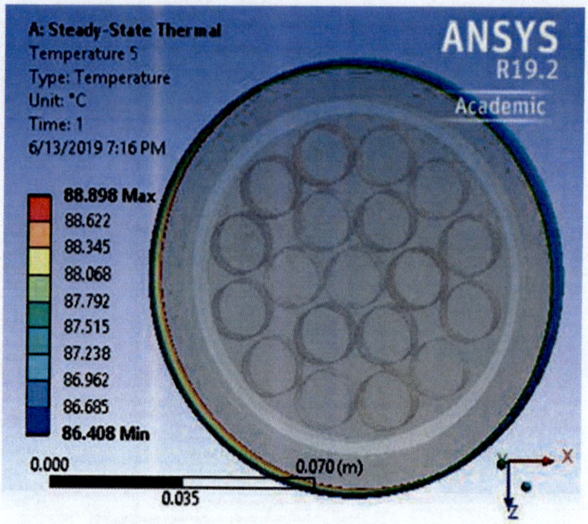

Fig. 7 Circumferential temperature variation over CT

Table 4 Calculation for net energy outgoing/gaining by the channel components

	Incident radiation (W)	Reflected radiation	Emitted radiation (W)	Net outgoing radiation (W)
Center clad tube surface	102.77	20.553	98	15.786
Middle clad tubes ring surfaces	465.09	93.019	399.33	17.258
Outer clad tubes ring surfaces	482.58	96.515	408.2	22.134
PT inner surface	160.08	32.016	73.616	-54.448
CT inner surface	71.34	14.268	2.602	-54.473

Table 5 Energy balance

Eccentricity PT with CT	Input energy (W) (all clad tubes)	Energy received by PT inner surface (W)	Energy received by CT inner surface (W)	Energy loss to PT inner surface	Energy loss to reach CT inner surface
e = 0	55.495	54.473	54.448	1.84%	1.89%

4 Conclusions

The following conclusions can be drawn from the steady-state thermal analysis in a 220 MWe, IPHWR channel using ANSYS Workbench:

1. The center pin is at the maximum temperature, which confirms the findings of Sharma et al. [2].
2. The maximum outgoing radiation is found in the outer clads ring.
3. The maximum temperature difference between center clad and middle clad and center clad outer clad are found 57.5 and 254.17 °C.

References

1. Bajaj SS, Gore AR (2006) The Indian PHWR. Nucl Eng Des 236(7–8):701–722
2. Sharma M, Kumar R, Majumdar P, Mukhopadhyay D (2018) Effect of the eccentricity of pressure tube on circumferential temperature distribution of PHWR fuel bundle under postulated accident condition. Nucl Eng Des 331:274–281
3. Nandan G, Sahoo PK, Kumar R, Chatterjee B, Mukhopadhyay D, Lele HG (2010) Thermo-mechanical behavior of pressure tube of Indian PHWR at 20 bar pressure. World Acad Sci Eng Technol 61:205–213
4. Ajay K, Kumar R, Gupta A (2021) Assessment of heat transfer in fuel channel of Indian Phwr under postulated large break loss of coolant accident: experimental and numerical study. In: EPJ web of conferences, vol 247. EDP Sciences
5. Kim HT, Rhee BW, Park JH (2006) Benchmark calculations of a radiation heat transfer for a CANDU fuel channel analysis using the CFD code. J Nuclear Sci Technol 43(11):1422–1430

Experimental Comparative Analysis of Heat Transfer Enhancement in Shell and Coiled Tube Heat Exchanger with Winged Insert and Annular Fin

R. Periyasamy and S. Santhia

Abstract Annular fins and winged insert are utilized in the helically coil to upgrade the shell and curled pipe heat exchanger, and furthermore, the frictional and heat attributes are introduced in this paper. The experimental examination is done in four methods of activity. Water is utilized as the liquid in the heat exchanger. In the primary mode, the winged insert is in the coiled tube, and the annular blades are on the external surface tube. In the subsequent mode, just the addition is set inside the coiled tube. In the third mode, just the annular fins are fitted above the external side of the curled tube. In the fourth mode, an empty tube is analyzed. From the previous study, this work differs in the use of the turbulator which has a novel design of thorn like projections present along the length of the insert which was not used before. Findings demonstrated that for the coil tube inner diameter of 10 mm, shell length of 700 mm, and annular fin surface are of 4 mm^2, the greatest estimation of effectiveness and the overall heat transfer coefficient (O-hTc) are acquired as 0.62 and 148.18 W/m^2K for the arrangement of coil with embed and annular blade at $Q_{shell} = 5$ LPM, $Q_{Coil} = 1$ L per minute (LPM).

Keywords Annular fins · Shell and coiled tube heat exchanger · Winged insert

1 Introduction

Heat transfer is a procedure wherein the heat is moved starting with one medium then onto the next medium because of temperature contrast [1]. With a similar mass transition and same pitch, at high fume characteristics, the most elevated curl embed accomplished greatest heat transfer improvement in a tube in tube heat exchanger [2]. Embedded cylinders had better P and D proportion and also better a proportion

R. Periyasamy (✉)
SANPAR Microfilters Pvt Ltd, Bommasandra, Bangalore, Karnataka 560099, India
e-mail: rperiyasamy13@gmail.com

S. Santhia
Government Polytechnic College, Krishnagiri, Tamil Nadu 635001, India

N. M. Sivaram et al. (eds.), *Advances in Manufacturing, Automation, Design and Energy Technologies*, Lecture Notes in Mechanical Engineering,
https://doi.org/10.1007/978-981-99-1288-9_77

of thickness [3]. The ideal condition for expanding generally speaking warmth move coefficient in looped tube heat exchanger is gotten by most elevated measure of the curl, cold, and hot water stream rates and curl pitch [4]. The O-hTc is improved by the increment of air flow rate [5]. The most significant parameters for design of transfer of heat and loss of exergy are the tube diameter and the drift of cold fluid, respectively. Furthermore, [6] with decrease in the inlet temperature along the shell and increase inlet temperature along the curled tube, the dimensionless exergy loss increases [7]. If the size of the pitch is multiplied, the Nusselt number in shell increments in high value and the loop Nusselt numbers also increases [8]. The embedded curled wires upgraded the warmth move coefficient up to 1.24 occasions contrasted with the plain cylinders [9]. With the rise of cold or hot water rates of flow, there is increase in exergy loss [10]. The heat transfer enlargements which use turbulators show varying values finer when compared to smooth cylinder for various pitch separations [11]. The significant Nusselt number is effected high by both hot and cold stream rates, and heat move coefficient is dropped down by expanding of curl of pitch just as by increasing the flow of hot fluid [12]. The amount of loss of exergy and effectiveness increases by increase in ratio of volume [13]. With the increase in pitch of the coil, the Nusselt number increases [14]. BI transition opened up two mechanisms such as accumulation of bubble in the direction of flow due to difference in bubble velocity and accumulation of bubble in peripheral and radial direction [15]. In normal, expansion in heat move coefficient and drop of pressure was seen at the most elevated [16]. Witchayanuwat tentatively examined heat move from the quality exhaust particulate of cleanser shower drying pinnacle to water by utilizing looped tube heat exchanger [17]. The performance of a double pipe heat exchanger of vertical type can be improved by the higher value of (n/d) [18]. Nu_{mis} is affected by the ratio of the pitch of the coiled tube, and with a low value of ratio of pitch by diameter, higher value of Nusselt number can be obtained [19]. The tools for detecting and studying the laminar and turbulent flow transition are given by IHCP approach [20]. Decrement in the intersection angle along the velocity vector increases the average Nusselt number under constant wall temperature boundary condition [21]. By varying the hot region temperature through the water flow, the cold region of thermoelectric module can be converted in to the right air cooling unit [22]. The performance of heat exchanger and the Nusselt number can be increased utilizing corrugated tubes [23]. Maximum thermal performance factor is obtained using corrugated tube outside the heat exchanger.

2 Test Setup

2.1 *Design of the Test Trial*

See Fig. 1.

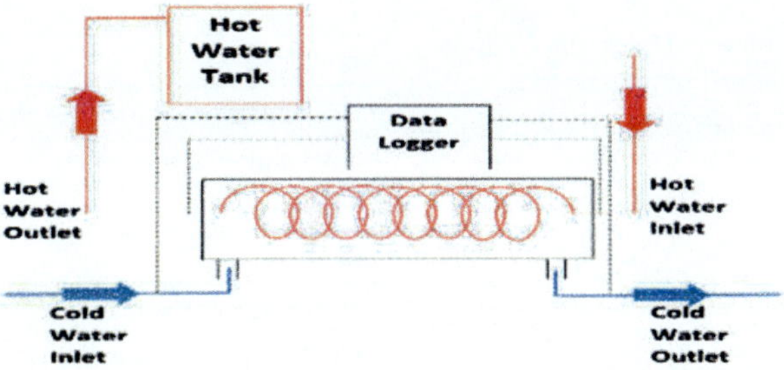

Fig. 1 Diagrammatic representation of the overall trail setup

2.2 *Measurement Method and Experimental Setup*

Figure 1 means the diagrammatic representation of the overall trail setup. Figure 2 shows a turbulator used in this experiment. It is a wire on which wing like structure is made on it. Figure 3 depicts the annular fin which increases the overall surface area of the coiled tube used in this experiment. As the overall surface area increases, the heat transfer increases in the copper coil which is more than that of the similar pin fin. In Fig. 4, the entire arrangement of the experiment is shown.

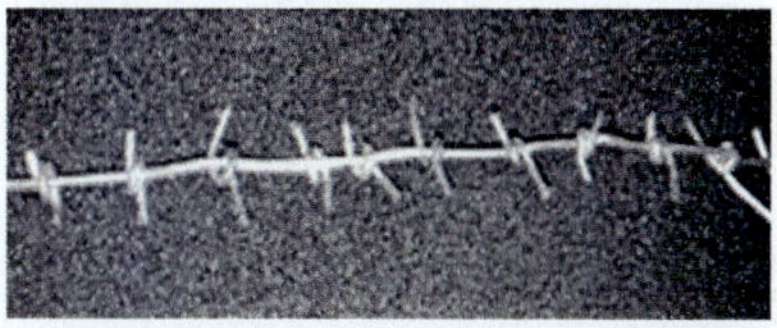

Fig. 2 Wire insert

Fig. 3 Annular fins

Fig. 4 Entire arrangement of the experiment

Table 1 Uncertainty values for various parameters considered

Variable	Unit	Value
Uncertainty value—temperature °C		
Outlet and inlet temperature	°C	±0.4
Uncertainty—volume flow rate measurement $m^3\ h^{-1}$		
Water	$m^3\ h^{-1}$	±0.03
Uncertainty in the result %		
O-hTc	%	±8.35
Drop in pressure	%	±2
Final effectiveness	%	±9.3

2.3 Measurement of Uncertainty

As portrayed in [21–23], the values of uncertainty in evaluating a outcome ($W_R{}^+$) due to various non-dependent parameter are determined (Table 1) using the following equation.

$$W_R^+ = \left(\left(\frac{\partial R^+}{\partial X_1}W_1\right)^2 + \left(\frac{\partial R^+}{\partial X_2}W_2\right)^2 + \cdots + \left(\frac{\partial R^+}{\partial X_n}W_n\right)^2\right)^{\frac{1}{2}} \tag{1}$$

2.4 Data Reduction

The data reduction of the observed results is outlined as follows.

The mean heat transfer rate (Q_{avg})

$$Q_{avg} = (Q_{hot} + Q_{cold})/2 \tag{2}$$

The quantity of heat interchanged to the cold water

$$Q_c = m_c c_{pc}\left(T_{c,\,out} - T_{c,\,in}\right) \tag{3}$$

The quantity of heat interchanged to the hot water

$$Q_h = m_h c_{ph}\left(T_{h,in} - T_{h,out}\right) \tag{4}$$

The O-hTc is calculated by

$$U_i = q_{ave}/A_i \Delta T_{LMTD} \tag{5}$$

$$\Delta T_{LMTD} = \left(\left(T_{h,out}-T_{c,out}\right) - \left(T_{h,in}-T_{c,in}\right)\right)/Ln\left(\left(T_{h,out}-T_{c,out}\right)/\left(T_{h,in}-T_{c,in}\right)\right) \tag{6}$$

ε = actual heat transfer/maximum possible heat transfer

$$\text{Nusselt number Nu} = 0.913\ De^{0.476}\ Pr^{0.2} \tag{7}$$

$$\text{Dean number De} = Re\left(\frac{d}{D}\right)^{0.5} \tag{8}$$

$$\text{Prandtl number, Pr} = \frac{c\mu}{K} \tag{9}$$

$$\text{Reynolds number Re} = \frac{4\rho Q}{\pi \mu d} \tag{10}$$

3 Results and Discussion

Trial examination to improve the exhibition of shell and curled tube heat exchanger is done, and the outcomes are talked about underneath.

Figure 5a shows the effectiveness for coil with insert and annular fins, and here, the maximum value of effectiveness is 0.62 for flow rate of $Q_{Coil} = 1$ LPM. Figure 5b shows the effectiveness for coil with insert and without annular fins, and here, the maximum value of effectiveness is 0.55 for flow rate of $Q_{Coil} = 3$ LPM. Figure 5c shows the effectiveness for coil without insert and with annular fins, and here, the maximum value of effectiveness is 0.54 for flow rate of $Q_{Coil} = 4$ LPM. Figure 5d shows the effectiveness for plain coil, and here, the maximum value of effectiveness is 0.53 for flow rate of $Q_{Coil} = 4$ LPM.

Figure 6a depicts the O-hTc for coil with insert and annular fins, and here, the top value is 148.18 W/m^{2}K for the flow rate of $Q_{Shell} = 5$ LPM and $Q_{Coil} = 1$ LPM. Thus at this point, the maximum value is achieved because that in the coil side, the flow rate is maximum, and due to the presence of the insert, more turbulence is created, and thus, more amount heat energy is transferred to the fluid along the shell side which is moving steadily. Further rise in rate of flow will increase the O-hTc because the film thickness near the pipe wall decreases and also the temperature difference is more. Figure 6b shows the O-hTc for the coil using insert and with no annular fins, here the maximum magnitude of O-hTc 118.06 W/m^{2}K for the flow rate of Q_{Shell} = 5 LPM and $Q_{Coil} = 2$ LPM. Figure 6c reveals the O-hTc for coil with no insert and with annular fins, and here, the maximum value of O-hTc 116.25 W/m^{2}K for the flow rate of $Q_{Shell} = 5$ LPM and $Q_{Coil} = 4$ LPM. Figure 6d shows the O-hTc for plain curl tube, and here, the maximum of the O-hTc is 117.57 W/m^{2}K for the flow rate of $Q_{Shell} = 5$ LPM and $Q_{Coil} = 4$ LPM.

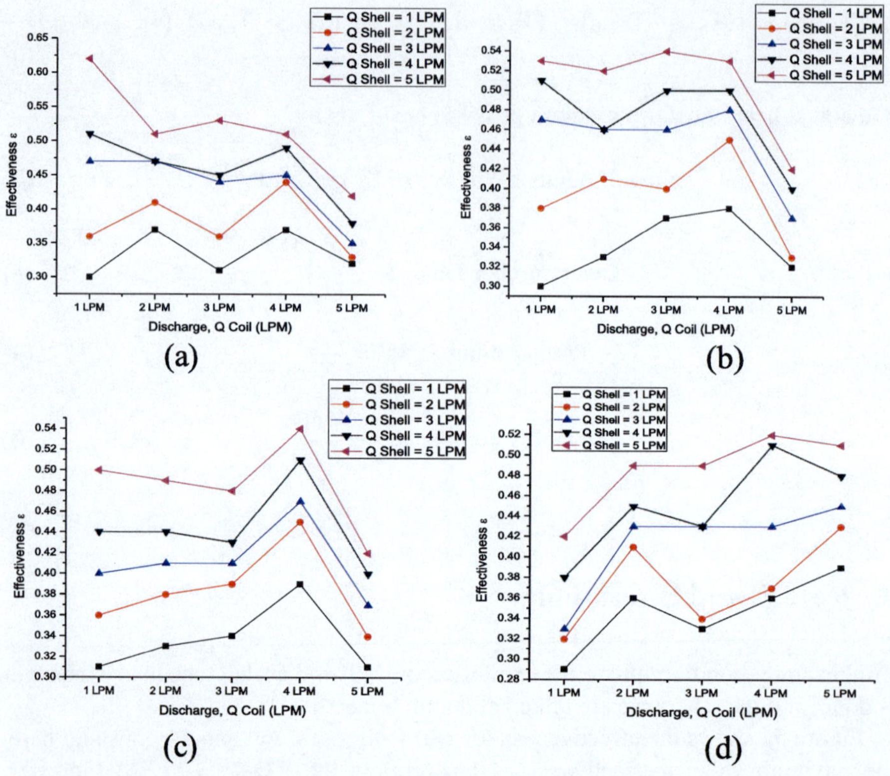

Fig. 5 Graphical representation—effectiveness for **a** coil with insert and annular fins, **b** coil with insert and without annular fins, **c** coil without insert and with annular fins, **d** plain coil

Figure 7a presents the graph plotted between Nusselt number and Dean number for coil without turbulator. The graph shows that the Nusselt number varies in a parabolic manner to the corresponding values of Dean number for the set up without using turbulator. Figure 7b presents the graph plotted between Nusselt number and Dean number for coil with turbulator. The graph shows that the Nusselt number varies in a parabolic manner to the corresponding values of Dean number for the set up using turbulator. Various effectiveness values are obtained for varying flow rates. From the effectiveness graph, it is seen that in the first mode for higher flow rates along the shell side, the effectiveness changes in the descending order. The effectiveness gradually rises with increase in the coil rate of flow. The maximum effectiveness, $\epsilon = 0.62$, is obtained for the setup of coil tube with insert and annular fin at $Q_{Shell} = 5$ LPM, $Q_{Coil} = 1$ LPM.

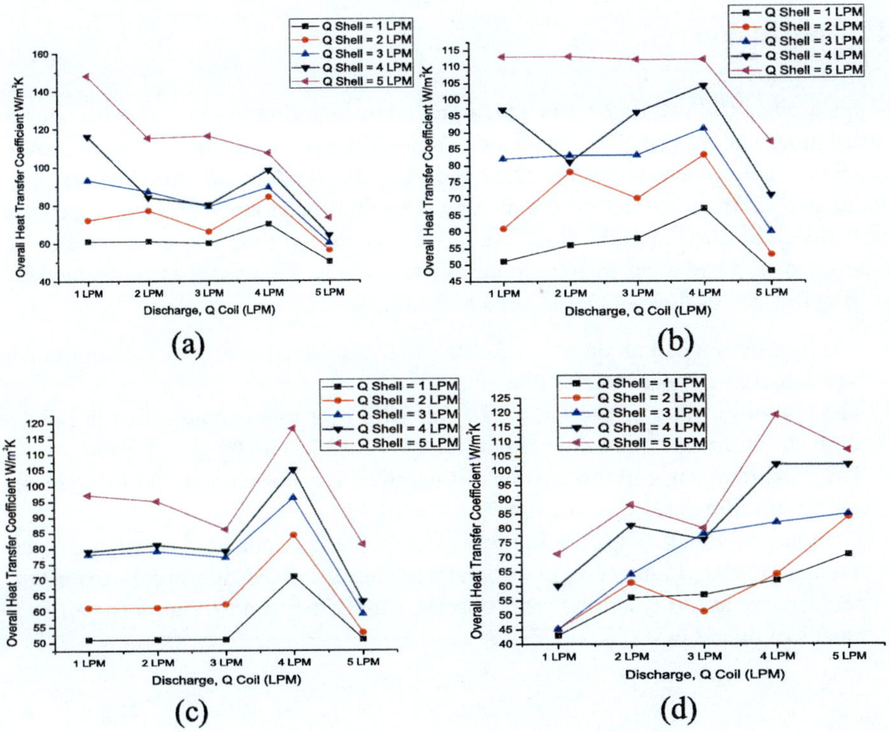

Fig. 6 Graphical representation—O-hTc for **a** coil with insert and annular fins, **b** coil with insert and without annular fins, **c** coil without insert and with annular fins, **d** plain coil

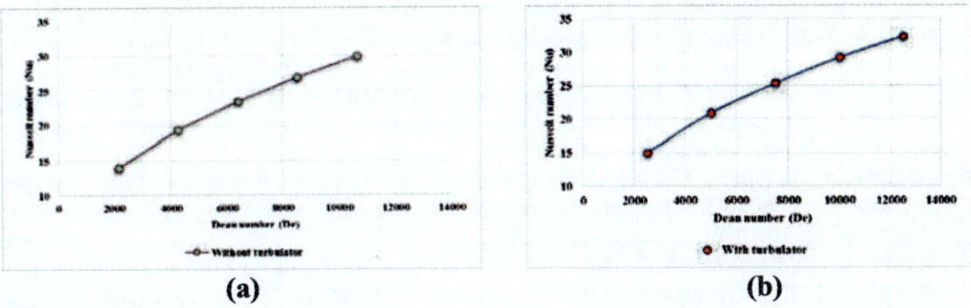

Fig. 7 **a** Nusselt number versus Dean number for coil without turbulator **b** Nusselt number versus Dean number for coil with turbulator

4 Conclusion

In this project, the heat transfer rate is improved by inserting a wire through a copper curled tube. The analysis has been done in four modes of operation, and the optimum arrangement has been suggested. Then, the analysis was done on effectiveness, frictional, and thermal characteristics of tube. The fluid used along the hot water was used along the shell, and the fluid used along the coiled tube was cold water. The experiment is conducted in four modes of operation. This work is performed by varying the rate of flow along the shell and coil.

- The installation of fins on the coil acts as an additional heat transfer equipment which increases heat transfer rate notably.
- The optimum value of O-hTc for the coil with insert and annular fins is 148.18 W/m^2K for the optimum flow rate of $Q_{Shell} = 5$ LPM and $Q_{Coil} = 1$ LPM.
- The maximum value of effectiveness is obtained for the setup of coil using insert and annular fin at $Q_{shell} = 5$ LPM, $Q_{Coil} = 1$ LPM.
- The reasons which cause the improvement in the frictional and thermal characteristics of coiled tube (due to turbulator) were discussed thoroughly from new perspective, and the actions were observed for effectiveness charts for all four modes of operation.

References

1. Agrawal KN, Anil Kumar MA, Behabadi A, Varma HK (1998) Heat transfer augmentation by coiled wire inserts during forced convection condensation of R-22 inside horizontal tubes. Int J Multiph Flow 24(4):635–650. https://doi.org/10.1016/S0301-9322(97)00061-X
2. Yun R, Hwang J, Chung JT, Kim Y (2007) Flow boiling heat transfer characteristics of nitrogen in plain and wire coil inserted tubes. Int J Heat Mass Transf 50(11–12):2339–2345. https://doi.org/10.1016/j.ijheatmasstransfer.2006.10.038
3. Jamshidi N, Farhadi M, Ganji DD, Sedighi K (2012) Experimental analysis of heat transfer enhancement in shell and helical tube heat exchangers. Appl Therm Eng 51(1–2):644–652. https://doi.org/10.1016/j.applthermaleng.2012.10.008
4. Moosavi A, Abbasalizadeh M, Dizaji HS (2016) Optimization of heat transfer and pressure drop characteristics via air bubble injection inside a shell and coiled tube heat exchanger. Exp Thermal Fluid Sci 78:1–9. https://doi.org/10.1016/j.expthermflusci.2016.05.011
5. Majid Etghani M, Amir Hosseini Baboli S (2017) Numerical investigation and optimization of heat transfer and exergy loss in shell and helical tube heat exchanger. Appl Therm Eng (121):294–301. https://doi.org/10.1016/j.applthermaleng.2017.04.074
6. Dizaji HS, Jafarmadar S, Hashemian M (2015) The effect of flow, thermodynamic and geometrical characteristics on exergy loss in shell and coiled tube heat exchangers. Energy 91:678–684. https://doi.org/10.1016/j.energy.2015.08.084
7. Alimoradi A, Veysi F (2016) Prediction of heat transfer coefficients of shell and coiled tube heat exchangers using numerical method and experimental validation. Int J Therm Sci 107:196–208. https://doi.org/10.1016/j.ijthermalsci.2016.04.010
8. Shafaee M, Alimardani F, Mohseni SG (2016) An empirical study on evaporation heat transfer characteristics and flow pattern visualization in tubes with coiled wire inserts. Int Commun Heat Mass Transf 76:301–307. https://doi.org/10.1016/j.icheatmasstransfer.2016.06.001

9. Dizaji HS, Khalilarya S, Jafarmadar S, Hashemian M, Khezri M (2016) A comprehensive second law analysis for tube-in-tube helically coiled heat exchangers. Exp Thermal Fluid Sci 76:118–125. https://doi.org/10.1016/j.expthermflusci.2016.03.012
10. Sahin HM, Baysal E, Dal AR, Sahin N (2015) Investigation of heat transfer enhancement in a new type heat exchanger using solar parabolic trough systems. Int J Hydrog Energy 40(44):15254–15266.https://doi.org/10.1016/j.ijhydene.2015.03.009
11. Etghani MM, Baboli SAH (2017) Numerical investigation and optimization of heat transfer and exergy loss in shell and helical tube heat exchanger. Appl Therm Eng 121:294–301. https://doi.org/10.1016/j.applthermaleng.2017.04.074
12. Khorasani S, Dadvand A (2017) Effect of air bubble injection on the performance of a horizontal helical shell and coiled tube heat exchanger: an experimental study. Appl Therm Eng 111:676–683. https://doi.org/10.1016/j.applthermaleng.2016.09.101
13. San J-Y, Huang W-C, Chen C-A (2015) Experimental investigation on heat transfer and fluid friction correlations for circular tubes with coiled-wire inserts. Int Commun Heat Mass Transf 65:8–14. https://doi.org/10.1016/j.expthermflusci.2016.05.011
14. Li Z, Jiang S, Yang X, Huang Y, Zhu G, Tu J, Zhu H (2017) Bubbly-intermittent flow transition in helically coiled tubes. Chem Eng J 323:96–104. https://doi.org/10.1016/j.cej.2017.04.029
15. Saeedinia M, Akhavan-Behabadi MA, Nasr M (2011) Experimental study on heat transfer and pressure drop of nanofluid flow in a horizontal coiled wire inserted tube under constant heat flux. Exp Thermal Fluid Sci 36:158–168. https://doi.org/10.1016/j.expthermflusci.2011.09.009
16. Witchayanuwat W, Kheawhom S (2017) Heat transfer coefficients for particulate airflow in shell and coiled tube heat exchangers. Int J Chem Biol Eng 3:1. https://doi.org/10.5281/zenodo.1058147
17. Dizaji HS, Jafarmadar S, Abbasalizadeh M, Khorasani S (2015) Experiments on air bubbles injection into a vertical shell and coiled tube heat exchanger; exergy and NTU analysis. Energy Convers Manage 103:973–980. https://doi.org/10.1016/j.enconman.2015.07.044
18. Moawed M (2011) Experimental study of forced convection from helical coiled tubes with different parameters. Energy Convers Manag 52(2):1150–1156. https://doi.org/10.1016/j.enconman.2010.09.009
19. Cattania L, Bozzoli F, Rainieri S (2017) Experimental study of the transitional flow regime in coiled tubes by the estimation of local convective heat transfer coefficient. Int J Heat Mass Transf 112:825–836. https://doi.org/10.1016/j.ijheatmasstransfer.2017.05.003
20. Guo J, Huai X (2016) Numerical investigation of helically coiled tube from the viewpoint of field synergy principle. Appl Therm Eng 98:137–143. https://doi.org/10.1016/j.applthermaleng.2015.12.012
21. Sadighi Dizaji H, Jafarmadar S, Khalilarya S, Moosavi A (2016) An exhaustive experimental study of a novel air-water based thermoelectric cooling unit. Appl Energy 181:357–366. https://doi.org/10.1016/j.apenergy.2016.08.074
22. Sadighi Dizaji H, Jafarmadar S, Mobadersani F (2015) Experimental studies on heat transfer and pressure drop characteristics for new arrangements of corrugated tubes in a double pipe heat exchanger. Int J Therm Sci 96:211–220. https://doi.org/10.1016/j.ijthermalsci.2015.05.009
23. Sadighi Dizaji H, Jafarmadar S (2016) Experiments on corrugated tubes through a double pipe heat exchanger. Exp Heat Transf 29(5):577–592. https://doi.org/10.1080/08916152.2015.1046015

Modeling of Solar Thermal-Based Adsorption Cooling System for Residential Building

Shubha Deep Paul and K. Ravi Kumar

Abstract The detrimental effect of mechanical vapor compression air conditioning system on the environment and their energy intensive nature are compelling the researchers to make the efforts to enhance the capability of solar thermal-based adsorption cooling system (STACS). In the current study, modelling of a STACS is performed under the weather condition of Delhi, India, in order to fulfill the thermal energy demand of a typical building. According to the cooling load of the building, 3 TR (Ton of Refrigeration) adsorption chiller has been considered. The performance parameter is considered as solar fraction (SF) for the system optimization. The performance of the STACS is found better for the major contributing parameters such as evacuated tube solar collector area and storage tank capacity for the specified mass flow rate of 100 kg/hr are 30 m^2, and 600 L, respectively. Based on the analysis, the optimized system could achieve coefficient of performance (COP) of the STACS is 0.49.

Keywords Solar energy · Evacuated tubular collector · Vapor adsorption · TRNSYS

1 Introduction

Energy is a critical component for the economic development of a country. Nearly 40% of worldwide energy consumption and more than 33% of global greenhouse gas emissions are attributed by the building sector [1]. Furthermore, heating, ventilation, and air conditioning (HVAC) systems account for the majority of a building's energy consumption [2]. This energy is mainly supplied by non-renewable energy sources. These energy resources are depleting day by day, and per capita energy consumption is increasing continuously. Electricity used to produce cooling and heating in

S. Deep Paul · K. Ravi Kumar (✉)
Department of Energy Science and Engineering, Indian Institute of Technology Delhi, Hauz Khas, Delhi 110016, India
e-mail: krk@ces.iitd.ac.in

N. M. Sivaram et al. (eds.), *Advances in Manufacturing, Automation, Design and Energy Technologies*, Lecture Notes in Mechanical Engineering,
https://doi.org/10.1007/978-981-99-1288-9_78

many household and commercial buildings, which is a thermodynamically inefficient operation. Instead, the cooling and heating may be provided using renewable energy particularly solar energy, and hence contribute to a decrease in CO_2 emissions. Adsorption cooling has been proposed as a viable alternative to typical vapor compression refrigeration devices for utilizing low-grade thermal energy [3]. Waste heat-driven cooling systems are readily available in the industrial sector; however, the utilization of solar thermal-based adsorption cooling system (STACS) is very rare in industrial as well as building sector. In the hot and arid climate of Egypt, Reda et al. analyzed, through TRNSYS simulation of a STACS for residential space [4]. It was found that the proposed system could attain a daily COP of 0.41. Ali et al. analyzed the performance of residential scale STACS employed in the upper region of Egypt. It is observed that by replacing cooling tower of capacity of 34 kW with 50 kW, the COP of adsorption cooling system (ACS) is improved [5]. Compound parabolic collector (CPC) integrated with ACS in France was designed for 100 m^2 building in Orly, France. It was found that CPC improved the solar COP, and the system achieved in-door temperature below 23 °C [6]. Alam et al. optimized the CPC-assisted ACS for the weather condition of Tokyo, Japan, on basis of cycle time. The cycle time is considered as the most significant parameter for the optimization ACS system [7].

Based on the literature review, it is observed that few researchers employed TRNSYS to optimize the solar water heating system (SWHS) integrated with ACS. In the present study, the performance assessment of a STACS has been performed using TRNSYS to address the comfort cooling needs of a musical room of a Shivalik Hostel in the Indian Institute of Technology (IIT) Delhi, India. The proposed system consists of evacuated tube collectors (ETC), hot water storage tank, adsorption chiller, cooling tower, heat exchanger, and blower. According to the cooling load of the room, 3 TR (Ton of Refrigeration) adsorption chiller has been considered. Also, the efforts have been made to simulate, and optimize the SWHS comprising of ETC integrated with hot water storage tank for adsorption chiller. The performance parameter SF is considered for the system optimization of SWHS.

2 Modeling of a Vapor Adsorption System

The cooling requirement of a musical room in Shivalik Hostel, IIT Delhi, is simulated using TRNSYS Building environment (TRNBuild). TRNSYS features TRNSYS3D, allows the user to plug in a SketchUp™ (3D design software) model to create multizone buildings and import the geometry into the TRNBuild directly from the SketchUp interface [8]. The cooling load of a building depends on various parameters such as area of building, occupancy, type of building, type of wall, material of construction, window type, location of the building, orientation of the building, weather conditions, etc. The thermophysical properties of all these parameters are presented in Table 1. The input data of building provided in the TRNBuild are given in Table 2. Figure 1 exhibits the weather data extracted from the Meteonorm weather

Table 1 Thermal properties of building constituents

Constituents	Thermal conductivity (W/m–k)	Density (kg/m^3)	Specific heat (kJ/kg-k)
Cement plaster	0.721	1648	0.80
Brick	0.980	1920	0.80
Asbestos cement sheet	0.245	1520	0.84
Reinforced concrete (RCC)	1.580	2288	0.88

Table 2 Input data for building required by TRNBuild

Floor area	7.47 m × 3.50 m
Building volume	79.785 m^3
Windows type Single clear glass	6 mm glass
Occupancy	10 people
Cooling set point temperature	24 °C
Operating hours	9 A.M. to 5 P.M

file for New Delhi. As shown in Fig. 1, monthly maximum global solar radiation incident on the earth surface and monthly maximum ambient temperature recorded are 1032.8.2 W/m^2 and 32.8 °C, respectively. The monthly cooling load of a room is illustrated in Fig. 2, where it exceeds 5 kWh during summer months March to November and it reaches its maximum of 29.8 kWh during June.

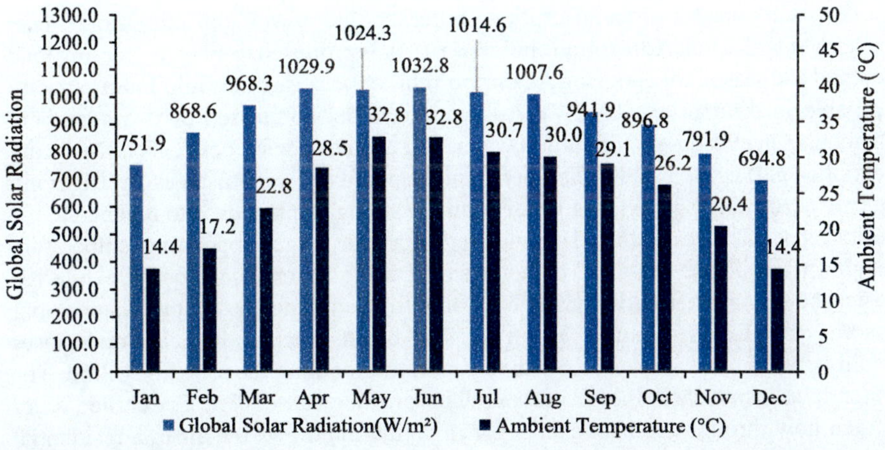

Fig. 1 Monthly maximum global solar radiation and monthly average ambient temperature for New Delhi, India

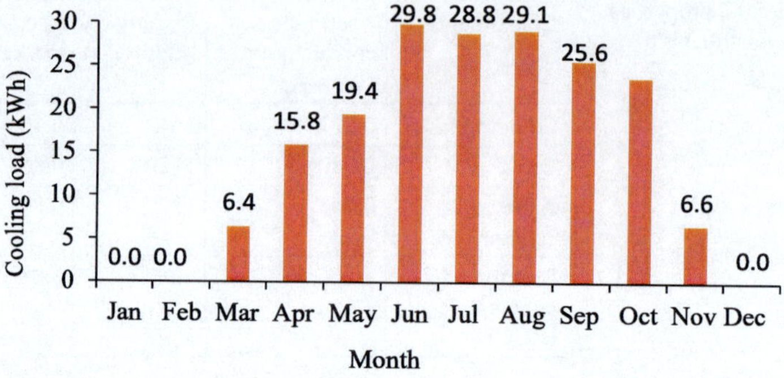

Fig. 2 Monthly cooling load of space throughout the year for operating hours from 9 AM to 5 PM

3 System Description

The system comprises of a solar field, an ACS, and a hot water storage tank as shown in Fig. 3. Two auxiliary heating elements are connected inside the hot storage tank to elevate the hot water temperature when it falls below the temperature of 80 °C during fluctuation of the solar radiation. The solar field is containing evacuated tube collector (ETC) to generate hot water greater than 95 °C. ETC is considered due to higher fluid delivery temperature (50–120 °C) and performance [9]. The water from the bottom of the hot water storage tank is pumped to the evacuated tube collector using pump powered by solar energy. The system has a differential controller, which activates the solar pump when the difference in temperature between the outlet collector temperature and at the bottom of the hot storage tank is more than 10 °C.

The ACS consists of an adsorption chiller, cooling tower, and liquid-to-air heat exchanger with a fan. Adsorption chiller of 10 kW is proposed to meet the cooling load demand and reason for choosing Adsorption chiller because of the minimum capacity available in the market (10 kW). Adsorption chiller comprised of two adsorbent chambers, an evaporator for chilled water, and a water-cooled condenser. Desiccant gel (silica gel) is used as the adsorbent, and standard tap water used as a refrigerant. During desorption process, hot water from the storage tank is used to regenerate the desiccant gel. The desiccant gel regenerated with wide range of operating temperature ranging from 50 °C to 96 °C of hot water. During adsorption process, while the desiccant gel is externally cooled by water from cooling tower, the water vapor adsorbs onto the surface of desiccant gel. This creates vacuum in chiller and causes liquid water in an evaporator to flash to vapor, thus cooling the remaining fluid. The water flows inside the evaporator tubes, which are therefore cooled. The chilled water is then flow through the heat exchanger in which fan draws the air in a residential space and then blows it over a coil for cooling of the air.

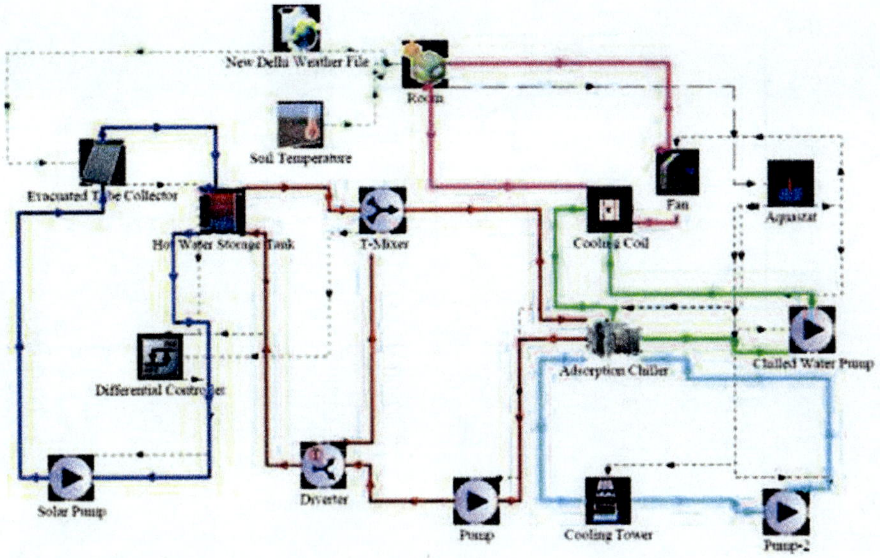

Fig. 3 Flow diagram of STACS modeled in TRNSYS

4 Results and Discussion

4.1 Validation

The present model is validated with Reda et al. in terms of COP of adsorption chiller for the month of June [4]. The parameters considered by the Reda et al. are given in Table 3.

Figure 4 shows the comparison of the COP of the present model with the Reda et al. as a function of time. It can be seen from the figure that the present model has good agreement with the Reda et al. data. As a result, it is possible to infer that the employed model is suitable for simulating the effective behavior of the STACS system.

Table 3 Parameters considered by the Reda et al. model

Parameters [4]	Values
Solar collector area	58 m^2
Storage tank capacity	5500 L
Tilt angle	5°

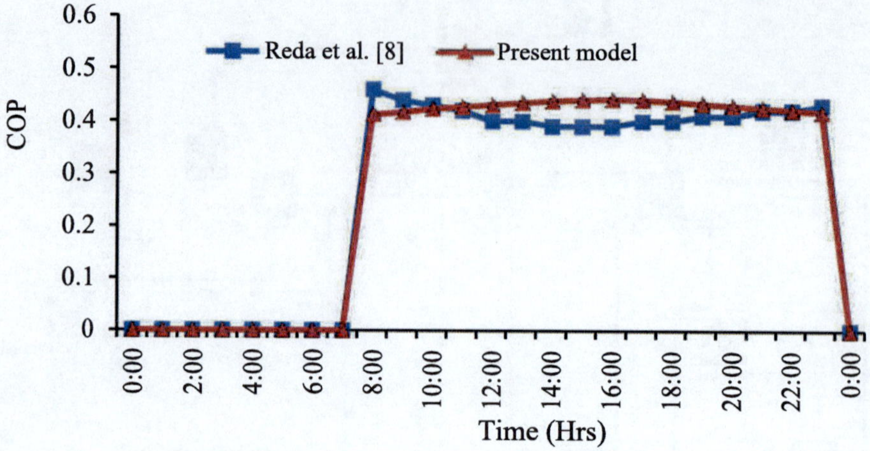

Fig. 4 COP of adsorption chiller for the typical day in month of June

4.2 Optimization of Evacuated Tube Collector Area

SF is defined as the fraction of thermal energy provided by solar thermal system to the thermal energy required for regeneration of desiccant gel in adsorption chiller. SF may be expressed as:

$$SF = QSWH/Qchiller \tag{1}$$

where QSWH = Energy delivered by the solar thermal system (kJ/hr) and Qchiller = Energy required in the chiller (kJ/hr).

The SF has been determined during the month of June, where the cooling load is more than 29 kWh and maximum ambient temp is 44.7 °C. Based on the analyzes, the variation in SF on various solar collector area for the water flow rate of 100 kg/hr, is as shown in Fig. 5. The SF increases as the solar collector area increases, although increasing the collector area by more than 30 m^2 for the month of June has no meaningful effect on the SF. As a result, demand coverage and heat supply from solar collectors are well matched. If the solar collector area exceeds 30 m^2, it is neither technically nor economically feasible to oversize the system. Therefore, collector area of 30 m^2 is regarded as optimum value for the proposed system. A maximum of 0.96 SF can be achieved when solar collector area is 30 m^2.

4.3 Optimization of Hot Water Storage Tank Volume

The default average tank loss coefficient of 2 kJ/hr m^2 K is considered for the thermal storage tank in the TRNSYS model. The variation in SF on various storage tank

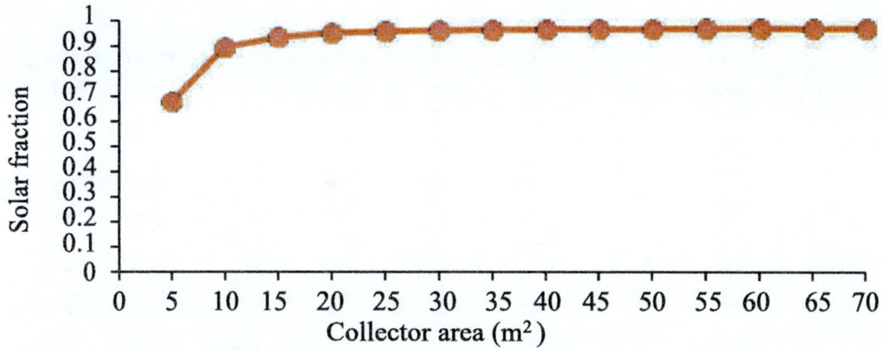

Fig. 5 Variation of SF for various solar collector area

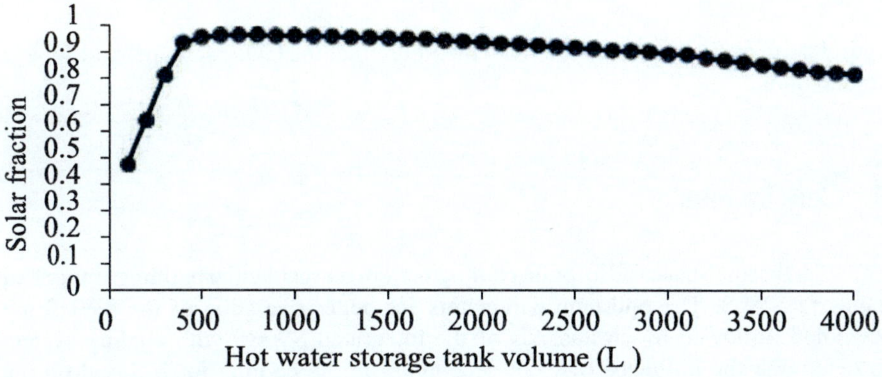

Fig. 6 Effect of SF for various storage tank volume

capacity is shown in Fig. 6. The commercially available storage tank capacity of 100–4000 L was chosen for optimum collector area of 30 m^2. The results portrayed that the SF improved with an increase in storage capacity from 100 to 600 L. The SF started to decline beyond storage volume of 600 L, because large storage tank has a large quantity of low-temperature hot water, and the tank can use more auxiliary power. Therefore, optimum value for the storage tank is set to 600 L for the proposed system.

4.4 Simulation of Adsorption Cooling System

Based on the analysis, the above-optimized SWHS is integrated with the ACS. The rated capacity and rated COP of ACS are considered to be 10 kW and 0.6, respectively. Fig. 7 depicts the COP of refrigeration unit over the course of a year using an

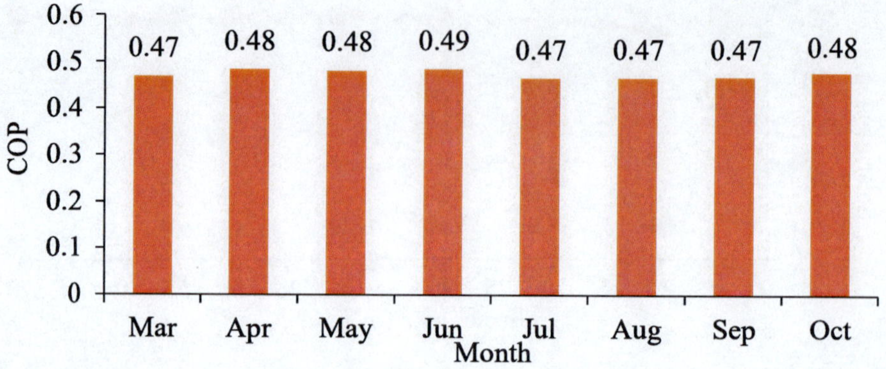

Fig. 7 COP of adsorption chiller integrated with optimized SWHS

optimized SWHS. It can be seen from the figure that the COP attain peak in June at around 0.49.

5 Conclusions

The solar thermal-based adsorption cooling system for residential building *i* modelled using TRNSYS. The optimum parameters for better operation of the SWHS are identified, followed by an analysis of the integrated SWHS with cooling system to determine the influence of COP depending on the cooling load. Based on the investigation, the following results are obtained:

- SWHS with evacuated tube collector with optimal design parameters such as a collector area of 30 m^2, and hot water storage capacity of 600 L for mass flow rate of 100 kg/hr, is determined to be efficient in delivering the energy required for the functioning of the adsorption chiller with capacity of 10 kW.
- The optimum SF of STACS is 0.96.
- The integrated system produced the maximum COP of 0.49 for the month of June and average COP of 0.47 throughout the summer (March to October).

Acknowledgements The financial support for the research work is provided by the "Yardi System, USA", through the research project titled "Design and Demonstration of Sustainable Building Infrastructure" (Grant Number-IITD/IRD/RP03743)) is duly acknowledged.

References

1. Buildings Are the Foundation of Our Energy-Efficient Future I World Economic Forum. https://www.weforum.org/agenda/2021/02/why-the-buildings-of-the-future-are-key-to-an-efficient-energy-ecosystem/. Accessed Oct. 20, 2021.
2. Jouhara H, Yang J (2018) Energy efficient HVAC systems. Energy Build 179:83–85. https://doi.org/10.1016/j.enbuild.2018.09.001
3. Elsheniti MB, Abd El-Hamid AT, El- Samni OA, Elsherbiny SM, Elsayed E (2021) Experimental evaluation of a solar two-bed lab-scale adsorption cooling system. Alex Eng J 60(3):2747–2757. https://doi.org/10.1016/j.aej.2021.01.024
4. Reda AM, Ali AHH, Morsy MG, Taha IS (2019) Optimization and performance of a small scale adsorption cooling system fully powered by solar energy for hot arid areas. SN Appl Sci 1(9):1–14. https://doi.org/10.1007/s42452-019-1150-8
5. Ali AHH (2017) Performance assessment and gained operational experiences of a residential scale solar thermal driven adsorption cooling system installed in hot arid area. Energy Build 138:271–279. https://doi.org/10.1016/j.enbuild.2016.12.062
6. Clausse M, Alam KCA, Meunier F (2008) Residential air conditioning and heating by means of enhanced solar collectors coupled to an adsorption system. Sol Energy 82(10):885–892. https://doi.org/10.1016/j.solener.2008.04.001
7. Alam KCA, Saha BB, Akisawa A (2013) Adsorption cooling driven by solar collector: A case study for Tokyo solar data. Appl Therm Eng 50(2):1603–1609. https://doi.org/10.1016/j.applthermaleng.2011.09.028
8. Kummert M (2007) Trnsys 16. Program 5:1–11
9. Kumar KR, Chaitanya NVVK, Natarajan SK (2021) "Solar thermal energy technologies and its applications for process heating and power generation – A review." J Clean Prod **282**. doi:https://doi.org/10.1016/j.jclepro.2020.125296.

A Review on Piezoelectric Vibration Energy Harvesters

Kirandeep Singh and Rohit Tamrakar

Abstract Piezoelectric energy harvesters are attractive for developing self-generating power. The combination of sparse energy and the huge weight of batteries poses the most censorious problem for individuals having high electricity demands in remote areas and who are already carrying hefty loads (such as explorers or field scientists). Piezoelectric energy harvesters are cost-effective energy harvesters, are used for many applications like recharging the batteries, and help move toward the technology of battery-less devices and sensors. In our environment, ambient energy is available in various forms like energy from the sun which is called solar energy, energy from heat which is called thermal energy, and the energy produced from vibratory motion of the body called vibration energy. Among the mentioned energies, vibration energy is constantly present in nature and man-made devices and structures. This paper reviews the work done in the area of piezoelectric vibration harvesters from the last ten years. This paper also reviews the recent advancements in piezoelectric materials that are used in piezoelectric vibration energy harvesters.

Keywords Piezoelectric · Energy harvester · Vibration

1 Introduction

A reduction in the size of sensors has mainly drawn the attention of many researchers to research battery-less power generating devices and sensors and how to replace battery devices with battery-less devices. In many applications such as the use of sensors in remote areas (geographically inaccessible regions), battery changing may be impractical and costly. In some harsh and hard terrain areas, battery replacement might be a risky and hard job. Piezoelectric vibration energy harvesters are the energy

K. Singh (✉) · R. Tamrakar
Mechanical Engineering Department, Sardar Vallabhbhai National Institute of Technology, Surat, Gujarat, India
e-mail: p20cc002@med.svnit.ac.in

N. M. Sivaram et al. (eds.), *Advances in Manufacturing, Automation, Design and Energy Technologies*, Lecture Notes in Mechanical Engineering,
https://doi.org/10.1007/978-981-99-1288-9_79

harvester devices used to eliminate batteries and are used in microelectronics for self-powered sensors and circuits and are used in many other applications. A lot many articles were published for review of work done in piezoelectric vibration energy harvester till 2010, and in this paper, the research that was done in last decade from 2010 is reviewed.

Piezoelectric vibration energy harvesting mainly works on the principle of conversion of liberal mechanical energy (from vibratory motion) into electrical energy. Piezoelectric vibration energy harvester mainly focuses on low-level energy harvesting mainly in milliwatts or microwatts. Piezoelectric vibration energy harvesting has certain advantages over solar or thermal energy. The major advantage is that ambient vibrations are always present, and we do not have to check environmental or weather conditions for energy harvesting, but in solar and thermal energy harvesting, the environmental conditions must be favorable for energy harvesting.

This paper tells about recent advancements in piezoelectric materials, types of piezoelectric transducers, and advancements in various transducers, thereafter the author's conclusion on work done in the area of piezoelectric energy harvesting through vibrations and piezoelectric materials.

2 Review of Piezoelectric Materials and Piezoelectric Vibration Energy Harvesting Through Piezoelectric Transducers

Piezoelectric vibration energy harvesting works on the basic principle of exciting the proof mass, which is attached, on a cantilever beam made from piezoelectric material. The proof mass is excited from environmental vibrations, this input energy on proof mass is then converted into kinetic energy, and this kinetic energy is stored as potential energy in a beam, which is made of piezoelectric material. The potential energy stored leads to strain in the piezoelectric material, which generates an electric charge and thus produces energy. This sort of energy harvester mainly has two types of structures that are unimorph structures and bimorph structures. A cantilever beam is mainly attached with seismic mass at its tip to adjust the resonance frequency of the structure with available environmental frequency. Piezoelectric vibration energy harvester uses Barium Titanate ($BaTiO_3$) and Lead Zirconate Titanate (PZT) materials, which are ferroelectric in nature. In this Sect. 2.1, piezoelectric material advancements are reviewed, and in the next Sect. 2.2, piezoelectric transducer types are reviewed with recent advancements.

2.1 *Piezoelectric Materials*

Over the past years, many piezoelectric materials are developed. The most commonly used piezoelectric material is named perovskite lead zirconate titanate also familiar as PZT, and this material is covered with niobium and lanthanum to get hard and soft structural properties. Piezoceramics are widely used in sensors and actuators due to their ability to produce high voltage outputs in the range of 50 V to 100 V. PZT is the most commonly used material, and there is ongoing research on PZT structure improvement for better output voltages [1–4]. Gao et al. in 2018 [5] informed about a new piezoelectric material, PNN-PZT that has a very high coupling coefficient of 1753Pc N^{-1}, which is up above the other piezoelectric materials. The coupling of piezoelectric materials is improved by texturing in grains of materials [6] and after adding epitaxial thin films on substrates [7]. High hydrostatic piezoelectric strain is accomplished by porous piezoelectric materials, which are used in hydrostatic sensors [8–10]. Piezoelectric polymers are also developed, and they contain polyvinyl fluoride also called PVDF. Harstad et al. in 2017 increased the beta (β) phase present in the material to improve the coupling of piezoelectric (PVDF) material [11]. Piezoelectric materials are mainly categorized into five groups named piezoelectric single crystals, piezoelectric foams, piezoelectric nanocomposites, lead-free piezoelectric materials, and high-temperature piezoelectric materials.

2.1.1 Piezoelectric Materials with Single-Crystal Microstructure

Piezoelectric single crystals are mainly evolved for the achievement of good coupling in piezoelectric materials. The piezoelectric materials with single crystals have a lofty electromechanical coupling coefficient [12–15]. The main drawback of these materials is that they are very costly and their damping is high [16].

In 2010, Ren et al. [17] have done fabrication and testing on an unimorph cantilever consisting of PMN-PT material, and this cantilever was exposed to sinusoidal base excitations. The dimensions of unimorph beam were 50.0 mm length, 6 mm width, and 0.3 mm thick brass plate, which is bonded with 13 mm length, 6 mm width, and 1 mm thick PMN-PT material with mass at the tip of a cantilever beam, which was able to produce 4.16 mW of power under periodical excitation and 60 Hz frequency. The voltage output was 91.2 V. There is another research, which is done on materials like Lead Zirconate Titanate (PZT), which is a piezoelectric material, and Lead Magnesium Niobate (PMN), which is also a popular piezoelectric material. In 2017, Yue et al. [18] PMN-PZT nano powders were synthesized which contains MnO_2 is used to fabricate ceramics. Shahab et al. do another study in 2018 [19] on the performance of piezoelectric ceramics which are soft and hard in nature. The piezoelectric materials used in this study are piezoelectric ceramics named PZT-5H and PZT-5A, which are soft, and other piezoelectric ceramics named PZT-5H and PZT-5A, which are hard in nature. Piezoelectric crystals are also used that are PMN-PT and PMN-PZT. He found that soft materials perform better in off-resonance frequency than hard materials, and on the other hand, in wideband frequency excitation the hard materials perform better than soft ones. In 2015, Hwang et al. [20] evolved an energy harvester which is made from single-crystalline PMN-PZT thin film and it was set down on the heel of a boot that was capable of glowing LEDs while normal routine walks.

2.1.2 Piezoelectric Lead-Free Materials

Lead is a toxic material, and using lead may be risky for health. There are ecological restrictions on lead usage, and it cannot be used in the medical industry and thus motivated researchers for research on lead-free piezoelectric materials, their development, effectiveness, usage, etc. [2, 21–24]. Piezoelectric lead-free materials are of three different. The first one is titanate-based lead-free piezoelectric materials. Second, one is alkaline niobate perovskite-based lead-free piezoelectric materials, and the third one are bismuth perovskite-based lead-free piezoelectric materials [25].

In 2018, Wu et al. [26] illustrate lead-free piezoelectric material based on the material composition of KNN-BNZ-AS-Fe. A cantilever harvester was made of this material which is capable to glow 10 LEDs. BZT-BCT composition is a widely studied lead-free piezoelectric material due to its great piezoelectric properties [27].

2.1.3 High-Temperature Piezoelectric Materials

Piezoelectric materials can be used only where favorable temperature conditions are possible. At high-temperature conditions, piezoelectric materials have limited usage because of their phase sensitivity at high temperatures. Their phase gets changed and thus material properties are changed. To remove these limitations, researchers focused on developing high-temperature piezoelectric materials.

In 2018, Qaiser et al. [28] developed a piezoelectric composite that is embedded with $BiFeO_3$ grains which are present in the Bi3TaTiO9 material matrix to achieve both piezoelectric properties and high-temperature properties which can be used up to 500 °C temperature. In 2017, Li et al. [29] illustrated magnesium (Mn)-based modified $BiFeO_3$-$BiTiO_3$ lead-free piezoelectric ceramic and this ceramic has a high curie temperature which is above 500 °C.

2.1.4 Piezoelectric Foam Type Materials

In the past few years, researchers are focusing on the development of cellular-like foam material which acts as piezoelectric material and gives similar advantages as a piezoelectric material for energy harvesting [30]. Firstly, this development was begun in the 1980s in a country named Finland [31]. This category of materials is also called electret, which means a type of material that contains a permanent electric charge.

Anton et al. [32] 2014 have worked on vibration energy harvesting using ferromagnetic electret material. He created an energy harvester, which is pre-tensioned and have dimensions of 15.24 cm length, 15.24 cm width, and 85 μm of thickness. When this harvester is excited in the longitudinal direction with a 60 Hz excitation frequency, this harvester produced an 8 V output voltage. This energy harvester is capable of charging a 4.6 V capacitor in 30 min with a generated power output of 6.0 μW.

In past years, many studies have been performed on stacked piezoelectric foams to improve their performance. In 2014, Pondrom et al. [33] made a piezoelectric foam that contains 9 to 10 layers of piezoelectric material. When this foam is excited harmonically, it produced about 1.3 μWg^{-2} of power for resistance in a load of 100 MΩ. After that, he increased the number of layers to 20 in the stack to improve its performance and then excited it harmonically and piezoelectric electret which showed an output of about 3.8 mW g^{-2} for an optimal resistance on the load of 650 kΩ.

2.2 *Piezoelectric Transducer Types*

Piezoelectric materials are used in transducer devices to optimize their energy harvesting efficiency because these transducer devices are the devices that provide a stand to absorb mechanical form of energy ineffective way to generate more power output from piezoelectric materials. Among the piezoelectric transducers, the most common type of transducer is the piezoelectric cantilever beam type transducer in which the cantilever beam, which is made of piezoelectric material, is attached at its one end and loaded with a proof mass on its other end. In this section, different types of piezoelectric cantilever transducers are reviewed according to their configuration geometry, shape, size, structure, etc.

In a piezoelectric cantilever transducer, a piezoelectric layer is glued to the substrate beam. There are two types of cantilever piezoelectric transducers that are unimorph cantilever beam transducer and bimorph cantilever beam transducer. In the unimorph cantilever beam transducer, a single piezoelectric layer is glued on the substrate beam, and on the other hand, in the bimorph piezoelectric cantilever beam transducer two piezoelectric layers are glued with substrate beam on each side. The electromagnetic behavior of piezoelectric unimorph beam and bimorph beam piezoelectric cantilever beam type energy harvester is been studied in the literature [34, 35]. Piezoelectric cantilever beam type energy harvester provides great energy output when they are excited under resonance frequency. Some piezoelectric cantilever beam harvesters use triangular-shaped profiles or trapezoidal-shaped profiles (other than rectangular profiles) to increase their performance. Their energy output is improved by using different profiles and cross-sections of beams [36–38].

The main issue with piezoelectric vibration energy harvesters is their high frequency and low energy output from the beam when excited below or above the resonance frequency of the beam. Adding mass to the tip of the beam reduces its natural frequency. Different frequency up conversion mechanisms are used to convert to high-frequency oscillation from low-frequency vibrations. In 2013, Dhakar et al. [39] proposed a frequency up-conversion device that uses a spring and contains proof mass at the tip of the cantilever beam.

Piezoelectric patches can also be used to harvest energy when they are attached to some oscillating body such as vehicles and aircraft and are another form of energy harvesters [40, 41]. Diaphragm-type energy harvesters are also used for energy harvesting from fluids such as blood pressure and have future scope in the medical industry [42, 43].

2.2.1 Nonlinear and Broadband Transducers

Piezoelectric vibration energy harvester power output is weak when excited below or above their resonance frequency and this is a major issue in energy harvesting with piezoelectric energy harvesters [44]. In the past decade, wide research has been done on increasing the bandwidth of vibration energy harvesters to improve their performance. In 2016, Li et al. [45] proposed a piezoelectric bimorph cantilever structure that contains two bimorph beams with two proof masses at their tips and both of them are of different resonant frequencies. It was seen that this structure has a wide bandwidth frequency of 14 Hz with an improvement in output power by 80%. On the other hand, if beams are used separately, then power output comes out is less.

Wide research on nonlinear energy harvesters is also done in the last few years. With non-linear piezoelectric vibration energy harvesters, bandwidth is increased, and hence performance is increased. Piezoelectric magnetic structures are used, and the resonator is forced to swing back and forth around positions that are dynamically stable [46–49]. In 2009, Stanton et al. [50], Cottone et al. [51], and Erturk et al. [52] in the field of nonlinear vibration consisting of a piezoelectric cantilever beam type energy harvester that contains a tip magnet and two fixed magnets performed initial work. This arrangement of magnets provides one steady position that results in additional frequency bandwidth and increased power output.

2.2.2 Microelectromechanical Type Piezoelectric Transducers (MEMS)

With the development of small-scale microelectronics and decreased power, consumption in electronics chips requires only tens of milliwatt power. This decreased consumption of power and development of small-scale microelectronics encourages researchers to research microelectromechanical piezoelectric vibration energy harvesters. This will enable fully self-powered sensors and microelectronics, which are widely used in the electronics industry.

In 2012, Berdy et al. [53] presented a MEMS piezoelectric vibration energy harvester that contains a fixed–fixed design. This harvester was tested under a 49.7 Hz frequency and a power of about 120 μW was generated. In 2014, Yu et al. [54] proposed another MEMS piezoelectric vibration energy harvester. Firstly, a finite element simulation on this design is done and the device was designed and optimized. The final prototype consists of five piezoelectric beams and was tested under an excitation frequency of 234.5 Hz and with this device 66.75 μW of power was achieved.

3 Conclusion

Piezoelectric vibration energy harvesting is a wide research area for the last two decades. Piezoelectric vibration energy harvesting can be done from civil infrastructures [55, 56], human body [57, 58], and aerospace [40, 59–62]. The purpose of this paper is to tell the recent advancements that are being done in the last ten years in the area of piezoelectric vibration energy harvesting and piezoelectric materials. It is very difficult to capture all the recent advancements in one research paper, but this paper summarizes the main advancements, which has been done in the area of piezoelectric vibration energy harvesting.

Ceaseless development in piezoelectric materials with increased mechanical, thermal properties introduced new piezoelectric materials that include single-crystal piezoelectric materials, lead-free piezoelectric materials, high-temperature piezoelectric materials, and piezoelectric foams. On the behalf of this research, various types of piezoelectric transducers are introduced in the field of piezoelectric vibration energy harvesting that includes nonlinear broadband transducers to improve frequency bandwidth and microelectromechanical (MEMS) transducers to use in the electronics industry for self-powered circuits and sensors.

References

1. Panda P (2009) Review: environmental friendly lead-free piezoelectric materials. J Mater Sci 44:5049–5062
2. Priya S, Nahm S (2011) Lead-free piezoelectrics. Springer, New York
3. Uchino K (2017) The development of piezoelectric materials and the new perspective Advanced Piezoelectric Materials 2nd edn. pp 1–92, Elsevier, Amsterdam
4. Marino A et al (2017) Piezoelectric nano transducers. Futur Neural Stimul Nano Today 14:9–12
5. Gao X et al (2018) Giant piezoelectric coefficients in relaxor piezoelectric ceramic PNN-PZT for vibration energy harvesting. Adv Funct Mater 28:1706895
6. Messing GL et al (2004) Templated grain growth of textured piezoelectric ceramics. CritAl Rev Solid State Mater Sci 29:45–96
7. Isarakorn D et al (2010) Epitaxial piezoelectric MEMS on silicon. J Micromech Microeng 20:055008
8. Roscow J, Taylor J, Bowen C (2016) Manufacture and characterization of porous ferroelectrics for piezoelectric energy harvesting applications. Ferroelectrics 498:40–46
9. Martínez-Ayuso G et al (2017) Homogenization of porous piezoelectric materials. Int J Solids Struct 113:218–229
10. Roscow J et al (2017) Modelling and fabrication of porous sandwich layer barium titanate with improved piezoelectric energy harvesting figures of merit. Acta Mater 128:207–217

11. Harstad S et al (2017) Enhancement of β-phase in PVDF films embedded with ferromagnetic Gd5Si4 nanoparticles for piezoelectric energy harvesting. AIP Adv 7:056411
12. Erturk A, Bilgen O, Inman DJ (2008) Power generation and shunt damping performance of a single crystal lead magnesium niobate-lead zirconate titanate unimorph. Anal Exp Appl Phys Lett 93:224102
13. Palneedi H et al (2017) Strong and anisotropic magnetoelectricity in composites of magnetostrictive Ni and solid-state grown lead-free piezoelectric BZT–BCT single crystals. J Asian Ceram Soc 5:36–41
14. Ko S-Y et al (2017) Improved solid-state conversion and piezoelectric properties of 90Na1/2Bi1/2TiO3-5BaTiO3-5K1/2Na1/2NbO3 single crystals. J Eur Ceram Soc 37:407–411
15. Rao GB, Rajesh P, Ramasamy P (2017) Enhanced optical, thermal and piezoelectric behaviour in dye-doped potassium acid phthalate (KAP) single crystal. J Cryst Growth **468**: 411–415
16. Bernard J et al (1971) Piezoelectric ceramics. Elsevier, Amsterdam
17. Ren B et al. (2010) Piezoelectric energy harvesting using shear mode 0.71Pb(Mg1/3Nb2/3)O3-0.29PbTiO3 single crystal cantilever. Appl Phys Lett **96**: 083502
18. Yue Y et al (2017) High power density in a piezoelectric energy harvesting ceramic by optimizing the sintering temperature of nanocrystalline powders. J Eur Ceram Soc 37:4625–4630
19. Shahab S, Zhao S, Erturk A (2018) Soft and hard piezoelectric ceramics and single crystals for random vibration energy harvesting. Energ Technol 6:935–942
20. Hwang GT et al (2015) A reconfigurable rectified flexible energy harvester via solid-state single crystal grown PMN– PZT. Adv Energy Mater 5:1500051
21. Shibata K et al (2018) Applications of lead-free piezoelectric materials. MRS Bull 43:612–616
22. Zheng T et al (2018) Recent development in lead-free perovskite piezoelectric bulk materials. Prog Mater Sci 98:552–624
23. Zhang Y, Sun H, Chen WA. (2017) brief review of Ba (Ti0.8Zr0.2) O3-(Ba0.7Ca0.3)TiO3 based lead-free piezoelectric ceramics: Past, present and future perspectives. J Phys Chem Solids **114**: 207–219
24. Kang W-S, Koh J-H. (2015) (1−x) Bi0.5Na0.5TiO3–xBaTiO3 lead-free piezoelectric ceramics for energy-harvesting applications. J Eur Ceram Soc **35**: 2057–2064
25. Leontsev SO, Eitel RE (2010) Progress in engineering high strain lead-free piezoelectric ceramics. Sci Technol Adv Mater 11:044302
26. Wu M et al (2018) High performance piezoelectric energy harvester and self-powered mechanosensing using lead free potassium−sodium niobate flexible piezoelectric composites. J Mater Chem A 6:16439–16449
27. Yan X et al (2017) Composition-driven phase boundary and its energy harvesting performance of BCZT lead–free piezoelectric ceramic. J Eur Ceram Soc 37:2583–2589
28. Qaiser MA et al (2018) 0–3 type Bi3TaTiO9: 40 wt% BiFeO3 composites with improved high-temperature piezoelectric properties. J Alloy Compd 740:1–6
29. Li Q et al (2017) High-temperature dielectric, ferroelectric and piezoelectric properties of Mn-modified BiFeO3-BaTiO3 lead-free ceramics. J Mater Sci 52:229–237
30. Challagulla K, Venkatesh T (2012) Electromechanical response of piezoelectric foams. Acta Mater 60:2111–2127
31. Savolainen A, Kirjavainen K (1989) Electrothermomechanical film. Part I. Design and characteristics. J Macromol Sci A **26**: 583–591
32. Anton SR, Farinholt KM, Erturk A. (2014) Piezoelectret foam-based vibration energy harvesting. J Intell Mater Syst Struct **25**: 1681–1692
33. Pondrom P et al (2014) Vibration-based energy harvesting with stacked piezoelectrics. Appl Phys Lett 104:172901
34. Erturk A, Inman DJ (2008) A distributed parameter electromechanical model for cantilevered piezoelectric energy harvesters. J Vib Acoust 130:041002
35. Erturk A, Inman DJ (2008) On mechanical modelling of cantilevered piezoelectric vibration energy harvesters. J Intell Mater Syst Struct **19**: 1311–1325

36. Goldschmidtboeing F, Woias P (2008) Characterization of different beam shapes for piezoelectric energy harvesting. J Micromech Microeng 18:104013
37. Muthalif AG, Nordin ND (2015) Optimal piezoelectric beam shape for single and broadband vibration energy harvesting: modelling, simulation and experimental results. Mech Syst Signal Process **54**: 417–426
38. Roundy S et al (2005) Improving power output for vibration-based energy scavengers. IEEE Pervasive Comput 4:28–36
39. Dhakar L et al (2005) A new energy harvester design for high power output at low frequencies sensor. Actuators A-Phys 199:344–352
40. Lee S, Youn BD (2011) A new piezoelectric energy harvesting design concept: multimodal energy harvesting skin. IEEE Trans Ultrason, Ferroelectr Freq Control **58**: 629–645
41. Aridogan U, Basdogan I, Erturk A (2014) Analytical modelling and experimental validation of a structurally integrated piezoelectric energy harvester on a thin plate. Smart Mater Struct 23:045039
42. Chen X-R et al (2012) Vibration energy harvesting with a clamped piezoelectric circular diaphragm. Ceram Int 38:S271–S274
43. Wang W et al (2012) Vibration energy harvesting using a piezoelectric circular diaphragm array. IEEE Trans Ultrason Ferroelectr Freq Control 59:2022–2026
44. Stephen N (2006) On energy harvesting from ambient vibration. J Sound Vib 293:409–425
45. Li S et al (2016) Bi-resonant structure with piezoelectric PVDF films for energy harvesting from random vibration sources at low frequency Sensor. Actuators A Phys 247:547–554
46. Naseer R et al (2017) Piezomagnetoelastic energy harvesting from vortex-induced vibrations using monostable characteristics. Appl Energy 203:142–153
47. Fan K et al (2018) A monostable piezoelectric energy harvester for broadband low-level excitations. Appl Phys Lett 112:123901
48. Lan C, Tang L, Qin W (2017) Obtaining high-energy responses of nonlinear piezoelectric energy harvester by voltage impulse perturbations. Eur Phys J Appl Phys 79:20902
49. Wang W et al (2018) Comparison of harmonic balance and multi-scale method in characterizing the response of monostable energy harvesters. Mech Syst Signal Process 108:252–261
50. Stanton SC, McGehee CC, Mann BP (2009) Reversible hysteresis for broadband magnetopiezoelastic energy harvesting. Appl Phys Lett 95:174103
51. Cottone F, Vocca H, Gammaitoni L (2009) Nonlinear energy harvesting. Phys Rev Lett 102:080601
52. Erturk A, Hoffmann J, Inman DJ (2009) A piezomagnetoelastic structure for broadband vibration energy harvesting. Appl Phys Lett **94**(254102) p 3
53. Berdy DF et al (2012) Low-frequency meandering piezoelectric vibration energy harvester. IEEE Trans Ultrason Ferroelectr Freq Control 59:846–858
54. Yu H et al (2014) A vibration-based MEMS piezoelectric energy harvester and power conditioning circuit. Sensors 14:3323–3341
55. Ali SF, Friswell MI, Adhikari S (2011) Analysis of energy harvesters for highway bridges. J Intell Mater Syst Struct **22**: 1929–1938
56. Erturk A (2011) Piezoelectric energy harvesting for civil infrastructure system applications: moving loads and surface strain fluctuations. J Intell Mater Syst Struct 22:1959–1973
57. Karami MA, Inman DJ (2012) Powering pacemakers from heartbeat vibrations using linear and nonlinear energy harvesters. Appl Phys Lett 100:042901
58. Qi Y et al (2010) Piezoelectric ribbons printed onto the rubber for flexible energy conversion. Nano Lett 10:524–528
59. Anton SR, Erturk A, Inman DJ (2012) Multifunctional unmanned aerial vehicle wing spar for low-power generation and storage. J Aircr **49**: 292–301
60. Erturk A, Renno JM, Inman DJ (2009) Modeling of piezoelectric energy harvesting from an l-shaped beam-mass structure with an application to UAVs. J Intell Mater Syst Struct **20**: 529–544
61. Featherston CA, Holford KM, Greaves B (2009) Harvesting vibration energy for structural health monitoring in aircraft. Key Eng Mater **413–414**: 439–446

62. Le MQ et al (2015) Review on energy harvesting for structural health monitoring in aeronautical applications. Prog Aerosp Sci 79:147–157

Simulation Studies on Harmonic Analysis of Lighting Loads

Madhu Palati and M. C. Madhu

Abstract In recent days, the demand for electrical power is increasing. To cater the need, generation has to be raised. Since the conventional resources are depleting, the focus was made on renewable energy generation. Also, it is aimed to reduce the electrical power consumption by switching the home lighting system to CFL or LED lamps. By using these lamps, the power consumption is reduced and efficiency is increased, at the same time, these lamps are posing a threat to the distribution system by injecting harmonics into the source leading to poor power quality. In this paper, emphasis is made in studying the harmonic analysis of different lamps, and methods are proposed to mitigate the harmonics. Models are developed in PSPICE, and FFT analysis is carried out on the current waveform. Design of filter circuit is proposed to reduce the total harmonic distortion in the system.

Keywords CFL · LED · Harmonics · Filter · THD · FFT · PSPICE

1 Introduction

The demand for electrical energy is increasing in India due to rapid growth in the urbanization, increasing industrial loads, agriculture load, and domestic load. The load on the plant is always dynamic in nature, increase in load demand can be met with an increase in generation. Most of the electrical power generation in India is happening through coal-based plants. As coal is depleting year by year, as an alternative to these thermal plants, the government is focused on the generation of electrical energy through renewable energy sources to meet this increase in demand. At the same time, the increase in demand can be met by decreasing the transmission and distribution losses. A lot of research is done to reduce the transmission losses, and the methods are already implemented to reduce the losses. Government is encouraging the firms to manufacture the electrical appliances that consume less power compared

M. Palati (✉) · M. C. Madhu
BMS Institute of Technology and Management, Bangalore, India
e-mail: mfmadhu@gmail.com

N. M. Sivaram et al. (eds.), *Advances in Manufacturing, Automation, Design and Energy Technologies*, Lecture Notes in Mechanical Engineering,
https://doi.org/10.1007/978-981-99-1288-9_80

to conventional appliances. The decrease in the power consumption helps to cater the need of increase in demand and also reduces the electricity bill of the customer which is a cost-effective solution. All the domestic appliances run on AC supply. Lighting system forms the major part of domestic loads; different types of lamps are used based on category of lighting system such as ambient, task, accent, and decorative lighting system. At home, nowadays commonly used lamps are incandescent lamps, fluorescent Lamps, compact fluorescent lamps (CFL) and light emitting diode (LED) lamps. Apart from this, for street lighting sodium vapor lamp, mercury vapor lamp, and nowadays, LED street light is used. One of the methods of reducing the power consumption of lighting system is using the energy-efficient lamps in lieu of incandescent lamp with fluorescent, compact fluorescent lamp, or an LED lamp.

LED lamps have better efficiency than other lamps, longer life, and environmentally friendly when compared to other lamps. Hence, demand for LED lamps is growing, and many manufacturers are showing interest in producing LED lamps in large scale. In the future, this may ramp up exponentially due to advancement in technology, making it compact and the overall cost will come down. The disadvantage of these LED lamps is heat dissipation, producing distorted current which penetrates into the power system and create power quality issues. Most of the researchers have concentrated on the electronic ballast design, making it compact and improving the performance. Few researchers have carried out on studying the harmonic effect due to these LED lamps and mainly focused on low rating LED lamps of rating less than 25 watts and even the power factor of these lamps was low in the range from 0.6 to 0.8. As per standards, LED lamps of wattage less than 25 should comply that the third harmonic component should not exceed 86% and 5th harmonic should not exceed 61%.

Every customer expects a good quality power supply from the electrical distribution power supply companies. LED light is a nonlinear load which produces harmonics and has an impact on distribution grid. In general, it is assumed that the supply voltage is free from harmonics and has only fundamental frequency, and the waveform is sinusoidal in nature. In reality, due to the introduction of power electronic converters in these lamps leads to production of harmonics and the supply voltage get distorted, which gets added with other frequency waveforms along with fundamental frequency waveforms. This reduces the power quality of the supply voltage. Hence, there is a need to analyze the harmonic content in the supply voltage when CFL or LED are connected as loads. Non-sinusoidal waveforms are generated when circuits with switching transformers, iron-cored inductors, rectifiers, inverters, CFL which has an electronic ballast. The distortion in sinusoidal waveform is often termed as harmonics and are multiples of the fundamental frequency. The frequency of the harmonic waveforms will be 2f, 3f, 4f, 5f, etc., These can be classified into even harmonics (2f, 4f, 6f, etc.,) and odd harmonics (3f, 5f, 7f, etc.,). Also, these can be classified into positive-sequence harmonics and negative-sequence harmonics. Positive-sequence harmonics are responsible for causing overheating in conductors. Negative-sequence harmonics rotate in opposite to the rotating magnetic field in induction motors causes weakening of flux leading to reduction in mechanical torque. Another type is zero-sequence harmonics which have frequency multiples of three. These causes heating in the neutral wire. Therefore, the presence of harmonics causes a potential problem in conductors, rotating machines, and electrical appliances which are inductive in nature, leading to increase in losses and thereby resulting in decreased efficiency of the distribution system.

2 Literature Review

Ivan Ramljak et al., presented LED bulbs and street light modeling using EMTP-ATP software, Experiments were conducted on single LED lamp and examined the Total Harmonic Distortion (THD) values, it was observed that the values are as high as 150% especially 3rd harmonic is predominant and the values are more for larger systems. It is also observed that in phase conductors 5 and 7th harmonics are of higher magnitude [1]. Lom GM et al., conducted experiments on different LEDS of different ratings, CFL and GIL bulb types. To study the effect of voltage fluctuation, experiments were conducted to obtain the active, reactive power, and harmonic distortion by subjecting the lamps to different under voltage, rated voltage, and over voltage. The results obtained were compared with the standard values [2]. Pedro VELOSO presented three case studies, replacing the mercury vapor and high-pressure sodium vapor lamps with LED lamps. PQ parameters were analyzed, and the tests were performed as per international standards. The voltage harmonics were plotted for all the three cases before replacement and after replacement of lamps with LED. It was observed that voltage harmonics there was no much difference before replacement and after replacement, whereas there was reduction in current harmonics after replacing the lamps with LED lamps [3].

Akila et al., carried out experiments on different LED lamps of different ratings starting from 6 to 18W ratings. Parameters like actual power, frequency, voltage, current, power factor, THD of the voltage waveform and THD of the current waveform were obtained through experiments. Harmonic profiles of each lamp were studied and the cause of the increase in THD value of 42–70.3% is due to implementation of 10 million LEDs in Egypt was explained [4]. Amir Tokic presented a new model of fluorescent lamp and simulation was performed using EMTP/ATP software and this can also be performed using MATLAB software. The analysis is based on the extended curve fitting method of voltage versus current characteristics. Harmonic studies were done on these models and the simulation results were compared with experimental results [5]. Piyush Verma et al., presented the harmonic analysis of different branded LED lamp driver circuits and explained the importance of phase angle of each harmonic component in reducing the harmonic emissions. Also presented the variation of harmonics levels from 12 to 102% for lamps of the same wattage but of different company products. Also explained the method to select the LED lamps with less harmonic emission [6].

Molina developed several CFL models to analyze the harmonics based on the admittance matrix method which reduces the computational time. Four models were developed, and suitable application of each model was also presented based on THD values [7]. Uddin et al., conducted experiments on 37 numbers of different LED lamps and CFL with rating 4–8W of different brands. The current waveforms were observed without filtering, with passive filtering, and active filtering based on company brand as each company has different filtering mechanism and few brands do not use any filtering mechanism. Some lamps produced the THD more than acceptable range, this happened in lamps without filtering circuit. Diversity factor was estimated and it decreases with increase in number of lamps. LED and CFL generates lots of heat inside the tube which will be in the order of 60–85 and up to 100° respectively [8]. Carillo et al., developed a fluorescent lamp model considering nonlinearities of the lamp and neglecting the nonlinearities of the reactance. The voltage, current and power waveforms were obtained. The behavior of the lamp under different voltage fluctuations was not studied, and there was more error in the obtained results [9].

From the literature, it is very much clear that fluorescent, CFL, and LED lamps are advantages in reducing the power consumption when compared to incandescent lamps, but on the other end these cause harmonic distortion in the main supply which is a major concern. Therefore, the impact of these lamps on distribution grid has to be studied, and method to reduce the harmonic distortion has to be analyzed. Since it is a time-consuming process, approach of analyzing the harmonics in different types of lamps can be done easily by developing the models and performing the simulation using software like PSPICE, MATLAB, ORIGIN, etc., in this work, the models were developed using PSPICE software. This saves a lot of time and cost for a design engineer.

3 Simulation Model

Three types of models for CFL, spiral CFL, and LED were developed in PSPICE. Each lamp has to be represented by its equivalent circuit. Exact circuit implementation becomes difficult and complex; hence, the equivalent circuit will be simplified to make the model development easy in PSPICE. The inverter circuit of a CFL can be represented using a capacitor of 6.52 microfarads in parallel with equivalent resistance of 3996 ohms [10] and is shown in Fig. 1.

Since any lamp with rectifier or inverter circuit or with electronic ballast generates harmonics. To mitigate these harmonics, digital filter has to be added. In the present work, the digital filter is tuned to the fifth order frequency ($5 * 50 = 250$ Hz), standard values of resistor and capacitance are taken as 100 ohms and 470 microfarads. From literature, the researchers have tuned for higher order frequency; in this paper, the filter is tuned to fifth order to get better results in mitigating the harmonics. The inductance of the filter circuit is estimated using Eq. (1). The equivalent circuit of CFL with filter circuit is shown in Fig. 2.

$$L = \frac{1}{2\pi f \sqrt{C}} = \frac{1}{2\pi * 250\sqrt{470 * 10^{-6}}} = 0.862\, mH \tag{1}$$

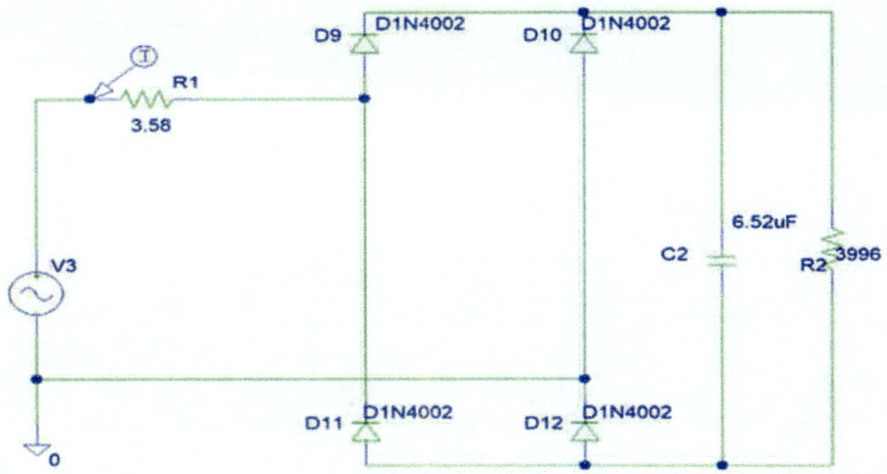

Fig. 1 Equivalent circuit of CFL without filter

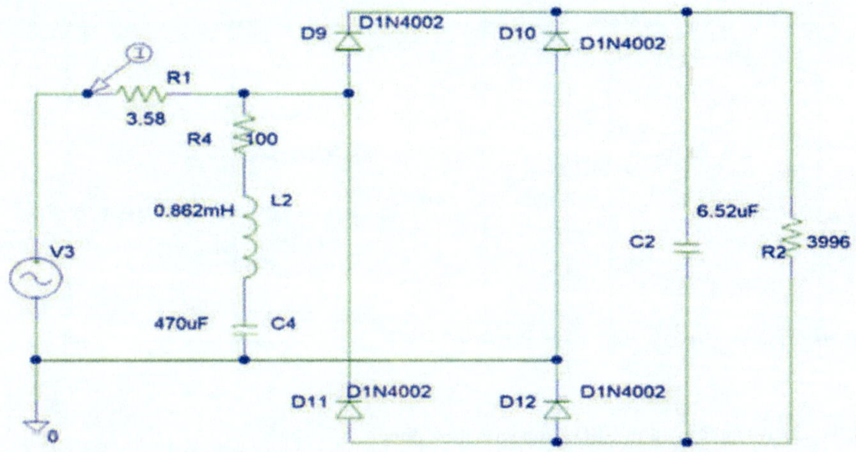

Fig. 2 Equivalent circuit of CFL with filter circuit

An equivalent circuit for spiral CFL is developed in PSPICE with capacitance and equivalent resistance of the inverter circuit as 22 μF and 22 kΩ, respectively. Figure 3 shows an equivalent circuit of spiral CFL along with filter circuit. Equivalent circuit for LED lamp which includes the digital filter is shown in Fig. 4 with I_1 and I_2 values as 34.53 mA and 24.99 mA, respectively. R and L values are 14.3 ohms and 164.5 H respectively [11].

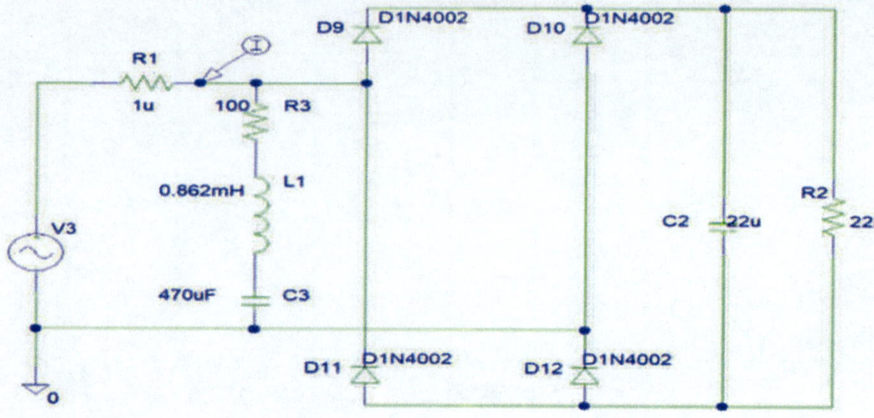

Fig. 3 Equivalent circuit of Spiral CFL with filter circuit

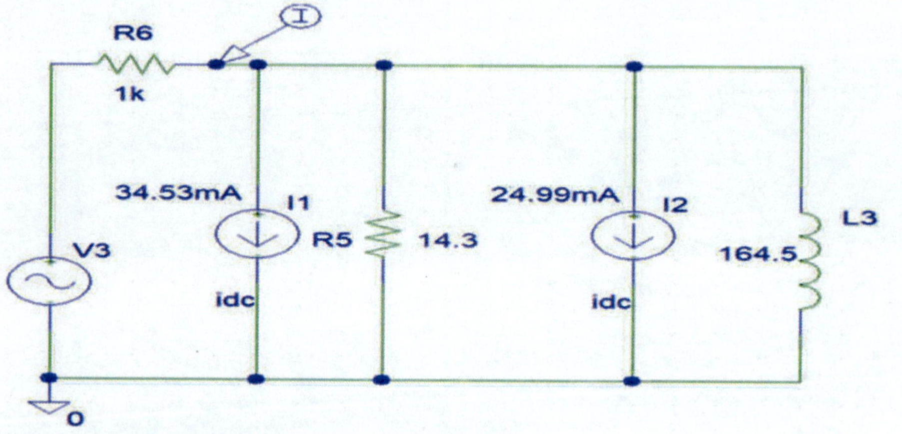

Fig.4 Equivalent circuit of LED lamp with filter circuit

4 Results and Discussions

Total harmonic distortion (THD) of current and voltage waveforms is expressed as

$$THD_i = \sqrt{\frac{\Sigma_{n=2} i_n^2}{i_1^2}} \quad (2)$$

$$THD_v = \sqrt{\frac{\Sigma_{n=2} v_n^2}{v_1^2}} \quad (3)$$

where i_n and v_n are the rms values of nth fourier components of current and voltage waveforms $i(t)$ and $v(t)$, respectively.

$$THD_{even} = \sqrt{\frac{i_2^2 + i_4^2 + i_6^2 + \cdots + i_{2n+2}^2}{i_1^2}} \tag{4}$$

$$THD_{odd} = \sqrt{\frac{i_3^2 + i_5^2 + i_7^2 + \cdots + i_{2n+1}^2}{i_1^2}} \tag{5}$$

$$THD = \sqrt{\frac{i_2^2 + i_3^2 + i_4^2 + \cdots + i_n^2}{i_1^2}} \tag{6}$$

A fast fourier transform (FFT) waveform of CFL without and with filter circuit is shown in Fig. 5. The harmonic distortion percentage of odd harmonics, even harmonics, and total harmonic distortion is estimated using Eq. (4), (5), and (6), respectively. The values are 51.36%, 74.5%, and 90.48%, respectively. From Fig. 5 with filter, the estimated value of total harmonic distortion is 3.53%. The voltage waveforms are not shown in this paper as the harmonics in voltage waveform will be almost minimum. An FFT wave form of spiral CFL without filter and with filter circuit is shown in Fig. 6. The FFT waveform for LED lamp is shown in Fig. 7.

From Fig. 6, for spiral CFL without filter circuit, the percentage of harmonic distortion of odd harmonics, even harmonics, and total harmonic distortion are 42.2%, 58.57%, and 72.19%, respectively. From Fig. 6 spiral CFL with filter circuit the total harmonic distortion is 2.73%. Figure 7 shows the FFT analysis of LED lamp with filter circuit and the total harmonic distortion is 2.2%. It is observed that in the CFL after inclusion of filter circuit the THD has come down to 3.53% from 90.48%. Similarly, in spiral CFL by including filter circuit, the THD value has reduced from

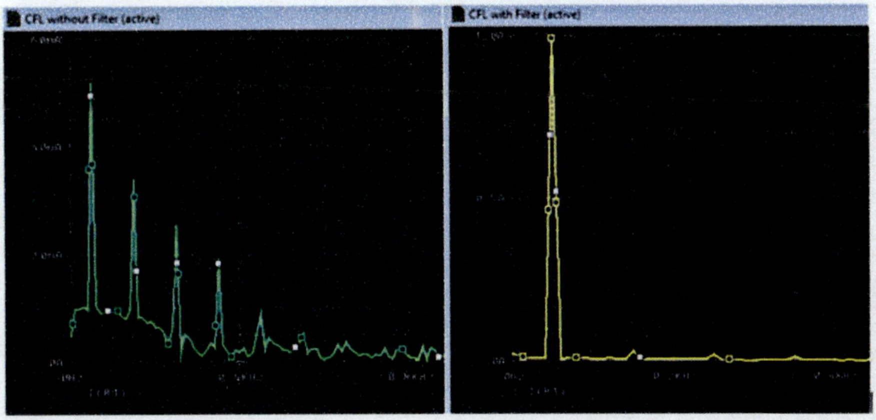

Fig. 5 FFT waveform of CFL without and with filter circuit

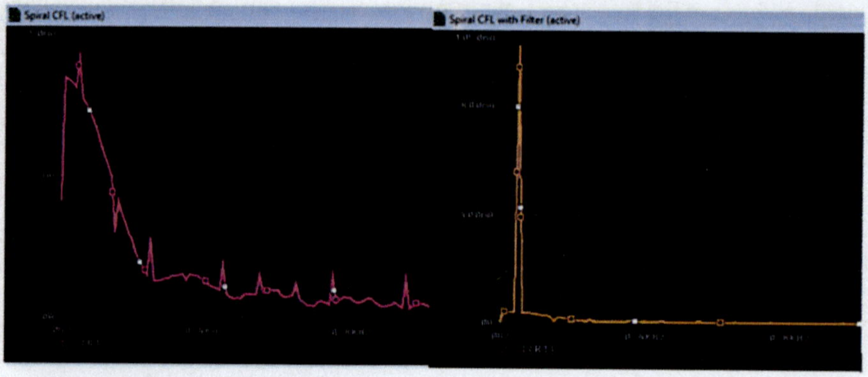

Fig. 6 FFT waveform of Spiral CFL without and with filter circuit

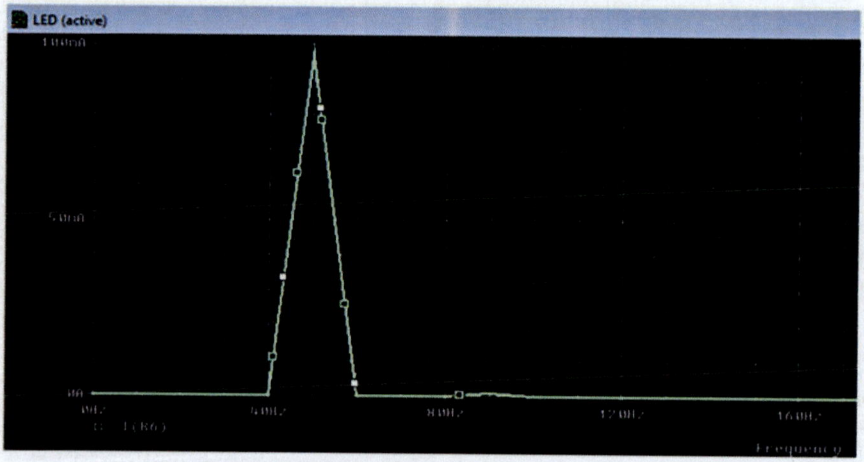

Fig. 7 FFT waveform of LED lamp with filter circuit

72.19 to 2.73%. In LED lamp with filter circuit, the THD is 2.2%. By including filter circuit, most of the harmonics can be eliminated, which reduces the burden on the grid and the power quality of the supply will be improved.

5 Conclusion

In this paper, attempt was made to study the harmonic analysis of CFL, spiral CFL, LED and simulation was carried out in PSPICE software. The THD value is very high in case of lamps without filter circuit and is very high and it was brought down to a very low value by adding a filter circuit with optimized values. It was observed that the THD value of the single lamp seems to be of less importance, but considering the future where all the existing lamps will be replaced with LED lamps will definitely lead to huge harmonic penetration into the distribution grid. Simulation was performed by combining different driver topologies to reduce the THD level of lamps. Attention was kept especially on 3rd and 5th harmonic level as this will have a huge impact on the grid, and hence always the THD values has to be kept well within the limits. The LED lamps will have an impact on the voltage quality of grid system; neutral carries high current and compensatory circuits are required in addition to mitigate the harmonics. Filter circuit was proposed for each model which reduced the THD to a significant value. The diversity factor has to be maintained low so that, lower will be the harmonic injection into the system. Combination of CFL and LED lamps can be implemented to reduce the diversity factor.

References

1. Ivan R, Amir T (2020) Harmonic emission of LED lighting. AIMS Energy 8(1):1–26
2. Lom GM, Laverty D, Morrow DJ et al (2019) Load and harmonic distortion characterization of modern low-energy lighting under applied voltage variation. Electr Power Syst Res 169:124–138
3. Veloso P, Lebre A, Bastiao F (2019) LED technology in public lighting—analysis of the impact in power quality in the low voltage grid distribution. In: 25th International conference on electricity distribution (CIRED), pp 1–5
4. Akila A, Etman M, Youssed K (2019) Passive mitigation technique for the harmonics caused by LED lamps. In: 25th International conference on electricity distribution (CIRED), pp 1–5
5. Tokić A, Jukan A, Uglešić I et al (2018) Nonlinear model of fluorescent lamp in harmonic studies. In: IEEE PES innovative smart grid technologies conference, pp 1–6
6. Verma P, Patel N, Nair NC (2017) LED electronics, harmonic issues and augmenting Standards, In: IEEE international conference on consumer electronics (ICCE), pp 190–191
7. Molina J, Sainz M (2015) Compact fluorescent lamp modeling for large-scale harmonic penetration studies. IEEE Trans Power Deliv 30:1523–1531
8. Uddin S, Shareef H, Mohamed A, Hannan MA (2012) Harmonics and thermal characteristics of low wattage LED lamps. Przegląd Elektrotechniczny 88:266–271
9. Carrillo C, Cidras J (2001) Fluorescent lamp modelling for voltage fluctuations. Eur Trans Energy Power-ETEP 11(2):119–127
10. Hanan M, Hashmi K, Ai X, Salman S, Masood A (2018) A shunt active filter for extenuation of harmonics by compact fluorescent lamp drives. In: 2nd IEEE conference on energy internet and energy system integration (EI2), pp 1–7
11. Mauro SD, Raciti A, Rizzo SA, Susinni G (2018) LED lamp PSpice circuit emulating first harmonic current absorption as a function of the main voltage. In: International symposium on power electronics, electrical drives, automation and motion (SPEEDAM), pp 1101–1106

Effects of Hot and Cold EGR in CRDI Diesel Engine Fuelled with B20 Blend of Rice Bran Oil Biodiesel

S. Aravinthan, Davis Lazar, M. A. Habeebullah, and J. Jayaprabakar

Abstract The rice bran oil biodiesel is prepared; properties are measured and tested in a CRDI diesel engine. B20 blend of rice bran biodiesel and normal diesel is tested without exhaust gas recirculation, and in the next phase both hot and cold exhaust gas recirculation are introduced. Four sets of EGR are varied, viz. 10% hot EGR (EGR 10 H), 20% hot EGR (EGR 20 H), 10% cold EGR (EGR 10 C), and 20% cold EGR (EGR 20 C). The diesel recorded high brake thermal efficiency compared to biodiesel blend either with EGR or without EGR. At full condition, the specific fuel consumption for diesel was less compared to the biodiesel blend. Diesel produces fewer amounts CO at higher loads, whereas EGR 20 C produces more. EGR 20 C produces low NOx rate at higher loads. Diesel has recorded less hydrocarbon emissions, while EGR 20H blend produced more. B20 blend tested at without EGR condition has produced high smoke opacity values, but when EGR is introduced, the smoke opacity reduced considerably.

Keywords Common rail direct injection · Rice bran oil · Exhaust gas recirculation · Emission

1 Introduction

Economic progress has been heavily reliant on the use of oil as a source of energy in recent decades. Meanwhile, natural gas and coal remain the world's primary energy sources [1, 2]. Over the last few decades, there has been a rise in environmental consciousness amongst the general public. As a result, people are increasingly turning to alternate fuels. Biodiesel is a green alternative fuel created from vegetable oils, animal fats, and greases that is chemically synthesised. Vegetable oils are transesterified into fatty acid alkyl esters, which can be utilised as a low-viscosity diesel fuel

S. Aravinthan · D. Lazar · M. A. Habeebullah · J. Jayaprabakar (✉)
Department of Mechanical Engineering, Sathyabama Institute of Science and Technology, Chennai, India
e-mail: jp21tn@gmail.com

N. M. Sivaram et al. (eds.), *Advances in Manufacturing, Automation, Design and Energy Technologies*, Lecture Notes in Mechanical Engineering,
https://doi.org/10.1007/978-981-99-1288-9_81

substitute. There are a variety of alternative fuel sources for IC engines. One of these types is biodiesel, which is a diesel engine alternative fuel. It possesses a number of advantageous characteristics, including high lubricity, biodegradability, minimal carbon emissions, and a low-cost feedstock. Transesterification is a common method for producing it. Biodiesel is made up of long-chain fatty acid mono alkyl esters generated from green lipid sources like vegetable oil or animal fat, in a nutshell. Rice bran oil, algal oil, and waste cooking oil are all used in the production of biodiesel [3, 4]. Different biodiesel combinations, such as alcohol blended biodiesel, di-ether blended biodiesel, and others, were also tested in IC engines to improve biofuel operating performance [5, 6]. According to various publications, rice bran and algae are considered to be low-cost biodiesel sources. Rice bran is a waste product, while algae are a non-edible source [7, 8]. Diesel engines use compression ignition and are referred to as "versatile" because of their vast range of applications. Currently, the majority of diesel engines are direct injection models with a low-pressure fuel injection pump. Common rail direct fuel injection system, in contrast to a low-pressure fuel pump feeding unit injectors, is built around a high-pressure (over 2,000 bar) fuel rail feeding unit. By injecting fuel as a larger number of tiny droplets with a much higher surface area to volume ratio than conventional lower-pressure fuel injection, high-pressure fuel injection improves strength and fuel economy. This allows for better vaporisation, resulting in a more effective combination of ambient oxygen and evaporated fuel, resulting in a more complete combustion. The usage of a CRDI engine has undeniably improved the overall efficiency of diesel engine operating. Because biodiesel has a significantly higher density and viscosity, the pressure required to break the molecules will be higher, to have better mixing and vaporisation. Although biodiesel has the potential to be used in CRDI engines, it has yet to be validated. It's essential to understand the blended fuel's maximum possible performance, which might provide significant comparison parameters with the base engine. Furthermore, the impact of changes in load and blends are important considerations [9]. The use of biodiesel in CRDI engines will undoubtedly improve the performance and carbon emission characteristics of diesel engines, as biodiesel is known for its low-carbon emissions. However, NOx emissions continue to be a major issue in biodiesel engines. The generation of NOx increases exponentially as the temperature rises. Using low-nitrogen fuels is one way to reduce NOx emissions. Another option is to adjust combustion conditions to produce significantly less NOx. Exhaust Gas Recirculation (EGR), Selective Catalytic Reduction (SCR), Two-Stage Turbocharger, High Scavenge Pressure, Compression Ratio, Water Injection and Water Emulsion, and Humid Air Method are all options for reducing NOx emissions. Cooled EGR improves engine volumetric efficiency by increasing consumption density. At the same time, the lower temperature can reduce NOx emissions; nevertheless, when compared to warm EGR, HC emissions and cycle-by-cycle versions are higher. Injection of hot EGR does not necessitate the use of an extra cooler to cool the outgoing exhaust gas. This allows for better surface vaporisation, resulting in a more effective combination of ambient oxygen and evaporated fuel, resulting in a more complete combustion. However, cold EGR method requires a cooler, and in comparison to hot EGR, cool EGR has higher thermal efficiency, volumetric efficiency, brake specific energy consumption, and braking power. With all of the information presented above, the present study's goal is to examine the effect of hot and cold EGR methods in a CRDI engine operated with B20 blend of rice bran biodiesel.

2 Materials and Methods

2.1 Materials Required

Methanol, Rice Bran Oil, Catalyst–NaOH (Sodium hydroxide), Diesel, and CRDI Engine with Cooled and Hot EGR provisions.

2.2 Preparation of Fuel Samples

With low temperature and low pressure involvement, transesterification is a common process for producing biodiesel [10, 11]. Transesterification is a reversible reaction in which fatty acids, alcohol, and a catalyst are combined together. In this study, the reaction is held between rice bran oil, methanol, and NaOH pellets. The rice bran oil has been collected, and the transesterification procedure has been completed. These dry caustic balls were mixed with methanol. Approximately 5 grammes (0.5 weight per cent of rice bran oil) of NaOH pellets are mixed with 182 ml of methanol (molar ratio of 1:6 to rice bran oil) for 90 min using a magnetic stirrer. It was ensured that the dry NaOH pellets and methanol were appropriately mixed. The reacted mixture is poured into the two flasks with openings and allowed to settle for 15 min. Finally, the two layers are clearly identified and separated as glycerin and methyl ester. In the next phase, the fuel properties of the prepared biodiesel blends are measured as per ASTM standards. The measured properties are presented in Table 1.

2.3 Engine Setup

The test rig used in this is a single cylinder CRDI engine connected to an eddy current dynamometer. It is provided with necessary instruments for crank angle, airflow, fuel flow, load and temperature measurements, and combustion pressure. The signals are interfaced to personal computer through a high-speed data acquisition device.

Table 1 Properties of Fuel samples

Parameters	D100	B100	B20
Acid value mg NaOH/g	0.11	1.3	0.42
Density g/cc	0.845	0.882	0.8524
Kinematic viscosity @ 40 °C	0.84 cSt	8.4 cSt	1.69 cSt
Copper strip corrosion @ 60 °C for 3 h	1 a	1 a	1 a
Flash point by PMCC method	52 °C	221 °C	153 °C
Fire point by PMCC method	67 °C	234 °C	161 °C
Cloud point	−5 °C	2 °C	20 °C
Pour point	−12 °C	−10 °C	13 °C
Moisture content	0.010%	0.13%	0.06%
Calorific value Cal/g	10,143.916	8642.63	9842.6

3 Results and Discussions

The performance and emission characteristics of the prepared fuel blend in CRDI engine with hot and cold EGR methods are presented and discussed in this section of the study.

3.1 *Brake Thermal Efficiency*

The brake thermal efficiency of a heat engine is a measurement of how efficiently an engine transforms heat from a fuel into mechanical energy. It is mainly affected by fuel properties such density and heating values. Figure 1 shows that variation in brake thermal efficiency against different loads. It clearly shows that the diesel has highest brake thermal efficiency, and it is followed by B20, B20 EGR 10C, B20 EGR 20 C, B20 EGR 10H, and B20 EGR 20H. The loss in brake thermal efficiency with greater EGR flow rates is due to a decrease in the availability of oxygen in the combustion chamber, which is replaced by higher percentages of EGR [12].

3.2 *Specific Fuel Consumption*

With reference to Fig. 2, the changes in specific fuel consumption decreases with increase in loads. At 25% load, EGR 10 H has more specific fuel consumption than others. SFC values for all the fuel samples are decreases with increase in load. Finally, at 100% load, EGR 10 C & 20 C and EGR 10 H & 20 H shows the highest specific fuel consumption, and they are similar to each other at the same load, diesel shows the least specific fuel consumption.

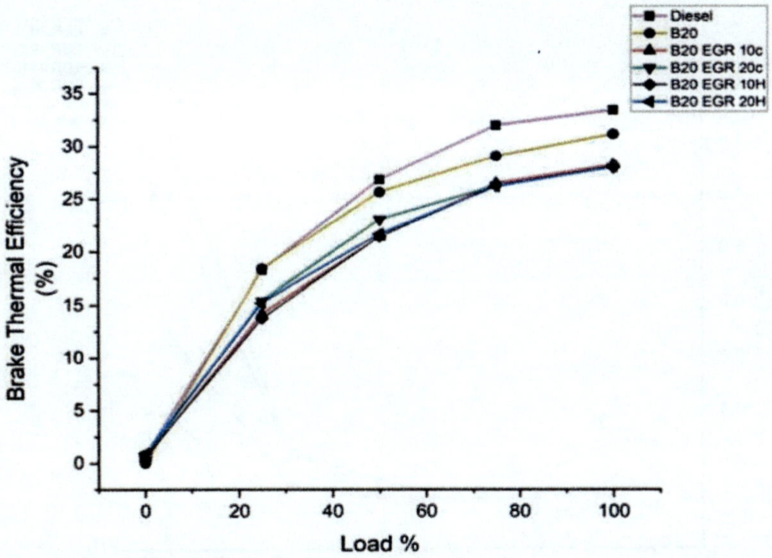

Fig. 1 Brake thermal efficiency

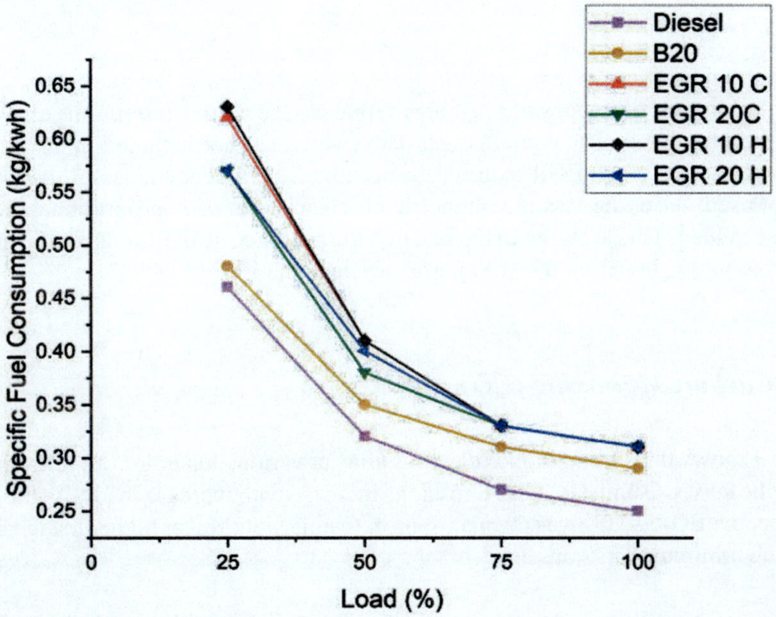

Fig. 2 Specific fuel consumption

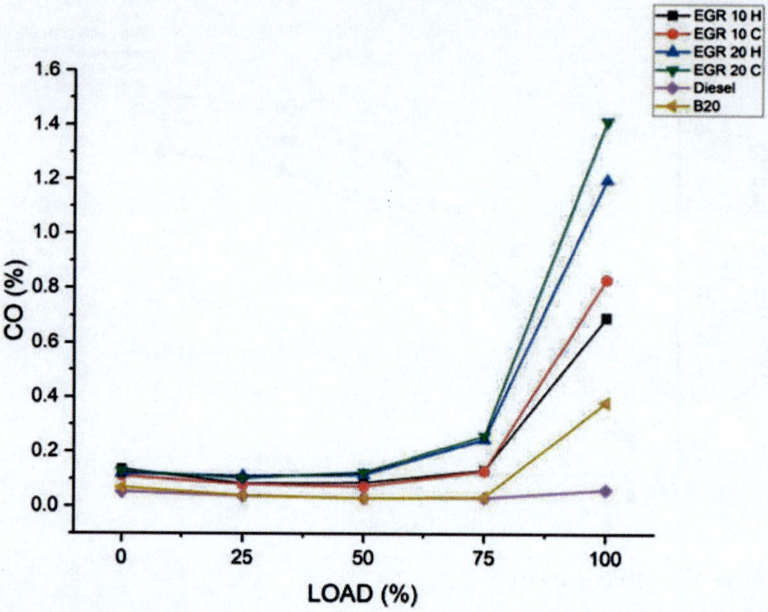

Fig. 3 CO emissions

The volumetric efficiency of a piston is defined as the ratio of the volume displaced to the swept volume. The experimental data (which are not included in the article) clearly show that when EGR is used, the volumetric efficiency is lesser than when it is not used. Also, the loss in volumetric efficiency is exactly proportionate to the increase in load. This is owing to the fact that the presence of EGR reduces the intake of air mass flow, lowering the volumetric efficiency [13].

3.3 *Carbon Monoxide (CO)*

Figure 3 shows the carbon monoxide emissions at various loads. At 0% load, diesel shows the least CO emission. CO emissions increase with increase loads. As the load increase, the EGR 20 C gives the maximum CO emission. But at higher loads, diesel gives the minimum CO emission.

3.4 *Nitrogen Oxides (NOx)*

The NOx emissions are depicted in Fig. 4. As the load increases, NOx emission also increases for all the fuel samples, and the highest NOx emission was recorded by

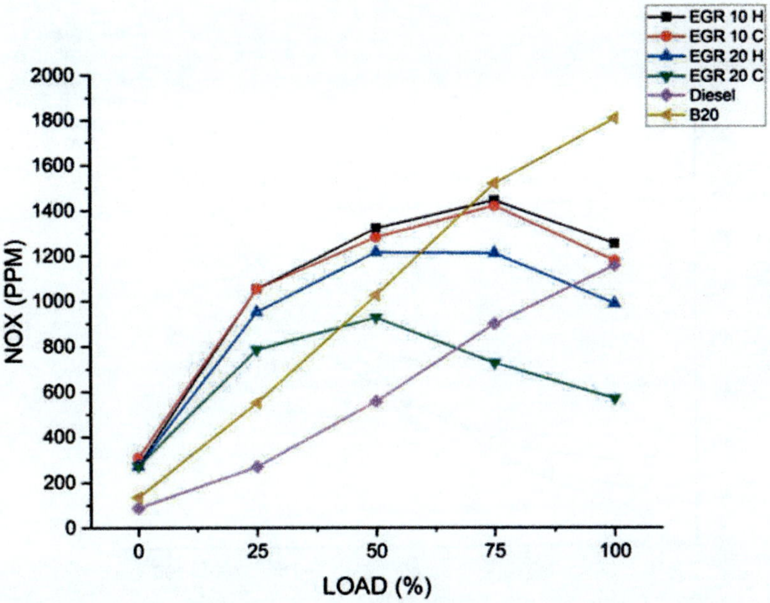

Fig. 4 Nitrogen oxide emissions

B20. Where all EGR initially contains maximum NOx, but when we increase loads it slightly decreases NOx level. EGR 20 C has a minimum amount of NOx. The amount of NOx produced reduces as the EGR rates increase under all loads. The results also show that when the engine load increases, the NOx reduction effect at a fixed EGR rate increases. The varied exhaust gas compositions and hence variances in the dilution effect at different loads account for a major part of the load dependence. When NOx reduction is plotted versus intake manifold oxygen for various loads, all points fall on the same curve. This indicates that at high loads, a given NOx reduction will require less EGR than at low loads [14].

3.5 Hydrocarbon Emission (HC)

The incomplete combustion of fuel in the combustion chamber causes hydrocarbon emissions. This is due to the fuel's composition or structure, the engine's design, and the operating conditions. The hydrocarbon emissions at varying loads are depicted in Fig. 5. The amount of HC released increases as the load increases. At 0% load, B20 has highest value, but at 100% load it moved to 3rd highest HC emission. EGR 20 H became maximum HC emission value at 100% load, whereas diesel has lowest HC emission value. The findings revealed that when the load is increased without

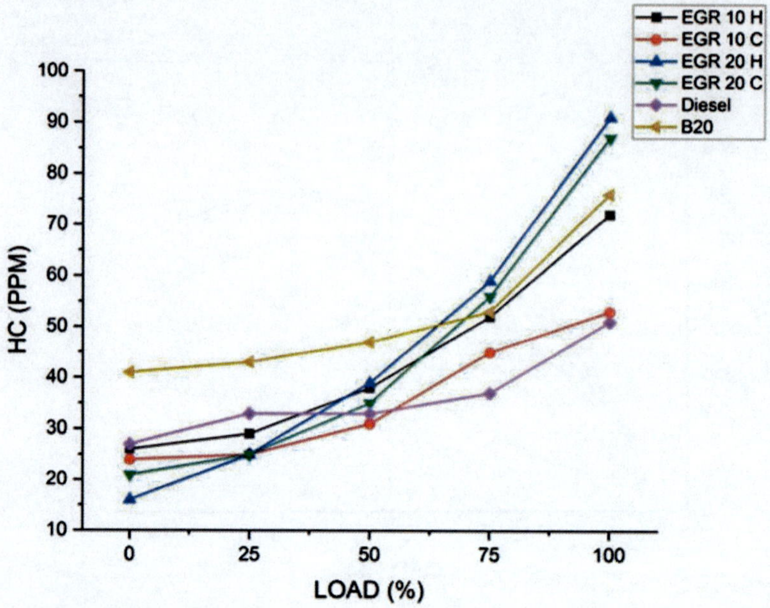

Fig. 5 Hydro carbon emissions

EGR, hydrocarbon emissions are exceptionally high in comparison to hydrocarbon emissions when EGR is present [15].

4 Conclusion

The following conclusions were drawn at end of the study,

- The diesel has high brake thermal efficiency compared to B20 with EGR. But all EGR shows low BTE in high load.
- For specific fuel consumption for diesel has lower rate at 100% load, but EGR 20H has high rate.
- In CO emissions, diesel produces less amount of CO at higher loads, where EGR 20 C produces more CO.
- For NOx EGR 20 C produces low NOx rate at higher loads.
- HC emission has low in diesel where EGR 20H has more.

To conclude, NOx reduction without significant penalty on the brake thermal efficiency is not possible, but the combination of B20 blend of rice bran oil biodiesel at CRDI technology would positively reduce this penalty.

References

1. Abedin MJ, Masjuki HH, Kalam MA, Sanjid A, Ashrafur Rahman SM, Masum BM (2013) Energy balance of internal combustion engines using alternative fuels. Renew Sustain Energy Rev 26:20–33
2. Canakci M, Hosoz M (2006) Energy and exergy analyses of a diesel engine fuelled with various biodiesels. Energy Sources Part B 1(4):379–394
3. Jayaprabakar J, Karthikeyan A (2014) Analysis on the performance, combustion and emission characteristics of a CI engine fuelled with algae biodiesel. Appl Mech Mater 591:33–37
4. Karthikeyan A, Jayaprabakar J (2019) Energy and exergy analysis of compression ignition engine fuelled with rice bran biodiesel blends. Int J Ambient Energy 40(4):381–387
5. Saikrishnan V, Karthikeyan A, Jayaprabakar J (2018) Analysis of ethanol blends on spark ignition engines. Int J Ambient Energy 39(2):103–107. https://doi.org/10.1080/01430750.2016.1269678
6. Jayaprabakar J, Anish M, Beemkumar N, Mathew A, Alex George A (2019) Effect of diethyl ether blended with neem oil methyl esters in CI engine. Int J Ambient Energy 40(2):116–118
7. Lopez I, Quintana C, Ruiz J, Peragon F, Dorado M (2014) Effect of use of olive-pomace oil biodiesel/diesel fuel blends in a compression ignition engine: preliminary exergy analysis. Energy Convers Manag 85:227–233
8. Jayaprabakar J, Karthikeyan A, Josiah A, Shajan A (2015) Experimental investigation on the performance and emission characteristics of a CI engine with rice bran and micro algae biodiesel blends. J Chem Pharm Sci 7:19–22
9. Jayaprabakar J, Karthikeyan A, Gokula Kannan K, Ganesh A (2015) Combustion characteristics of a CI engine fuelled with macro and micro algae biodiesel blends. J Chem Pharm Sci 7:68–71
10. Rajalingam A, Jani SP, SenthilKumar A, Adam Khan M (2016) Production methods of biodiesel. J Chem Pharm Res 8(3):170–173
11. Verma D, Raj J, Pal A, Jain M (2016) A critical review on production of biodiesel from various feedstocks. J Sci Innov Res 5(2):51–58. http://www.jsirjournal.com/Vol5_Issue2_05.pdf
12. Edara G, Murthy YVVS, Srinivas P, Nayar J, Ramesh M (2018) Effect of cooled EGR on modified light duty diesel engine for combustion, performance and emissions under high pressure split injection strategies. Case Stud Therm Eng 12:188–202. https://doi.org/10.1016/j.csite.2018.03.004
13. Agarwal D, Singh SK, Agarwal AK (2011) Effect of exhaust gas recirculation (EGR) on performance, emissions, deposits and durability of a constant speed compression ignition engine. Appl Energy 88(8):2900–2907
14. Yokomura H, Kohketsu S, Mori K (2003) EGR system in a turbocharged and intercooled heavy-duty diesel engine: expansion of EGR area with venturi EGR System. Mitsubishi Tech Rev 15
15. Jayaraman J, Pradip K, Kalaiselvan MP, Joy N, Mariadhas A, Appavu P (2020) Effect of EGR in a CI engine operated with high performance blend of diesel. In: AIP conference proceedings, vol 2311, p 020008. https://doi.org/10.1063/5.0034498

Critical Review on Optimization of Star Propellant Grain Design

Aditya Sharma and Amit Kumar Thakur

Abstract The alignment of the star grain design has shown that it is very attractive for the efficient use of strong power-generating rockets. Among the many benefits of star shape of grain is to reduce the natural weight of a rocket motor structure; protection of the walls of the rooms from extreme heat; and the opportunity to get a wider range of that fixed pressure. The effects of changes on various design parameters are investigated in this paper; each design parameter is treated differently while other parameters. It is always thought. Different types of grains are compared, about the same size as the propellant material. The investigated parameters include grain geometry, initial temperature range, grain erosion and critical width erosion, web burning, and sliver burning. The results are compiled to show variations in end-to-head pressure and time.

Keywords Star grain · Solid propellant grain design · Port geometry

1 Introduction

There are so many kinds of grain designs available for solid propellants, and it depends on the requirement to choose the grain design. Star-shaped grain geometry is found more efficient concerning other designs to use it for sounding rockets, missiles, and heavy payload-carrying rockets. The star-shaped grain configuration is vastly a two-dimensional configuration, and it does not necessarily depend on three-dimensional end effects for burning neutrality [1]. The star-shaped configuration can provide all three types of burnings such as neutral burning, progressive burning, and regressive burning, depending on the need of mission and some parameters. Although star-shaped propellant grain geometry has axial instability behavior in the solid rocket motor [2]. Seven parameters may be used to define a star such as grain radius (r_g), ray length (r_l), number of star points (n), an angular fraction (ε), star

A. Sharma · A. K. Thakur (✉)
Lovely Professional University, Phagwara 144411, Punjab, India
e-mail: amitthakur3177@gmail.com

N. M. Sivaram et al. (eds.), *Advances in Manufacturing, Automation, Design and Energy Technologies*, Lecture Notes in Mechanical Engineering,
https://doi.org/10.1007/978-981-99-1288-9_82

angle (α), fillet radius (r_f), and tip radius (r_t) [3]. When combustion takes place and propellant starts to burn radially, there will be two phenomena that occur in the case of star propellant. The first one is known as web burning, and the second one is sliver burning. These can be elaborated as follows:

Sliver fraction: Sliver fraction is the measure of the ratio of the remaining area of grain after the web burns out with the area of the grain. Sliver fraction leads to tail off, and the tail off ultimately provides a loss in thrust.

Neutrality: Neutrality is defined as the ratio of maximum burning area to the average burning area of the propellant grain.

2 Grain Burnings

The grain geometry of solid propellant is a very important factor to calculate the performance of the motor.

- We know that the volume of solid propellant is always equal to the multiplication of cross-sectional are and length of solid fuel.
- The burning at the propellant surface is always normal to the surface.
- The thrust developed by solid rocket motor can be written as follows [4]:-

$$F = C_f S_b^{\frac{1}{1-\sigma}} \left(C^* a\rho\right)^{\frac{1}{1-\sigma}} A_t^{-\frac{\sigma}{1-\sigma}} \tag{1}$$

The term F stands for thrust produced by a solid-propellant rocket. C_f indicates the thrust coefficient. Index σ signifies the effects of pressure based on the combustion process. Here, S_b refers to the burning surface area, and A_t refers to the nozzle throat area. The constant is physically based on the initial temperature and other characteristics of the propellant. The terms C^* and ρ refer to characteristic velocity and propellant density, respectively.

Types of grain burning:-

Neutral burning

When combustion takes place in such a way that the S_b does not vary with time, the thrust will also be constant. This kind of burning is said neutral burning.

Progressive burning

When combustion takes place in such a way that S_b increases, the thrust will also increase with time, this kind of burning is said progressive burning.

Regressive burning

In the regressive burning process, the value of S_b decreases with time as burning proceeds and this leads to a decrement in chamber pressure and thrust. This process is said regressive burning.

3 Pressure and Thrust

Star grain design is responsible for inducing a greater value of impulse at lift-off [5]. This section provides a basic understanding of pressure and thrust developed by grain burning and also introduces the effects of surface curvature on the pressure induced inside the rocket motor. Different parameters directly influence the burning velocity inside the motor case such as the temperature of propellant, pressure inside the motor, rate of change of pressure, and grain geometry. The pressure influences the burning velocity in such a way that it satisfies the following Eq. (2).

$$\dot{r}_0 = \hat{a} P^{\sigma} \tag{2}$$

Here, $\dot{r}_0$ refers to the burning rate of solid propellant, whereas $\hat{a}$ and σ are the constants that can be calculated experimentally. P is the internal pressure of the motor which depends upon the gas flowing inside the motor.

$$\dot{r} = \dot{r}_0\left(1 + \frac{M_a}{R_c}\right) \tag{3}$$

Here M_a is the Markstein number (having units of length) which is again experimentally calculated constant, along with it R_c is the radius of curvature of grain geometry and it is always positive for convex surfaces [6].

A_t refers to the area of the throat of the nozzle, A_b refers to the burning area of solid propellant, and V_c is the volume of the combustion chamber.

The pressure produced inside the combustion chamber can be determined by using the law of mass conservation in the form of the following Eq. (4)

$$\rho \dot{r} A_b = \frac{d}{dt}(\rho_c V_c) + c A_t P \tag{4}$$

Here, ρ refers to the density of solid propellant, and ρ_c refers to the density of the gas inside the combustion chamber. The Eq. (4) also depends on a constant 'c' which is called discharge coefficient, and it is determined by the following formula

$$c = k\sqrt{\frac{\left(\frac{2}{k+1}\right)^{\frac{k+1}{k-1}}}{kRT}} \tag{5}$$

Here, k is the ratio of specific heats, R refers to the gas constant, and T refers to the stagnation temperature of the produced gases inside the motor.

According to the ideal gas equation, we can write the term ρ_c as $\rho_c = \frac{P}{RT}$. When combustion takes place, the pressure inside the chamber rapidly increases which leads to the presence of choking at the throat of the nozzle on the Mach number of unity [6]. The rate of mass flow which is passing through the nozzle can be determined from the following Eq. (6).

$$\dot{m} = \rho \dot{r} A_b = c A_t P \tag{6}$$

Therefore, we can write the thrust produced by the motor in terms of mass flow rate and specific impulse in the following manner

$$F = \dot{m} I_s \tag{7}$$

4 Star Propellant Grain Burning Evaluation

In this section, the burning surface of star-shaped grain is evaluated. Any star-shaped grain design which has n vertices will always contain $2n$ flat surfaces in the geometry. Length of the flat surface is assumed as 's'. The angle subtended by each of the $2n$ surfaces of length s at the center would be π/n.

The evolution of one particular segment as it regresses by distance Δ normal to its initial surface (Δ = AE = AC). The evolution is between the bisector of two adjacent points in the star as shown to give a straight-line CD and a circular arc EC.

The angle subtended by the arc at A is $\left(\frac{\pi}{2} + \frac{\pi}{n} - \frac{\theta}{2}\right)$ giving the arc length EC as $\Delta\left(\frac{\pi}{2} + \frac{\pi}{n} - \frac{\theta}{2}\right)$. Here the θ is the angle between the star segments. The length CD is $s - \Delta\cot\left(\frac{\theta}{2}\right)$.

The length ECD is $\Delta\left(\frac{\pi}{2} + \frac{\pi}{n} - \frac{\theta}{2}\right) + \mathrm{s} - \Delta\cot(\theta/2)$. The total surface area, considering the $2n$ segments, when the burning regresses by Δ is, therefore:

$$2n\left[\Delta\left(\frac{\pi}{2} + \frac{\pi}{n} - \frac{\theta}{2}\right) + s - \Delta\cot\left(\frac{\theta}{2}\right)\right]L \tag{8}$$

Here, in Eq. (8), the term L refers to the length of the star propellant grain.

The value of θ affects the type of propellant burning. By altering the θ, we can get either progressive, regressive, or neutral burning. For neutral burning, the following condition must be satisfied.

$$\left(\frac{\pi}{2} + \frac{\pi}{n} - \frac{\theta}{2}\right) = \cot\left(\frac{\theta}{2}\right) \tag{9}$$

As the result of Eq. (9), we get $\theta = 67°$ for $n = 6$ and $\theta = 74.6°$ for $n = 8$ [4].

We can observe that progressive burning occurs when θ will greater than the above value since $\cot\left(\frac{\theta}{2}\right)$ is less then $\left(\frac{\pi}{2} + \frac{\pi}{n} - \frac{\theta}{2}\right)$. For the lower value of θ, regressive burning will occur.

In another way, grain regression can be achieved by equal web increment in all directions. Propellant mass can be determined by $m_p = \rho V_k$ where ρ is the density of solid propellant [7, 8].

Finally, we can conclude this section by stating that we can get any kind of burning by altering the value of θ for star propellant grain burning.

5 The Burn Rate of Propellant Grain, Pressure, and Temperature

Chemical reactions get affected by variations in the temperature of solid propellant which leads to effects on the burning rate. The temperature of the propellant can influence the pressure of the combustion chamber; therefore, it is vital to study this parameter then we can be able to optimize the grain geometry. The following Eq. (11) shows the influence of propellant grain temperature on the pressure of the chamber [9].

$$\Delta P = P_0^{\pi_k \Delta T} \tag{10}$$

Here, ΔP refers to the change in pressure, and P_0 refers to the initial pressure. Temperature variation is indicated by ΔT, and π_k is an experimentally determined coefficient named as temperature sensitivity of pressure. K shows the burning area of propellant grain.

Now according to Saint-Robert's or Vieille's law:

$$r_b = a_0 P_c^{\sigma} \tag{11}$$

In Eq. (11), a_0 and σ are experimentally measured constants. It is directly observable from Eq. (11) that a set of different values for a_0 and σ can provide the required relations between burning rate and pressure throughout the combustion. By determining the a_0 parameter for the correct reference temperature, and giving the entire propellant that uniform temperature, we can neglect the effect of a temperature difference [9].

6 Influence of Star Grain Geometric Parameters on the Pressure–Time Profile and Optimization

An overview of the parameters which influence the performance of star-shaped grain is proposed in this specific section.

6.1 Effect of Variation in the Angular Fraction

If we change the value of angular fraction in the interval of 0.5–1.0 in the case of six petals star-shaped grain, leads to a decrement in the thickness of the star port and increment in the propellant weight.

There will be a rise in the neutral pressure at a higher angular fraction. Further, it leads to a reduction in the neutrality factor at a higher value of the angular fraction.

Apart from that, the tail off duration will also increase along with total burn time at a high angular fraction.

6.2 *Outer Diameter of Star Grain*

The pressure–time curve remains close to neutral at the higher outer diameter of the star grain geometry, and increment will be occurring in the tail off duration simultaneously [10]. The increment in the outer diameter of the star leads to a reduction in neutrality factor and increment in the peak value of the pressure. Further, the tail off factor also increases with an increment in the outer diameter of the star. For optimization or better performance, we need to get high neutrality and less tail off factor [11, 12]. So, we can conclude this section by observing the contradiction between the needs of star outer diameter for optimization on two parameters (neutrality and tail off factor).

Optimized Number of Star Points

The star-shaped optimization is generally a trade-off among sliver mass of propellant, neutrality at the first phase, and volumetric loading. The Port area of propellant grain is directly proportional to the number of star points in the geometry [13]. Now we can say that an increase in star points will lead to increment in the port area, and further, it will because of the corresponding decrease in the burning parameter and final sliver after web burning [14]. An increase in volumetric loading results in an increment of sliver fraction. The volumetric loading fractions in the range of 0.7–0.9 can be obtained for 3–5-star points [15, 16].

7 Conclusion

Several alterations can be done to improve the existing efficiency of star-shaped propellant grain. By going through the burning evaluation of star-shaped propellant grain, it is investigated that by altering the value of θ, the type of burning can be changed according to the need. An increment in the star outer diameter causes a decrement in neutrality factor and a raise in a tail off factor.

References

1. Brooks WT (1982) Ballistic optimization of the star grain configuration. J Spacecr Rocket 19(1):54–59
2. Loncaric S, Greatrix DR, Fawaz Z (2004) Star-grain rocket motor–nonsteady internal ballistics. Aerosp Sci Technol 8(1):47–55

3. Shekhar H (2000) Modelling of burning surface regression of taper convex star propellant grain. Def Sci J 50(2):207
4. Ramamurthy K (2010) Rocket propulsion. Laxmi Publications Pvt Ltd
5. Zeller B (1993) Solid propellant grain design. In: Solid rocket propulsion technology. Pergamon, pp 35–84
6. Mashayek F, Farzad H, Ashgriz N (1996) A geometry-independent technique for solid propellant grain design. Proc Inst Mech Eng Part G: J Aerosp Eng 210(3):209–220
7. Kamran A, Guozhu L (2010) Design and optimization of 3D radial slot grain configuration. Chin J Aeronaut 23(4):409–414
8. Mukunda HS, Paul PJ (1997) Universal behaviour in erosive burning of solid propellants. Combust Flame 109(1–2):224–236
9. Reddy KO, Pandey KM (2013) Burnback analysis of 3-D star grain solid propellant. Int J Adv Trends Comput Sci Eng 2(1):215–223
10. Nawaz A, Kamran MA (2016) Parametric sensitivity analysis of convex star port propellant grain for solid rocket motor. J Eng Appl Sci (JEAS) 35(2)
11. Ricciardi A (1992) Generalized geometric analysis of right circular cylindrical star perforated and tapered grains. J Propul Power 8(1):51–58
12. Shekhar H (2005) Parametric studies on star port propellant grain for ballistic evaluation. Def Sci J 55(4):459
13. Krishnan S (1975) Design of neutral burning star grains. J Spacecr Rocket 12(1):60–62
14. Sutton GP, Biblarz O (2016) Rocket propulsion elements. Wiley
15. Barrere M, Jaumotte A, Fraeijs de Veubeke B, Vandenkerckhove J (1960) Rocket propulsion, No. OA-2. LTAS
16. Sciences OA (2003) Evaluation of design parameters of the solid propellant star grain configuration. In: Proceedings of the 10th ASAT conference, vol 13, p 15

PMSG-Based Wind Energy Conversion System with MPPT Controlled Boost Converter

Vivek Kisku, Sukanta Roga, and Subir Datta

Abstract A computational study is conducted to evaluate the performance of a WECS consisting of a VAWT and a PMSG. In the paper, the implementation of the WECS in a low wind speed region has been studied using a VAWT in the simulated environment. To derive the maximal power as output from the WECS, a power point tracking method is implemented. The technique used in the simulation utilizes the power and voltage of the PMSG to check the maximum power point by generating the reference values and comparing them with the DC output values. A control scheme is designed for the power converter to generate PWM for the IGBT switch. The input parameters for wind turbines, PMSG, and boost converters were chosen to produce the best-optimized output comprising of DC power, voltage, current, rotor angular speed, power coefficient, electromagnetic, and mechanical torque. The simulated wind energy conversion system's computational results were acquired, and its performance is examined.

Keywords Wind energy · Green energy · Energy system control · Modeling · Simulation

List of Abbreviations Used

WECS	Wind Energy Conversion System
VAWT	Vertical Axis Wind Turbine
PMSG	Permanent Magnet Synchronous Generator
IGBT	Insulated Gate Bipolar Transistor
MPPT	Maximum Power Point Tracking
P & O	Perturb and Observe

V. Kisku (✉) · S. Roga
Visvesvaraya National Institute of Technology, Nagpur 440010, Maharashtra, India
e-mail: vivekkisku5@gmail.com

S. Datta
Mizoram University, Aizawl 796004, India

N. M. Sivaram et al. (eds.), *Advances in Manufacturing, Automation, Design and Energy Technologies*, Lecture Notes in Mechanical Engineering,
https://doi.org/10.1007/978-981-99-1288-9_83

DC	Direct Current
AC	Alternating Current
PI Controller	Proportional Integral Controller
PWM	Pulse Width Modulation
TSR	Tip Speed Ratio
WPD	Wind Power Density
d-q axis	Direct-Quadrature axis

1 Introduction

Renewable energy has long been a significant source of electricity generation. With the increase in the need for power and the depletion of conventional sources of energy such as fossil fuels, alternative sources of energy are the need of the hour. Therefore, it requires a clean and green source of energy. Wind energy has set up a good foundation for itself as one of the most reliable wellsprings of power.

WECS is a system that aids in the efficient conversion of wind energy into electricity. It is primarily comprised of a wind turbine for the conversion of kinetic energy which is then converted to electrical energy using a generator. For a WECS to work efficiently and provide a DC load, a three-phase rectifier is utilized for the conversion of two-directional AC into single-directional DC, along with a boost converter for DC-to-DC conversion, and lastly, an MPPT controller [1].

The major challenge possessed by the WECS is the control of the equipment; therefore, effective control tactics must be utilized to catch the greatest amount of wind energy. Thus, numerous strategies are being explored to get the most power from the system [2]. In this paper, previously known control schemes from the literature such as the MPPT method and PI controllers were employed, and their performance was simulated and analyzed.

2 Literature Review

Several studies on wind energy conversion systems have been conducted. Fathabadi [1] implemented a novel MPPT method that extracts optimum power from WECS using a sensorless controller with a Savonius type wind turbines to produce 600 W of power. Nambiar et al. [2] have developed a tracking system for WECS to maximize the power by utilizing laboratory devices with a rated power output of 500 W for the base wind velocity of 10 m/s.

3 Mathematical Modeling and Methodology

3.1 Wind Turbine

The mechanical power (P_m) produced from the wind turbine is expressed as [1]:

$$P_m = 0.5C_P(\lambda, \beta)\rho AV^3 \tag{1}$$

where C_p (λ, β) signifies the turbine's performance coefficient, which is controlled by the pitch angle of turbine blades, given by β and the tip-speed ratio (λ). ρ signifies the density of air (kg/m^3), A is the blade swept area (m^2), and v is the wind velocity hitting the turbine rotor (m/s).

The tip-speed ratio (TSR) or λ is computed from the blade tip-speed and wind speed,

$$\lambda = \frac{\mathrm{R}\omega_\mathrm{r}}{\mathrm{v}} \tag{2}$$

3.2 Permanent Magnet Synchronous Generator

PMSG is modeled using a two-phase synchronous reference, known as the *dq* axis or direct and quadrature axis frame. *vd* is the voltage on the d-axis, and it is calculated using the relation given below [3]:

$$v_d = L_q\omega_e i_q - L_d\frac{di_d}{dt} - R_s i_d \tag{3}$$

Here, *d-q* inductances are given by L_d and L_q, Similarly, the d-q output currents are given by i_d and i_q, R_s is the stator's resistance, ω_e is the angular frequency produced by the PMSG.

V_q is the voltage on the q-axis, and it is calculated as follows:

$$v_q = \sqrt{\frac{3}{2}}\phi_{s_f}\omega_e - L_d\omega_e i_d - L_q\frac{di_q}{dt} - R_s i_q \tag{4}$$

where ϕ_{s_f} is the flux linkage created by the rotor of the PMSG.

3.3 Boost Converter

The switching time (T_s) is given as $T_s = 1/f_s$, where f_s signifies the boost converter's switching frequency. The boost converter's output voltage can be computed from [4]:

$$\frac{V_o}{V_{in}} = \frac{1}{(1-d)} \tag{5}$$

where V_o is the output voltage of the boost converter, V_{in} is the input voltage of the boost converter, and d signifies the duty cycle or the duty ratio. The values for inductance and capacitance of boost converter can be calculated using the equations:

$$L = \frac{(1-d)^2 dR}{I_r f_s} \tag{6}$$

$$C_{min} = \frac{dV_0}{V_r R f_s} \tag{7}$$

where R is the resistor load and V_r is the voltage ripple.

3.4 MPPT Technique

The system was designed using the Perturb and Observe MPPT approach, as its algorithm is more efficient compared to its counterparts. The Perturb and Observe MPPT method's working flowchart is depicted in Fig. 1. The boost converter gets input from the rectifier as DC voltage and current. The boost converter duty ratio is set to ensure that the maximum DC voltage is obtained. With the proper implementation of the MPPT algorithm, the maximum power point for each wind speed can be achieved, which also helps in reducing losses, and hence maximum possible power can be extracted for different wind speeds [5].

The main concept behind the Perturb and Observe method is to plot a curve between power and voltage, assuming there is only one maximum power point at which maximum voltage is obtained. It has two locales, one on the left side and one on the right. The rate of change of power concerning velocity is greater than zero on the LHS. The rate of change of power in relation to velocity is smaller than zero on the RHS. As a result, the voltage becomes inversely proportional to the power in this range. To make all this happen, an MPPT controller is employed in the system which keeps track of the voltage and power changes and depending upon the error obtained between the reference and DC output voltage, the duty ratio is set to generate a PWM signal [6, 7].

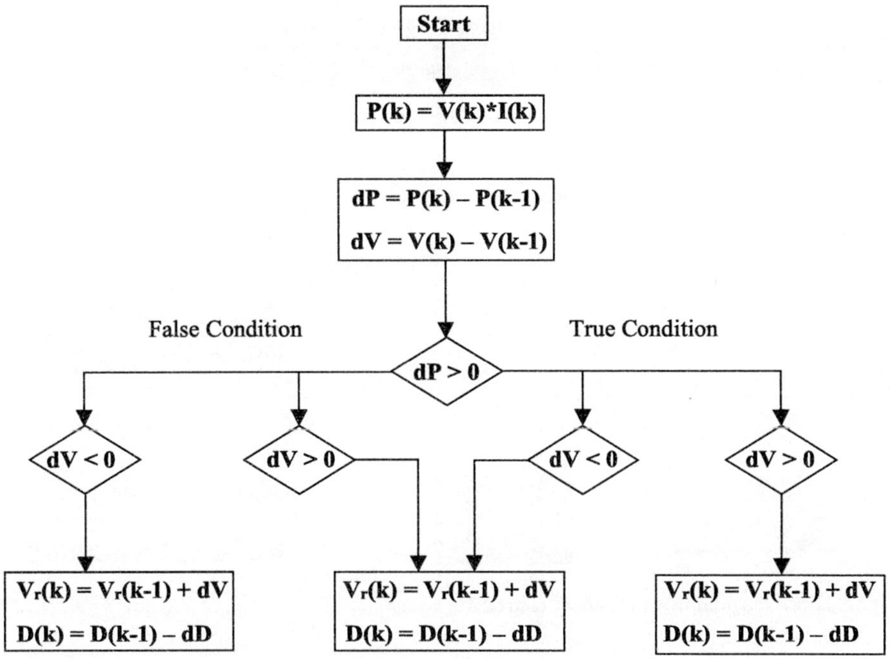

Fig. 1 Flowchart depicting the working of P and O MPPT technique

4 Working Model of Wind Energy Conversion System

Figure 2 shows the working model of the WECS. It is used for preparing a simulated model in the Simulink software. A 1 kW system was analyzed in the simulation. The mechanical torque produced by the wind turbine is used by PMSG to convert it into electrical energy. The output is an uncontrolled signal; hence, a rectifier is used which provides a stable output. These DC outputs are then fed to the MPPT controller, which generates the reference voltage. The control scheme for the power converter can also be seen in which error is controlled with the help of a PI controller. The PWM signals generated are given to an IGBT switch working at a particular switching frequency. The duty ratio for the MPPT controller is set by the boost converter and at the load resistor, DC output voltage is obtained.

The important parameters for different components considered in the simulation have been summed up in the table below:

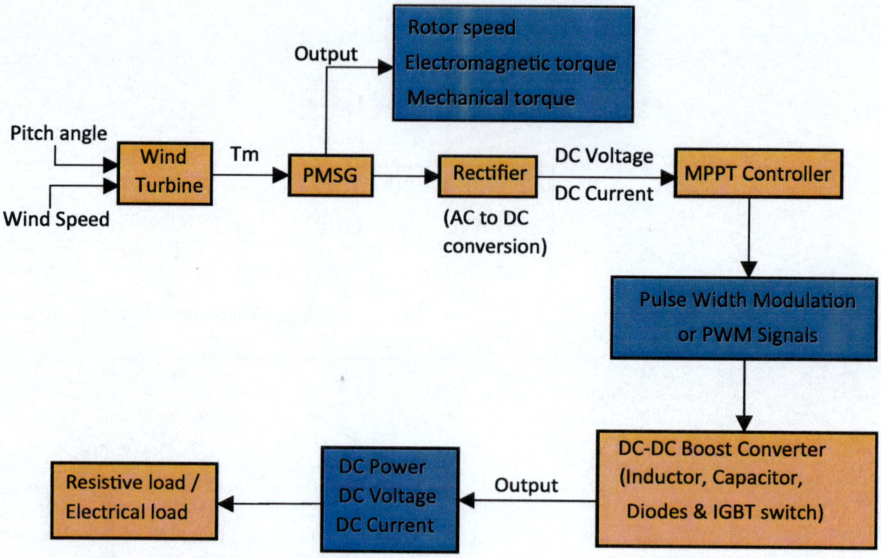

Fig. 2 Block diagram of wind energy conversion system

5 Results and Discussion

For the WECS model, simulation was done at different wind speeds. The input parameters used for different components of WECS are given in Table 1. The voltage, current, DC power output, and wind power density obtained for different wind speeds in the summer season are shown in Table 2. The rotor speed, power coefficient, mechanical, and electromagnetic torque obtained from simulation for different wind speeds are shown in Table 3.

Here, Darrius type VAWT has been considered with $D = 1.6$ m and $H = 1.8$ m, where $H =$ Height of the turbine and $D =$ Diameter of the turbine. Tip-speed ratio (λ) has been chosen as 6.5.

Table 1 Input parameters

Input parameters	Values
Rated output power (W)	1000
Rated wind speed (m/s)	6
Stator Phase Resistance, Rs (Ohm)	0.0485
Inductances, Ld and Lq (Milli-henry)	0.395
Pole Pairs	6
Switching Frequency (Hz)	1000
Inductance, L (Milli-henry)	3.125
Capacitance (Milli-farad)	0.5

Table 2 Voltage, current, DC power output, and wind power density were obtained from simulation for different wind speeds in the summer season

Wind speed (m/s)	Voltage (V)	Current (A)	DC Power (W)	WPD (W/m^2)
3.225	22	2.1	69.14	24
4.68	52.14	5.2	285.2	100.02
4.22	44.74	4.4	193.5	67.18

Table 3 Rotor speed, power coefficient, mechanical, and electromagnetic torque were obtained from simulation for different wind speeds

Wind speed (m/s)	Rotor speed (rad/sec)	Electromagnetic torque (Nm/A)	Mechanical torque (Nm)	*Cp*
3.225	21.48	2.017	2.83	0.19
4.68	41.47	5.935	6.055	0.28
4.22	34.01	5.079	5.092	0.26

The graphs in above figures demonstrate the variation of output parameters for wind speeds during the monsoon season. The variation of recorded wind speeds for the summer season is depicted in Fig. 3, and it also depicts the fluctuation in DC power output for different wind speeds. For 4.68 m/s, which is the highest recorded wind speed in the summer season, the maximum value obtained is 285.2 W. The obtained value for the DC power output is less than 1 kW WECS as the designed WECS has a rated wind speed of 6 m/s. The output produced by PMSG is uncontrolled three-phase signals. Therefore, a rectifier has to be used. They convert these uncontrolled AC signals into DC-controlled signals, which are then fed to the MPPT controller.

Figure 4 shows the output DC voltages and currents for various wind speeds. The value of voltage and current decreases as the wind speed decreases. The development of MPPT reference values is also obtained and compared with DC output voltage and then fed to a PI controller with appropriate constants values. The highest value of voltage yielded by the system for the summer season is 52.14 V and for current, it is 5.2 A.

Figure 5 shows the fluctuations of power coefficient caused by different wind speeds. The simulation results obtained show the maximum value of C_p to be around 0.3, i.e., 30% efficient. As stated by the Betz limit, the maximum possible value of C_p is limited to 0.59 which means the theoretical maximum efficiency of the wind turbine is limited to 59%.

The above graphical results were obtained using the optimized values for the input parameters after calculations to produce the best results for the given input parameters.

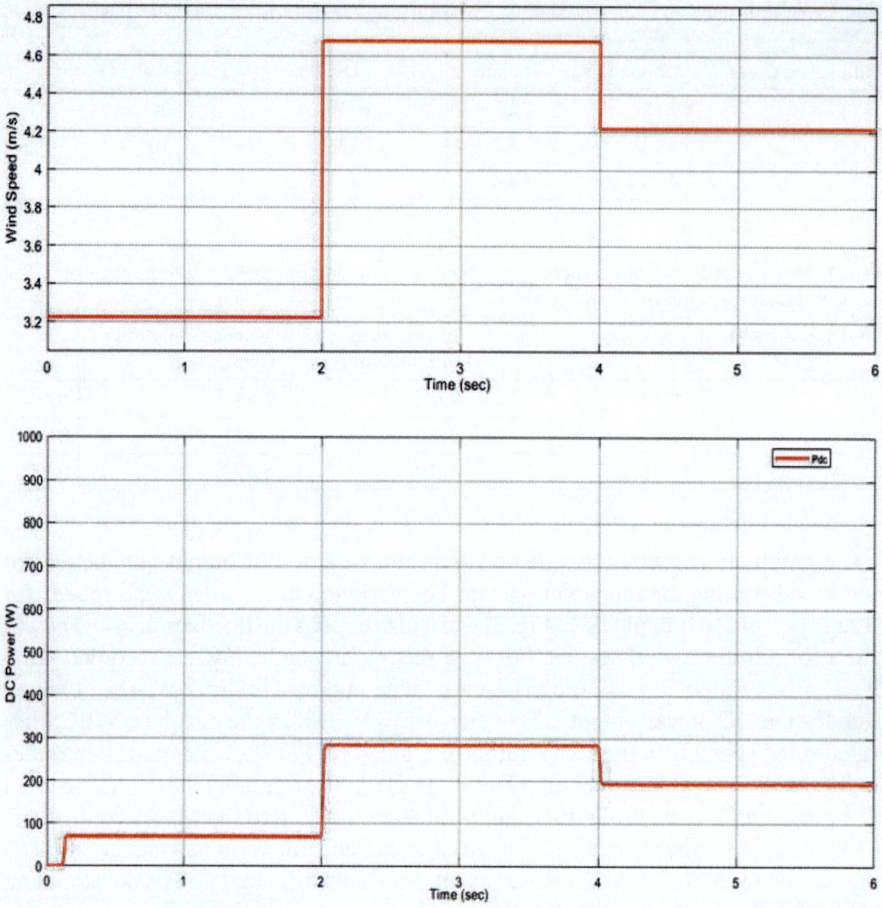

Fig. 3 Variation in wind speed and DC power output

6 Research Gaps and Novelty of the Research

The previous research work done in this field mainly comprises of modeling and simulation of WECS for medium- to high-speed regions comprising of 10 m/s or above. The rated power for developed WECS is restricted to below 1000 W, as explained in Sect. 2. Also, the wind turbine used for simulations is a horizontal axis wind turbine which is commercially preferred for electricity production from wind energy.

In the paper, a novel approach of modeling and simulation of WECS is done with the implementation of a vertical axis wind turbine. The VAWT is well suited for low wind speed regions; however, they are still in their initial phase. The study focuses on the generation of 1000 W from a PMSG-based WECS using a VAWT. The results

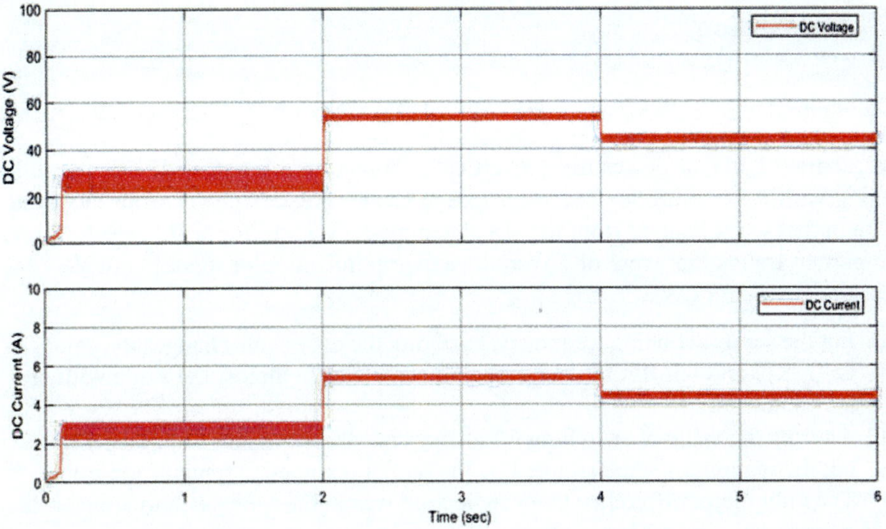

Fig. 4 Vdc and Idc variation

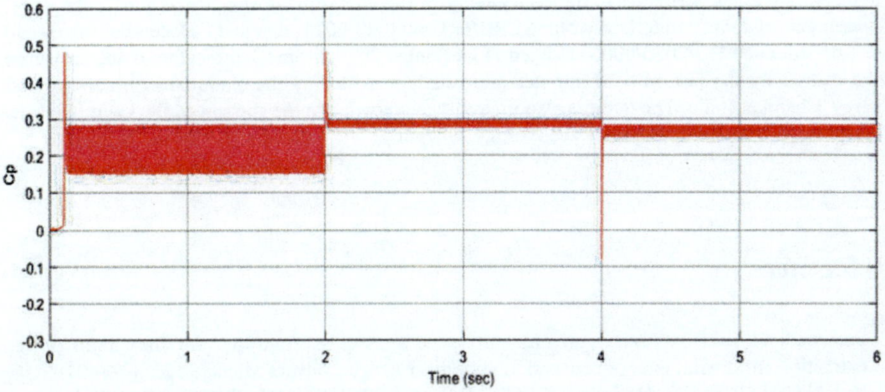

Fig. 5 Power coefficient variation

obtained in the study give a brief idea about the generation of power, voltage, and current.

This study can help to understand the challenges in the generation of power using a VAWT in a WECS, and further improvements can be made in the future as more research work is carried out in this field.

7 Conclusion

A simulation model of a wind energy conversion system was created, and its performance was evaluated in this paper. A PMSG and DC-DC boost converter were utilized to generate 1 kW of power from a vertical axis wind turbine. An MPPT algorithm along with a PI controller was used to extract the greatest power from the wind. The output was obtained from the simulated model for varying wind speeds. Some important trends that were observed after the optimization of input parameters for each component of WECS are:

a. For the same switching frequency, changing the duty cycle changes the values of L, C, Kp, and Ki. But the output parameters remain almost the same with only 2–3% deviations.
b. Altering the value of switching frequency changes the values of L, C, Kp and Ki, but the maximum value attained by the output parameters remains the same.
c. The only major difference that can be seen was in the time taken to attain stable output, which gets better with the most optimized value.

Acknowledgements This research funding for the research work is supported by the Science and Engineering Research Board (A stationary body of the Department of Science and Technology, Government of India), vide Diary No: SERB/F/5544/2020-2021, dated: 11 December 2020, and sanction order no EEQ/2020/000631, dated 11 December 2020. From the project fund, the measured wind dataset for 2 years with 10 minutes intervals is provided by the National Institute of Wind Energy, Chennai, India. The authors also gratefully acknowledge the support of the Core Institute, Visvesvaraya National Institute of Technology, Nagpur.

References

1. Fathabadi H (2016) Novel high efficient speed sensorless controller for maximum power extraction from wind energy conversion systems. Energy Convers Manage 123:392–401
2. Nambiar AS, Raja GM, Harindran K, Rajan P, Sankar SS, Sankar VJ (2017) Design and development of a maximum power point tracker and its testing setup for wind energy conversion system with laboratory level devices. In: International conference on technological advancements in power and energy (TAP Energy). IEEE, pp 1–5
3. Acharya P, Papadakis A, Shaikh MN (2016) Modelling and design of a 3-kW permanent magnet synchronous generator suitable for variable speed small wind turbines. In: MATEC web of conferences, vol 55. EDP Sciences, p 04001
4. Mesbahi A, Aljarhizi Y, Hassoune A, Khafallah M, Alibrahmi E (2020) Boost Converter implementation for Wind Generation System based on a variable speed PMSG. In: 1st international conference on innovative research in applied science, engineering and technology (IRASET). IEEE, pp 1–6
5. Verma N, Banerjee S, Gupta S, Goyal S, Sharma R (2019) PMSG based WECS with MPPT via modified P & O algorithm. In: 3rd international conference on recent developments in control, automation and power engineering (RDCAPE). IEEE, pp 646–650

6. Roga S, Dubey SK (2020) DMST approach for analysis of 2 and 3 bladed type Darrieus vertical axis wind turbine. EAI Endorsed Trans Energy Web 8(33):e2
7. Roga S (2020) Wind energy investigation of straight-bladed vertical axis wind turbine using computational analysis. EAI Endorsed Trans Energy Web 8(33):e5

Four-Port Converters with PV and Fuel Cell for Low-Voltage Bipolar DC Microgrid

S. Sharma and K. Rajambal

Abstract Renewable energy sources are quite easy to interface with a low voltage DC microgrid applications. It has one step of power conversion which is essential for power flowing into the grid. PV system is growing everywhere in this century, so the research on combining other sources with PV attracted in recent times. This article employs a four-port DC-DC converter which connected with the photovoltaic solar (PV) and hydrogen-powered fuel resources to a low-voltage Bipolar DC Microgrid. With only one inductor, this model contains lesser circuit elements. The performance of the four-port converter is investigated for varying input voltages and load conditions through Matlab simulation. It deals with the comparison of both dual-input dual-output converter and four-port proposed converters. The converter's capacity to provide dual output with a single input source as well as two sources is also investigated, and the results discussions were given.

Keywords Four-port converter (FPC) · Photovoltaic (PV) system · Fuel cell (FC) · Bipolar DC microgrid (BDCMG)

1 Introduction

Generally, the needs of energy increasing day by day in this world, so everyone concentrate on the regenerating and non-pollute energy sources [1, 2]. In this crisis, renewable energy sources (RES) places a vital role. PV system is growing everywhere in this century, so the research on combining other sources with PV attracted in recent times [3, 4]. In a pair of redox reactions the chemical energy of a fuel (hydrogen) and an oxidizing agent (oxygen) converted into electrochemical cell which is called as fuel cell [5, 6].

S. Sharma (✉) · K. Rajambal
Puducherry Technological University, Puducherry, India
e-mail: sharma.sha@pec.edu

K. Rajambal
e-mail: rajambalk@pec.edu

N. M. Sivaram et al. (eds.), *Advances in Manufacturing, Automation, Design and Energy Technologies*, Lecture Notes in Mechanical Engineering,
https://doi.org/10.1007/978-981-99-1288-9_84

DC-DC converter is an important device which leads to convert the steady-state DC to Variable DC voltage [7, 8]. This variable voltage is used to give the dc appliances or stores in energy storage system (ESS) or invert into AC and gives to ac load [9, 10]. So, the need of this variable DC voltage is mandatory to every system [11, 12]. DC-DC converters change the voltage level of a direct current source from one to another [13, 14]. It classified into two types: (i) non-isolated types. (ii) isolated types [15, 16].

This work investigates non-isolated DC-DC converter topologies such as buck-boost converters. Normally, the four-port converter can build by just adding another port in three port converter [17, 18]. In this type of converters cannot gives a satisfied efficiency to the system. So, the working on new converters is on process. In this conventional type, DC-DC converters are need a Bipolar DC Microgirds (BDCMG) to get high efficiency [19, 20].

In this paper shows the comparison of both dual-input dual-output converter and four port proposed converters. It shows the best effective converter which is suitable for the system. The calculation of capacitor and inductor parameters to the DC-DC converter also presented. In the absence of one source, the system should not disconnect to the load and need to give the continuous power supply that also discussed here. The simulation results of the four-port converter presented.

2 Four-Port DC-DC Converter

Reduced part count might improve the efficiency and simplicity of non-isolated unidirectional DC-DC converters. As a result, this study suggests a new dual-input dual-output (DIDO) unidirectional DC-DC converter to combining PV and FC sources into a low-voltage BDCMG. A general block diagram for the proposed system was mentioned in the Fig. 1. The following are the primary characteristics of the proposed four-port DC-DC converter:

1. These have fewer switches than any other DIDO converter investigated in the literature (two switches).

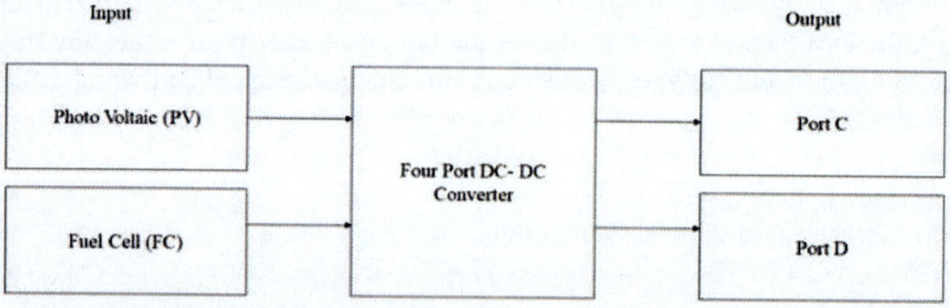

Fig. 1 General block diagram of proposed system

2. As it just uses one inductor, this converter is far more cost-effective, small, and efficient.
3. It could be used as a single-input dual-output (SIDO) converter, permitting it to regulate one of the BDCMG's pole voltages with just one power source.
4. It may operate in both continuous and discontinuous conduction modes. (DCM).
5. The converter has two degrees of freedom because there are two switches. This converter can control one of the DC micro grids' pole voltages as well as the inductor currents.
6. MPPT operation of such PV source is conceivable with the proposed converter, and this may be accomplished by easily observing the inductor current and PV voltage. To put it another way, no extra PV current sensing is necessary.
7. Regardless of an operating conditions, the converter's dynamic model leftovers intact.

The suggested converter's controller design is quite simple. The above-mentioned facts show that the proposed topology can enhance the total efficiency of BDCMG system.

3 Converter Configuration and Modes of Operation

Figure 2 depicts the proposed converter's circuit diagram. The converter is made up of two switches (S1 and S2), one inductor (L1), and four diodes (Da–Dd), as shown. However, only three diodes (Db–Dd) are necessary for the converter to function properly. The diode Da is solely used to safeguard PVs (against negative currents) [21, 22]. The load resistance as sensed by the terminals "O–N" and "P–O" in Fig. 2 is R1 and R2, respectively. Put another way, as shown in Fig. 2, the loads connected to the converter's output terminals might be represented. The load correlations shown have been computed. The proposed topology is a unidirectional converter, as can be seen in Fig. 2, and this is especially relevant to combining any two RESs, which including solar PV, FC, and wind. The converter is connected to dc loads (or dc micro grid) in this paper via PV and FC sources [as illustrated in Fig. 2].

The following conditions must be met for the converter to function properly.

1. To establish volt–second balance across inductor L1, $Vin2 > VT(=V1+V2) > Vin1$
2. R2 is greater than $R1$ (Sect. 3 explains why this would be the case).

If such duty ratios of switches $S1$ & $S2$ are $d1$ & $d2$, accordingly, yet these duty cycles have satisfy one of its two requirements listed below for the converter to work as intended: (1) $d1 > d2$ (i.e., switch S2's turn-on time is longer than switch S1), or (2) $d1 > d2$ (i.e., Switch S1 takes longer to turn on than switch $S2$). Those figures below depict just about all the possible working modes of the proposed converter, considering diverse duty ratio situations and the type of inductor current (i.e., CCM and DCM). If indeed the converter is in CCMmode & $d1 > d2$, the functioning

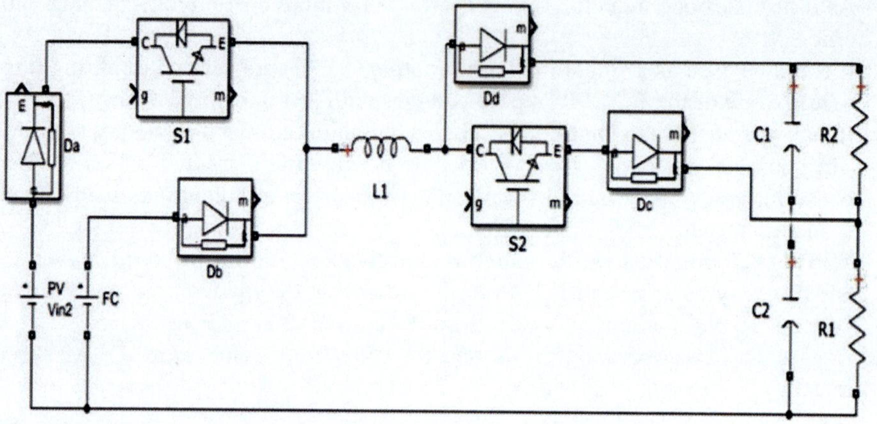

Fig. 2 Circuit diagram of four-port converter

modes will be as shown in CCM mode. The same set of waveforms is depicted in Fig. 7 in the DCM mode of operation.

In Figs. 3, 4, and 5 depicts the average PV current increases when switch S1 is "ON" for a longer length of time. If the "ON" period shortens, the opposite is true. As a result, it can be deduced that the average PV current can be regulated by modifying S1. Similarly, regulating the average voltage across capacitor C1 can be accomplished by modulating S2.

The suggested converter has the benefit of being able to operate as a SIDO converter, that is, in the lack of a PV or FC source, as long as the following conditions are met: $V1 \leq (Vin1\ or\ Vin2) \leq VT.$. Switch S1 is endlessly ON (i.e., $d1 = 1$) in the nonexistence of the FC source, while switch S2 is regulated to regulate $V1$. Switch $S1$ is continuously OFF (i.e., $d1 = 0$) in the nonappearance of the PV source,

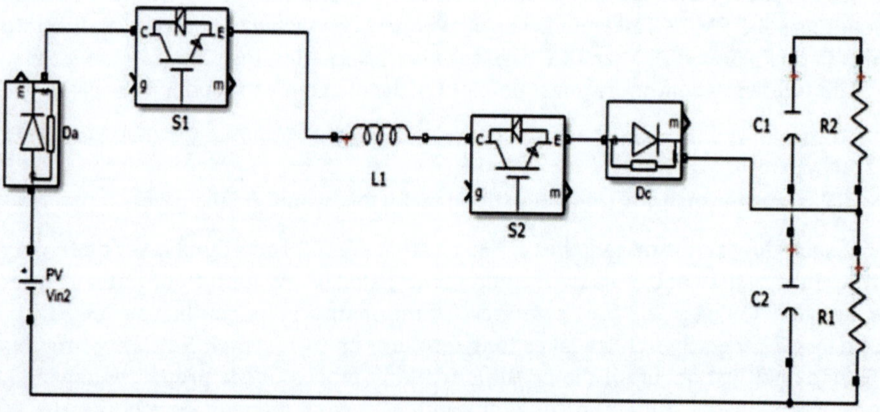

Fig. 3 MODE 1 (S1 and S2 ON)

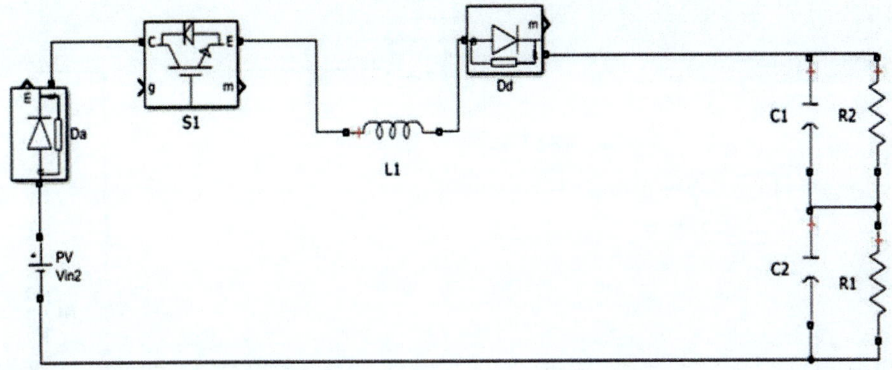

Fig. 4 MODE 2 (S1 and S2 OFF)

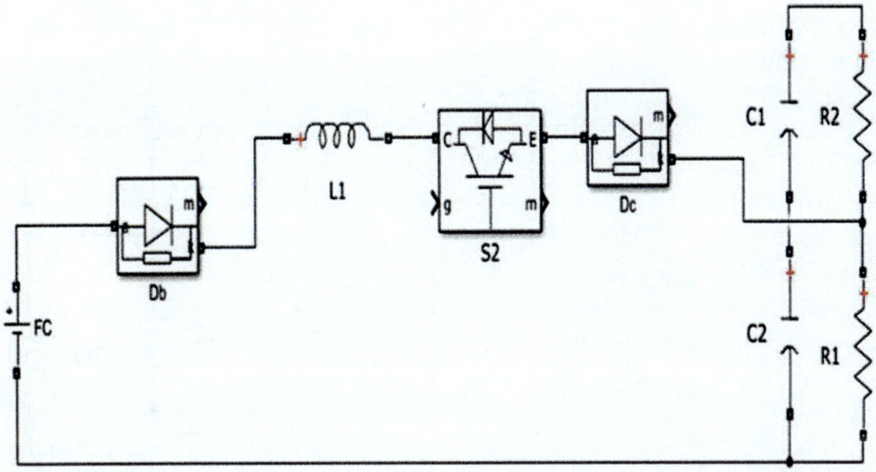

Fig. 5 MODE 3 (S1 OFF and S2 ON)

while switch $S2$ is regulated to regulate $V1$. However, it should be emphasized that regulating $V1$ only with a PV source is only viable if the PV source has sufficient power. During this circumstance, the operating modes would be as indicated in Figs. 6, 7, and 8.

Steady-state equations:

$$I_{s2}, \text{avg} = \frac{V1}{R1} - \frac{V2}{R2} \quad (1)$$

$$I_{s2}, \text{avg} = \frac{V1}{Ra} \quad (2)$$

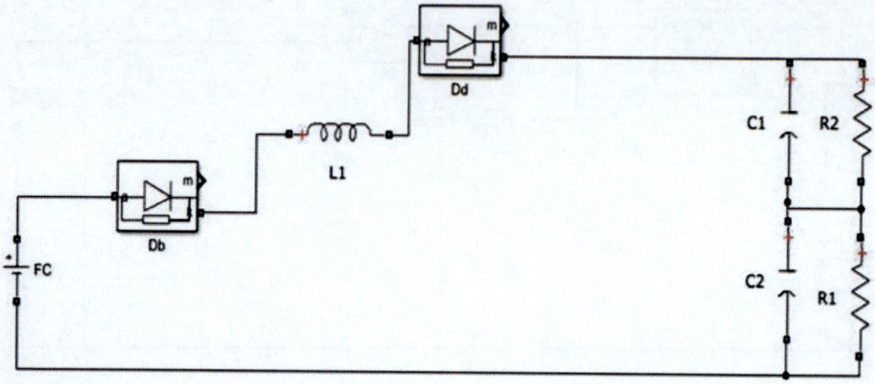

Fig. 6 MODE 4 (S1 and S2 OFF in CCM)

Fig. 7 MODE 5 (S1 and S2 OFF in DCM)

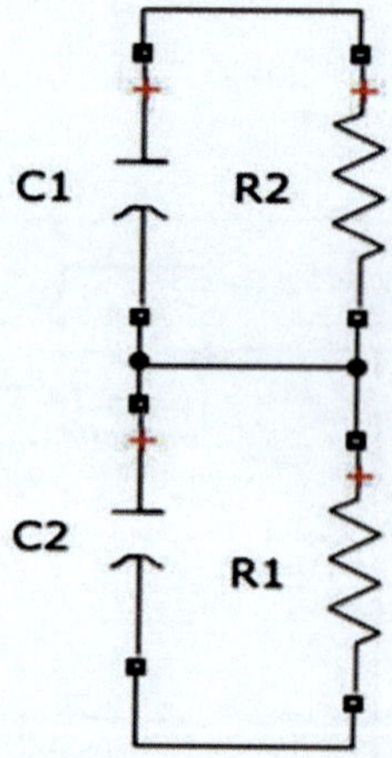

$$R_a = \frac{R1 \times R2}{V1R2 - V2R1} \times V1 = \frac{R1 \times R2}{R2 - R1}(\text{if}\, V1 = V2) \tag{3}$$

$$I_{Dd}, \text{avg} = \frac{V2}{R2} \tag{4}$$

$$I_{Dd}, \text{avg} = \frac{V1 + V2}{Rb} = \frac{VT}{Rb} \tag{5}$$

$$Rb = \frac{VTR2}{V2};\ Rb = 2R2(\text{if}\, V1 = V2) \tag{6}$$

$$I_L, \text{avg} = I_{s2}, \text{avg} + I_{Dd}, \text{avg} = \frac{V1}{R1} = \frac{VT}{RB} + \frac{V1}{Ra} \tag{7}$$

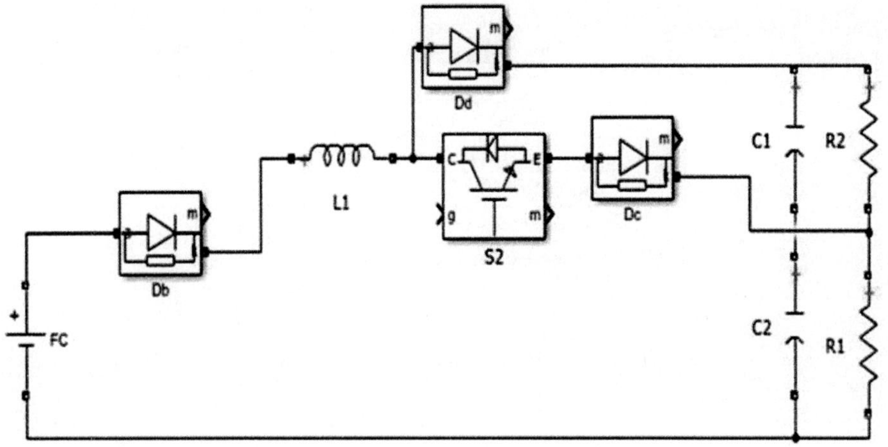

Fig. 8 MODE 6 (S2 is ON (in SIDO mode))

$$I_L, \text{avg} = I_{PV}, \text{avg} + I_{FC}, \text{avg} \tag{8}$$

$$I_{S2}, \text{avg} = \frac{V1}{R1} - \frac{V2}{R2} \geq 0;\ R2 \geq R1 (\text{if}\, V1 = V2) \tag{9}$$

$$V_{Da} = 0V (\text{when } S1 \text{ is off}),\ V_{Ds1} = V_{in2-}V_{in1}$$

$$V_{Db} = V_{in1-}V_{in2} (\text{when } S1 \text{ is on}),\ V_{Ds2} = V_2$$

$$V_{Dc} = 0V (\text{when } S2 \text{ is off}),$$

$$V_{Dd} = -V_2 (\text{when } s2 \text{ is on}) \tag{10}$$

$$P_{T=}\frac{V_1^2}{R1} + \frac{V_2^2}{R2} = V_x, \text{avg} \times I_L, \text{avg} = \frac{V_1^2}{Ra} + \frac{V_T^2}{Rb} \tag{11}$$

$$V_x, \text{avg} = V_{in2} \times D1 + V_{in1} \times (1 - D1) \tag{12}$$

$$D1 = \frac{\frac{PtXR1}{V1} - Vin1}{(Vin2 - Vin1)} \tag{13}$$

$$D2 = \frac{(Vin2 - Vin1)XD1 + (Vin1 - Vt)}{(V1 - Vt)} \tag{14}$$

Table 1 Parameters for simulation

Components	Ratedvalue
InputDCVoltage($Vin1$, $Vin2$)	$30V$, $70V$
Switchingfrequency	30kHz
LoadResistorrL, $R1$ & $R2$	0.1Ω, 2.02Ω, 2.69Ω
Inductor	250μH
Capacitor$C1$and$C2$	1000μF
Powerswitches	MOSFET

$$\text{L1, min} = \frac{R1}{2f} \times \frac{1}{V1}((Vin2 - V1) \times D1D2 + (Vt - Vin1)(1 - D1)(1 - D2)) \tag{15}$$

$$C2 = \frac{D2}{R2X\frac{\Delta vc2Xf}{Vc2}} \tag{16}$$

with $d1 = 0$ and $Vin2 = 0$, just the steady-state equations for the FC source (in CCM). This situation is more common and more realistic since FC is a dispatchable source that is available 24/7 of the day and night [23]. Corresponding steady-state equations may be constructed when just the PV source is evaluated. However, in this case, $d1 = 1$ and $Vin1 = 0$ would be the conditions.

The circuit is built for two input voltages of 30 and 70 V, as well as the computed parameters for DC-DC converter design are listed in Table 1.

4 Simulation Study of Four-Port DC-DC Converter Topology

Figure 9 shows a Simulink model of a four-port DC-DC converter. The converter's performance is investigated for varying solar intensity and fuel cell voltages. As shown in Fig. 10, the continuous conduction mode (CCM) of operation is obtained for D1D2. During the circuit's operation at CCM, the capacitor voltage and current be there shown in Fig. 11.

4.1 Various Simulated Results for the Converters

PV and FC serve as input sources in this circuit. By keeping one source constant while varying the other source from minimum to maximum voltage, the converter can be examined at different voltages for each source [24]. Tables 2 and 3 show how the converter's performance changes when the PV and FC values change.

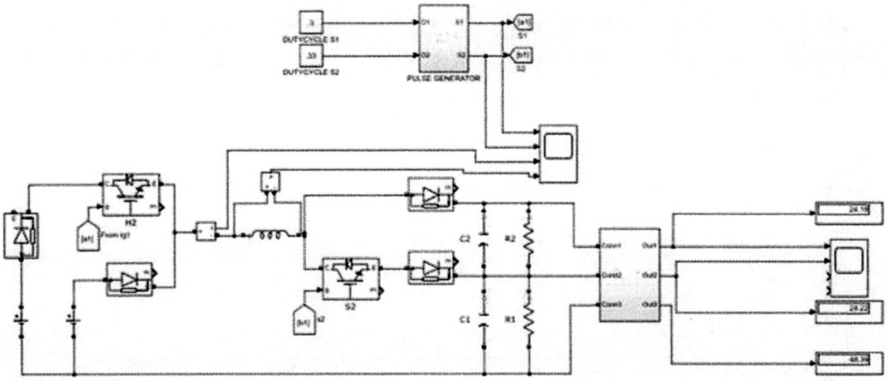

Fig. 9 Simulink model of four-port DC-DC converter

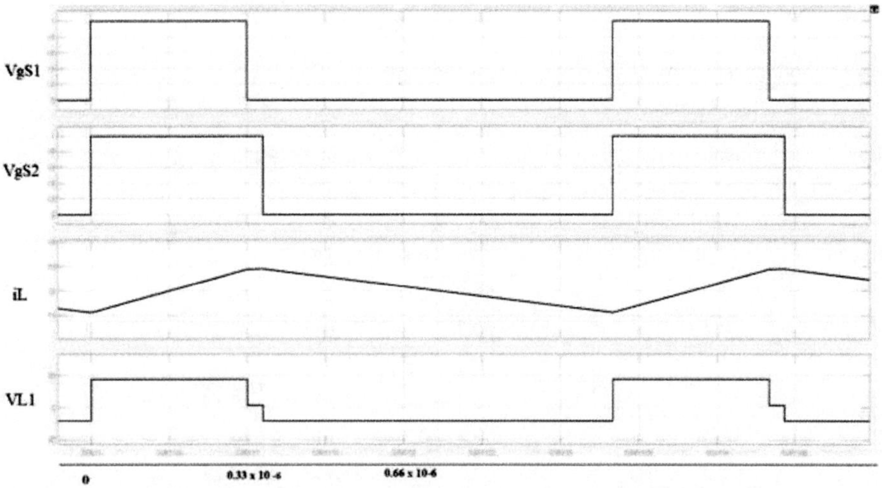

Fig. 10 Continuous conduction mode obtained under the condition ($D1 < D2$)

SIDO and DIDO are the two modes in which the converter can be used. Below are the simulation findings for these two modes of operation.

4.2 Four Port DC-DC Converter in SIDO Mode

The proposed converter really does have the benefit of being a SIDO converter, which implies it really can run in the absent of a PV/FC source.

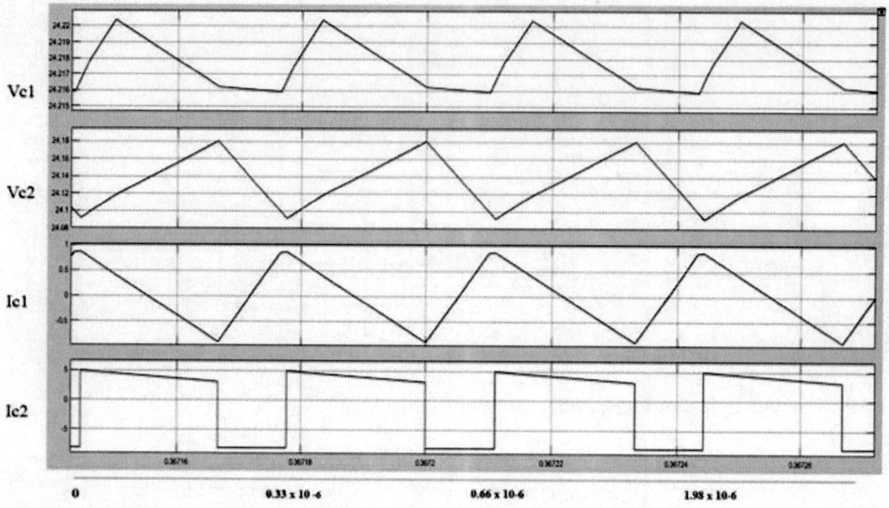

Fig. 11 Voltage and current waveforms across the capacitor

Table 2 Performance of the converter under varying photovoltaic (PV) voltage

PV voltage (port A)	Fuel cell voltage (port B)	D1	D2	Voltage 1 (port C)	Output voltage (Vt) (port D)
10	30	0.3	0.5	20.68	36.14
20	30	0.3	0.6	22.92	36.63
30	30	0.3	0.7	25.01	36.21
40	30	0.3	0.6	25.35	40.35
50	30	0.3	0.4	22.39	42.12
60	30	0.3	0.3	21.61	44.19
65	30	0.3	0.42	25.91	48.27
67	30	0.3	0.37	24.84	48.05
70	30	0.3	0.33	24.22	48
71	30	0.3	0.3	23.51	48.02
75	30	0.3	0.26	23.03	48.54

SIDO operation (in the absence of FC)

1. When the FC source is not present, switch $S1$ remains on and switch $S2$ is modulated to regulate $V1$.
2. However, it should be emphasized that regulating $V1$ only with a PV source is only achievable if the PV source has sufficient power.

Table 3 Simulation Results under varying Fuel Cell (FC) voltage

PV (V) (port A)	FC (V) (port B)	D1	D2	V1 (port C)	Output voltage (Vt) (port D)
70	25	0.38	0.33	24.21	48.54
70	27	0.35	0.33	24.41	48.46
70	28	0.33	0.33	24.12	48.24
70	29	0.32	0.33	24.27	48.54
70	31	0.28	0.33	24.18	48.22
70	32	0.26	0.33	24.16	48.24
70	33	0.24	0.33	24.24	48.17

SIDO operation (in the absence of PV)

- When the PV source is not present (Fig. 12), switch $S1$ is turned off (i.e., $d1 = 0$), and switch $S2$ is modulated to control $V1$.
- The steady-state equations are valid only using the FC source (in CCM)

$$D2 = ((Vin2 - Vin1)XD1 + (Vin1 - Vt))/((V1 - Vt))$$

- With the FC source alone (in CCM), the steady-state equations are $V1 = ((Vin2D1 - Vt(1 - D2) + Vin1(1 - D1))/D2$

$T/(2L1, \min)((Vin2 - V1) \times D1D2 + (Vt - Vin1) \times (1 - D1)(1 - D2))ILIfD1 > D2orD1 < D2,$

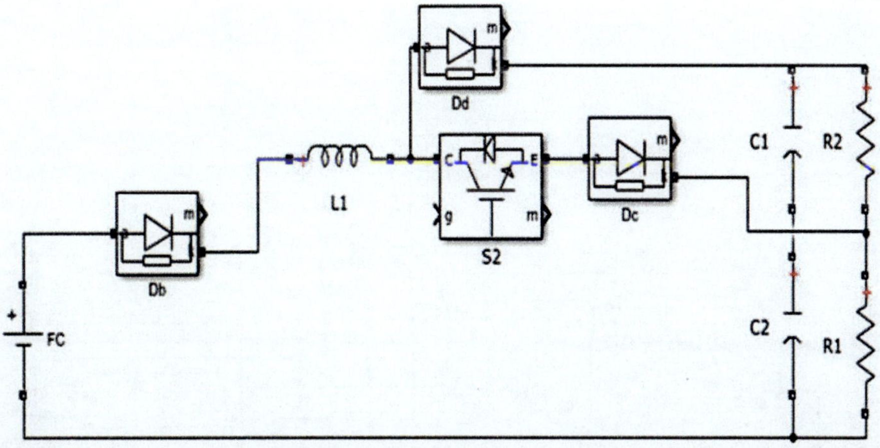

Fig. 12 Operation of four- port DC-DC converter in single-input dual-output mode (in the absence of PV)

Table 4 SIDO simulated values

PV(*V*)(Port*A*)	*FC*(*V*)(Port*B*)	*D*1	*D*2	OutputVoltage(*Vt*)(Port*D*)
0	27	0	0.7	48
0	30	0	0.8	47.6
0	33	0	0.5	48
0	34	0	0.5	48

$$L1, \min = R1/2f \times 1/V1((Vin2 - V1) \times D1D2 \\ + (Vt - Vin1)(1 - D1)(1 - D2))If D1 > D2or D1 < D2$$

- Because FC is accessible late at night and may be dispatched, the condition with $d1 = 0$ and $Vin2 = 0$ is more realistic. Table 4 displays the SIDO Simulated values in the absence of PV

4.3 Four-Port DC-DC Converter in DIDO Mode

WORKING OF DIDO-DC CONVERTER:

Combining two typical positive output buck-boost converters with a common freewheeling diode D yields the recommended converter (Fig. 13).

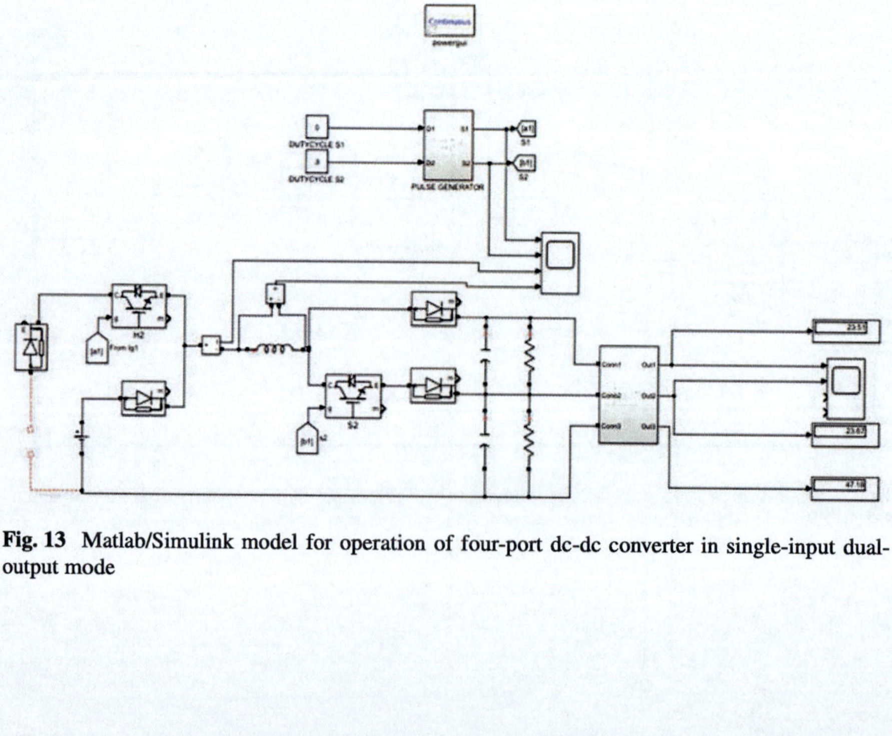

Fig. 13 Matlab/Simulink model for operation of four-port dc-dc converter in single-input dual-output mode

- The input sources output power is controlled by the source side switches $Ss1$ and $Ss2$, which allow unidirectional current.
- The switches $S1$ and $S2$ regulate the power generated in the inductors.
- To operate the DIDO-DC converter in buck-boost mode, the duty ratios of switches $Ss1$, $Ss2$, $S1$ and $S2$ are set correctly [25, 26].
- The converter's input sources are two dissimilar voltage sources, $VS1$ and $VS2$. Despite the fact that the switches Ss1 and Ss2 are both switched on. Either of the input sources with a high voltage value may power the inductors at the same time [27].
- The DIDO-DC converter has four states of operation based on switching pattern.

  Output waveforms of the DIDO converter in boost operation is shown in Fig. 14

Prominent features:

- Simple circuit and compact topology.
- Easy control
- Low cost
- Converter can be powered even if one of the sources is not available.

Problems faced by the DIDO DC-DC converter for DC microgrid topology are:

- Less reliable

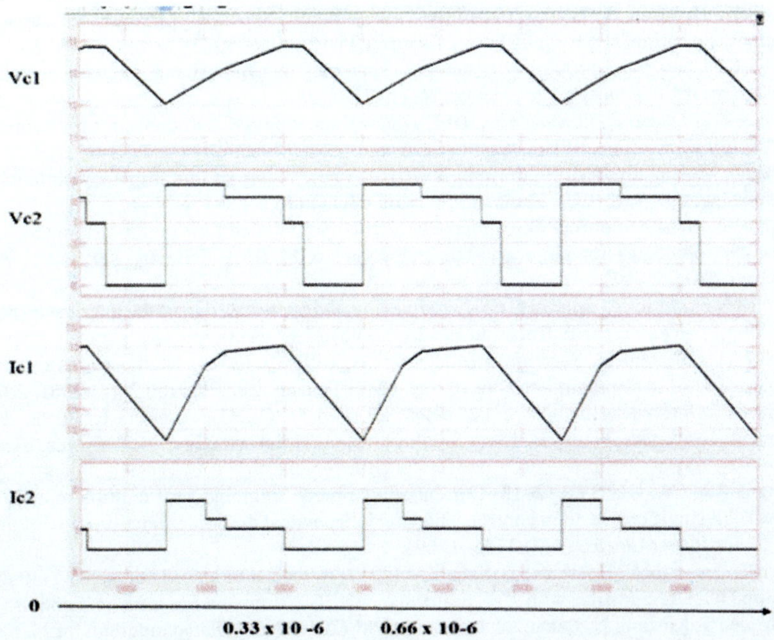

Fig. 14 Simulation Results for DIDO converter

- Although the converters are unidirectional, they employ a high number of switches, reducing overall efficiency.

5 Conclusion

This study proposes a four-port DC-DC converter that may be utilized to link PV & FC sources to a low-voltage BDCMG system. The proposed converter employs fewer switches and just one inductor in comparison to standard multi-converter rivals. As a result, the converter is small, efficient, and economical. Furthermore, it was noticed for the examination of converters in various voltage sources, SIDO mode operation, and DIDO mode operation, and their simulation results were reported.

References

1. Strunz K, Abbasi E, Huu DN (2014) DC microgrid for wind and solar power integration. IEEE J Emerg Sel Topics Power Electron 2(1):115–126
2. Madduri PA, Poon J, Rosa J, Podolsky M, Brewer EA, Sanders SR (2016) Scalable DC micro-grids for rural electrification in emerging regions. IEEE J Emerg Sel Topics Power Electron 4(4):1195–1205
3. Kakigano H, Miura Y, Ise T (2010) Low-voltage bipolar-type DC microgrid for super high quality distribution. IEEE Trans Power Electron 25(12):3066–3075
4. Liu Y-C, Chen Y-M (2009) A systematic approach to synthesizing multi-input DC–DC converters. IEEE Trans Power Electron 24(1):116–127-
5. Kardan F, Alizadeh R, Banaei MR (2017) A new three input DC/DC converter for hybrid PV/FC/battery applications. IEEE J Emerg Sel Topics Power Electron 5(4):1771–1778
6. Azizi M, Mohamadian M, Beiranvand R (2016) A new family of multiinput converters based on three switches leg. IEEE Trans Ind Electron 63(11):6812–6822
7. Hawke JT, Krishnamoorthy HS, Enjeti PN (2014) A family of new multiport power-sharing converter topologies for large grid-connected fuel cells. IEEE J Emerg Sel Topics Power Electron 2(4):962–971
8. Garcia FS, Pomilio JA, Spiazzi G (2013) Modeling and control design of the interleaved double dual boost converter. IEEE Trans Ind Electron 60(8):3283–3290
9. Dasika JD, Bahrani B, Saeedifard M, Karimi A, Rufer A (2014) Multivariable control of single-inductor dual-output buck converters. IEEE Trans Power Electron 29(4):2061–2070
10. Nami FZ, Ghosh A, Blaabjerg F (2010) Multi-output DC-DC converters based on diode-clamped converters configuration: Topology and control strategy. IET Power Electron 3(2):197–208
11. Dong Z, Tse CK, Hui SYR (2018) Current-source-mode singleinductor multiple-output LED driver with single closed-loop control achieving independent dimming function. IEEE J Emerg Sel Topics Power Electron 6(3):1198–1209
12. Nouri T, Vosoughi N, Hosseini SH, Sabahi M (2017) A novel interleaved nonisolated ultrahigh-step-up DC–DC converter with ZVS performance. IEEE Trans Ind Electron 64(5):3650–3661
13. Jalilzadeh T, Rostami N, Babaei N, Maalandish M (2018) Nonisolated topology for high step-up DC-DC converters. IEEE J Emerg Sel Topics Power Electron. https://doi.org/10.1109/JESTPE.2018.2849096
14. Li W, He X (2011) Review of nonisolated high-step-up DC/DC converters in photovoltaic grid-connected applications. IEEE Trans Ind Electron 58(4):1239–1250

15. Jafari M, Hunter G, Zhu JG (2012) A new topology of multi-input multi-output buck-boost DC-DC converter for microgrid applications. In: Proceedings of the PECon, Kota Kinabalu, Malaysia, pp 286–291
16. Nahavandi MT, Sharifian MBB, Danyali S (2015) A nonisolated multiinput multioutput DC–DC boost converter for electric vehicle applications. IEEE Trans Power Electron 30(4):1818–1835
17. Behjati H, Davoudi A (2013) A multiple-input multiple-output converter. IEEE Trans Ind Appl 49(3):1464–1479
18. Babaei E, Abbasi O (2016) Structure for multi-input multi-output DC–DC boost converter. IET Power Electron 9(1):9–19
19. Chen M, Rincón-Mora GA (2007) Single inductor, multiple input, multiple output (SIMIMO) power mixer-charger-supply system. In: Proceedings of the ACM/IEEE international symposium low power electronics and design, pp 310–315
20. AA de Melo Bento, da Silva ERC (2016) Dual input single switch DC-DC converter for renewable energy applications. In: Proceedings of the IEEE energy conversion congress and exposition (ECCE), Milwaukee, WI, USA, pp 1–8
21. Prajof P, Agarwal V (2015) Novel solar PV-fuel cell fed dual-input-dual output DC-DC converter for DC microgrid applications. In: Proceedings of the IEEE 42nd photovoltaic specialist conference (PVSC), New Orleans, LA, USA, pp 1–6
22. Kazimierczuk MK (2008) Pulse-width modulated DC-DC power converters. Wiley, Hoboken, NJ, USA
23. Hong J, Yin J, Liu Y, Peng J, Jiang H (2019) Energy management and control strategy of photovoltaic/battery hybrid distributed power generation systems with an integrated three-port power converter. IEEE Access. 17(7):82838–82847
24. Zhang G, Wang Z, Iu HH, Chen SZ, Zhang B, Qiu D, Fernando T, Zhang Y (2018) A five-terminal impedance network based three-port converter. IEEE Access 25(6):29474–29485
25. De Francesco M, Arato E (2002) Start-up analysis for automotive PEM fuel cell systems. J Power Sources 108:41–52
26. Saxena N, Hussain I, Singh B, Vyas AL (2017) Implementation of a grid-integrated PV-battery system for residential and electrical vehicle applications. IEEE Trans Indus Electron 65(8):6592–6601
27. Singh SA, Carli G, Azeez NA, Williamson SS (2018) Modeling, design, control, and implementation of a modified Z-source integrated PV/grid/EV dc charger/inverter. IEEE Trans Ind Electron 65(6):5213–5220

Evaluation of Combustion Characteristics of Fuel Derived from the Waste Lubricating Oil with N-Pentanol Additives in Diesel Engine

S. P. Venkatesan, Subbiah Ganesan, S. Lakshmisankar, P. L. Leonard Ignatius, and R. Naveen

Abstract An experiment is carried out with three blends consisting of used engine oil, n-Pentanol and diesel. This study provides an alternate fuel to diesel. Engine fuel is extracted from used engine oil by the method of pyrolysis distillation. Zeolite, sodium carbonate (Na_2Co_3) and lime (CaO) are used as a catalyst in the pyrolysis process. The blend has a volume percentage of 65% diesel, 25% pyrolysis waste engine oil and 10% n-Pentanol has shown best performance compared to diesel and other blends. The brake thermal efficiency of this blend had noticed a decrease of about 1% compared to pure diesel at full load, and the NO_X, CO, HC and CO_2 emissions were found to be 20, 10, 15 and 10% lower than that of pure diesel at maximum load.

Keywords Diesel engine · n-Pentanol · Waste engine lubrication oil · Emissions · Efficiency

1 Introduction

Fossil fuel is a combination of carbon and hydrogen. When it is burned, carbon dioxide is produced, which is a greenhouse gas that is trapped in the atmosphere for an unknown amount of time and contributes significantly to global warming. In response to the rapid increase in energy consumption, a variety of methods for using alternative energy sources have emerged. Biodiesel is a home-made, clean-burning, renewable alternative to petroleum diesel. Biodiesel as an automotive fuel improves energy security, environment, and the economy by allowing diesel engines to run

S. P. Venkatesan (✉) · S. Ganesan · S. Lakshmisankar · P. L. L. Ignatius · R. Naveen
School of Mechanical Engineering, Sathyabama Institute of Science and Technology, Chennai 600119, India
e-mail: spvenkatesan.mech@sathyabama.ac.in

N. M. Sivaram et al. (eds.), *Advances in Manufacturing, Automation, Design and Energy Technologies*, Lecture Notes in Mechanical Engineering,
https://doi.org/10.1007/978-981-99-1288-9_85

without modification. Our country with the highest per capita consumption of fossil fuels in the world. In India, Jatropha bio-diesel is used to meet the diesel fuel needs of remote areas. Jatropha oil is used directly in diesel generators on the rural side. There haven't been enough literary works written about the use of a large amount of alcohol as an additive in a diesel engine. Alcohols with long carbon chains, such as hexanol, had extremely low mixing with aqua. As the number of carbon atoms in the alcohol links increases, the proclivity to mix with water decreases due to increased molecular stability, increasing alcohol's proclivity to mix with water, which causes an increase in alcohol's proclivity to mix with gasoline-based fuels. In comparison to ethanol and butanol, hexanol has greater strength, which increases automotive fuel economy since the volatile form is less volatile, preventing unnecessary gaps and evaporative ejections. Venkatesan and Kadiresh [1] identified an optimum dosage of aluminum oxide nanofluid in diesel to produce improvement in performance and emissions. Blends of waste plastic oil, diethyl ether and diesel give good results compared to diesel [2]. The WPO blends have increased viscosity and lower calorific value than diesel. To improve the combustion characteristics, brake thermal efficiency (BTE), brake specific fuel consumption (BSFC) and emissions, the WPO can be used as a diesel alternative at higher compression ratios without modifying the compression ignition engine. Babu and Anand [3] recommending alcohol-diesel blends, n-hexanol-biodiesel, and n-Pentanol-biodiesel to meet the world's future energy requirement. Kaimal and Vijayabalan [4], identified the blends of diethyl ether and plastic oil that can be used in CI engines without modification. Venkatesan et al. [5] developed a fuel consist of ZnO nanofluid and diesel to improve engine working performance. As the waste plastic oil vaporize quickly inside the cylinder, the cylinder combustion pressure is more than petrol [6]. Due to better combustion, waste plastic oil has a higher heat release percentage than diesel. NOX levels rise as the percentage of oil involved rises. In a diesel engine, Ganesan et al. [7] submitted an extensive study using n-hexanol as a diesel additive and Calophyllum Inophyllum Methyl Ester as ternary blends at different concentrations in a diesel engine under various loads. The good performance of diesel engine on using WPO and n-pentanol is accounted by Damodharan et al. [8]. Adding alcohol to blends, decrease the viscosity and density of oil in the blends and improve blends properties [9, 10]. Nanofluid blends can be an effective alternate for fossil diesel [11, 12] and which have seen a significant increase in engine efficiency. Increasing the content of pentanol and propanol as alcohols in diesel and vegetable oil biodiesel helps to explore good engine performance. The blends of diesel, pyrolysis waste engine oil and n-Pentanol are our choice of study on alternate fuel for diesel engine.

2 Bio-diesel Preparation

This study is to extract biodiesel from scrap lubrication engine oil by the process of pyrolysis refining (or) distillation. Testing of this biofuel is carried out in an internal combustion engine. In this process, waste oil is collected and it is purified from contaminants like carbon black, dirt, metal granule, some gum-like substance, and various impurities will be refined in this process of purification. In this process to convert waste engine oil to biodiesel by the pyrolysis distillation method to add some catalyst in the process like zeolite, sodium carbonate, and lime. These catalysts are mixed in the fixed-bed reactor with the help of a mixture. The mixture will stir the oil continuously, and the heating source is independent of the reactor with the help of wood, coal, etc. After that, the steam will be passed through the condenser and converted into 80% of biodiesel. The remaining 20% is converted into the asphalt.

3 Fuel Blends

The three blends are made by the composition of 25, 35 and 45% of PWEO and the remaining diesel. The 10% vol of pentanol mixed with the blends of PWEO and diesel. The blends are called as D65PWEO25P10 (Diesel(65%) + PWEO(25%) + Pentanol(10%) by volume), D55PWEO35P10 (Diesel(55%) + PWEO(35%) + Pentanol(10%) by volume) and D45PWEO45P10 (Diesel(45%) + PWEO(45%) + Pentanol(10%) by volume). The fuel properties of blends are presented in Table 1.

4 Experimental Setup of Engine

This setup is made with a computerized variable compression ratio (VCR) four stroke, single-cylinder CI engine. Provision is made in the engine to change the compression ratio of the engine while the engine is running. Engine combustion pressures and crank angles are measured using instruments attached in the engine. A engine indicator is used to send the signal to the computer to draw the PV diagrams. Measurement of fuel flow, airflow, engine temperature and load acting on the engine are interfaced and sent to computer by means of sensors. The air box with orifice meter connected to manometer is fixed on the top of the panel of the setup. The fuel tank, manometer, fuel consumption measuring instrument, speed, load and temperature indicators are fixed on the side of panel configuration. Two rotameters are fixed at

Table 1 Properties of blends

Properties	Diesel	D65PWEO25P10	D55PWEO35P10	D45PWEO45P10
Density (g/cc)	0.824	0.801	0.794	0.742
Kinematic viscosity @40 °C	2.426	0.78	0.720	0.812
Calorific value (kJ/kg)	45273.35	43797.6	42899.28	37517.066
Flashpoint by PMCC METHOD	55	51	51	52

Fig. 1 Engine setup

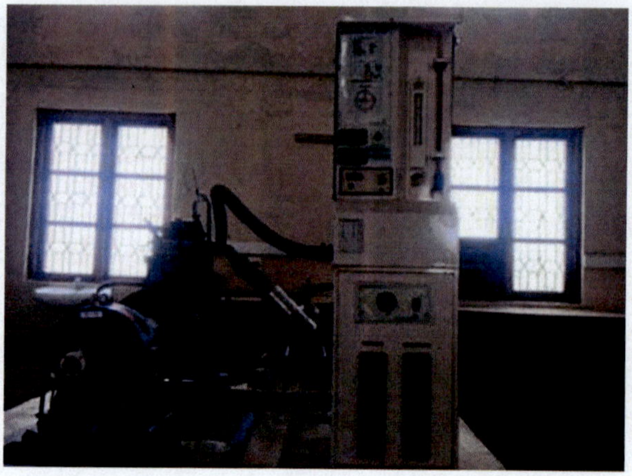

the bottom of the panel to measure the engine cooling water and calorimeter water flow rates. Engine torque is measured by eddy current dynamometer which is coupled to an engine crank shaft and sent as a signal to a computer. Provision is made to read the performance parameters such as cylinder pressure, rate of heat release, indicated, brake and friction efficiencies, brake fuel consumption, indicated and brake mean effective pressures, and the mechanical efficiency, etc., directly from the computer at the end of the test run. The above-mentioned components are the major parts that are necessary for the working of the test rig. The major components present serve various use in the different operations conducted in the test rig. Proper maintenance of these components in this engine test setup is necessary for the efficient working of the system.

The gas analyzer (AVL Make) is used to measure exhaust gas constituents such as nitrogen oxide, unburnt hydrocarbon, excess oxygen, carbon monoxide and carbon dioxide. A smoke meter (AVL Make) is used to measure engine smoke. The photographic view of engine setup and five gas analyzer are shown in Figs. 1 and 2, respectively.

5 Results and Discussion

5.1 Brake Thermal Efficiency

Figure 3 shows the BTE of blends for different loads. From low load to full load, BTE varies from 15 to 31% for diesel, 14.06–30.77% for D65-PWEO25-P20, 13.17–28.45% for D55-PWEO35-P30, and 14–30.05% for D45-PWEO45-P10, respectively. As the load increases, the fuel supplied to the engine also increases, so thermal efficiency increases. As the calorific value of PWEO is lower than diesel, the BTE of

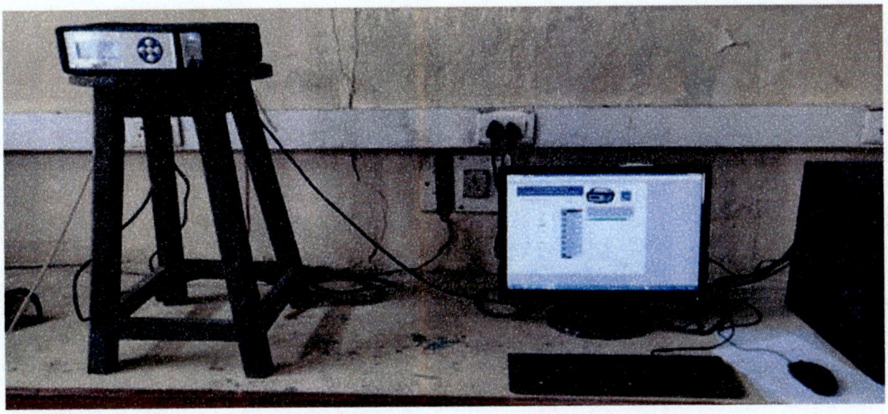

Fig. 2 Gas analyzer

Fig. 3 Thermal efficiency

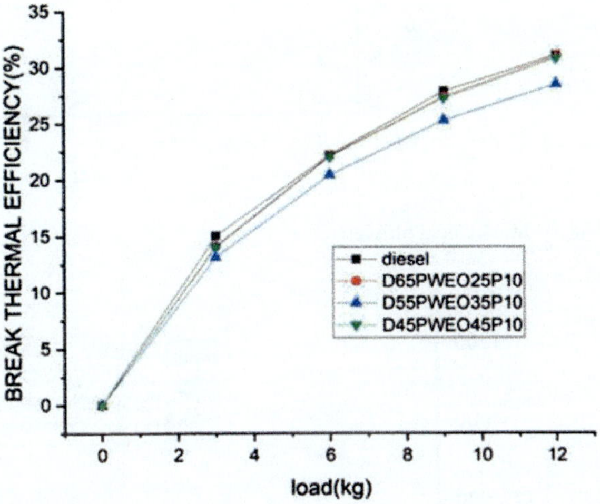

blends is slightly lower than diesel. Among the blends, D65-PWEO25-P10 showed more percentage increase in brake thermal efficiency.

5.2 *Mechanical Efficiency*

Figure 4 presents the variation of mechanical efficiency for blends. From no load to maximum load, mechanical efficiency varies from 41.85 to 74.05% for D65PWEO25P10, 40–73.25% for D55PWEO35P10, and 40–73.21% for D45PWEO45P10, respectively. Figure 5 displays that as load increases, there is an increase in mechanical efficiency for all fuels. The n-Pentanol assists for complete burning of fuel, so that more heat energy converted in to work energy, and hence, mechanical energy increases. Among the blends, D65-PWEO25-P10 showed more percentage increase in mechanical efficiency.

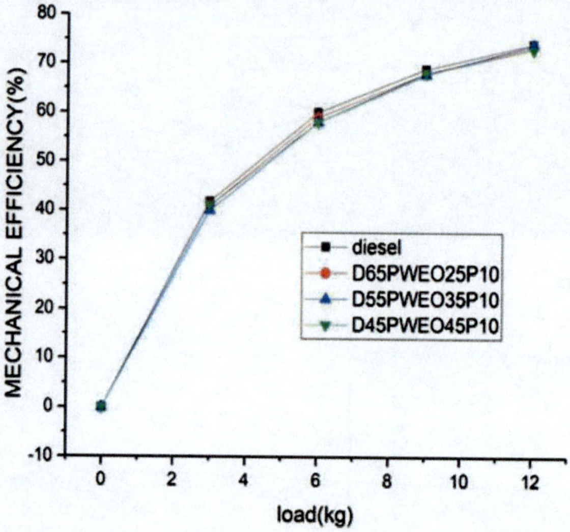

Fig. 4 Mechanical efficiency

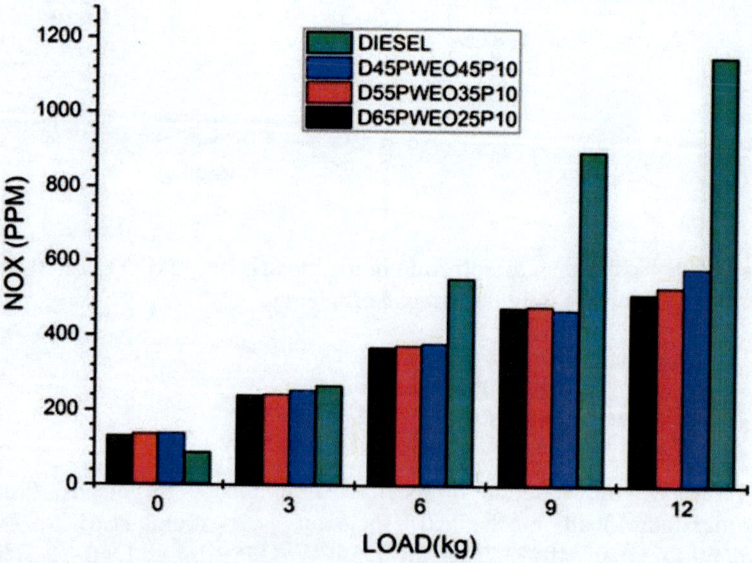

Fig. 5 NO_X emission against various loads

5.3 Nitrogen Oxides

NO_X formation is depends on the engine combustion temperature, availability of oxygen and reaction between nitrogen and oxygen in the combustion chamber. Figure 5 shows the nitrogen oxides of blends for different loads. The NO_X emission variation from no load to top load is 195-1180PPM for diesel, 180-450PPM for D65-PWEO25-P10, 190-470PPM for D55-PWEO35-P10, and 185-480PPM for D45-PWEO45-P10, respectively. Among the blends, D65-PWEO25-P10 showed lower percentage increase in NO_X emission, but compared to diesel, NO_X emissions increase slightly.

5.4 Carbon Monoxide

Figure 6 presents the variation of carbon monoxide for blends. CO emission varies from zero to top load from 0.057 to 0.06% for diesel, 0.04–0.12% for D65-PWEO25-P10, 0.06–0.11% for D55-PWEO35-P10 and 0.04–0.95% for D45-PWEO45-P10, respectively. This may be due to the presence of PWEO which worsens the normal ignition process leading to a slightly lower percentage of H_2O compared to diesel with increased CO emission. Among the blends, D65-PWEO25-P10 showed lower percentage increase in CO emission.

Fig. 6 Carbon monoxide emissions at varying loads

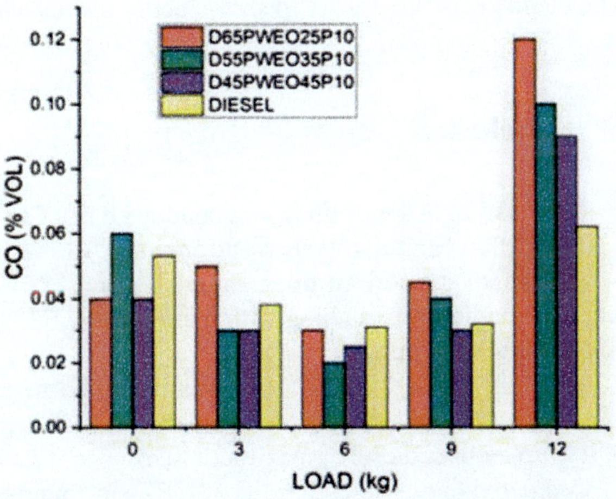

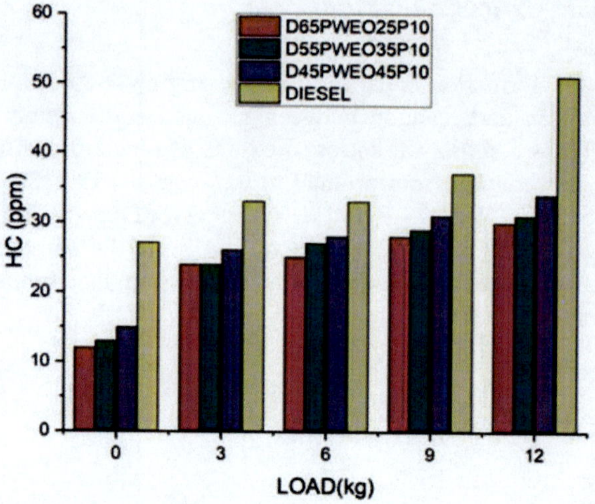

Fig. 7 Hydrocarbon emissions with an increase in varying load

5.5 *Hydrocarbon Emissions*

HC emission depends on ignition quality of fuel. HC emission of blends for various load are given in Fig. 7. For the engine test loads from zero to maximum, hydrocarbon emission varies from 27 to 52 ppm for diesel, 30–62 ppm for D65-PWEO25-P10, 36-65 ppm for D55-PWEO35-P10, and 38-70 ppm for D45-PWEO45P10, respectively. The blend D65-PWEO25-P10 shows better results when compared to the diesel.

6 Conclusion

In this work, an in depth study was conducted in a CI engine using three blends of n-pentanol, diesel and Pyrolysis waste engine at various loads. The 10% concentration of n-Pentanol used in all three samples blends. No abnormalities were observed when the engine was running with these blends. From the test results the following conclusions were drawn.

The D65-PWEO25-P10 blend was identified to produce the best performance among the three blends, but the brake thermal efficiency of this blend was found to be 1% lower than diesel at maximum load.

Compared to diesel, emission of NOX was found to be 20% lower when engine was operated with the D65PWEO25P10 blend at maximum load.

Compared to diesel, emission of CO was found to be 10% lower when engine was operated with the D65PWEO25P10 blend at maximum load.

Compared to diesel, emission of HC was found to be 15% lower when engine was operated with the D65PWEO25P10 blend at maximum load.

References

1. Venkatesan SP, Kadiresh PN (2018) Combustion performance study of aqueous aluminum oxide nanofluid blends in compression ignition engine. J Energy Resour Technol 141(4). https://doi.org/10.1115/1.4042086
2. Ananthakumar S, Jayabal S, Thirumal P (2016) Investigation on performance, emission and combustion characteristics of variable compression engine fuelled with diesel, waste plastics oil blends. J Braz Soc Mech Sci Eng 39(1):19–28. https://doi.org/10.1007/s40430-016-0518-6
3. Babu D, Anand R (2017) Effect of biodiesel-diesel-n-pentanol and biodiesel-diesel-n-hexanol blends on diesel engine emission and combustion characteristics. Energy 133:761–776. https://doi.org/10.1016/j.energy.2017.05.103
4. Kaimal VK, Vijayabalan P (2016) An investigation on the effects of using DEE additive in a DI diesel engine fuelled with waste plastic oil. Fuel 180:90–96. https://doi.org/10.1016/j.fuel.2016.04.030
5. Venkatesan SP, Kadiresh PN, Beemkumar N, Jeevahan J (2019) Combustion, performances, and emissions characteristics of diesel engine fuelled with diesel-aqueous zinc oxide nanofluid blends. Energy Sour Part A Recovery Utilization Environ Effects 1–15. https://doi.org/10.1080/15567036.2019.1666933
6. Senthilkumar P, Sankaranarayanan G (2016) Effect of Jatropha methyl ester on waste plastic oil fueled DI diesel engine. J Energy Inst 89(4):504–512. https://doi.org/10.1016/j.joei.2015.07.006
7. Ganesan S, Hemanandh J, Venkatesan SP, Mahalingam S, SenthilKumar J, Padmanabhan S, Mohan Raj M (2021) Experimental analysis on DI diesel engine characteristics using watermelon seed oil with Magnesium oxide. Mater Today Proc. https://doi.org/10.1016/j.matpr.2021.05.326
8. Damodharan D, Sathiyagnanam AP, Rajesh Kumar B, Ganesh KC (2018) Cleaner emissions from a DI diesel engine fueled with waste plastic oil derived from municipal solid waste under the influence of n-pentanol addition, cold EGR, and injection timing. Environ Sci Pollut Res 25(14):13611–13625. https://doi.org/10.1007/s11356-018-1558-5
9. Ganesan S, Mahalingam S, SenthilKumar J, Hemanandh J, Venkatesan SP, Padmanabhan S, Lakshmi Sankar S (2021) Experimental analysis on DI diesel engine characteristics using watermelon seed oil with Magnesium oxide. Mater Today Proc. https://doi.org/10.1016/j.matpr.2021.04.521
10. Ganesan S, Venkatesan SP, Kumar NB, Purushothaman M, Hemanandh J, Senthilkumar J, Senthilkumar G (2020) Taguchi based optimization of diesel engine parameters with blends of lemon grass oil with ZnO. AIP Conf Proc 2311. https://doi.org/10.1063/5.0034021
11. Venkatesan SP, Kadiresh PN (2018) Thermodynamic analysis of diesel engine fuelled with aqueous nanofluid blends. In: Innovative design, analysis and development practices in aerospace and automotive engineering (I-DAD 2018), pp 445–456 (2018). https://doi.org/10.1007/978-981-13-2718-6_43
12. Venkatesan SP, Kadiresh PN, Kumar KS (2014) Experimental investigation of aqueous cerium oxide nanofluid blend in diesel engine. Adv Mater Res 938:286–291. https://doi.org/10.4028/www.scientific.net/amr.938.286